T0133391

On Sea Ice

On Sea Ice

by

W. F. Weeks

Professor of Geophysics, Emeritus
Geophysical Institute
University of Alaska Fairbanks

with (Chapter 16)

W. D. Hibler III

International Arctic Research Center
University of Alaska Fairbanks

University of Alaska Press
Fairbanks, Alaska

University of Alaska Press
P.O. Box 756240
Fairbanks, AK 99775-6240

Library of Congress Cataloging-in-Publication Data

Weeks, W. F.
 On sea ice / by W.F. Weeks ; with W.D. Hibler III.
   p. cm.
 Includes bibliographical references.
 ISBN 978-1-60223-079-8 (lithocase : alk. paper)
 1. Sea ice. I. Hibler, W. D., III. II. Title.
 GB2403.2.W44 2010
 551.34'3—dc22
            2009035399

Interior design and layout by Rachel Fudge
Cover design by Dixon Jones, Rasmuson Library Graphics

Contents

Preface

An interest in sea ice is an odd sort of an addiction. I acquired the habit over 50 years ago when I volunteered to spend a winter on the Labrador Coast investigating the properties of this somewhat odd natural material. At the time I was a lowly second lieutenant in the United States Air Force serving out a two-year ROTC commitment. In that the Air Force had deferred calling me to active duty for several years in order to let me finish my Ph.D. in the petrology/geochemistry area of the earth sciences, they presumably thought that they could find some use for my assumed talents. When I was ordered to active duty, I was shipped off to the AF Cambridge Research Center (AFCRC) in Boston and assigned to a project concerned with soil mechanics. Now I knew nothing whatsoever about soil mechanics. As I was totally uninterested in the subject, I immediately began to look for an escape route. Lo and behold, I had no sooner started looking than it arrived in the form of an operational requirement from the Northeast Air Command. This command had the task of supplying the radar stations located along the coasts of Labrador and Baffinland: remote mountainous regions in northern Canada, then almost totally lacking in aircraft runways. Resupply during the summer was no problem as ships could be used. However, winter was another story in that the presence of sea ice along these coasts prohibited the use of normal shipping. In those preturbine days, helicopters had only limited payloads and were difficult to operate in cold weather. Also, the winter weather along the Labrador Coast was not only cold and snowy, it was truly vile. One apparent answer to this problem was, of course, to land fixed-wing aircraft on the sea ice, starting at the earliest possible date in the winter and continuing until the last minute in the spring. However, sea ice along the Labrador Coast was both thin and warm, and existing bearing capacity tables were of doubtful applicability. As research requirements from operational commands were not a common occurrence, AFCRC immediately sprang into action by inquiring, "Who would like to volunteer to spend an exciting winter living in an isolated Eskimo community on the Labrador Coast?" Volunteers were so limited that I got the job and was placed in charge of the project, as I was the only Air Force officer involved. Looking back on my elevated status, I have always thought that my real responsibility was to serve as the person to blame when an aircraft broke through the ice and sank.

Now, my lack of interest in soil mechanics was not the only reason that I volunteered. While thinking about the decision, I had looked in the literature available to me at the time and, to my surprise, I found very few references on sea ice in general and almost no references on its properties. Now in 1955 the petrology/geochemistry area of geology was a very active subject, with many players with varying views on how the world worked. Whenever a research paper was written in this field, it was expected to display an in-depth knowledge of a seemingly endless number of earlier papers, refuting all previous misguided interpretations. This not only took time, it was hard work. Finding a research field such as sea ice that was "unplowed" was most appealing. A field where everything you did would be brand new and you only needed to reference your own papers. Wonderful! Now, as the reader will discover in Chapter 2, there were earlier references. I just had not discovered where they were hiding. However, I was right in that their numbers were comparatively small.

It also dawned on me that sea ice covered a very significant percentage of the surface of the Earth. It therefore seemed reasonable to assume that this material and its behavior must be of some significance, even though this possibility was rarely mentioned in the textbooks of the day. In fact, the only text at the time that devoted a significant number of pages to sea ice and its properties was *The Oceans*, by Sverdrup, Johnson, and Fleming (1942). As will be seen, there was a good reason for this. It was also clear that human presence in the polar regions was expanding with the building of the DEW (Distant Early Warning) Line, the impending start of the International Geophysical Year, and the increasing capabilities of air transportation. These activities involved operations on the polar oceans, where sea ice invariably caused problems. In my head this translated into the conclusion that studies of sea ice and its properties would definitely be needed in the future. I thought about these matters and decided that I knew a good deal when I saw one. As a result of my decision, I spent the winter of 1955–1956 at Hopedale, Labrador, and of 1956–1957 at Thule, Greenland. Although in 1957 I returned to teaching conventional earth sciences for five years, I had become addicted to the polar regions and the world of snow and ice. Fortunately, I was able to arrange to spend the ensuing summers simulating the growth of sea ice by freezing NaCl solutions in the cold rooms at the Snow, Ice and Permafrost Research Establishment (SIPRE). By 1962 I had to admit to myself that I could not stay on top of both the petrology/geochemistry area and also study sea ice at the same time. To add to the difficulty, typical sea ice field operations occurred in the winter at times when I was expected to teach. Besides, I missed the excitement of field operations in the polar regions. As a result, when in 1962 the Cold Regions Research and Engineering Laboratory (CRREL) offered me a position with their Snow and Ice Branch, I accepted. This was a decision that I have never regretted as it allowed me to enter the study of glaciology full time. Here by *glaciology* I refer to the study of snow and ice in all its varied forms (I have heard that some glaciologists even study glaciers). In my particular case I have tried to focus as much as possible on sea ice, although I have also been involved in research programs on lake ice, river ice, alpine snow, avalanches, icebergs, and large ice caps at different times in my career.

I have enjoyed writing this book as it deals with a subject that has fascinated me these many years. I hope that in this book I have been able to convert this fascination into a communicable disease. At the least, I hope that this book provides a window on a portion of the Earth's surface that will appear as foreign to most people as the surface of the moon and where humans are not necessarily the top of the food chain.

W. F. "Willy" Weeks
Portland, Oregon
w-f-weeks@comcast.net

About This Book

A book should serve as the ax for the frozen sea within us.
Franz Kafka

Now that the reader knows how and why I became involved with the study of sea ice, it is reasonable to inquire as to what is to be found in this book. My purpose in writing this book is twofold: one personal and one hopefully more generally educational. The personal reason is to bring my efforts to understand sea ice to some sort of closure. Over the years I have worked on a variety of research projects investigating a number of different aspects of sea ice properties and behavior. As anyone who is involved in research can attest, one effective way to deal with a research problem is to develop a sort of tunnel vision. You read all the papers you can locate that relate to your specific problem and then perform the appropriate experiments or go into the field to take the pertinent observations. Finally you write up your results. At this stage you hope to find that you know more about the problem than you thought you did. While all this is going on you commonly neglect everything else. This allows the broad-based overarching view of your field that you would like to possess to gradually deteriorate, assuming that it ever existed. Then you move on to the next research problem, which because of variations in funding support is commonly focused on a quite different aspect of the field or even on some quite different field. In my case I have, over the years, been involved in research on many of the different aspects of sea ice. In some cases my involvement has been quite intense, in others I have been a mere dabbler on the margins of the subject. What I hope to gain for myself by writing this book is an overall feel for where the many different aspects of the study of sea ice are at the approximate time of the International Polar Year during 2007–2009, including what we know, what we may know, what we should know but don't, and what this all may mean. At best, I hope to be able to write with some insight into the specific topics at hand. On topics where I have less experience, I will simply try to do a yeomanlike job of providing a reasonably thorough guide concerning the status of the subject.

What I hope to accomplish for readers is that by putting this down on paper, I will assist them in rapidly coming to grasp with the pre-2009 literature concerning the study of sea ice. As one will soon become aware, the study of sea ice has expanded at an amazing rate since the end of World War II. At the present, a seemingly endless (to a reviewer) stream of papers are being published. Because of contract pressure, there is an unfortunate tendency for investigators to publish many short, less detailed papers instead of one integrated, high-quality one. This is the old "If you can't dazzle them with quality, bury them under quantity" approach, and it makes a reviewer's task difficult. Although there are a few journals such as the *Journal of Geophysical Research*, *Geophysical Research Letters*, the

Journal of Glaciology, and *Cold Regions Science and Technology* where papers on sea ice can routinely be found, many interesting papers on the subject are hidden away in specialty conferences or in even darker recesses of the "gray" literature. I have tried to present a reasonable amount of this material to the reader. There are a number of difficulties here: first, gray literature publications can be difficult to find as most libraries do not acquire this type of material; second, this material is frequently too expensive for most individuals to acquire. This, of course, also makes matters difficult for persons attempting to write reviews, particularly when working in the basement of their homes at considerable distance from a specialized polar library.

I will also attempt to provide the reader with what insights I have concerning cracks in the sea ice research edifice. By this I mean places where things simply do not appear to add up correctly, where something appears to be missing, or even where the whole approach appears to have gone awry. The reader should note that just because I may be picking at some perceived flaw in a paper does not mean that the paper is poor. In fact, it is commonly the better papers that provide enough detail to allow one to suggest alternative explanations. In examining flaws and in suggesting new approaches and explanations I will attempt to emulate two of my personal heroes: the geochemist V. M. Goldschmidt and the oceanographer N. N. Zubov. Both of these individuals produced books that were for their time full of useful information and, perhaps more important to me as a neophyte researcher, contained numerous useful insights. It was said that Goldschmidt's book suggested a Ph.D. thesis topic on every page. How successful I am in this endeavor will depend on the interests of each individual reader. At least I will have tried. Incidentally, the English translation of Zubov's classic book is a good example of the gray literature in that it was published in 1943 in Russian and translated into English in 1945 on a contract for the U.S. Navy Electronics Laboratory. Although a number of copies were prepared, the translation was never formally published.

One advantage that the people of my generation had was that we were able to spend a considerable amount of time in the field studying sea ice in situ. That was certainly true in my case. I have always considered myself to be the glaciological equivalent of a field geologist: a person who goes in the field, looks around to see what's happening, takes some measurements and samples, and goes home and tries to figure out why it's happening. It is my impression that currently field opportunities are less frequent (they clearly are expensive) and that the emphasis has shifted more to remote sensing and modeling. This is fine, but logging time on the ice clearly gives one a sense of the relative importance of different processes. I hope that this book will, in some small way, help to fill this perceived gap. Besides, certain sea ice types are becoming increasingly rare and difficult to access—old ice for example.

A word of warning is also due to graduate students reading this book. In my experience, new graduate students commonly make the mistake of believing what they are assigned to read. This is a fundamental error. A graduate student's job is to find the cracks in the walls of the research edifice and to remove them by providing new interpretations, insights, and data. This cannot be accomplished by being a believer. Always think that even if the original author finally was able to get something right, the present reviewer (such as myself) undoubtedly either misunderstood it or hopelessly confused things. To help the reader in these matters I will commonly write in the first person when I am pontificating on sea-ice-ological matters. I realize that it is common to write technical tomes in the sleep-inducing third person with the liberal use of the royal *we*, implying that not only the author thinks so but the local Kings and Gods agree with him. In this book the use of the word *I* should be considered as a flag signifying *caveat emptor*.

As can be seen by examining the table of contents, after an introductory discussion of the historical aspects of sea ice research, the book gradually works its way from the small scale (component properties, phase relations, internal structure) to a slightly larger scale (salinity, growth, and decay). This, in turn, leads to a discussion of the physical basis for interpreting the considerable variations in sea ice properties and to a yet larger scale where polynyas and leads are described. This is followed by a discussion of the deformation of natural ice covers and of ice-induced gouging of the seafloor, of conditions near the boundary between ice and open ocean, and of the distribution and nature of snow on the ice. Next, there is a discussion of the drift of pack ice and of models that have attempted to simulate this behavior. This is followed by a chapter that explores the very interesting and comparatively new subject of the formation of the different types of underwater ice. The book concludes with an examination of recent changes in ice extent and thickness. In short, the book gradually progresses from considering sea ice on a microscopic scale to examining its effects on a more global scale. As it is commonly easier to investigate matters on a small scale than on the very large, this allows us to use more firmly established smaller-scale information as building blocks in assessing larger-scale problems. However, as the reader will soon discover, there are still large numbers of missing bricks in the sea ice edifice considered on every scale. Appendices include a listing of symbols, acronyms, a glossary of sea ice terminology, and brief discussions of sampling procedures, thin section preparation, and remote sensing.

Although this book is written more as a monograph, it can easily serve as a text for a graduate course on sea ice if supplementary problem sets are developed. It can also serve as a source of ancillary reading on selected topics in courses on glaciology, oceanography, meteorology, and polar engineering. One of the more perplexing problems that I had to face in writing this book is that of deciding exactly how much detail should go into the discussion on specific subjects. Clearly a book such as this cannot be everything for everyone. In general, I have decided to provide the reader with considerable background information on a variety of subjects. This decision was largely based on my experiences gained at the University of Alaska where for several years I taught a graduate course on sea ice. I invariably found that, for every subject that I discussed, there was someone in the class who knew as much as I did about that subject, or more. More importantly, there was also a student who knew next to nothing about that subject. This book tries to help the latter students gain confidence in areas that are outside their specialty. The cost to the already knowledgeable is small in that skipping over familiar material is both easy and fast. It also makes you feel so superior!

The reader should also know what this book is not about. First, it is not a totally new treatment of the subject. I have written reviews on several aspects of sea ice in the past and I have used some of this material in the present book. The most obvious example is the chapter on the history of research on sea ice. However, even here the material has been tweaked a bit. Second, the book is certainly not an adequate treatment of the large and diffuse Russian literature on sea ice. As I have never mastered that language, this goal is clearly beyond my grasp. However, I have tried to reference some of their more important contributors and papers as an aid to further investigations by more linguistically gifted readers. As I have recently assisted several AARI scientists in preparing English versions of their reviews on different sea ice subjects, I have had the advantage of being exposed to their views (Frolov and Gavrilo 1997). Unfortunately, an English version of this material has not, as yet, been published.

The present book also does not present an adequate link to the very large literature on the material sciences that bears on the study of sea ice. In addition, although there is a short appendix (F) on remote sensing, and the results gained by the use of a variety of these techniques are discussed in the book, no attempt is made to explore this subject in detail other than to provide a few critical references. Clearly this would require another tome. In that such tomes are available (Carsey 1992; Lubin and Massom 2006; Martin 2004; Rees 2001), I see no reason to add to this list. Next, although the subject of the biological aspects of sea ice is of considerable importance, it is barely mentioned in the present book. My excuse here is one of ignorance. It is highly recommended that readers interested in sea ice as a biological habitat examine copies of Melnikov (1997), Lizotte and Arrigo (1998) and the latter part of Thomas and Dieckmann (2003) as useful gateways into this literature. Also, the book makes no attempt to cover the engineering aspects of sea ice although it does contain considerable material that an individual interested in ice engineering would find useful.

I started to work on the book at about the time that I retired in 1996, as something to keep me amused during retirement. One reason that I thought that I might pull this off is that the subject of sea ice tends to be very faddish. As one works one's way through the book one may notice that there will be a flurry of activity on a particular subject for a few years and then little or no activity on that subject. For example, at present everyone appears focused on the interrelations between sea ice and climate. This unevenness can be blamed on the funding agencies' tendency to throw money at subjects that they perceive to be "hot." I frequently felt that just when I knew enough about a particular aspect of sea ice to be able to make real progress was when funding on that topic dried up. It is clearly not the case that the abandoned subjects are necessarily well understood. Hopefully the present book will provide the reader with a feel for currently neglected areas worthy of further investigative effort.

The fact that I have been writing this book for over 10 years attests to both my leisurely approach and the fact that I appreciably underestimated the amount of work involved. A particular difficulty at the present time is that field observations indicate that both the extent and the thickness of the Earth's sea ice covers are changing as the result of changes in the Earth's environment. These trends will be discussed in Chapter 18. In the earlier portions of the book, I have simply tried to provide a description of the status of the ice cover prior to noticeable recent changes. As such it is hoped that the book will, at the very least, provide a useful picture of how ice conditions appeared prior to ~1980.

Under favorable lighting conditions sea ice can occasionally provide the photographer with the possibility of striking color photographs. The reader will note that I have avoided color completely in the body of the book. My reasons are simple. Essentially all sea ice features of scientific interest can be clearly illustrated using grayscale images. In fact, this is how such features frequently appear in the field. My purpose here is to make the book as inexpensive as possible so that interested students might possibly afford a copy.

One other important point: as I neared completion of the book with just two chapters left (Chapter 16, "Ice Dynamics," and Chapter 18, "Trends"), I became aware of a 2004 review paper by W. D. Hibler III entitled "Modelling the Dynamic Response of Sea Ice." Now, sea ice dynamics is a subject worth discussing in considerable detail because understanding it is essential to understanding estimates of both future sea ice extent and current behavior. It dawned on me that if I could talk Bill into letting me use his paper as the basis for Chapter 16, it would greatly speed the completion of the book. Now Bill and I have known each other for years in that we are both former members of CRREL's Snow and Ice Branch and coauthored several papers during the 1970s. Fortunately, enlisting his assistance proved to be possible in that Bill's contributions to sea ice dynamics are major while mine are quite minor. Although much of Chapter 16 is a direct insert from Hibler (2004), in an effort to make the present book as homogeneous as possible some sections have been extensively modified. Other sections have been deleted in that their material is either covered elsewhere in the book or deemed to be more advanced than needed here. More importantly, I take responsibility for any inappropriate changes or misinterpretations.

Finally, an examination of the dates in the reference list will show that while *On Sea Ice* attempts a reasonably complete survey of the sea ice literature through 2008, journal articles published in 2009 are generally not included. The only exceptions are a very few papers and books in press that I was able to examine during 2008. Also note that there are a few references in the bibliography that are not cited in the text. Most of these concern studies that add detail to subjects discussed in the book. I have found it particularly useful to examine several sequential papers by a given author. With luck you can find out not only what the author thinks and the path he or she took in arriving at this conclusion—if you are really lucky, you may see that there is a better path that will take you even further.

Remember,

When you copy from one author, it's plagiarism;
When you copy from many, it's research.
Paraphrased from Wilson Mizner (1876–1933)

Acknowledgments

*I've learned that the easiest way for me to grow as a person
is to surround myself with people smarter than I am.*

Andy Rooney

I have asked a number of friends and former associates of mine whom I consider to be particularly knowledgeable in specific areas of sea ice research to review different chapters of this book with an eye toward removing foolish statements and identifying subjects and interpretations that I have missed. These are Hajo Eicken, Bill Hibler, Martin Jeffries, Austin Kovacs, Pat Langhorne, Gary Maykut, Don Perovich, Dan Pringle, Jackie Richter-Menge, Ian Stone, Terry Tucker and Norbert Untersteiner. Inadequacies that remain are clearly my responsibility.

I would also like to thank the following people and organizations for contributing and/or assisting me in locating a number of the figures in the book: Reinhard Krause, Alfred Wegener Institute, American Geophysical Union (Washington, DC), Arctic and Antarctic Research Institute (St. Petersburg), Arctic Submarine Laboratory (San Diego, CA), U.S. Army Cold Regions Research and Engineering Laboratory (Hanover, NH), Geological Survey of Canada, International Glaciological Society (Cambridge, UK), Jens-Ove Näslund (Stockholm), Russian State Museum of the Arctic and Antarctic (St. Petersburg), Scott Polar Research Institute and *Polar Record* (Cambridge, UK) and the Whitby Museum (Yorkshire, UK).

Thanks are also due to the late Jim Bender, former head of the Research Division at CRREL; Ron McGregor and Chuck Luther of ONR; Bernie Lettau of the Division of Polar Programs at NSF; and Gunter Weller of the OCSEAP program at the University of Alaska. These individuals always seemed to find a way to support sea ice research, even when it was very out of favor.

I would like to dedicate this book to the memory of four close friends (Andrew Assur, Vladislav Gavrilo, Arnie Hanson, and Tadashi Tabata) who were dedicated students of sea ice as well as to my many associates with whom I have had the pleasure of working in, and thinking about, the polar regions.

Figure Sources

The question is not what you look at, but what you see.
Henry David Thoreau

In that this book is primarily a review of the published sea ice literature, it is hardly surprising that most of the figures are from this same literature. Fortunately, I have been able to obtain permission to use these figures courtesy of the original publishers. The following is a listing of these organizations followed by the name of the journal and the specific figures contributed.

American Association for the Advancement of Science
   Science: 8.34, 12.25, 16.24
American Geophysical Union
   Geophysical Research Letters: 8.30, 9.20, 9.21, 18.3, 18.4, 18.5, 18.6
   Journal of Geophysical Research: 7.21, 7.36, 7.37, 8.5, 8.9, 8.10, 8.20, 8.21, 8.29, 8.32, 8.37, 8.38, 8.39, 8.40, 8.41, 9.5, 9.6, 9.7, 9.13, 9.18, 9.19, 11.6, 11.7, 11.9, 11.10, 11.11, 11.13, 11.14, 11.16, 11.17, 11.18, 11.19, 11.25, 11.26, 11.27, 11.28, 11.29, 11.30, 11.31, 11.45, 11.48, 12.45, 12.48, 12.49, 12.50, 12.52, 12.53, 12.54, 12.55, 12.59, 12.60, 14.4, 15.4, 15.6, 15.7, 16.3, 16.5, 16.11, 16.13, 16.25, 16.29, 16.32, 17.3, 17.4, 17.5, 17.6, 17.8, 17.11, 17.12, 18.7, 18.8, 18.9
   Monographs and Books: 3.3, 8.12, 10.35, 10.36, 10.38, 11.8, 16.24, 17.10
   Reviews of Geophysics: 15.8
American Meteorological Society
   Journal of Climate: 15.3, 16.24, 16.30, 16.31
   Journal of Physical Oceanography: 16.9, 16.17, 16.23, 16.24, 16.26, 16.27
   Monthly Weather Review: 16.28
Arctic and Antarctic Research Institute
   Files 12.22, 12.29, 17.1
Bildarchiv der Osterreichischen National Bibliothek, Wein
   Photograph: 2.3
Cambridge University Press
   Rees, W. G. 2001. *Physical Principles of Remote Sensing* (2nd ed.): E-1, E-2
Cold Regions Research and Engineering Laboratory
   Monographs: 5.6, 5.7
   Special Report: 6.9
Elsevier
   Cold Regions Science and Technology: 6.10, 6.11, 6.12, 7.38, 8.24, 10.19, 10.22, 10.23, 10.26, 12.33, 12.34, 12.34, 12.36
Institute of Low Temperature Science
   Science Contributions: 4.2, 7.17, 7.18, 7.19, 7.20
International Glaciological Society
   Annals of Glaciology: 7.42, 8.33, 8.35, 10.21, 17.9
   Journal of Glaciology: 4.1, 6.6, 7.11, 7.32, 8.2, 8.8, 8.9. 8.15, 8.36, 9.1, 9.10, 9.11, 9.12, 10.5, 10.13, 10.15, 10.16, 12.46, 16.6, 17.2, 17.7

Optical Society of America
    Applied Optics: 5.4
Russian State Museum of the Arctic and Antarctic
    Files: 2.7
Scott Polar Research Institute
    Polar Record: 2.4
Sigma Xi
    The American Scientist: 3.4

1 Introduction

Here was a crystalline world of azure and emerald, indigo and alabaster—
dazzling to the eye, disturbing to the soul.
P. Berton, describing the Ross expedition's first
encounter with sea ice, Davis Strait, 1818

The reasons that individuals living in the Arctic need to understand the behavior of sea ice are both numerous and varied. Here, by the term *sea ice*, I refer to any type of ice that forms in or on the surface of the sea by the freezing of seawater. This definition excludes both icebergs, which are pieces of glacier ice adrift in the sea, and spray icing formed on either natural or man-made objects. Although at first glance this definition may seem to be unnecessarily restrictive, the reader will soon find that sea ice is surprisingly varied, reflecting the local environmental state during ice formation. However, as the great majority of readers of this book are probably not residents of the shores of the polar oceans and some will have never seen a piece of sea ice, they might well inquire why they should be interested in this seemingly obscure and physically remote material. Why should they not simply conclude that out of sight should equate with out of mind, and proceed on to other less esoteric matters?

To answer this question one needs to examine both the amount of snow and ice on the Earth and how it is distributed. Table 1.1 presents some of this type of information compiled from a variety of sources. When the maximum volume of sea ice that exists in the World Ocean during a given year ($\sim75 \times 10^3$ km^3) is compared with the present volume of glacier ice ($\sim24 \times 10^6$ km^3), we find that the volume of sea ice is roughly 0.3% of that of glacier ice. A more realistic comparison would be to compare the volume of sea ice existing at any given time, recalling that maximum sea ice extent in one hemisphere roughly corresponds to a minimum extent in the other hemisphere. In this case the result would be even less impressive, with the volume of sea ice amounting to only $\sim0.2\%$ of the volume of glacier ice. In making these approximate calculations I have arbitrarily assumed that all northern hemisphere sea ice has a thickness of 3 m and that all southern hemisphere sea ice is 1.5 m thick. Needless to say this is far from the case. However, the exact values are immaterial in that any reasonable estimate of average sea ice thicknesses would lead to the same general result: volumetrically, sea ice does not appear to be an important geophysical entity. When compared to the total volume of water in the World Ocean ($\sim1370 \times 10^6$ km^3), the volume of sea ice shrinks to total insignificance (0.004%). Even the total volume of all types of ice on the surface of the Earth comprises only $\sim1.7\%$ of the total water volume on the Earth's surface. Approximate values for the volumes and surface areas of the Earth's sea and glacier ice can be found in Table 1.1.

There are two reasons why many scientists are currently interested in sea ice. The first reason is shared with ice in all its forms except for the small amount of ice suspended in the atmosphere. When ice occurs it forms the surface layer of the Earth. When sea ice exists it forms the surface of the sea, since sea ice forms from seawater by definition and ice floats in its own melt. Sea ice is also

Table 1.1. Information concerning the various types of ice occurring on the Earth's surface. *These values can be taken as reasonable estimates for the time period 1950 to ~1980. Decreases in ice extent and volume occurring since ~1990 will be discussed in some detail in Chapter 18.*

Approximate volumes	($\times 10^6$ km^3)		
Earth's surface waters	1600		
Oceans and seas	1370		
Lakes and rivers	230		
Glacier ice (total)	24		
Sea ice	0.075		

Surface areas			
Earth	509×10^6 km^2		
World Ocean	$\sim 361 \times 10^6$ km^2		
Arctic Basin	12.2×10^6 km^2		
Southern Ocean	$\sim 35 \times 10^6$ km^2 (ocean areas south of the antarctic convergence)		

	Ice covered ($\times 10^6$ km^2)	% of continental area	% of total global glacier area
Antarctic continent	13.91	99.93	85.81
Greenland Ice Cap	1.80	82.45	11.11
N. America	0.22	0.98	1.34
Asia	0.14	0.33	0.85
Europe	0.12	1.01	0.72
S. America	0.03	0.14	0.16

Total area covered by glaciers	16.22×10^6 km^2

Sea ice areas	($\times 10^6$ km^2)
N. Hemisphere	
Maximum extent	~15
Minimum extent	~8
S. Hemisphere	
Maximum extent	~18
Minimum extent	~3

a good thermal insulator and serves as a platform for the deposition of snow, which is an even better insulator. In addition, the albedo of bare sea ice is ~0.8 and that of newly fallen snow is 0.85 to 0.9, meaning that during the polar summer only 10 to 20% of the incoming shortwave radiation is absorbed at the surface of snow- and ice-covered areas, with the rest being reflected back into either space or the atmosphere. Furthermore, during the winter the snow cover on the sea ice acts as a nearly perfect black body, radiating heat into space. In short, sea ice acts as an efficient lid on the surface of the polar oceans controlling the exchange of heat and water vapor between the comparatively warm ocean and the frigid atmosphere. As a result, sea ice is an important player in the puzzle of the world's climate system.

By now it is also probably obvious that the areal extent of the world's sea ice cover is vastly more impressive than its volume. At maximum extent sea ice covers ~35×10^6 km^2 during some time of the year. This amounts to ~7.3% of the Earth's surface or, more importantly, ~11.8% of the surface of the World Ocean. To put these numbers into some perspective, I note that the contiguous United States has an area of ~7.825×10^6 km^2 which means that in the northern hemisphere at maximum extent sea ice covers an area equal to slightly less than twice (192%) that of the contiguous United States. In the late summer the sea ice area shrinks to just slightly more (102%) than this value. A map showing the general distribution of sea ice in the northern hemisphere during both its maximum and minimum extent is given in Figure 1.1. As is obvious from this figure, at maximum extent sea ice does not advance parallel to lines of latitude. In both the Atlantic

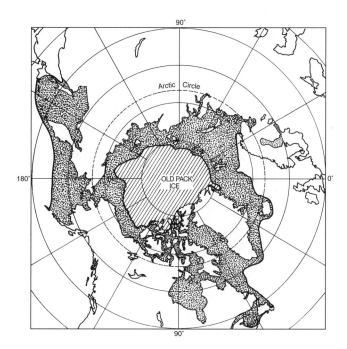

Figure 1.1. Sketch map of a portion of the northern hemisphere showing the approximate maximum extent of sea ice during late winter and early spring. The most southerly location where an extensive sea ice cover forms during the winter is located on the coast of China in the Gulf of Bo Hai (Yellow Sea) at ~40°N and is slightly off the edge of this map. The occurrence of multiyear (MY) ice is the result of the fact that presently some ice survives the summer melt period and continues into the following winter. The amount and location of this ice varies considerably from year to year. The sea ice boundaries shown are schematic and are generally representative of the time period 1950 to 1990. Recent changes are discussed in Chapter 18.

and Pacific Oceans the advance down the western sides of these ocean basins far exceeds the advance down the eastern sides. Also, the presence of shallower water greatly favors the occurrence of sea ice. In the summer the ice primarily retreats to locations within the Arctic Basin, plus locations between the islands of the northern part of the Canadian Arctic Archipelago and a southern extension off the east coast of Greenland referred to as the East Greenland Drift Stream.

In the southern hemisphere (Figure 1.2) at maximum extent sea ice covers an area equal to 2.6 times that of mainland Australia $(7.633 \times 10^6 \ km^2)$ and shrinks to ~39% of the Australian area at the end of the austral summer. At maximum extent the ice advances farthest from the continent in the South Atlantic and South Pacific sectors, with the least advance occurring in the Indian Ocean sector. At minimum extent sea ice occurs in a fairly thin belt around the continent. However, there are several so-called ice massifs where consistently heavy sea ice can be found even at the end of summer. Examples include the extremely heavy ice in the western portion of the Weddell Sea, the Pine Island Bay area of the Amundsen Sea, and the Victoria Land–Balleny Islands region of the western Ross Sea. Considerably more will be said later in this book (Chapter 18) about recent variations in the amount and in the distribution of sea ice and in the significance of these changes.

The presence of sea ice is also associated with other grand-scale effects. For instance the freezing of seawater is a somewhat inefficient desalination process which concentrates water in the sea ice in the form of pure ice and rejects salt back into the underlying ocean in the form of cold, dense brine. Brine rejection is especially intense in regions such as the coastal margins of Antarctica, where the intense katabatic winds that flow off the continent continuously strip the newly formed sea ice away from the coast, forcing it to the north. As a result, open water and thin ice areas called polynyas are constantly being formed and refrozen in locales where thick ice covers would be expected considering the local climate. The resulting rapid growth of thin ice causes large amounts of cold, dense brine to form. This brine sinks to the bottom of the continental shelf and ultimately flows off the shelf contributing to the cold saline bottom water of the World Ocean.

Figure 1.2. Sketch map of the Antarctic and its surrounding seas giving the approximate locations of the absolute maximum, the average maximum, the average minimum, and the absolute minimum extents of the southern hemisphere sea ice cover. The numbers identify three of the more important ice shelves located along the continental margins: (1) Ross, (2) Ronne-Filchner, and (3) Amery.

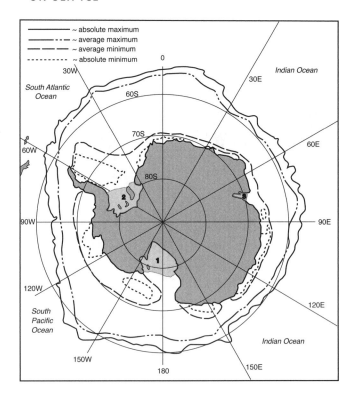

In the Arctic, the other half of the sea ice desalination process also has an interesting oceanographic role to play. Every year some ~10% of the ice in the Arctic Basin exits the basin via the East Greenland Drift Stream, the name used to describe the cold, ice-laden current that flows toward the south off the east coast of Greenland. In that much of this ice has been desalinating for two or more years, the Drift Stream can be thought of as a river of freshwater flowing into the Greenland Sea. The scale of this freshwater flux is very surprising (~2366 km^3 yr^{-1}). It is over twice the total annual flow of North America's four largest rivers (the Mississippi, St. Lawrence, Columbia, and Mackenzie). When compared to the world's largest rivers, it proves to be exceeded only by that of the Amazon (Aagaard and Carmack 1989). The fresh, surface water layer resulting from the melting of this ice is transported with little dispersion at least as far south as the Denmark Strait and in all probability can be followed completely around the subpolar gyre of the North Atlantic. Even more interesting is the speculation that in the past this freshwater flux has been sufficient to alter or even stop the convective regimes of the Greenland, Iceland and Norwegian Seas and perhaps also of the Labrador Sea. This is a sea ice–driven, small-scale analog of the so-called halocline catastrophe that has been proposed for past deglaciations when it has been argued that large freshwater runoff from melting glaciers severely limited convective regimes in portions of the World Ocean. The difference is that, in the present instance, the increase in the freshwater flux that is required is not dramatic because at near-freezing temperatures the salinity of the seawater is appreciably more important than the water temperature in controlling its density. It has also been proposed that this process has contributed to the low near-surface salinities and heavy winter ice conditions observed north of Iceland between 1965 and 1971, the decrease in convection described for the Labrador Sea during 1968–1971, and perhaps to the so-called "great salinity anomaly" which has freshened much of the upper North Atlantic during the last 25 years. Is this all true? Perhaps. What is certain is that these speculations are currently actively being investigated.

For people who live in the polar regions, there are more obvious reasons for being interested in sea ice. It directly affects daily life by totally changing the nature of the sea. Like most things in nature, there are both good and bad aspects to these changes. For instance, sea ice makes a good hunting platform that is used by humans as well as other top-level predators. Its presence helps prevent coastal erosion by directly protecting the coast through the formation of a fast ice belt and by limiting the fetch available to generate waves during storms. It also provides a vertically stable platform from which to take geophysical observations in that the presence of sea ice damps out all but very long-period ocean waves.

Sea ice can also be used as a platform for a wide variety of operations. For instance, the United States Antarctic Program recently has operated winter resupply flights from fast sea ice runways using extremely large, heavy C-5 aircraft. On the downside, sea ice can be very unstable in a horizontal sense in that pack ice commonly drifts 1 to 3 km/day and can drift as much as 30 to 40 km/day during storms. During much of the year the presence of sea ice closes desirable shipping routes such as the Northwest Passage through the Canadian Arctic Archipelago and the Northeast Passage off the northern coast of Russia. Even in the summer, expensive icebreaker escorts are required along these routes although recently there have been summers when these routes were ice free for short periods of time. If offshore structures are to be built at locations on the continental shelves of the Arctic, the maximum forces, which they must be designed to withstand, are caused by ice moving against the structures. Furthermore, the seafloor of the polar continental shelves at water depths of less than ~60 m is regularly plowed by grounded sea ice masses that are pushed along by the surrounding moving pack. Gouges of up to 6–8 m are known, a fact that adds considerable uncertainty and expense to any engineering scheme that utilizes buried cables, pipelines, or other types of seafloor structures in the offshore Arctic. The presence of sea ice also changes the acoustic environment of the Arctic Ocean, a fact that has received considerable attention from the British, Russian, and United States navies, who have operated submarines in the Arctic since the cruise of the USS *Nautilus* across the Arctic Basin in 1957.

The properties and behavior of sea ice encompass a natural geophysical system that is fascinating in its own right. Sea ice is the one igneous rock that crystallizes from its melt at temperatures that humans can almost tolerate. Furthermore, the crystallization process commonly occurs at the surface where it can be observed. The ice is then overlain by snow, a sedimentary rock that is deposited in complex patterns. Added to this is the fact that both sea ice and snow undergo metamorphic transformations at rates that can be measured in terms of days. The whole system then races around undergoing high-speed plate tectonics and mountain building during deformation events whose time span can be measured in terms of hours. Yet none of this is ever mentioned in texts on geology. As my academic training was in geology, I have always found this to be somewhat amazing!

The study of sea ice is also able to profit from the extensive studies that have been made on crystal growth and solidification theory as applied to metals and ceramics. In fact, when the structure of sea ice is examined, it is found to bear a striking resemblance to structures formed during the solidification of impure melts of hexagonal metals such as zinc. Fortunately, sea ice studies also have something to contribute back to metals and ceramics, as ice is transparent, allowing one to optically examine its internal characteristics at temperatures within a few degrees of the initial freezing temperature.

Finally, in using this book the reader should keep the following in mind. The scientific study of sea ice has essentially all been carried out during the last 100 years, with most of the effort concentrated on the period after 1950. During most of this period and perhaps for a period of several thousand years, the polar sea ice covers appear to have remained remarkably stable in extent and presumably in thickness. However, this does not mean that every year was the same; clearly there were changes in extent from year to year and presumably there were associated changes in thickness. What this does mean is that there were no obvious strong trends indicating systematic changes in ice conditions. Much of the general description of ice types and ice conditions distributed throughout this book are based on observations taken during this apparently static period of historically "normal" ice conditions. Today it will be a rare reader who does not realize that since the late 1980s *static* would be a very poor term to use in describing the current state of sea ice. Currently the world's sea ice covers are undergoing rapid changes in extent and thickness, changes that are likely to have far-reaching repercussions. These recent trends are described in some detail in Chapter 18.

I hope that the above discussion will encourage the reader to delve further into this book as well as provide a general idea of a few of the topics that will be discussed.

2 Historical Background

A science that... forgets its founders is lost.
Alfred North Whitehead

2.1 Introduction

The early literature on sea ice is both sparse and scattered in that there were no specialized journals at that time in which such material typically appeared. Fortunately there are a few books that are to be particularly recommended as guides. Here I refer to the books by Nansen and Zukriegel. In 1911, Fridtjof Nansen published *Nebelheim*. This is an exhaustive examination of the early literature on polar exploration written by a master of the subject who was one of the great heroes of both polar exploration and science. As might be expected, it contains extensive information concerning early references to and studies of sea ice. *Nebelheim* or its English translation, *Northern Mists: Arctic Exploration in Early Times,* can usually be found in any large reference library. The second book, which is more of a literary curiosity than Nansen's work, was published in 1935 in Prague by the Geographical Institute of Charles IV University. Its author, Josef Zukriegel, as well as the Institute director, Professor Svambera, were both interested in the polar regions and apparently assembled a unique scientific literature on this subject at their Institute. More to the point, Zukriegel was specifically fascinated by the subject of sea ice and his book, *Cryologia Maris,* focuses on this. Even more fortunate for the present author, this book is in neither Greek, nor Latin, nor Czech, but English. It is a veritable gold mine of exotic references and florid phrases and, although it may be difficult to locate, is to be highly recommended. An impression of Zukriegel's writing style can be gained from his description of the study of sea ice where he noted, "The ice has to be studied far away, in the north or south, in white deserts of ice, where there is nothing, nothing else, no shelter no help; and where, on those vast plains in which the chasms of the sea keep tearing open and unsurmountable obstacles in the shape of mountains of ice keep piling up, the Lord alone is with man." Wow!!

2.2 Some Ancient History

In the northern hemisphere, people have been living in the subarctic and even the Arctic for very long periods of time. However this fact appears to have been lost on the learned world of early antiquity that was centered on the Mediterranean. As Nansen (1911) points out, amber beads obtained from Greek graves from the 14th to 12th century B.C. are believed to indicate that even then trading routes extended to locations across the Black Sea and up the river systems of Russia.

Written records of the world beyond and to the north of the Gates of Hercules (i.e., the Strait of Gibraltar) are fanciful to the extreme, with descriptions ranging from an endless, hostile ocean to a region so cold that survival was impossible. A sense of this view of the world's northern regions can be gained from the following quote: "Outside, and beyond the region inhabited by man there is a realm of darkness and fog, where the sea, land and heaven melt into one curdled liquid mass ending in a bottomless abyss and empty space of horror." The other view was that the Arctic, or at least the region to the north of the roaring Boreas (the cold north wind), possessed a splendid climate and was populated by the Hyperboreans. These were people who lived in the woods and groves, were untouched by sickness and old age, and appeared to lead an outstanding existence occupied with song, dancing maidens, and sacrificial feasts. Not only did they live to immense ages, they could even fly.

Credit for inserting some small sense of reality into these realms of fancy goes to the traveler Pythias of Massilia who, in about 350 to 320 B.C., passed through the Gates of Hercules and headed north ultimately reaching Thule. Nansen devotes several pages to discussing the varying views concerning exactly what Pythias did and did not do. One of the things that he did was to either encounter sea ice or, at the least, hear it described by other individuals as a viscid (lunglike) material (not a bad description for grease ice in a marginal sea ice zone). Now, exactly where Thule was located is a matter of some debate. Pythias described it as six days sailing to the north of Britannia. Some authors have favored Iceland. Another view, and one strongly favored by Nansen, is that Thule was actually somewhere on the coast of Norway where frazil ice could be encountered in the fjords. Another possibility noted by Zukriegel (1935) is the Skagerrak, where ice such as described as Pythias can occur during cold winters. One thing is certain: if Pythias recorded a description of his travels, it has been lost. The description of his journey comes down to us through other writers such as Strabo, a geographer from Asia Minor (63 B.C. to A.D. 25), and Pliny the Elder (23 to 79 A.D.), who referred to the frozen sea as *Mare Concretum*, an apt description in any language. As much of this information appears to be either second- or even thirdhand, it is not too surprising that there was some skepticism as to its veracity. Strabo also refers to a solid or motionless sea, a *Mare Immotum*. Other descriptions date from Ptolemy who, in approximately A.D. 150, referred to the sea of the Hyperboreans as congealed or dead, *Mare Kronios* or *Mare Nekros*.

As exploration gradually expanded over time, brief descriptions of sea ice occasionally appeared as fragments in the travel literature. For instance, in about A.D. 825 Irish monks described encountering *Mare Concretum* during voyages to Iceland that occurred in about A.D. 795. Reports also exist of wolves crossing the ice to islands in the Baltic Sea. In about 1070, Adam of Bremmen described both Iceland and Greenland as well as sea ice and became the first to mention the mythical Vineland, a term which is believed to refer to the east coast of Canada. In about 1250 the book *The King's Mirror* was written, presumably by the priest Ivarr Bodde. This was a notable event, both because of the general high quality of the book, and more importantly because the book contained detailed, reliable, nonfantastic descriptions of both sea ice and the Greenland Ice Cap. References to additional brief descriptions of sea ice during this period can be found in Zukriegel (1935). One thing is clear: although these mariners saw the ice and noted it in their logs, their goal was not to study it, but to avoid it. In a recent compilation of ice edge positions in

Figure 2.1. A portion of Olaus Magnus Gothus' *Carta Marina* map, initially produced in 1539. Note the varied over-ice activities.

the Nordic seas, although the earliest observations date from 1553 it was not until the 1850s that fairly consistent observations became available (Divine and Dick 2006).

This situation continued essentially unchanged through the 1700s, although the accumulated knowledge concerning the occurrence and behavior of sea ice as well as of the geography of the Arctic gradually increased. A testimony to the knowledge of sea ice of the peoples who actually lived in regions where such ice occurred can be seen in Figure 2.1, which is a section of the map *Carta Marina*, of the region around the Baltic, prepared by Olaus Magnus Gothus in 1539. A variety of activities are seen to be taking place on the ice, including crossings made via horse-drawn sledges between Finland and Sweden near the southern end of the Bay of Bothnia and people poling on skis in the Gulf of Finland. Individuals tempted to repeat these trips during the present day would be wise to remember that 1539 was near to the peak of the Little Ice Age. It has recently been noted that this same map gives a quite accurate representation of the currents to the south of Iceland. Expedition reports that contained general descriptions of the ice conditions encountered during arctic explorations include those of Martin Frobisher (1576–1578), who discovered Baffin Island; John Davis (1585), who rediscovered Greenland and whose earlier occupation by Norse colonists appears to have been forgotten; and Willem Barents (1596–1597), who sailed in the eastern Arctic, discovering Bear Island and Svalbard. The fact that Mercator's map of 1569 contained seven islands presumably representing Svalbard suggests that these islands were actually discovered at some earlier time by Russian navigators (Holland 1994).

Because of the general expansion of ocean trade routes during the later part of the 1700s, there was increased interest in any route that offered faster passage between Europe and the Orient. Foremost on the list of possibilities were two fabled but unexplored candidates, the Northeast Passage across the north coast of Siberia, and the Northwest Passage, presumed to be located somewhere either through or to the north of Canada. The economic advantages that could ensue for the country that discovered and controlled such routes could be enormous. The visions of such a prize beckoned. Martin Frobisher's sense of the importance of this prize is clearly summarized in his statement that "It is still the only thing left undone, whereby a notable mind might be made famous and remarkable."

It should also be added here that during the late 1700s the geographic pictures of the Northeast and Northwest Passages were in very different stages of development (Starokadomskiy 1976). During the period between roughly 1550 and 1616, Russian traders, starting at the White Sea and portaging the Yamal Peninsula, had reached and proceeded up the Ob and Taz Rivers as part of the expanding fur trade. The termination of this sea trade in 1616 was for political rather than for geographic reasons: overland and river routes were easier to control and tax. Then, during the period 1633 to 1680, traders reached the Lena Delta by going north downriver, where they proceeded to explore over 1400 km of the Siberian coast between the Olenek and Kolyma Rivers. Also in 1648, Semen Deznev sailed east from the mouth of the Kolyma rounding the cape (Mys Dezneva) on the western side of the Bering Strait that was much later (1878) named in his honor by Nordenskiöld. Deznev then proceeded into the Bering Sea, ultimately coming ashore in the vicinity of the mouth of the Anadyr River. This was definitely not a pleasure trip. Of the 90 men that set out on this expedition, only 16 survived. In the same time frame, at least one attempt appears to have been made to sail around the Taymyr Peninsula, documentation of which is limited. This is not surprising in that clandestine passage of this portion of the Russian coast could prove to be very profitable to persons in the fur trade and had been forbidden by the Tsar. Whether or not some attempts were successful is not known as what little is known comes from shipwrecks that obviously did not complete the passage.

The credit for clarifying the missing portions of the Northeast Passage goes to the Great Northern Expedition that was initiated in 1733 under the command of Vitus Bering. This was a major production by any standard, being comprised of slightly less than 1000 men, not including the members of the science support team. It operated via seven separate detachments, each of which explored different segments of the Northern Sea route, including portions of the Alaskan coast and the route south to Japan. The remaining "impenetrable" portion of the route around the Taymyr Peninsula was reconnoitered by Chelyuskin using a dog team as part of this expedition. At the conclusion of the expedition (1743), the fact that a Northeast Passage existed was known, coupled with the belief that its use as a navigable sea route was not then practical because of year-round sea ice massifs at several choke points.

Another notable but much smaller British expedition initiated in 1773 was under the command of Constantine John Phipps (Phipps 1774). Its task was to explore the possibility of reaching India via a northern route, presumably over the Pole. Although based on current knowledge this was a totally impossible task, it must be remembered that at the time it was a commonly held belief that if a ship could penetrate the ring of ice surrounding the Pole, the polar ocean itself would be found to be ice free. The two ships sailed up the eastern coast of Svalbard, mapped the northern coasts of these islands, and ultimately reached 80° 48'N before becoming beset in the ice for a period and turning back at the onset of winter. While the ships were beset, a map was prepared describing the local ice conditions, presumably a first. A member of the crew was the 14-year-old midshipman Horatio Nelson, who showed early signs of later heroics and more than a little foolhardiness during a one-on-one encounter with a large polar bear. Only a warning shot from the ship's gun frightening off the bear saved Horatio from the possibility of becoming lunch. The Captain was not amused (Holland 1999).

2.3 The 19th Century

Foremost among the several nations that understood the advantages that could follow from the discovery of a new sea route between Europe and the Orient were the English, who at this time also possessed the world's largest navy. Furthermore, the 1820s were a period when England was at peace (Napoleon had been shipped off to exile in 1815). This meant that there was an overabundance of naval officers and that commands and promotions were nearly impossible to obtain. What to do? Why, discover the Northwest Passage, of course, while in the process filling in the vast region of *terra incognita* existing on the maps of the North American Arctic. The first problem might appear to be the difficulty of convincing the Admiralty that this was the right thing to do. In fact, one might reasonably suspect that the Admiralty bureaucracy would nip such potentially heroic ventures in the bud before they became expensive or even embarrassing if failure ensued. Although this might generally be true, in this particular case the cause of arctic exploration and the search for the Passage had powerful friends upstairs, in particular John Barrow, who was the Second Secretary of the Admiralty, a position far more powerful than the title might imply. Barrow was a believer in arctic exploration (Barrow 1846) and supported an active English role therein. All this was good for the cause of polar exploration. However, Barrow had some offsetting traits. First, he was a firm believer in the legends of an ice-free polar sea and had little use for individuals who suggested that this might not be true. He apparently did not even believe that there was a Baffin Bay, although it had been discovered by Bylot in 1616. Second, he exerted undue control over who would be in charge of British naval expeditions. In addition to his interest in finding the Northwest Passage, Barrow's other enthusiasm was focused on discovering the origin of Africa's Niger River. Although succumbing to frostbite, scurvy, and lead poisoning in the north was hardly a pleasant fate, it was definitely preferable to the even less healthy options awaiting Barrow's boys in equatorial Africa (Fleming 1998).

What was needed here was clearly a touch of reality: a pragmatic, realistic captain who knew both the Arctic and in particular the ways of the sea ice. Amazingly, in England there was such a man and his name was William Scoresby (Figure 2.2). More specifically he was William Scoresby the younger, in that his father had been a distinguished whaling captain. Scoresby's credentials were exemplary (Stamp and Stamp 1976). Although still relatively young, he was an experienced captain with an excellent record. Furthermore, he understood the ice and the general environmental conditions of the Arctic. In fact, at this same time he was in the process of completing a paper, "On the Greenland or Polar Ice" (1818, republished 1980) and later a book, *An Account of the Arctic Regions* (1820). He already knew what it would take the bureaucracy of the British Navy decades to learn. In my view, his understanding of sea ice and the arctic environment qualifies him for the title of the first sea ice scientist. Unfortunately, Scoresby's views were not welcomed, as he was non-Navy. Even worse, his experience had been gained as a commercial whaler. Very nonaristocratic! Not surprisingly, Scoresby's efforts to obtain a command were rejected, as was the possibility of his even serving as a pilot. Scoresby's views—that there was no open ocean beyond the ice barrier and that even if ice conditions proved to be light one year, there was no guarantee that conditions would generally be favorable—were prophetic and are still considered to be true today. Unfortunately, he did not prove to be a prophet in his own time.

This set the stage for a series of explorations: by John Ross (1818, 1829–1833) as well as William Edward Parry (1819–1820, 1821–1822, 1824–1825, and 1827), not to mention four overland expeditions which in 1845 led to the Franklin Expedition, on which 134 men and two ships vanished into the vast unknown of the Canadian Arctic Archipelago. The disappearance of Franklin spawned a veritable flurry of expeditions, starting in 1848 with James Ross. This culminated in the actual discovery of the final part of the western portion of the Northwest Passage as well as the transit of the Passage by Robert McClure (1850–1854), partially on foot after abandoning his ship on the north coast of Banks Island; in the discovery by Dr. John Rae of relicts from the Franklin Expedition in the hands of Eskimos (1854); and in the verification of Rae's discovery by M'Clintock (1857–1859).

For individuals unacquainted with the lore associated with the attempts to discover and transit the Northwest Passage, two recent books are to be recommended. *The Arctic Grail,* written by the Canadian historian Pierre Berton (1988), is the story of the exploration of the western Arctic. It describes the sagas of the search for the Northwest Passage as well as for the missing Franklin Expedition. It also includes the attempts of Nansen, Cook and Peary to attain the North Pole. Although this book is not specifically about sea ice, it is filled with references to this material and to the problems associated with its occurrence. Berton's descriptions of the steadfast refusal by most Royal Navy commanders to learn survival skills from the native population in the Arctic, and of their tendency to ascribe the successful attempts of the few explorers such as Parry who profited from this cultural exchange as resulting from the demeaning phrase "going native," can best be described as withering. The second book, *Across the Top of the World: The Quest for the Northwest Passage* (Delgado 1999), also provides a useful survey of the large literature on the subject starting at the end of the 15th century and concluding with both the transit of the passage by the tanker *Manhattan* in 1969 and with modern tourist transits. The book also contains excellent maps and photographs as well as a number of reproductions of historic paintings and line drawings, many of which show the early explorers or their ships in dire straits as the result of the presence of ice.

The importance of these facts here is that as both the Northwest and the Northeast Passages gradually became known and as probes were initiated toward the North Pole, the general picture of the distribution of sea ice improved. For instance, Ferdinand von Wrangell (1840) published quite detailed descriptions of ice conditions along the Northeast Passage to the east of the Kolyma Delta based on expeditionary work carried out during 1821–1824. Incidentally, although he completed several hazardous trips over the sea ice, he never actually sighted the island that bears his name, marking it on his charts based on descriptions by native people that he encountered. (It might also be noted that Wrangell was the second governor of Russian Alaska.) Documented sightings of Ostrov Vrangelya did not occur until over 40 years later by the German explorer Dallman in 1866 and the American whaler Thomas Long in 1867. It is also believed by some that Henry Kellett, an Englishman, saw the island in 1849. The collection of ice information during such trips ultimately culminated in the publication of good maps of the general sea ice distribution in the eastern Arctic by

the Russians in 1884, and in both the western Arctic and Antarctic by the British Admiralty in 1866 and 1875. Incidentally, the first complete transit of the Northeast Passage was completed in 1879 by Adolf Nordenskiöld on the ship *Vega*. The journey/voyage took two years, a fact supporting those who doubted that the route would ever become an operational passage.

The next event of particular note in the context of the present book occurred in 1871, when Payer and Weyprecht (Figure 2.3) led an expedition to Svalbard and then in 1872–1874 organized the Austro-Hungarian Arctic Expedition (Weyprecht 1875). After their

Figure 2.3. Karl Weyprecht (1838–1881). Image courtesy Alfred Wegener Institute. Source: Bildarchiv der Osterreichischen National Bibliothek, Wien.

ship, the *Tegethoff*, was beset in the ice, they discovered and wintered off Franz-Josef Land. Ultimately, they were forced to abandon both their ship as well as their scientific results and to proceed, first over the sea ice and then by small boat, to the west coast of Novaya Zemlya. Although this expedition had more than its share of disaster, it led to two very important results. First among these was the publication of the book entitled *Die Metamorphosen des Polareises* by Weyprecht (1879). This book focused on the ice itself and dispatched the prevalent idea of the existence of perpetual or paleocrystic sea ice with the statement, "There is absolutely no perpetual ice, if we have in mind one and the same piece of ice. Such ice cannot exist, just as there are no perpetual trees, and as there are no people thousands of years old." The book devoted a major section to describing the natural history of the different forms of pressure ridges and hummocks, providing one of the rare factual descriptions of these often romanticized features. In fact, with the exception of papers by Russian workers such as Burke (1940), who was an icebreaker captain, good descriptions of pressure ridges only became common after the 1960s. Even more important, Weyprecht's experiences in the Arctic convinced him that a better understanding of the ocean, weather, and climate could only become possible through international cooperation. His advocacy of this cause ultimately led to the organization of the first International Polar Year in 1882–1883 (unfortunately Weyprecht did not live to see this occur). This was followed in 1932–1933 by the less successful second International Polar Year, which probably owed its outcome to the generally depressed economic conditions at that time. Nevertheless, these events were milestones in international scientific cooperation (Bretterbauer 1983), and served as the model for the highly successful International Geophysical Year (IGY) during 1957–1958.

It should also be mentioned that by the 1870s papers had begun to appear in the scientific literature that discussed the properties of sea ice as well as the yearly variations in ice conditions (Chavanne 1875; Pettersson 1883; Tomlinson 1871). In addition, papers describing experiments on both the freezing and melting of seawater as well as the properties of the resulting ice were published (Buchanan 1874, 1887, 1911). Particularly important here are two papers by famous scientists whose names are not primarily associated with sea ice; specifically, Hertz and Stefan. In 1884, Heinrich Hertz (the "Hz," cycles per second, Hertz), while still a student, pondered whether he could safely cross over a thin river ice cover to visit some friends. To resolve this problem he solved the bearing capacity problem by considering the ice to behave elastically and to rest on an elastic foundation whose resistance is proportional to the deflection by the ice (a so-called Winkler foundation). Although I do not know whether he tested his theory by crossing the river, his solution still stands today and with some modification can be applied to short-term loading problems concerning both lake and sea ice.

For his part, Stefan (1889, 1891) examined the classic freezing problem that is now known under his name, using polar ice as a model. It has generally been assumed that his interest in this problem was sparked by Weyprecht's field observations that the ice thickness was a function of the accumulated number of freezing degree-days where this number is obtained by adding the average air temperatures for the days in question as corrected relative to the freezing temperature of seawater. For instance, the accumulated freezing degree-days for a three-day period in which the average air temperatures were –10, –15, and –12°C, respectively, would be

31.6 if the freezing point of the seawater was constant at −1.8°C (i.e., [10 − 1.8] + [15 − 1.8] + [12 − 1.8] = [10 + 15 + 12] − [3 × 1.8] = 31.6). However, as pointed out by Wettlaufer (2001), Stefan's 1889 paper contains no references to field observations and his 1891 paper, although referencing field observations from a number of other sources and locales (Alaska, Canada, Greenland), contains no reference to Weyprecht's field observations made on the Austro-Hungarian Arctic Expedition of 1872–1874. It is difficult to believe that Stefan was unaware of this data set in that the data are archived at the Austrian Academy of Science of which Stefan was a member. Why these data were not included remains a mystery, although Wettlaufer offers several possible explanations. Perhaps Stefan actually was unaware of Weyprecht's data, or the data were in a form unsuitable for analysis. Another possibility is that the data were ignored because the fit between the theory and the data was poor. It is not difficult to imagine that this was a possibility if there was a large oceanic heat flux in the portion of the Barents Sea where the drift of the *Tegethoff* took place. Wettlaufer suggests that these questions could possibly be resolved via a reanalysis of Weyprecht's archived data. To date such a study has not occurred. It is also interesting to note that according to Carslaw and Jaeger (1959), an even more general solution to Stefan's problem was actually worked out by Neumann in the 1860s but was never published until it appeared in a text by Rieman-Weber in 1912 (Carslaw and Jaeger 1946). Doronin and Kheisin (1977) also mention that results similar to those obtained by Stefan were obtained by members of the Russian Academy in 1831. Be that as it may, it is Stefan's name that is usually associated with one-dimensional freezing problems of the sea and lake ice type.

The 1800s closed with a singular event in the history of polar science: the voyage of the ship *Fram*, whose name appropriately means 'forward' in Norwegian. This ice-strengthened ship with a rounded hull was built by Colin Archer for Nansen (1897) (Figure 2.4) to test his theory that there was a general circulation of the pack ice of the Arctic Basin and that if the *Fram* was inserted into the ice at the right place, it would gradually drift across the Basin and ultimately exit the ice somewhere to the south of Fram Strait (now named for the ship) via the East Greenland Drift Stream. This idea

was based on Nansen's observation that identifiable relicts from the *Jeannette*, which had sunk in 1881 to the north of the New Siberian Islands, were discovered in southwest Greenland in 1884 after drifting across the Arctic Ocean. The four-year cruise of the *Fram*, and the associated attempt by Nansen and Johansen to reach the Pole by dogsled when it became apparent that the drift of the ship would not cross the Pole, is one of the great polar adventure stories, only equaled by Shackleton's *Endurance* cruise in the Weddell Sea, Antarctica. The scientific data collected during the drift of the *Fram* (Nansen

Figure 2.4. Fridtjof Nansen (1861–1930). Photograph courtesy of the *Polar Record*.

1900–1906) vindicated Nansen's view of the general circulation of the ice, proved that the Arctic Ocean both was deep and contained several subbasins, served as the basis for the simple ice drift model that still proves useful today (Thorndike and Colony 1982; Zubov 1945), and demonstrated the importance of the Coriolis force in treating the wind drift of the ice (Nansen 1900). Nansen's data also stimulated the development of the Ekman spiral analysis that is so important in boundary layer theory as applied in both oceanography and meteorology (Ekman 1905). With the cruise of the *Fram*, sea ice ceased to be an odd aside noted in books of polar exploration as a hazard to be overcome and became part of the world of modern geophysics.

2.4 The 20th Century

Sea ice research in the 20th century can be divided into two periods separated by the Second World War, with the prewar period being described as somewhat limited and leisurely and the postwar period as frequent and frenetic. The reasons for these differences were political and logistical as well as scientific.

2.4.1 1900 to 1945

A report published in 1901 by Makarov indicated that not only had sea ice entered the world of geophysics, it was also entering the world of engineering. As was mentioned earlier, at the end of the Great Northern Expedition in the 1740s it was generally concluded that the Northeast Passage did not offer a viable sea route. The problem was, of course, that portions of the route were clogged with ice even during the summer. During the time of sail, there was little that could be done about this difficulty. However, as ship construction changed from wood to metal and propulsion from sail to steam, a solution to the problem of transiting the difficult portions of the route appeared in the form of the icebreaker. Considering that the desirability of an operational sea route along the northern coast of Russia had, if anything, become more acute, it is not surprising that the construction of the first polar icebreaker, the *Yermak*, was completed in 1898. An individual extremely influential in this project was Admiral S. O. Makarov, who intended to use the ship for both research and route development. In the summer of 1901 Makarov carried out sea ice studies for the purpose of improving icebreaker design using the *Yermak*. This appears to be the first sea ice program that was both specifically engineering oriented and not tied to a trip of exploration. The outbreak of the Russo-Japanese War in 1904 provided additional motivation for the further development of the Northern Sea Route in that this route could provide relief to the overloaded single-track Trans-Siberian Railway in moving men and equipment to the war zone. After the Russian defeat at the naval battle of Tsushima, in which Makarov was killed, it was argued that a fully operational Northern Sea Route would have allowed the Baltic Squadron to arrive in the Far East in better condition, hopefully catching the Japanese off guard resulting in a Russian victory in both the battle and the war. Although, as pointed out by Barr (1991), this conclusion is more than a bit suspect, it appears to have resulted in the Arctic Ocean Hydrographic Expedition of 1910–1915, which was extremely successful in improving the state of knowledge of the hydrography of the overall route. It is also reasonable to assume that the decision to form the Arctic Research Institute (later to become the Arctic and Antarctic Research Institute [AARI] in

1920 in Petrograd (now St. Petersburg) was to some degree affected by such strategic considerations. AARI is currently the largest organization in the world specializing in polar ocean and sea ice–related studies and has a long and proud history.

Skipping briefly to the Southern Hemisphere, several studies occurring during this general time period are particularly worthy of mention. For example, excellent descriptions of antarctic sea ice conditions can be found in papers by Arctowski (1903, 1908, 1909) based on observations made during the *Belgica* Expedition of 1897–1899 and by von Drygalski (1921) based on his work on the German South Pole Expedition of 1901–1904. Also, in 1921 Wordie published his observations on the natural history of pack ice as observed in the Weddell Sea on the Shackleton Expedition of 1914–1917. Here he documented many aspects of antarctic pack ice, such as its frequently flooded upper surface, that have only recently been "rediscovered." Next, in 1922 Wright and Priestley, who were members of Scott's last expedition (the British Terra Nova Expedition of 1910–1913), published their results under the title *Glaciology*. This book is a classic that, although not primarily about sea ice, contains a number of interesting observations relative to this material. Specifically, it provides a good description of the geometric selection process occurring during crystal growth, the first observations on platelet ice, including the fact that in the McMurdo Sound region this ice type does not appear to form until some appreciable time (several weeks to months) after the initial fast ice sheet has developed. They also noted that there appeared to be consistent differences in the nature of pressure ridge formation in the Antarctic as contrasted with the Arctic. Finally in 1923 Dobrowolski, who had also been a member of the *Belgica* Antarctic Expedition, published *Historja naturalna lodu* (*The Natural History of Ice*), a 940-page tome that dealt with all aspects of ice, including sea ice. As was the case with Weyprecht, Dobrowolski's polar experience had convinced him of the importance of cooperative international studies in this portion of the world (Dobrowolski 1933).

Returning to the north, in 1927 Finn Malmgren published his doctoral thesis, which was based on his sea ice growth and property observations carried out during the drift of the research vessel *Maud* between 1922 and 1925, observations that were completed while the *Maud* was drifting across the Arctic Basin. The quality and focus of this work causes me to consider

Malmgren (Figure 2.5) to be the first true student of the geophysics of sea ice. As a scientific experiment, the cruise of the *Maud* was extremely successful, in no small part due to the studies of another member of the science party, H. U. Sverdrup. Sverdrup's observations, published in 1928, in a sense completed the "ring" initiated by Nansen by verifying Ekman's theory of the spiral in the boundary layers. Although Malmgren met a tragic death on the Nobile Expedition in 1928 while attempting to reach Svalbard after the crash of the airship *Italia* (Cross 2000), Sverdrup had a long

Figure 2.5. Finn Malmgren (1895–1928). Source: Behounek 1928.

Figure 2.6. Photograph of the first North Pole Station (*Severny Polius* 1), or NP-1. Source: AARI files.

and outstanding career in oceanography. For many years the only physical oceanography text containing anything approximating an adequate treatment of sea ice was *The Oceans* by Sverdrup, Johnson, and Fleming (1942), in that it incorporated many of Malmgren's results. (For a fascinating movie about the crash of the *Italia* and the death of Malmgren, try to locate *The Red Tent* at a video store. The movie makers—an Italian–Russian combine—have, as usual, gussied up the facts a bit, including adding opening and closing scenes in which the ghosts of individuals who perished in the expedition come back to visit Nobile, who amazingly was still alive at the time the movie was made. Love interest was also added by having Malmgren, while still on Svalbard, roll around in the snow with Claudia Cardinale. Cool! Be this as it may, the basic story presented by the movie is true and the sea ice photography and the views of Svalbard are spectacular. Claudia isn't hard to look at either.)

In 1937, the first of the Russian North Pole stations, NP-1 (Figure 2.6), was established under the leadership of Ivan Papanin. This successful expedition proved that research stations could be established on the drifting pack ice and operated over significant periods of time (months to even years). Its success was a testament both to the Russian understanding of sea ice conditions in the Polar Basin and to their confidence in their operational capabilities. A very useful summary of the timing and start and termination locations of the 35 NP stations to date (as of July 2008) can be found in Wikipedia under the heading *Drift_ice_station*. Although major Arctic Basin activities such as the NP program were terminated as a result of World War II, I have been told by Russian scientists that even during the war limited sea ice studies were continued by the USSR. One of these efforts is of particular importance. In 1943, N. N. Zubov completed his book *Arctic Ice* (1945). Zubov (Figure 2.7) was a physical oceanographer who had considerable experience with sea ice. His scientific skills were of the highest level and he was a truly inventive individual. His book served to close an era by providing an innovative synthesis of widely scattered work from both the Russian and the Western literature. Although an English translation exists that was published in the "gray" literature by the U.S. Navy Electronics Laboratory, it is not easy to locate and can probably only be found in libraries specializing in polar subjects. Every serious student of sea ice should examine a copy, as there are insights on every page. Table 2.1 presents a listing of some of the Russian scientists who were carrying out sea ice studies in the period between the First and Second World Wars, as well as their general areas of interest. The dates given indicate the years of their first and last references in my sea ice library and only serve as a rough guide

Figure 2.7. N. N. Zubov (1885–1960). Photograph courtesy of the Russian State Museum of the Arctic and Antarctic.

to the period when a particular individual was actively involved in the study of sea ice. I have also included references to a few of the papers written by these men in the bibliography at the end of this book.

A few additional comments on some of these individuals follow. Arnol'd-Aliab'ev (1929b) carried out studies from the icebreakers *Krasin* and *Malygin* in the Gulf of Finland as well as in the Barents and Kara Seas completing the first direct determinations of the volume of air included in sea ice. Vize, along with Zubov, was particularly interested in long-term ice forecasting. Their approaches were quite different, with Vize favoring a more historical–climatological approach (1923, 1924, 1944) and Zubov a more theoretical one. Burke (1940) was an icebreaker captain whose monograph on sea ice contains careful observations of the ridging process and particularly of the internal structure of the ridges. Golovkov (1936) appears to have made the first thin sections of sea ice and also developed a rather elaborate genetic classification of the material. Unfortunately, the quality of the reproduction of his photomicrographs in his papers was extremely poor, which limited the impact of his observations.

V. L. Tsurikov, who appeared to be the only individual interested in sea ice at the Institute of Oceanology in Moscow, had a long and productive career. During the 1940s he developed the first geometric model for the variation in the strength of sea ice with changes in the gas and brine volumes (1940, 1947a, 1947b). The flaws with the model were caused by the fact that he did not, at the time, have a realistic picture of the internal distribution of the phases within the ice. During

Table 2.1. Russian scientists active in sea ice research during the period 1920–1940. Also given are their areas of specialization and approximate dates bracketing the time when they were active in these research areas.

Scientist	Primary area of interest	Approximate period of activity
V. I. Arnol'd-Aliab'ev	Strength, friction, gas content	1924–1939
V. Iu. Vize	Air/ice/ocean interactions	1924–1944
A. K. Burke	Pressure ridges	1932–1940
M. P. Golovkov	Ice structure	1936–1937
D. B. Karelin	Air/ice interactions	1937–1946
V. S. Nazarov	Sea ice properties	1938–1963
N. T. Chernigovskii	Radiation balance	1936–1963
M. M. Somov	Ice/ocean interactions	1939–1941
V. L. Tsurikov	Strength, growth, composition	1938–1979
N. N. Zubov	Varied	1934–1959

the later part of his career, Tsurikov (1976) published a monograph as well as several papers on the sea ice phase diagram (Tsurikov 1974; Tsurikov and Tsurikova 1972). Nazarov, who was on the staff at AARI, carried out work on both ice properties and forecasting, and also published a review of the current knowledge of antarctic sea ice conditions including a useful map showing Russian iceberg sightings in that region (Nazarov 1962). Incidentally, an excellent review of the multifaceted research program carried out during this general period in support of the development of the Northern Sea Route can be found in Armstrong (1952). Later systematic updates of Russian activity in this area were also published in the journal *Polar Record*.

Research activities in the West during this same period of time could be described as scattered and sparse. Table 2.2 lists a few of the workers active during this period. Ringer was a Dutch chemist who worked on phase relations in seawater brines. Although the work appears to have been completed by 1906, his results were also published in German in 1928. Whitman (1926) was the first person to propose a mechanism (brine pocket migration) for removing the included brine from sea ice. Although the current view of his proposed mechanism is that it is too slow to be of much importance in sea ice, the same mechanism is believed to be important in other materials. Crary is a particularly interesting individual who, with Maurice Ewing, did experimental work on wave propagation speeds in ice in the 1930s (Ewing et al. 1934); carried out fieldwork in the Arctic Ocean on the ice island T-3 and the Ellesmere Ice Shelf in the 1950s; and in 1957 became the chief scientist for the U.S. IGY Program in the Antarctic. He also had the distinction of being the first man to stand at both the North and the South Poles. I had the pleasure of sharing an office with "Bert" during 1955–1956 while I was assigned to the Arctic Group at the Air Force Cambridge Research Center. He was a wonderfully pleasant, low-keyed individual of legendary toughness in the field. Koch was a Dane who carried out research on and completed a major book about the sea ice in what is now referred to as the East Greenland Drift Stream (Koch 1945).

Table 2.2. Western scientists active in sea ice research during the period 1920–1940. Also given are their areas of specialization and approximate dates bracketing the time when they were active in these research areas.

Scientist	Primary area of interest	Approximate period of activity
W. E. Ringer	Phase relations	1906
W. G. Whitman	Brine migration	1926
A. P. Crary	Elastic waves in ice	1934–1960
Lauge Koch	East Greenland pack ice	1945

2.4.2 1950 to the Present

After World War II there was a marked increase in sea ice research. This increase started slowly and gradually increased until the investigation of sea ice became a major research specialty involving a large community of scientists. As the present book is specifically focused on this period, I will not refer to the contributions of specific scientists here as their names will be mentioned in conjunction with their contributions later. However, I will comment briefly on the political considerations that were influential in setting the funding priorities that made possible an expanded research effort on sea ice. Understanding these considerations helps to clarify why particular major projects occurred at certain times. At least in the West, the nature of the factors affecting the funding for

Table 2.3. Factors affecting sea ice funding since World War II (at least in the West). The use of *present* implies an activity continuing as of 2008.

The Cold War (1945–1992)
    The DEW and Pine Tree Lines and the Strategic Air Command's dispersal strategy
    The Soviet North Pole (NP) Stations and the Sever Expeditions
    The missile race
    The Arctic Surface Effects Vehicle (ASEV)
    The Soviet Arctic submarine strategy
Arctic Offshore Oil and Gas (1968–present)
    The *Manhattan* cruises
    Ice forces on offshore structures
    Oil spills in ice-infested waters
    Gouging and subsea pipelines
    Outer Continental Shelf Environmental Assessment Program (OCSEAP)
Satellite-based remote sensing (1979–present)
The global greenhouse and climate change (1990–present)
The transfer of pollutants by sea ice (1990–present)
The easing of ice conditions in the Arctic (1995–present)

---

research on sea ice changed drastically after the war (Table 2.3). As a generality, it might be said that prior to 1940 most sea ice investigations were tied to the fact that sea ice was an obstacle to marine transportation. Although there were some limited concerns relative to national security, such as the speculation that a fully functional Northern Sea Route might have allowed the Russians to win the Russo-Japanese War, these considerations were always secondary to the transportation issue.

When the Cold War started after the end of World War II, this changed. The two most powerful nations in the world glared at each other over a frozen ocean. It was a very nervous time on all sides. Prior to 1940, the Arctic Ocean would have been considered a barrier separating the contestants in this continuing battle of national purpose. After the war, the effectiveness of this barrier had vastly decreased because of improved logistic capabilities that had been honed by the war and were possessed by both sides. Also, the fact that the aircraft had come of age as a logistic tool in the Arctic had been clearly demonstrated in 1934, when the Russians evacuated by air 104 passengers and crew from the ship *Chelyuskin*, which had been trapped in the ice and crushed (Smith 1971), and in 1937, when they established the research station NP-1 on the drifting pack. In addition, missiles and submarines were soon to further complicate the picture. Table 2.3 lists a number of factors that I believe were important in setting sea ice funding agendas during the period between the 1950s and the 1990s.

Sea ice studies related to the establishment of the DEW and Pine Tree Lines (Fletcher 1990) were primarily focused on improving the understanding of the bearing capacity of sea ice so that ice runways could be utilized as early as possible in the fall and as late as possible in the spring. Another related consideration here was the Strategic Air Command (SAC) requirement to have available a large number of alternative runways that could be used under emergency conditions.

The Soviets reestablished the series of North Pole Stations in 1950–1951 with NP-2. This was to be a long-continuing effort that at times included several stations operating simultaneously. It should be noted that recently a useful summary of the results of the ice studies carried out on these stations has been published by Romanov (1993). (The last drift station [NP-31] in this essentially continuous series was closed in August 1992 because of fiscal reasons associated with the demise of the Cold War. NP-32 was not established until almost 10 years later [April 2003] and was abandoned during March 2004 as the result of the breakup of the floe on which the station was sited.) Not only were the NP stations able to collect valuable data, they also served as jumping-off points for what was referred to as the Sever Expedition. By this I refer to aircraft operations wherein a large number of ice floe sites were temporarily occupied while geophysical data were collected. A published map showing large numbers of such stations located in the Chukchi and Beaufort Seas (Figure 2.8) appears to have greatly impressed planners in Washington, particularly as U.S. and Canadian information concerning these locations off their arctic coasts was essentially nonexistent. As a result, funding appeared for operations on the ice pack such as the ARLIS Stations, Ice Station Charlie, the Ice Island T-3, AIDJEX (Arctic Ice Dynamics Joint Experiment) and the Canadian Polar Continental Shelf Project. A few years later, after the U.S. Army Cold Regions Research and Engineering Laboratory (CRREL) had just completed constructing a new laboratory, we received a request from AARI for a photograph of the new facility. A very nice photo, in color too, was promptly dispatched. I always thought that this was the very least that we could do in assisting them in their effort to also obtain a new building (they were successful), as they had done so much indirectly to advance our research program.

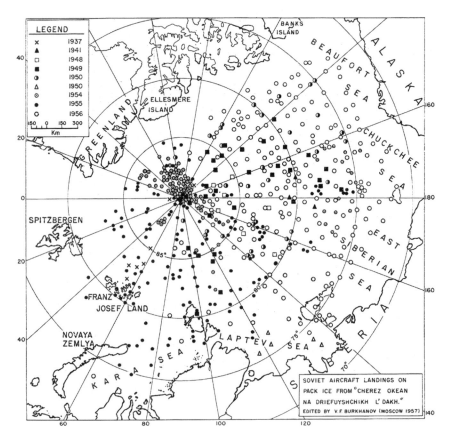

Figure 2.8. A Soviet map published in ~1951 showing Arctic Ocean locations visited by the Sever Expedition.

Although I have listed the missile race, its effects on sea ice research were largely indirect through keeping a general focus on the importance of the Arctic. A much more direct impact was exerted by the Arctic Surface Effect Vehicle (ASEV) program, which was a brainchild of an organization in the U.S. Defense Department called ARPA (Advanced Research Projects Agency). I have heard it said by some wag that if an idea was sufficiently realistic to ever be a possibility, ARPA would not fund it. The ASEV program was a case in point. (The Internet was not.) The idea was to construct air-cushioned vehicles that had lateral dimensions of roughly 200 m and would be capable of moving across the Arctic Ocean at very high speeds (200 knots). Clearly, the individual who spawned this idea had never seen the surface of the Arctic Ocean. Of course, the problem with developing such an ASEV was the presence of pressure ridges, which were hard, high, and all over the place. Anyway the ASEV project caused funding to become available to study pressure ridges, features that are clearly important in influencing the behavior of pack ice and that had largely been ignored since Weyprecht's book in the 1870s. By the time that the ASEV program was shelved as being impractical, the database on pressure ridges had increased considerably, a fact that has proved to be quite important for subsequent scientific developments.

The last item listed under the Cold War is the Soviet arctic submarine strategy. This refers to the decision by the Soviet Navy to hide missile-launching submarines under the pack ice after increases in the range of their missiles placed the continental U.S. within range of locations such as the Barents Sea. Considering that certain regions of the pack ice (particularly the marginal ice zones) are acoustically quite noisy, locating such hidden submarines is, at best, difficult. To do so, one must be able to separate the noise of the target from the environmental noise. This, in turn, requires that one thoroughly understand the relations between environmental factors and noise. Portions of the funding support of a large number of projects were related to this general purpose. Although many of these projects were unclassified and their results were published in the open literature, it was nevertheless hoped by the funding agencies, in particular the Office of Naval Research (ONR), that the increased geophysical understanding of the environment of the Arctic Basin and its surrounding seas would give NATO forces an edge in the arctic acoustic game. The same might be said for similar Russian research programs carried out during the same period. For instance, it was well known that much of the research carried out by the AARI group headed by Bogorodskii (e.g. Bogorodskii and Gavrilo 1980) was of interest to the Russian Navy.

In 1968 an extremely large oil field was discovered at Prudhoe Bay located on the shore of the Arctic Ocean on Alaska's North Slope. This immediately brought a variety of other sea ice problems to the forefront. These are also listed in Table 2.3. First was the question of getting the oil south to market. Two basic candidates were considered: tankers and pipelines. To explore the tanker option, the SS *Manhattan* was structurally modified and sent north to attempt the Northwest Passage route between the east coast of the U.S. and Canada to Prudhoe Bay. The purpose of the two *Manhattan* cruises was to obtain test data on the resistance of ice to the passage of the ship that would lead to the design of specialized vessels that would be able to make the transit even during the Arctic winter. As might be expected, such tests involved extensive characterization of sea ice properties along the route. The tests carried out on the first cruise were unsuccessful primarily because of the timing. The ship came out of the shipyard later than scheduled. This placed the ship in the ice during the late summer and early fall. This meant that in the late summer the ship encountered rotten multiyear ice whose properties were next to unspecifiable and in the fall thin first year (FY) ice whose resistance was so slight as to be nearly unmeasurable. The second cruise occurred the following spring, even though the decision had been made in the interim to select the pipeline alternative. This cruise was very successful in providing the data needed for tanker design. To the best of my knowledge, the reports resulting from this cruise are still not generally available. There is little question that although the tanker route was a viable alternative, the pipeline was much more politically attractive and contained fewer imponderables. For instance, the lengths of periods of time when ice conditions could be so severe as to prevent ship arrivals at an offshore loading facility were, and still are, poorly known. Also, at that time there was considerable uncertainty concerning how to design such offshore facilities that, because of shallow water over the continental shelf, would have to be far offshore (~40–60 km) in regions of highly mobile heavy pack ice. Of course, the problem of designing fixed offshore structures is not tied to the tanker option but is a necessity for offshore development even if the oil is pumped ashore and sent south via a pipeline. Therefore, extensive studies have been completed related to ice forces on fixed structures. Excellent reviews of recent work on this subject have been published by Sanderson (1988) and Timco et al. (2000). Interesting topics that are still poorly understood are the effect of scale on the apparent strength of sea ice; sea ice–induced gouging of the seafloor (see Chapter 13), which can rupture improperly designed subsea pipelines; and the cleanup and control of oil spills in areas of pack ice. Major resources have been devoted to these problems by oil companies. In addition, both the U.S. and Canadian governments have supported related, coordinated programs (e.g., the Outer Continental Shelf Environmental Assessment Program [OCSEAP]).

To date, the only significant production from the offshore Arctic is from sites that are near shore, in relatively shallow water and somewhat protected by offshore barrier islands. Although a number of additional offshore discoveries have been made, they have not proved to be economic until the recent increases in oil prices. As this book is being edited, consideration is being given to the construction of offshore production facilities at more exposed locations. Even if further exploration activity in the U.S. and Canadian offshore regions were to prove to be unsuccessful, research on these types of subjects will undoubtedly be continued, with the geographic area of interest shifting to the vast continental shelf of Russia. In fact, this shift is already underway.

I have listed the development of satellite-based remote sensing in Table 2.3 for several reasons. First it opened up new sources of sea ice funding from agencies such as NASA, NOAA, and ESA. However, more importantly, it gradually provided the sea ice community with increasingly high-resolution imagery of the ice and ultimately with the capability to obtain such imagery under all-weather conditions. These enhanced capabilities have gradually changed the manner by which sea ice research is carried out, as remote sensing is now commonly used as an integral part of field operations both to characterize ice features and types and to monitor drift and deformation. This allows in situ field observations to focus on other aspects of the problem under study as well as on improving and verifying the interpretation of the imagery.

The next-to-last two items in Table 2.3, the role of the polar regions and of sea ice in particular in (a) climate change and (b) the transport of pollutants, are currently "hot" topics under whose banners a wide variety of sea ice research either has or has the potential for funding. Major progress has been made relative to the climate change problem (Solomon et al. 2007). The results to date, which particularly include satellite-based remote sensing and coupled air–sea ice modeling, clearly support Weyprecht's belief that progress on a subject such as the role of sea ice on climate requires both interdisciplinary and international efforts. How important sea ice is in the transport of pollutants remains somewhat moot. Discussions of possible effects and related references can be found in papers by Weeks (1994) and Pfirman et al. (1995).

The final item, the recently documented (see Chapter 18) easing of ice conditions in the Arctic, has resulted in increased interest in a variety of open-ocean economic activities in the arctic offshore regions. Obvious examples are the expanded use of both the Northwest and the Northeast Passages as shipping routes during portions of the year and oil development at sites exposed to more difficult environmental conditions. Another factor here is the possibility of expanded territorial claims. At present it is too early to have a clear picture of how these matters will play out. Discussions of both the historical details and the current and proposed usages of the Northern Sea Route are to be found in Armstrong (1952, 1996), Nielsen (1996), and Bulatov (1997). Unfortunately these reports were written before it was generally realized how drastic the changes in the ice conditions along the route might be.

Listings of individual investigators associated with particular topics and with major research programs occurring since 1950 can be found in the paper by Weeks (1998a), from which some of the material presented in this section was taken. In that research during this time period will be discussed in detail in the remainder of this book, this material is not included here.

3 The Ocean Setting

The pack ice is a country without a name.
It is the emptiest country in the world.
The most dangereous.
The least explored.
Bond and Siegfried

Before starting to discuss the details of sea ice formation, I believe that it is helpful for the reader to have a general description of the ocean settings in which sea ice occurs. As will be seen, the two principal sea ice regions, the Arctic Ocean and the Southern Ocean, are very different. There is a large and rapidly developing literature on both of these areas. In that the current book is primarily focused on the ice instead of on the underlying water column, only a general outline of some of the important aspects of the oceanography of these regions, as they relate to the ice, will be discussed. I will also only provide a few of the many references on these subjects. For an exhaustive listing of polar oceanography sources, the reader should refer to papers in Smith (1990) and in Jacobs and Weiss (1998) and in particular to the reference by Carmack (1990). I will not discuss the distribution of sea ice here in that this has been treated in a general way in Chapter 1 and will be discussed in detail later in this book. It is also hoped that this chapter will help individuals unacquainted with the polar regions to start to get a general feeling for their geographic settings.

3.1 Topography

A quick examination of a globe quickly reveals one fundamental difference between the two polar regions. In the Arctic the primary ocean area containing sea ice occurs to the north of ~70°N latitude. Furthermore, the Arctic Ocean is largely landlocked in that it has only one deep exit with a sill depth of ~2600 m, the 600-km-wide Fram Strait located between Svalbard (Spitzbergen) and Greenland. As a result, one occasionally sees the Arctic Ocean referred to as a Mediterranean-type sea. The two principal areas of this system are the Arctic Ocean per se to the north of Fram Strait, and the Greenland/Iceland/Norwegian Seas located more to the south with Fram Strait serving as the primary connection. A look at any map of the northern polar regions clearly shows that there are several other ways that water can move in and out of the Arctic Basin: for instance, the Bering Strait between Alaska and Siberia, the Robson Channel and Nares Strait between Ellesmere Island and northwest Greenland, the complex channel systems through the Queen Elizabeth Islands of Arctic Canada, and finally, the Barents Sea to the east of Svalbard. Although

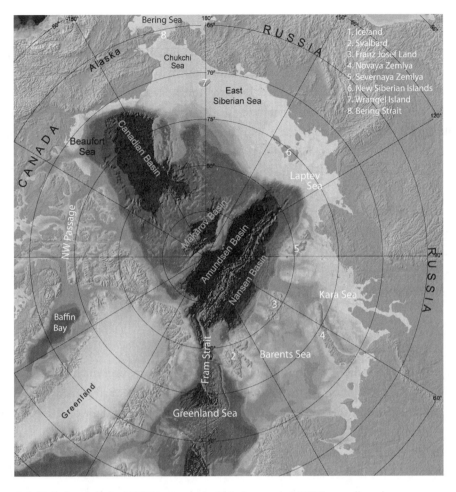

Figure 3.1. Map showing the IBCAO topography (Jakobsson et al. 2008) as well as place names applied to different locations in or near the Arctic Ocean.

these passages may appear attractive on a map their effectiveness is limited by the fact that the water in these areas is generally shallow.

The Arctic Basin (Figure 3.1) is itself divided into two primary subbasins, the Eurasian and the Canadian, with depths of 4200 and 3800 m, respectively. The Eurasian Basin is comprised of several significant subbasins: the Amundsen, the Nansen, and possibly the Makarov. In many discussions, the Lomonosov Ridge, with a sill depth of ~1400 m, located between the Makarov and Amundsen Basins, is considered to be a feature separating the Canadian from the Eurasian Basin. The apparent spreading center separating the Amundsen and Nansen Basins is referred to as the Gakkel Ridge. Clearly, the tectonics of this portion of the globe are complex. Furthermore, the tectonics are only partially understood, largely owing to the difficulty of access resulting from the presence of sea ice. The continental shelves around these deep basins are of two different types. From Alaska across the north coast of the Canadian Archipelago and Greenland, the shelf is quite narrow. On the other hand, starting from Svalbard and continuing east through the Barents, Kara, Laptev, East Siberian, and Chukchi Seas, the shelf is very wide, with typical values in the range of 600 to 800 km.

The Antarctic, on the other hand, is primarily land to the south of ~70°S, with the primary exceptions being the embayments of the Ross and Filchner Ice Shelves extending to the south and the Antarctic Peninsula extending to the north. To the north the Southern Ocean is bounded by the Polar Front, the contact between the polar and subpolar water masses. In a sense, the Southern Ocean can be thought of as the mother of all oceans in that it is truly circumpolar and directly connects to the South Atlantic, Indian, and South Pacific Oceans. This gives the Southern Ocean an area of ~38 × 10^6 km^2 and a volume of ~140 × 10^6 km^3, values that are 2.2 and 8.2 times that of their arctic equivalents. The nominal seaward edge of a significant portion (~45%) of the Antarctic Continent is comprised of ice shelves. At first glance it might seem that this fact would have only a slight effect on the Southern Ocean. However, as will be seen later, this is not the case. The continental shelves of the Southern Ocean are generally quite narrow, with two notable exceptions: the broad (~400 km) and deep (~400 m) shelves of the Ross and Weddell Seas. North of the shelves are four basins with depths exceeding 4500 m that are separated by broad submarine ridges and plateaus.

3.2 Hydrology

Not surprisingly, the hydrologic inputs to the two polar oceans are very different. In the Arctic the total annual amount of freshwater entering the Arctic Ocean from rivers is estimated to be ~3500 km^3 yr^{-1}. Contributing to this are several of the great rivers of the world, including the Yenisei, Ob, Lena, and Pechora in Russia and the Mackenzie in Canada. Another 2000 km^3 yr^{-1} of fresh water are estimated to enter the Arctic Basin as the freshwater fraction of the inflow through the Bering Strait. The flow of these arctic rivers varies considerably from river to river and is highly seasonal. Carmack estimates that the overall interannual variability is about 5 to 20% of the mean annual flow.

The situation in the Southern Ocean is different in that the freshwater input into the ocean is in the form of freshwater glacier ice deposited in the ocean by glacier and ice shelf calving and by processes such as the basal melting of ice shelves. Jacobs (1985) has reviewed estimates of the volume of ice calved from antarctic ice shelves and glaciers, estimates that vary from 500 to 2400 km^3 yr^{-1}. He recommends 2000 km^3 yr^{-1}. Clearly, this is a difficult number to estimate. Readers of the popular press will note that in the last decade there have been several articles discussing the calving of very large icebergs from locations such as the Ross Ice Shelf. In most cases the increase in reportage is believed to be more the result of the increase in surveillance capabilities offered by satellite systems as opposed to an actual pronounced increase in the calving rates (Long et al. 2002). An exception to this statement would be the breakup of the Prince Gustav Channel and the Larsen Ice Shelves located on the east side of the Antarctic Peninsula, in that this area has shown particularly rapid regional warming. An estimate of the 100-year warming rate based on station observations located along the peninsula weighted by the length of the observing period gives a value of +5.2°C per 100 years (Vaughn et al. 2001). Also, observations using satellite radar interferometry support the conclusion that bottom melting and the retreat of the grounding lines of some ice shelves are more rapid than previously estimated (Rignot and Jacobs 2002).

3.3 Currents

3.3.1 Arctic

In the Arctic there are two main currents that exchange water between the Arctic Ocean and water bodies located further to the south. Geographically both are located in Fram Strait between Svalbard and northeast Greenland, the one deep exit from the Arctic Ocean. Specifically, these are the West Spitzbergen Current (WSC), which flows off the west coast of Svalbard and carries warm, relatively salty water northward into the basin, and the East Greenland Current (EGC), located off the east coast of Greenland, that carries less saline water and ice southward out of the basin.

A number of studies using different approaches have provided values for the volume of water transported by the WSC. Estimates range between 2 and 8 Sv (1 Sv [one Sverdrup] = 10^6 m^3/s). There also appears to be temporal variability related to the observation that the current appears to be banded containing a number of eddies that follow topographic contours. Considerable recirculation also appears to occur in the Fram Strait area. Rudels (1987) has argued that only about half of the water within the WSC actually enters the Arctic Ocean, with the remaining half becoming entwined with the EGC returning southward to the Greenland and Iceland Seas. One interesting aspect of the WSC is the fact that as its water moves to the north, surface cooling continuously increases the density and the instability of the surface layer until it enters the Arctic Ocean and encounters the southward moving ice pack. There, the sensible heat carried by the WSC causes ice to melt. This, in turn, decreases the salinity and the density and increases the stability of the surface layer. As noted by Carmack (1990), the dynamical consequences of a current crossing a density extremum are not well understood.

The EGC extends for roughly 2500 km along the Greenland coast. In fact, south of Fram Strait it is simply the southward extension of the Transpolar Drift Stream that originates within the Arctic Basin. The EGC serves as a conveyor belt carrying between 4000 and 5000 km^3 of ice south each year into the North Atlantic, where it melts. There appear to be two branches of this current, the smaller of which branches to the southeast at approximately 77°N and the larger, referred to as the Jan Mayen Current, which branches east at about 73° forming the southern edge of the gyre in the Greenland Sea. Transport estimates for the EGC vary ranging from 2 to 5 Sv. Recent work based on both both moored current meter and drifting buoy data suggests that, for the upper 700 m, a reasonable transport estimate is roughly 3 Sv. The current appears to be quite stable and apparently does not serve as a source of eddies.

The other exchange current that should be mentioned generally flows northward through the Bering Strait between Alaska and Russia. This entrance to the Arctic Ocean is both narrow (~85 km) and shallow (~50 m) and has been studied for a number of years in spite of the operational difficulties imposed by the Cold War (Coachman et al. 1975). An examination of available data by Coachman and Aagaard (1988) suggests a long-term mean transport of about 0.8 Sv, with higher flows occurring in the summer than in the winter, when strong winds from the north can occasionally drive significant amounts of ice southward through the Strait. Whether this is only a near-surface effect or whether the flow thoughout the complete water column reverses during these events appears to be unknown. North of the Strait the flow field typically splits into two different regimes, with a broad western regime with weak flows (< 0.2 m/s) and a narrow eastern regime with stronger flows (~0.7 m/s). In spite of these differences, the total transports of both flows appear to be approximately equal.

Within the Arctic Basin itself, there are two main surface currents that show themselves quite clearly in the movement of the ice. In the first of these, the Transpolar Drift Stream, surface water and ice move across the Eurasian Basin, over the North Pole, and on through Fram Strait where, as noted, the current undergoes a name change to EGC. The other current is the anticyclonic, clockwise flow referred to as the Beaufort Gyre. In the past the Gyre was believed to be centered over the Beaufort Sea. Present information indicates that the speeds of these intrabasin currents are low, averaging about 0.02 m/s. Observations during 1950 to 1970 on the drift of identifiable objects such as ice islands within the Gyre gave rotation times on the order of 10 years. Within the Gyre the major portion (80%) of the transport occurs within the upper 300 m in the region of the pycnocline. Of the 3 Sv transport of the Gyre, 2.3 Sv is believed to be surface water, whereas only 0.7 Sv occurs within the Atlantic layer. The Gyre is particularly strong along the Alaskan coast of the Beaufort Sea, presumably as a result of the topography. Although in this region the near-surface ice and water move toward the west, the average subsurface motion (the Beaufort Undercurrent) above the continental slope is toward the east. The presence of the Undercurrent is indicated by a subsurface temperature maximum that is associated with the movement across the shelf of warmer water that originates in the inflow through the Bering Strait.

3.3.2 Antarctic

Just from considerations of topography alone one would guess that the situation in the Antarctic would be very different. Particularly missing elements are the broad shelf seas found in the Arctic. The exceptions are, of course, the embayments of the Ross and Weddell Seas. More importantly, between 56°S and 63°S there is no significant blocking land mass in a complete latitudinal circle around the Antarctic Continent. This, coupled with the fact that this same latitudinal region is known for its gale-force winds coming from the west, would suggest that a strong current should be present. It is! This current, termed the Antarctic Circumpolar Current (AACC), has a mean annual flow estimated at approximately 134 Sv around the continent in a clockwise direction, and displays little attenuation with depth. The AACC is of great geophysical importance as it serves as the connecting link between the Atlantic, Pacific, and Indian Oceans.

South of the AACC there is a narrow, nearshore countercurrent, referred to either as the Antarctic Coastal Current or the East Wind Drift, that flows toward the west following the continental margin. It lies to the south of the belt of low atmospheric pressure that surrounds the continent at approximately 65°S and shows current speeds of ~0.1 m/s. As might be anticipated by looking at a map of the Antarctic, the Antarctic Circumpolar Current is not completely circumpolar as a result of the topography of the Antarctic continent, in particular the blocking effect of the Antarctic Peninsula. Considering that to the north of the Antarctic continent the AACC flows toward the east, whereas along the coast of the continent the Antarctic Coastal Current flows toward the west, it is not surprising to find that there are weak baroclinic clockwise gyres in the western Weddell Basin and in the northern part of the Ross Sea. Estimates of transport within the Weddell Gyre are in the range of 75 to 100 Sv.

3.4 Water Masses

3.4.1 Arctic

There are several main classes of surface water that need to be considered in describing the Arctic Ocean. Within the basin the uppermost water layer is referred to as Polar Water (PW). PW typically shows temperatures below 0°C and salinities below 34.4‰, and it comprises the upper layer of the ocean to a depth of about 200 m. The upper 30 to 50 m of PW represents a seasonal mixed layer that in the winter shows vertically uniform temperature and salinity values. The prime factor driving the mixing is the strong vertical convection owing to salt rejection during the growth of sea ice. In the summer the melting of ice at the sea surface causes a pronounced salt stratification, resulting in a stable vertical density profile. Except in areas that become completely ice free in the summer, this water always remains near the freezing point. It also makes up the surface outflow of the East Greenland Current and within the Canadian Arctic Archipelago.

(The oceanographically inclined reader will note that in the present book I have chosen to use the parts per thousand symbol [‰] in describing salinity values instead of the more in vogue practical salinity units [psu]. This is purely a matter of personal artistic taste, as the values expressed are identical independent of which units are used.)

The main halocline, which shows temperatures of less than −1°C and salinities between 30.4 and 34.4‰, is found in the lower portion of the surface layer. This cold, relatively saline layer is believed to be maintained by drainage from the continental shelves during the winter. The importance of the halocline comes from two factors. First, it lies between the cold, upper mixed layer that contains the ice and the warmer underlying Atlantic Water (AW) layer. Second, the halocline both is at near-freezing temperatures and shows a strong, stable, vertical stratification as a result of its salinity profile. This suppresses vertical convective motions and limits the upward heat flux from the warmer underlying water. Figure 3.2 shows representative salinity and potential temperature and density profiles from the Arctic Ocean as reported by Perkin and Lewis (1984). Studies have generally found that the arctic halocline is not horizontally uniform and is possibly composed of more than a single water mass derived from one region (Jones and Anderson 1986). For instance, in the Eurasian Basin salinity increases rapidly with depth, reaching 34.9 to 35.0‰ at approximately 200 m, whereas the water temperature remains below −1.5°C to a depth of 150 m before increasing. In contrast, in the Canadian Basin the halocline is deeper and the salinity increase with depth is more gradual. Also, there are two temperature minima near the

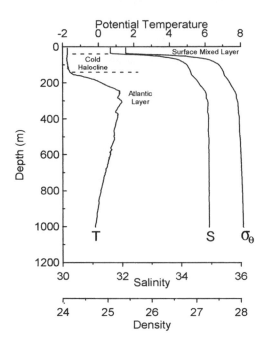

Figure 3.2. Representative salinity, potential temperature, and density profiles from the Arctic Ocean (84.485°N, 17.212°E). The observations were obtained from EUBEX (Perkin and Lewis 1984) and are reprinted in Padman (1995).

depths where the salinities are 31.6 and 33.1‰. There is also a temperature maximum near the depth where the salinity is 32.4‰. The minimum near 31.6‰ is thought to be a remnant of winter cooling. On the other hand, the maximum near 32.4‰ and the minimum near 33.1‰ may be the result of the influx of warmer water through the Bering Strait. Whatever the interpretation of these details, it seems clear that the origin of the halocline in the Arctic Ocean is far from simple.

Until ~1990 there was no data that suggested that the halocline might become unstable and even disappear. This impression of constancy of conditions in the near-surface layers of the Arctic Ocean has now changed as the result of scientific cruises using both nuclear submarines and ice-breakers as sampling platforms. These observations indicate that in the Amundsen Basin during the period 1991–1995, the salinity gradient from 60–90 m depth vanished. More will be said about these changes in Chapter 18 in the discussion of trends.

The second water mass is termed Atlantic Water (AW), in that it is transported into the system as a part of the West Spitzbergen Current (WSC). In the vicinity of Fram Strait it is initially characterized by water showing temperatures above 3°C and salinities greater than 35‰. Upon entering the Arctic Basin north of Svalbard, the AW is rapidly cooled by the atmosphere, by mixing with colder water and by melting ice at the margins of the pack. The resulting modified water then dives beneath the less dense PW and spreads out across the Arctic Basin where it occurs at depths between 200 and 800 m. Some studies have identified three water types occurring within this layer, each with a different temperature/salinity (T/S) signature (Swift and Aagaard 1981). AW also enters the Arctic Ocean through the Barents Sea as part of the Norwegian Current.

The third class of surface water is referred to as Arctic Surface Water (ASW). This water is warmer and saltier than PW but cooler (0 to 3°C) and fresher (34.4 to 34.9‰) than AW. ASW occurs mainly within the gyres of the Greenland and Iceland Seas. In that ASW is appreciably denser than either AW or PW, it cannot form by simple mixing, implying significant modification as the result of local air–sea exchange.

The deep water of the Arctic Basin is to be found at depths below ~800 m. Several varieties having different T/S characteristics are currently recognized. These range from the warm (ca. −0.5°C) and saline (34.95‰) deep water of the Canadian Basin to the cold (−1.2°C) and fresher (<34.90‰) Greenland Sea deep water. More will be said about deep water formation in the chapters on polynyas and underwater ice.

Finally, it should be noted that the waters overlying the broad continental shelves of the Arctic Polar Basin show more compositional variation (2 to 4‰) than the surface waters within the main basin (0.5‰). During summers, shelf water is typically more dilute than comparable water offshore as the result of the large inflow of water from the large northward-flowing arctic rivers. This is particularly true on the Russian Shelf. During the winters the reverse is possible, with salinities reaching values in excess of 34.5‰ as the result of the release of brine during freezing and the possible upwelling of saline water onto the shelf.

3.4.2 Southern Ocean

The water masses of the Antarctic are more complex than those of the Arctic because of the unconfined nature of the Southern Ocean. Even so, the general distribution of the water masses is similar to that found in the Arctic. On the top to the north of the continental shelf there is a layer of cold, lower-salinity Antarctic surface water (AASW) occurring to depths of ~200 m. The lower salinity of this water comes not from the negligible direct runoff from the Antarctic continent, but from dilution resulting from the melting of icebergs and from the general excess of precipitation over evaporation in this region. It is interesting to note that the thickness of the AASW layer is approximately the same as the draft of tabular antarctic shelf icebergs. The release during melting of gas bubbles trapped within the ice and their subsequent rise to the sea surface have been postulated as a mechanism contributing to vertical mixing within this surface layer (Jacobs et al. 1979).

The greatest volume of water in the Southern Ocean is the Circumpolar Deep Water (CDW) that lies between the AASW and the Antarctic Bottom Water (AABW). Although it is generally considered to be the least spatially variable of the antarctic water masses, its temperature does change from about 0.5°C in the Weddell Sea to +1.5°C in the Ross Sea. This nutrient-rich water mass is both warmer and more saline than the overlying AASW. Its point of origin is not in the Antarctic, but instead is believed to be the North Atlantic. Off the Antarctic continent, CDW generally moves from west to east as part of the AAC except near the coast, where it moves east to west as part of the Antarctic Coastal Current.

At the base of the water column is the AABW. This water mass is defined as being colder than 0°C. Its exact properties also vary somewhat with location, with the bottom water in the Ross Sea being somewhat warmer and more saline than that of the Weddell Sea. The origin of this water is local in that it is fed by the cold saline water that forms on the antarctic continental shelves as the result of rapid sea ice growth in polynyas and leads and ocean–ice shelf interactions (see Chapters 11 and 17). Figure 3.3 is a cartoon by Laws (1985) that shows an interpretation of the interrelations between the different water masses off the Antarctic continent.

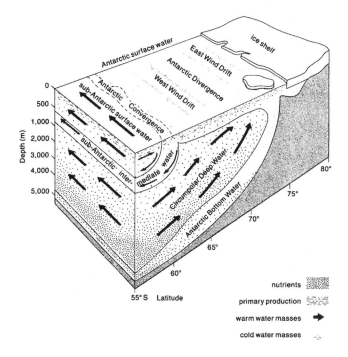

Figure 3.3. A cartoon summarizing the many different water masses and currents that occur in the Southern Ocean off the coast of the Antarctic (Laws 1985).

The water that develops over the Antarctic continental shelves shows similarities to the water found over shelves in the Arctic. For instance, its properties through most of the water column are highly variable, being warmer and less saline in the summer and colder and more saline in the winter. The exception is the water at the base of the water column, which is generally cold and saline on a year-round basis. It is this water that feeds the AABW and is produced by a variety of processes associated with the growth of sea ice. It is referred to as High Salinity Shelf Water (HSSW) and is usually taken to be more saline than 34.6‰ in that this is the salinity required for shelf water to drive deep convection. Its temperature is essentially at or near the surface freezing point.

The final water mass occurring over the continental shelves is Ice Shelf Water (ISW). This water type is unique to the Antarctic and forms as the result of interactions between HSSW and the meteoric ice found at the base of ice shelves. It is the only water type in the World Ocean whose potential temperature is below its surface freezing temperature. One of the end products of these interactions is so-called marine ice, which occasionally can be sighted in green icebergs. ISW will be discussed in some detail in Chapter 17.

In considering water mass development over the shelf areas of both the Arctic and the Antarctic, Carmack (1990) points out an interesting fact. During the summers the shelves act as positive estuaries with a net outflow of low-density water at the surface, whereas during the winters they act as negative estuaries with a net output of high-density saline water at depth.

However, even considering the above, in general the arctic and the antarctic ocean settings are not at all similar. Even so one might guess that arctic and antarctic sea ice might prove to be essentially identical in that freezing seawater up north should produce the same end product as freezing seawater down south. However, as will be demonstrated, because the nature of the ice that forms is a reflection of the environmental conditions under which it forms, the resulting mix of ice types can be very different.

4 An Introduction to Sea Ice Growth

I have reached these lands but newly
From a wild weird clime that lieth, sublime,
Out of SPACE—out of TIME.
Edgar Allan Poe

4.1 A Growth Model

Before examining the details of the freezing of seawater, it is useful to examine freezing in a more general way. By this I mean that the following discussion could be applied to the solidification of a large variety of materials. The material "constants" would change but the basic heat conduction physics would remain the same. As noted in Chapter 2, the study of sea ice played a seminal role in the development of an analytic approach to freezing problems. Here I refer to the field observations carried out in the vicinity of Franz Josef Land as part of the Austro-Hungarian North Pole Expedition of 1872–1874. While his ship, the *Tegethoff*, was beset in the ice, Weyprecht (1875) obtained a time series of first-year ice thicknesses and associated meteorological observations. On analyzing these observations, he noted that there appeared to be a simple relation between the cumulative number of freezing degree-days θ

$$\theta = \int_0^t \left(T_f - T_a \right) dt, \tag{4.1}$$

and the resulting ice thickness H of the general form

$$H = \text{constant} \, (\theta)^n \tag{4.2}$$

where the constant n had a value of approximately $1/2$. In equation 4.1 T_f is the freezing point of the water and T_a is the air temperature usually measured at a set distance such as 2 m above the ice surface. Fortunately, the problem of sea ice growth in the polar regions gained the attention of the physicist Stefan (1891), who then solved the one-dimensional freezing problem that is still referred to as the Stefan problem. In doing this he made the reasonable first assumption that the temperature of the upper surface of the ice was identical to the temperature of the air.

A representative example of this type of ice growth data is shown in Figure 4.1 (Anderson 1961), which presents a log-log plot of ice thickness vs. accumulated degree-day observations. This data was collected in 1956 at Thule, Greenland, during a study of the flexural strength of sea ice (Anderson and Weeks 1958; Weeks and Anderson 1958b). In order to test ice of a variety of thicknesses, "ponds" were cut and cleared in the ice sheet and the ice that formed in them was harvested at different times. As a result, over the winter a wide variety of ice thicknesses

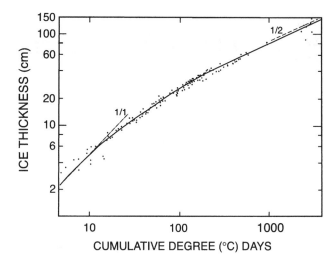

Figure 4.1. A plot of ice thickness vs. accumulated degree-days based on data collected at North Star Bay, Thule, Greenland, during the winter of 1956–1957 (Anderson 1961).

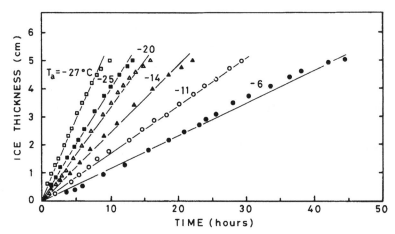

Figure 4.2. Changes in the growth rate of thin sea ice with air temperature, Lützow-Holm Bay, Antarctica (Wakatsuchi 1983).

resulted and were sampled in the thickness range of 20 to 150 cm. In that new ice sheets were continuously being produced, there was commonly only a very thin snow cover on the ice. This was particularly true of the thinner categories of ice ($H < 20$ cm). Note that the curve shows a slope of 1/2 at ice thicknesses greater than ~50 cm, as should be the case if Stefan's analysis were to strictly hold. However, at smaller ice thicknesses the slope of the curve appears to approach a value of 1. An excellent field documentation of the fact that in thin ice, ice thickness is a linear function of time for a constant air temperature is shown in Figure 4.2, based on observations in Lützow-Holm Bay located near Syowa Station, Antarctica (Wakatsuchi 1983). A reasonable first guess as to the cause for the deviation from the square root relation, of course, would be that in the thinner classes of ice, the surface temperature of the ice is no longer equal to the air temperature. Anderson described his observations using an empirical relation suggested earlier by Zubov (1938, 1945) that contains both H and H^2 terms and fits the observations quite well:

$$H^2 + a_1 H = a_2 \theta \qquad (4.3)$$

Here a_1 and a_2 are empirically determined constants. In the case of the Thule data a_1 and a_2 were found to equal 5.1 and 6.7, respectively, where H is in centimeters and θ is the air temperature in

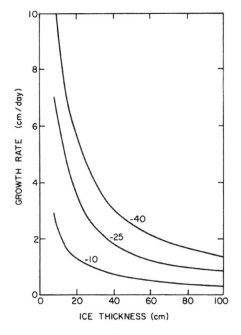

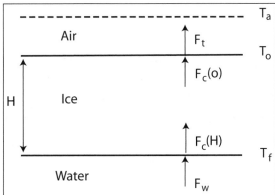

Figure 4.3. Variations in the growth rate of young sea ice with thickness for three constant air temperatures (Maykut 1986). The curves are based on the thickness measurements of Anderson (1961) at Thule, Greenland.

Figure 4.4. Schematic of a slab of snow-free sea ice showing the pertinent heat fluxes.

°C. This simple relation tells us a great deal about sea ice growth as shown in Figure 4.3, which plots growth rate against ice thickness for three different constant air temperatures. Clearly, as sea ice thickens there is a drastic decrease in the growth rate.

Now let us examine a simplified but extremely informative physical model for ice growth as developed by Maykut (1986). The general physical setup is as shown in Figure 4.4, with ice with thickness H, density ρ_i and thermal conductivity k_i resting on seawater that is at its freezing point T_f. The ice–seawater interface temperature is assumed to be fixed at T_f. The ice surface temperature T_o is controlled by the surface heat balance and is not necessarily equal to the near-surface air temperature T_a. As noted, if the ice is thin, the assumption that the temperature gradient within the ice is linear is known from field observations to be a good approximation to reality. Therefore the conductive heat flux through the ice is $F_c(z=0) = F_c(z=H)$ and

$$F_c(H) = k_i \left(\frac{T_o - T_f}{H} \right). \tag{4.4}$$

As noted by Maykut, the amount of growth or ablation at the bottom of the ice $(z = H)$ is determined by the sum of $F_c + F_w$ where F_w is the heat flux from the underlying water column. If this sum is positive, ice melts; if it is negative, ice growth occurs, releasing enough latent heat to balance the heat lost by conduction:

$$-\rho_i L \frac{dH}{dt} = F_c(H) + F_w. \tag{4.5}$$

If the simplest possible case is considered $F_w = 0$ and $T_o = T_a$ and equation 4.5 becomes

$$\rho_i L \frac{dH}{dt} = k_i \left(\frac{T_f - T_a}{H} \right).$$ (4.6)

If it is then assumed that $H(t = 0) = 0$, integration of equation 4.6 gives

$$H^2 = \frac{2k_i}{\rho_i L} \int_0^t \left(T_f - T_a \right) dt = \frac{2k_i}{\rho_i L} \theta.$$ (4.7)

As already suggested, the problem with this equation, which is the simple form of Stefan's solution, is that T_o can be appreciably warmer than T_a when H is small. To deal with this difficulty Maykut makes the simple assumption that the rate of heat exchange between the ice and the atmosphere F_t is proportional to the difference between T_o and T_a

$$F_t = C_t(T_a - T_o).$$ (4.8)

Here the term C_t can be considered to be an averaged surface heat transfer coefficient. In that $F_c = F_t$, we can write

$$k_i \left(\frac{T_o - T_f}{H} \right) = C_t \left(T_a - T_o \right),$$ (4.9)

which, when solved for T_o, gives

$$T_o = \frac{k_i T_f + C_t H T_a}{k_i + C_t H}.$$ (4.10)

Substituting this relation into equation 4.4 gives

$$F_c = \frac{k_i C_t}{k_i + C_t H} \left(T_f - T_a \right).$$ (4.11)

Recalling that we have assumed that $0 \sim F_w << F_c$ and that at time $t = 0$, $H = 0$, integration of equation 4.5 results in

$$H^2 + \frac{2k_i}{C_t} H = \frac{2k_i}{\rho_i L} \theta.$$ (4.12)

Maykut evaluated this relation by letting $C_t = 50$ cal / (cm day °C) (a value which is a bit larger than the nominal value for the sensible heat flux alone), $k_i = 419.9$ cal / (cm day °C), and $\rho_i L = 65$ cal cm^{-3} (a value that corresponds to an average brine volume in the ice of $\sim 10\%$). He found that the resulting equation

$$H^2 + 16.8 H = 12.9 \theta$$ (4.13)

resulted in ice thicknesses which were larger than observed by Anderson. To correct for this, he added a snow cover of thickness H_s, with a thermal conductivity k_s that, if the conductive heat flux in the snow equals that in the ice, gives

$$H^2 + \left[\frac{2k_i}{k_s}H_s + \frac{2k_i}{C_t}\right]H = \frac{2k_i}{\rho_i L}\theta \qquad (4.14)$$

or

$$H^2 + (13.1H_s + 16.8)H = 12.9\theta. \qquad (4.15)$$

If snow thicknesses of between 5 and 10 cm are assumed, this relation roughly brackets the empirical relations of Anderson, of Zubov, and of Lebedev (1938) that were fitted to field observations.

What have we learned here? First, that a simple physical analysis allows one to develop an equation that provides a reasonable fit to field data on the growth of first-year sea ice. Second, to fit data on the growth of both thin and thick sea ice, we need an H^2 and an H term in our relation. Even more important is the fact that as the ice thickens the H^2 term becomes increasingly important relative to H. This is the same as saying that as ice thickens H gradually becomes a linear function of $\sqrt{\theta}$. Furthermore, the significance of the H term relative to the H^2 term increases with an increase in the thickness of the snow (the bracketed term in equation 4.14). For example, if $H_s = 0$ the crossover point where the H^2 and H terms are equal occurs at $H \approx 17$ cm. However if $H_s = 5$, the crossover point becomes ≈ 80 cm. Also sea ice growth rates are clearly very sensitive to the amount of snow present on the ice. Finally, again pointed out by Maykut, matters are not really this simple, as we have either ignored or glossed over all sorts of potentially important matters. For instance, we have not considered either longwave or solar radiation, which are major components of the surface heat balance. Likewise, the thermal properties of both the sea ice and the overlying snow cover do not remain constant and will change with time. Also we will undoubtedly have to consider the upward heat flux from the underlying ocean.

Some sense of the presence of these additional factors can be gained by examining Figure 4.1 more carefully. Note that up to ice thicknesses of ~80 cm, the growth of snow-free sea ice appears to be well described as a simple function of freezing degree-days $\Sigma\theta$. However, once the ice becomes thicker in the $\Sigma\theta > 1000$ range, there are significant thickness variations in the range of 90 to 150 cm for roughly the same $\Sigma\theta$ values. Presumably these variations are the result of factors that have not, as yet, been considered here. We will return to these matters in Chapter 9 when we consider ice growth in more detail.

4.2 Multiyear Ice

Next consider conditions at the end of the summer. There are two possible situations: either (a) the first-year (FY) ice at a particular location melts completely, or (b) some of the ice layer survives the summer. If it has all melted, ice growth starts from zero in the fall, the situation that we have just described. In the northern hemisphere roughly half of the area covered by sea ice at its maximum extent melts away by the end of the summer (recently more like two-thirds; see Chapter 18). In the southern hemisphere ~85% melts away. If some ice survives, then these areas start the freezing season with a "head start" in that at time $t = 0$, ice thickness $H > 0$. However, the surviving ice does not grow as fast as at nearby areas where there was initially no ice. As we have already shown in the previous section, the thicker the ice layer, the lower the growth rate. Nevertheless, at the end of the winter, the second-year (SY) ice will be thicker than neighboring FY ice.

Now consider an idealized case where the annual climate cycle does not change from year to year. Every summer experiences exactly the same amount of total ablation from the top and the bottom of the ice, and every winter has exactly the same climate forcing. At the end of the second summer, the ice surviving will be thicker than at the end of the first summer. This means that during the third winter the thickness of new ice accreted at the bottom of the floe will be less than was accreted during the second winter. The result of such a process is shown in Figure 4.5, which is an extension of a schematic initially published by Maykut and Untersteiner (1971). Note that the net amount of ice added every winter decreases exponentially while the ice removed every summer stays the same. In the figure I have assumed that the initial winter's ice growth is 2 m and that every summer the floe loses 0.5 m owing to surface ablation. As a result the ice ultimately reaches a thickness where the winter's ice growth exactly equals the summer's ice ablation. When this thickness is reached, the multiyear (MY) ice floe has achieved a steady state in which the thickness cycle remains the same from year to year. What is this thickness and what are the factors that affect it? Depending on the climatic regime, equilibrium thicknesses are surprisingly thin in the range of 2.5 to 5 m and, based on past climatic data, have been estimated to be attained in 10 to 15 years (Maykut and Untersteiner 1971). Not surprisingly these numbers are affected by a large number of factors including snow cover and the oceanic heat flux F_w, matters that will also be discussed in Chapter 9.

Do steady-state MY ice floes really exist? They do, or at least they did. As will be seen later in the discussion of the structure of old sea ice, prior to 1965 several such old floes were described in some detail by field workers operating in the central Arctic Basin (Cherepanov 1964a; Schwarzacher 1959; Shumskiy 1955a). Even then the percentage of the pack ice in the Arctic Basin that was composed of such steady-state ice floes was not well known. My guess is that the

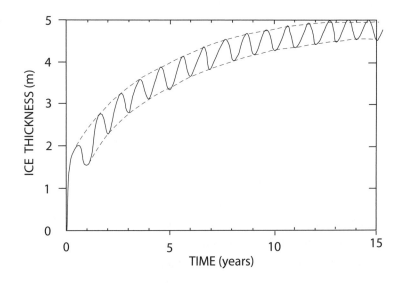

Figure 4.5. General pattern of ice growth in the central Arctic Basin (based on Maykut and Untersteiner 1971).

areal value was probably less than 50%. This impression is based on the observation that many cores taken during that period from apparently old sea ice that appeared to be undeformed, on close examination showed obvious signs of deformation in contrast to the layering characteristic of annual growth increments (Bennington, pers. comm.). Factors affecting this value are that even prior to 1980 a significant percentage (>10%) of the ice in the basin was either swept out of the basin every year or melted within the basin. Also, during the 10 to 15 years required for a floe to reach a steady-state thickness, there is, as noted, a good chance that the floe will become so deformed that the telltale layering associated with old MY floes will be destroyed. Another possibility is that in recent decades the amount of total ablation during the summer has increased sufficiently to result in a significant decrease in the equilibrium thickness as well as in the number of annual layers present in a steady-state floe.

One thing is clear, the thickest sea ice in the Arctic Basin (~60 m) is not the result of thermodynamics but of deformation. Is the 3.5 to 5 m range of thicknesses the thickest ice that can be formed in the Arctic as the result of thermodynamic processes? The answer here is, or at least was, no, in that at sites along the northern coasts of Ellesmere Island and Greenland fast ice with thicknesses of 12 m has been reported (Walker and Wadhams 1979). This ice even has its own special name, *sikussak*. Presumably these sites are locales where the oceanic heat flux F_w is near zero and the winter snow cover is thin enough to allow appreciable ice growth during the winter and thick enough to minimize surface ablation during the summer. Such extremely thick floes may also be relics of a time when the climate was colder than at the present. How much of this ice still exists is not well known. I would guess that the amount existing at present (2008) is small.

What is the situation in the Antarctic? Only some 15% of antarctic sea ice survives the summer. Therefore, the chances of steady-state floes developing within the antarctic pack are slight. If such floes do exist, they would be expected to occur in the perennial, heavy pack ice areas that are referred to as ice massifs. Possible locations would include the western portion of the Weddell Sea. Other likely locations are at sites where the development of very old fast ice is possible, for instance, in the more protected environments of some of the fjord systems along the antarctic coast. In the past there were areas of very thick sea ice to the southwest of McMurdo Station. However, there have been no recent studies at these locations and I do not know how much of this ice still exists. It is, however, safe to say that there is appreciably less very old sea ice in the Antarctic than in the Arctic.

I would also like to generally support the suggestion made by Weyprecht, in the late 1800s, that there is no perennial sea ice in either the Arctic or the Antarctic. Here I use the term *perennial* to refer to ice that would be hundreds to thousands of years old. Let us assume that there is a protected site somewhere in the polar regions where a very thick, stable fast ice cover has been in existence for the last 10,000 years. Here one would expect to find a floe of some considerable thickness, say ~15 to 20 m. It would be a layer cake of annual growth increments because once the steady-state thickness is reached, every layer that is added on the bottom during the winter by growth is balanced by an equivalent amount of ice removed during the summer by ablation. If we were to assume that the annual growth increment at the equilibrium thickness was 25 cm, the 20 m floe would only contain ice formed during the last 80 years. There would be no record remaining of the ice that had formed during the previous 9920 years. As will be seen later, the best-described steady-state floe observed to date in the Arctic contained only approximately 10 annual layers. Knowledgeable individuals might well ask, how does the sea ice in the Ellesmere Ice Shelf fit into this discussion in that it appears to be a "fossil"? However, I note that the Ellesmere Shelf is clearly not at an equilibrium thickness relative to recent climate and that it is currently rapidly disintegrating via breakouts that send ice islands out into the Arctic Ocean (Jeffries 1991, 1992; Mueller et al. 2003), where they will drift for a few years before exiting into the North Atlantic via the East Greenland Drift Stream. That said, it appears that the sea ice that comprises the lower portion of a part of the Ellesmere Shelf is the exception that proves Weyprecht's "rule." Current thinking concerning the age of this shelf is that it has almost certainly occupied the mouth of Disraeli Fjord since about 3000 B.C. (Crary 1960; Jeffries 1991, 1992).

This concludes our initial examination of the growth of sea ice, a subject that will be considered in more detail in Chapter 9, after the structure and properties of sea ice have been discussed.

5 Components

Don't worry about people stealing an idea.
If it's original, you will have to
ram it down their throats.
Howard Aiken

In studying the interrelations between growth conditions and the properties of the ice masses that form from the freezing of the water in the world's oceans, it is essential to remember that the phases involved are relatively few and simple. This is particularly true when the characteristics of the phases comprising sea ice are compared with the complex compositions and structures that are found in either naturally occurring rocks or in industrially produced ceramics and metals. What we have to deal with in sea ice is the everyday hexagonal form of ice referred to as ice I(h) plus a few simple hydrated salts, specifically, $NaCl \cdot 2\ H_2O$, $Na_2SO_4 \cdot 10\ H_2O$, and $CaCO_3 \cdot 6\ H_2O$, in order of decreasing importance and increasing temperature of precipitation (−22.9, −8.2, and −2.2°C), respectively based on the experimental observations of Nelson (1953) and Nelson and Thompson (1954). Three of the other four solid salts that are usually mentioned as occurring in sea ice (KCl, $MgCl_2 \cdot 12\ H_2O$, and $CaCl_2 \cdot 6\ H_2O$) only crystallize at temperatures below −36°C and, therefore, rarely occur under natural conditions. The effect of the remaining salt, $MgCl_2 \cdot 8\ H_2O$, has remained indiscernible. Even $CaCO_3 \cdot 6\ H_2O$, which starts to crystallize at temperatures just below the freezing point of seawater, has not as yet been shown to have a measurable effect on sea ice properties, although its formation has been postulated to result in a significant introduction of CO_2 into polar waters (Jones and Coote 1981). Therefore, in essence, the study of sea ice is the study of the solidification of solutions comprised primarily of water with some NaCl and Na_2SO_4 added. This would appear to be simplicity itself: ice + brine + no, one, or two solid salts depending on the temperature of the sea ice. However, as is common with many natural phenomena, matters are rarely as simple as they first appear. For example, these phases are arranged in several different geometric configurations that are related to the meteorological and oceanographic conditions prevailing during freezing. In the following sections the structures and a few general properties of these different phases will be discussed. More details will be provided in later chapters when applications in which these properties become important are discussed. In the next chapter the "equilibrium" relations between the phases as summarized in the so-called sea ice phase diagram will be described.

In that this book makes no attempt to provide a through review of the chemical physics of water, ice, seawater, or of the solid salts existing in sea ice, the number of references provided in the present chapter has been kept to a minimum. Extensive references on these matters can be found in Fletcher 1970; Glen 1974; Hobbs 1974; Petrenko 1993a, 1993b, 1994, 1996; Petrenko and Whitworth 1994a, 1994b, 1999; and in other references included with the text.

5.1 Water

There is general agreement among the scientific community as to the origins of the hydrogen and oxygen that comprise water (Waller and Hodge 2003). Hydrogen is believed to be a condensate from the hot Big Bang itself, whereas oxygen formed later in the thermonuclear cores of massive stars during their final supergiant phase. There is far less agreement concerning the origin of water itself on Earth. Scenarios include wayward comets and asteroids, an impact by an unusually wet body roughly the size of the moon, and the possible retention of water from the inner portion of the solar system despite intense solar heating (Drake 2005).

Although questions of its origin remain, pure water is clearly an atypical material. In fact the development of life on Earth would presumably be considerably different if the properties of water were not so unusual. Table 5.1 summarizes a number of its anomalous characteristics. A partial reason for these unusual properties can be found in the molecular structure of water, which is

Table 5.1. Anomalous physical properties of liquid water (Sverdrup et al. 1942).

Property	Comparison with other substances	Importance
Specific heat $(= 4.18 \times 10^3 \mathrm{Jkg^{-1}{}^\circ C^{-1}})$	Highest of all solids and liquids except liquid NH_3.	Prevents temperature extremes; heat transfer by water movement is very large; tends to maintain uniform body temperatures.
Latent heat of fusion $(= 3.33 \times 10^5 \mathrm{Jkg^{-1}{}^\circ C^{-1}})$	Highest except NH_3.	Thermostatic effect at the freezing point as the result of absorption or release of latent heat.
Latent heat of evaporation $(= 2.25 \times 10^6 \mathrm{Jkg^{-1}})$	Highest of all substances.	Extremely important in heat and water transfer within the atmosphere.
Thermal expansion	Temperature of maximum density decreases with increasing salinity; for pure water it occurs at +4°C.	Brackish water and seawater have their maximum density at temperatures above the freezing point whereas the maximum density of normal seawater is at the freezing point.
Surface tension $(= 7.2 \times 10^9 \mathrm{Nm^{-1}})$	Highest of all liquids.	Controls certain surface phenomena and the behavior of drops, important in cell physiology.
Dissolving power	Dissolves more substances and in greater quantities than any other liquid.	Obvious physical and biological implications.
Dielectric constant (87.9 at 0°C, 80 at 20°C)	Pure water is the highest of all liquids except H_2O_2 and HCN.	Affects the behavior of inorganic dissolved substances and the ice characteristics as viewed by EM remote sensing.
Electrolytic dissociation	Very small.	A neutral substance but contains both H+ and OH– ions.
Transparency	Relatively great.	Large absorption in IR and UV; selective absorption in the visible range is small; important in both physical and biological phenomena.
Conduction of heat	Highest of all liquids.	Primarily important on a small scale as natural heat transfer is primarily by convective processes.
Molecular viscosity	Less than most other liquids at similar temperatures.	Flows readily to equalize pressure differences.

comprised of an oxygen atom bonded to two hydrogen atoms. Hydrogen has three isotopes, ^1H (or H), ^2H (or D, deuterium), and ^3H (tritium), whereas oxygen has six: ^{14}O, ^{15}O, ^{16}O (or O), ^{17}O, ^{18}O, and ^{19}O. Therefore the exact isotopic composition of natural water varies somewhat depending upon its origin, a fact that allows one to distinguish ice formed from seawater from ice formed in the atmosphere. Within these limits natural water is comprised of 99.73% $^1H_2^{16}O$. As a result, for many purposes water is taken to be 100% $^1H_2^{16}O$ or H_2O.

Water molecules are polar (i.e., they possess an electric dipole moment), a fact which indicates that they cannot be linear, with the two hydrogen atoms located at equal distances on either side of the oxygen atom. In fact, studies of the infrared spectrum of water reveal that the bond angle is 104.5°, that the hydrogen–oxygen distance is 0.957 Å, and that an isolated water molecule has a triangular shape. However, it is generally accepted that the structure of bulk water is more the result of the highly directional hydrogen bonds between the molecules than of the largely covalent bonds within the molecules (Eisenberg and Kauzmann 1969; Stanley and Ostrowsky 1990; Tokmakoff 2007). These hydrogen bonds arise from the fact that the electrical properties of hydrogen and oxygen are quite different. In each of the hydrogen atoms there is one electron occupying a spherically symmetrical 1s orbit. In the oxygen atom there are two inner 1s electrons, two 2s electrons also in spherical orbits, and four external 2p electrons. When the two hydrogen atoms attach themselves to the oxygen atom, eight electrons take part in forming the external molecular orbitals. Two of these come from the two hydrogens while six come from the 2s and 2p orbitals of the oxygen. The process involved is referred to as s–p hybridization and the result is that the electron cloud of the water molecule consists of four orbitals each having two electrons.

The two orbitals directed toward the two hydrogen nuclei are referred to as the bonding orbitals, whereas the other two orbitals, which are on the opposite side of the oxygen nuclei from the hydrogen nuclei, are referred to as the lone-pair orbitals. The exterior shape of the water molecule, which is determined by the electron density in the region around the hydrogen and oxygen nuclei, therefore takes the shape of a slightly distorted regular tetrahedron, as shown schematically in Figure 5.1. (The H–O–H angle in a regular tetrahedron is 109.467° as compared to the value of 104.523° observed in water.) As noted by Petrenko (1993a), even though the hydrogen–oxygen bond results from electron sharing and can therefore be considered to be covalent, the fact that a part of the electron cloud is "deferred" from the hydrogen to the oxygen causes them to assume opposite

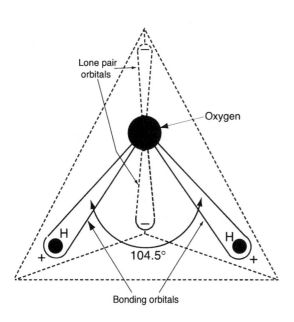

Figure 5.1. Schematic showing the angular relations in a water molecule.

electrical charges (+ at the hydrogen, – at the oxygen) which results in the electrostatic attraction typical of ionic bonding.

Although monatomic fluids can successfully be treated as disordered aggregates of atoms with no orientational preference in a unit volume greater than the coordination number, molecular fluids such as water have, to date, defied simple structural descriptions (Oxtoby 1990). However, as noted, it is clear that the bulk structure of water is related to the highly directional intermolecular hydrogen bonding between the H_2O molecules instead of to the intramolecular bonds within the water molecules. That the intramolecular bonds are significantly stronger than the bonding between the molecules can be seen by comparing the energies required to dissociate a water molecule into its constituent atoms (9.510 eV to remove the first hydrogen atom and 5.11 eV to remove the second) with the energies required to melt ice I(h) at 0°C and to vaporize liquid water (0.06 and 0.39 eV per molecule, respectively).

Even though the intermolecular bonds are weaker, they are nevertheless effective at longer ranges than the covalent bonds that control the directionality within the molecule.

This causes the water molecules to have an attraction for one another at temperatures near the freezing point, where the H_2O molecules tend to arrange themselves into partly ordered groups or clusters. Not surprisingly these clusters show a tetrahedral coordination similar to that observed in ice I(h). At temperatures significantly above the freezing temperature, thermal vibrations result in both increasing distances between the H_2O molecules and increasing randomness in their motions, both factors that decrease ordering in liquid water. On the other hand, as water cools polymerization gradually develops as more and more partly ordered groups form. If this ordering did not occur, the decreasing distances between the water molecules would result in the density of liquid water gradually increasing until the freezing point was reached, as is the case in almost all pure materials. In water the structures of the clusters that form in the liquid are icelike in that they have a more open structure that anticipates the impending formation of ice. As the freezing point nears, the openness of the increasing number of clusters first balances the tendency for the H_2O molecules to become more closely packed with decreasing temperature. As a result a density maximum occurs at +4°C. At temperatures below +4°C, the openness of the increasing number of clusters becomes more and more dominant, causing the density of the liquid to continue to decrease until the freezing point is reached (In_2Te_3 appears to be the only other compound that shows a similar behavior). Figure 5.2 shows these density changes (note that the density scales for the water and ice portions of the diagram are quite different). Attempts to explain the density maximum in water by the use of computational models have to date not been successful. For instance Cho et al. (1996) obtained behavior somewhat similar to that observed in water by the use of a two-dimensional model that considers interactions between second nearest neighbors. However, attempts to extend this treatment to more realistic three-dimensional models have not been able to achieve a density maximum and instead show the usual monotonous increase in density with decreasing temperature (Velasco et al. 1997).

Two parameters that characterize the strength of intermolecular hydrogen bonds in a series of compounds are their melting T_m and boiling temperatures T_b. When these values for H_2O are compared with values that would be obtained by extrapolating values in the chemical series H_2Te, H_2Se, H_2S, one finds that the actual T_m and T_b values are ca. 90° and 180°K higher than

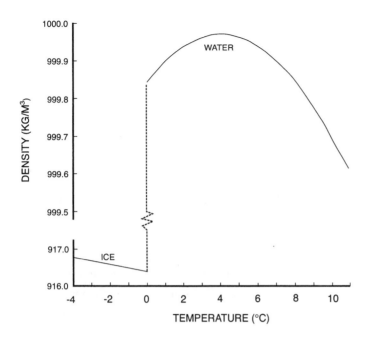

would be anticipated. If this were not so, the only state in which H_2O would occur on the Earth's surface would be water vapor and life as we know it would certainly not exist.

Another of water's atypical characteristics is that in the laboratory it can easily be super-cooled below its equilibrium freezing temperature. In fact, the temperature of homogeneous nucleation for pure water is commonly given as −40°C. Here the term *homogeneous nucleation* refers to the formation of ice in a system in which the nucleation of the ice crystals is not aided by the presence of foreign particles that serve as nucleation sites. Homogeneous nucleation almost never occurs in natural settings, in that foreign nuclei are invariably present in nature to aid in nucleation. Nevertheless small supercoolings appear to be common in nature and are essential to several processes that occur during sea ice formation and growth. Of particular interest here is the formation of the different types of underwater ice, as will be discussed in Chapter 17. An interesting aspect of supercooled water is that it continues to become less dense as it supercools. In the same manner, its diffusion constant decreases and appears by extrapolation to go to zero at temperatures just below the homogeneous nucleation temperature. Although this behavior has led to the suggestion that there is a stability limit to supercooling in water, this limit does not ap-pear to be the glass transition temperature as in many other liquids. It appears to be the start of a region in which disordered water cannot exist. For water the glass transition temperature appears to be much lower, at roughly 136°K (Soper 2002).

5.2 Seawater and Brine

When a second material is dissolved in a liquid, it has two effects: it increases the density of the liquid and depresses its freezing point. In both cases the effect increases as the amount of dissolved material increases. For instance, freshwater has a density of approximately 1.00×10^3 kg m^{-3} while typical seawater has an average density of 1.03×10^3 kg m^{-3}. In addition typical seawater freezes

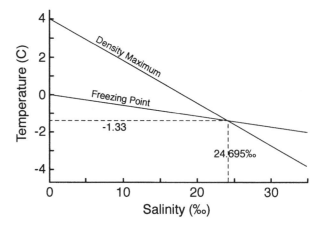

Figure 5.3. The temperatures of the density maximum and of the freezing point of brackish and seawater as a function of salinity.

at ca. −1.8°C as compared to 0.0°C for freshwater. Although seawater contains on the average ~35‰ salt (parts per thousand by weight), near-surface salinities in the polar regions are typically less than this, with values in the 31–34‰ range. As increasing amounts of salts are added to freshwater, the salt ions interfere with the tendency of the water molecules to form ordered polymerized groups as the freezing temperature is approached, as was just discussed. As a result, as the salinity increases the temperature at which the density maximum occurs gradually decreases to lower and lower temperatures. This effect is shown in Figure 5.3, which plots both the freezing point of seawater and the temperature of the density maximum against salinity. As can be seen, at salinities of greater than 24.7‰ freezing occurs before the cluster formation occurring in the water becomes sufficiently pronounced to result in the occurrence of a density maximum. This does not mean that the density maximum cannot form, but only that if it does form, it must occur in water that has been supercooled. In that large supercoolings are rare in natural situations, if salinities are in excess of ~24.7‰ ice formation invariably occurs before supercoolings become large enough for the density maximum to occur.

Therefore, the salinity of 24.7‰ separates two very different cooling regimes as far as fluid behavior is concerned. In the first regime (S_w > 24.7‰), as atmospheric cooling proceeds, the surface layer of water is consistently being cooled to densities that are higher than those of the underlying warmer fluid. This is an unstable situation that results in the development of convective mixing in the upper portion of the water column. As cooling continues convection also continues until the complete thickness of the convecting layer has been cooled to the freezing temperature and ice starts to form. The thickness of this mixed layer is typically in the range of 10 to 40 m. However, if the salinity of the water is < 24.7‰, a similar convective situation exists only until the temperature of the surface convecting layer is cooled to +4°C. As cooling proceeds further, it now results in the surface water becoming less dense. This is a stable situation in that it produces a less dense (colder) liquid layer sitting above a denser (warmer) liquid layer. Heat transfer in this stable upper layer now takes place by conduction, a process that is less efficient than convection by a factor of 10^2 to 10^3. Now only the thin upper conducting layer is cooled to the freezing point, whereas the much thicker layer that was initially undergoing convection stays at +4°C. Water that has a salinity of < 24.7‰ is therefore referred to as brackish water, whereas water with a salinity of > 24.7‰ is referred to as seawater. Because of the incredibly large volume of water in the ocean having salinities in excess of 30‰, brackish water is actually fairly rare, primarily occurring in estuaries and near the mouths of large rivers such as the Yenisei, Ob, and Lena Rivers, which flow into the Arctic Ocean along the north coast of Russia. As will be seen later, ice that forms from brackish water has not been well described to date.

Some readers may now be wondering exactly what is meant by *salinity*. Formally it is defined as "the total amount of solid material in grams in 1 kilogram of seawater when all the carbonate has been converted to oxide, the bromine and iodine replaced by chlorine, and all organic matter completely oxidized." Prior to the mid-1950s, salinity was commonly determined by measuring the amount of chloride ion in the seawater by a silver nitrate titration. The relation between this value and the salinity was then calculated by the use of an empirical formula. Having performed a large number of these titrations I can attest that this procedure was a "pain." Currently most salinity determinations are made by measuring both the electrical conductivity and the temperature of the seawater sample. In studying problems of "wet" oceanography these measurements must be made very precisely, as an uncertainty of ±0.01°C in temperature results in an uncertainty of ±0.01‰ in salinity. According to Knauss (1978), typical salinity measurement accuracies are ±0.003‰, with the ultimate precision of the technique at ±0.001‰. Here the limit is more the result of possible variations in the ionic ratios of the different salts present in the sample than in the test procedures themselves. The reason for this concern about accurate salinity measurement is the fact that the total salinity range of much of the ocean is between 34.50 and 35.00‰. Investigators interested in sea ice do not have to meet such high standards because sea ice salinities can vary between 0 and 15‰. Furthermore samples taken a few centimeters apart from apparently homogeneous ice sheets can vary by as much as ±1‰. This subject will be discussed in some detail later. Sea ice salinity measurements are commonly made by first melting the sample in an enclosed container. Then the conductivity and temperature of the resulting solution is determined by the use of a small portable electrical conductivity meter. As instruments currently available have the temperature corrections incorporated into their circuits it is only a matter of dialing in the appropriate values and reading the resulting salinity from the display. I am not aware of specific studies of the accuracy of this procedure but based on my observations I would estimate it to be on the order of ±0.1‰. (Since 1978 it has become common in the oceanographic literature to express salinity values in terms of a practical salinity scale that is based on the ratio K_{15} of the conductivity of the sample to that of a KCl solution at 15°C and 1 atmosphere pressure, in which the mass fraction of KCl is 0.0324356. A K_{15} value of exactly 1 is then taken by definition to have a practical salinity equal to 35. In that ratios have no units, this scale is dimensionless and 35 on the psu scale no longer exactly equals 35 grams of salt per liter of solution. However considering the accuracy of typical sea ice salinity measurements and the scale of the observed natural variations, for the purposes of this book values on the psu scale can be considered to be identical to the old parts per thousand (‰) values. In that the salinities discussed here have been determined by a variety of different procedures, I have decided to use ‰ throughout this book.)

The main constituents of seawater are of course hydrogen and oxygen, in that H_2O comprises between ~965 and 975.3 kg of every 1000 kg of seawater, with the exact value depending on the salinity. When the compositions of the other components are examined, values presented in Tables 5.2 and 5.3 are obtained.

Table 5.2. Main components of seawater in the form of salts per m³ of seawater of 35‰ at 20°C (Neumann and Pierson 1966).

NaCl	28.014
$MgCl_2$	3.812
$MgSO_4$	1.752
$CaSO_4$	1.283
K_2SO_4	0.816
$CaCO_3$	0.122
KBr	0.101
$SrSO_4$	0.028
H_2BO_3	0.028

Table 5.3. Major composition of seawater for a total salinity of 35‰.

Cations	g/kg	Milliequivalents per kg
Sodium	10.752	467.56
Potassium	0.39	9.98
Magnesium	1.295	106.50
Calcium	0.416	20.76
Strontium	0.013	0.30
Total		605.10
Anions	**g/kg**	**Milliequivalents per kg**
Chlorine	19.345	545.59
Bromine	0.066	0.83
Fluorine	0.0013	0.07
Sulfate	2.701	56.23
Bicarbonate	0.145	---
Boric acid	0.027	
Total		602.72

Another fact that should be noted is what I have heard referred to as the first law of marine geochemistry. This states that although the total amount of salt in seawater can change from location to location, the ratios of the different components are constant (i.e., ratios such as the SO_4/Cl, Ca/Cl, and Mg/Cl values are constant at 0.1396, 0.02150, and 0.06694, respectively). Although this is not precisely true, it is a surprisingly good approximation for most seawater. If it were not a good approximation it would not be possible to determine salinity values from electrical conductivity measurements in that conductivity is not only a function of the temperature and the total amount of salt in solution (i.e., the salinity); it also varies with the ratios of the different ions in solution. Where this "law" breaks down is in nearshore regions where riverine input has resulted in the formation of brackish water. In that the main constituents of most river and lake waters are carbonate, sulfate, and calcium, with sodium and chloride comprising only small fractions of the total salts, brackish water can show large variations in the ion ratios of the salts.

Because the thickest sea ice known only has a draft of ~50 m and seawater is relatively incompressible, we will normally not be concerned with the variation in the density of seawater as a function of pressure. The one exception here relates to the formation of underwater ice in locations near large ice shelves. The same is true of the variation in the density of seawater as a function of temperature, in that seawater temperature variations in regions where sea ice is forming would be expected to be quite small. The same is not true of the salinity of the seawater where melting and freezing effects can result in significant changes both locally and regionally. The relation between seawater density and salinity at 0°C can be expressed as

$$\sigma_0 = -0.093 + 0.8149 S_{sw} - 0.000482 S^2_{sw} + 0.0000068 S^2_{sw}. \qquad (5.1)$$

Here σ_0 is the σ_t value of water at 0°C and at atmospheric pressure, and S_{sw} is the salinity in ‰. (Sigma-t values are commonly used in oceanography to express the density of seawater. For instance, the numerical values for the density of seawater always start with 1.0 . . . with the exception of very low salinities and high temperatures, conditions that will rarely be discussed in this book. Therefore, by convention one can discard these values and express a density of, say, 1.01765 by its σ_t value of 17.65 (i.e., $\sigma_t = (\rho_{s,t,p} - 1) \times 10^3$). Here the subscripts s, t, and p indicate salinity, temperature, and pressure, respectively.]

Another useful equation gives the relation between the freezing point of seawater (T_f) at zero pressure and its salinity (S_{sw}) (Millero 1979):

$$T_f = -0.0575 \times S_{sw} + 1.1710523 \times 10^{-3} S^{3/2}_{sw} - 2.154996 \times 10^{-4} S^2_{sw} + 7.53 \times 10^{-4} p \quad (5.2)$$

where T_f is in °C, S_{sw} is in ‰, and p is the pressure in dbars. In studies of unusual sea ice types occurring under and in the vicinity of antarctic ice shelves, the last term in equation 5.2 becomes of considerable importance, as will be seen in Chapter 17.

In many sea ice problems one is more interested in the temperature dependence of the physical properties of the brine that is descended from standard seawater and is in equilibrium with ice, than in the seawater itself. In the past there does not appear to have been a great amount of information available on such matters. The reason for this data void was that most investigators who have studied brine properties were not interested in sea ice and therefore were not particularly concerned with the equilibrium state between ice and seawater brine as a function of temperature. Even Nelson and Thompson (Nelson 1953; Nelson and Thompson 1954), on whose experimental work the sea ice phase diagram has been based, were not specifically interested in sea ice but were investigating freezing as a possible process contributing to the origin of selected sulfate deposits. To avoid problems associated with the presence of ice, they first extracted brine samples from freezing solutions but then warmed the samples to +4°C before determining their densities. The relations typically used to estimate the properties of sea ice brine are commonly linear extrapolations to the freezing point. Although this may be adequate for many purposes, it is hardly desirable. For instance, consider brine density, which is an extremely important property that enters into the calculation of both the brine and gas volumes in sea ice samples. In the past this dependence has been estimated using a simple linear relation suggested by Zubov (1945):

$$\rho_b = 1 + 0.0008 S_b \tag{5.3}$$

where ρ_b is in Mg m^{-3}.

However, more exact measurements by Maykut and Light (1995) made at the freezing point on brines produced from a synthetic sea salt mix show that the relation is nonlinear and can be separated into two different temperature regimes (−2 to −8°C and −8 to −32°C):

$$\rho_b(T) = 0.997978 - 0.01658912T - 5.126629 \times 10^{-4}T^2 \tag{5.4}$$

$$(-2 \geq T \geq -8°C)$$

and

$$\rho_b(T) = 1.024326 - 0.01039362T - 1.307606 \times 10^{-4}T^2 \tag{5.5}$$

$$(-8 \geq T \geq -32°C).$$

Here T is in °C and ρ_b is in Mg m^{-3}. The r^2 values for equations 5.4 and 5.5 are 0.95 and 0.99, respectively. As noted by Maykut and Light, as the densities of brine solutions with salinities in excess of 24.7‰ increase as temperature decreases, it would be expected that the density values measured by Nelson and Thompson should increasingly underestimate ρ_b as the brine becomes colder. As shown in Figure 5.4 the Maykut and Light observations are consistently higher that those of Nelson and Thompson at temperatures below −8°C. That these relations should be both nonlinear and exhibit a break at −8°C is not surprising, in that the precipitation temperature of Na$_2$SO$_4$•10 H$_2$O from seawater is commonly taken to occur at −8.2°C. What is surprising here is that there is no clear break in the slope of the data at −22.9°C, where NaCl•2 H$_2$O precipitates. Using similar procedures Maykut and Light have also determined the density variations in brines produced by the freezing of simple NaCl–H$_2$O solutions. In that NaCl ice is frequently utilized as a surrogate for sea ice in experimental studies, the $\rho_{NaCl}(T)$ for this system is also given here:

$$\rho_{NaCl}(T) = 1.003618 - 0.01306119T - 2.147406 \times 10^{-4}T^2 \tag{5.6}$$

$$(-2 \geq T \geq -21.2).$$

In this case there is no break in the curve near −8°C as there is no Na$_2$SO$_4$ in the system and brine does not exist at temperatures below −21.2°C, which is the eutectic point. The fitted curve describes the data extremely well ($r^2 = 0.997$).

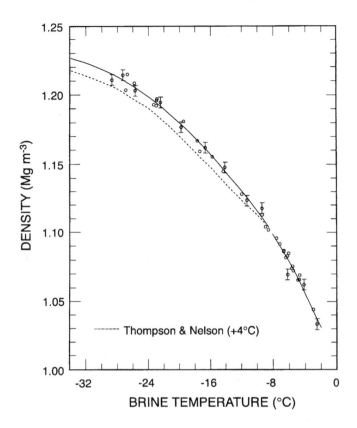

Figure 5.4. Density vs. temperature plots of the freezing point of synthetic sea ice as determined by Maykut and Light (1995). Also shown is the curve as determined by Nelson and Thompson (1954) at +4°C.

5.3 Ice

When one considers the fact that ice I(h) floats in its own melt as the result of an increase in specific volume of 9.05% on freezing (a characteristic shared by Bi, Ge, and Si), it is not surprising that there is more than one polymorph of ice. Here the term *polymorph* refers to another solid phase having the same composition but with the atoms arranged in a different structural pattern. What is surprising is the number of polymorphs. There are currently 13 known polymorphs of ice, plus three additional amorphous forms (Petrenko 1993a; Petrenko and Whitworth 1999; Zubavicus and Grunze 2004). It has been suggested that the changes between these amorphous forms occur through a series of intermediate metastable forms (Tulk et al. 2002). There is also an additional polymorph (ice XII) that has been predicted via the use of computer simulations (Szpir 1996). In fact, ice has more polymorphs than any other known material. It has also been discovered that at very high pressures (60 GPa) ice changes from a molecular solid, in which each hydrogen atom is bonded strongly to the oxygen atom in its H_2O molecule and is weakly bonded to the oxygen in the neighboring molecule, to an ionic solid in which symmetrical hydrogen bonding occurs (Goncharov et al. 1996). In short, the phase diagram for H_2O is far from simple. A discussion of work leading to the synthesis of ice VIII and ice XIV can be found in Salzmann et al. (2006).

Just consider the exciting geophysical possibilities and complexities that might occur if all of the ice polymorphs were to occur naturally on Earth. Alas, this is not to be, in that for ice I(h) to change to ice III by crossing the nearest phase boundary would require a pressure of roughly 2 kbar. This is a value equivalent to the pressure encountered at a depth of ca. 22.6 km in an ice sheet. As the thickest real ice sheet in East Antarctica only has a thickness of slightly over 4 km, this is clearly not a possibility. In fact, the only other ice polymorph that might occur naturally is cubic ice I(c), which forms when water vapor sublimates onto a substrate in the temperature interval from −80 to −150°C. If this were to occur naturally, ice I(c) would only be found in the upper atmosphere. Although arguments and experiments have been presented that support the transient occurrence of ice I(c) in the upper atmosphere (Murray and Bertram 2007; Shilling et al. 2006), to date I know of no direct observations that prove such an occurrence. This leaves us with ice I(h) as the only ice polymorph that is stable under the physical conditions existing at or near the surface of the Earth. This is the ice of sea ice, snowflakes, glaciers, skating rinks, and ice cubes. In the remainder of this book, when the term *ice* is used the ice polymorph will always be ice I(h), unless specifically noted otherwise. In fact, in most sea ice problems the variations in pressure are small enough (< 2.5 atmosphere) that their effect can be neglected and the density of pure ice can be expressed as a simple function of temperature

$$\rho_i = 0.917 - 1.403 \times 10^{-4}T \qquad\qquad (5.7)$$

where ρ_i is in Mg/m³ and T is in °C (Pounder 1965).

That ice is undoubtedly a member of the hexagonal crystal system has probably occurred to anyone who has examined the shapes of snowflakes (Nakaya 1954). This is the crystal system that is characterized by four axes: the three equivalent *a*-axes, which lie in a plane separated by angles of 120°, and the *c*-axis, which is oriented perpendicular to this plane. It is also important to note that these axes are not vectors in that they are not directional. The positions of the three *a*-axes and the *c*-axis of the hexagonal system, as well as the orientations of several directions and planes in an ice crystal, are given in Figure 5.5. The notations used here are the so-called Miller or Miller–Bravais indices of the directions and faces. These indices are defined as the reciprocals of the intercepts of the planes on the coordinate axes where the units of the intercepts are those

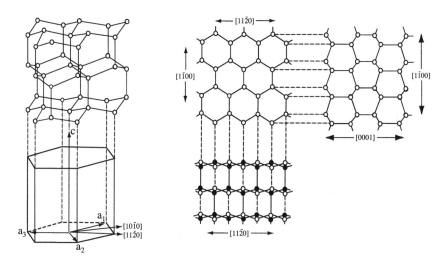

Figure 5.5. Positions of the oxygen atoms in ice I(h) as viewed from several different angles. In the middle diagram the view is parallel to the *c*-axis and in the diagram on the right it is perpendicular to the *c*-axis.

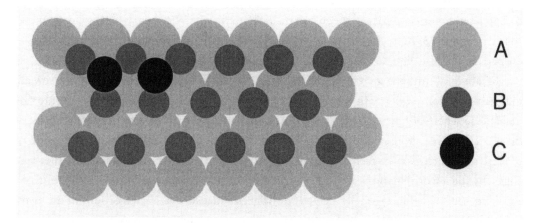

Figure 5.6. Dense packing of identical, ideal balls on a level surface (after Petrenko 1993a).

of the appropriate lattice parameter. More detailed descriptions of Miller indices can be found in introductory crystallography texts such as Phillips (1946), Hobbs (1974, 725–26), or Glen (1975, 43). One should note that when parentheses are placed around the indices this indicates that they apply to either a single plane or a set of planes. Indices of a direction such as $[11\bar{2}0]$ are given in brackets while the full set of equivalent directions of a form are indicated by carets $\langle11\bar{2}0\rangle$. As an example consider the indices for planes that are oriented normal to the c-axis. The a_1, a_2, a_3, and c intercepts of these planes are respectively ∞, ∞, ∞, and 1. The reciprocals of these values are then 0, 0, 0, 1 and the Miller indices of the basal planes are (0001). In the same way it can be shown that $(11\bar{2}0)$ refers to the planes normal to the a-axes while $\langle11\bar{2}0\rangle$ refers to the directions normal to these planes (i.e., parallel to the a-axes). The c-axis direction, which is perpendicular to the (0001) plane, is also frequently referred to as the optic axis direction. Therefore if one says that an ice crystal has its c-axis (or optic axis) oriented exactly horizontally and in the N–S direction, this would indicate that the (0001) planes of the crystal are aligned vertically and in the E–W direction. This description does not, however, exactly specify the orientations of the a-axes, other than that they will obviously lie in the (0001) plane.

The structural arrangement of the oxygen in ice as shown in Figure 5.5 has been known from the early days of x-ray crystallography (Bragg 1922), in that the oxygen atoms scatter x-rays strongly as they are each surrounded by 10 electrons (8 of their own and 2 from the hydrogen). On the other hand, hydrogen atoms do not scatter x-rays well as they have been virtually stripped of their electrons. Petrenko (1993a) has suggested that a useful way to think about the arrangement of the oxygen atoms in ice is to consider the close packing of a layer of spheres of identical size resting on a plane as shown by the spheres marked A in Figure 5.5. Next, place an identical close-packed layer on top of level A in the positions marked B. In the figure the sizes of the spheres in the different layers are changed to aid in the viewing. In ice they would all be the same size. After layers A and B are in place, there are two alternatives for the next (C) layer. The first would be to position the third layer exactly above the A layer. This would result in the hexagonal close-packing sequence … ABABAB… Now if the B balls are replaced by dumbbells that consist of two oxygens connected by a hydrogen bond normal to the plane of the layers, one obtains almost exactly the arrangement of the oxygen in ice I(h). The primary difference is that the ice lattice is slightly squashed along the *c*-axis relative to the close-packed spheres, in that for ice $c/a = 1.6293$ whereas for the close-packed spheres $c/a = 1.633$. If the third layer were placed in the positions marked C, resulting in the sequence ...ABCABCABC..., the structure of cubic ice I(c) would result.

Figure 5.5 shows the general position of the oxygen atoms in ice as viewed from several different directions. If the positions of the O atoms are projected parallel to the *c*-axis onto the basal plane, a perfect series of hexagons is obtained. Note however that this hexagonal arrangement is composed of three puckered close-packed rows of oxygen atoms, with the directions of these rows corresponding exactly to the directions of the three equivalent *a*-axes. Other reasons why the *a*-axis directions are important are because they coincide with the directions of the dendritic arms of hexagonally shaped snowflakes and also with the directions of the arms of Tyndall figures (internal snowflakelike melt figures that can form inside ice crystals as the result of melting induced by absorbed solar radiation; Nakaya 1956). Particularly important in view of the subjects discussed in the present book are the fact that the hexagonal arms of the dendritic sea ice crystals that form during the initial growth of ice crystals from fresh water and seawater are oriented parallel to the *a*-axes. Also the basal (0001) plane of an ice crystal is the plane of easy glide or fracture, as rupture along this plane results in the breakage of fewer bonds than ruptures along other directions in an ice crystal. This can be seen by examining the unit cell of ice (Figure 5.7), in which one finds that fracture along the (0001) plane requires breaking fewer bonds than a fracture along a prism face. The (0001) plane is also the plane of easy growth in geometrically unconstrained ice crystals growing into supercooled water. In addition, as might be expected, the properties of ice crystals are invariably different when measured parallel to the *c*-axis than when measured parallel to the basal plane. Also, as all three *a*-axes are identical, properties measured in the (0001) plane are not dependent on the exact orientation of the measurement. The important point made by all the above items is that in ice, as in many other materials, the macroscopic physical properties reflect the internal structural arrangement of the atoms comprising the material.

The highly regular net of oxygen atoms shown in Figure 5.5 is held together by a series of hydrogen bonds whose distribution ideally obeys the so-called Bernal–Fowler rules: (1) two hydrogen are near each oxygen atom and (2) only one hydrogen atom can be on or near the line connecting two neighboring oxygen atoms. The first rule requires the electroneutrality of a water molecule in the ice crystal lattice whereas the second rule fulfills the main characteristic of the hydrogen bond in that a single proton is required. Within the constraints of these rules all configurations are considered to be equally probable. The average structure that results can be specified by assigning hypothetical half-hydrogen atoms to each of the 4N sites contained in any array of N oxygen atoms. As a result each oxygen atom is tetrahedrally surrounded by four half-hydrogens. This arrange-

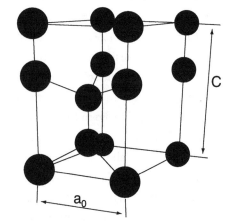

Figure 5.7. The unit cell of ice I(h) (after Petrenko 1993a).

ment of half hydrogen atoms is in good agreement with both the observed zero-point entropy of ice (Nagle 1966; Pauling 1935; Suzuki 1967) and with the results of single-crystal neutron diffraction studies (Peterson and Levy 1957). Any violation of the Bernal–Fowler rules can be considered to produce a defect in the ice structure. A violation of rule 1 results in an ionic defect in that an oxygen atom surrounded by three protons produces a positive ion $(H_3O)^+$ whereas one with only one proton produces a negative ion $(OH)^-$. When rule 2 is violated a Bjerrum defect results; when two protons occur on the bond a so-called D-defect results, whereas a bond with no protons gives an L-defect.

Another important point is that the ice in sea ice is not only ice I(h), it is *pure* ice I(h) without containing any appreciable amount of impurity in solid solution, where this term refers to the impurity atoms actually occupying lattice sites in the host crystals, a situation common in many other solids. Again, the reason is related to the characteristics of the O–H bond. For a solid solution to occur, the foreign atom must not only be of the right size to fit into the site in the host crystal structure, it also must have an appropriate charge to maintain electrostatic neutrality, and tend to form a similar type of chemical bond. For ice these tend to be rather stringent requirements that are not met by materials present in significant quantities in seawater. This is rather fortunate, in that the chemical compounds that do form limited solid solutions in ice are not particularly pleasant and include HF, F^-, NH^{+4}, and NH_4F, as well as the other hydrohalogen acids (Hobbs 1974).

5.4 Solid Salts

Our knowledge of the characteristics of the solid salts that occur in sea ice is somewhat limited. As noted earlier, the main salts which have been reported as precipitating from seawater brine are $CaCO_3 \cdot 6\,H_2O$, $Na_2SO_4 \cdot 10\,H_2O$, and $NaCl \cdot 2\,H_2O$ at –2.2, –8.2 and –22.9°C, respectively. The other salts that precipitate (KCl, $MgCl_2 \cdot 12\,H_2O$, and $CaCl_2 \cdot 6\,H_2O$) do so only at temperatures below –36.8°C, conditions that are rarely encountered in sea ice in nature. Some properties of these phases are listed in Table 5.4. Only in a few cases are the effects of the presence of these salts in sea ice reasonably well understood. What is clear is that the precipitation of $NaCl \cdot 2\,H_2O$ and in some cases $Na_2SO_4 \cdot 10\,H_2O$ are associated with breaks in slopes on plots of some sea ice properties against either temperature or, more importantly, against brine volume. In addition, Maykut and Light (1995) have shown quite conclusively that the presence of solid salts results in a large increase in the observed scattering from sea ice at optical frequencies. In fact, when one visually examines cold sea ice, one can observe a clear color change associated with the temperature of ~22.9°C, with gray ice occurring at the warmer locations below this temperature boundary (closer to the seawater–ice interface) and whitish, chalky-colored ice occurring at the colder temperatures above it. There can be little doubt that this color change is the result of the precipitation of $NaCl \cdot 2\,H_2O$. It is clear that adequate simulation of these variations by current radiative transfer models such as those used by Grenfell (1991) and Jin et al. (1994) will require hard numbers on the size distributions, shapes, and spatial distributions of the salt crystals as a function of temperature. Collections of this type of data can be found in the papers of Dieckmann et al. (2008), Light et al. (2003a, 2009), and Perovich and Gow (1991, 1992), and will be discussed in Chapters 7 and 8. Photographs as well as sizes of mirabilite and ikaite crystals can be found

Table 5.4. Properties of the solid salts occurring in sea ice (for additional discussion see Chapter 6).

Composition	Mineral name	Crystal system	Density (Mg/m^3) and [index of refraction]	Eutectic temp. (aqueous solution)	Ppt. temp. in seawater (experimental)	Ppt. temp. in seawater (theory)
H_2O	ice	hexagonal	0.9167 (0°C) [1.309, 1.313]		−1.921	−1.924
$CaCO_3 \cdot 6 H_2O$	ikaite	monoclinic	1.771 [1.460, 1.535, 1.545]	?	−2.2	?
$Na_2SO_4 \cdot 10 H_2O$	mirabilite	monoclinic	1.464 [1.394, 1.396, 1.398]	−3.6°C	−8.2	−6.3
$MgCl_2 \cdot 8 H_2O$				−33.6	−18.0	
$NaCl \cdot 2 H_2O$	hydrohalite	monoclinic	1.630 (0°C) [~1.43] (Light et al. 2009)	−21.1	−22.9	−22.84
KCl	sylvite	cubic	1.984 [1.490]	−11.1	−36.8	−34.25
$MgCl_2 \cdot 12 H_2O$		monoclinic	(1.24)	−33.6	−43.2 (erratic)	−36.82
$CaCl_2 \cdot 6 H_2O$	antarcticite	hexagonal	1.718 (4°C) [1.417 1.392]	−55.0	<−55.0	−53.64

in Light et al. (2003a) and Dieckmann et al. (2008), respectively. Attempts to determine crystal sizes for hydrohalite have, to date, not been successful. The dangers of arbitrarily introducing descriptions of the state of the solid salts in sea ice are attested to by the salt reinforcement theories advanced in the 1950s and 1960s by Assur (1958) and Peyton (1966), in that the predictions of their models were not supported by further measurements. In fact, the effect of solid salts on sea ice strength is still far from clear.

6 The Phase Diagram

If you don't know where you are going, you will end up somewhere else.

L. Peter

Wherever you go, there you are.

Buckaroo Banzai

6.1 Fundamentals

In order to understand sea ice, one must have a general understanding of the so-called sea ice phase diagram, which specifies the amounts and compositions of the phases that exist at different temperatures when one freezes seawater to form sea ice. The fact that the development of a phase diagram was an essential step in understanding the variations in the properties of sea ice has been realized for a considerable time (Gitterman 1937; Malmgren 1927; Ringer 1906, 1928; Zubov 1945), for without such a diagram it would be impossible to specify the phases and their relative volumes in a sea ice sample having a known salinity and temperature.

As was noted in the previous chapter, the ice that forms when seawater freezes is composed of extremely pure H_2O if one only considers impurities that occupy lattice sites within the ice crystals, that is, impurities that are in solid solution. Even impurities such as F^-, HF, NH_4^+, and NH_3, which fulfill the requirements of possessing an appropriate charge to maintain electrical neutrality, having the right size to fit into the ice structure, and forming a similar type of chemical bond, only show limited substitution in ice. Maximum substitution of these materials corresponds to a molar ratio of about 1 in 5000. If the materials that actually occur in seawater are considered, equilibrium substitution values are not well known and are extremely small. Harrison and Tiller (1963a) estimate that for most solutes the equilibrium solute partition coefficient k_o (where k_o is given by

$$k_o = \frac{x_A^s}{x_A^\ell} \qquad (6.1)$$

and x_A^s and x_A^ℓ are the mole fractions of solute A in the solid and liquid, respectively) is less than 10^{-4}. This says that, even though thermodynamic arguments can be made that imply that at equilibrium a finite amount of impurity should exist as a solid solution in the ice structure, this amount is so small that for our purposes the ice in sea ice can be considered a pure phase. As a result when ice forms during the freezing of an aqueous salt solution, solute is rejected back into the remaining solution. As the temperature decreases, resulting in the formation of more and more ice, the solution (melt) that coexists in equilibrium with the ice contains more and more solute (NaCl).

The phase relations simply specify the number and composition of the different phases (ice, brine, and solid salts) that coexist at different temperatures and pressures. The results of the

Figure 6.1. A portion of the NaCl–H₂O phase diagram.

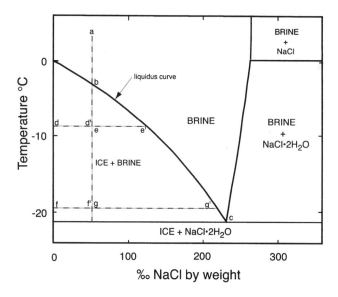

experimental observations made to determine such values can be displayed graphically in the form of a phase diagram. In that the water-rich portion of the NaCl–H$_2$O system is the simplest system that approximates the formation of sea ice, let us examine it more closely. Figure 6.1 shows this diagram for the high water–low temperature portion of this system. This diagram tells us what will happen if we take pseudo-seawater (a fairly dilute NaCl solution) and freeze it. Consider a solution with a composition of ca. 50‰ NaCl by weight (a little saltier that typical seawater) that is initially at a temperature of +8°C, as indicated by the location of letter a in the figure. What does the phase rule tell us about the state of the solution a? The general statement of the phase rule is $P + F = C + 2$ where P, F, and C are the number of phases, the number of degrees of freedom, and the number of components, respectively. A phase is simply a homogeneous portion of a system. For instance ice, brine, and NaCl are all phases. The number of components of a system is the smallest number of independently variable constituents by means of which the composition of each phase involved in the equilibrium can be expressed. For instance the two-component system NaCl–H$_2$O can contain the phases NaCl, NaCl·2 H$_2$O, ice, and brine. Finally the degrees of freedom F of a system can be thought of as the number of variables to which values must be assigned in order to completely specify the state of the system. Actually, for the situation we are considering the phase rule should be modified to $P + F = C + 1$, in that we have arbitrarily fixed one degree of freedom by the fact that all our experiments on salt solutions have been carried out at a fixed pressure of ~ 1 atmosphere pressure. For individuals who have escaped from college *sans* physical chemistry, I strongly recommend that they obtain a suitable textbook (texts on physical chemistry for students of ceramics and metallurgy invariably treat the phase rule thoroughly) and read up on this deceptively simple yet powerful relation, whose author, J. Willard Gibbs, was clearly the premier American scientist of the late 1800s (Gibbs 1957).

In our case, at point a, $P = 1$ (the salt solution), $C = 2$ (NaCl and H$_2$O), and therefore $F = C + 1 - P = 2$. Two degrees of freedom means that if the pressure is fixed at 1 atmosphere, we must also fix two variables (temperature and composition) to specify the state of the system. One might well ask why composition only counts as one variable, yet we are examining a two-component system. The answer is that in a two-component system the sum of the two components must equal 100%. Therefore, if we know the amount of NaCl in our initial solution, we also know the amount of H$_2$O and vice versa. As the system cools, F remains at 2 until point b is reached at a temperature of ca. −3°C, when ice starts to freeze out of the salt solution. Once ice starts to form, $P = 2$ in that both ice and brine now coexist, C remains at 2, and from the phase rule $F = 1$, indicating that the system is now univariate. In our case this means that as long as ice and brine coexist and the pressure is constant, it is only necessary to specify one parameter (either the temperature or the bulk composition) to fix the state of the system. As we initially specified the bulk composition of our solution (~50‰ NaCl), if we fix the temperature we have also specified the composition of the brine that can coexist in equilibrium with ice. In fact, the liquidus curve (b to c) gives the composition of the brine that can coexist with ice at equilibrium as a function of temperature. As can be seen from this curve, as cooling continues and more and more ice is formed, the remaining brine becomes saltier and saltier in that the growth of ice removes water from the solution. For instance, at –8°C a brine composition of ~120‰ (pt. e') coexists with pure ice (pt. d), whereas at –19°C the brine composition becomes ~220‰ (pt. g') while the ice composition remains the same (pt. f = pt. d = 0‰ NaCl by weight). Note that at each temperature the coexisting compositions are given by the two ends of lines such as d–e' and f–g'. The fact that the compositions of points d and f are the same (e.g. pure H_2O in the form of ice) merely attests to the fact that in these systems ice does not accept a measurable amount of impurity into its structure as a solid solution. Also, at each temperature the amount of ice separated (say between points b and e') can be determined from ratios such as $100 \times (e-e')/(d-e')$ percent of H_2O. In a similar manner the amount of remaining brine of composition e' is given by $100 \times (d-d')/(d-e')$ percent of composition a. Note also that these values are dependent on the initial composition of the solution (the location of point a on the composition axis). Historically, although liquidus curves could in principle be determined from the thermodynamic properties of the participating phases, in fact they have usually been determined by careful experiments in which the initial freezing and complete melting points of a number of samples having different bulk compositions are determined. More will be said about this later.

If cooling continues to –21.2°C a third phase forms (the solid salt NaCl•2 H_2O) so that ice + brine + NaCl•2 H_2O now coexist. This means that $P = 3$ and $C = 2$ in that the bulk composition of our sample has not changed and therefore $F = 0$. Such conditions are referred to as invariate (i.e., further cooling cannot continue until one of the phases has disappeared). As heat continues to be removed from the sample, solidification continues at the constant eutectic temperature of –21.2°C, with both ice and NaCl•2 H_2O forming until the last of the brine disappears. Only then can the sample, which is now composed of only two solid phases (ice and NaCl•2 H_2O), cool further in that the disappearance of the brine has caused the system to gain a degree of freedom. Note that the eutectic point (c) is the intersection of the liquidus curve that gives the composition of brines that are in equilibrium with ice, on the one hand, with the liquidus curve that gives the composition of brines that are in equilibrium with NaCl•2 H_2O, on the other. It is the lowest temperature attainable in the NaCl–H_2O system in which a liquid phase, the brine, is present. In fact, careful nuclear magnetic resonance (NMR) measurements indicate that in the H_2O–NaCl system there is a small but finite amount of liquid present below the eutectic temperature (in the temperature range between –22 to at least –40°C; Cho et al. 2002). It is currently believed that this is not a violation of the phase rule but instead is a manifestation of the quasi-liquid layer that is observed in pure ice.

Now consider a somewhat more realistic representation of seawater by examining the phase relations in the system H_2O–NaCl–Na_2SO_4. Figure 6.2 is a three-dimensional representation of the

Figure 6.2. A three-dimensional schematic drawing of a portion of the NaCl–Na$_2$SO$_4$–H$_2$O phase diagram.

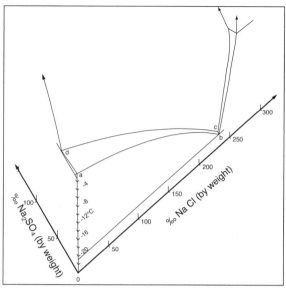

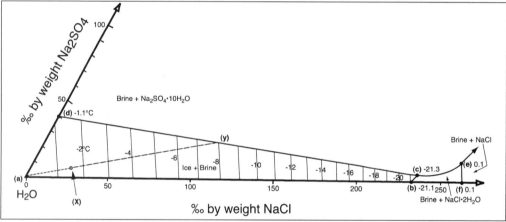

Figure 6.3. A plan view of a portion of the NaCl–Na$_2$SO$_4$–H$_2$O phase diagram.

phase diagram for this system, and Figure 6.3 is a plan view of the H$_2$O-rich, low-temperature portion of the system. Let us use Figure 6.3 to examine the sequence of events that occurs when such a system freezes. Note that the H$_2$O–Na$_2$SO$_4$ binary system is also a simple eutectic with a eutectic temperature of –1.1°C where the phases ice + brine + Na$_2$SO$_4$•10 H$_2$O coexist. Now examine what happens when a solution with a composition x is cooled. Starting at temperatures high enough that solid phases have not started to crystallize, we find that $P = 1$, $C = 3$, and therefore $F = 3$. Again this means that to fix the state of the system we need to specify the temperature and the composition, but fixing the composition now requires us to specify the amounts of any two of the three components. As our initial composition x lies in the ice + brine field, the first solid phase to form is ice. As ice continues to form, the composition of the remaining liquid changes along the x–y portion of the longer straight line a–y, as lower and lower temperatures are reached. During this portion of the freezing cycle $P = 2$ (ice and brine) and $F = 2$. If the composition of the brine reaches point y an additional phase (Na$_2$SO$_4$•10 H$_2$O) forms. We now have three phases coexisting (ice + brine + Na$_2$SO$_4$•10 H$_2$O), so that $P = 3$ causes F to become 1. As before in the simple NaCl–H$_2$O system we only need to fix one parameter, such as the temperature, to specify the state of the system. Curves such as the

d–y–c curve are called cotectic curves. The composition of the remaining brine gradually changes along the curve y–c until it reaches point c where NaCl·2 H$_2$O also forms. We now have brine + ice + NaCl·2 H$_2$O + Na$_2$SO$_4$·10 H$_2$O coexisting ($P = 4$) causing the system to again become invariate, meaning that the sample will remain at a constant temperature (a ternary eutectic at –21.3°C) until all the brine disappears, resulting in a sample of three solids, the two solid salts and ice. Note that if the Na$_2$SO$_4$ content of the initial solution is sufficiently small so that the extension of the a–"bulk composition" line encounters the b–c line instead of the d–c line, a eutectic is achieved only after the compositional path of the liquid proceeds down the b–c cotectic line to point c. Along the b–c line the phases in equilibrium would be ice + brine + NaCl·2 H$_2$O. Again a eutectic is only achieved when the liquid composition reaches point c at a temperature of –21.3°C and Na$_2$SO$_4$·10 H$_2$O forms, causing $F = 0$.

If we were to add another component, the problem of showing graphically exactly what is happening when a salt solution is frozen would become difficult in that the curves would have to be plotted inside a tetrahedron. When seawater is considered, the problem is even worse in that simple models for sea ice consider it to be at least an eight- to ten-component system. This means that for a true eutectic to exist requires brine plus eight to ten solid phases to achieve a chemical equilibrium. This appears to me to be very improbable. To this should be added the fact that nuclear magnetic resonance studies of the amount of brine present in real sea ice have indicated the presence of a liquid phase at a temperature of –51°C (Richardson 1976) and possibly even at a temperature of –70°C (Richardson and Keller 1966). However, experimental uncertainties place this latter result in some doubt. Even if only the –51°C result is true, it shows that brine exists at temperatures far colder than temperatures experienced by real sea ice. The only time that one would possibly find sea ice that is colder than –50°C would be in a block in the snow-free upper part of a pressure ridge, and this ice typically has a salinity near zero anyway. I mention this because it is not unusual in the literature to see the freezing of seawater referred to as a process in which eutectics occur. Strictly speaking this is probably not the case. Certainly, as we have shown, many of the salts present in seawater show eutectic relations in simple hydrous binary systems. In addition, eutectics can be attained at very low temperatures when, in experiments, artificial seawater containing only a few components is used. In nature the freezing of seawater can be considered to be a series of slides down connected cotectic curves. When new phases appear, the slopes of the curves and their compositional directions change. However the system never becomes invariant, staying at a constant temperature waiting for a phase to disappear so that a degree of freedom can be gained.

6.2 Experiments and Analysis

So what is one to do when confronted with working out the phase relations for real sea ice? The first thing is to find a chemist who has determined the compositions and the amounts of the phases that coexist in the seawater system at different temperatures below 0°C. In the mid-1950s there were three such sources: Ringer (1906, 1928), Gitterman (1937), and Nelson and Thompson (Nelson 1953; Nelson and Thompson 1954). The results and procedures of Ringer and of Nelson and Thompson were in reasonable agreement, whereas the procedures and results of Gitterman were slightly different and were also not readily available in the West. However, as is

Table 6.1. Phase relations for standard sea ice as developed by Assur (1958). The information presented includes (in g/kg soln.) the amounts of water in the brine of the different solid salts and of ice as a function of temperature (°C). The bulk salinity is assumed to have a constant value of 34.325 (chlorinity = 19.0). In the original table Assur also presents the amounts of the K^+, Ca^{++}, Mg^{++}, $SO_4^=$, Na^+, and Cl^- ions and the apparent amount of unidentified salts. The second column, labeled H_2O (NMR), gives the H_2O values as determined by nuclear magnetic resonance techniques (Richardson 1976).

T–°C	H_2O (NMR)	H_2O	Soln.	$MgCl_2 \cdot 12\,H_2O$	KCl	$NaCl \cdot 2\,H_2O$	$MgCl_2 \cdot 8\,H_2O$	$NaSO_4 \cdot 10\,H_2O$	$CaCO_3 \cdot 6\,H_2O$	Ice
0	965.00	965.68	1000.	0	0	0	0	0	0	0
2	860.83	878.91	913.2	0	0	0	0	0	0	86.77
4	488.71	451.61	485.9	0	0	0	0	0	0.07	515.01
6	335.00	309.30	343.6	0	0	0	0	0	0.13	656.23
8	247.53	236.76	271.0	0	0	0	0	0	0.18	728.78
10	196.9	195.29	227.8	0	0	0	0	3.95	0.21	768.02
12	173.41	169.17	200.8	0	0	0	0	5.92	0.21	793.04
14	154.77	151.08	182.4	0	0	0	0	6.80	0.21	810.65
16	139.78	137.26	168.3	0	0	0	0	7.36	0.21	824.15
18	127.04	125.78	156.6	0	0	0	0	7.75	0.21	835.41
20	116.23	115.56	146.3	0	0	0	0.10	7.99	0.21	845.43
22	106.67	106.72	137.3	0	0	0	0.24	8.17	0.21	854.07
24	77.48	75.91	98.7	0	0	12.40	0.39	8.35	0.28	879.93
26	52.35	48.93	63.8	0	0	24.90	0.55	8.51	0.29	901.94
28	39.18	36.64	47.8	0	0	30.56	0.73	8.66	0.31	911.88
30	32.10	30.22	39.5	0	0	33.38	0.93	8.78	0.31	917.03
32	27.40	26.54	34.8	0	0	34.88	1.12	8.86	0.31	919.97
34	23.39	23.84	31.4	0	0	35.89	1.36	8.88	0.31	922.14
36	19.72	21.94	29.0	0	0	36.49	1.58	8.88	0.31	923.67
38	17.01	20.39	27.1	0	0.02	36.92	1.81	8.88	0.31	924.93
40	14.89	18.97	25.3	0	0.09	37.25	2.05	8.88	0.31	926.08
42	13.51	17.80	24.0	0	0.15	37.49	2.28	8.88	0.31	927.01
44	10.99	13.39	18.1	3.75	0.23	37.66	2.28	8.88	0.31	928.75
46	7.80	8.12	11.1	8.59	0.34	37.79	2.28	8.88	0.31	930.60
48	5.15	5.38	7.5	10.91	0.50	37.89	2.28	8.88	0.31	931.69
50	3.40	3.90	5.4	12.09	0.63	37.96	2.28	8.88	0.31	932.34
52	2.65	3.00	4.2	12.84	0.69	38.00	2.28	8.88	0.31	932.70
54		2.44	3.5	13.28	0.73	38.03	2.28	8.88	0.31	932.94

shown in Figure 6.4, which presents results from all three investigations on brine concentration as a function of temperature, the differences between the studies are not large when compared to the total variation. This figure, which was taken from Doronin and Kheisin (1977), has been changed slightly in that the curve through the data has been linearized with breaks in slope occurring at the crystallization temperatures of $Na_2SO_4 \cdot 10\,H_2O$ and $NaCl \cdot 2\,H_2O$. Therefore, in 1957, two individuals who were both exploring the problem of developing a phase diagram for sea ice, Anderson (1958a) and Assur (1958), made the same reasonable decision, which was to primarily base their investigations on the Nelson and Thompson experimental observations, which were reported in considerable detail in Nelson's (1953) Ph.D. thesis. Not surprisingly, in that they both started with the same numbers, they came up with similar results.

The phase relations as developed by Assur are tabulated in Table 6.1 and graphically summarized in Figure 6.5 by consecutively adding and plotting the values from the table. Note that there are pronounced breaks in the slopes of the curves at temperatures where the different solid salts are assumed to start to crystallize. Assur used this somewhat unusual semilog presentation so that the presence of components only present in small amounts could be clearly seen. In the diagram he considers 1000 g of "standard" sea ice having a bulk salinity of 34.325‰. By comparing the table and the figure one can see that at −10°C such ice is composed of 768 g of ice, 4.2 g of solid salt $(4.0 \text{ g of Na}_2\text{SO}_4 \cdot 10 \text{ H}_2\text{O} + 0.2 \text{ g CaCO}_3 \cdot 6 \text{ H}_2\text{O})$, and 228 g of brine composed of 195 g H_2O and 32 g of other ions. At −30°C the number of grams of ice and of solid salt present has risen to 917.0 and 43.4, respectively, with 76% of the solid salt being $\text{NaCl} \cdot 2 \text{ H}_2\text{O}$. At this temperature the amount of brine has decreased to 39.5 g.

The question that one must now ask is how good the experiments were, particularly the Nelson and Thompson experiments on which the Anderson and Assur calculations were based. First, it is essential to note that Nelson and Thompson were not performing experiments focused on determining a phase diagram for sea ice. The primary problem they were considering was the possibility that certain sulfate mineral deposits formed by freezing seawater. As a result, their experimental observations have some shortcomings when applied to the sea ice phase diagram problem. The experimental procedures were as follows. An initial volume of natural seawater was cooled to a series of temperatures in a freezing tube with insulated sides so that ice growth would essentially be one-dimensional from the top down. When the tube had thermally equilibrated at the desired temperature, the brine and solid salts beneath the ice were drained off and analyzed. As the temperatures became lower the problems of obtaining enough brine for the chemical analysis procedures used at the time became increasingly difficult. To overcome this problem they resorted to two different procedures. In the first they went through a preliminary freeze concentration procedure, cooling 8 and 50 liters of seawater to about −8°C. They then separated the brine before proceeding on to lower temperatures. In the second, they initially concentrated 50

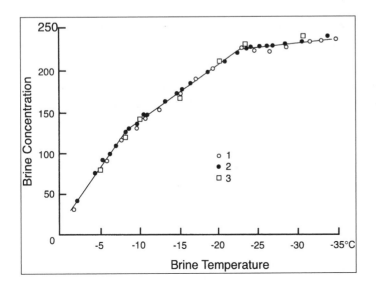

Figure 6.4. The brine concentration during the freezing of seawater plotted vs. temperature according to the experimental results of (1) Gitterman, (2) Nelson and Thompson, and (3) Ringer as shown by Doronin and Kheisin (1977). Their original figure has been modified slightly by breaking their fitted curve into a series of straight lines.

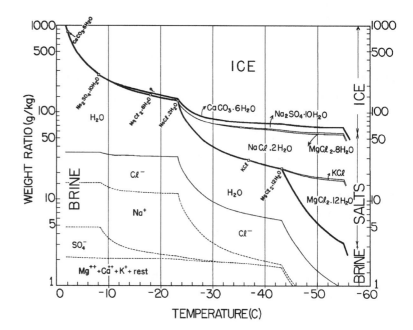

Figure 6.5. The phase diagram for "standard" sea ice according to Assur (1958). The open squares on the brine–salt line indicate temperatures at which different solid salts were observed to precipitate in the experiments of Nelson (1953) and Nelson and Thompson (1954).

liters of seawater, first at –17°C and then, after separating the brine, at –25°C, before finally transferring the residual brine to the initial freezing tube.

When these procedures are examined relative to the sea ice phase diagram problem, several deficiencies have been noted (Savel'yev 1963; Tsurikov and Tsurikova 1972; Weeks and Ackley 1986). They are as follows:

1. There was never a direct determination of the compositions of the solid salts that formed. Compositions were inferred by changes in the composition of the brine and in other cases by chemical procedures that enabled one to determine the presence of specific cations or anions.

2. No attempt was made to verify that the phases had achieved equilibrium by bracketing the temperatures where specific salts crystallized. Past experience with other systems shows that apparent equilibrium relations based only on cooling curves can be seriously in error.

3. No attempt was made to assure that the liquid phase was both compositionally homogeneous and able to react with all the solid phases in the system. As noted above, the procedures used by Nelson and Thompson almost certainly guarantee that this was not true from two different points of view. First, the brine isolated in brine pockets (liquid inclusions) within the ice could no longer react with either the brine in other inclusions or with the main body of residual brine located in the lower part of the container. Second, the separation of the brine at –8, –17 and –25°C during the concentration procedures essentially prohibits the possibility of reactions between the brine and previously formed solid salts. For instance, both Savel'yev (1963) and Gitterman have suggested

$$CaSO_4 \cdot 2\,H_2O + 2\,NaCl \cdot 2\,H_2O \leftrightarrow CaCl_2 \cdot 6\,H_2O + Na_2SO_4 \cdot 10\,H_2O$$

as a possible reaction. Another more likely possibility is

$$Na_2SO_4 \cdot 10\,H_2O + CaCl_2\,(aq) \leftrightarrow CaSO_4 \cdot 2\,H_2O + NaCl\,(aq) \leftrightarrow NaCl \cdot 2\,H_2O$$

(Marion, pers. comm.). Prior to the precipitation of $NaCl \cdot 2\,H_2O$ that creates a Na sink resulting in the reaction shifting to the right, the high Na/Ca ratio in seawater brine would favor the left side of this relation.

4. Finally, all experimental work to date has ignored the carbonate content of the brine, even though $CaCO_3 \cdot 6\,H_2O$ appears to be the first solid salt to form during freezing. However, this deficiency may not be as significant as it appears in that $CaCO_3 \cdot 6H_2O$, would be only a minor constituent of the freezing of seawater and also appears to be metastable relative to calcite. Marion (pers. comm.) has simulated the freezing of seawater both with and without the presence of $CaCO_3$ using a thermodynamic model that will be discussed later. There was no significant difference.

I will return to these apparent deficiencies later but now I will discuss the typical application of the phase diagram as applied to sea ice a bit further.

In developing his version of the phase diagram, Assur made the following assumptions:

1. that the freezing point of the residual brine in the Nelson and Thompson experiments is the same as that of the liquid brine in the brine inclusions in sea ice, and
2. that the relative concentration of the different ions, relative to each other, is also the same as measured in the residual brine.

Assur also pointed out that in a strict sense his calculations apply to what he termed "standard sea ice" (i.e., sea ice of such a composition that its meltwater will have the same relative concentration of ions to each other as in normal seawater). He also noted that at its freezing point, the salt content of the brine produced from seawater and in presumed equilibrium with ice can be expressed as a linear function of temperature if breaks in slopes are permitted at temperatures where solid salts form. As noted earlier, his tabular results giving the phase relations for standard sea ice derived from seawater with an initial salinity of 34.325‰ are given in Table 6.1. A table giving the interrelations between temperature, salinity, and brine volume for standard sea ice having a salinity of 1‰ can be found in Assur (1958). To obtain the brine volumes V_b for sea ice samples having other salinities, simply multiply the values in the table by the appropriate salinity value (for example, a sea ice sample with a salinity of 10.0‰ at a temperature of $-12.4°C$ has a brine volume of 47.1‰). For most applications where extreme accuracy is not essential, the fact that brine volume V_b can be linearized as a function of $(1/T)$ and S_i where T is the temperature (°C) and S_i is the salinity of the ice in ‰ can be utilized using the equations developed by Frankenstein and Garner (1967):

$$V_b = S_i\left(\frac{45.971}{T} + 0.930\right) \quad -8.2 \le T \le -2.0°C \qquad (6.2)$$

$$V_b = S_i\left(\frac{43.795}{T} + 1.189\right) \quad -22.9 \le T \le -8.2°C \qquad (6.3)$$

Note that different relations are used on the opposite sides of the $Na_2SO_4 \cdot 10\,H_2O$ crystallization temperature of –8.2°C. At temperatures above and below the stated range, it is necessary to refer directly to the table. Also note that both Assur's table and the above equations assume that sea ice has a constant density of 0.926 Mg/m³. Therefore, to determine the applicable brine volume for sea ice having a known bulk density of ρ, the calculated brine volume should be multiplied by $(\rho/0.926)$. Equations incorporating this density change and allowing the calculation of both the brine and gas volumes in sea ice of known densities will be developed later in this section.

The best check on the adequacy of the Assur/Anderson phase relations can be found in the studies by Richardson and Keller (1966) and Richardson (1976), who directly determined the amount of liquid water present in an artificially frozen seawater sample via the use of nuclear magnetic resonance (NMR) techniques. Their values were then compared with values for the same parameter as given by Assur, based on Nelson and Thompson's measurements. As can be seen in Figure 6.6, the agreement is very good at temperatures above –31°C and below –45°C. In between these two temperatures Assur's liquid water contents are consistently somewhat higher than the NMR values. Geophysically these differences are probably of little import in that sea ice rarely exists at temperatures below –30°C.

An additional uncertainty about the adequacy of the Anderson/Assur phase diagram relates to the "standard sea ice" assumption that its meltwater will have the same relative concentration of ions to each other as does normal seawater. One possible mechanism that would "nonstandardize" sea ice would be a situation in which as the ice ages and brine gradually drains out of the ice, the solid salts remain fixed in the ice. This would result in the ice gradually becoming enriched in the solid salts that form at the higher temperatures. For instance, if the upper surface of sea ice were to be maintained at –20°C for a considerable period of time, this would cause $CaCO_3 \cdot 6\,H_2O$ and $Na_2SO_4 \cdot 10\,H_2O$ to form while all the Cl^- ions would remain in the brine. Under such conditions brine drainage could conceivably produce increases in the $SO_4^=/Cl^-$ and $CO_3^=/Cl^-$ ratios. If such changes were large enough, the standard sea ice phase diagram of Assur would no longer be applicable. The question is, do such changes occur? In newly formed ice and in FY ice that has not been subjected

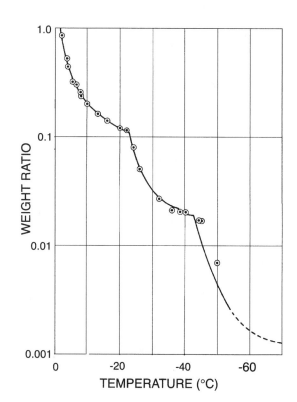

Figure 6.6. A plot of the observed brine volumes as determined by Richardson and Keller (1966) using nuclear magnetic resonance techniques (circles) vs. values (the smooth curve) as calculated by Assur (1958), based on the experimental studies of Nelson (1953) and Nelson and Thompson (1954).

to long periods of low temperatures, most observations suggest that any changes in the ion ratios are small (Addison 1977; Bennington 1963b; Blinov 1965). For MY ice, where changes in the ion ratios would be expected to be the largest, observational results are quite varied (Anderson and Jones 1985; Reeburgh and Springer-Young 1983; Tsurikov 1974, 1976). However there does not appear to be a clear trend or a consistent difference even between FY and MY ice (Anderson 1998; Meese 1989). This has a negative consequence in that it is not possible to use parameters such as $SO_4^=/Cl^-$ ratios as indicators of water masses that have been affected by sea ice. Nevertheless, the absence of consistent ratio changes is quite fortunate in that if this were to occur the standard phase diagram would no longer be generally applicable, with somewhat different phase relations being exhibited by ices that had different thermal histories. Such a situation would require the field worker to either perform a complete chemical analysis on every sample or to utilize some technique such as NMR that would allow the direct estimation of brine volume values. In that both of these procedures would be expensive and time-consuming, characterizing sea ice properties would become even more difficult than it is at present.

What we need now are a series of equations that build on and formalize our knowledge of the phase relations and that can be utilized later in estimating a variety of sea ice properties. Such relations have been worked out by several different authors (Cox and Weeks 1983; Doronin and Kheisin 1977). An excellent review of Russian work on this subject can be found in the paper by Nazintsev (1997). Here I will follow the Cox and Weeks development because I am better acquainted with it.

In the derivations that follow, m_ℓ, ρ_ℓ, and V_ℓ are the mass, density, and volume of component ℓ where the subscripts a, b, i, s, and ss denote air, brine, pure ice, salt, and solid salts, respectively. Also M, V, and ρ represent the bulk mass, the bulk volume, and the bulk density, respectively. These values should not be confused with the gas-free theoretical mass, volume, and density. Also the terms m_s^b and m_s^{ss} indicate the mass of salt in the brine and the mass of salt in the solid salts, respectively. The mass of gas in the ice is assumed to be negligible. By definition the salinity of the ice S_i and the salinity of the brine S_b are given by

$$S_i \equiv \frac{m_s}{M} = \frac{m_s^b + m_s^{ss}}{m_b + m_{ss} + m_i} \tag{6.4}$$

and

$$S_b \equiv \frac{m_s^b}{m_b} . \tag{6.5}$$

Recalling that the brine salinity and the relative amounts of salt present in the brine and as solid salts are unique functions of the ice temperature via the phase relations, we can use the above two equations to derive an equation for the brine volume of sea ice. From equation 6.4

$$m_s^b = M S_i - k m_s^b \tag{6.6}$$

where

$$k = \frac{m_s^{ss}}{m_s^b} . \tag{6.7}$$

Now solving for m_s^b from equation 6.7 and noting from equation 6.5 that

$$m_s^b = \rho_b \, V_b \, S_b \,, \tag{6.8}$$

it follows that

$$V_b = \frac{M \, S_i}{\rho_b \, S_b} \left(\frac{1}{1+k} \right). \tag{6.9}$$

In that $M = \rho \, V$, equation 6.9 can also be written as

$$\frac{V_b}{V} = \frac{\rho \, S_i}{\rho_b \, S_b} \left(\frac{1}{1+k} \right) \tag{6.10}$$

where V_b/V is the relative brine volume.

In that the brine density can be approximated as a function of brine salinity (see equations 5.3, 5.4, and 5.5) and because S_b and k are unique functions of temperature, the relative brine volume can be expressed as

$$\frac{V_b}{V} = \frac{\rho \, S_i}{F_1(T)} \tag{6.11}$$

where S_i is in ‰, ρ is in Mg/m³, and

$$F_1(T) = \rho_b \, S_b \, (1+k). \tag{6.12}$$

The function $F_1(T)$ is plotted against temperature in Figure 6.7 and the parameters for the least squares curves fitted to the data are presented in Table 6.2. A tabulation of the $F_1(T)$ and $F_2(T)$ values can be found in Cox and Weeks (1983). Note that although there is a pronounced break in the slope at the temperature where NaCl•2 H$_2$O starts to precipitate, a similar feature is not apparent in the vicinity of the crystallization temperature of NaSO$_4$•10 H$_2$O. A comparison between the brine volume estimates given by equation 6.12 based on the Assur tabulations and the results of Nazintsev (1997) are shown in Figure 6.8. As can be seen, their results are in good agreement when the magnitudes are compared directly. However, when viewed as % differences, significant

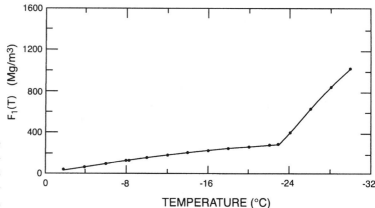

Figure 6.7. A plot of the function $F_1(T)$ vs. temperature (Cox and Weeks 1983). The curve was determined by least squares.

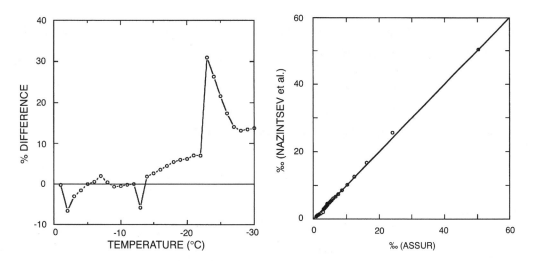

Figure 6.8. Two different presentations showing the dependence of the amount of brine present in sea ice having a salinity of 1‰ as estimated by Assur (1958) and by Nazintsev (1997).

deviations (up to 30%) are apparent, with the largest deviations ($> 5\%$) occurring at low temperatures ($< -20°C$) where the brine volume values are quite small. The temperatures where the larger deviations occur are interesting ($-2, -8, -13,$ and $-23°C$) in that the $-2, -8,$ and $-23°C$ temperatures correspond roughly with temperatures where $CaCO_3 \cdot 6\,H_2O$, $Na_2SO_4 \cdot 10\,H_2O$ and $NaCl \cdot 2\,H_2O$ are known to precipitate. On the other hand the deviation (6%) at $-13°C$ does not correspond to the precipitation temperature of any known salt. (According to Gitterman $CaSO_4 \cdot 2H_2O$ started to precipitate at $-15°C$ although the thermodynamic studies of Marion suggests a lower temperature around -22 to $-23°C$.) I have discussed these differences with Nazintsev and neither of us has an adequate explanation. At the present time if I were to have to decide between using Assur's tables and Nazintsev's, I would choose Assur's based on the good agreement between his estimates and Richardson's (1976) NMR measurements through most of the studied temperature range. However, as the absolute differences are small, for most purposes either set of results would be adequate.

One can build upon the above relations to obtain a variety of other useful information. For instance the relative volume of pure ice is given by

$$\frac{V_i}{V} = \frac{\rho}{\rho_i} - \left(1 + C\right)\frac{\rho_b}{\rho_i}\frac{V_b}{V} \tag{6.13}$$

where C is the proportionality constant at each temperature between the mass of brine and the mass of solid salts assuming that solid salts are present

$$m_{ss} = Cm_b \tag{6.14}$$

where C is a function of T. In a similar way the relative volume of solid salts is given by

$$\frac{V_{ss}}{V} = C\,\frac{\rho_b}{\rho_{ss}}\frac{V_b}{V} \tag{6.15}$$

where ρ_{ss} is taken to be the average solid salt density assumed to be constant at 1.5 Mg/m³. The relative volume of gas (air?) in the ice is equal to

$$\frac{V_a}{V} = 1 - \frac{V_b}{V} - \frac{V_i}{V} - \frac{V_{ss}}{V}.$$
(6.16)

If we substitute from (6.13) and (6.15) into (6.16), we obtain

$$\frac{V_a}{V} = 1 - \frac{\rho}{\rho_i} + \frac{V_b}{V}\left[(1+C)\frac{\rho_b}{\rho_i} - C\frac{\rho_b}{\rho_{ss}} - 1\right],$$
(6.17)

a relation that can be simplified by defining

$$F_2(T) = \left[(1+C)\frac{\rho_b}{\rho_i} - C\frac{\rho_b}{\rho_{ss}} - 1\right].$$
(6.18)

Now if one substitutes for $\frac{V_b}{V}$ from (6.9) and recalls (6.12), the following is obtained

$$\frac{V_a}{V} = 1 - \frac{\rho}{\rho_i} + \rho\, S_i \frac{F_2(T)}{F_1(T)}.$$
(6.19)

Values for F_2 are also included in Table 6.2. Finally, the density of the sea ice considering both the presence of air and solid salts can be found from (6.19) by solving for ρ:

$$\rho = \left(1 - \frac{V_a}{V}\right)\frac{\rho_i\, F_1(T)}{F_1(T) - \rho_i\, S_i\, F_2(T)}.$$
(6.20)

Cox and Weeks compare the density estimates given by equation 6.20 with estimates made by Zubov (1945), Anderson (1960), and Schwerdtfeger (1963a). The agreement with Anderson's results is excellent, which is not surprising in that both sets of calculations were based on the Nelson and Thompson experiments. The largest difference was between the Cox and Weeks results and those of Schwerdtfeger (+0.0058 Mg/m³), who did not consider the presence of solid salts and assumed that the volume of brine was equal to the volume of pure water. Comparisons between the Cox and Weeks results and the very careful direct determinations of the air volumes in sea ice samples made by Nakawo (1983) have also shown excellent agreement.

Table 6.2 Coefficients for the functions $F_1(T)$ and $F_2(T)$ as determined by least squares from the Nelson and Thompson experimental data.

T(°C)	a_o	a_1	a_2	a_3	Correlation coefficient
$F_1(T)$					
$-2 \geq T \geq -22.9$	-4.732	-2.245×10^1	-6.397×10^{-1}	-1.074×10^{-2}	0.9999
$-22.9 \geq T \geq -30$	9.899×10^3	1.309×10^3	5.527×10^1	7.160×10^{-1}	0.9999
$F_2(T)$					
$-2 \geq T \geq -22.9$	8.903×10^{-2}	-1.763×10^{-2}	-5.330×10^{-4}	-8.801×10^{-6}	0.9999
$-22.9 \geq T \geq -30$	8.547	1.089	4.518×10^{-2}	5.819×10^{-4}	0.9999

Useful tabulations of a variety of different bulk sea ice properties based on the above type of approach can be found in Nazintsev (1997). These include brine salinity, brine mass, brine volume, and density of air-free sea ice, as well as the effective heat capacity, the latent heat of sea ice, and its thermal conductivity, thermal diffusivity, and coefficient of volumetric expansion all expressed as functions of temperature and salinity. Also included are tabulations of the phase composition of sea ice with a composition of 10‰ and changes in the mass of the solid salts and a weighted mean of their heat of melting, both as a function of temperature.

The reader will note that the Cox and Weeks relations cut off at temperatures warmer than –2°C. However individuals working on sea ice during the late spring or early summer many times find themselves dealing with near-melting ice temperatures in the –0.1 to –1.9°C interval. This problem has been addressed by Leppäranta and Manninen (1988) by using the relation between salinity and the freezing point suggested by Millero (equation 5.2) to recalculate $F_1(T)$ and $F_2(T)$. Their suggested values are given in Table 6.3, where we will refer to these parameters as $F_3(T)$ and $F_4(T)$ to avoid confusion with the values presented in Table 6.2. Their paper also includes a useful discussion of the possible errors in such procedures.

Table 6.3. Coefficients for the third-degree polynomials $F_3(T)$ and $F_4(T)$ for use in the temperature interval 0° to –2°C based on the UNESCO equations for $S_b(T)$ and $\rho_b(S_b, T)$ (Leppäranta and Manninen 1988).

	a_0	a_1	a_2	a_3
$F_3(T)$	-4.1221×10^{-2}	-1.8407×10^{1}	5.8402×10^{-1}	2.1454×10^{-1}
$F_4(T)$	9.0312×10^{-2}	-1.6111×10^{-2}	1.2291×10^{-4}	1.3603×10^{-4}

6.3 Questions

Now let us return to the question of whether or not the Nelson and Thompson measurements represent a series of observations on a system that had achieved chemical equilibrium. This subject has been addressed via the application of a model which calculates the equilibrium states in multicomponent saltwater systems at below freezing temperatures (Marion and Grant 1994). The model, which has been referred to as FREZCHEM, is designed to deal explicitly with the type of low-temperature brine systems that occur in sea ice (i.e., solutions from which the chloride and sulfate salts of sodium, potassium, calcium, and magnesium precipitate in the temperature range between –60 and +25°C). The initial version of the model was described in a paper by Spencer et al. (1990). A specific description of FREZCHEM, including a FORTRAN program, can also be found in the Marion and Grant paper. A detailed description of the model is beyond the scope of the present book. As the authors stress, FREZCHEM is a chemical thermodynamic equilibrium model that calculates the equilibrium compositions of aqueous solutions at specified temperatures. What it does not provide, as is typical of such models, is any information on the time required to reach, or the path of change to, the equilibrium state. For readers who are not acquainted with such considerations, let me note that if we were to examine the thermodynamics of one of the building blocks of modern construction, we would find that steel is thermodynamically unstable relative to graphite and alpha iron. However, this fact does not prevent us from using steel in a great number of applications without the least worry that it will revert before our eyes into its fundamentally more stable components. The minerals whose thermodynamic parameters are utilized in the FREZCHEM seawater freezing simulations are given in Table 6.4.

Table 6.4. Minerals currently considered in the FREZCHEM models as applied to the freezing of seawater, their calculated initial crystallization temperatures, and the ionic species considered in solution (Marion et al. 1999; Mironenko et al. 1997).

Solid phases	Estimated temperature of first appearance (°C)		Ionic species
	N/T Path	Gitterman Path	
Ice I(h) (ice)	−1.92	−1.92	H_2O
$Na_2SO_4 \cdot 10\,H_2O$ (mirabilite)	−5.87	−6.3	Na^+
$NaCl \cdot 2\,H_2O$ (hydrohalite)	−22.87	−22.9	K^+
KCl (sylvite)	−34.30	−33.6	Ca^{++}
$MgCl_2 \cdot 12\,H_2O$	−36.82	−36.2	Mg^{++}
$CaCl_2 \cdot 6\,H_2O$ (antarcticite)	−53.73		Cl^-
$CaSO_4 \cdot 2\,H_2O$ (gypsum)		−22.2	

The initial FREZCHEM simulations of seawater freezing (Mironenko et al. 1997) only considered the thermodynamics of the first five solid phases listed in Table 6.4. The reasons given for excluding the calcium sulfate salts and calcium carbonate were several. For one, all of these salts were known to be comparatively insoluble at the temperatures considered. As a result adequate solubility data were felt to be lacking. Also, there was reason to believe that seawater was supersaturated relative to calcite at 0°C. In fact, minor amounts of $CaCO_3$ or $CaCO_3 \cdot 6\,H_2O$ were observed to precipitate during the Nelson and Thompson (N/T) experiments. There was also excellent agreement between the Ca concentrations in the N/T experimental data and the FREZCHEM simulations (see Figure 6.9).

However when $CaSO_4 \cdot 2\,H_2O$ was added to the FREZCHEM simulation the crystallization sequence changed appreciably (Marion et al. 1999). Mirabilite formed at −6.3°C, which is an appreciably higher temperature than was observed in the N/T experiments (−8.2°C) and

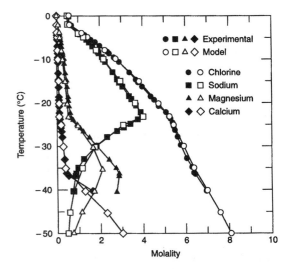

Figure 6.9. The concentrations of major seawater constituents during freezing as calculated using the FREZCHEM model (Marion and Grant 1994). The experimental results are those of Nelson and Thompson (1954).

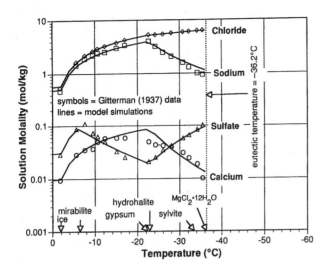

Figure 6.10. Experimental data and model simulations of chloride, sodium, sulfate, and calcium molalities during the freezing of seawater along the Gitterman pathway (Marion et al. 1999).

closer to the temperature (−7.3°C) observed by Gitterman. The model also predicted the precipitation of gypsum beginning at −22.2°C via a reversible reaction with mirabilite that works as follows.

First NaCl•2 H₂O starts to form at −22.9°C, lowering the Na concentration in the solution, and leading to the dissolution of mirabilite. The increasing sulfate concentration, in turn, causes gypsum to form (Figure 6.10). At temperatures of < −23°C, the SO_4^{-2} concentration gain in solution resulting from the mirabilite dissolution coupled with the general concentration gain in the solution produced by the removal of ice is greater than the SO_4^{-2} lost from solution by gypsum precipitation. This results in an increase in the soluble sulfate concentration at temperatures < −23°C (Figure 6.10). The model predicts that the last salt to form in this sequence is $MgCl_2 • 12\ H_2O$ at −36.2°C, also in agreement with Gitterman's results. The only minor discrepancy between Gitterman's results and the model predictions is the temperature at which gypsum begins to form, with Gitterman reporting −15°C in contrast to the FREZCHEM's prediction of −22°C. Marion et al. (1999) explored this difference with a series of lengthy experiments (43 and 366 days), which indicated that the seawater samples were undersaturated with gypsum at −15 and −20°C in agreement with the model predictions.

Marion et al. (1999) offer a very reasonable explanation for this seemingly confusing state of affairs: that there are two alternative compositional pathways that freezing seawater can take, the equilibrium Gitterman G-pathway (Figure 6.10) and the nonequilibrium Ringer, Nelson, and Thompson (R/N/T) pathway (Figure 6.11). The G-pathway occurs when there is adequate time for mirabilite dissolution and gypsum precipitation, which are both processes that are known to be sluggish even at +25°C. In his experiments Gitterman assisted the attainment of equilibrium by making certain that there was an adequate amount of precipitated salts present, in addition to the amounts of salt present in solution. He also ran his experiments for periods up to four weeks. On the other hand R/N/T have essentially focused on the concentrated solution phase from which, according to the theoretical calculations, 90% of the sulfate has been removed as mirabilite by the time that the solution has reached −20°C. As the sulfate necessary for gypsum precipitation appears to come primarily from the dissolution of mirabilite (Figure 6.12), any process such as fractional crystallization that would hinder effective reactions between the brine and the mirabilite would favor the R/N/T pathway because at temperatures < −20°C the brine by itself does not provide sufficient sulfate for gypsum to precipitate (Marion et al. 1999).

Figure 6.11. Experimental data and model simulations of chloride, sodium, sulfate, and calcium molalities during the freezing of seawater along the Ringer–Nelson–Thompson pathway (Marion et al. 1999).

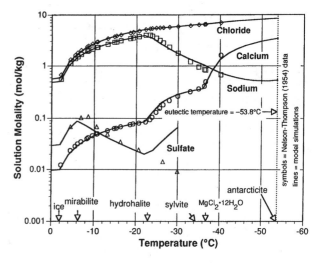

Figure 6.12. Distribution of sulfate during the freezing of seawater along the Gitterman pathway (Marion et al. 1999).

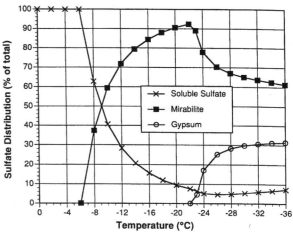

To summarize, the difference between the two pathways is the result of the poor kinetics of the mirabilite dissolution–gypsum precipitation reaction and fractional crystallization. When this reaction occurs an approximation to chemical equilibrium is achieved, gypsum forms, and the final low-temperature salt in the sequence is $MgCl_2 \cdot 12\,H_2O$ at $-36.2°C$. When the mirabilite dissolution–gypsum precipitation reaction does not occur, a situation favored by inadequate mixing in the brine pockets and channels within the ice and by any process that would result in the separation of solid salts and brine, gypsum does not form and antarcticite becomes the final low-temperature solid phase at $-53.8°C$.

This brings us to the question that is by now probably on the reader's mind: "Look, I am going to go in the field tomorrow to study sea ice. What am I supposed to do about the phase relations?" My response at this time is as follows. If you just need to know brine volumes, then measure ice temperature and salinity and use the Frankenstein and Garner equations 6.2 and 6.3. If you intend to measure ice temperature, salinity, and bulk density in order to also estimate the volume of gas in the sample, then use the Cox and Weeks equations, a procedure that can be facilitated by writing a small program for a pocket calculator that can be carried into the field. An example of such a program written for the HP-21C can be found in the 1982 CRREL version of the Cox and Weeks (1983) paper. These methods are built on Nelson and Thompson's experiments and the Anderson–Assur tables. The reason that I believe that this should be adequate is that most sea ice studies use ice that has probably never been colder than −30°C. Also if warm (> −20°C) samples are suddenly removed from a sheet of sea ice and subjected to very cold ambient temperatures, cooling across the temperature range during which the mirabilite–gypsum reaction occurs will be rapid, a situation that will favor the R/N/T pathway. If I had the misfortune of having to study sea ice that was in the temperature range between −30 and −45°C, then my recommendations become less confident and I would also consider using the values given by Richardson (1976). I guess that I would still stay with the R/N/T pathway as that choice would at the least allow me to readily compare my results with those of previous investigators. However, I would keep my eyes on the literature and look for further studies relative to the appropriate choice of pathways as a function of environmental conditions during sampling and of the thermal history of the sample.

As the reader may recall, in Chapter 5 it was mentioned that there was some question as to whether $CaCO_3 \cdot 6\,H_2O$ (the mineral ikaite) actually occurs in sea ice. Indirect evidence based on laboratory and test tank studies definitely indicated that a $CaCO_3$-rich phase of some type was forming (Killawee et al. 1998; Tison et al. 2002). Fortunately this question has now been answered: $CaCO_3 \cdot 6\,H_2O$ does form. Noting that $CaCO_3 \cdot 6\,H_2O$, once formed, is stable at temperatures <+4°C, Dieckmann et al. (2008) placed samples of antarctic sea ice in containers at +2°C for a period of two days allowing the ice phase to melt. The meltwater was then filtered and the residue subjected to microscopic and x-ray examination. Crystal shapes varied from xenomorphic to idiomorphic, with sizes varying from 0.005 to 0.6 mm. The diffraction patterns for these crystals and for those obtained from artificial ikiate were identical. These results indicate that the brine in the sea ice samples was supersaturated with respect to ikiate and that the formation of less soluble phases (e.g., calcite, aragonite, and vaterite) appears to be kinetically inhibited.

Finally, it should be noted that organisms play an important role in determining the chemistry of seawater, which in turn affects the sea ice phase diagram as well as the composition of the gas present in the ice. Individuals interested in exploring this subject should start by examining the review of the biogeochemistry of sea ice by Thomas and Papadimitriou (2003). Clearly the last word on sea ice phase relations has not been written.

7 Sea Ice Structure

The beauty of crystals lies in the planeness of their faces.

A. E. Tutton

We have just seen that an understanding of the phase diagram and of the properties of the different pure phases is an essential first step in the path leading to understanding the large natural variation in sea ice properties. The other essential ingredient is an understanding of the different types of structures that occur in sea ice, and of how these structures are affected by the environmental conditions that prevail during its formation. As a result of these linkages, the study of the structure of sea ice has received considerable attention during the last 50 years.

7.1 Environmental Pathways and Terminology

There are two environmental pathways that usually prevail during the initial formation of a sea ice cover. The first of these, and the least common, occurs during calm, cold conditions and in the absence of an appreciable swell. The initial fine spicules or platelets of ice that form on the water surface are called *frazil* until they start to freeze together forming a continuous initial ice skim on the sea surface. The thin (< 5 cm) elastic crust of ice that develops from subsequent downward growth is called *dark nilas*. It is both very dark in color and quite flexible under conditions where a slight swell is present. The reason for its dark color is that it is sufficiently thin to be somewhat transparent so that one sees the dark underlying ocean. Sometimes you will also see dark nilas referred to as *black ice*, a term that is better applied to thin layers of fresh or brackish ice which, when bubble free, can be very transparent. The flexibility of nilas is the result of the fact that its temperature is only slightly below freezing and its salinity is quite high (12–20‰). When both of these conditions occur, the result is a large brine volume within the ice which allows the platelets of ice that make up the ice crystals to flex and slip relative to one another. A special variety of dark nilas is *ice rind*, a term used to describe brittle, shiny crusts of ice that form either by direct freezing or from grease ice. Ice rind is commonly found in areas where the surface water salinity is low—for instance, coastal regions affected by the input of river water or areas within the pack where melt during the day has reduced the salinity of the surface layer. Ice rind is easily broken into rectangular pieces by either differential movements within the pack or by swell. The brittleness of this ice type is the result of its low salinity, which results in a low brine volume within the ice. As nilas thickens (5–10 cm) it becomes somewhat lighter in color and is termed *light nilas*. Incidentally, *nilas* is one of several sea ice terms that originated with the Pomor people who inhabited the coastal regions of the White Sea in northern Russia (Badigin 1956). All of the above ice types are subsets of *new ice*, which is a general term used to describe these initial ice types. Sea ice types possessing thicknesses between 10 and 30 cm are referred to as *young ice*. Young ice, which is the transition stage between nilas and first-year ice, is in turn subdivided into *gray ice* (10–15 cm) and *gray-white ice* (15–30 cm). The behavioral differences here are that gray ice is less elastic than nilas, commonly is broken by swell, and usually rafts when subjected to pressure. Gray-white ice, in turn, is even less flexible than gray ice and when deformed more often forms ridges than rafts.

Once the ice thickness has entered the range 30 to ~ 200 cm it is referred to as *first-year* (FY) *ice*. As will be discussed in the chapter on salinity variations, there is a good reason for having a terminology break at 30 cm in that a significant change in the rate at which the mean salinity of the ice decreases with time occurs at about this ice thickness. FY ice is referred to as *thin FY* (30–70 cm), *medium FY* (70–120 cm) and *thick FY* (>120 cm). Thin FY has also been subdivided at 50 cm into two additional stages of growth in the WMO terminology (1989). These breaks may be useful for individuals involved in ice reconnaissance; however, as I know of no consistent changes in the properties of sea ice that occur at 50 or at 70 and 120 cm, these terms will rarely be used in this book.

The other ice growth pathway develops under the conditions of appreciable swell, waves, wind, and blowing snow. In short, it is the common pathway occurring under environmental conditions more representative of the open ocean. Again, the initial crystals that form suspended in seawater are termed *frazil*. However, instead of freezing together into a thin skim, the continual motion of the water churns the frazil crystals into a soupy layer on the sea surface that becomes increasingly viscous as continued freezing proceeds to form additional frazil crystals within the layer. This soupy layer is referred to as *grease ice*. Figure 7.1 shows the initial stages of the formation of a layer of such ice. It is obviously very flexible. It is also very weak and if removed from the water would be unable to support itself. Figure 7.2 is an oblique aerial photograph that shows streamers of grease ice, aligned roughly parallel to the wind direction, being blown from the lower-right-hand side toward the upper-left-hand side of the photograph. It is possible to tell the direction of the drift from the shape of the frazil accumulations. For instance, note that several of the frazil plumes have broad heads and narrow tails. Field observations have shown that in such features, which have been referred to as "tadpoles," the tails point upwind (Dunbar and Weeks 1975; Martin and Kauffmann 1981).

At the same time it is common for both falling and blowing snow to be deposited on the sea surface. If all the crystals in the sea surface layer are snow

Figure 7.1. Initial stages in the formation of grease ice (Bellingshausen Sea).

Figure 7.2. Oblique aerial photograph showing streamers of grease ice being blown from the lower right toward the upper-left-hand portion of the picture (Beaufort Sea). See text for details of the interpretation.

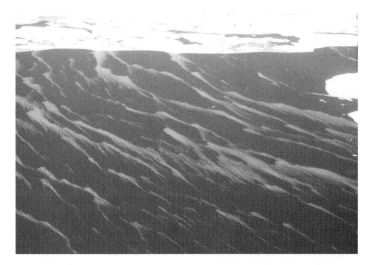

Figure 7.3. An oblique aerial photograph showing grease ice being formed as the result of snow being blown off of the surface of the older fast ice in the upper portion of the picture and subsequent deposition of blowing snow on the sea surface (Beaufort Sea).

crystals, then the layer is strictly termed *slush*: a term that WMO (1989) defines as "snow which is saturated and mixed with water on land or ice surfaces," or as "a viscous floating mass in water after a heavy snowfall." Figure 7.3 shows an aerial photograph of an area in which drifting snow, which is being blown off of the fast ice area in the upper-left-hand portion of the photograph, is being deposited on the surface of the sea forming a layer of grease ice. Unfortunately there is no simple way to visually examine current or former grease ice to tell whether its crystals were originally frazil or snow. Although there are isotopic methods for making such determinations, they require specialized laboratory procedures and the results are commonly only available after the completion of one's fieldwork. In this book we will consider grease ice to be a slurry comprised of seawater, frazil crystals initially formed by freezing seawater, and/or snow crystals initially formed either in the atmosphere or as the result of wind erosion from existing snow covers on nearby land and/or sea ice surfaces. In only a few cases will we know the relative inputs of snow and frazil.

An additional ice type which can form from grease ice and apparently even from masses of anchor ice rising from the seafloor is referred to as *shuga*, another Pomor term. This expression refers to accumulations of spongy, white ice lumps a few centimeters across. In all my years of working on sea ice I have never seen shuga. For one, I do not believe shuga to be particularly common, but more than that, it is an ice feature that only can be seen for a short time in that even a thin cover of snow would obscure the fact that shuga had formed during the initial stages of the development of the ice cover. It also only forms under windy conditions and while the ice is very thin and of dubious bearing capacity. In short, to optimize one's chances of seeing shuga and living to tell about it, it is best to work off an icebreaker operating during the ice growth season. A photograph showing shuga can be found in Armstrong et al. (1966, Fig. 43).

Although shuga may be comparatively rare, there is another ice "arrangement" which frequently forms during development of new ice that is far from rare. Here I refer to *pancake ice*. According to the WMO classification the term *pancake* refers to predominately circular pieces of ice 30 cm to 3 m in diameter and up to ca. 10 cm in thickness. They commonly show slightly raised rims that form as the pancakes continuously bump into one another as the result of the motion produced by the passage of swell through the ice field. Pancakes can form from grease ice, from shuga, or by the breaking of ice rind. In the Southern Ocean, in particular, and in marginal seas in general where significant swell is common, pancake formation is a major factor in the initial stages of the formation of an ice cover, a process that has been referred to as the "pancake cycle" (Lange et al.1989; Lange and Eicken 1991). Figure 7.4 offers a diagrammatic summary of the sea ice types mentioned in terms of ice thickness and environmental factors.

It should be noted that other types of ice classifications do exist. For instance, Golokov (1936, 1951) suggested a quite detailed classification that applies petrographic terminology (*porphyroblastic, gneissic*, etc.) to delineate the different structural types of naturally occurring ice masses. This approach has never "caught on" for several reasons, foremost of which being that it requires one to examine thin sections of the ice, a time-consuming and for many purposes an impractical process.

The formation of new ice will now be examined in more detail.

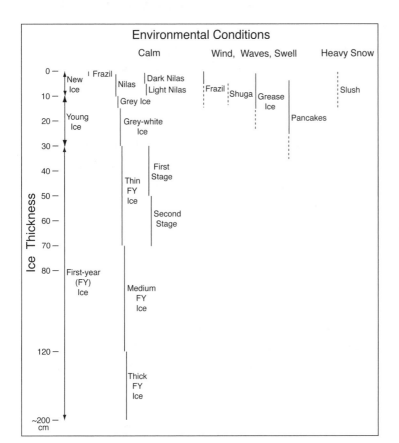

Figure 7.4. A schematic diagram showing the thickness–environmental space in which different sea ice types form. Note that the thickness ranges of the ice types that form under poor weather conditions are not exactly specified and can vary over a wide range.

7.2   First-Year Ice

7.2.1   The Initial Layer

7.2.1.1 Calm Conditions   The fact that frazil crystals start to form in the near-surface layer of the sea indicates that this layer is supercooled, in that some supercooling is a prerequisite for the nucleation and growth of ice crystals. As noted earlier, nucleation of ice crystals in nature invariably occurs with the aid of an existing particle (i.e., via heterogeneous nucleation). This has been shown to be the case in the atmosphere even at sites, such as the summit of Mauna Kea volcano in Hawaii, that have been selected for their distance from sources of atmospheric pollution. During initial freezing in the sea, heterogeneous nuclei also appear to be readily available and supercoolings are commonly believed to be both small ($<<1°C$) and transient (Katsaros, 1973), with the supercooled layer being very thin. One might expect to find that measurements of supercoolings in the polar oceans would abound in the literature. However, this is not the case. Although supercooling measurements would appear to be quite easy to perform, they are, in fact, difficult. The problem is that in order to measure supercooling one must simultaneously observe the salinity, temperature, and pressure of a volume of seawater. The difficulty comes from the salinity measurement in that if there are any tiny unnoticed ice crystals present in the sample in the field, they will have melted by the time the salinity measurement is performed, thereby diluting the water. The theoretical freezing point of the diluted seawater is then compared with the measured in-situ temperature and it is concluded that supercooling has been present. On the other hand, if it had been possible to measure the nondiluted salinity and compare it with the in-situ temperature, one would have concluded that there was either no supercooling present or that the value was smaller than previously assumed. For this reason many, if not most, early oceanographic measurements indicating supercooling are now believed to be questionable (Lewis and Lake 1971). The first careful measurements were made in rough sea ice by Untersteiner and Sommerfeld (1964), who determined the temperature differential between iced and noniced faces of a thermopile in the vicinity of the ice island T-3. Their results suggested that supercoolings of 0.004°C were occasionally present. Later careful measurements were made within the upper 10 m of the water column north of Svalbard by Lewis and Perkin (1983), who observed occasional supercoolings of up to 0.008°C. Considering that the measurements were made in a region where heavy ice was present they suggested that the supercoolings were transitory and not the result of surface cooling. Their explanation was based on a mechanism initially proposed by Foldvik and Kvinge (1974) to explain supercoolings observed in the vicinity of Antarctic ice shelves. In the Svalbard case, the supercooling was posited to have been the result of the occurrence of nearby pressure ridges and the pressure dependence of the freezing temperature of seawater that had achieved equilibrium at depth in the presence of ice. Lewis and Perkin referred to the mechanism as the *ice pump*. It will be described in considerable detail later in Chapter 17 ("Underwater Ice") and will also be discussed in Chapter 12 ("Deformation").

Russian experience, as summarized by Doronin and Kheisin (1977), also suggests that in-situ supercoolings rarely exceed a few hundredths of a degree C. The largest supercoolings Russian investigators have observed in the Arctic were measured off the coast of Greenland during the winter, where supercooling values varying between 0.20 and 0.29°C were found. Even larger values have been reported off the coast of Antarctica, where values as large as –2.0°C have been reported. Although these trends are reasonable, the supercoolings reported would appear to be unrealistically large. As noted, any errors make real supercoolings less than the apparent values. Experimental studies suggest that during the freezing of freshwater, an estimate of the amount of supercooling might be obtained by examining the angular relations between the crystals. Unfortunately, during the freezing of aqueous solutions with salt concentrations in the range encountered in seawater, there are fewer changes in growth mode in the temperature range of interest than in fresh water (see Pruppacher 1967 for details).

There appears to have been very little work completed on the nucleation of ice crystals from seawater. Even so, it is possible to piece together the general outlines of what happens, based on studies of the freezing of pure water and a knowledge of nucleation in other materials. The initial embryo of ice that forms will be only a few molecular units in size and will tend to be spherical in shape, thereby minimizing the surface free energy per unit volume (Hobbs 1974). However, by the time the ice embryo is of a sufficient size to be visible using a conventional microscope, the shape of the growing crystal has usually changed from spherical to discoidal (Arakawa 1954). The assumption that the initial shape was spherical is supported by the fact that the shapes of these discoids as observed in the (0001) plane are circular. Figure 7.5 shows the diagrammatic changes in the shapes of such crystals as observed both parallel and perpendicular to the c-axis of the crystal. The discoidal growth shape results from both the equilibrium and kinetic properties of ice owing to its highly anisotropic surface free energy. Estimates of the differences in surface free energy values for pure ice between the basal plane and a prism plane have been obtained by studying the shape anisotropy of water-filled negative ice crystals (Koo et al. 1991). A very large anisotropy ratio of 10^2 was obtained. Although this high ratio specifies a planar form, it does not specify any particular favored growth direction within the (0001) plane that is the plane of maximum reticular atomic density in an ice crystal. Therefore, ice growth is well described by Bravais' law: the smaller the reticular density of a crystal surface, the faster it grows normal to itself. In fact, for ice the (0001) surface is a facet at temperatures up to the melting point. Such molecularly smooth flat surfaces require undercoolings that are larger than the threshold value sufficient to allow the formation of new two-dimensional nuclei on the surface of the plane. The other possible surfaces of an ice crystal possess a structure that results in a gradual interphase variation over the scale of a number of molecular planes (Wettlaufer 1998). Such surfaces are referred to as molecularly rough or diffuse. In ice the surfaces perpendicular to the basal plane take such a structure at near-melting temperatures. Rough surface growth appears not to require a threshold supercooling and is able to proceed at any supercooling (Fernandez and Barduhn 1967). Measurements of ice growth velocities by Hillig (1958) have supported the idea that growth parallel to the c-axis is controlled by an activated nucleation process. More importantly,

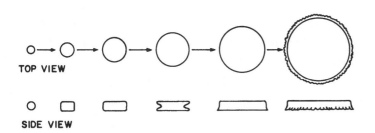

TOP VIEW

SIDE VIEW

Figure 7.5. The growth sequence of ice crystals in bulk water as observed by Arakawa (1954).

he also observed that at supercoolings of less than 0.03°C, growth in the c-axis direction was zero. At small supercoolings he was able to grow ice crystals that had dimensions of 25 cm parallel to the basal plane and 5 µm parallel to the c-axis: a ratio of 50,000 to 1. The concept that ice growth perpendicular to the basal plane occurs layer by layer, and is rate-limited by the necessary two-dimensional nucleation on a molecularly smooth interface, whereas prismatic interfaces are rough and do not require significant supercooling, has also received additional support via the use of molecular dynamics simulations (Nada and Furukawa 1996).

Experiments on the growth of ice crystals from pure water in a closed chamber have shown that if the amount of supercooling is low (< 0.18°C), the initial crystals that form are always smooth-surfaced disks. At higher supercoolings dendrites appear to form directly. The maximum diameter to which such discs grow, at least in freshwater, is on the order of 2 to 3 mm, and again is a function of the degree of supercooling, with larger circular discs resulting (in time) from smaller supercoolings. Figure 7.6 shows a number of such discs formed by the freezing of seawater. Note the characteristic notched edges of some of the disks when viewed on edge (i.e., perpendicular to the c-axis). A discussion of the morphological stability of such disc crystals growing in pure water can be found in Fujioka and Sekerka (1974). Although the order of magnitude of their calculated critical radius for instability is similar to that observed experimentally, available experimental studies have not proved sufficiently detailed to serve as a test of their theoretical treatment. Limited studies by Kumai and Itagaki (1953) indicate that under comparable growth conditions the maximum disc size is depressed by the presence of an appreciable amount of solute in the water. However, they observed the formation of discs in all cases when inorganic solutes were used. The limiting disc size occurs when the disc transitions to a dendritic hexagonal star. That the growth of circular discoids would be limited by a transition to a dendritic growth mode is reasonable, in that it is necessary for an ice crystal growing in supercooled seawater to dissipate both heat and solute into the surrounding liquid. As the radius of curvature of the disc increases, the ease of heat and solute dissipation decreases, until a critical radius value is reached. At this value, which is determined by the thermal conductivity of the melt, the diffusion coefficient of the solute in the melt, and the growth velocity (Glen 1955a), the discoidal growth form becomes unstable, breaking up into a hexagonal dendritic star which again offers a much smaller tip radius of curvature. Although the change to a stellar form results in an increase in the relative amount of surface, this is apparently compensated for by the fact that the crystal is now more readily able to dispose of heat and solute at the advancing interface. The disc-to-star transition is therefore marked by an appreciable increase in the growth velocity (Kumai and Itagaki 1953). In their experiments most if not all stellar crystals appear to have gone through an initial discoidal stage. However, studies discussed by Tirmizi and Gill (1989) suggest that if supercooling is greater than the value of 0.18°C suggested by Fujioka, the initial crystals will be dendrites. Once the dendritic star morphology develops, the arms of the stars always form parallel to the *a*-axis directions in the ice crystal, which as noted earlier is the shortest lattice vector in the (0001) plane.

On lakes, the first crystals to form on calm, cold, clear nights are dendritic needles that extend rapidly in the thin supercooled surface water layer (Fujino and Suzuki 1959; Hallett 1960). Such needles can have their *c*-axes inclined at various angles to the vertical including horizontal (Gow 1986). They also frequently are curved in the horizontal plane (Knight 1962c). Once an interlocking network of needles is established, the space between the needles is filled by platelike tabular crystals with their *c*-axes approximately vertical. As the spaces between the needles can be quite large, the *c*-axis vertical crystals in lakes can also be quite large, achieving cross-sectional areas in excess of 1 m^2 (Fedotov and Cherepanov 1991). Although surface needle formation occurs in sea ice, it appears to be less common than in freshwater. This can be explained by the fact that in seawater convective mixing will have lowered an appreciable thickness of the water layer to the freezing point, making it possible for growth to proceed both along the upper water surface and downward into the water column. An appreciable downward growth component results in the ice crystal developing as a planar dendritic star as opposed to a needle. Such growth also results in a moment owing to buoyancy forces, which tends to rotate the stellar crystal until its (0001) plane is parallel to the upper water surface. It seems possible that needlelike growth in seawater primarily occurs when part of the needle has frozen to or is partially restricted by the surrounding crystals so that it is prevented from rotating as downward growth proceeds. When needle crystals occur in sea ice they do not appear to show the pronounced curvatures observed in freshwater ice. This is in keeping with the experimental observations of Knight (1962c), who found that an increase in solute content tends to eliminate the curvature. In freshwater, complex morphologies have been noted under conditions of more pronounced supercooling: simple double pyramids between –2.7 and –5.5°C and, at yet lower temperatures, complex double pyramids showing secondary and higher-order nonrational growth directions. Photographs of such crystals can be found either in Macklin and Ryan (1966) or in Hobbs (1974). To date, I am not aware of such morphologies being observed during the freezing of seawater. This is probably because comparable supercoolings are never reached in nature, even though experimental studies indicate that the presence of solutes appears to reduce the supercooling necessary for crystal growth to occur in nonrational crystallographic directions (Ryan 1969; Ryan and Macklin 1968).

Figure 7.7 is a photomicrograph of the upper surface of a thin sea ice skim collected off the coast of Labrador. Several needle crystals are present. The most prominent of these needles is surrounded by crystals with their *c*-axes vertical (normal to the plane of the photograph). Figure 7.8 is an inked copy of a rubbing of the upper surface of a similar skim. As rubbings will be presented at several different locations in this chapter, a few words on the technique are in order. In essence, ice rubbings are similar to the more common rubbings of gravestones and were initially applied

Figure 7.7. A photomicrograph of the upper surface of a sea ice skim showing a needle surrounded by crystals with their *c*-axes vertical (perpendicular to the plane of the photograph).

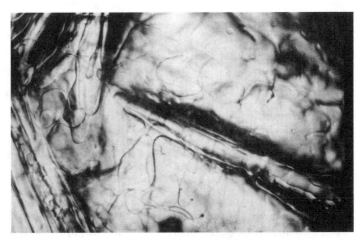

Figure 7.8. Inked copy of a surface rubbing taken from an ice skim (Hopedale, Labrador). The diameter of the rubbing is 7.62 cm.

to glacier and lake ice for studies of grain size variations and textural analysis (Ragle 1962; Seligman 1949). Usually, a flat ice surface is prepared, either by melting using a heated glass plate or by using the flat underside of a teakettle or a similar flat-bottomed metallic container containing warm water. Then, an infrared lamp is held over the ice surface to induce localized melting along the grain boundaries and other structural defects. Any excess surface water is poured off the ice. Next, a piece of thin vellum paper is placed on the ice surface and rubbed using either a soft lead pencil or a block of graphite. The end result is a permanent record of the topography of the surface with projections appearing as lines or "ticks" on the rubbing, whereas "smooth" (0001) surfaces appear as white or uniformly shaded areas. Rubbings of the upper and lower surfaces of ice skims are a bit different in that it is not necessary to prepare the surface prior to applying the paper because natural crystal growth processes produce a sufficiently textured surface to produce useful rubbings. Studies of a number of such rubbings of the surfaces of ice skims formed during calm conditions have shown that typically more than 50% of the surface area is composed of crystals with their *c*-axes near vertical.

When one considers the initial formation of ice nuclei during the formation of a sea ice skim, one soon realizes that in many if not most cases, the initial ice nuclei do not form in or on the surface of the water but in the surrounding air. For instance, during investigations of the freezing of the surface layer of calm water in streams, it has been found that ice crystal concentrations ranging from 6×10^1 to 6×10^4 m^3 existed in the air near the surface of the stream (Osterkamp 1977, 1978). These crystals had an average size of 180 μm and were in the form of hexagonal plates, a fact that clearly

Figure 7.9. Frost flowers growing on the surface of dark nilas occurring in a recently refrozen lead. Considering the color of the nilas, it could properly be referred to as black ice. Individual frost flower clumps are ~3 to 5 cm in diameter. Such ice is very weak and would not support a person's weight.

distinguished them from the discoidal crystals characteristic of ice formation in or on a water surface. Once the stream had skimmed over with a new ice cover, the airborne crystals were no longer observed, suggesting that the moisture source was the water surface of the stream. In addition, airborne crystals were observed falling into the stream. In all probability the ice crystals nucleated in the water droplets that comprise the vapor cloud that typically forms over open water during freezing conditions. The freezing of droplets in such vapor clouds is a situation that has received considerable attention by cloud physicists. See for instance the paper by Vali (1971) and the general reviews by Fletcher (1970) and Hobbs (1974). There apparently have been no comparable studies in the ocean. The sea ice that forms under such conditions is, as noted earlier, referred to as *nilas* or *ice rind* while it is still thin. Another expression that has occasionally been utilized to describe such ice occurring on lakes is "sheet ice," in that the resulting ice cover "presents a smooth unbroken surface on which there are no highly evident horizontal changes in the structure of the ice layer" (Wilson et al. 1954).

Another common feature occurring during the growth of nilas is the development of frost flowers on the upper ice surface. Frost flowers are dendritic "flowers" of ice (Figure 7.9) that can develop on the upper surface of thin ice if conditions are favorable. Particularly favorable conditions for the growth of these flowers occur when leads open during calm periods with very low air temperatures (Perovich and Richter-Menge 1994). During such periods frost flower growth can be quite rapid, with extensive fields of flowers forming overnight. Initial flower nucleation appears to occur at many isolated sites followed by lateral spreading. As the flowers form, a 1- to 4-mm-thick, highly saline slush layer develops on the surface of the ice beneath the flowers. Although it is generally believed, based on a mass balance argument, that the source of the surface brine is the ice interior, the detailed mechanisms are still under discussion. A current favorite is that a pressure gradient that develops as the result of the strong temperature gradient in the near surface ice produces the upward transport of the brine (Dash et al. 1995; Wettlaufer and Worster 1995). It is clear, however, that the presence of the saline slush layer is essential to frost flower development. To date, two different crystal habits (dendrites and needles) have been noted in the field (Perovich and Richter-Menge 1994). Laboratory studies of frost flower development carried out by Martin et al. (1995) and by Martin et al. (1996) using NaCl solutions have produced results similar to the field observations, with dendritic flowers growing between $-12°$ and $-16°C$ and rodlike needles between $-16°$ and $-25°C$. These temperature ranges are in general agreement with the temperatures associated with similar morphological changes during crystal growth from the vapor in pure H_2O systems (Hobbs 1974). Current observational data collected by Steffen (reported in Martin et al. 1996) indicate that frost flowers do not form if wind speeds exceed 3–5 m/s. There appear to be two reasons for this. First, high winds destroy the near-surface micro-meteorological environment required for the growth of the flowers, and second, as frost flowers provide a rough upper ice surface, drifting snow can obliterate them in minutes. Although at first glance frost flowers may appear to be a purely surficial curiosity, their presence appreciably changes the surface heat balance of the ice, causing the ice surface temperature to be $1–2°C$ warmer than bare ice of an identical thickness. At the same time, the radiative (IR) temperature of the flower-covered ice will appear 4 to $6°C$ colder than that of bare ice (Martin et al. 1995).

It is important to note that frost flowers, if well developed, can conceal the dark color that tells one that the underlying ice is very thin. During the AIDJEX project I was traveling using two ski-doos across a series of refrozen leads that were covered with frost flowers. The ice was thin but thick enough for safe travel. After traveling on these leads for some distance, we traversed an MY floe and came to another lead, also covered with similar-appearing frost flowers. As everything looked the same, we assumed that all the leads formed during the same divergence event and that the underlying ice was of the same thickness. It was not! As the first skidoo was moving fairly rapidly, it traveled ~5 m onto the refrozen lead before it broke through, followed by the equipment sled that it was towing. These items proceeded to travel rapidly to the bottom of the Arctic Ocean (water depth ~3000 m). Fortunately the driver (Steve Ackley), being highly motivated, was able to dive to the side and avoid being snagged by the rapidly disappearing equipment. His freestyle swimming and icebreaking technique was so impressive that he was able to reach the safety of the MY floe without getting the back of his anorak wet. Although this whole event occurred in a few seconds, it left a lasting impression on both the participant and the observers.

7.2.1.2 Turbulent Conditions   Under windy conditions typical of the open ocean, the physical situation would appear to be even more conducive to the formation of ice crystals in the atmosphere in that splashing, wind spray, and the bursting of air bubbles in the water would all tend to transfer water into the very cold air above the sea surface. Also common of the open ocean is wave-induced turbulence in the upper portion of the water column during initial ice formation. Such mixing results in an appreciable thickness of the upper ocean being reduced to the freezing temperature and probably slightly supercooled. When ice nuclei are introduced onto the sea surface from the atmosphere or by in-situ heterogeneous nucleation, vertical mixing provides the energy necessary to overcome the buoyancy forces so that the initial crystals can be stirred throughout the upper portion of the water column to depths of up to several meters (Martin and Kauffman 1981; Savel'yev 1958, 1963). As a result, effective supercooling is presumably reduced, more crystals form per unit volume, and the abrasive action between crystals is increased, favoring extensive discoidal growth. Even if stars or needles form they will rapidly be broken, resulting in smaller and more equigranular

Figure 7.10. A photograph showing the formation of marine frazil as observed in a series of flume experiments carried out by Ushio and Wakatsuchi (1989) to simulate frazil growth in a polynya. The side view shows an accumulated upper layer composed of frazil crystals as well as numerous free-floating crystals distributed within the water column. The photograph is 40 cm across.

Figure 7.11. Grease ice forming on the surface of a Chukchi Sea lead and being driven by the wind into its downwind end (Martin and Kauffman 1981).

grains. It also should be noted that under such conditions it is not necessary for each new crystal to nucleate separately. It has been found in studies of frazil ice formation in freshwater that new crystals can form by what is referred to as collision breeding (Chalmers and Williamson 1965; Garabedian and Strickland-Constable 1974). The process is believed to involve collisions with the other ice particles that are present, thereby mechanically forming new nuclei. In that this is a cascade-type process, the rapid accumulation of crystals is possible. A useful review discussing these matters as they apply to the generation of frazil crystals in freshwater can be found in Daly (1984). The end result is a sort of upside-down snowstorm of ice crystals that are then herded by the wind, waves, and current into a variety of different forms. Some sense of what occurs can be obtained from Figure 7.10, a side view of frazil ice accumulating during a series of flume experiments.

As noted, grease ice is a suspension of such frazil crystals into a soupy surface layer with typical thicknesses in the range of 0.1 to 0.3 m. This layer is distinguished by the low-reflectivity, matte-like appearance of its upper surface and by its viscous, fluidlike properties in contrast to the more typical "solid" behavior of nilas and other thicker ice types. Based on wave-damping experiments, Martin and Kauffman (1981) have found that in general a concentration of about 40% by volume of frazil crystals is necessary before the viscous characteristics of a frazil layer begin the transition to more representative solid characteristics. While still in the viscous, grease ice stage, the frazil crystals characteristically accumulate on the downwind sides of leads (Figure 7.11) where they can also be swept underneath the blocking ice sheet. In cases where the blocking ice is thick or when the frazil is being swept onto a coast, frazil ice thicknesses of up to 1 m have been observed.

As a frazil layer thickens, its color gradually changes from a dark shade of gray through increasingly lighter shades, ultimately becoming white. Presumably this change is the result of two factors. First, when the ice is very thin the apparent ice color is affected by the presence of the water beneath the ice. An extreme example of this occurs during calm conditions when so-called "black" ice forms on a lake. In fact, black ice is not black at all but is a layer of nearly bubble-free ice that is transparent similar to a plate of glass. Its blackness comes from its large optical depth and the apparent color of the underlying water. As sea ice is never as transparent as bubble-free lake ice, it usually appears to be gray instead of black except when it is very thin. However, the principles are the same. The second factor particularly occurring in frazil ice is that it initially contains large amounts of included brine within the suspension of ice crystals. As the ice thickens and cools, the volume of brine decreases rapidly as the result of brine drainage. This is particularly true of the upper portion of the frazil layer that gradually rises higher above sea level as the ice thickens, allowing the brine to be replaced by air.

There are several processes that are presumed to occur in a developing frazil layer. I say *presumed* because these matters have not yet been carefully investigated. Fieldwork during periods when frazil is developing is difficult at best and can be quite dangerous. First, it is invariably windy, and frequently a large swell is running. In addition, frazil crystals are typically small, and detailed observations require the use of a microscope. Add to this the facts that the boat is rocking and pitching and water is sloshing around freezing to you and to your equipment. Messy and miserable! It doesn't take much thought for most people to decide to wait until the developmental stage of frazil formation is over before walking out to collect a sample of the end product of the process. Fortunately some direct observations are available (Bauer and Martin 1983; Martin 1981; Martin and Kauffman 1981).

One process possibly occurring in frazil after the initial formation of dispersed crystals is somewhat analogous to changes observed in precipitate particles in fluids, where it is referred to as Ostwald ripening. For instance, in water-saturated snow, grain coarsening is believed to be determined by this process in which the smaller crystals shrink and eventually disappear while the larger crystals grow. Once a steady relative size distribution is established, the rate of volume change of typical grains increases linearly with grain volume, from the characteristic negative rate for the smallest grains, through zero for grains of average volume, to positive values for the larger grains (Raymond and Tusima 1979). The process can be explained by considering the relation between the sizes of the crystals and their freezing points (Colbeck 1982). As water-saturated snow is a two-phase, single-component system, the jth phase must obey the Gibbs–Duhem relation

$$d\mu_j = v_j dp_j - s_j dT \tag{7.1}$$

where μ is the chemical potential, v the mole volume, s the entropy, and T and p the temperature and pressure, respectively. At equilibrium any changes in the chemical potential of the liquid (l) and solid (s) phases must be equal so that

$$v_\ell dp_\ell - s_\ell dT = v_s dp_s - s_s dT. \tag{7.2}$$

The difference in entropy is

$$s_l - s_s = L / T \tag{7.3}$$

and the difference in pressure is given by Laplace's equation

$$p_i - p_j = 2\sigma_{ij} / r_{ij} \tag{7.4}$$

where σ is the surface tension. The mean radius of curvature, r_{ij}, is given by

$$\frac{2}{r_{ij}} = \frac{1}{r_1} + \frac{1}{r_2} \tag{7.5}$$

where r_1 and r_2 represent the two principal radii of curvature (or radii in any two mutually orthogonal planes). It then follows that the melting temperature of an unstrained ice particle in pure water is given by

$$\ln\left(T / T_0\right) = -\frac{2}{L p_s} \frac{\sigma_{sl}}{r_{sl}} \tag{7.6}$$

which is approximately equal to

$$T_m = -\frac{2 T_0}{L p_s} \frac{\sigma_{sl}}{r_{sl}} . \tag{7.7}$$

Here T_m is the melting temperature in °C and the reference temperature T_0 is the melting temperature of a flat surface. In that the smaller grains are at lower temperatures in water-saturated snow, heat flows to them, causing them to melt while, at the same time, the larger grains grow.

The above analysis of Ostwald ripening definitely would appear to apply to water-saturated snow occurring on the surface of MY ice during the summer. Unfortunately, observations of the efficiency of this process as observed in water-saturated snow cannot be directly applied to frazil ice in the sea because the process is known to be slowed by the presence of impurities such as salt. The explanation for this tendency is as follows. Melting by the smaller crystals causes the local solute concentration to be lowered as the result of dilution by the meltwater (thereby raising the local freezing temperature). This is in contrast to the situation at locations where the larger grains are growing, in that impurities (salts in the case of seawater) are being excluded by the growing solid phase (ice), thereby lowering the local freezing temperature. The end result is a lowering of the temperature differences that result from the surface curvatures of the grains. A quantitative discussion of this general effect can be found in Raymond and Tusima (1979). It also should be noted that several of the assumptions that they use in analyzing their saturated snow data, although realistic for snow, are of dubious applicability to the frazil ice situation in the ocean. Here I particularly refer to the assumptions that heat is transferred only by diffusion and that the liquid between the grains is motionless.

Bonding between frazil particles starts to occur as soon as they are brought into contact. This favors the formation of flocs. There appear to be several factors active here. For one, actively growing frazil crystals are known to be very sticky in that they readily adhere to any foreign substances that they contact, including other ice crystals. In addition to active growth favoring floc formation, the development of intercrystalline bonds is favored both by deformation and by surface energy forces (Colbeck 1979; Hobbs 1974). Two effects favor the welding of the frazil particles together. In the first, any force pressing ice particles together will lead to local depressions of the freezing point (regelation). This leads to the flow of heat toward these locations and freezing at other contacts. In addition, when the stress is removed, the unfrozen contact will change its freezing point and refreeze. The second mechanism depends on the radius of curvature of the frazil particles and is generally believed to occur in the unsaturated, above-sea-level portions of the flocs. As noted earlier, small particles in contact with the water and vapor that is present are thermodynamically unstable relative to the equilibrium pressures between solid–liquid, solid–vapor, and liquid–vapor at the surfaces of larger particles. This pressure imbalance tends to deform the ice particles, leading to a flattened solid interface between them and causing them to be welded together at a solid boundary. A distinction between grease ice and pancake ice during the formation process may be related to the ability of the pancake floes to drain increasing the likelihood of freezing contacts developing. Generally, in the completely saturated case when the ice crystals are immersed in seawater, bond development is not as dramatic as in unsaturated compacts.

During the initial stages of frazil development there are clearly significant grinding motions occurring between crystals. The situation could be likened to being in a ball mill where the grinding balls are the other crystals. As a result, once frazil has solidified sufficiently to be removed from the water column and thin sectioned, it is commonly found to be composed of very small crystals of ice generally less than 5 mm in diameter. Because of the assumed random motions between the crystals in frazil flocks, the c-axis orientations of the crystals are usually assumed to be random. As noted in Appendix E, standard glaciological thin section techniques are difficult to apply to frazil ice because of its typical small grain sizes. Therefore, detailed petrographic observations are largely lacking for this type of ice, other than noting the generally small grain size and the fact that strong preferred crystal orientations are not obvious.

Because of the noted operational difficulties in obtaining detailed field observations during periods of active frazil growth, the initial approaches to analyzing grease ice behavior were theoretical studies by Weitz and Keller (1950) and Peters (1950). They approximated grease ice as a layer of noninteracting point masses and solved for the two-dimensional velocity potential in an inviscid fluid with mixed surface boundary conditions. In this mass-loading model, waves were assumed to propagate from open water across a distinct ice edge into ice-covered water. Both papers predicted that both phase speed and wavelength should decrease as the waves entered the ice. Weber (1987) has modeled the behavior of brash ice by assuming a Lagrangian formulation for waves propagating in a two-layer fluid with an inviscid, infinitely deep lower layer and an upper layer having a Newtonian viscosity. If the upper layer is thin and highly viscous, the model leads to an exponential wave decay rate and an increase in wave damping with results that agree well with field observations (Hunkins 1962; Liu, Holt, et al. 1991; Squire et al. 1995).

Comparisons of these model results with experimental results have been carried out by Newyear and Martin (1997, 1999), who froze NaCl solutions in a thermally insulated flume. A range of wave frequencies and amplitudes were artificially generated and measured using five independent strain gauge probes. It was found that in a field of grease ice with an effective ice thickness ch, where c is the volume concentration of ice in the frazil layer and h is the layer thickness, wave damping decays exponentially with distance. It is also frequency dependent in that the dimensionless damping coefficient increases with dimensionless frequency. In addition, the derived bulk viscosity of the grease ice is at least four orders of magnitude larger than that of the underlying water and increases with frequency. This suggests that grease ice behaves as a non-Newtonian fluid. After rejecting the mass loading model, which predicted a significant wavelength shortening relative to open water as opposed to the observed wave lengthening, Newyear and Martin (1999) made direct comparisons with two viscous fluid models. They first examined the classic infinite depth, constant viscosity one-layer model of Stokes (1851) and Lamb (1932), and then the finite depth two-layer model of Keller (1997), which assumes a constant viscosity, an immiscible upper layer (the grease ice), and an inviscid lower layer (the underlying ocean). Not surprisingly, the more physically realistic Keller model provided the best agreement with the laboratory data, provided that the viscosity of the grease ice was assumed to be $2\text{–}3 \times 10^2$ that of the seawater. Newyear and Martin also suggest that the reason that the constant viscosity model is successful is that the scale of the incident waves is much greater than the scale of the individual ice crystals in the frazil layer. As a result, grease ice behaves as a sufficiently continuous medium for the bulk viscosity treatment to be meaningful. Of the two different experiments that were analyzed, the second, in which the frazil layer was both slightly thicker and a day older, gave somewhat higher viscosity estimates. This is reasonable following the arguments of Martin and Kauffman (1981), who suggested that the viscous nature of grease ice is the result of interactions between frazil crystals. In their quite realistic laboratory simulations of grease ice formation, they initially observed the formation of isolated small discoids of ice having diameters of ~1 mm and thicknesses ranging between 1 and 10 μm. These individual platelets then bonded together, forming clusters as large as 5 mm in diameter. This sintering process occurred because two ice spheres in contact at a point are thermodynamically unstable in that the transfer of ice to the contact point, thereby forming a neck between the spheres, results in a decrease in the surface free energy (Hobbs 1974). Martin and Kauffman point out that this sintering occurs quite rapidly (0.1–6 s), a time that is comparable with the inverse of the wave frequency (0.06–0.1 s). They also suggest that within a wave, bonds between crystals are constantly forming and being broken, one effect of which is to produce an increase in viscosity at low rates of shear. They also note that although at low wave amplitudes the grease ice crystals appear to be randomly oriented, at larger amplitudes the crystals tend to line up parallel to one another as they approach the zone (the dead zone) where the properties of the grease ice change from liquid to solid. This orientation change results in a more tightly packed arrangement than would be possible with randomly oriented platelets. To date, this structural change has not been documented in the field. However, considering the limited amount of structural studies that have been focused on grease ice, this is not surprising. Considerably more will be said about frazil later in the discussion of leads and polynyas, where frazil is herded by the wind and the local circulation of the water. Also, in Chapter 17 in the present book one will find a discussion of very peculiar types of sea ice referred to as marine ice, platelet ice, underwater ice, and anchor ice. They can also be considered to be forms of frazil in that they form in the water column by growth into supercooled water.

In concluding this section on the initial formation of an ice cover, it is important to note that the initial stages of ice formation are largely the result of ice growth into supercooled water. Therefore the cooling pathway is not from the atmosphere to the ice to the water but from the atmosphere to the water to the ice.

7.2.2 The Transition Zone

Once a continuous layer of ice has formed on the sea surface, the possibilities for crystal growth change. After an initial skim or frazil layer forms, the latent heat of freezing is extracted through the overlying ice sheet which separates the melt (seawater) from the cold source (the atmosphere) as shown by equation 4.5 in Chapter 4. The growth rate is determined by the temperature gradient in the sheet and the effective thermal conductivity of the sheet. Ice growth is also subjected to another important constraint in that a degree of growth freedom is lost. This means that only if the grain boundaries separating the growing crystals are exactly vertical (perpendicular to the ice–water interface) can crystal growth proceed without one grain interfering with the growth of the adjoining crystal. Therefore, any tendency for anisotropic growth will result in geometric selection, with the crystals in the favored growth orientation eliminating crystals in less favored orientations by cutting them off from the melt. As anisotropic growth is the common growth mode observed in ice crystals growing into supercooled seawater, it is hardly surprising that geometric selection is invariably observed at the base of the initial skim or frazil ice layer. The layer in which this occurs has been called the transition layer by Perey and Pounder (1958), who first described it. Figure 7.12 presents their idealization of the process. As will be seen, geometric selection in sea ice is very effective in that major orientation changes resulting from anisotropic crystal growth commonly occur in vertical distances of less than 10 cm. When a thin initial ice skim forms, the base of the transition layer is commonly within 15 cm of the upper ice surface. However, the base of the transition layer can be located at distances of over a meter below the upper ice surface at locations where the initial ice sheet is comprised of a thick layer of frazil (Pounder and Little 1959; Weeks and Gow 1980). In marginal seas where frazil crystals are frequently swept beneath existing ice layers, an ice sheet may contain several transition layers.

The following summarizes the observational evidence for the existence of transition layers. Figure 7.13 shows a rubbing of the bottom surface of a 2.5-cm-thick sea ice layer in which geometric selection is clearly occurring. In this figure, as in Figure 7.8, the tick lines are made by the raised edges of the basal (0001) platelets of the ice crystals. Local areas where all the platelets are

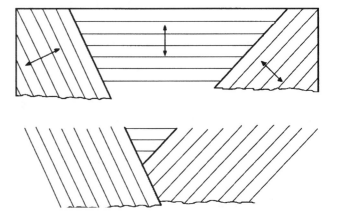

Figure 7.12. A schematic diagram showing the process of geometric selection in sea ice according to Perey and Pounder (1958). The lines within the crystals represent the orientations of the (0001) basal planes.

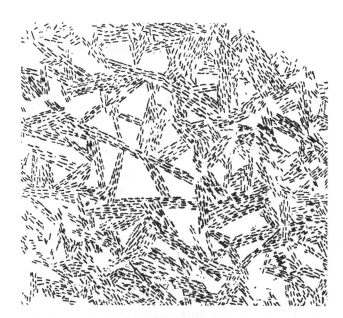

Figure 7.13. A rubbing of the bottom of a 2.5-cm-thick ice skim in which the crystals are undergoing geometric selection (Hopedale, Labrador).

Figure 7.14. Photograph of the lower surface of sea ice showing a triangular crystal located in the center of the photograph being eliminated by geometric selection. The diameter of the triangular crystal is ~3 mm.

parallel represent individual crystals. The blank plane-sided polygonal areas are locations where crystals occur with *c*-axes either vertical or nearly vertical. The reason that these areas appear blank in the rubbing is not that the (0001) planes are comparatively smooth (which they are). It is the result of the fact that these areas are recessed by distances of up to 2 to 5 mm behind the ice–seawater interface represented by the crystals that make the ticks. In short, the "tick" crystals, which have their *c*-axes essentially horizontal, are growing faster than the crystals with their *c*-axes essentially vertical. A more visual representation of this can be seen in Figure 7.14, which is a photomicrograph of the bottom surface of an ice skim showing in its center such a depressed area due to the presence of a *c*-axis ~vertical crystal. If this ice were to continue growing this crystal would be rapidly eliminated by the surrounding crystals as the result of preferential growth parallel to the (0001) plane.

A more quantitative demonstration of geometric selection can be obtained by first making a series of thin sections of the ice at different vertical depths and then determining the *c*-axis orientations of the crystals using a three-axis universal stage. The data obtained by this process are then plotted on a Schmidt equal-area net. A description of these procedures and a discussion of

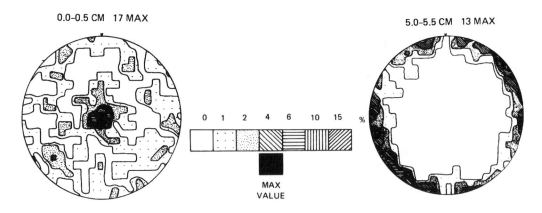

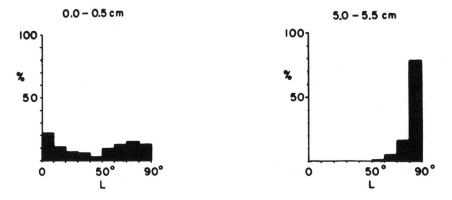

Figure 7.15. Two contoured fabric diagrams of the orientation of the *c*-axes of crystals as observed at different depths in a layer of sea ice that initially formed as an ice skim (Thule, Greenland). The diagrams are in the horizontal plane. The contour intervals (per a 1% circle) are indicated.

Figure 7.16. Two histograms showing the relative percentage of the different *c*-axis orientations (0° = vertical, 90° = horizontal) at different levels in the same ice studied in Figure 7.15.

how to interpret the resulting plots, which are referred to as fabric diagrams, is given in Appendix E. Figure 7.15 shows two fabric diagrams prepared at the 0.0–0.5 and 5.0–5.5 cm levels of ice that formed initially as a skim under relatively calm conditions in North Star Bay at Thule, Greenland. As can be seen, although *c*-axis vertical orientations are those most frequently observed at the uppermost level in the ice sheet, by the time the ice reaches a thickness of 5.0–5.5 cm almost all of the crystals with *c*-axis orientations of greater than a few degrees from horizontal have been eliminated. A more detailed presentation of the results from this site that included seven levels of measurements can be found in Weeks and Ackley (1986). Figure 7.16 is a somewhat different presentation taken from the same data set, in which only the percentage of the crystals having *c*-axis orientations inclined within different class intervals from the vertical are noted. This would appear to be a reasonable simplification in that there does not appear to be any appreciable preferred orientation direction within the horizontal plane in these particular samples. Here the measurements are from the 0.5–1.0 and 5.0–5.5 cm levels. Note that in the initial skim crystals show all different orientations, ranging from vertical to horizontal. As a result, it is not necessary to nucleate new grains having *c*-axis horizontal orientations. They are already present in the initial ice skim (or frazil layer). The new orientation forms by the survival of those crystals that have the favored orientation (*c*-axes horizontal). Orientation changes such as shown in Figures 7.15 and 7.16 are characteristic of transition zone ice. In short, the transition zone is the vertical region within the ice sheet in which a strong *c*-axis horizontal crystal orientation develops. Although in an ice skim or frazil layer any *c*-axis orientation is possible, below the transition layer almost all the crystals have their *c*-axes oriented within a few degrees of horizontal.

How does this work? There have been, to date, two papers written specifically on this subject as it occurs in sea ice. The first was by Kvajic et al. (1973), who used bicrystals to study selective growth from dilute NaCl solutions. Elimination rate was shown to depend on impurity concentration and growth rate, as well as on both the relative orientation of the two crystals and their absolute orientation relative to the growth direction. This was followed by an even more detailed study by Kawamura (1986, 1987), who used somewhat similar procedures initiating one-dimensional growth from artificial freshwater ice seeds, which were produced by welding together rectangular blocks of single crystals that had known crystallographic orientations. The growth direction was oriented parallel to the grain boundary of the seed (or stated in another way, the initial interface between the two crystals in the seed was oriented perpendicular to the ice–seawater interface). Experiments were carried out with growth rates varying from 0.5 to 2.0 mm/hr and with salinities ranging from 8 to 32‰. These growth conditions were all in the range in which the advancing sea ice interface developed a typical cellular structure. A schematic diagram of the first of the three types of initial *c*-axis arrangements studied is presented in Figure 7.17. Here the *c*-axes were arranged in a vertical plane that was oriented perpendicular to the grain boundary. For orientations where both crystals expose their basal planes in the grain boundary groove (zones I and II as shown in Figure 7.18), the encroachment angle β between the grain boundary and the vertical axis z is given by the mean value of the inclination angles a_1 and a_2 of the two grains as shown in Figure 7.19. This is the result to be expected if the two exposed planes both advance at the same growth rate as measured parallel to the respective *c*-axes.

When the basal planes of both crystals were not exposed (zones III and IV in Figure 7.18), the situation is a bit different. Assume that the preferred crystal is the one with its *c*-axis nearest to horizontal. It then follows that grain 1 will grow slightly ahead of grain

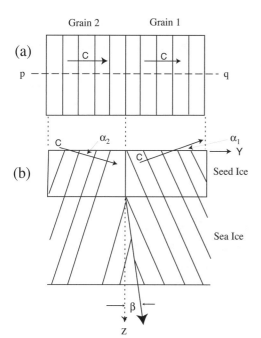

Figure 7.17. Grain boundary encroachment as seen in sea ice (Kawamura 1987). Here (a) shows a top-down view of the seed and (b) shows the cross-sectional view of both the seed and the subsequent sea ice. The thin sets of parallel lines can be taken to represent either the traces of the (0001) planes of the ice crystals or the platelike substructure characteristic of sea ice. The inclination angles between the c-axes and the horizontal plane and between the vertical and the resulting grain boundary in the sea ice are given by the αs and βs, respectively.

Figure 7.18. The four possible types of intragrain contact referred to by Kawamura (1987) as zones I to IV and the associated direction of grain growth, as indicated by the arrows, based on the values of both the angles a_1 and a_2. As in Figure 7.17 the parallel lines represent the basal planes as projected onto the vertical cross section.

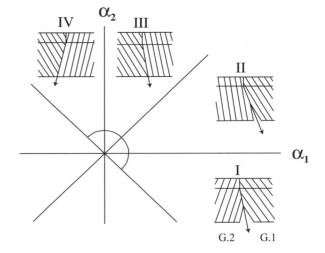

Figure 7.19. A more detailed schematic of the grain boundary groove in zones I and II in which the basal planes of both crystals can be exposed. If it is assumed that the basal planes of both crystals advance at the same rate ($V_1 = V_2$) in the direction parallel to their c-axes, then with subsequent ice growth the grain boundary will advance in the direction of the arrow (Kawamura 1987).

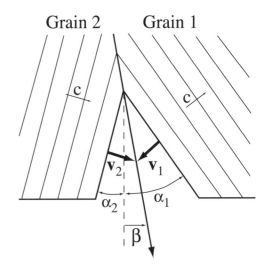

2, thereby limiting its lateral growth. Furthermore, if the lateral edge of the growth step is the basal plane, then the grain boundary direction is specified by the c-axis inclination angle of the preferred grain or simply $\beta = a_1$. Note that as shown in Figure 7.20, the grain boundary is inclined toward grain 1 even though its c-axis is nearest to horizontal. At first glance this observation appears to contradict the common field observation noted two paragraphs earlier, that the crystals that survive to the bottom of the ice sheet are those with their c-axes closest to horizontal. However, there are cooperative effects as also shown in Figure 7.20. If a crystal having the same orientation as grain 1 is also placed to the left of grain 2, thereby forming a zone II–type grain boundary with an encroachment angle $\beta = (a_1 + a_2) / 2$, the fact that the β for this zone II grain boundary is always greater than for the zone III boundary ($\beta = a_1$) ensures that grain 2 will be cut out.

Kawamura also carried out experiments in which the c-axes of the adjoining crystals were both horizontal and at right angles to each other. In these cases, geometric selection occurred with the central crystal being cut out. Even when the platelets of the neighboring crystals were in contact with each other at angles other than 90°, the β values observed were similar to those obtained when the c-axes were at right angles to each other. Within the growth rate and salinity range studied, which was always sufficient to produce a sea ice substructure, changes in the environmental conditions did not affect the geometric rules.

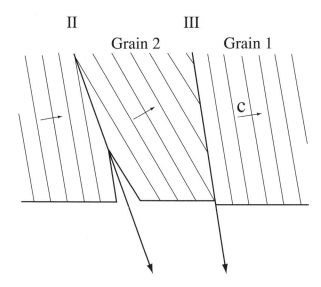

Figure 7.20. The right side of this figure shows the intragrain contact features in zone III. Here it is assumed that grain 1 is able to advance very slightly ahead of grain 2. Note that grain 2 is encroaching on grain 1 even though grain 1's c-axis is the closest to horizontal. The left side of the figure also shows a grain with the orientation of grain 1 in contact with grain 2 at what is now a type II boundary. Note that grain 2 is being cut out even though it is able to encroach on grain 1 at the type III boundary (Kawamura 1987).

Now that we know what happens within the transition zone, we can ask why this occurs. There are two reasons that are usually given.

1. *The favored crystal orientation has its direction of maximum thermal conductivity oriented parallel to the direction of maximum heat loss.*

At first glance this explanation would appear to be inapplicable to sea ice, in that the available information on directional differences in the values of the thermal conductivity for pure ice suggests that the differences between thermal conductivity values measured parallel to the c-axis and parallel to the basal plane are very small. Available data (Landauer and Plumb 1956) suggest that the value parallel to the c-axis is larger by 5% and that the difference in the two values is certainly < 8%. However, as will be discussed in detail later in the section on the thermal properties of sea ice, a sea ice crystal is not just ice but is comprised of a layered composite of ice and brine with the brine layers oriented parallel to the (0001) planes within each ice crystal. In that brine has a lower thermal conductivity than ice, the end result is that in sea ice the effective thermal conductivity parallel to the basal plane of the ice crystal is larger than the value parallel to the c-axis (Anderson 1958a, 1960). Not surprisingly, these values also vary considerably depending on the salinity and temperature of the ice.

2. *The favored crystal orientation has its easy growth direction oriented parallel to the direction of maximum heat loss.*

Because of the unidirectional nature of heat flow in sea ice during the development of the transition layer and the fact that, as noted, ice crystals show a pronounced growth anisotropy, with the easy growth direction being in the (0001) plane, the above statement predicts that the favored crystal orientation in sea ice should be c-axis horizontal. As noted, this is exactly what is observed.

In the sea ice transition zone, the process of geometric selection, in which c-axes horizontal crystals selectively cut out crystals with other orientations, satisfies both of the above reasons. The process as described provides no reason why one c-axis horizontal direction should have a growth advantage over another, provided that the heat loss is purely unidirectional. In situations in which the heat loss is bidirectional, additional constraints are placed on the resulting fabrics. For instance, during the refreezing of narrow cracks that form in cold ice, heat is lost both to the overlying air and also to the cold ice that comprises the sides of the crack. Therefore one would expect that

Figure 7.21. A cartoon of the alignment of the crystal grain boundaries in a freezing crack (Petrich et al. 2007). The crystals growing in from the sides experience a growth advantage upstream due to the salinity gradient (density of dots) and freezing point in the downwelling (arrows) liquid at the ice–water interface.

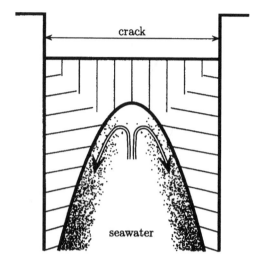

the crystal orientations of the ice in refrozen cracks would be different from sea ice formed by unidirectional freezing from above. That this was the case has been known since the mid-1950s when refrozen cracks containing ice with c-axes both horizontal and parallel to the axis of the crack were reported at Hopedale, Labrador (Weeks and Ackley 1986). Fortunately, this phenomenon has now been examined in the detail that it deserves (Petrich et al. 2007). Figure 7.21 is a cartoon showing the typical grain boundary alignments observed in cracks formed in antarctic fast ice. The fact that the crystals growing in from the sides of the crack grow slightly upward is believed to result from buoyant convective flow within the crack, a process that they have modeled. The generally archlike alignment of the brine pockets is also attributed to this convective flow. They also observed that in wide cracks, the salinity near the air–ice interface is higher at the side of the crack than in the center. However, deeper, and in narrow cracks in general, the salinity is higher at the center of the crack than at its sides.

One would think that once one understood what happens in the transition zone in sea ice, similar processes occurring in lake ice would be obvious. This is not the case. Although c-axis horizontal orientations always occur at the bottom of the transition zone in sea ice, this is not true in lake ice, where either c-axes horizontal or c-axes vertical orientations can result. In some lakes both orientation types have been reported (not in the same sample). In both cases the favored orientation is able to survive by apparently cutting the less favored orientations off from the underlying melt. Furthermore, field observations by Gow (1986) indicate that in freshwater bodies certain types of initial ice skims uniquely result in certain favored orientations. For example, c-axis vertical orientations result via the formation of ice skims on calm, cold, clear nights, whereas c-axis horizontal orientations result when ice forms during windy conditions resulting in waves and an initial layer consisting of frazil crystals. Although I never doubted the reality of this correlation, it appeared to me that this was not an explanation but an observational rule that must follow from any true explanation. I have always felt that these factors must be secondary, in that everything that we know about ice skims and frazil layers indicates that they provide a variety of crystal orientations at the top of the transition layer. If so, then it is not the initial ice layer but differences within the transition layer that establish which crystal orientation is favored and dominates.

There was also an alternate explanation that appears to have been initially advanced by Shumskiy (1955b), and then later developed by Cherepanov and his coworkers at AARI (Cherepanov 1968, 1976; Cherepanov and Kamyshnikova 1973; Fedotov and Cherepanov 1991). A similar explanation was also favored by Knight (1962b) in his studies of arctic lake ice. This is that the favored crystal orientation is controlled by the thermal characteristics of the underlying water body. Simply put, this theory suggested that once constrained crystal growth is initiated, c-axis vertical orientations form if the near-surface portion of the water possesses a stable stratification, with the water temperature gradually increasing with depth. Under such conditions the ice–water interface is assumed to be both flat and smooth, as it is fixed by the position of the 0°C isotherm. Such a control removes any growth advantage associated with horizontal c-axis orientations, as such growth would cause the more rapidly growing crystal to protrude into water that is above the freezing point. The grain boundary groove between the crystals of differing orientations will then be asymmetric, with the crystal that exposes the lower energy surface (the basal plane) to the solid–liquid interface having a lateral growth advantage leading to c-axis vertical orientations. This is similar to an argument advanced by Bolling and Tiller (1960) to explain preferred orientations in metals.

On the other hand, if the water temperature is exactly at the freezing point or slightly supercooled, then the thermal restraint on the direction of easy growth is removed and c-axis horizontal crystals with their basal planes oriented parallel to the direction of heat flow are able to gain the growth advantage. One interesting aspect of this theory is that it is consistent with the fact that sea ice below the transition zone always shows c-axis horizontal orientations. The reason is that as discussed in Chapter 5, the density–composition variation for water with salinities > 24.7‰ always results in water temperature profiles that are either at freezing or slightly supercooled.

In spite of all its apparent advantages, this explanation has not been widely accepted, in that the Russian results have only been presented in a very general way without detailed documentation of either their temperature measurements or their resulting crystallographic observations. Also, experiments demonstrating that the postulated temperature changes do actually result in the appropriate crystallographic changes have not, to date, been reported. Another apparent roadblock to acceptance is the additional work of Gow (1986), who froze several tanks of freshwater having different initial temperatures, observing no effect on the resulting orientations other than that caused by seeding. However, it has been argued (Weeks and Wettlaufer 1996) that these experiments are equally ambiguous, in that the water temperature when the tanks are filled is of little importance. What is important are the temperature profiles in the water column during the time that the initial skim forms and the preferred orientation develops. In short, Gow's observations do not necessarily disprove the Russian explanation.

Fortunately, Müller-Stoffels (2006) has completed a careful set of controlled freshwater freezing experiments in an attempt to resolve these questions. He found that if the ice sheet is seeded, c-axis horizontal orientations *always* develop regardless of the temperature gradient in the water column. In short, his experiments are in complete agreement with the observations of Gow. Müller-Stoffels' explanation is as follows. First consider the case where the temperature gradient downward into the water is positive and conditions are calm (no vertical mixing). In such cases ice crystals can only form when the uppermost surface layer of the water becomes supercooled. As discussed in Section 7.2.1.1, the first crystals that occur are needles that form a very open two-dimensional interlocking surface grid. This is followed by the formation of c-axis vertical crystals in the polygonal spaces between the needles. Once this has occurred, the skim is now laterally continuous, containing both c-axis vertical and c-axis horizontal crystals. In addition, the horizontal area occupied by the platelike c-axis vertical crystals is appreciably larger than that occupied by the needlelike c-axis horizontal crystals. In this situation, the c-axis vertical crystals have the growth advantage in that their easy growth direction is parallel to the interface where the coldest water occurs. The growth of crystals with their c-axes horizontal is limited by the fact that the temperatures ahead of the interface are above freezing.

If the initial crystals are introduced by seeding, the situation is quite different: the initial crystals are much smaller (frequently less than a millimeter in diameter), and the liquid at the ice–air interface is commonly above freezing. As a result an appreciable amount of the early seed crystals melt, as evidenced by an initial drop in the near-surface water temperature associated with the introduction of the seeds. This is followed by a temperature rise associated with the melting and then a gradual monotonic surface cooling to the freezing point. During the partial melting, the coldest water is to be found in a layer at the ice–water interface as the result of buoyancy effects. Although the thickness of the cold layer is not, at present, well specified, it is thought to be at least the thickness of the roughness of the seeded layer. If seeding occurs under windy conditions, its thickness could be several centimeters. When growth occurs into this cold layer, it is the c-axis horizontal crystals that have the growth advantage. Also, because the initial grain sizes resulting from seeding are typically small, c-axis vertical crystals are eliminated from the ice–water interface within vertical growth distances as small as a few millimeters.[*]

If one wishes to study processes occurring in the sea ice transition zone in the field, it is essential to be present during freeze-up. This can be achieved either by being present when the initial ice cover forms, by utilizing newly formed ice in a lead, or more easily (and safely) by forming a pond in an existing ice sheet by cutting and removing blocks of ice. Attempting to study the transition zone several months after freeze-up is not recommended, in that the zone usually occurs either at or very near the ice–atmosphere interface. Therefore it is frequently subjected to significant temperature variations including above- and near-freezing temperatures. Having to deal with the results of changes resulting from these additional environmental conditions in addition to the changes normally occurring within the transition zone is not desirable.

The percentage of a winter's ice cover that is composed of transition zone ice is small because the thickness of this zone is also small. Nevertheless, it is the selective growth processes within this zone that establish the fundamental c-axis horizontal fabric of the underlying ice. Ice showing this easily distinguishable fabric is commonly referred to as *congelation ice*. In the Western literature the use of this term implies that the ice being discussed grew as the result of essentially one-dimensional heat loss upward through previously formed ice. In short, congelation ice is ice whose growth is well described by the classical Stefan ice growth analysis. The term *congelation* was first used by Russian investigators who initially applied it to all types of ice that formed

---

[*]   I would like to conclude this aside into the world of lake ice with a personal comment. The lake ice orientation problem has interested me since 1955 when I first read *SIPRE Report 5* (Wilson et al. 1954). At that time everything seemed to be so simple, as the New England lakes under study developed c-axis vertical orientations, and current experiments (Landauer and Plumb 1956) suggested that the thermal conductivity of ice was slightly greater parallel to c than perpendicular to c. I also would like to congratulate Cherepanov, Gow, Müller-Stoffels, and Shumskiy for their contributions to this problem. First-class work. However, I still think that we are missing something. Something that's right in front of us and I can't seem to see it. Of course, maybe I'm wrong. It wouldn't be the first time.

from the direct freezing of bulk seawater. Currently their usage has also narrowed to the sense described here (Borodachev et al. 2000). Under this definition, layers of frazil ice within a sea ice sheet are not considered to be congelation ice in that the immediate cold source causing the formation of these individual crystals is the supercooled surrounding water as opposed to the loss of heat upward through an overlying ice layer.

7.2.3 The Columnar Zone

Sea ice below the transition layer has all the characteristics associated with the so-called columnar zone in metal ingots. That is, it possesses a strong crystal elongation parallel to the direction of heat loss, a pronounced crystal orientation, and a gradual increase in grain size over crystals closer to the cold source (Walton and Chalmers 1959). Therefore, this term is also useful in discussing sea ice. Because the transition zone is both thin and located at or near the ice–air interface, it frequently sublimates away or recrystallizes into the snow cover as the ice sheet grows. Therefore, essentially the complete thickness of a first-year ice sheet can, to a good approximation, be considered to be in the columnar zone. Compared with the pronounced changes in the transition layer, the changes in the columnar zone are considerably more subtle. The fabric diagrams all show predominately c-axis horizontal orientations. In some cases the c-axes appear to be randomly distributed in the horizontal plane, giving fabric diagrams similar to that from the 5.0–5.5 cm level as shown in Figure 7.15. In other cases a strong c-axis alignment develops in the horizontal plane. More will be said about this later. Typical horizontal thin sections from the columnar zone are shown in Figure 7.22. The individual crystals are indicated by thin section regions where the interference colors show a constant shade of gray and the substructure shows a common

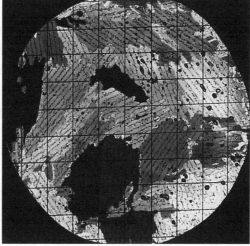

Figure 7.22. Typical photographs taken using crossed polaroids of thin sections of columnar zone FY sea ice. The grid spacing is 1 cm. The lefthand photo shows ice from offshore Barrow, Alaska (Weeks and Hamilton 1962). The ice on the right was from the Russian Arctic (Cherepanov et al. 1997).

Figure 7.23. Plot of the change in average grain size with distance below the upper ice surface. The grain size curves are shifted to zero at the upper ice surface.

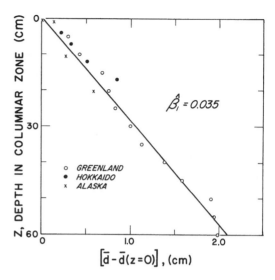

orientation. As can be seen, each crystal is comprised of a number of subcrystals that are separated either by layers of brine or layers of isolated inclusions of liquid brine that are referred to as brine pockets. The edges of the crystals are frequently serrated, reflecting the boundaries of the subcrystals that make up the crystals.

One interesting aspect of ice in both the transition and columnar zones is the vertical change in grain size as observed in the horizontal plane. This subject has been studied by several investigators (Tabata and Ono 1962; Weeks 1958; Weeks and Hamilton 1962). As the depth in the ice sheet increases, the number of crystals per unit area decreases and their average diameter increases. Here, mean diameters for individual crystals were calculated from $\bar{d} = \sqrt{\ell \times w}$ where the length measurements ℓ were taken parallel to the (0001) plane while the width w was measured normal to this (parallel to the c-axis). As might be expected, considering the growth anisotropy so characteristic of ice, and as can be seen in Figure 7.22, diameters measured parallel to the basal plane are invariably equal to or greater than the diameters parallel to the c-axis. Figure 7.23 is a plot of data from Greenland, Hokkaido, and Alaska that shows mean grain size $\bar{d}$ shifted to zero at the ice's upper surface plotted against depth z. A pronounced linear increase in $\bar{d}$ with increasing z is apparent. I used to think that the slope of this curve might be a constant for different sheets of sea ice. This would mean that if the average grain size was determined at one level in an ice sheet, it would specify the mean grain size distribution for the whole sheet (see the discussion in Weeks and Ackley 1986). Although this would be convenient, particularly in studies where the grain size is an important parameter, this no longer appears to be generally true. Perhaps it would be better to say that if such a relation exists it is no longer obvious (to me).

One problem occurring in real ice covers is that the steady growth of congelation ice is frequently interrupted by the introduction of layers of frazil ice at the base of the ice sheet. These introduced layers interrupt any systematic changes in grain size with depth by essentially causing the growing ice to form a new transition layer with its associated changes in grain size. As will be seen later in this chapter, an additional difficulty occurs in ice sheets that develop strong c-axis alignments in the horizontal plane. The problem is simply that once all the c-axes are pointed in essentially the same direction, it becomes very difficult to separate the aligned pattern of platelets as seen in thin sections into individual grains whose lengths and widths can be measured. Another poorly understood feature of vertical variations in grain size is that in FY ice the lowest layers that form just before the cessation of growth (resulting from the onset of the melt season) show an appreciably smaller grain size than the overlying ice in the columnar zone. One explanation for this might be that the development of biological activity in the spring leads to nucleation of many new grains, resulting in a decrease in the average grain size. Another possibility is that the competitive process that results in the vertical increase in grain size in the columnar zone ceases to be effective at growth rates below some minimum value.

Based on his study of sea ice cores collected from Jones Sound off of Devon Island in the Canadian Arctic, Koerner (1963) has suggested that grain size differences observed in cores collected at differing distances from shore could possibly be explained by differing growth rates (nearshore ice showed a larger initial grain size and a less rapid increase in grain size with depth than ice collected farther off the coast). Although this is conceivable, I doubt very much that it will prove to be a useful generality. If I were to explore this possibility further, I would first survey the metals literature, where grain size control is an important economic consideration, then proceed to some simple cold room experiments, and finally test my results with field observations.

Although c-axis orientations in sea ice have received considerable attention in the literature, very limited attention has been given to whether or not a preferred orientation also develops in the a-axis directions. The reason is, of course, that a-axis orientations are more difficult to measure. While the c-axis orientation of a given sea ice crystal can readily be determined by using the optical procedures described in Appendix E, a-axis orientations are usually determined by producing etch-pits on the (0001) surface on an ice crystal. These pits have a hexagonal outline that in turn reveals the a-axis directions as shown by x-ray investigations of crystals of pure ice. The only study of possible a-axis orientations to date is by Kawamura and Ono (1980), who examined both c- and a-axis directions in sheets of thin sea ice at Barrow, Alaska. Although there were large differences in the c-axis orientations between their upper (4.0–7.7 cm) and lower (17–22 cm) samples, there was little change in the a-axis orientations with depth. Even so, all a-axis orientations were not equally represented, with the most frequent observations being either horizontal or nearly horizontal (38%) or at 30° (17%). This suggests that the favored a-axis growth direction is close or equal to either $\langle 11\bar{2}0 \rangle$ or $\langle 10\bar{1}0 \rangle$. If there is a preferred direction I would guess that it would be $\langle 11\bar{2}0 \rangle$, an orientation that would place an a-axis vertical and parallel to the direction of heat loss, as opposed to placing the normal to the first-order hexagonal prism $\langle 10\bar{1}0 \rangle$ in this orientation. It is also reasonable to assume that the growth advantage of $\langle 11\bar{2}0 \rangle$ would be small. This would suggest that it might take a long period of ice growth before the dominance of $\langle 11\bar{2}0 \rangle$ would begin to show. In the future it would be interesting to check a-axis orientations in much thicker FY and MY ice to see if pronounced a-axis orientations do develop.

Occasionally, isolated individual crystals found in the columnar zone do not have c-axis horizontal orientations. In my experience these crystals all seem to share the following characteristics. They are small (1–4 mm) and have roughly circular outlines in the horizontal plane. When examined under crossed polaroids they remain dark during a complete rotation of the stage indicating that the c-axis is vertical or nearly vertical. Finally, they do not appear to have an appreciable vertical extent, indicating that they are rapidly "cut out" by the surrounding c-axis horizontal crystals. A reasonable explanation for the occurrence of these crystals is that they form as small discoids within the underlying water column and then float upward until they contact the base of the overlying ice sheet. There, their discoidal shape causes them to come to rest with their c-axes vertical. This however places them at a growth disadvantage relative to the surrounding crystals, resulting in their rapid termination as the ice sheet continues to grow.

There are some columnar zone features that are only clearly seen in vertical sections. For instance, it is common to observe small-scale horizontal banding in sea ice (Bennington 1963a, Fig. 3; Cole et al. 2004; Langleben 1959, Fig. 3; Paige 1966; Shumskiy 1955a; Tabata and Ono 1957, Plate XVI, 1962; Tison et al. 2002). Shumskiy observed a total of 58 secondary layers in a 291-cm section of old pack ice, giving an average layer thickness of 5 cm. The layering appeared to be the result of variations in the amount of impurities (gas and brine) trapped in the ice and was assumed to be the result of variations in the rate of ice growth. Photographs showing horizontal bands that also result from porosity variations can be found in Cole and Shapiro (1998) and in Cole et al. (2004). These bands do not appear to be associated with the nucleation of new crystals. Air bubble layering in lake ice is well known (Swinzow 1966; Taylor and Lyons 1959, 16–19). At times it can be very striking and has been used to map time horizons in ice covers in regions where the weather is laterally relatively homogeneous (Gow and Langston 1977). They also used bubble layers in lake ice as a reference level in studying the relative amounts of ablation from the top and bottom of a sheet. Langleben (1959) has also used such layering in sea ice as a reference mark in studying the location of the transition layer throughout the growth season. Theoretical analyses of the effect of changes in the growth velocity on the amount of impurity incorporated in solid solution in the solid phase have been developed by Tiller et al. (1953) and Smith et al. (1955), assuming that solute transfer in the liquid phase takes place by diffusion only. Attempts to apply this theory to the formation of bubble layers in freshwater ice have been published by Carte (1961), Bari and Hallett (1974), and Gow and Langston (1977). To date, no related work has been carried out on sea ice, although similar relations would be expected to hold. It is also possible to obtain multilayered structures as the result of the rafting of several ice layers over one another while the ice is still quite thin. This will be discussed later in the section on deformation.

The most detailed examination of banding in FY ice to date is by Cole et al. (2004). Although the study is thorough, it raises more questions than it answers. For instance, the authors identify several different types of bands that occur both offshore and in a nearby lagoon near Barrow, Alaska. These include corrosion bands first identified by Bennington (1963a, 1963b) as well as bands resulting from fine-grained ice layers that characteristically occur in the upper part of the ice sheet. The most common bands in the columnar portion of the ice sheet appear to be the result of variations in the amount of gas and brine in the ice. Both high-porosity and low-porosity bands were noted. It was found that band characteristics did not vary systematically with depth but varied significantly from year to year. Furthermore, although the two study sites were only a few kilometers apart and were apparently subjected to the same weather, the banding patterns were not similar. This would appear to reject explanations that depend on variations in the local weather. The best correlations appear to be with changes in local sea level, a parameter that, in turn, is assumed to correlate with variations in the under-ice current velocity. Current suggestions are that a band occurs when the current velocity drops below the level necessary to maintain a c-axis alignment in the columnar crystals. At present it appears that banding patterns are telling us more about changes in oceanographic conditions under the ice than about meteorological conditions above the ice.

7.2.4 Substructure

One characteristic distinguishing congelation sea ice from freshwater or glacier ice is the fact that its ice crystals invariably show the presence of a cellular substructure consisting of reasonably evenly spaced ice platelets or cells, separated by small-angle grain boundaries. It is along these boundaries that the salt in sea ice is present in the form of liquid and solid inclusions. This cellular substructure is clearly shown in Figure 7.22. To understand sea ice, one must understand why this structure forms and how it varies. Although, as we will see, there have been several studies of substructure variations in sea and NaCl ice, the amount of this work pales in comparison to the number of related studies carried out on metals and transparent organic compounds. The reason for such intense interest is that in most materials the development of the substructure, the entrapment of impurities, and the properties of the resulting material are interrelated. Therefore, understanding these interrelations allows one to manufacture materials that have desirable properties. Although in the case of natural sea ice the manufacturing process is hardly under one's control, a similar understanding assists one in interpreting the large ice property variations that are observed in the field. It is also good to remember that although the general principles governing the solidification of metals such as lead, cadmium, and zinc; organics such as succinonitrile and pivalic acid; and seawater are similar, the details can be quite different.

7.2.4.1 Constitutional Supercooling

The earlier discussion of the NaCl–H$_2$O phase diagram could lead one to believe that the ice that forms when seawater freezes should be pure ice. If this were strictly true it would be simple to desalinate seawater: simply freeze a layer of ice and separate it from the underlying brine. The reason this is not done is that when one determines the composition of the resulting ice, one finds that it contains salt in the form of liquid inclusions located along the boundaries between the groups of platelets of pure ice that comprise the individual sea ice crystals. What the phase diagram told us is still correct: the ice phase in sea ice is pure ice. The confusion here is partially semantic in that bulk samples of sea ice are composed of pure ice + entrapped brine + solid salts (if the temperature is sufficiently cold).

The question is, why does sea ice form with a nonplanar, cellular interface, resulting in brine entrapment, whereas lake ice forms with a planar interface, resulting in essentially complete impurity rejection back into the melt? Both sea ice and lake ice result from the freezing of impure melts, with the difference being only a matter of degree in that seawater contains ~35‰ salts as compared with < 1.0‰ in lake or river water. The answer is that sea ice is a classic example of a material that invariably forms under growth conditions that result in a nonplanar solid–liquid interface becoming stable. This is a condition that is essential to brine entrapment and is referred to as *constitutional supercooling*. What is constitutional supercooling and how does it work? There have been two different approaches to this problem. I will first discuss the original classical analysis (Elbaum 1959; Rutter and Chalmers 1953; Tiller et al. 1953) in some detail, in that I believe that understanding this phenomenon is essential to understanding many aspects of sea ice. Later I will outline the results of other approaches.

Consider an NaCl solution freezing as the result of one-dimensional cooling. For such a system the distribution coefficient k_o (defined as the ratio of the solute content of the solid to the solute content of the liquid when the two phases are at equilibrium) has a value of 10^{-4} or less

(Harrison and Tiller 1963a). As noted, the ice phase that forms is essentially pure. Next assume that (a) diffusion in the solid is negligible; (b) convection and mixing in the liquid due to causes other than diffusion are negligible (for salt solutions this is a good assumption only if we freeze from the bottom up); and (c) k_o is constant. If the origin of the coordinate system is taken to be at the freezing interface, then freezing can be viewed as moving the liquid toward the interface at a velocity v producing a net flow out of a unit volume of $v\,(dC/dx)$ where C is the concentration and x is the distance normal to the interface. At the same time there is a diffusive flow of impurity into the volume element of $D(d^2C/dx^2)$, where D is the diffusion coefficient. If a steady state is achieved these two terms must be equal (the amount of impurity flowing in equals the amount of impurity flowing out), leading to the differential equation

$$D\frac{d^2C}{dx^2}+v\frac{dC}{dx}=0. \tag{7.8}$$

The boundary conditions are

$$C_L=C_0 \text{ at } x=\infty$$

$$C_L=\frac{C_0}{k} \text{ at } x=0.$$

Here C_L is the concentration in the liquid and C_o is the initial concentration in the liquid (the composition far removed from the interface). The steady-state solution is then

$$C_L=C_i\exp\left(-\frac{v\,x}{D}\right)+C_0 \tag{7.9}$$

where C_i is the composition of the liquid at the interface. This ultimately leads to

$$C_L=C_0\left[1+\left(\frac{1-k}{k}\right)\exp\left(-\frac{v\,x}{D}\right)\right], \tag{7.10}$$

showing that there should be a thin layer of liquid at the interface that exhibits an exponential decrease in C_L as one moves away from the interface. This layer has a characteristic distance D/v in which the excess concentration (above C_0) falls to $1/e$ of its initial value. This is shown schematically in Figure 7.24. Because of these compositional variations, there are associated differences in the equilibrium freezing temperature of the liquid ahead of the interface (Figure 7.25) as specified by the phase diagram, inasmuch as

$$T_E=T_0-mC_L \tag{7.11}$$

where T_E is the equilibrium freezing temperature, T_o is the melting point of the pure solvent (ice), and m is the slope of the liquidus line from the phase diagram. This results in

$$T_E=T_0-mC_0\left[1+\frac{1-k}{k}\exp\left(-\frac{v\,x}{D}\right)\right]. \tag{7.12}$$

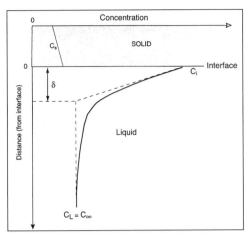

Figure 7.24. A schematic diagram showing the solute enriched layer that develops in front of an advancing liquid–solid interface during the freezing of a salt solution.

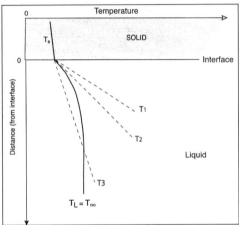

Figure 7.25. A schematic diagram showing the equilibrium freezing temperature T_E of the liquid in front of an advancing liquid–solid interface that results from the compositional profile shown in the previous figure. Also shown are three possible temperature profiles in the liquid ahead of the interface. Note that profile T_2 is exactly tangent to the T_E curve at the interface. In profile T_1 the temperature below the advancing interface is always above freezing. However, for profile T_3 there is a region where the temperature is below the equilibrium freezing temperature as the result of the compositional profile.

Next, take the actual temperature distribution in the liquid to be given by

$$T = T_0 - m\left(\frac{C_0}{k}\right) + G_L x \tag{7.13}$$

where G_L is the temperature gradient in the liquid at the interface and the first two terms on the right side of the equation give the temperature at the interface. For some compositional profiles and temperature profiles the composition is such that the liquid ahead of the interface becomes supercooled (see Figure 7.25). Hence the term *compositional supercooling*.

If the slope of the actual temperature profile is more than the slope of the T_E curve at the interface, a planar interface remains stable. If, on the other hand, the slope is less than the slope of the T_E curve (profile T_3 in Figure 7.25), *constitutional supercooling* (CS) occurs. The criterion for this condition

$$\frac{G_L}{v} < \frac{mC_o}{D}\left(\frac{1-k}{k}\right) \tag{7.14}$$

is a relation obtained by differentiating equations 7.12 and 7.13 with respect to x and evaluating them at $x = 0$. This relation has been found to be in good agreement with experimental observations on metal systems and has been used extensively in the materials science literature (Chalmers 1964; Flemings 1974; Tiller 1991; Walton et al. 1955). When equation 7.14 is applied to the freezing of both salt solutions and seawater, it is found that when optimum conditions for the maintenance of a planar interface occur ($v \sim 10^{-6}$ cm/s, $G_L \sim 1.0$ C/cm, and $k_0 = 10^{-4}$), the ratio $G_L/v = 10^6$. For this ratio CS should occur (see Fig. 43 in Weeks and Ackley 1986), even for salinities representative of typical lake water (~ 100 ppm or 0.1‰), not to mention the salt contents studied in the book (> 24.7‰). Flemings points out that equation 7.14 should also be applicable even if convection is occurring in the liquid in that a laminar sublayer with its small but finite concentration gradient is generally believed to exist at the solid–liquid interface regardless of the degree of convection. In this case, equation 7.14 becomes

$$\frac{G_L}{v} < \frac{-mC_\infty\left(1-k\right)}{D}.$$ (7.15)

Here C_∞, the composition of the bulk liquid, is equal to C_0 for a small amount of solidification from a large volume of melt, a situation that is certainly applicable to the natural freezing of seawater.

Although the existence of a CS layer ahead of the advancing solid–liquid interface is necessary for cell formation, it is not necessarily sufficient. In addition, there is considerable uncertainty in the preceding calculation as the result of possible variations in G_L. This difficulty can be avoided by calculating if the growth conditions are such that the steep-walled cell boundary grooves associated with the entrapment of brine along plate boundaries are stable. The stability criterion for the formation of these grooves (Tiller 1962) is

$$\frac{G_s}{v} < -\frac{mC_i\left(1-k\right)}{D}$$ (7.16)

where G_s is the temperature gradient in the solid and C_i is the composition of the liquid at the interface. If transfer in the liquid is by diffusion only, $C_i = C_0/k$ and equation 7.16 becomes similar to equation 7.15, with G_s substituted for G_L. Equation 7.15 is the more stringent criterion in that for ice–solute systems $G_s \gg G_L$. Now, if

$$t = \frac{\rho L h^2}{2\kappa \Delta T_0}$$ (7.17)

(Carslaw and Jaeger 1959), where t = time, ρ = density of the solid, L = latent heat of fusion, ΔT = difference between the upper surface (i.e. cold plate–ice interface) temperature of the ice and the freezing temperature of the seawater, h = thickness of the ice, and κ = thermal conductivity of the ice, then

$$v = \frac{dh}{dt} = \frac{\kappa \Delta T_0}{\rho L h}$$ (7.18)

and

$$G_s = \frac{\Delta T_o}{h}.$$ (7.19)

Therefore, equation 7.15 becomes

$$\frac{\rho L}{\kappa} < \frac{-mC_0\left(1-k\right)}{D k}.$$ (7.20)

In this relation C_0 is the only parameter under the control of the experimenter. Substitution of numerical values in equation 7.20 shows that for all sea and brackish water salinities cell boundary grooves are stable if the solute transfer in the liquid is by diffusion only. In actuality, because natural sea ice freezes from the top down, the primary mechanism of solute transfer is free convection (Farhadieh and Tankin 1972; Wakatsuchi 1977), which causes C_i to approach C_0. Convection also results in an increase in the effective value of D by as much as a factor of 100 over the value for pure diffusion (for NaCl solutions $D \sim 10^{-5}$ cm^2/sec; Harned and Owen 1950). This explains the observations of Weeks and Lofgren (1967), who recorded transitions from a nonplanar to a planar interface during the freezing of 1‰ (unstirred) and 3‰ (stirred) NaCl solutions. In these experiments the transitions occurred as the growth rate decreased even though salt rejection caused an increase in the salinity of the underlying freezing solution, thereby favoring cellular growth.

7.2.4.2 Diffusive Instabilities   The constitutional supercooling approach discussed above examines which state, solid or liquid, is thermodynamically stable in front of a planar interface. If this state is liquid, the interface is assumed to remain planar; if the state is solid, the planar interface is assumed to become unstable and evolve into one of the other possible interface shapes. Like all purely thermodynamic approaches, it tells us what should happen but does not tell us if it will happen. As pointed out by Wettlaufer (1998), CS also tells us nothing about heat transfer in the solid, the effects of latent heat release, surface tension, or interfacial attachment kinetics. Clearly, a more detailed examination of this problem was warranted.

As a result, an alternate approach, initially developed in papers by Mullins and Sekerka (1963, 1964), has focused on the dynamics of the development of the interface shape. This approach has been referred to as a *Mullins–Sekerka instability*, a *morphological instability*, or a *diffusive instability*. Here I will refer to it using the latter designation and the letters (DI). In this approach, a small perturbation δ is assumed to form on a growing planar interface. If the perturbation decays, the planar interface remains stable; if it grows, the planar interface breaks down. The fundamental assumptions of the theory are similar to those of constitutional supercooling in that equilibrium is assumed to hold at the interface, convection is typically not considered, and the surface energy is taken to be isotropic (crystallographic factors are neglected). The advantage of the approach is that it is capable of describing the time evolution of a perturbed interface and the accompanying temperature and concentration fields. The growth velocity of the perturbation, a function of the wavelength of the perturbation λ and several thermal parameters, is then calculated. If δ remains greater than zero, the instability grows and develops into cells. Although the resulting equations are appreciably more complex than the relations resulting from the CS approach, the important terms in the CS theory also occur in the DI theory.

Although an examination of the details of this theory are beyond the scope of the present book, a few results relating to ice should be mentioned. Sekerka et al. (1967), who utilized the theory to examine the effect of the initial buildup of salt on interface stability during the freezing of seawater, concluded that a seawater–ice interface is effectively instantaneously unstable with respect to the rise time associated with the buildup of salt at the advancing ice interface. In addition, Hardy and Coriell (1968, 1973) have used DI theory to study the interface breakdown observed on ice cylinders in order to obtain an estimate of the ice–water interfacial free energy. The

Figure 7.26. The neutral stability diagram in the (v, λ) plane where v is growth velocity and λ is the wavelength of the disturbed interface. Conditions inside the closed curve represent those under which a planar interface is unstable (Wettlaufer 1992a).

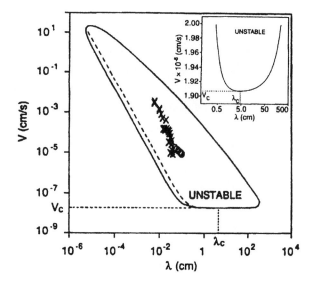

resulting estimate was in good agreement with values obtained by other methods. The most detailed application of the theory to sea ice has been by Wettlaufer (1992a, 1992b, 1998), who has utilized the H_2O–NaCl system as a model for the solidification of seawater. The model he considers is thermally nonsymmetric in that the thermal conductivity of ice and brine are different. It is also chemically one-sided in that solute diffusion in the solid is taken as negligible. The effect of the latent heat liberated at the growing interface is also considered. However, the advective/convective transport of heat and solute is initially neglected. Figure 7.26 shows a plot in v–λ space of the neutral stability curve separating the region where a planar interface is stable from the region (interior to the curve) where it is unstable and a cellular interface should develop (for a 35‰ salt solution). The inset is a blowup of the curve near the critical values of v and λ. The short wavelength truncation is due to the Gibbs–Thompson effect, which limits the finite nucleation size of an interfacial disturbance, whereas the long-wave cutoff results from the finite interaction range of the solute field as the result of the slow diffusion of impurities. Note the flatness of the neutral curve near the critical point (i.e., the weak dependence of λ on v), resulting in weak wavelength selection. Wettlaufer's analysis (1992a, 1992b) also indicates that for typical seawater compositions and growth rates a cellular interface will always be the stable natural form. He concludes that in cases where a planar interface transitions to a cellular interface, the transition will not be gradual but sharp, a situation that he refers to as a subcritical bifurcation. This conclusion is in good agreement with the limited experimental observations on such transitions observed during the freezing of NaCl solutions (Weeks and Lofgren 1967). Details on the above analysis can be found in Wettlaufer (1992a, 1992b). His general review of many additional crystal growth aspects of sea ice is also highly recommended (Wettlaufer 1998). Readers who might be interested in expanding Wettlaufer's study of interface stability in the NaCl–H_2O system to a six- to eight-component system more representative of the composition of seawater are advised to scan the papers by Coates et al. (1968) and Coriell et al. (1987) on interface stability in ternary systems to gain a sense of the additional difficulties.

7.2.4.3 Interface Shape   In the above discussion I have tried to convince the reader that when seawater freezes a planar solid–liquid interface is unstable both from a theoretical as well as from an observational point of view. Assuming that this effort has been successful, one might well inquire as to the exact shape of the so-called cellular interface that results. Surprisingly, there is

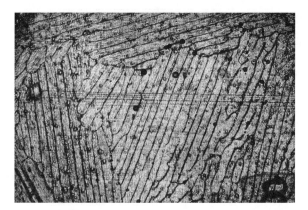

Figure 7.27. A photomicrograph taken in the horizontal plane showing the substructure of NaCl ice. The initial solution salinity was ~3‰ and the speed of the advancing interface was 1.8×10^{-4} cm/s (Lofgren and Weeks 1969). Note the straightness and regularity of the cell boundaries.

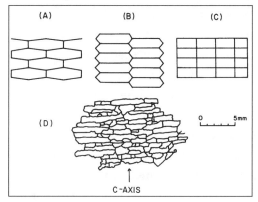

Figure 7.28. Idealized cell packings as observed in zinc (*A, B,* and *C*) by Damiano and Herman (1959) and typical cellular structure as observed in sea ice (*D*) at Point Barrow, Alaska (Weeks and Hamilton 1962).

very little detailed information available on this subject, particularly in the salinity and growth rate range commonly encountered in nature. Nevertheless, it is possible to piece together a general description. First consider the horizontal thin sections shown in Figure 7.22. They show that each individual ice crystal is comprised of a number of cells of pure ice that are separated by arrays of brine inclusions or films of brine. Within a given crystal all of the cells have essentially the same orientation, as determined optically under crossed polaroids. Here the word *essentially* is used as a qualifier in that small differences in alignment of up to 3° are not unusual within such crystals, indicating the presence of small-angle grain boundaries. In some situations described in the metals literature, the boundaries are extremely regular (Weinberg 1963). Also, it has been noted that when freezing NaCl solutions the cells are commonly very straight and regular and extend the complete length of the crystals. For instance, see Figure 7.27. Extremely straight cell boundaries appear to be particularly common in rapidly solidified materials. In zinc, also a hexagonal material, the orientation of the (0001) plane is clearly revealed by an elongation of the individual cells parallel to the basal plane, and the individual cells are packed together in what appears to be three different patterns (Figure 7.28). In natural sea ice, matters do not appear to be nearly as clear-cut. In a limited study of the horizontal cell geometry of sea ice at Barrow, Alaska, Weeks and Hamilton (1962) found that although an elongation parallel to the basal plane is invariably present, it is much less pronounced and precise. The packing of the cells was also less regular than in metals and appeared to be best described by packing type (B) as shown in Figure 7.28. In sea ice and salt ices there is considerable variation in the regularity of the geometry of the individual cells. In metals, the radius of curvature of the tips of the cells, as shown in Figure

Figure 7.29. Schematic illustration of variations in cell profiles as observed in metals (modified from Tiller 1991). The parameter λ is the cell size (or brine layer spacing) and the position of the number 5 gives the location of the brine-filled intercellular groove.

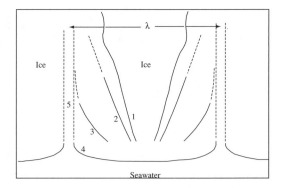

7.29, is known to sharpen as the value of vC_∞ / G_L increases (Tiller 1991). Here v is the growth velocity, C_∞ is the bulk composition of the liquid, and G_L is the temperature gradient in the liquid at the interface. To date, in sea ice studies the value of G_L has not been determined. It would appear that the variations in cell geometry are telling us something about the environmental history experienced at that specific level in the ice sheet. This appears to be a subject that has not been thoroughly explored.

Consider the shape of the ice–seawater interface as it would look as examined in a vertical cut oriented in the plane normal to the c–axis. Each cell is separated from its neighbors by brine-filled channels (Figure 7.29, number 5) that are very deep compared to the horizontal spacings, λ, between the brine channels, which typically have dimensions of ~1 mm. The observational basis for this is as follows. When one examines the strength of the bottom layer of a growing sheet of sea ice one finds that the lower 2 to 3 cm of the ice can be scraped away using just one's finger. This weak layer is believed to result from a lack of lateral ice–ice contacts between the individual platelets of ice. The upper boundary of this so-called "skeletal" layer is fairly well defined and the layer appears to have a fairly constant thickness. To date no detailed studies are available. Fortunately, the entrapped brine leaves a permanent record of the positions of the intercellular grooves at each level in the ice sheet, allowing measurements of λ values to be made at the convenience of the investigator. However, measurements of skeletal layer thicknesses would have to be made immediately upon removing the sample from the lower portion of the ice sheet. In a study of Pb–Sn alloys, Stewart and Weinberger (1972) used radioactively tagged Pb to determine how far the melt flowing past the growing interface penetrated the interface. Penetration occurred up to the point where the solid volume fraction achieved values greater than 0.12 to 0.22. They then used the value of 0.20 to define the thickness of what has been referred to here as the skeletal layer.

The exact shapes of the cell tips have not been studied in either seawater or NaCl–H_2O systems. However, in other systems cell tips are generally rounded and blunt. In sea ice, typical profiles would probably be between curves 3 and 4 in Figure 7.29. A Russian view of the probable cell tip shape can be seen in Figure 7.30 (Cherepanov et al. 1997). A review of different approaches to the cell shape problem can be found in a paper by J. D. Weeks et al. (1991). There are several reasons why the detailed shapes of the cells are of interest. If the tips are blunt the solute is primarily rejected back into the melt with only a small amount trapped in the intercellular grooves. If the cells are more parabolic in shape, more impurity will presumably

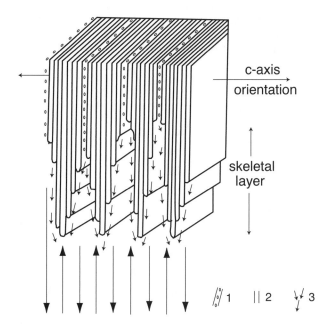

c-axis
orientation

skeletal
layer

||1 ||2   ↓3

Figure 7.30. An alternate view of the skeletal layer at the bottom of a growing sheet of sea ice (Cherepanov et al. 1997). The numbered symbols indicate the following: (1) layers of brine and gaseous inclusions, (2) the orientation of the (0001) planes in the individual cells of pure ice, and (3) the general directions of brine drainage.

be trapped. Also, the prediction of the exact shape of the cell tips provides a check on the adequacy of theoretical simulations. In metals, studies of cell shapes are made difficult by several factors. First, the high melting temperatures of most pure metals make observations difficult. Second, to examine the solid–liquid interface one must decant the surface, as both solid and molten metals are opaque. The problem here is that during the decanting procedure surface tension may cause liquid to be retained both on the interface and in the grain boundary grooves, thereby modifying the shape of the decanted interface. Certainly these problems would be expected to be less in studies of the freezing of seawater in that the melt is transparent, making observations possible during active growth. In addition, measurement systems can be placed in the melt if necessary and the temperature range encountered facilitates ready, if slightly uncomfortable, handling of the specimens. It is therefore surprising to me that so little work has been done along these lines, a point also noted by Wettlaufer (1998). The primary exception here (Harrison and Tiller 1963a, 1963b) is quite interesting in that they both obtained cells developed in the (0001) direction as observed in nature and also in the c-axis direction. Although they primarily discuss cellular development, their photographs are of interfaces comprised of arrays of "cells" that show appreciable ornamentation more characteristic of dendritic growth. Although several studies have been completed in metal and organic systems of the detailed changes in interface morphology during the transition from planar to cellular, only one related study exists using NaCl solutions that were frozen from the bottom up (Kvajic et al. 1971). Several morphological changes were noted during the interface breakdown on both c-axis vertical and c-axis horizontal crystals. The authors attempt to analyze these changes in terms of diffusive interface theory as formulated at that time was, in my view, only partially successful.

7.2.4.4 Cell Size Variations   It has been known for some time that the spacing of the cellular substructure in sea ice (λ) changes with vertical location in an ice sheet (Fukutomi et al. 1953). In an attempt to understand these variations, observations have been carried out on natural sea ice (Gow and Weeks 1977; Nakawo and Sinha 1984; Paige 1966; Tabata and Ono 1962; Weeks

Figure 7.31. Schematic illustration of possible mechanisms for changing cell spacings suggested in the materials science literature (Lu and Hunt 1992): (A) to (B) = overgrowth resulting in an increase in the cell spacing; (B) to (C) = tip splitting, resulting in a decrease in the cell spacing.

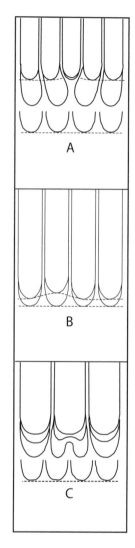

and Hamilton 1962) and on a variety of artificial sea ices (Lofgren and Weeks 1969; Rohatgi and Adams 1967a, 1967b, 1967c; Weeks and Assur 1963). Here, the important parameter in characterizing the substructure is the brine layer spacing, λ, a parameter that has also been referred to as the plate width in the sea ice literature and as the cell size in metals. In this book I will use the terms *brine layer spacing* and *cell size* interchangeably. This is the distance between the centers of adjacent layers of brine pockets measured parallel to the c-axis (see Figures 7.29 or 7.30; there the double-headed arrows indicate both the c-axis direction and the width of the particular cell). Laboratory experiments have shown that in NaCl ice formed under conditions of unidirectional freezing, there is a gradual increase in the value of λ with increasing distance from the upper ice surface or from a constant temperature cold plate. There are no observations on exactly how these size adjustments occur in sea ice. Based on observations on other materials the change mechanisms are probably overgrowth (A to B) and tip splitting (B to C), as shown schematically in Figure 7.31. Considering the fact that λ values generally increase as the ice thickens and growth rates decrease, the mechanism in sea ice is presumably primarily one of overgrowth. The important parameter here, of course, is not the position in the ice sheet or the distance from the cold plate but the fact that these locations correspond to specific growth velocities with the velocity commonly decreasing as the ice thickens. Similar results have been obtained in studies of the solidification of metals and transparent organic compounds. In that work, the experimentalists have been able to compare cell sizes from freezing runs performed with different constant growth velocities (Jesse and Giller 1970; Okamoto and Kishitake 1975; Okamoto et al. 1975; Palacio et al. 1985; Rutter and Chalmers 1953). Individuals interested in exploring this general subject further in the materials science literature should look at the results of the symposium on the *Establishment of Microstructural Spacing during Dendritic and Cooperative Growth* (Perepezko and Shiflet 1984).

An analysis of the factors controlling the variation in λ has been published by several different authors including Bolling and Tiller (1960), Lu and Hunt (1992), Rohatgi and Adams (1967c), Tiller (1991), and Wettlaufer (1992a). Although the details of these different approaches vary, the physical reasoning is generally similar. The exact solution to the steady-state solute distribution in the liquid ahead of an advancing cell consists of both plane and nonplane wave terms. These latter terms cause lateral diffusion and can be considered to extend some effective distance into the liquid. Once lateral diffusion starts it continues until it is terminated by the advancing interface. The time allowed for diffusion is a function of interface shape and the interface velocity. The distance x that the solute can diffuse in time t is $x \approx \sqrt{Dt}$ where D is the diffusion coefficient. The relations developed by Bolling and Tiller suggest that λ should increase with an increase in C_0, the bulk composition of the freezing solution, and that the functional form of the relation between λ and v is $\lambda v = A$ for small values of v and $\lambda \sqrt{v} = A$ for large values of v (here A is a constant). The theoretical treatment developed by Rohatgi and Adams is specifically focused on extremely rapid freezing such as occurs in splat metallurgy. When their theoretical treatment is modified so that it more closely corresponds to the sea ice situation (Lofgren and Weeks 1969), it gives λ vs. v relations of a form that are in reasonable agreement with the Bolling and Tiller analysis for all but very high-growth velocities where the Rohatgi and Adams treatment predicts larger λ values. The DI approach used by Wettlaufer (1992a) indicates that if the observed λ values correspond to the fastest-growing disturbance and to the maximum growth rate, then the expected relation would be $\lambda \propto v^{-1/2}$, a relation similar to the rapid growth rate suggestion of Bolling and Tiller (i.e., that $\lambda^2 v = A$). Incidentally, this relation has also been found to hold for the spacing of substructure elements such as rods in directionally solidified eutectics. Other relations that have been used in fitting metallurgical data include $\lambda = A / v\, G_L + b$ and $\lambda = A v^{-a} G_L^{-b}$ where a, b and A are experimentally determined constants and G_L is the temperature gradient in the liquid. However, these latter relations have not been utilized in studies of the freezing of aqueous solutions as values for G_L are typically not well known.

One distinct difference between the predictions of these models is in the compositional dependence, where Rohatgi and Adams conclude that $\lambda^2 \propto C_0^{-1}$, in contrast to the direct dependence suggested by the Bolling and Tiller analysis. DI analysis, although not specifically focused on compositional variations as the formation of sea ice, commonly does not result in a significant change in the average salinity of the underlying seawater, which suggests that λ should decrease with increases in C_0 if other factors remain essentially constant.

Existing experimental studies of cell size variations in solidified aqueous systems are a bit confusing. The experiments that most closely correspond to normal freezing are those of Lofgren and Weeks (1969), who unidirectionally froze insulated cylindrical columns filled with NaCl solutions with varied initial compositions. They found that when convection was predominant, $\lambda \sqrt{v} \approx A$ at high values of v whereas at low values of v, λ became roughly constant. Figure 7.32 shows the test data. However, in the one freezing run in which the bulk salinity was sufficiently low ($1.05‰$) to prevent convection from occurring, $\lambda v \approx A$ at lower freezing velocities. The most detailed experiments to date are by Rohatgi and Adams (1967a, 1967b, 1967c). The "problem" with some of their experiments is that their freezing geometry was very different, in that spherical droplets were frozen while slowly sinking through a refrigerated organic liquid that was held at a constant temperature. In their unidirectional experiments λ increased linearly with distance from the chill, indicating that $\lambda \sqrt{v} \approx A$. They also found that the value of λ at a fixed distance from the chill increased linearly with the diffusivity of the solute.

In addition, Rohatgi et al. (1969) also investigated the effect of composition on λ in freezing runs in the system NaCl–KCl–LiCl–H$_2$O. They concluded the following:

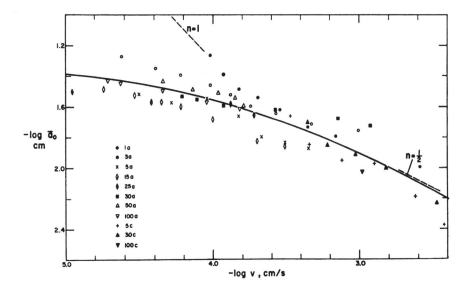

Figure 7.32. A plot of $-\log \lambda$ v vs. $-\log v$ (Lofgren and Weeks 1969). The fitted curve is of the form $\log(\bar{\lambda}v) = \log A + B\left(\log\dfrac{1}{v}\right)^2$ where $A = -4.069$ and $B = -0.093$. The numbers in front of the symbols indicate the approximate initial NaCl content in ‰ of each freezing solution.

1. If two solutes are present in equal amounts, λ will lie between the values obtained with individual solutions in which each solute is present at its same concentration.
2. The solute present in the larger concentration tends to govern the cell size, with small additions of other solutes causing relatively small shifts in λ. If the secondary solute has a higher diffusivity, mean cell size will increase and vice versa.
3. In solutions in which the constituent solutes are in a fixed ratio, λ increases with total solute concentration.

Field observations are in general agreement with the above laboratory observations and have generally shown an increase in λ with increasing depth z in sea ice sheets (Paige 1966; Tabata and Ono 1962; Weeks and Hamilton 1962). However, at Narwhal Island, Alaska, a decrease in λ with increasing z was noted by Gow and Weeks (1977), and at Eclipse Sound, Baffin Island, a rather complex variation was observed by Nakawo and Sinha (1984). In first-year sea ice, typical λ size variations range from 0.4 mm near the upper ice surface to 1.0 mm at a depth of 2.0 m. At a given level in the ice sheet, λ values are approximately normally distributed (Weeks and Hamilton 1962). Clearly, the best field study to date of the relation between λ and v in a natural ice cover was undertaken at Eclipse Sound by Nakawo and Sinha (1984). Curve (a) in Figure 7.33 shows the calculated growth rate plotted as a function of the vertical position (depth) in the ice sheet (z). Curve (b) is the mean of the growth rate for an interval of $+50$ mm centered on the 25-mm segment for which it is plotted. There is a pronounced variation in growth rate with z, with maxima at about 0.2, 0.5, and 1.0 m corresponding to the observed cold periods in November, December, and January. Curve (c) presents the salinity profile obtained from an adjacent core. There is clearly a positive correlation between growth rate and salinity. More will be said about this in the next chapter. The curve on the right shows the corresponding values of λ and suggests

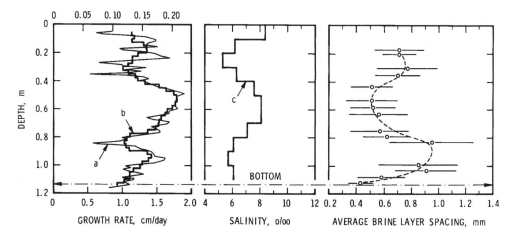

Figure 7.33. Profiles of growth rate, salinity, and average brine layer spacing ($\bar{\lambda}$) as a function of depth in the FY ice at Eclipse Sound, Nunavut (Nakawo and Sinha 1984).

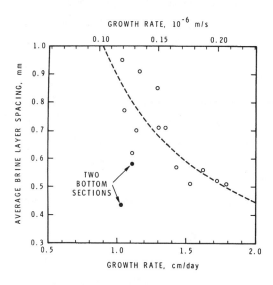

Figure 7.34. Plot of the average brine layer spacing vs. the corresponding rate of growth from curve b in Figure 7.33. The broken line shows a least-squares fit (neglecting the two bottom values) assuming that *l* is inversely proportional to λ (Nakawo and Sinha 1984).

an inverse proportionality, with growth rate in agreement with the work of other investigators. Figure 7.34 shows the same data presented as a λ versus v plot. The dashed line is a least-squares fit of a relation suggested by Bolling and Tiller (1960). The agreement is reasonable if the two λ values from near the bottom of the ice sheet are disregarded. Whether or not this omission is legitimate is apparently moot. One justification that I have mentioned elsewhere is that limited observations suggest that when sea ice sheets are sampled near the end of the growth season, when growth velocities are very low, the relation between v and λ become particularly enigmatic. The ice studied here showed a pronounced c-axis alignment not only in the horizontal plane but specifically in the NE–SW direction in that plane. At a given level in the ice sheet (i.e., at a constant growth velocity), limited data suggest that λ is a maximum value when the c-axis is parallel to the favored direction, and that λ decreases when the angle to this direction increases. This strikes me as an important observation, but unfortunately I do not know why it should be so.

The hypothesis that slower growth corresponds to larger λ values is strongly supported by field observations on very thick sea ice—ice that in all probability grew extremely slowly. For

instance, Cherepanov (1964a) obtained λ values of 1.5 mm on old 10- to 12-m-thick sea ice observed on drift station NP-6. Gow (pers. comm.) obtained λ values of 1 to 1.5 mm on thick sea
ice that formed as part of the Koettlitz Ice Tongue, McMurdo Sound, Antarctica. Finally, Zotikov
et al. (1980) obtained λ values of 5 mm from sea ice formed on the base of the 416-m-thick Ross
Ice Shelf. The growth velocity at the sample site (Camp J-9) was estimated to be 2 cm/year, far
slower than could easily be obtained in the laboratory. Incidentally, this ice is true congelation ice
and should not be confused with marine ice as discussed in Chapter 17.

One possibility that might occur to the reader is that there might be some coupling relation between the quasi-hexagonal patterns of ice and brine, as seen in the horizontal plane within sea ice, and
convective patterns within the upper part of the water column, because convective patterns many
times show a roughly hexagonal symmetry. Current thinking on this subject (Wettlaufer 1998) suggests that there is little if any such relation in that the characteristic scale of convection in the upper
portion of the ocean and presumably that at the ice–ocean interface is sufficiently larger (1–10 m)
than the scale of the sea ice substructure (0.1–5 mm), making such coupling improbable.

In that simple freezing experiments are both much less expensive and easier to control than
observations in the field, someone should definitely undertake additional controlled studies of
cell size and shape variations.

7.2.4.5 Alignments   In the earlier portions of this book I pointed out that the crystals in the
columnar zone invariably have their c-axes oriented in the horizontal plane. I have also shown an
early fabric diagram (Figure 7.15) in which the c-axis alignments within the horizontal plane were
either random or near-random. Initially, it was reasonable to expect that this might be the case, as
the heat flow was apparently one-dimensional. Also, such random radial c-axis alignments were
common in the solidification of metals. However, in the early 1960s Peyton (1963, 1966, 1968),
while examining 3×3 m blocks of 1.6-m-thick sea ice at Barrow, found that the bottom meter
exhibited a near-constant c-axis orientation over the entire 9-m^2 cross section. In his studies of
mechanical properties, Peyton then utilized this ice to investigate the effect of changes in the orientation of the uniaxial stress to the c-axis direction on the failure strength. At roughly the same
time, Cherepanov (1964a) and Smith (1964) observed that thick old sea ice incorporated in the
ice islands NP-6 and ARLIS II showed near-perfect c-axis alignments over large areas (on NP-6
the entire 80-km^2 area of the "island" showed a similar c-axis alignment). Clearly such alignments
were not statistical fluctuations, and specific efforts to study this phenomenon (or these phenomena) began. In 1971 a quite remarkable set of c-axis observations (Figure 7.35) were published by Cherepanov (1971a), who observed nearly constant c-axis alignments over thousands
of square kilometers in the Kara Sea. This was followed by studies of the directional dependence
of the electrical properties of sea ice as observed at widely separated sites in the Canadian Arctic.
These observations also supported the contention that c-axis alignments might well extend over
lateral distances of the order of tens of kilometers (Campbell and Orange 1974; Kohnen 1976).
This work was followed by a series of several papers that provided a detailed look at c-axis alignments along the coast of arctic Alaska (Gow and Weeks 1977; Weeks and Gow 1978, 1980);
for the Mackenzie Delta portion of the Beaufort Sea (Langhorne 1980); for the ice surrounding artificial gravel islands that are used for drilling platforms (Vittoratos 1979); for channels in

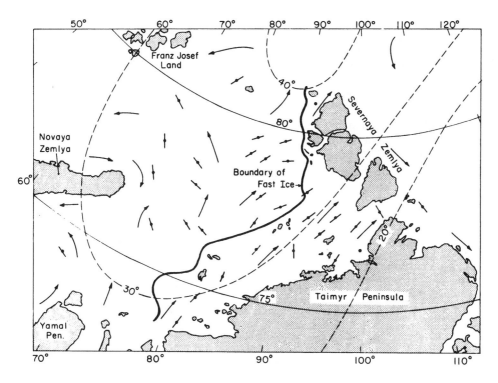

Figure 7.35. C-axis alignments in the horizontal plane for the Kara Sea as determined by Cherepanov (1971a).

the Canadian Arctic Islands (Nakawo and Sinha 1984); for the Russian Arctic (Strakhov 1989b, 1991); and for the Antarctic (Gow et al. 1982, 1998; Serikov 1963; Strakhov 1987, 1989a).

These observations suggest the following:

1. In all sea ice formed by unidirectional freezing, c-axis horizontal orientations develop rapidly after an initial ice skim forms, and they dominate the rest of the ice growth.

2. In much of the ice examined (95% of the sample sites along the Alaskan coast, Weeks and Gow 1980; 94% of the sites sampled in the Kara Sea, Cherepanov 1971a; Strakhov 1991), strong c-axis alignments also develop within the horizontal plane. Figure 7.36 shows a sampling of published fabric diagrams. Standard deviations around the mean are usually < 10° for ice collected from near the bottom of ice sheets in excess of 1.5 m thick.

3. Although alignments are occasionally present in ice less than 20 cm thick, more typically they are observed at distances of > 50 cm below the upper ice surface. Also alignments typically become stronger (show less scatter around the mean) with increasing depth in the ice.

4. Although the mean c-axis direction is not constant at different levels in the ice sheet, the variations are commonly < 20°.

5. Mean c-axis directions at the same site commonly appear to be similar from year to year although the strength of the alignment may vary. The best example of this can be found in the work of Strakhov (1991), who revisited many of the sites sampled by Cherepanov (1971a) in the Kara Sea. He noted that the insignificant alignment differences that were

Figure 7.36. Representative fabric diagrams (Schmidt net plots, see Appendix E) showing strong horizontal alignments observed at sampling sites along the coast of the Beaufort Sea (Weeks and Gow 1978). Each data point indicates the *c*-axis orientation of a single crystal. The centimeter distances specify depth in the ice sheet.

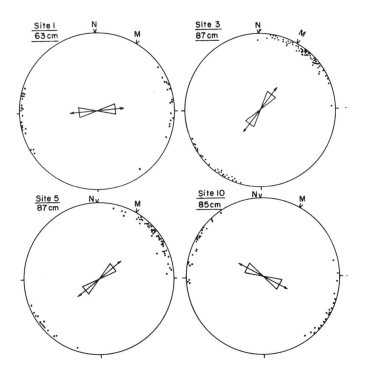

observed suggest that the under-ice current directions in this area are essentially constant. The one area where alignments did not form was also the same in both data sets.

6. In the nearshore regime, the *c*-axes are generally aligned parallel to the coast; they swing around both natural and artificial islands in a streamlike pattern and are aligned parallel to the axis (throats) of passes and inlets. (When Tony Gow and I proposed the current direction explanation for alignments, we noted that there was one site in Cherepanov's data that definitely did not fit our theory. In the strait between the northern tip of the Tamyr Penninsula, Mys Chelyuskin, and the southernmost island of Severnaya Zemlya, the one site sampled gave a *c*-axis alignment normal to the axis of the strait instead of parallel. Our only explanation was that perhaps Cherepanov had made a 90° rotation error during the collection of the sample as this can easily occur. Although we will never know if this is true, I am pleased to note that in Strakhov's sample collected at the same location, the *c*-axis alignment had come to its senses and was oriented parallel to the axis of the strait.)

7. Limited information indicates that strong crystal alignments can develop in areas of heavy pack ice (Cherepanov 1971a; Kovacs and Morey 1978, 1980; Strakhov 1991).

There have been several different suggestions concerning the cause or causes of these alignments. Initially, Cherepanov (1971a) suggested that they might be the result of an interaction between the earth's magnetic field and the electric potential produced at the freezing interface. Later, Stander and Gidney (1980) suggested that the alignments were possibly the result of stress-activated mechanisms such as grain boundary sliding and dislocation generation. The generally favored theory (Weeks and Gow 1978) is that the mechanism most likely to produce the aligned crystals works through the control that ocean currents have on the composition of the liquid at the cell tips, and thus on the interface temperature. As such, this is a less effective selective growth mechanism than anisotropic growth, which, as has been described, controls geometric selection in the transition layer. This is demonstrated by the fact that strong c-axis horizontal orientations develop in a few centimeters of vertical growth whereas strong c-axis alignments frequently only develop in the lower portions of ice sheets at depths of at least 50 cm. It is important to realize that this is not a recrystallization phenomenon; it is a competition between preexisting crystals over space domination at the growing solid–liquid interface.

The explanation appears to be as follows. If the current is parallel to the basal ice plates, each plate presents a smooth form to the flow, allowing a stable boundary layer to build up along the tip within which solute transfer is largely a diffusive process. If, however, the flow is perpendicular to the plate tips (i.e., parallel to the c-axis), then mixing will be enhanced at each plate tip, and the thickness of the diffusion-limited solute boundary layer reduced. A reduced boundary-layer thickness should give the crystals in this orientation a very slight growth advantage, allowing them to extend further into the melt and permitting them to grow sideways at the expense of neighboring crystals that have less favored orientations. Although approximate calculations suggest that the Weeks and Gow hypothesis is plausible, these mechanisms are far from proved and to date no detailed model has been developed.

Fortunately, it is possible to test the suggested correlation between the mean current direction and the direction of the c-axis alignment by field measurements (Kovacs and Morey 1978; Stander and Michel 1989b; Strakhov 1991; Weeks and Gow 1978, 1980). For example, Figure 7.37 shows the relative frequency of different angular deviations between the observed instantaneous current direction determined just under the ice and the mean c-axis direction. The agreement is quite good considering that short-term current measurements are known to commonly show significant deviations from the long-term average current direction because of eddies. For instance, Weeks and Gow measured a 7-hour average of the current vector at a site near Barrow and obtained an average value that differed from the mean c-axis alignment by only 4 degrees. Although generally similar results were reported by Stander and Michel, they also documented a site where a c-axis alignment had developed even though the currents appeared weak and non-directional.

The best support for the correlation between the c-axis alignment direction and the current direction at the growing interface comes from a variety of flume experiments (Cherepanov and Strakhov 1989; Langhorne 1982, 1983; Langhorne and Robinson 1986; Stander and Michel 1989a; Strakhov 1991). All of these studies show that c-axis alignments develop parallel to the current direction. Stander and Michel's results are particularly interesting for several different reasons. These experiments were initially designed to test the alternative

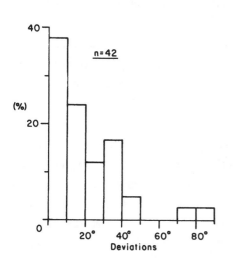

Figure 7.37. Histogram showing the relative frequency of different deviations between the observed instantaneous current direction and the mean c-axis direction (Weeks and Gow 1980). Chukchi coast, Barrow, Alaska.

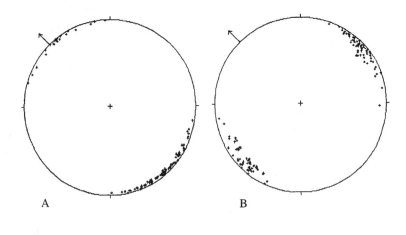

Figure 7.38. *C*-axis plots
(Schmidt net, lower
hemisphere) of the crys-
tal orientations obtained
in two freezing runs in a
cylindrical flume. The arrows indicate the direc-
tion of fluid flow. The ice
grown in plot (a) formed
from an NaCl solution
with a salinity of 25‰
whereas the ice in plot
(b) formed from a solu-
tion with a salinity of
0.7‰ (Stander and Mi-
chel 1989a).

hypothesis that the alignments were due to deformation (Stander and Gidney 1980; Stander and Michel 1991). They concluded that this was not the case in that their freezing runs usually resulted in strong *c*-axis alignments when there was a significant current, whereas alignments were missing when there was no current. In addition, when the initial NaCl content of their freezing solution was > 1‰ the *c*-axes were always aligned parallel to the current direction, as would be expected from the Weeks and Gow theory. However, when initial solution salinities were < 1‰, something unexpected was observed: *c*-axis alignments developed that were perpendicular to the current direction (Figure 7.38). I say relatively unexpected, as one case of similar observations had been reported on undeformed river ice (Lasca 1971). Stander and Michel explained this 90° change in orientation by noting that experiments by Vlahakis and Barduhn (1974) and Simpson et al. (1973) had shown that the growth velocity of ice crystals growing into a current of salt water showed a maximum value at ~5‰ NaCl. They then suggest that if the initial salinity of the solution is >1‰, salt rejection during the initial growth of the ice sheet plus salt buildup in their closed flume system is sufficient to raise the salinity of the boundary layer at the base of the ice sheet to values in excess of 5‰, causing those crystals whose basal planes lie perpendicular to the current to grow faster. If, on the other hand, initial salinities are < 1‰ (actually 0.6 to 0.8‰), then boundary-layer salinities will remain below 5‰, in the range where salinities favor crystals whose basal planes lie perpendicular to the flow.

Although these experiments are very definitive in supporting the view that sea ice alignments are controlled by the current directions, I am not totally convinced that the proposed explanation for the orientation change at ~1‰ is correct. I note that the transition from a planar (lake ice) interface to a cellular (sea ice) interface occurs at ~1‰ (Weeks and Lofgren 1967). I suggest that the observed orientation change is somehow associated with the change in the interface shape. I think that I understand why a sea ice crystal that has a cellular interface and its *c*-axis oriented parallel to the current has a growth advantage over cellular crystals in other orientations. What I obviously don't understand is why an ice crystal with a planar interface would have a growth advantage if its *c*-axis is oriented perpendicular to the current. In Strakhov's experiments alignments readily developed in ice formed from brackish water, but in contrast to the results of Stander and Michel, he reports that when the salinity of the freezing solution is zero and presumably the interface is planar, alignments do not develop. Clearly, additional experiments in which interface morphology and salinity profiles in the fluid are monitored with time would be very helpful. Once the mechanisms involved in these changes are understood we may have a better understanding of brackish ice types as well as of the differences between sea and lake ice. As will be shown later, the development of strong preferred c-axis orientations is not merely a crystallographic curiosity but has an important controlling effect on the directionality of ice sheet properties.

If the c-axis alignment direction is a measure of the mean current direction, it might at first glance be concluded that alignments would only be found in ice that was fast when the alignment developed. Several pieces of information suggest that this is not always true. Cherepanov (1971a) found alignments in first-year ice in areas of the Kara Sea that he identified as pack ice (Figure 7.35), and Kovacs and Morey (1980) obtained similar results for offshore areas of the Beaufort Sea that were definitely composed of drifting ice. This can be explained in the context of the present theory if the current velocity at the ice–water interface remains reasonably constant relative to the ice long enough for an alignment to develop. As both these sets of observations were made in the later winter when the ice pack in the Beaufort and Kara Seas was extremely tight, making floe rotations difficult, this would appear to be a plausible explanation.

Another interesting result from Strakhov's (1991) studies is that the oblateness of the crystals as measured in the horizontal plane was appreciably higher for the flume studies than for ice grown under natural conditions in the ocean, presumably representing smaller fluctuations in the current direction. Also, the degree of alignment and the rate at which the alignment develops appears to increase as the speed of the under-ice current increases.

Present observations in both the Arctic and the Antarctic suggest that very significant portions of the world's fast ice show strong c-axis alignments. The percentage of pack ice showing such alignments in unknown. Strakhov notes MY floes that show a change in c-axis alignment from layer to layer, as would be expected from either floe rotations or regional shifts in current directions. He has also carried out field experiments in which he rotated blocks of aligned ice to an unfavorable orientation with respect to the current and then examined orientations in the ice that formed below the block after 1.5 to 2.5 months of additional growth. Based on these observations he suggests that realignments to changes in relative current directions are slow, occurring over time periods of about a month. I feel that this value is probably on the high side in that the rotated block no longer contains crystals in the new favored orientation, whereas in most natural situations some crystals in the new favored direction would invariably be present.

I would guess that pack ice that forms in locations such as the Bering Sea, Baffin Bay, and much of the Antarctic, where the ice is highly mobile and the floes are free to rotate, would show random to highly variable c-axis alignments in the horizontal plane or at least highly variable alignment directions at different levels in the floes. When, in locations such as the central Arctic, the pack ice becomes very tight during the ice growth season, thereby restricting the possible rotation of floes, then aligned ice may become the dominant form. Systematic field observations should be made to resolve these questions. It would also be desirable to have detailed time series of current velocity observations at locations where alignments have not been observed to develop.

7.3 Old Ice

There is surprisingly little information available on the structural details of old ice, considering the fact that at least in the central Arctic this ice "type" has historically covered a large percentage of the area of the ocean. I do not expect this lack of information to change rapidly in the near future in that decreases in the extent of the pack at the end of the summer (Chapter 18) have made access to old ice even more difficult as it is now typically found farther from shore. For instance, until the late 1980s one could invariably find old ice within a few miles of Barrow, Alaska. Sometimes old floes were even located within the fast ice. Since 1990 this has not been the case. Also, even when old ice was logistically accessible near the coast, the condition of this ice was usually poor in that such ice, being located at the margin of the pack during the late summer, had invariably been subjected to extensive wave action and was frequently highly deteriorated and/or deformed.

In the following I will follow the WMO usage by using the term *old ice* to refer to sea ice that has survived one or more summer melt seasons. If the ice has survived one melt season but not two it is referred to as *second-year ice*, and if it has survived two or more melt seasons it is called *multiyear* (MY) *ice*. The reader should be warned that at least in the ice properties and remote sensing literature, ice that is referred to as MY may very well be second-year and more aptly described as old, in that its exact age may not be known, only that it has survived at least one summer. Once sea ice survives a summer its salinity profile and physical properties are drastically changed. As will be discussed in the chapter on the salinity profile, this change largely occurs during the first melt season, with changes during subsequent melt seasons being much less pronounced.

In the Arctic, old ice is usually identified by two main characteristics. First, it has a rolling, hummocky surface that is produced by differential melting. On second-year ice the melt-induced vertical relief is a few tens of centimeters and the surface drainage pattern is either disorganized or at best poorly organized. Also, the blocks of ice in pressure ridges, although rounded, still possess a blocky appearance, and fresh fractures appear greenish to light blue. More importantly, the surface salinities have decreased to nearly zero, a change which reveals itself through changes in the passive microwave and synthetic aperture radar (SAR) signatures of the ice. After this ice passes through a second summer, the surface relief becomes higher still, reaching values of over a meter. In addition, the surface drainage becomes more integrated and the melt ponds more distinct. Old ridges become completely rounded, causing their initially blocky nature to become indistinct. In addition, the essentially complete desalination of the upper part of the ice above the waterline causes fractures in the ice to have a bluish tint. In the Antarctic, where summer surface melt is less common, the presence of old ice is indicated by the occurrence of high freeboards and thick snow covers.

The point here is that if one has the opportunity of viewing a region of sea ice several times during a year, it is possible to clearly differentiate between regions of sea ice that have seen one, two, or even three or more summers. If, on the other hand, one does not have such sequential observations, which is the common case, it is easy to distinguish FY ice from old ice, but the separation between second-year and MY ice becomes more tenuous (Johnston and Timco 2008). One should also remember that even if it can be established that the top layer of an old floe is,

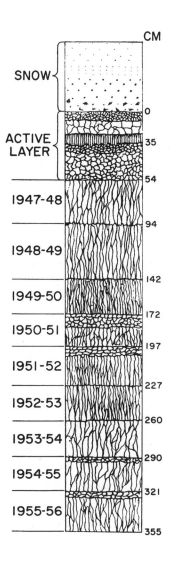

SNOW

ACTIVE LAYER

CM

0

35

54

1947-48

94

1948-49

142

1949-50

172

1950-51

197

1951-52

227

1952-53

260

1953-54

290

1954-55

321

1955-56

355

Figure 7.39. Schematic drawing showing the cross-sectional structure of an MY flow, NP-4, Arctic Ocean (Cherepanov 1957).

say, four years old, the layer below it is only three years old, and the lowest part of the ice is less than one year old and is probably still growing.

In Chapter 4, I discussed the ideas that the ideal steady-state MY floe should be a layer cake of annual growth layers and that perennial sea ice of great age in a geological sense does not exist. These ideas, which originated with Weyprecht (1875, 1879), were not based on the discovery of a discernable layering in the ice, but instead on his ice growth observations in Franz Josef Land. There, he found that 209 cm of FY ice formed when the average number of freezing degree-days was 5,625. He then calculated that if 100 cm of ice melted during the summer, and if the number of freezing degree-days remained the same, the ice should be 234 cm thick at the end of the second summer. He also estimated that when the ice thickness reached 260 cm as much ice would melt in the summer as grew in the winter (see also Zubov 1938, 1945). To date, the number of observational checks on the assumption that old pack ice is typically a simple sequence of growth layers are few. The first did not occur until 1954, when Shumskiy (1955a) examined a 291-cm-thick floe during a brief visit to the drift station NP-3. Unfortunately, because of above-freezing temperatures only a cursory examination was made, with the ice being divided into two main types: infiltered (0–38 cm) and normal (38–291 cm) sea ice. As the infiltered ice resembled the firn of polar glaciers, Shumskiy concluded that the snow line in the central Arctic was presently at sea level. Although layering in the 253-cm portion of the floe was rather ambiguous, he also concluded that the floe was not less than six to seven years old. Clearly, this floe was not a sterling example of Weyprecht's MY ice.

Soon after this preliminary study was completed, other studies were carried out on NP-4 (Cherepanov 1957) and on IGY Drift Station Alpha (Schwarzacher 1959). These floes were more "textbook" examples of what MY ice might be expected to look like. Both authors reported that annual layering was readily observable in the lower part of the ice and was less clearly defined nearer the upper surface, where the effects of the summer melt were more pronounced. Cherepanov found 10 recognizable layers in a 335-cm floe (Figure 7.39), and Schwarzacher found either 7 or 8 layers in a 345-cm-thick floe.

The most common boundary between the annual layers is a thin (2–5 mm) layer of milky-white ice with a sharp upper boundary and a somewhat more irregular lower boundary. Although the details of the formation of this layer are not clear, the layer obviously forms when the ice growth rate is either zero or close to this value. One possibility is that this layer forms during a period when the combination of the low growth rate and a surface melt–induced lowered salinity at the ice–ocean interface have been such as to result in a planar growth interface becoming stable, with the associated complete rejection of the salt back into the underlying ocean. For example, studies on the freezing of NaCl solutions by Weeks and Lofgren (1967) have shown that clear ice will form from a solution if the solution salinity is < 1‰, and that a clear layer can form below a layer possessing the typical sea ice substructure if the growth rate falls to a sufficiently low value. They also observed that the boundaries between these layers were sharp, as would be expected from diffusive interface stability theory (Wettlaufer 1992a, 1992b). One thing that is clear is that the formation of the milky layer is not associated with either recrystallization or the nucleation of new grains, because when ice growth starts in the fall the new crystals resume growth with the same crystallographic orientation as those crystals on the underside of the overlying layer (Schwarzacher 1959). In some cases the milky layer has a yellowish or brownish tinge. It has been suggested, but not proven, that this may be the result of biological activity during the summer.

As can be seen from Figure 7.39, the other type of summer layer is characteristically thicker (1–10 cm) and shows a sharp decrease in grain size and shape when compared to the overlying ice. In the ice studied by Schwarzacher, the c-axis orientations in this type of summer layer were horizontal but with deviations of up to 30°, far greater than found in normal columnar-zone sea ice. In the floe studied by Cherepanov, the summer layer crystals were equiaxed with random c-axis orientations. The summer layers also did not show the typical sea ice substructure and had a much lower salinity (< 1‰) than the surrounding ice. The formation of these types of annual layers is clearly the result of the establishment during the summer of a low-salinity layer of surface-derived meltwater between the lower surface of the ice and the underlying more saline and therefore denser seawater. The formation of such freshwater layers has been known for some time (Malmgren 1927; Nansen 1897) and was described in some detail on Drift Station Alpha by Untersteiner and Badgley (1958). There, they observed that surface meltwater pouring off the margins of floes, or flowing through melt ponds that had completely perforated the floe, formed a stable freshwater layer between the bottom of the ice and the underlying normal seawater. In most cases these freshwater lenses, which have been termed *under-ice melt ponds* by Hanson (1965), appear to be trapped under thinner ice areas or in depressions in the bottoms of

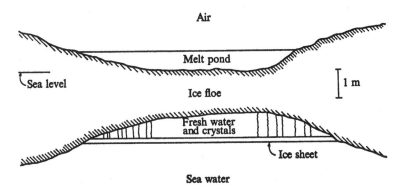

Figure 7.40. Sketch of the formation of an under-ice melt pond, modified after Hanson's 1965 description by Martin and Kauffman (1974).

thicker ice. A sketch showing the geometry of the situation can be found in Figure 7.40. It should be noted that typically a number of boreholes are made to drain surface meltwater from the vicinity of manned camps in order to keep surface albedo values high and minimize surface melt rates. This obviously assists in the transfer of surface meltwater to the underside of an ice floe.

At the base of this fresh layer, water with a freezing point of 0°C is in direct contact with seawater with a freezing point of −1.8°C. As the thermal diffusivity is roughly 5×10^3 times larger than the diffusion coefficient for salt in water, crystals of pure ice start to grow at the boundary and eventually float upward until the freshwater layer is filled with a mesh of fragile crystals. The hydrodynamics of this situation was initially examined, both theoretically and experimentally, by Martin and Kauffman (1974). They found that heat transfer rates at the seawater–freshwater interface were 5 to 10 times the values that would be expected if all the heat transfer had been by thermal diffusion. What happens is as follows. Because of the rapid diffusion of heat relative to that of salt, and the fact that the density of water with a salinity of < 24.7‰ decreases on cooling, a zone of water forms that is both supercooled and less dense than the overlying water (see Figure 7.41). The resulting vertical density distribution is unstable, with a Rayleigh number of 10^3 to 10^4, resulting in free convection. When this supercooled water rises it is nucleated by the overlying ice layer and a mesh of thin vertical interlocking ice crystals forms that ultimately grows down to the freshwater–seawater interface. At this time, the development of ice crystals throughout the complete thickness of the freshwater layer eliminates the supercooling. It also constrains both the temperature and the salinity to lie on the freezing curve and permits them to diffuse in the vertical direction until both quantities become horizontally uniform. Once ice crystals are nucleated at the freshwater–seawater interface they also grow laterally along the interface, resulting in the formation of solid horizontal sheets of freshwater ice across the bottoms of melt ponds. Such sheets are referred to as false bottoms and have been observed by a number of authors. Field observations made during the Surface Heat Budget of the Arctic Ocean (SHEBA) project indicate that the lateral growth of false bottoms can occur sufficiently rapidly to trap arctic cod within the ice cover (Notz et al. 2003). Excellent photographs of the growth of ice crystals during a laboratory simulation of this process can be found in the Martin and Kauffman paper. Field observations also indicate that the horizontal sheet both thickens and migrates upward with time (Hanson 1965; Untersteiner and Badgley 1958). The vertical migration of the sheet can be explained as follows. As the ice sheet grows upward, the heat released by the formation of ice on the top of

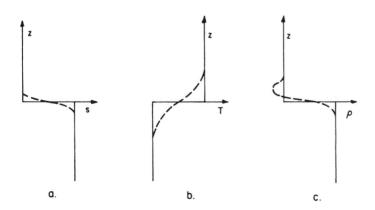

Figure 7.41. The initial (solid lines) and subsequent (dashed lines) profiles of a) salinity, b) temperature, and c) density occurring when a freshwater layer overlies a saltwater layer, with both layers at their respective freezing points. The size of the density inversion is exaggerated (Martin and Kauffman 1974).

the sheet flows downward through the ice toward the cold source, which is the underlying sea-water. There it warms the lower ice–seawater interface. To restore equilibrium the lower surface of the ice dissolves, thereby cooling and diluting the adjacent water back to the freezing curve. With growth on the top and dissolution on the bottom, the sheet migrates upward. Equations for estimating the migration rates have been developed by Martin and Kauffman; however, as accurate measurements of temperature and salinity gradients under summer pack ice did not exist at that time, they did not estimate these rates. The theoretical treatment initiated by Martin and Kauffman has been extended by Notz et al. (2003), who developed analytical solutions describing the diffusional transport of heat and salt during the simultaneous growth and dissolution of false bottoms. In the later treatment the fact is stressed that the ablative process occurring at the interface between the lower surface of the false bottom and the underlying seawater is not the result of melting (the melting temperature of pure ice is, of course, 0°C, whereas the temperature of the underlying seawater is appreciably colder than this at –1.8°C), but is one of dissolution (see Woods 1992 for a discussion of the differences). As a result, the process is limited by the rate of transport of salt. Notz et al. also extend the model to account for more realistic turbulent heat and salt transports in the ocean system. Their model simulations are in reasonable agreement with field data collected earlier during AIDJEX.

If the fresh ice layer survives until the growth season starts again, it becomes incorporated in the growing ice sheet and serves as a summer marker. In that the lower surfaces of old floes typically show some relief, the thickest freshwater layers would be expected to form where the ice is thinnest. As a result, I would expect summer ice layers to show appreciable lateral variations in thickness and from year to year. Also, even if such freshwater layers always form, they do not always survive. Cherepanov found that only four out of the nine annual layers on the NP-4 floe were marked by "fresh" ice layers, while Schwarzacher concluded that such layers did not contribute significantly to the total growth of the ice that he studied. Schwarzacher also examined a set of 150 cores that he obtained from what appeared to be undeformed MY ice located in the vicinity of the Alpha camp. He found that only 25% of this ice was actually undeformed and that only 2% showed annual layering throughout the complete thickness of the floes. In studies in the Eurasian Arctic, Eicken (1994) found that 8 of 52 locations showed false bottoms whereas Jeffries et al. (1995) observed false-bottom ice in 22 of 57 cores collected in the Beaufort Sea. The conclusion that the occurrence of false bottoms is not a rare event is also supported by submarine sonar measurements (Wadhams and Martin 1990).

Although it will also be mentioned later in a discussion of the mass balance of pack ice floes, the importance of drainage integration and of melt pond perforation on the mass balance cannot be overstated. On a second-year floe, drainage leading to the edge of the floe is frequently poorly integrated. Provided that the melt ponds do not perforate completely through the floe opening a pathway to the underlying ocean, most of the surface meltwater will still reside on the upper surface of the floe at the end of the melt period. Once freezing sets in, the pools and ponds will refreeze rapidly and little, if any, mass will be lost. On the other hand, if a well-integrated drainage system exists, as is frequently the case on older floes, and if the melt ponds perforate completely through the floes, the meltwater is rapidly transferred to the ocean where much of it may be "lost" via mixing. I know of no information on the amount of surface melt that is "saved" by refreezing the freshwater layer beneath the ice as has just been discussed. I would guess both that the percentage is small and that the values are highly variable. One thing certain, however, is that false-bottom ice can relatively easily be distinguished from normal sea ice on the basis of its salinity ($< 1‰$), the shapes of its crystals, their lack of substructure, and their isotopic composition, indicating a significant input of meteoric water presumably originating from melting snow (Eicken 1994).

The top 50 cm of the Station Alpha floe examined by Schwarzacher was comprised of ice that formed at or near the surface of the floe. As can be seen in Figure 7.39, a similar thickness of ice (denoted as an active layer) that has been modified by surface effects was found on NP-4. This ice appears to be of several types, of which the most predominant results from the freezing of the meltwater pools that form on the ice surface. On Alpha, this ice was characterized by c-axis vertical orientations, frequently an indicator of freshwater ice from refrozen melt ponds. A fine-grained granular ice can also be formed by snow either falling or drifting into melt ponds, where it forms slush that subsequently freezes. Whatever the process, the melting and refreezing experienced by the upper part of the ice tends to obliterate the annual layering and to make the estimation of the total age of the floe difficult.

An additional process affecting the upper levels of old ice floes has been called *retexturing*, i.e., the rounding of originally sharp, serrated grain boundaries and the significant modification or even the elimination of the platy substructure within the sea ice crystals. Although such effects have been known for some time (Knight 1962a), additional studies did not occur until almost 1990 (Gow, Tucker, and Weeks 1987; Jeffries 1991; Tucker et al. 1987, 1992). It seems that at least two factors are necessary for retexturing to occur. First, retexturing only appears to occur in sea ice that has been extensively desalinated by the freshwater flushing that occurs during the summer melt season. Second, temperatures at or near 0°C also seem to be necessary. It would appear that desalination is the more essential factor in that ice in the lower portions of MY floes, which has not been completely desalinated, does not show retexturing even though it has undoubtedly been exposed to near-melting temperatures during significant periods of time during successive summers. As long as appreciable numbers of brine pockets still exist along the small-angle grain boundaries that occur within the sea ice crystals, the defect arrays comprising these boundaries appear unable to break away from the included material and migrate to the large-angle intergrain boundaries, even though this would decrease the total surface free energy of the system. Incidentally, such retexturing, driven by the presence of substructure boundaries similar to those present in sea ice, is frequently observed in metals (Gottstein and Shvindlerman 1999). Photographs of several horizontal thin sections of retextured ice from the upper layers of old arctic floes can be found in Gow, Tucker, and Weeks (1987) and in Tucker et al. (1992). Note that although the substructure within the crystals has essentially been removed, the crystals still retain their c-axis horizontal orientation. Other changes noted included the substantial smoothing of the characteristically angular, interpenetrating outlines of the crystals.

Observations on the structure of very thick old sea ice are very limited. One of the best is by Cherepanov (1964a, 1964b) who in 1955 examined the fabric of the floe on which the Russian drift station NP-6 was located. The floe was 10 to 12 m thick with a horizontal area of 80 km². In addition, its ice showed strong horizontal alignments, suggesting a fast ice origin. Cherepanov commented that such floes are quite often encountered. Although Wadhams (1981b) noted a few other examples of observations of very thick floes in the arctic pack, he concluded that such floes are rare. Current information would tend to support Wadhams' conclusion. However, this does not mean that Cherepanov was incorrect in 1964. Very thick floes that have formed by congelation growth require quite long periods of time to reach an equilibrium thickness. For example, Walker and Wadhams (1979), using the Maykut and Untersteiner multiyear ice growth model, estimated that it would take roughly 200 to 300 years for a floe to reach a thickness of 20 m even if favorable thermodynamic conditions are assumed. Considering what is now known about the circulation of the pack in the Arctic Ocean, the chances of a floe remaining in the basin for that period of time would appear to be very small. In the Antarctic the chances of an ice floe surviving in the pack for a long period of time would appear to be even less than in the Arctic. More will be said about these matters in Chapter 18.

Where might one expect to find extremely thick sea ice? Favorable sites should provide a protected location for continuing fast ice growth, coupled with a low oceanic heat flux and just the right amount of snowfall to protect the ice from appreciable surface melt during the summer but not thick enough to limit the ice growth during the winter. A good guess would be the fjord systems of northern Greenland, the Canadian archipelago, the Antarctic, and locations such as Severnaya Zemlya and Franz Josef Land. In fact, such ice has been reported to occur in fjords in northern Ellesmere Island (Lyons et al. 1971; Serson 1972); in fjords located in northern Greenland (Koch 1945; Wadhams 1981b); and in the sea ice portion of the Koettlitz Glacier Tongue near McMurdo Sound, Antarctica (Gow et al. 1965; Gow and Epstein 1972). In the Arctic the Iñupiaq name *sikussak* has been applied to such ice. Sikussak typically shows thicknesses in the range of 6 to 15 m, well-developed surface drainage, and pronounced surface melt–induced topography with > 2 m of relief. In his review of studies of such ice, Wadhams (1981b) suggests that sikussak should supply an excellent platform for studies of ice structure developed under conditions of very slow growth. I agree. Unfortunately, how much of this ice survives is not known, as the locations mentioned are rarely visited. The best-studied location and also the thickest sea ice (44.5 m) studied to date is the Ward Hunt Ice Shelf located on the northern coast of Ellesmere Island. There, Jeffries and Sackinger (1989) and Jeffries (1991) found that the shelf was made up of four distinct layers, two of which were congelation layers of sea ice 9 and 24.5 m thick. This later layer is in itself thicker than the thickest ice studied to date within the polar pack (Cherepanov 1964a). The identification of the congelation layers was accomplished by oxygen isotope measurements (Figure 7.42). Although the elongated interlocking crystals and the platy substructure typical of congelation sea ice are identifiable, these features are not nearly as clearly developed as in younger sea ice. In that it is reasonable to assume that initially these features were similar to substructures observed in newly formed sea ice, Jeffries suggests that this modification of the cellular substructure is another example of retexturing, as described earlier. He also notes that the causes frequently assumed to result in retexturing (solar radiation and near-melting temperatures) do not appear to be applicable to this ice, in that it has probably never been exposed at the surface. He suggests that the extreme age of the ice, which he estimates to be as much as 3000 years, is the important factor here. I have no reason to doubt that this is important but I also note that the slow growth of such thick ice results in newly formed sea ice remaining at near-melting temperatures for years. In short, near-melting temperatures do not necessarily require near-

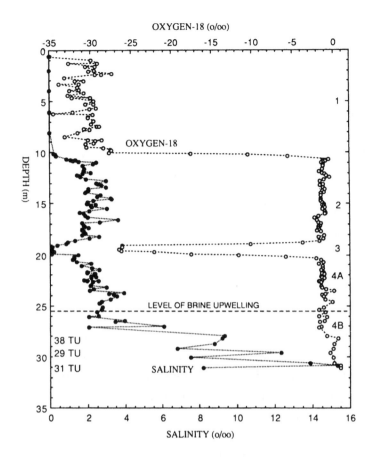

Figure 7.42. Salinity (solid circles) and $\delta^{18}O$ (open circles) profiles for ice core 83-1 obtained from the west Ward Hunt Ice Shelf. Anthropogenic tritium values are shown at the lower far-left-hand side at the depths at which they occurred. A four-layer stratigraphy is apparent and the number of each stratum is shown at the right side (Jeffries 1991).

surface conditions. I feel that considering the age of this ice, its surprising feature is that it has retained as much of the original substructure as it has. This possibly can be explained by the fact that the salinities within the sea ice layers are in the range of 2 to 3‰, values typical of the lower (below sea level) portions of normal MY floes. Locations where pronounced retexturing has been observed in more typical floes are usually above sea level, where desalination is essentially complete.

If one were to locate such a floe one might think that the best time to obtain samples would be in the summer. Based on my experience this is not the case in that, during the summer, surface meltwater (~fresh) tends to run down the core hole. As the ice in the deeper portions of such floes is still commonly at below-freezing temperatures, the meltwater rapidly freezes, cementing the core barrel into the hole. When in 1990 Martin Jeffries and I attempted to obtain additional ice cores from the sea ice portion of the Koettlitz Glacier Tongue, we experienced considerable difficulty because of this problem and were lucky to escape with our corer intact. Working just prior to the melt season would have been preferable. I note that by 1990 a considerable part of the Koettlitz as originally studied in 1963 by Gow et al. (1965) had broken up and gone out to sea (Gow and Govoni 1994), and that additional portions of the Ward Hunt Ice Shelf located along the northern coast of Ellesmere Island were also in process of failing and drifting into the Arctic Ocean (Mueller et al. 2003). Once these ice islands enter the pack, their life expectancy is at the most a few tens of years, based on past observations of the drift of ice islands such as T-3.

7.4 Reality

What I have discussed earlier in this chapter, although frequently based on field observations, does not really give an accurate picture of what is "out there" in the polar ice packs. The reason is, of course, that when studying a naturally occurring material such as sea ice, scientists tend to pick locations that appear to be "clean-cut" in the hope that that they will be able to unravel the different factors influencing the state of the material. One result of this approach is that it is only since the 1970s that we have started to understand pressure ridge formation. Why study something that is obviously a mess when you can select undeformed areas within the pack ice for study? What follows is an attempt to give the reader some impression of what is currently known about more representative ice conditions within the polar ice packs. As will be seen, the structure of the sea ice in the Arctic is somewhat different from that in the Antarctic. At first glance this result might appear to be a bit absurd, in that it is reasonable that if one freezes compositionally identical seawater under identical environmental conditions, one should obtain essentially the same material independent of the location. Although this is true, it is a simple observational fact that the environmental conditions in the Southern Ocean are on the average sufficiently different from those in the Arctic Ocean to produce systematic differences in the internal structure of the ice that is formed.

7.4.1 Arctic

As far as ice structure is concerned the important feature of the Arctic Ocean is that it is an essentially landlocked basin with only one principal exit, Fram Strait. It is not only landlocked, it has been ice-clogged, even in summer, with major ice-free areas only occuring around its margins. In the winter when new ice is forming the amount of open water and thin ice present at any given time are a few percent, largely present in narrow, ever-changing leads that have limited fetches. Because of the very low air temperatures common during the arctic winter, once open water forms it is rapidly replaced by a layer of thin ice. With the exception of the period of initial ice formation over the extensive arctic continental shelf, the combination of wind, waves, and fetch that favors the formation of large amounts of frazil ice are largely missing. This means that in the Arctic one would expect to find the largest amounts of frazil in thin ice types with the percentage of congelation ice gradually increasing as the ice becomes thicker. This is exactly what has been reported. For instance, Gow, Tucker, and Weeks (1987), in their study of ice exiting the Arctic Basin via Fram Strait, found that with few exceptions the undeformed ice sampled typically contained < 20% frazil. This estimate is also supported by related studies of undeformed floes in the Eurasian sector of the Arctic Ocean by Eicken, Lensum et al. (1995) and in the Beaufort Sea by Meese (1989). The one prominent exception (85% frazil) to this trend was thin (0.44 m) lead ice. However, if the ice was collected in the vicinity of a pressure ridge, frazil amounts ranging between 30 and 70% were observed. This is a particularly useful reference in that it contains numerous structural profiles (58) and thin-section photographs (242), allowing the reader to gain a good impression of the complexity of pack ice in the field. In their study of lead ice in the eastern Arctic, Gow et al. (1990) found that frazil ice primarily occurred in the uppermost part of the ice cores (i.e., the ice that formed while a portion of the lead was presumably still open to the atmosphere; see their Fig. 4). Working in the Beaufort Sea in 1992, Jeffries (pers. comm. 2002) also found that in second- and multiyear ice the most common ice type was congelation, provided that the ice was undeformed. This was particularly true for thicker MY ice. Exact ice-type percentages separated on a regional basis do not appear to be available, as in most cases the exact provenance of the sampled ice is unknown. The only regional map showing the spatial distribution of structural types of which I am aware was prepared in late 1959 for the Russian coast by Cherepanov (1971b). In it he divides the ice into six different structural types based on the primary divisions of prismatic, fibrous, and granular. Although he does not provide detailed definitions of these terms, I presume that the term *prismatic* refers to straight grain boundaries as seen in horizontal thin sections of freshwater ice, whereas *fibrous* refers to a more typical irregular grain boundary pattern as seen in congelation sea ice. The sampling points are quite scattered. Although the map is crude, some trends are obvious: (1) prismatic ice only forms near the coast where the large influx of freshwater from the Russian rivers causes very low near-surface salinities, and (2) ice that has a granular structure (presumably frazil) is widely developed in the region of the continental slope where polynyas commonly form.

Another feature of the arctic ice pack is the nature of the snow cover, whose characteristics are associated with the common occurrence of quite low temperatures, the very limited amount of open water during the winter, and the occurrence of MY ice in excess of 2 m thick. This snow is typically thin (10–30 cm) and hard-packed in areas where the ice is undeformed. The thickest accumulations are the result of drifting snow and are found in areas where the ice surface is rough, primarily because of ridging. Snow primarily accumulates in the late fall and early winter, when local open-water areas still exist. Most arctic snow remains dry and serves as an insulating layer throughout most of the year. Even so, snow thicknesses are rarely sufficient to load the ice sheets until their upper surfaces are depressed below sea level. As a result, infiltrated snow ice rarely forms on the upper surface of arctic floes.

Although most extensive occurrences of platelet ice are from antarctic sites and appeared to be primarily associated with the presence of ice shelves (discussed in section 7.4.2 and in more detail in Chapter 17), one might nevertheless ask if platelet ice has been reported in the Arctic. It has. For instance, during their 1992 and 1993 field programs in the central and western Beaufort Sea, Jeffries et al. (1995) found a moderate amount of platelet ice having similar characteristics to antarctic platelet ice. Platelets occurred in 22 out of a total of 57 cores, which were widely distributed throughout the study area. The platelets were invariably associated with congelation ice, as was also found to be the case in the Antarctic. In analyzing the arctic data, if platelets occurred at a given level in an ice sheet, that level was considered to be a platelet layer even though the platelets typically comprised only a small percentage by volume of the layer. Using this definition, platelet ice comprised 9% of the total ice core collected. Clearly the volumetric percentage of platelet ice was much smaller than this value.

Although the origins of arctic platelet ice are still somewhat speculative, Jeffries et al. (1995) have appraised the several possibilities as follows:

Ice Shelf Sources: Although platelet ice formation undoubtedly does occur in association with ice shelves in the Arctic, the limited sizes of these shelves and their lesser thicknesses as compared with antarctic ice shelves make them improbable sources of the widely distributed supercooled water necessary to produce the observed platelets.

Nearshore Processes: Nearshore processes do not appear to be a probable explanation in that the platelets observed in the Beaufort Sea are better developed and grain sizes are significantly larger than would be expected for anchor or normal frazil ice. Also, the $\delta^{18}O$ values obtained from near-coastal ice, where anchor ice might be expected to form, is two to three times more negative than values obtained from the platelet ice observed in typical Beaufort Sea floes. In addition the significant amount of silt frequently found in nearshore ice is missing.

Under-Ice Melt Ponds: Structurally platelet ice appears to be reasonably similar to ice that has been observed growing in the low-salinity water in under-ice melt ponds (Eicken 1994). However, this melt pond ice is less saline and has less positive $\delta^{18}O$ values than the platelet ice. If under-ice melt ponds are the primary source of the platelets, this suggests that ice growth in such melt ponds is not a cryological oddity but instead is a fairly common phenomenon.

Ice Pumps: As envisioned by Lewis and Perkin (1983, 1986), an ice pump also operates by the mechanisms discussed by Foldvik and Kvinge (1974, 1977), but in this case the water starts at the surface of a lead where it is at the freezing point. As it moves downward, passing under a pressure ridge or a thick old floe, it warms, allowing it to melt a small amount of ice from the bottom of the floe or from the ridge keel. This melting produces a slightly lower-salinity layer of seawater that is now exactly at its freezing point at depth. When this water subsequently moves upward the decrease in pressure then results in supercooling and platelet growth. This is certainly a real mechanism, but the difference is that in the Arctic the maximum observed depth of pressure ridge keels is ~50 m and most ridges and old floes are far less thick than that. When this is contrasted with antarctic shelf ice that can be several hundred meters thick at the grounding line, it is clear that arctic ice pumps are certainly less effective than antarctic ones. The relatively high salinities and the fact that the platelet ice has $\delta^{18}O$ values close to zero also favor this mechanism.

What have we concluded here? Clearly, platelet formation is a real phenomenon that may make a moderate contribution to the growth of undeformed congelation sea ice in the Arctic. However, the mechanistic details of the processes involved are still not totally clear. My best guess at the present is that both under-ice melt ponds and the ice pump mechanism are contributing factors.

Individuals interested in fine-tuning their ability to operationally distinguish between the varieties of old ice and FY ice during the summer in the Arctic should examine Johnston and Timco (2008), a 236-page booklet that contains numerous color shipboard and aerial photographs as well as satellite imagery of old ice. Accompanying each photograph or image are the identifiers used in classifying the ice as well as an estimate of the authors' confidence in the suggested classification.

7.4.2 Antarctic

Although most of our basic understanding of sea ice structure comes from work carried out in the Arctic, I believe that we currently know more about the nature of the ice in the antarctic pack than we do about the ice in the arctic pack. The reason for this somewhat surprising status is logistical. Since the 1980s the scientific community has had frequent access to the antarctic pack, even during the winter, via research icebreakers such as the *Polarstern* and the *Nathaniel B. Palmer*. The same has not been true in the Arctic, where the presence of much thicker ice has hindered access even during the summer.

In considering antarctic sea ice one must pay closer attention to the geographic setting of the formation site than is typically the case in the Arctic. The possible exception here would be near-coastal sites along the north coast of Russia, where the input of large quantities of river water produces low salinities. In the Antarctic there are several reasons for distinguishing sea ice formed at near-coastal locations from ice formed farther offshore. For one, because a significant portion of the antarctic coastline is comprised of ice shelves, based on our previous discussion one would expect to find appreciable amounts of platelet ice near such locations (see also Chapter 17). At some sites this ice type may, in fact, make a major contribution to the total ice thickness. Near-coastal locations where platelet ice has been observed include Mirny (Serikov 1963); Prydz Bay near the Amery Ice Shelf (Kozlovskii 1976a); Leningrad-skaya Station on the Cates Coast (Cherepanov and Kozlovskii 1973b); Terra Nova Bay on the Victoria Land coast (Tison et al. 1993, 1998); sites in the Weddell Sea near the Filchner Ice Shelf (Dieckmann et al. 1986; Eicken and Lange 1989a; Kipfstuhl 1991); and locations in McMurdo Sound which are located near the Ross Ice Shelf (Gow et al. 1982, 1998; Jeffries et al. 1993; Jones and Hill 2001; Leonard et al. 2006; Paige 1966; Wright and Priestly 1922). Because of the numerous studies carried out in McMurdo Sound, this area has become essentially the type location for the occurrence of such ice. It is interesting to note that in their sampling transit from the Balleny Islands down the east coast of Victoria Land, Jeffries and Weeks (1992b) only observed significant amounts of platelet ice in samples collected near the Drygalski Ice Tongue and further south. In ice that had formed away from the coast and its ice shelves, significant quantities of platelet ice were not observed.

There are several other common characteristics of sea ice that has formed near the antarctic coast. For one, there are large amounts of frazil ice generated in the polynyas that frequently occur along the coast. Therefore frazil commonly forms the initial layer of ice that has developed in this zone. At times these initial frazil layers can be several tens of centimeters thick. In that frazil can be swept beneath previously formed ice layers, complex floes existing of alternating layers of frazil and congelation ice can form. A particularly well-documented example of such layered ice located near Syowa Station on the coast of East Antarctica has been described by Tison and Haren (1989). The 1.64-m-thick floe consisted of 77% frazil ice, with the remaining congelation ice occurring in five different layers. Of these layers only the lower 18-cm-thick layer showed the classical features characteristic of congelation ice in the Arctic. Figures 52 through 56 in Weeks and Ackley (1986) also show examples of such complex floes.

Adding to the frazil crystals, which by definition form in the sea, are large amounts of snow that are blown off the continent by the intense katabatic winds that are common features of the antarctic coastal regime. In that these drainage winds are topographically steered, large variations in wind speed can occur as one moves laterally along the coast, with the highest wind speeds occurring at the mouths of valleys and ice shelves. The amount of snow moved onto the coastal sea ice and, if they exist, into the coastal lead and polynyna systems can be very large. For instance, at Mirny significant amounts of snow drift from the land onto the ocean starting in February, with wind speeds of 6 to 7 m/s. By March, when the average wind speed reaches 15 m/s, it is estimated that nearly 800 kg of snow per linear meter crosses the coast per hour (Tauber 1957). Over the course of a year the transport of snow across one kilometer of coastline in this region is estimated at 1.5 million tons. At other locations such as the notoriously windy Adélie Coast snow transport estimates are as high as 50 million tons per year per kilometer of coastline. Although katabatic winds do occur in Greenland, there clearly is no equivalent of this phenomenon on this scale in the Arctic. Furthermore, in that katabatic winds decrease rapidly in speed upon reaching the ocean, where the mean surface slope becomes effectively zero, this large mass of snow is dumped within a few tens of kilometers of the coast. If an open polynya exists the snow is deposited in the water, where it mixes with the frazil crystals that are forming, thereby contributing to the formation of a thick slush layer that will ultimately form the surface layer of the sea ice.

If a stable ice sheet has formed, a thick layer of snow can build up on the ice surface. Depending on the thickness of the ice and the thickness and density of the deposited snow, the weight of the snow is frequently sufficient to depress the upper surface of the ice to below sea level. The appropriate equations here, assuming hydrostatic balance, are as follows (Ackley et al. 1990; Spichkin 1966):

$$\rho_s h_s + \rho_i h_i = \rho_w d \qquad (7.21)$$

where ρ_s, ρ_i, and ρ_w are the average densities of snow, sea ice, and seawater, and h_s and d are the sea ice thickness and draft (depth below sea level of the bottom of the ice). Recalling that $h_i = f + d$ where f_b is the freeboard, we can see that for flooding to occur $f_b = 0$ and $d = h_i$ or

$$\frac{h_s}{h_i} \geq \left(\frac{\rho_w - \rho_i}{\rho_s} \right). \qquad (7.22)$$

This situation, in which seawater infiltrates the snow pack lying on the top of the ice, is frequently observed not only in near-coastal ice but in the main antarctic pack as well. For instance, for $\rho_s = 0.2, 0.3$, or $0.4\ \mathrm{g/cm^3}$, the critical ratio $h_s/h_i = 1/2, 1/3$, and $1/4$, respectively. Spichkin found that in the fast ice area of the Davis Sea, h_s/h_i values were ~0.5 as compared to h_s/h_i values necessary for flooding of ≥ 0.25. He also noted that the formation of infiltrated snow-ice noticeably increased the effective ice thickness. Examinations of the results of this process can be found in the papers by Eicken, Fischer, and Lemke (1995), Haas et al. (2001), and Kawamura et al. (2004). It appears that such ice can form in at least two different ways: (1) by the infiltration of seawater as discussed above, and (2) by the collection and subsequent refreezing of snow meltwater on the top of the sea ice. The later process certainly would appear to be dominant in the Ross Sea ice studied by Kawamura et al., based on the isotopic signature of the ice, which was similar to that of snow. Also its polygonal structure suggests Ostwald ripening. The ice formed by these processes has been termed both *infiltration ice* and *superimposed ice*. *Superimposed ice* is perhaps the better term in that it does not imply a specific mechanism.

As was noted in Chapter 1, as the percentage of the antarctic pack that survives the summer is small, the amount of thick MY ice present is also small. In that the antarctic pack consistently moves toward the north, where it encounters much warmer water and melts, for an ice floe to exist for several years some protection from typical conditions is required. In some cases such protective conditions are found in the near-coastal regime, with old ice existing near the heads of narrow bays or fjords or in shoal areas, where accumulations of grounded icebergs serve to prevent the drift of the ice to the north. Another region where heavy ice is invariably observed is off the east coast of the Antarctic Peninsula, where the clockwise circulation in the Weddell Sea presses the pack against the peninsula. This ice is also commonly topped by a thick snow pack.

Based on observations during trips aboard the icebreaker *Nathaniel B. Palmer*, it has generally been noted that when the ship encounters thick ice, a thick snow cover is invariably present. This is double trouble in that the snow adds considerable friction to ice thicknesses that ships such as the *Palmer* would find difficulty transiting even if no snow were present. A frequent result is that the ship is required to back and ram, a process that eats time and fuel and stresses the equipment. At worst, a ship can become beset with the possibility of sustaining structural damage and running out of fuel. Wise captains frequently ask the scientists aboard to modify their sampling plans to avoid such ice.

As a result, the structure of antarctic MY ice has not received much attention. As might be anticipated, some of this material is very different from arctic MY ice. For instance, Cherepanov and Kozlovskii (1973b) have described 3.6-m old fast ice in the vicinity of Leningradskaya Station. This ice was topped by a snow–firn layer that ranged from 60 to 120 cm and became 160 cm thick in an area where shelf ice occurred. At the base of the firn layer there was a 30- to 45-cm-thick brine + ice layer, with brine salinities running as high as 58‰. Beneath this, 260 cm of granular ice comprised the remainder of the floe. This ice showed considerable variations in both grain size (1–15 mm) and in texture but without any obvious pattern. However, there were several distinct layers. The structure of the floe was interpreted as indicating approximately five years of growth without the formation of congelation ice. Instead, the floe grows as the result of thermal metamorphism occurring at the boundary between the snow–firn layer and the underlying fine-grained ice. If this interpretation is correct, this is an MY floe that accretes on the top and ablates on the bottom. In the Arctic, MY ice ablates on the top and accretes on the bottom, just the opposite. Another interesting feature of the antarctic floe was that the lowest-salinity ice was on the bottom of the floe. In the Arctic it would be on the top. Although this scenario might at first glance seem odd, considering the varied processes that clearly affect the growth of antarctic sea ice it clearly is a plausible explanation. There is no way to know at present how typical floes of this type might be. I would guess that such floes are not common. One thing is for certain: such floes have not, as yet, been described in the Arctic.

Before we examine the structure of the interior of the antarctic pack, let us examine what occurs at the northern edge of the pack during the winter when the ice edge is gradually expanding to the north. Understanding these conditions is crucial to understanding how the structure of the pack develops, in that the initial development of much of antarctic sea ice occurs in such locales. As will be described in more detail in the chapter on the marginal ice zone, the ice edge is commonly quite a violent locale, with the frequent occurrence of high winds and associated waves and swell. Such conditions are ideal for the formation of frazil crystals in the upper layer of the ocean, ultimately leading to the development of a thickening layer of grease ice. Initially, this layer is unconsolidated with a very low viscosity. However, as the addition of more and more frazil thickens the layer of grease ice with the associated cooling of the layer, the development of intergranular bonding, and the development of more solidlike characteristics, pancakelike bodies begin to develop. As waves pass through the fields of pancakes, the contact regions between adjacent pancakes open and close. During these moments when open water is exposed, new frazil crystals form. When the pancakes come back into contact these crystals are both pushed upward, where they form rims on the edges of the pancakes, and pushed downward, where they

Figure 7.43. A schematic showing the pancake cycle as envisioned by Lange et al. (1989).

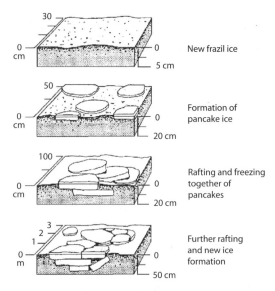

New frazil ice

Formation of pancake ice

Rafting and freezing together of pancakes

Further rafting and new ice formation

Figure 7.44. An early stage in the pancake cycle, Bellingshausen Sea, Antarctica.

Figure 7.45. A late stage in the pancake cycle showing several different ages and sizes of pancakes, Bellingshausen Sea, Antarctica. The diameters of the larger pancakes are ~2 m.

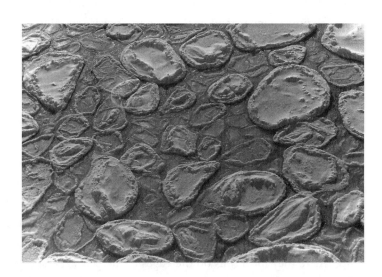

contribute to the thickening of the pancakes. Frequently, numerous small pancakes will adhere together, forming larger composite pancakes. This is such a striking and common phenomenon in the Southern Ocean that it has been termed the pancake cycle (Lange and Eicken 1991; Lange et al. 1989). Figure 7.43 shows a schematic of this cycle as envisioned by the above authors. Figures 7.44 and 7.45 show two different views of this general process as it occurs in the Bellingshausen Sea, with the first photograph showing the initial stages of pancake formation. Note that the spaces between the pancakes are filled with a mixture of frazil crystals and open water. The second figure shows a group of pancakes of varied sizes, and presumably ages, at roughly the stage when they freeze together, forming a continuous sheet. As shown in Figure 7.43 the extensive rafting of pancakes over one another is another common aspect of this cycle. Lange et al. (1989) speculate that when pancake thicknesses reach 0.4 to 0.7 m, the damping effects of the ice will be sufficient to terminate the cycle. In that long-period swell can penetrate hundreds of kilometers into the pack, the pancake cycle has the potential of occurring essentially anywhere within the antarctic pack. However, it is most prevalent near the its northern edge. It should be stressed that in the Southern Ocean, as elsewhere, when conditions return to calm the type of sea ice forming returns to congelation, both via the development and thickening of nilas into the various young ice types, and by the growth of congelation ice on the bottoms of floes whose upper portions are comprised of pancakes.

Very little attention appears to have been paid to the internal structure of pancakes. Part of the problem has been, of course, the fact that grain sizes in pancakes are commonly small, thereby limiting the use of standard glaciological petrographic techniques designed for studies of larger crystals. In addition, pancakes examined in the Odden region of the Greenland Sea (Wadhams et al. 1996; Wadhams and Wilkinson 1999) showed reasonably homogeneous structures and salinity profiles that were consistent with classic ideas of pancake growth via bottom accretion. Except for the extreme surface layer, salinity values were lowest near the surface of the ice and gradually increased with depth. A similar trend was found for porosity, with the older near-surface ice being the most compact.

However, work in the Antarctic by Doble et al. (2003) suggests that a much more complex picture is justified as they found a variety of morphologies, which could be separated into two main types. The first type was similar to pancakes found in the Arctic in that they were single layers. The other main types of pancakes observed typically showed a two-layered structure. In some of these, both layers were similar, showing porosities and salinities that increased downward but with a clear step in values between the layers. In other cases the lower layer had a lower porosity than the overlying layer and also a lower salinity (3.2 versus 5.6‰). Finally, in some cases the lower layer was isothermal columnar ice that did not appear to be growing. Here again, the lower layer showed an appreciably lower average salinity (3.7 versus 6.3‰). The apparent explanation of these variations is that many of these pancakes form by a multistage process in which the lower "platform" layer can be either a pancake of the same age, a slightly older pancake, or even a piece of congelation ice. In some cases the upper layer appears to have formed by frazil crystals washing over the top of the earlier layer and subsequently freezing. In that ice growth via the top layer process was roughly twice as fast as congelation growth, Doble et al. (2003) suggest that a parameterization of this process is important for models that need to simulate the rapid advance and thickening of wave-influenced ice covers such as occur in the Antarctic.

One would generally expect to find that the highest percentages of frazil occur in the thinner ice classes located near the northern edge of the antarctic pack and that the relative amount of congelation would increase in ice formed at increasing distances south of the ice edge (Lange et al. 1989). In fact, this is exactly what has been observed by Jeffries and Adolphs (1997) working in the Ross Sea during the autumn. Clearly the floe-to-floe variations in frazil–congelation ratios can be large, depending on the sea conditions during a particular year. For instance, Gow, Ackley, et al. (1987) found floes containing between 6 and 90% congelation ice and between 3 and 100% frazil in their Weddell Sea samples collected during the austral summer of 1980. The average values were 43% congelation and 57% frazil. During the winter of 1986 Lange et al. (1989) found that the samples that they obtained in the same general region produced similar average values. In the western Ross Sea in 1990, Jeffries and Weeks (1992b) observed frazil ice amounts varying from 2% to 85%, with an average of 39%, and congelation varying from 14 to 100%, with an average of 64%. The same cores were estimated to contain an average of 12% snow ice based on its typically larger grain size than frazil ice. In a fall cruise during 1992 in the Bellingshausen Sea and a late winter cruise during 1993 in both the Bellingshausen and Amundsen Seas located to the west of the Antarctic Peninsula, Jeffries et al. (1997) obtained an average of 44% frazil and 26% congelation. This study is particularly interesting in that stable isotope procedures were combined with the textural studies in order to classify the granular ice portions of the cores into frazil ice and snow ice. Based on the particular isotope fractionation model that was assumed, snow ice averages were found to vary between 13 and 42%, whereas frazil ice averaged between 26 and 55%. These general values are similar to values reported from the Indian Ocean region by Allison and Qian (1985), and by Jacka et al. (1988). In their summary of the ANARE observations, Worby et al. (1998) give average values for the pack of 39% columnar, 47% frazil, and 13% snow ice, with other ice types only comprising 1%. Undeformed area-weighted mean ice thicknesses, which included the open water fraction, varied seasonally from 0.31 m in December to 0.52 m in August. The above papers also contain frequent reference to ice core structures, indicating varying amounts of deformation by rafting and ridging.

To summarize the above bipolar observations, one can say the following:

1. Antarctic sea ice appears to show appreciably more textural and structural variability than is typical of arctic sea ice.

2. Frazil ice makes a significant contribution to the growth of antarctic sea ice and is the major ice type in many floes, whereas in the Arctic congelation ice is the major ice type.

3. The relative amounts of frazil and congelation ice vary from year to year and from location to location, presumably depending on changes in the meteorological and oceanographic environment.

4. The highly variable layering observed in antarctic sea ice cores indicates that ice growth in the Southern Ocean is frequently affected by factors such as the penetration of swell into the pack, the opening and closing of leads, and rafting and ridging events. Although such factors can also be important in the Arctic, the resulting variability as seen in ice cores is less.

5. Superimposed ice formed on top of a preexisting layer of congelation and/or frazil ice can be a significant component in the growth of antarctic FY ice floes. Although such ice has been described in the subarctic, it appears to be rare in the Arctic.

6. Extensive development of platelet ice only occurs at sites located in the vicinity of antarctic ice shelves. To date, similar development has not been reported in the Arctic.

7. Undeformed antarctic FY pack ice has thicknesses that rarely exceed 1 m and is commonly appreciably thinner than arctic FY ice when measured at equivalent times of the year.

In concluding this section, one should note that all the above differences are either directly or indirectly related to differences in the climates and the resulting oceanographic states of the two regions. An additional factor is the latitude in which field observations have been made. An excellent example is the frequent reporting of superimposed ice in the Antarctic and its comparative absence in the Arctic. Typical arctic field studies have been carried out at latitudes greater than 70°N, where very cold temperatures favor the growth of thick ice and limit the amount of open water resulting in relatively thin snow covers. In the Antarctic, on the other hand, observations of superimposed ice have frequently been made in the pack ice at locations well to the north of the coast, where the FY ice is thinner as the result of warmer temperatures and the snowfall is heavier as the result of appreciable areas of open water. To find such conditions in the northern hemisphere one needs to go to the subarctic such as the Labrador Coast where, in fact, superimposed ice formation resulting from the infiltration of seawater into a heavy snow pack would appear to be a frequent event (Weeks and Lee 1958).

8 Sea Ice Salinity

If God had consulted me before embarking on the Creation,
I would have suggested something simpler.
Alfonso of Castile

8.1 Introduction

The fact that bulk sea ice contains sea salts and that the amounts of these salts vary with the age of the ice has been known for a very long time by the native peoples living in the far north. The reason that this is important is that the water produced by melting the upper portions of hummocks and ridges in old ice types is quite potable, whereas the water produced by melting most FY sea ice is usually not suitable for drinking. In fact, the earliest study of a possible mechanism for the natural desalination of sea ice was undertaken as the result of a suggestion by the explorer Stefansson that this was a scientific problem worthy of investigation. He had undoubtedly used the above knowledge to obtain drinking water in his travels over sea ice. The resulting paper (Whitman 1926) is excellent for its time, and although the mechanism suggested therein does occur in sea ice, it is currently not believed to be a major contributor to natural desalination. As will be seen, the factors controlling the salinity of sea ice are surprisingly varied, complex, and still not completely understood.

8.2 Observations

There are ice salinity measurements scattered throughout the sea ice literature. In fact, of all the possible measurements that have been made directly on sea ice samples, ice temperature and ice salinity are undoubtedly the most common. The reason for this is obvious. If one wishes to measure almost any other sea ice property and to relate its value to the state of the ice, one must at least know the ice temperature and the volumes of gas, brine, and solid salts present in the ice. To obtain these latter two values, one must also know the salinity of the ice. What this means is that although salinity and temperature values are numerous, they have commonly been collected for purposes other than studying the salinity profile and its changes with time. As a result there are still only a few studies where reasonably systematic salinity observations have been made over a complete growth season on ice whose history was known.

A few words on obtaining salinity samples could be useful here for the novice. Salinity samples are commonly obtained via coring, a procedure that produces a cylindrical sample having a diameter of 7 to 10 cm (see Appendix D). The sample should immediately be placed in some variety of insulated core container designed to minimize change from the in-situ temperature. The core temperature profile should then be determined via the use of some variety of probe thermometer or thermistor inserted into a sequence of drill holes. A convenient spacing of the temperature measurements is as follows: obtain a value close to the upper ice surface and then every 10 cm (20 cm for thick cores) down the core. As temperature profiles are typically smooth curves that are nearly linear, the use of tighter spacings is usually not necessary. During the growth season there is no need to obtain a measurement in the bottom few centimeters, as the temperature of the ice–seawater interface is known provided that one knows the salinity of the underlying seawater. Once this procedure is completed, the core can be cut into segments, frequently on a spacing of 10 cm, and placed in sealed containers for subsequent melting. I have found that plastic containers of the type used to store food in refrigerators work well. Once the ice has melted the water salinity can easily be determined using a conductivity cell. It is also possible to obtain sequential salinity profiles via the use of nondestructive systems that are frozen into the ice early in the growth season and remain in place during the winter. Descriptions of such systems can be found in Morey et al. (1984) and Notz et al. (2005).

8.2.1 Arctic

The first set of salinity values taken over a whole season, and still one of the few sets made on ice in the central Arctic Basin, was obtained by Malmgren (1927) during the drift of the *Maud*. Although only five generalized profiles were presented (Figure 8.1), they illustrate several aspects of sea ice salinity profiles that have been borne out by later studies. For example, note

1. the decrease in the average salinity with increasing ice age,
2. the fact that the salinity profile reported in thinner FY ice types is often C-shaped,
3. the sharp decrease in the surface salinity during the start of the melt period, and
4. the pronounced overall decrease in salinity by the end of the first summer.

The best set of sequential salinity observations published to date has been obtained by Nakawo and Sinha (1981) with the assistance of the native population of Pond Inlet in the Canadian Arctic. From November 1977 to April 1978, salinity profiles were made on a biweekly basis on ice in Eclipse Sound. As can be seen in Figure 8.2 the profiles were quite detailed, with salinities being determined every few vertical centimeters. Note also that throughout the complete thickness range studied, the near-surface and the near-bottom salinity values are consistently higher than the rest of the profile. This fact, as noted above, results in the C-shaped profiles observed in the thinner ice cores. However, once ice thicknesses reach values

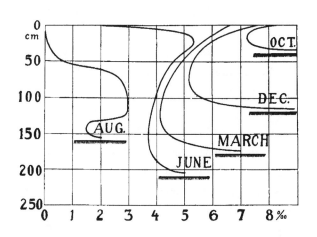

Figure 8.1. Curves showing the variation in sea ice salinity profiles for five different ice thicknesses based on data obtained during the drift of the *Maud* across the Arctic Ocean (Malmgren 1927).

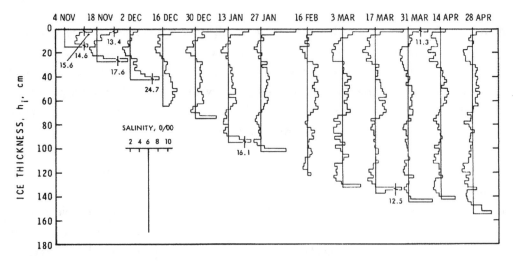

Figure 8.2. Sequential salinity profiles of ice in Eclipse Sound near Pond Inlet, Canada, measured at two-week intervals during the winter of 1977–1978. The scale for the salinity values is shown as an insert. The vertical solid lines represent a value of 6.0‰ and are given as a reference (Nakawo and Sinha 1981).

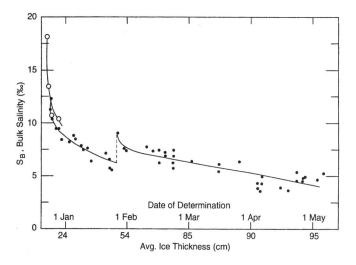

Figure 8.3. The average salinity of sea ice located in the bay at Hopedale, Labrador, plotted against ice thickness (cm) and the date of measurement during the winter of 1955–1956 (Weeks and Lee 1958). The dashed vertical parts of the curves indicate storms during which appreciable thicknesses of snow were deposited on top of the ice, resulting in an increase in thickness via the formation of infiltration ice.

in excess of ~0.6 m there is considerable variation in the salinity values in the middle portions of the profiles, and these patterns are not simple. There is also a general decrease in the average salinity with time, as was suggested by Malmgren's data, with the more rapid decrease occurring while the ice was still quite thin.

The other set of reasonably complete salinity measurements, which are far less detailed, was obtained by Weeks and Lee (1958) at Hopedale, Labrador, during the winter of 1955–1956. Although the winter climate at Pond Inlet was characterized by temperatures that were consistently well below freezing, at Hopedale there were several thaw periods when air temperatures were above –10°C and at times even above freezing. As a result periods of rapid desalination were observed (Figure 8.3) even during late December and January. The desalination was presumably related to the fact that the warm temperatures would definitely result in an increase in the volume of brine in the ice as well as in its permeability. The general increase in the average salinity of the ice sheet that occurred during late January was the result of a heavy snowfall that loaded the ice until its upper surface was below sea level. A salty surface slush layer then formed and subsequently froze rapidly when the temperature dropped to more normal levels, as this surface layer lacked the thermal protection of an overlying insulating thick snow and ice layer. The snow-loading event in effect re-created growth conditions that were similar to those encountered when the initial ice sheet formed in that ice was, once again, added by growth starting downward from the air–ocean interface. Prior to this event the thickest ice had been found in areas where the snow cover was the thinnest, thereby offering the least insulation. After the snow-loading event, the thickest ice was found at locations where the snow had been the thickest prior to flooding.

These observations show that one assumption that is usually made when computing ice thickness in the Arctic can be quite invalid in the subarctic or in the Antarctic. This is the assumption that if two different areas are subjected to the same general temperature regime, the ice thickness will vary inversely with the snow cover thickness. In truly arctic regions the formation of seawater-infiltered snow-ice does not typically occur; the snow retains its insulating capabilities throughout the winter and the above assumption holds (see Holtsmark 1955 and equation 4.15 in the present book). As will be seen later, many areas of the antarctic pack frequently show surface flooding similar to that found in subarctic regions.

Several studies have investigated small-scale lateral variations in the salinity of sea ice that appeared to be structurally homogeneous and that were known to have formed under quite stable freezing conditions. The fact that there is considerable lateral as well as vertical variation in an apparently uniform sheet of sea ice was first demonstrated by Kusunoki (1955). This was followed by a study of the salinity variations in the ice located in North Star Bay, Thule, Greenland (Weeks and Lee 1962), which utilized a large number of samples and analysis of variance techniques. It was found that samples collected at the same level from a visually uniform sheet of sea ice gave standard deviations varying from 0.51 to 0.62‰. Not surprisingly, similar sampling from a nearby area where the initial ice sheet was composed of rather heterogeneous pancakes gave higher values ranging between 0.77 and 1.06‰. In both ice types the largest standard deviations always occurred in the samples taken from the bottoms of the cores, where larger brine volumes would be expected to produce a larger scatter associated with brine drainage during sampling. Later Lake and Lewis (1970) determined salinities on several sets of samples obtained by coring a 1.6 m block that they had removed from a thick sheet of FY ice. They noted similar variations in samples obtained only a few centimeters apart. A more recent study (Tucker, Gow, et al. 1984) examined salinities obtained from 5 cores located at a center spacing of 38 cm in FY ice located north of Prudhoe Bay, Alaska. The core lengths varied from 2.00 to 2.05 m. An appreciable scatter was observed at all levels, with standard deviations ranging from 0.20 to 0.78‰. Again the largest variation was observed in samples obtained from the lower part of the ice sheet where higher in-situ brine volumes were present as the result of higher temperatures. If the upper 25 cm of these cores are neglected in that they contain bands of granular ice, again a similar pattern is revealed, with higher salinities occurring at the tops and the bottoms of the cores. The middle sections of these cores show a large local scatter with no obvious pattern.

The most recent work on this subject is a study by Cottier et al. (1999) using sea ice grown in an ice test basin. This allowed them to control the air temperature and to use procedures that were designed to minimize brine drainage during sampling and storage. They found that the presence of large brine drainage channels in the sea ice significantly altered the salt distribution on a centimeter scale. More will be said about this aspect of their results later. They also examined the differences in salinity patterns in ice that had continually been kept cold and ice that had been warmed to near-melting temperatures. Not surprisingly, the salt distribution in the "warm" ice was more homogeneous and less dependent on the distribution of brine drainage structures than that in the colder ice. This trend is believed to be related to the presumed increase in the lateral permeability of the ice associated with the increase in brine volume resulting from the warming. Whatever the mechanisms, warming the ice clearly has a significant effect on its salinity structure.

Bulk salinity trends in both FY and MY ice were initially examined by Cox and Weeks (1974), who collected data from a wide variety of sources and locations. Their results were encouraging in that the amount of scatter observed was surprisingly small and the trends were strong. They found that while the ice was still quite thin there was a rapid decrease in salinity from mean values that were initially as high as 16‰. At an ice thickness of ~35 cm the rate of the salinity falloff changed rapidly to a more gradual decrease. They were able to fit their data with good agreement by using two straight lines. This subject has since been reexamined by Kovacs (1996), who appreciably expanded the data set utilized. Somewhat surprisingly he found that there was little difference between the mean salinities of arctic and antarctic FY ice, even though average antarctic seawater salinities would be expected to be 2 to 3‰ higher than representative arctic values. His combined results for FY ice are shown in Figure 8.4. In that the additional data points smoothed out the break in the Cox and Weeks data set, Kovacs found that all of his FY data set could be well described by the simple curve

$$S_B = 4.606 + \frac{91.603}{H} . \tag{8.1}$$

Here S_B is the bulk salinity in ‰ and H is the total thickness of the ice in cm. Again considering the diversity of the sources, the amount of scatter in the data set is surprisingly small ($r^2 = 0.73$). It should be noted that relations such as equation 8.1 only are applicable if the ice is still growing.

Documentation of the salinity changes that occur during the spring and early summer is rare, as once surface melting sets in working conditions invariably become difficult. Exceptions to this generality are to be found in the paper by Holt and Digby (1985), who studied such changes during the start of the melt period at Moild Bay, a location in the western Canadian high Arctic. The purpose of their study was to

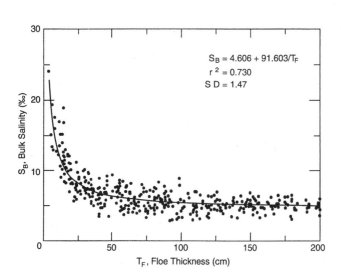

$S_B = 4.606 + 91.603/T_F$
$r^2 = 0.730$
$SD = 1.47$

Figure 8.4. Arctic and antarctic FY bulk sea ice salinities plotted vs. ice thickness (Kovacs 1996)

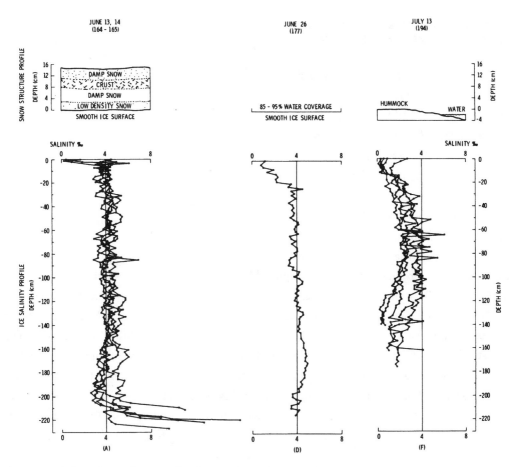

Figure 8.5. Selected ice salinity profiles from FY fast ice at Mould Bay, NWT, Canada, showing changes associated with surface melting (Holt and Digby 1985).

relate observed changes in the ice to changes in aircraft and satellite-based remote sensing imagery. The primary focus of the field effort was on the changes from the onset of melt through the period of drainage as observed in FY ice (Figure 8.5). The first set of profiles was obtained while there was still damp snow on the surface. Near-surface salinities are generally constant with an average value of ~4‰. Notice that in some of the profiles, near-surface values have dropped to values in the 1 to 2‰ range, indicating that at some locations near-surface melting was already underway. The second single profile was taken between 12 and 13 days later at a time when the smooth upper ice surface was 85 to 95% covered with water. Although there had been a pronounced decrease in the salinity of the upper 20 cm of the ice, the salinities at depths exceeding 25 cm remain centered on 4‰. The final set of profiles, which was taken a month after the first set, shows a general decrease in the salinity values, with values above 4‰ becoming the exception. The most pronounced decreases occurred in the near-surface portion of the cores. At the time that these cores were obtained, the upper surface of the ice had started to develop a slight hummocky topography. Melt ponds had also developed, and freshwater was flowing off the upper portion of the ice sheet and forming a low-salinity layer between the base of the ice and the underlying colder, more saline, and denser seawater. The effect of the formation of this low-salinity layer can also be seen in the salinity profiles that show a pronounced dropoff at ice thicknesses of greater than 120 cm.

Figure 8.6. Bulk salinities of MY sea ice floes from the Arctic and Antarctic (Kovacs 1996).

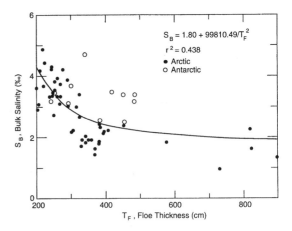

Although the actual FY→MY transition has received little attention, more information is available on typical MY profiles, although hardly the amount of information available on FY ice. There are several reasons for this difference. First, it is commonly difficult to reach MY ice via surface transport from shore stations. Second, MY salinity profiles are believed to be relatively static and therefore not as interesting as FY profiles. Although this is probably generally true, there undoubtedly are seasonal changes in MY profiles. For instance, the salinity of the FY ice layer accreting during a winter on the bottom of an MY floe must be different than its salinity at the end of the summer. These changes have, to date, not been specifically studied. Trends in the bulk salinity in MY ice have also been investigated by Kovacs (1996a), who used data from a wide variety of sources. In dividing FY from second-year ice, he arbitrarily chose 200 cm as the separation point. His results, shown in Figure 8.6, include data from the Antarctic as well as from the Beaufort Sea. Considering that 200 cm does not always give a clear-cut separation between FY and older ice types, it is not surprising that the scatter has increased, with the r^2 value decreasing to 0.44 for the fitted equation

$$S_B = 1.80 + \frac{99810.5}{H^2} \qquad (8.2)$$

as compared with the value of 0.73 obtained for FY ice. Nevertheless, it is apparent that the salinities of old ice types are appreciably less than for FY ice, with average values occasionally dropping to < 2‰ for very thick MY ice ($h < 5$ m). When MY data from Fram Strait are included in the analysis (Kovacs, his Fig. 17), the clear-cut decrease shown to occur between ice thicknesses of 2 and 4 m in Figure 8.6 becomes less obvious, with the r^2 value dropping to 0.22. Nevertheless, it is quite clear that MY ice is typically appreciably less saline than FY ice. There are exceptions in that Eicken (pers. comm.) has documented level second-year (SY) ice in the Weddell Sea that has mean salinities of 5.1‰ and thicknesses between 1.2 and 1.6 m. The reader should note that these thicknesses are outside of the range studied by Kovacs.

As far as detailed profiles are concerned, although information on MY salinity profiles can be found in Malmgren (1927), the classic reference is Schwarzacher (1959), who carried out a careful investigation of the MY ice located in the vicinity of IGY Drift Station Alpha in the Beaufort Sea. His final averaged salinity profile was based on 40 cores and over 2060 salinity measurements and is shown in Figure 8.7 (curve C). Further work by Cox and Weeks (1974) revealed that MY salinity profiles could be separated into two types, depending upon whether the cores were taken from hummocks (Figure 8.7, curve B) or from depressions, likely former melt ponds (Figure 8.7, curve A). Based on these observations Schwarzacher's data appear to be representative of hummocky ice or at least of ice with a higher freeboard. Some sense of the variability

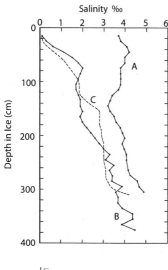

Figure 8.7. Average salinity profiles as obtained from depressions (curve a) in the surface of a MY floe as well as from hummocked areas (curve b) of an MY floe located in the central Beaufort Sea during one of the AIDJEX pilot experiments (Cox and Weeks 1974). Also shown is the average MY salinity profile (curve c) obtained by Schwarzacher (1959) during the drift of Ice Station Alpha.

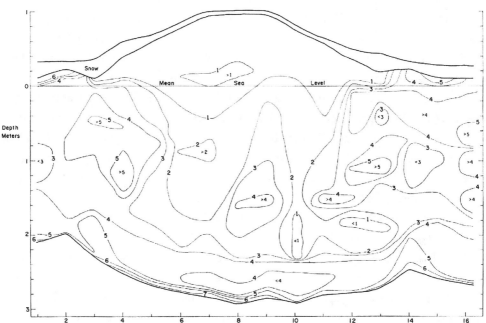

Figure 8.8. A cross section of an area of MY ice showing the variation of salinity with topography (Cox and Weeks 1974). The hummock appears to be the result of differential melting.

of salinity in MY ice can be gained by examining Figure 8.8, which is based on a series of cores taken in a line across a low hummock on an MY floe then located in the Beaufort Sea. There are several features of interest in the figure. Note the very low salinities ($\leq 1‰$) in the upper, above-sea-level portion of the hummock. This is clearly a site where ice could be mined for drinking water. Second, note that the highest salinities ($\sim 5–6‰$) only occur in the very bottom of the ice sheet. As this hummock was sampled in the late winter/early spring when it was still quite cold, it appears that these higher salinities are from a FY ice layer that has formed on the bottom of the floe during the current winter's growth period. Although the depressed areas in Figure 8.8 show slightly higher salinities than do hummocks in the spring, it is good to remember that these are the areas where melt ponds will form during the summer. As the water in the ponds largely comes from higher hummocked areas, it will be of a very low salinity and will tend to flush out any salt remaining in the near-surface ice. If the salinity of a melt pond is < 1‰, when the pond refreezes the resulting ice will effectively be lake ice.

During the late 1980s I was able to collect a number of MY cores from two sites located between 84 and 85° N to the north of Greenland. In attempting to analyze this data set, I plotted the salinity values against normalized thicknesses, $z(n)$, relative to the total thickness of each individual core. For instance, if a salinity value was obtained at a depth of 1 m in a core with a total thickness of 4 m, the $z(n)$ value would be 0.25, with all $z(n)$ values lying between 0 and 1. This approach was adopted in order to accommodate the fact that MY ice frequently exhibits appreciable variations in thickness. The results, which are shown in Figure 8.9, include 769 individual salinity measurements from ice with thicknesses varying from 1.99 to 5.45 m. Also included here are the MY salinities measured by Cox and Weeks (1974) in the Beaufort Sea, as well as the average MY salinity profile obtained by Schwarzacher (1959). However, I have excluded profile data that were clearly obtained from depressions and former melt ponds. Therefore, Figure 8.9 can be taken to represent regions of better-drained hummocked ice. There is considerable scatter. There is also a general linear increase in S_i values from 0‰ at $z(n) = 0$ to 1.5‰ at $z(n) = 0.5$. For $z(n)$ values greater than 0.5 there is a gradual nonlinear increase in S_i values as $z(n) = 1$ is approached. Although there is no clear separation between the values collected north of Greenland and the data from the Beaufort Sea, it is apparent that all the high salinity values (> 6‰) are from ice located at the very bottom of the ice sheets. This, plus the fact that it was still quite cold when these cores were obtained, indicates that the FY layer on the bottom of these floes had not started to deteriorate. Additional salinity and structural data collected during the summer melt period from level MY ice located in the Eurasian sector of the Arctic Ocean can be found in Eicken, Lensu, et al. (1995).

Considering the large area of MY ice present within the Arctic Ocean, it obviously would be useful to have a better characterization of its properties. The first step in accomplishing this would be to better characterize any regional and seasonal variations in the salinity profiles of this ice. In any such program the topographic setting of the cores should be noted and if possible the total age of the ice in the floe determined.

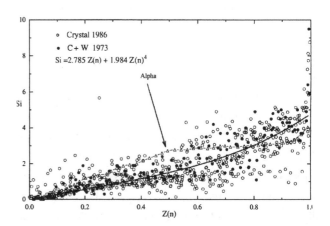

Figure 8.9. MY salinity values obtained on ice stations located between 84 and 85°N north of Nord, Greenland. Also included in the figure are MY salinities determined by Cox and Weeks (1974) from ice in the Beaufort Sea and the average MY profile as determined by Schwarzacher (1959; see also curve c, Figure 8.7).

8.2.2 Antarctic

As might be expected, the amount of information available on salinity changes in antarctic sea ice is more limited than in the Arctic. Although there are studies that have focused on the growth of FY ice during a complete austral winter (Crocker and Wadhams 1989; Weller 1968a, 1968b) and that obtained salinity values, I am unaware of any studies that specifically studied salinity changes throughout a complete winter. Information on MY profiles is also very limited. Available information largely comes from shore stations that allow access to FY fast ice sites. Sampling within the pack has largely been done from icebreakers that were typically in transit precluding investigators from following sequential changes on the same ice. To date there has been only one manned scientific drift station off the antarctic coast. However, the nature of the ice available in the near vicinity of this station (the Russian–U.S. effort in the western Weddell Sea) did not result in a major study of salinity variations. Nevertheless, it is still possible to piece together patterns and trends in the salinity of antarctic sea ice, some of which differ from trends observed in the Arctic.

As discussed in the chapter on ice structure, when seawater freezes you would expect to obtain similar salinity profiles independent of the hemisphere of occurrence provided that the environmental conditions driving the freezing process are similar. I have already made the case that structural studies show that conditions in the Arctic Basin and in the Antarctic are frequently quite different. Nevertheless one might expect that salinity profiles from near-shore locations protected from the turbulence of the Southern Ocean would be similar to profiles obtained from Arctic Basin FY ice. For instance it has generally been held that antarctic FY ice was on the average more saline by 0.5 to 1‰ than arctic ice (Doronin and Kheisin 1977). The reasons for this belief are several starting with the fact that the water off the Antarctic is slightly more saline by 1 to 2‰ than in the Arctic. Yet Kovacs, in his study of bulk salinities versus ice thickness for FY ice, found that there was no apparent difference in the salinity trends between the two hemispheres. As he did not offer an explanation for the similarity, I will suggest a possibility. Although one generally tends to think of the Antarctic as being colder than the Arctic, this is only true if one compares the climate of the Arctic Ocean with that of the extremely cold high plateau of East Antarctica. When one compares the climate of the sea ice areas off the antarctic coast with that of the Arctic Basin, this trend is reversed with the Arctic being the colder. (The Arctic Basin is located at a higher latitude and the thermal contribution of the ocean in warming the atmosphere is lessened by the nearly continuous ice cover.) Therefore, the lower seawater salinity and the faster ice growth in the Arctic possibly results in ice covers that have similar bulk salinity versus thickness profiles to antarctic ice that grows slower from more saline seawater. An interesting speculation; proving it is another matter. Considering that the amount of frazil ice found in the antarctic ice pack is highly variable, it is not surprising that there appears to be considerably more scatter in antarctic bulk salinity versus thickness plots than in similar plots of arctic data. For instance, see Fig. 27 in Weeks (1998b).

When individual salinity profiles are considered, antarctic profiles are clearly more complex than arctic profiles. For example, Eicken (1992), in his study of 129 profiles collected in the Weddell Sea, has found it useful to separate the profiles into the four different types shown in Figure 8.10. These are as follows: C-shaped profiles similar to profiles found in the Arctic during

Figure 8.10. Examples of the four types of antarctic FY salinity profiles as distinguished by Eicken (1992). In each figure the data points and the continuous curves (fitted third-degree polynomials) represent a specific example of each type whereas the dashed curves represent calculated average profiles for the type.

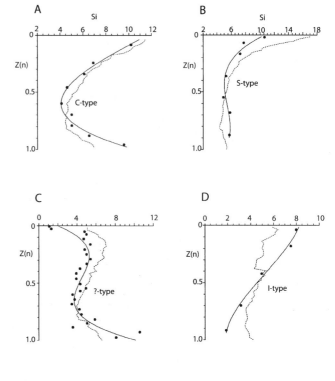

the ice growth season (33% of the samples); S-profiles, which, although similar to C-profiles, show a slight decrease in salinity near the bottom of the sheet and values higher than typical of C-profiles near the top (51%); inverted S-types that Eicken refers to as ?-profiles (16%); and I-profiles, which show either a relatively constant salinity or a linear decrease toward the bottom of the sheet (7%). An excellent example of a composite ?-profile can be found in Jeffries and Weeks (1992a), based on cores collected from the FY fast ice at McMurdo Sound. A sense of the variability in salinity profiles obtained from FY fast ice roughly formed at the same time in McMurdo Sound can be obtained by examining the 22 profiles shown in Fig. 4 in Gow et al. (1998).

The study by Eicken (1992) is particularly interesting in that it attempts to expand the factors influencing the salinity profile beyond the more standard meteorological factors that influence ice growth rates. For instance, flooding and upward brine expulsion are identified as factors contributing to high salinities in the near-surface portions of ice sheets. Low near-surface salinities, on the other hand, are frequently associated with textural changes produced by partial melting and refreezing and differential desalination, which commonly results from surface melting. One interesting aspect of this study is the utilization of the Cox and Weeks (1988a) numerical scheme for calculating salinity profiles (discussed later in this chapter) to provide information on the effect of changes in environmental parameters on the salinity profile. Of all the parameters studied, the most important factor in controlling the salinity profile at a given ice thickness appears to be the timing of initial ice sheet formation. Ice that starts to grow in a lead that opens in midwinter, when both air temperatures and incoming shortwave radiation are at a minimum, will typically grow faster and contain more salt than ice that starts to grow in the spring or fall. The effect of increasing rates of snow accumulation on top of the sea ice and the effect of increases in the oceanic heat flux are somewhat similar, in that both result in lowered salinities in the lower parts of the ice sheet. The mechanism is, of course, that increases in both snow accumulation rates and in ocean heat fluxes result in lower growth rates or even melting in the lower part of the ice sheet. Figure 8.11 summarizes the processes that Eicken suggests is important in determining the evolution of salinity profiles in antarctic ice sheets.

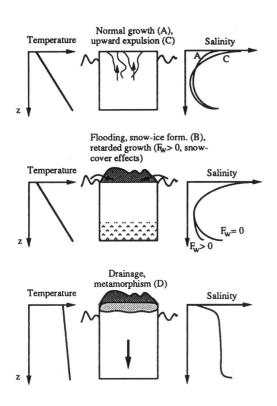

Figure 8.11. A schematic depiction of the processes believed by Eicken (1992) to be important in determining the evolution of salinity profiles within antarctic ice floes.

As was discussed in the chapter on structure, the amount of frazil ice present in antarctic cores is appreciably larger than in comparable samples obtained in the Arctic. Does this cause differences in the salinity profiles from these two very different regions? Eicken (1992) has tentatively explored this important subject, concluding that the evolution of the porosity of sea ice may be largely independent of the details of ice growth. If this is so, then the brine drainage mechanisms that function in congelation ice would also appear adequate to explain variations in the salinity profiles of ice sheets comprised primarily of frazil ice. Let us hope that this is the case, as it certainly makes life simpler. I think that all we can say at the present is that there is no strong evidence that salinity profiles in frazil ice have to be treated differently from those in congelation ice.

In a later paper, Eicken (1998) expanded his study of Southern Ocean sea ice, focusing on ice cores collected in the Weddell Sea. The particular emphasis of this work was to expand capabilities for using ice core data to make quantitative statements about the growth history of the ice sampled. The reason for this focus was, of course, the fact that, particularly in the Antarctic, many sea ice studies have been carried out within the pack ice, where ice samples cannot easily be associated with environmental histories measured at coastal stations. In such cases it would be quite useful to be able to utilize observations on ice structure and composition to infer assessments of regional environmental conditions. Specific observations carried out by Eicken included identification of different ice types (frazil, congelation) and their associated ages (FY and SY), as well as profiles of grain size variations. Measured compositional profiles included salinity and measurements of the stable isotopes $\delta^{18}O$ and deuterium. Here $\delta^{18}O$ refers to the ratio of $^{18}O/^{16}O$ as measured against laboratory standards. It was found that the mean salinities and $\delta^{18}O$ values were 6.0 ± 1.1 and 1.04 ± 1.0‰ for the FY ice, and 4.8 ± 1.4 and 0.40 ± 0.71‰ for the second-year ice. Deviations between ice of different textural characteristics and between averaged cores could largely be explained by differences in the amount of snow and superimposed ice. For instance, in Figure 8.12 note that without exception all congelation (columnar) ice has $\delta^{18}O$ values > 0, while all samples with $\delta^{18}O$ values < 0 have textures that are at least partially granular. It also proved easy to differentiate FY from SY ice on the basis of ice thickness (0.73 vs. 1.55 m, respectively, provided that the ice was undeformed); snow thickness and structure; the presence of clear ice layers on the surface of SY; and the occurrence of an internal chlorophyll maximum which forms

Figure 8.12. A plot of $\delta^{18}O$ vs. salinity for a variety of samples showing different structural types obtained from 31 ice cores collected from level ice in the Weddell Sea (Eicken 1998). The majority of the samples showing $\delta^{18}O$ values < 0 are composed, at least partially, of granular ice. The thin solid line is a least-squares fit to all data points with $\delta^{18}O$ values > -0.28.

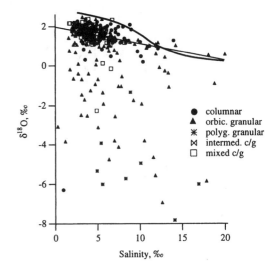

during the summer within the SY ice. Other related papers that use stable isotope compositions and textural variations to study environmental variations during sea ice formation in different regions of the Southern Ocean include those by Jeffries, Shaw, et al. (1994), Jeffries et al. (1997), Jeffries and Weeks (1992b), Jeffries, Li, et al. (1998), Lange et al. (1990), Souchez et al. (1988, 1995), Tison and Haren (1989), Worby and Massom (1995), and Worby et al. (1998). The reader will note that the amount of work involved in such studies is significant.

Now I will return to a consideration of specific mechanisms for removing brine from sea ice.

8.3 Mechanisms

8.3.1 Initial Entrapment

To understand just how the salinity of sea ice changes with time, one must first understand initial entrapment, as this sets the initial salinity values at each new level in a growing ice sheet. The first work on this subject was by Johnson (1943), who carried out simple freezing experiments in which he measured the salinity of the different thin ice layers that resulted. His results suggested a linear relation between the salinity of the ice S_i and the salinity of the water from which the ice formed S_w. On the other hand the growth velocity of the ice v did not appear to have a significant effect on the effective solute distribution coefficient k, where

$$k \equiv \frac{S_i}{S_w} \, . \tag{8.3}$$

This was followed 20 years later by a related study in which Adams et al. (1960) obtained results that seemed to demonstrate quite conclusively that S_i was approximately a linear function of v.

In order to resolve these differences Weeks and Lofgren (1967), and later Cox and Weeks (1975), carried out a number of freezing experiments using simple NaCl solutions as a proxy for seawater. Although these two studies were similar in purpose, the experimental aspects were quite different. In the Weeks and Lofgren study, ice sheets up to 25 cm thick were formed during freezing runs that lasted up to 110 hours. At the completion of each run the ice was removed and cut into a series of 1-cm slices; these were then melted and the salinity of the resulting solution determined. As the freezing runs were short, brine drainage during the runs was neglected. As is clearly shown by the data presented in Figure 8.4, this was not a very realistic assumption in that the sa-

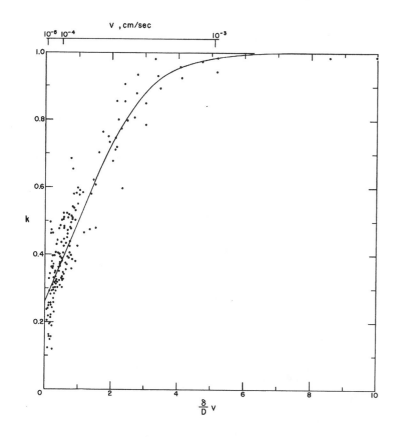

Figure 8.13. A plot of k vs. dv/D based on the salinity entrapment experiments of Weeks and Lofgren (1967). The curve is a least-squares fit of equation 8.8.

linity of sea ice changes rapidly during its initial growth. The salinity profiles were then used to calculate S_w values during the different growth stages. These values were then combined with the S_i values obtained at the same level in the ice in order to calculate the associated k values. One clear result from these experiments was

to verify the Adams et al. (1960) conclusion that the amount of entrapped salt was essentially a linear function of growth velocity.

In order to avoid the obvious problem associated with assuming no brine drainage, Cox and Weeks (1975) later used the radioactive isotope ^{22}Na as a tracer and performed "salinity" measurements by measuring the gamma ray emission from different levels in the ice sample and in the underlying water. As these freezing runs lasted up to 900 hours, brine drainage could clearly no longer be neglected. As a result, only the salinity values from the ice located just above the skeletal layer were used to study initial entrapment. As noted earlier the skeletal layer is the low-strength layer at the bottom of a growing ice sheet, where limited lateral bonding occurs between the ice platelets that comprise the individual sea ice crystals.

In fitting the data it was decided to use a relation initially developed by Burton, Prim, and Slichter (1953), henceforth called BPS theory, to fit similar experimental results in metallurgy. As will be seen, the fit obtained was excellent at all but the very slowest growth rates (Figure 8.13). The development of the BPS theory is as follows. First it is assumed that because of mixing, the concentration C approaches a constant value of C_ℓ at some distance δ ahead of the advancing solid–liquid interface. For distances less than δ the velocity component normal to the interface approaches that due to the crystal growth v. For a steady state at $x < \delta$, the continuity equation is

$$D\frac{d^2C}{dx^2}+v\frac{dC}{dx}=0 \qquad (8.4)$$

and the boundary condition is

$$(C_i - C_s)v + D\frac{dC}{dx} = 0 \quad \text{at } x = 0. \tag{8.5}$$

Here C_i and C_s are the concentrations in the liquid at the interface $(x = 0)$ and in the solid, respectively, and

$$C = C_\ell \text{ at } x \geq \delta.$$

Solutions to equation 8.4 with these boundary conditions are

$$\frac{C - C_s}{C_\ell - C_s} = \exp\left[\frac{v}{D}(\delta - x)\right]. \tag{8.6}$$

The concentration C_i of the liquid at the interface is

$$\frac{C_i - C_s}{C_\ell - C_s} = \exp\left[\frac{\delta v}{D}\right]. \tag{8.7}$$

By choosing δ correctly, this relation can be forced to give the correct value of C_i. Because $C_s/C_i = k_o$, equation 8.7 can be rewritten as

$$k = \frac{k_o}{k_o + (1 - k_o)\exp\left[-\dfrac{\delta v}{D}\right]}. \tag{8.8}$$

Here k_o should be considered to be the value of k at $v = 0$, provided that a cellular interface remains stable. In that $k_o < 1$ for NaCl and seawater systems, equation 8.8 can be rearranged as

$$\ln\left(\frac{1}{k} - 1\right) = \ln\left(\frac{1}{k_o} - 1\right) - \frac{dv}{D}. \tag{8.9}$$

This is a straight line with a slope of $-\delta/D$ and a zero intercept of $\ln(1/k_o - 1)$, if $\ln(1/k - 1)$ is plotted against v. Figure 8.14 is a plot of the Cox and Weeks data presented in this format. The presentation is linear except at very low growth velocities, when the observations gradually drift away from the straight line. Based on Weeks and Lofgren's observations, this drift toward higher $\ln(1/k_o - 1)$ values is undoubtedly related to changes in the morphology of the solid–liquid interface as conditions are approached where a planar interface becomes stable. Ignoring the values at low v, least squares gives $k_o = 0.26$ and $\delta/D = 7243$ s/cm. Note that at the highest growth velocities achieved essentially all the NaCl is entrapped within the ice. These runs used a mixture of dry ice and alcohol as a coolant, which was placed in a copper container in direct physical contact with the upper surface of the water column. The growth rates achieved far exceed maximum values observed in nature. More recent experiments using a slightly different geometry (freezing on a vertical plate) but similar high growth rates report similar high entrapment rates ($k_{eff} = 0.95$; Smedsrud et al. 2003). If growth rates are low, corresponding to temperature gradients in the ice of less than 1.3°C/cm, then the empirical equation

$$k = 0.1 + 0.293\frac{dt}{dz} \tag{8.10}$$

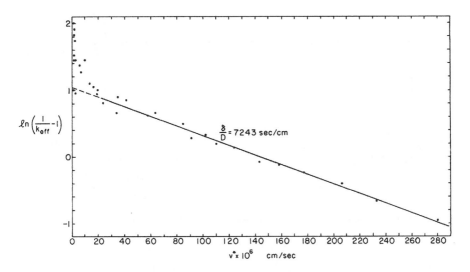

Figure 8.14. A plot of ln(1/k_{eff}–1) vs. v using the salinity entrapment data of Cox and Weeks (1975).

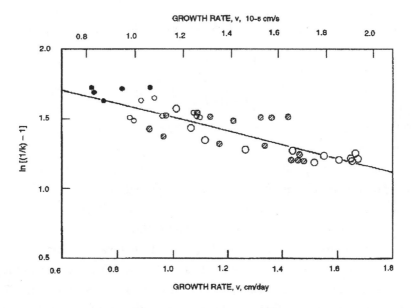

Figure 8.15. A plot of ln(1/k–1) vs. v for columnar-grained sea ice from depths of between 25 and 125 cm, Eclipse Sound, Nunavut, Canada (Nakawo and Sinha 1981). The solid line is a least-squares fit of equation 8.11.

developed by Feruck et al. (1972) might prove useful. To date this relation has not been used in the sea ice literature.

The comparisons of the BPS theory with field observations are also excellent. Figure 8.15 shows a similar ln(1/k − 1) plot using the salinity data from Eclipse Sound that were discussed earlier (Nakawo and Sinha 1981). Again the solid line is a fit of equation 8.9. Finally, Figure 8.16 shows a more straightforward plot of S_i versus v, based on the same data set. Note that although the laboratory and field data show similar trends, real sea ice appears to generally entrap less salt than the NaCl ice by ~2‰. One of several possible reasons for this difference is that because the NaCl salinities were determined by nondestructive testing, there was no opportunity for the brine losses that invariably occur during sampling of real sea ice.

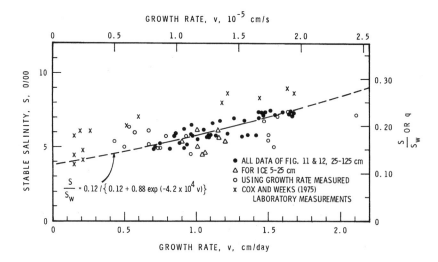

Figure 8.16. Stable salinity, or *k*, vs. growth rate *v* (Nakawo and Sinha 1981). The figure numbers refer to figures in the original paper.

In conclusion, it currently appears that the best way to estimate initial entrapment for normal sea ice is to use

$$S_i = S_w \left[\frac{k^*}{k^* + \left(1 - k^*\right) \exp\left(-\dfrac{\delta v}{D}\right)} \right] \tag{8.11}$$

with $k^* = 0.12$ and $\delta/D = 4.2 \times 10^4$ s/cm based on Nakawo and Sinha's results. This is the aspect of the salinity of sea ice where we have the best predictive capacity.

However, the BPS theory, although obviously practically useful, contains one fundamental flaw. It is based on a series of assumptions that strictly apply only to impurities that are incorporated into the solid phase as solid solutions (where the incorporated ions occupy lattice sites in the ice crystal). As discussed in some detail in Chapter 6, this does not occur in sea ice, where we are dealing instead with bulk entrapment of the melt at a nonplanar cellular interface. However, there are models in the materials science literature that specifically deal with the physical entrapment phenomenon (Edie and Kirwan 1973; Myerson and Kirwan 1977a, 1977b). It is even known that the correlations that they develop can be successfully applied to the freezing of aqueous NaCl solutions (Ozüm and Kirwan 1976). I will not describe the details of this approach here as it has not as yet been specifically applied to sea ice. However, I suggest that this approach would serve as a useful starting point in future investigations of initial entrapment in that it requires an accurate description of the geometry of the growing interface. As I pointed out in sections 7.2.4.3 and 7.2.4.4 in the chapter on structure, our current knowledge of the interface geometry of sea ice is limited. Attempting to expand a physical entrapment model would undoubtedly result in further understanding of the geometric changes in the growing interface with growth velocity and composition. Another possibility might be to apply the model developed by Ostrogorsky and Müller (1992) to sea ice, as this model considers the relation between the effective distribution coefficient and controlling factors such as the actual thickness of the solute layer, the convective velocity in the solute boundary layer, the equilibrium distribution coefficient, the characteristic length of the growth interface, and the macroscopic growth velocity.

8.3.2 Brine Pocket Migration

The first mechanism suggested for removing brine from sea ice was brine pocket migration or, as it is now referred to in the materials science literature, temperature gradient zone melting (TGZM; Pfann 1958; Tiller 1963; Wernick 1956). When in 1926 Whitman considered possible brine removal mechanisms, he realized that if a vertical temperature gradient existed in the ice, it would establish concentration gradients within the brine pockets if phase equilibrium was maintained. In the winter this would result in the cold upper end of each brine pocket being more saline than the warm lower end. This concentration gradient would, in turn, cause the diffusion of salt from the top of the brine pocket to the bottom. As a result, ice would form in the top of the brine pocket and dissolve in the bottom, causing the brine pocket to migrate downward toward the bottom of the ice.

The theory for this process is as follows. First note that for salt–water systems the equilibrium value of k (i.e., $k_o \sim 10^{-4}$) is very small. This means that for the migration of a small inclusion such as a brine pocket, all the salts can be considered to be excluded from the ice phase. As a result the flux of solute at the freezing interface of a brine pocket is

$$J = Cv(1-k) \approx Cv.$$ (8.12)

Therefore, for the steady state, equation 8.12 can be written in terms of the concentration gradient dC/dx as

$$Cv = -D\frac{dC}{dx}$$ (8.13)

where D is the diffusion coefficient of the salt (or salts) in water. Then substituting

$$\frac{dC}{dx} = \frac{dC}{dT} \bullet \frac{dT}{dx}$$ (8.14)

where T is the temperature and (dC/dT) is specified by the phase diagram, it follows that

$$v = -\frac{D}{C}\left[\frac{dC}{dT} \bullet \frac{dT}{dx}\right].$$ (8.15)

In this equation the appropriate temperature gradient is the gradient in the liquid G_ℓ. For a spherical droplet the relation between G_ℓ and the temperature gradient in the solid G_s is specified by the solution to Laplace's equation

$$\frac{G_\ell}{G_s} = \frac{3\kappa_s}{(2\kappa_s + \kappa_\ell)}$$ (8.16)

where κ is the thermal conductivity. In that κ_{ice} is approximately 4 times κ_{brine}, the gradient ratio is fairly insensitive to small changes in κ_{brine} resulting from changes in the composition of the brine. In addition, as the values of κ change slowly with temperature, the G_ℓ/G_s ratio is roughly constant. By substituting appropriate thermal conductivity values Seidensticker (1965) obtained $G_\ell = 1.34 G_s$. When he also added an additional correction resulting from the density (r) differences between the ice and the brine, he obtained

$$v=\left(\frac{3\kappa_s}{2\kappa_s+\kappa_\ell}\right)\left(\frac{\rho_\ell}{\rho_s}\right)\left(\frac{D}{mC}\right)=1.46\frac{D}{mC} \tag{8.17}$$

as the relation giving the brine pocket migration velocity in a unit thermal gradient. Here $m = dT/dC$ is the slope of the liquidus curve as specified by the phase diagram. The units of C do not matter as long as m and C are in comparable units. A more general derivation of equation 8.17 can be found in Shreve (1967).

Fortunately, the predictions of this equation can be checked as there have been several studies of the migration of artificially produced brine inclusions through ice crystals (Harrison 1965; Hoekstra et al. 1965; Jones 1973, 1974; Kingery and Goodnow 1963). Figure 8.17 shows the relation between predictions of equation 8.17 and the migration velocity as observed at different ice temperatures at a constant temperature gradient of 1°C/cm by Hoekstra et al. (1965). The general shapes of the experimental and theoretical curves are in good agreement although the theory predicts velocities that are ~20% higher than observed. A very recent discussion of these results plus additional migration rate determinations can be found in Light et al. (2009). Figure 8.18 shows the relation between the temperature gradient in the ice and the migration rate as observed by Jones

Figure 8.17. Relation between the migration velocity of spherical brine pockets and ice temperature at a constant temperature gradient of 1°C/cm. The curves are calculated using equation 8.19. Note that the solidly frozen KCl pockets migrate at an appreciable speed (see Hoekstra et al. 1965; Seidensticker 1965; Shreve 1967).

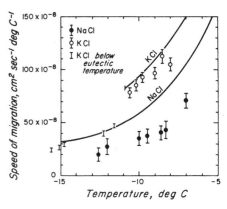

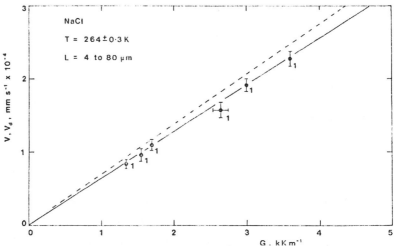

Figure 8.18. A plot of the observed brine pocket migration rate vs. the temperature gradient in the ice (Jones 1973). Each brace gives the mean and the mean deviation of the velocities of about 10 separate droplets. The dashed line indicates to an accuracy of ±10% the maximum diffusion limited values of v.

(1973). That v is a linear function of the temperature gradient in the ice, G_s, is quite clear. The experimental data clearly indicate that the rate of transport of the solute through the liquid is a major factor in controlling the migration rate. Jones also suggested that the presence of crystal defects in the ice ahead of the high temperature interface is also a factor affecting the migration rate.

Unfortunately, there is one main problem in applying these results to sea ice. This is the simple fact that both the theoretical and experimentally observed migration rates are extremely slow. As an example of an extreme case consider a 30°C temperature difference across a 1-m-thick piece of FY ice resulting in a temperature gradient of 30°/m. If the temperature at the brine pocket is –6°C, equation 8.17 gives a migration rate of 14 μm/hr, which corresponds to roughly a migration of 1 cm/month. Untersteiner (1968) has calculated that if more typical values for the temperature gradient in MY ice are used, a brine pocket located between 10 and 20 cm below the upper surface of an ice floe would migrate downward 2 cm between August and April and then migrate almost the same distance upward between May and July for an annual net travel of essentially zero.

It should be noted that all the brine pockets that have been studied have been very small (4 to 80 μm). In such small inclusions there is little chance of convective motions occurring in the brine. However, in larger inclusions with diameters > 1 mm, convection may become possible. It has also been argued that thermal convection should always take place in tilted brine pockets (Woods and Linz 1992). Convection is, of course, a much more efficient process that is associated with an increase in the "effective" diffusion coefficient to values of $\sim 10^{-3}$, as contrasted to 10^{-6} for pure diffusion. Migration rates would be accordingly enhanced. It is also important to note that a convection-driven process is essentially one-way, in that it will only occur when the temperature increases downward (during the winter). During the summer, when the temperature is reversed, the salt that is produced at the bottom of a brine pocket causes a stable density distribution in the liquid, turning the convection off. If convection plays a significant role in natural brine pocket migration, it presumably will be most effective in the lower portions of an ice sheet as the result of the increase in brine pocket size as the temperature approaches the melting point. One must conclude that although brine-pocket migration is a real process that does occur in sea ice, there is no indication to date that it is a significant contributor to the large salinity changes observed in nature.

8.3.3 Brine Expulsion

In early studies of both FY and MY ice by Knight (1962a) and Bennington (1963a), it was noted that if the liquid portion of a brine pocket separated from the vapor bubble, cooling with its associated ice formation would result in a pressure buildup in the pocket. If this buildup was sufficient, it could result in the failure of the surrounding ice, typically along vertically oriented (0001) planes. This would allow the brine to escape from the pocket and migrate toward the warm side of the ice sheet. This general process has been called *brine expulsion*. The contribution of brine expulsion to the overall removal of brine from sea ice has been considered by several authors. The initial model of this process, which was developed by Untersteiner (1968a), gave S, the decrease in the bulk salinity at a given level in an ice sheet, as

$$S = S_o \left(T_o / T \right)^{\Delta\rho/1-\Delta\rho} \tag{8.18}$$

Figure 8.19. A comparison between experimental salinity curves and theoretical salinity curves determined from a brine expulsion model. R2–3 and R2–6 are the initial and final observed curves (Cox and Weeks 1975). The other two curves were calculated. The dashed curve considers the effect of brine velocity; the dotted curve does not.

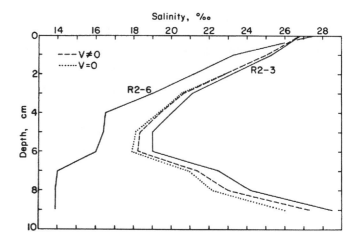

where $\Delta\rho \approx 0.1$ is the difference in density between water and ice. This relation only holds when the temperature is continuously dropping from T_0 to T. As a result the salinity of the ice at any time > 0 will depend upon the initial salinity entrapment as given by equation 8.9 and on all subsequent periods of cooling that the ice has experienced. When this equation is used to estimate the salinity profile of a MY floe, it gives near-surface salinities of a few tenths of a part per thousand, as observed. However, below the near-surface region the salinity profile is slightly concave upwards in all but the lowest portion of the ice. This is a characteristic that is not observed naturally. A more involved model developed by Cox and Weeks (1975) has been used in their studies of the salinity profiles in NaCl ice, where it was applied via the use of a finite difference scheme. Some of their results are shown in Figure 8.19, where the solid curves give the initial and final observed salinity profiles and the broken curves are calculated from the model. They conclude that brine expulsion plays only a small role in the desalination of FY ice. However, its effect is large enough that it cannot be neglected. This is particularly true during periods of rapid ice growth, when the rate of change of temperature at each level in the ice sheet is large. Somewhat later Cox and Weeks (1986) derived equations to predict the amount of brine expelled from sea ice during sampling and storage. It is these relations that have been used in estimating brine expulsion in numerical simulations.

Brine expulsion is clearly a real process and it is not difficult to apply the above equations. However, it is not simple to know what the results of such calculations tell us, in that one must now consider the movement of brine within interconnected channels within the ice. Consider a multilayered model in which the effects of brine expulsion are calculated during a cooling cycle. Depending on the amount of cooling, which of course will vary depending on the exact vertical location in the ice sheet, each layer will expulse a different amount of brine. Where does this brine go? It has typically been assumed that it exits immediately from the complete ice sheet. The reason why this may not be too outrageous an assumption lies in the existence of brine drainage tubes, which operate via a totally different mechanism and have the capability of removing brine rapidly from the ice sheet. All that brine expulsion must do is to move the brine to the nearest drainage tube, and efficient processes within the tube will take over. More will be said about this later. However, perhaps the brine stays within the ice and just moves to the next-lower (warmer) layer.

We do not even know if the brine always moves in the warmer direction. For instance, the surface layer of rapidly growing (and cooling) nilas characteristically shows very high salinities far in excess of the salinity of the underlying seawater. I used to think that this might be the result of brine expulsion forcing the expulsed brine upward to the ice–air interface. Why might the brine move upward instead of downward? A possibility could be that it moves in the direction of the greater permeability, which would also be in the direction of the increasing brine volume. Although in most ice sheets the brine volume increases downward, in very thin ice types such as nilas very high near-surface salinities are common and could counter the general trend. However, these speculations do not appear supported by the field observations of Ono and Kasai (1985), who studied surface layer salinities of very young sea ice grown in a pool cut in the natural ice cover of Saroma Lagoon, located on the north coast of Hokkaido. They observed that when there was no snow on the ice, the 1-mm-thick liquid surface layer had salinities of up to 42.4‰, far exceeding the salinity of the underlying seawater (31‰). They also observed that the salinity of the surface layer decreases when the air temperature drops, and increases when the air temperature rises. This is just the opposite of what would be expected if brine expulsion were an important factor. One should also keep in mind the fact that when mean ice sheet salinity values are plotted against ice thickness (Figure 8.4), there is a pronounced change in slope that occurs in the thickness range of 0.2 to 0.3 m. Possibly this change reflects the decreasing effectiveness of brine expulsion as temperature gradients decrease as the ice thickens. I mention these considerations to emphasize that there is undoubtedly more work ahead before the possible role of brine expulsion in determining the salinity profile of sea ice can be adequately evaluated.

Figure 8.20. Brine drainage channels as observed at Barrow by Cole and Shapiro (1998). The markings on the ice are at spacings of 10 cm.

8.3.4 Gravity Drainage

The term *gravity drainage* as defined by Eide and Martin (1975) refers to all processes in which the greater weight of the brine relative to seawater drives the brine down and out of the ice and into the underlying seawater. Any such mechanism is similar to brine expulsion and different from brine pocket migration, in the sense that it requires interconnected tubes and channels through which brine can drain. The existence of such features was first noted by Knight (1962a) and Bennington (1963a), and was described in some detail by Lake and Lewis (1970) who referred to these features as brine drainage tubes. Although these features are frequently difficult to see, good photographs can be obtained by using thick vertical slabs of ice specially cut for this purpose (Cole and Shapiro 1998). The tubes can be quite large, with lengths only slightly less than the total thickness of the ice sheet; see for instance Figure 8.20. They typically are vertical structures that are attended by smaller tributary tubes, much as in a vertically oriented, radially symmetric river system.

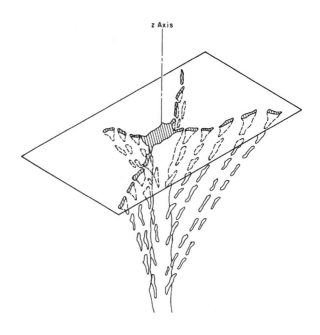

Figure 8.21. Schematic drawing of a cut through a brine drainage channel (Lake and Lewis 1970).

Figure 8.21 shows a schematic drawing of a cut through such a drainage feature as observed by Lake and Lewis, who found that representative tube diameters at the bottom of a 1.55-m-thick ice sheet were 0.5 cm and that there was one large channel every 180 cm². They also noted that the large channels showed starburst patterns in the horizontal plane, with the arms of the stars generally following the crystal boundaries. Figure 8.22 shows a photograph of a typical brine drainage channel as viewed in the horizontal plane. Although channels can occur both within and between sea ice crystals, they most commonly appear to occur in the boundary areas between two or more crystals. It has also been found that in ice thinner than 10 cm the number of channels is related to the average growth rate but not to the average grain size as measured in the horizontal plane (Saito and Ono 1980). They also observed the more frequent occurrence of channels (one channel per 33 cm²) than noted in thicker ice, again a similarity with river systems in which frequent smaller streams merge, forming more widely spaced master streams.

As the cold, dense, and very saline brine flows down the brine drainage tubes from the colder, upper part of the ice sheet to the warmer, lower portion of the sheet, it is warmed by the lateral flow of heat from the ice to the brine channel. This warms the brine above its phase equilibrium temperature. To reestablish the equilibrium or Nernst condition at the new lower level, the channel wall melts, cooling the ice and diluting the brine. As a result, the movement of brine downward through the ice causes both the cooling of the ice adjacent to the channel and an increase in the diameter of the channel. Furthermore, the increase in the lateral temperature gradients around the tubes assists in drawing brine into the tubes. The effects of this process were clearly shown in the recent study by Cottier et al. (1999) of lateral salinity distributions in relation to the location of tubes. These investigators observed that when the ice was cold, areas of higher salinity corresponded directly to the positions of the brine drainage channels and that these areas were surrounded directly by regions that showed brine depletion. When the ice sheets were warmed, thereby removing the driving force producing brine drainage down the tubes, these trends disappeared and the ice salinity distribution became laterally homogeneous.

There are also distinct differences between river systems and brine drainage channels, in addition to the differences in symmetry. For instance, river channels usually continuously widen as the mouth of the river is approached, whereas brine channels neck just above the ice–water interface. Even more surprising is the fact that rivers show a continuous downstream flow whereas brine channels show an oscillating flow with a periodic emptying and filling. Observations

Figure 8.22. Photograph of a typical brine channel as viewed in the horizontal plane, Barrow, Alaska (photo provided by David Cole).

suggest a periodicity of roughly one hour consisting of an 8- to 15-minute inflow of seawater followed by a 45-minute outflow of brine (Martin 1970). It turns out that in brine drainage channels the necking of the channel and the oscillating flow are related.

There are at least two factors at work here. First, not only is the seawater at the base of the channel at its freezing point, it is also less dense than the brine flowing from the ice. This density difference $\Delta\rho$ creates a buoyancy force $\Delta\rho g$ that acts upward and opposes the pressure gradient force F_p that drives the brine out of the ice. Martin (1974) calculated the value of F_p by assuming that the volume flux q in the neck of a brine drainage tube is imposed by the characteristics of the upper portion of the drainage channel and that the flow through the neck of the channel is Poiseuille flow. This results in

$$F_p = \frac{8\rho_b v_b q}{\pi a^4} \qquad (8.19)$$

where a is the radius of the channel neck and v_b is the brine volume. If a is sufficiently small then $F_p > \Delta\rho g$ and brine flows downward out of the neck. If the neck enlarges, then $F_p > \Delta\rho g$ and the more buoyant seawater can flow upward into the drainage tube. When this occurs, as a result of the fact that the seawater is exactly at the freezing point and the ice is colder than the seawater, ice forms on the inside of the neck, reducing the value of a. Martin then considered the ratio of these two forces which he refers to as the entrainment number E

$$E = \frac{\pi\,\Delta\rho\,g\,a^4}{8\,\rho_b\,v_b\,q} = \frac{buoyancy\ force}{pressure\ gradient\ force}. \qquad (8.20)$$

For sea ice the sequence of adjustments in the tip radius is as follows. Initially the brine flows uniformly out of the tip. This causes the value of a to increase so that $E > 1$. As a result, flow decreases until seawater begins to flow upward into the neck where it freezes on to the walls. This, in turn, causes a to decrease reducing the value of E to a subcritical value, at which time the whole process starts over again. When appropriate values are substituted into equation

8.20, a value of $a \sim 0.3$ mm is obtained, indicating that in the limit of $q = 0$, a channel should remain open.

For very slow ($q \rightarrow 0$) flows the diffusion of salt becomes important and the above criterion becomes invalid. In this case Lake and Lewis (1970) have argued that a more useful criterion to determine the minimum tube radius is the Rayleigh number criterion given by

$$\frac{g \dfrac{1}{\rho} \dfrac{\partial \rho}{\partial z} a^4}{D \mu} \approx 68 . \tag{8.21}$$

Here $\partial \rho / \partial z$ is the vertical density gradient within a drainage tube resulting from the temperature gradient plus the Nernst condition, μ is the viscosity, and D is the salt diffusivity.

The reason for the oscillations that occur in the fluid flow in and out of brine drainage channels is that there are two positions of hydrostatic equilibrium. When cold, dense, saline brine from the upper levels of the ice fills a brine channel, the equilibrium level is lower than when warmer, less saline seawater fills the channel. When the tip radius has increased until $E > 1$ then any small mass perturbation in the tube results in a large pressure imbalance, which accelerates seawater up the tube until the second, higher equilibrium is reached. While this is occurring ice growth associated with the influx of seawater is decreasing a, resulting in $E < 1$. Once the seawater has been cooled and reduced to brine, the process repeats.

A physical model for gravity drainage has been developed by Eide and Martin (1975), starting with earlier work on the formation of drainage tubes in temperate glaciers by Röthlisberger (1972) and Shreve (1972). In their treatment a linear temperature gradient is assumed to hold in the ice and the density of the brine within the ice is taken to be a linear function of temperature. They show that as the ice thickness increases, the equilibrium brine level relative to the free surface decreases. As a result, as the ice thickens the pressure gradient force drives the cold brine located in the smaller intergranular tubes in the upper part of the ice down into warmer ice. They also examine the thermal effects of this process, which has two components. First, there is the heat flux caused by the temperature difference F_t between the cold brine in the tube and its surroundings. There is also a flux F_s resulting from the melting of the wall of the tube necessary to maintain the Nernst condition, that the composition of the brine remain on the eutectic curve of the phase diagram. Their analysis indicates that melting resulting from the eutectic boundary condition results in a much larger heat flux by a factor of ~20 than that resulting from the warming of the cold brine.

The work of Martin (1974) and of Eide and Martin (1975) suggests that other aspects of brine channel formation that could assist in the desalination of sea ice are

1. increases in the diameter of brine channels, resulting in enhanced entrainment of brine pockets and reduced viscous drag on downward-flowing brine,
2. decreases in pressure associated with channel formation, facilitating the flow of brine into channels, and
3. increases in the local temperature gradient within the ice that surrounds channels, resulting in the expulsion of brine into the channel (Martin 1972, 1974).

In his study of different aspects of brine drainage, Wakatsuchi (1983) and Wakatsuchi and Ono (1983) attempted to obtain samples of brine as it leaves the base of a growing ice sheet. This is not an easy task; refer to the original paper for a description of the procedures used and the related experimental problems. Starting with seawater with a salinity of 33.0‰ and obtaining ice growth rates between 1.7×10^{-5} and 1.4×10^{-4} cm s^{-1}, brine salinities ranged between 42.3 and 92.7‰, and salt fluxes varied between 6.4×10^{-7} and 1.5×10^{-6} g cm^{-2} s^{-1}. He found that the total volume of brine excluded during a single growth cycle increases with an increase in both growth rate and the total growth time. Therefore, the longer it takes an ice sheet to reach a given thickness, the lower its salinity. Conversely, thicker sea ice grown during a fixed period of time will have a higher salinity in that a smaller amount of brine with a lower salinity has been excluded. These results indicate that the primary effect controlling both the salinity and the volume of brine excluded during the formation of a given thickness of ice is the time of year when ice growth is initiated. For instance during the fall, when air temperatures are relatively warm, the time required to reach a specified thickness is appreciably longer than during the dead of winter when air temperatures are much lower. This result is also supported by the model calculations of Cox and Weeks (1988a).

Observations of the details of brine drainage from laboratory-grown NaCl ice have also been carried out by Wakatsuchi (1983) via the use of a schlieren optical system. For very thin ice he observed a large number of fine streamers falling in bunches. As growth slowed, the streamers became thicker and fewer. He then attempted to obtain a volume flux for the brine leaving the ice by measuring the number of individual filaments and the associated falling velocity. He concluded that at higher growth rates, the volume flux of an individual filament is less as it has both a smaller diameter and a slower falling velocity. Nevertheless the volume flux of brine out of a unit area of ice is somewhat higher at higher growth rates as the result of the significantly larger number of brine filaments. Figure 8.23 shows a plot of Wakatsuchi's estimated salt flux values both as determined by brine collection and as calculated from the bulk salinity of the ice versus the growth rate of the ice.

The only experimental work to date that has attempted to isolate brine drainage as a function of parameters such as temperature gradient (G_{si}) and brine volume (v) is the study by Cox and Weeks (1975) that used radioactive ^{22}Na as a tracer. In it, they presented linear plots of the rate of change of salinity versus G_{si} and v. Recently a log-log reanalysis of this data has

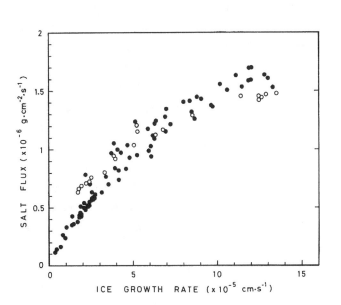

Figure 8.23. A comparison of salt flux values obtained by brine collection (open circles) and values estimated from sea ice salinity measurements (black circles) (Wakatsuchi 1983).

Figure 8.24. The gravity drainage data of Cox and Weeks (1975; their figures 30 and 31) as reanalyzed by Petrich et al. (2006). Temperature gradients in °K m⁻¹: (•)10 to 60; (○) 60 to 120; (×) 120 to 180. The broken line represents the linear fit of Cox and Weeks (1988a) with $f_c = 0.050$ and the solid line represents the fitted power law of Petrich et al. with $f_c = 0.054$ and $g = 1.2$.

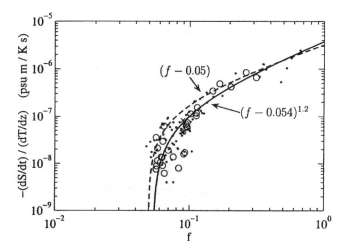

been undertaken by Petrich et al. (2006). The results are shown in Figure 8.24. It appears that the best fit to the data can be expressed as

$$\frac{DS_{si}}{Dt} = -A(f_t - f_c)^\gamma \frac{DT}{Dz} \tag{8.22}$$

where $A = 4.2 \times 10^{-6}$ ‰ m s⁻¹ K⁻¹, $f_c = 0.054$, $\gamma = 1.2$, $f_t > f_c$, and $\Delta T/\Delta z > 0$. Here f_t is the total brine porosity and f_c and γ are constants.

8.3.5 Stalactites

From the above discussion it should be clear that most of the brine that exits a sheet of sea ice does so from the mouths of the brine drainage channels once they have developed. Therefore, on a small-scale brine, sources are not reasonably evenly distributed at the ice–ocean interface, as would be the case if all the brine drainage uniformly occurred in the grooves between the individual platelets that make up the crystals of sea ice. Instead, the brine drainage channels serve as concentrated localized sources of brine in which the distance between channels is large in comparison to the diameter of the channel (15 cm/0.4 cm ≈ 37.5). In addition, during the ice growth season, the brine exiting a channel is both quite saline and cold, in that its condition can be assumed to be roughly an average salinity and temperature representative of the vertical temperature and salinity profiles through the overlying ice sheet. Considering that during the ice growth season the seawater at the ice–ocean interface is at its freezing point relative to its salinity, this localized brine drainage results in the formation of stalactitelike structures of ice that can extend several meters below the lower surface of the ice. Although stalactites would appear to be reasonably common, they were apparently not noticed until the late 1950s by divers performing biological studies on Drift Station Alpha in the Beaufort Sea. The first detailed published descriptions were from McMurdo Sound, Antarctica, an area where stalactites are particularly well developed (Dayton and Martin 1971; Paige 1970). There are several reasons for this. First, to observe the stalactites one must either dive under the ice or operate a remote system that provides imagery of the state of the ice–ocean interface. Secondly, stalactites are typically quite fragile, meaning that they would not survive under conditions of strong currents or remain attached to the bottom of the sheet during normal sampling procedures. Note that Lake and Lewis (1970) do not mention stalactites in their initial description of brine drainage channels, although in hindsight it is easy to believe that they were present on the bottom of the Cambridge Bay ice. I also must admit that in the period between 1955 and 1970, I did not note their presence on the bottom of the numerous sea ice samples I collected. One particular problem is associated with the manner in which large ice blocks, which quite possibly could have stalactites extending from their lower surface, are removed from the ice sheet. As they are very heavy, they are usually tilted and slid onto the top of the ice sheet, ultimately resting on their lower surface. It would be hard to design a more effective procedure for destroying entities as fragile as stalactites.

Paige (1970), who first referred to the tubes of ice as stalactites, reported lengths of 0.1 to 1 m, and Dayton and Martin (1971) observed and presented photographs of even larger features with lengths extending to 6 m. Lengths of 0.2 to 0.5 m appear to be more representative. Although difficult to observe in the field, stalactites are comparatively easy to grow in the laboratory by simply injecting cold NaCl brine downward at a constant flow rate from a glass tube into a salt solution. A particularly interesting field study of stalactite development was carried out during the LeadEx experiment (Perovich et al. 1995). While studying the refreezing of newly formed leads, it was observed that rapid stalactite development was associated with the rafting of thin (~7 cm) newly formed lead ice over older thicker (~90 cm) FY ice at the edge of the lead. About eight hours after the rafting event, an array of large stalactites was noted growing under the lead ice in a line parallel to the thick ice. The stalactites were placed approximately 5 m apart along the line, and their maximum length and diameter were 2.0 and 0.1 m, respectively. At the same time, at locations removed from the edge of the lead stalactite development was much less impressive. The authors suggest that the source of the cold, saline brine necessary for the enhanced stalactite development was the portion of the lead ice that was rafted over the thicker FY ice. This would favor brine drainage by more rapid cooling, by increasing the effective freeboard by placing the lead ice on top of the FY ice, and possibly by shearing off the skeletal layer during the rafting event.

In his classic paper on this subject Martin (1974) both presents a number of photographs of stalactites and also develops a theory for their formation. The following description is based on his work. Figure 8.25 shows a schematic diagram of a developing stalactite during two stages of its growth. As the cold brine flows downward out of the brine drainage tube, a stalactite of ice starts to grow down from the base of the ice sheet. Once this has occurred, horizontal transfer of heat between the brine within the stalactite and the seawater outside of the stalactite occurs through the wall of the tube. However, the transfer of salt through the lateral walls of the tube is either totally restricted or, at the least, significantly reduced. As a result the brine at a given level within the tube is warmer than when it left the actual ice sheet but is still at the same high salinity. For phase equilibrium to be maintained the brine within the tube must be both cooled and diluted back to the composition specified by the eutectic curve. This occurs through the melting of the inner wall of the tube, with the heat of melting coming from ice accretion on the outer wall of the tube. The result is a stalactite that both accretes on the outside and ablates on the inside as it lengthens. This is exactly the same process that occurs within brine drainage channels. The difference is that the stalactite is surrounded by seawater whereas the brine drainage tube within the ice sheet is primarily surrounded by ice. Therefore, ice growth occurs on the outer walls of the stalactites,

Figure 8.25. A schematic diagram of a stalactite of ice growing from the base of a sheet of sea ice (Martin 1974).

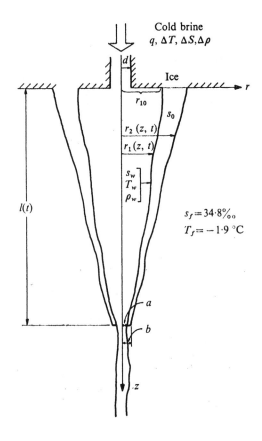

resulting in increased wall thickness, whereas the walls of the drainage tubes primarily cool, increasing the local lateral temperature gradient. The only ice growth is limited to changes in the small amounts of brine remaining in the ice. Another minor difference between drainage channels with or without attached stalactites is that if there is no stalactite the changes in the value of the tube diameter a controlling the flow in and out of the ice will occur near the mouth of the channel at the ice–ocean interface. If a stalactite is present these changes will presumably occur at the mouth of the stalactite.

8.3.6 Flooding and Flushing

There are two different aspects to the flushing phenomenon. The first occurs when the depression of the ice sheet, typically by snow loading, causes brine that is within the ice to be displaced upward until it produces flooding on the surface of the ice sheet. As was mentioned earlier, this is a common event in regions such as the subarctic and the offshore Antarctic, where it is favored by thin ice covers and heavy snowfalls. The only detailed investigation of the upward flushing of water through a sea ice cover is that of Hudier et al. (1995), carried out in Saroma-ko Lagoon on the north coast of Hokkaido. The site is well protected and typically develops 30 to 50 cm of sea ice during the winter. The experiment consisted of removing a 1×1 m piece of 40-cm-thick sea ice from the fast ice and emplacing on its underside a $60 \times 50 \times 40$ cm box that contained temperature sensors and conductivity probes at 2.5, 10, and 25 cm below the lower ice surface. A rubber gasket sealed the contact between the box and the ice to prevent interactions with the surrounding seawater. Then the ice block with its attached box was replaced in its original position in the ice sheet. Once a day, side panels on the box were removed and the interior of the box flushed until thermal and compositional equilibrium with the surrounding seawater was achieved. Additional temperature sensors were also placed both within and under the ice.

During the study period a storm deposited 25 cm of snow on the ice, resulting in the formation of a 10-cm layer of salty slush on the upper ice surface. The authors were able to document that the brine that produced this surface flooding passed through the ice, as opposed to traveling through cracks or saw holes, by observing that areas far from existing cracks were flooded and that shallow near-surface holes drilled in the sheet rapidly filled with water even at locations where surface flooding had been removed. Analysis of the resulting data revealed an initial rapid spikelike increase in the near-surface ice temperature from a preflooding value of –1.7°C up to

−1.47°C. This increase is believed to be the result of the upward movement of seawater in that this was the only heat source in the system. This was followed by a gradual (five-hour) decrease in the ice temperature back to the original temperature.

The most detailed look to date at the upward migration of brine through an ice sheet resulting in the production of infiltrated snow ice is by Maksym and Jeffries (2000). Their interest in the subject was piqued by Jeffries' extensive work on the ice pack of the Southern Ocean, where snow ice formation appears to be as important in thickening the ice cover as is the growth of congelation and pancake ice (Jeffries and Adolphs 1997; Jeffries, Hurst-Cushing, et al. 1998; Jeffries, Li, et al. 1998; Jeffries et al. 1997). That there are several ways by which surface flooding can occur (infiltration from the edges of floes, passage through poorly consolidated areas of brash and ridged ice, migration through cracks, and finally percolation through the ice sheet itself) has been pointed out by several different investigators (Ackley 1986; Ackley and Sullivan 1994; Massom et al. 1998). Although presumably all of these mechanisms contribute to surface flooding and their importance varies from site to site and from season to season, it is generally held that percolation through the sheet is, at least, an important mechanism and probably the dominent mechanism. Based on the material that has already been discussed in this chapter, it is clear that natural sea ice must be permeable much of the time. Otherwise the general decrease in the salinity of FY ice during the winter would not occur unless brine pocket migration and exclusion were significantly more effective processes than currently believed. A better phrasing of this question would be, Is natural sea ice always permeable? This query has been considered in some detail by Golden et al. (1998), who examined existing laboratory and field studies on both natural and artificial sea ice as well as the status of percolation theory as applied to other materials. Here percolation theory is a formalism that has been developed to analyze the properties of materials where the connectedness of a given component, in this case brine, determines the bulk behavior. They conclude that sea ice is not always permeable and that there is a critical brine volume fraction $V_C \approx 5\%$ such that if $V < V_C$, columnar sea ice becomes effectively impermeable. They then suggest that the relation between brine volume and temperature T and salinity S (Frankenstein and Garner 1967) implies that V_C corresponds to a critical temperature $T_c \approx -5°C$ for $S = 5‰$. They initially dubbed this the "Law of Fives." I personally feel that that although this result is correct and certainly useful, it can hardly be considered a law. My reasoning is as follows: it is neither T nor S per se that are important but the parameter controlled by these two factors, V_C, the critical brine volume fraction. However, a given value of V_C can be obtained by a variety of T and S combinations as shown by the Frankenstein and Garner equations. Perhaps the expression the "Rule of Fives" would be more apt. Clearly, zeroing in on the critical conditions for impermeability is important. Another important result from this study is that fluid permeability variations in columnar sea ice, as determined by Ono and Kasai (1985), show temperature trends similar to those observed in compressed plastic/nickel powders. More will be said about percolation theory later in this chapter.

One might well ask, in that the permeability of sea ice is so important, why doesn't someone simply measure it in that permeability measurements on porous materials can easily be made? However, in most materials where permeability measurements are made, the matrix material does not react with the fluid passing through it. This is definitely not the case with sea ice, where slight differences between the temperature of the ice matrix and of the temperature and salinity of

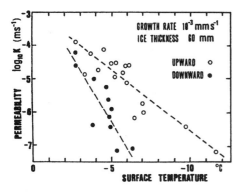

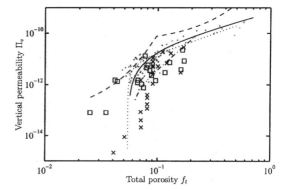

Figure 8.26. The dependence of brine permeability of sea ice on surface temperature (Ono and Kasai 1985). The open circles represent upward permeability and the solid circles represent downward permeability.

Figure 8.27. Vertical permeability as a function of the total porosity assumed to be equivalent to the brine volume (Petrich et al. 2006). Data from seven different sources are included.

brine can result in either the addition or the subtraction of ice from the matrix during the experimental procedure. Current laboratory results obtained by Ono and Kasai (1985) on a very thin (6-cm) sheet of sea ice are somewhat surprising in that the upward permeability was appreciably greater than the downward permeability, with the difference between the two values increasing as the surface temperature of the sheet was lowered; see Figure 8.26. I would have guessed that if there had been a difference in permeability with direction, it would have been the reverse of the observed trend. A somewhat different approach to this problem has been utilized by Saito and Ono (1978), who froze NaCl solutions and then used kerosene to determine the permeability. They found that NaCl ice grown from NaCl solutions with compositions lower than 0.2 mol/ liter is impervious. Maksym and Jeffries (2000) present a plot of all available data bearing on the correlation between permeability and brine volume (their Fig. 3). The scatter is large with one obvious trend: as brine volume increases permeability also increases. In fitting the data, they chose the Kozeny–Carman equation

$$K = \frac{d^2}{180} \frac{v_b^3}{\left(1 - v_b^2\right)} \tag{8.23}$$

where d is a representative grain size. As the authors point out, this relation is of doubtful representation for columnar ice, although it may be suitable for granular materials such as frazil or snow.

A more recent approach to obtaining useful values for the vertical permeability can be found in Petrich et al. (2006), who numerically obtained estimates for the isotropic permeability Π that was assumed to have the form $\Pi \propto (f_t - f_c)^\gamma$ (see equation 8.22). The proportionality constant ($\propto$) was then obtained via the use of a two-dimensional computational fluid dynamics model so that realistic salinity profiles resulted. The vertical permeability was estimated to be $\Pi_v = 7 \times 10^{-10}$ m $(f_t - 0.054)^{1.2}$ assuming an anisotropy factor of $\kappa = 10$. Figure 8.27 presents the vertical permeability $= f$ (total porosity; assumed to equal the total brine volume) resulting from this work using the field data of Nakawo and Sinha (1981). Also included are data from several other authors.

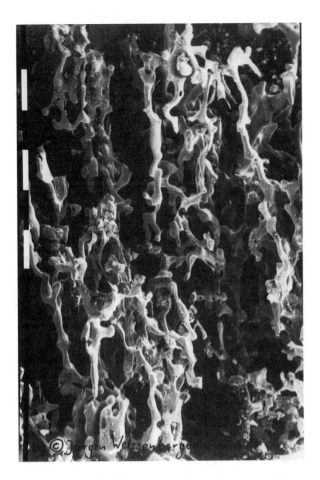

Figure 8.28. A photograph of a plastic cast of the interconnected pore space in a sea ice sample (Weissenberger et al. 1992).

I note that recent work has provided us with techniques for obtaining casts of brine channels (Figure 8.28, Weissenberger et al. 1992) and also for determining the geometry of pore microstructure via the use of magnetic resonance techniques (Eicken et al. 2000). Although it is my sense that we still do not quite know how to extract useful numerical parameters from these observations, these papers are clearly probes in the right direction. What is still lacking is a comprehensive data set that combines pore microstructural information in terms of quantitative stereometric parameters with brine volume and permeability measurements. Recent efforts attempting to tie microstructural data together with percolation theory are described in Golden et al. (2007). Once all this is worked out, we should know if, for given structural types such as frazil and congelation ice, a knowledge of brine volume uniquely specifies permeability. The Eicken et al. (2000) paper gives one a sense of the amount of work that will be involved.

Even taking the difficulty of adequately specifying permeability into account, the results of the Maksym and Jeffries modeling study are extremely informative. They considered two cases designed to bracket natural flooding occurrences. These are

1. A "*standard*" model, in which brine advection through the ice is considered. Here the bulk of the pore space is assumed to be effectively connected and the ice is taken to be reasonably laterally homogeneous.
2. A "*simple*" model, in which the brine within the ice is isolated from the network of pathways (cracks, floe edges, etc.) through which flooding is assumed to take place.

The results, which are driven by a representative temperature sequence for the Southern Ocean, are surprising in that it is the simple model that gives reasonable agreement with field data. Specifically, congelation and snow ice thicknesses are realistic and salinities in both the snow ice and the congelation ice layers compare well with field observations. On the other hand, the standard model produces only a small amount of snow ice even when highly negative

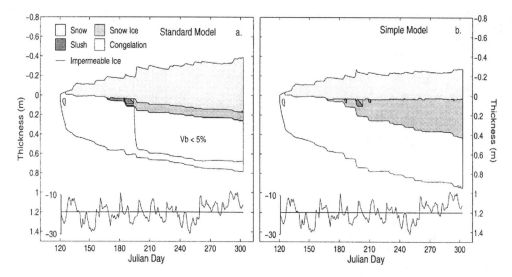

Figure 8.29. Simulated time series of sea ice growth in the Ross Sea: (a) standard model, which includes internal brine flow within the ice; (b) simple model, which does not consider internal brine flow. A region of impermeable ice is indicated by the solid contours within the congelation ice layer (Maksym and Jeffries 2000). Also shown is a trace of the air temperature used in the simulation (Possession Island).

freeboards are considered. Also, salinities within the congelation ice become unrealistically low, resulting in low brine volumes and causing the ice to become impermeable even when ice temperatures are high, assuming that the Rule of Fives holds.

The ice mass balance is primarily controlled by the presence of a snow cover. Although initial ice growth is rapid, once an appreciable snow cover develops further congelation ice growth is greatly diminished and is strongly dependent on the oceanic heat flux. In both models further ice growth is primarily through flooding and snow ice formation. In that the standard model in general produces little snow ice, total ice thickness is heavily dependent on environmental conditions during initial ice growth. This results in ice thickness being nearly independent of age. Although early ice growth is also very important in the simple model, in that snow ice thickness is strongly related to snow load, total ice thickness is much more predictable and ice thickness generally increases with age. Figure 8.29 shows two representative simulations.

Let me describe what happens in the standard model in a bit more detail. When the upper ice surface is depressed below sea level by the development of a snow cover, seawater will begin to move upward through the ice, assuming that this is initially permeable. This displaces the colder, denser brine within the ice upward into the snow cover, thereby modifying the salinity structure of the ice. As the seawater entering the ice is already at its freezing point, it will freeze against the walls of the conduits and as a result decrease the permeability, as the latent heat released in the process warms the surrounding ice. The overall result is a reduced brine volume and a decreased permeability. In the simulations brine volumes within the congelation ice frequently dropped well below the 5% value necessary for impermeability according to the Rule of Fives. In such cases this ice remains impermeable even when ice temperatures are quite warm. As a result, even if a negative freeboard is maintained, further flooding cannot occur as the result of the impermeable congelation layer. In the simple model brine flushing does not occur and the water entering the snow pack is normal seawater. During each flooding event the ice is warmed, high brine volumes are maintained, and the ice remains permeable. The salinity profiles generated are generally S-shaped or slightly C-shaped, similar to profile shapes frequently observed in Southern Ocean ice floes (Eicken 1992).

I find the Maksym and Jeffries paper to be particularly interesting in that it strongly suggests that the previous assumption that "the brine involved in the formation of infiltrated snow ice follows a pathway primarily through the underlying congelation ice" is largely incorrect. It also supplies a useful tool for looking at this surprisingly complex process. I also note that essentially the same simulation could be used to examine the different factors affecting the striking salinity change that occurs during the spring/summer period, when FY ice transitions to second-year and second-year changes to MY. In this case the displacing fluid would be low-salinity surface meltwater percolating downward into the ice. The paper also points out the importance of obtaining a better assessment of the factors controlling permeability in different structural types of sea ice, and also of obtaining a better assessment of the adequacy of the Rule of Fives.

Clearly the most detailed field study of what might be called the hydrology of the summer melt period occurred during the summer of 1998 as part of the SHEBA experiment (Eicken et al. 2002). In this study a variety of different techniques were used including fluorescent tracers, as well as stable ($\delta^{18}O$ and D) and unstable (^{7}Be) isotopes. This information was then coupled with in-situ permeability measurements to estimate the hydraulic gradients driving the lateral and vertical movement of water within the ice cover. Other interesting aspects of this study include both the retention of surface water in melt ponds and the life cycle of under-ice meltwater ponds. Based on the observational data set, the summer melt is divided into four different stages. The first is characterized by low ice permeability ($< 10^{-11} m^2$), widespread ponding largely hidden beneath the snow cover, and the lateral transport of meltwater draining into cracks and flaws. Toward the latter portion of this stage under-ice ponds develop and gradually become significant reservoirs of snow meltwater. During stage 2, melt continues with both permeability and hydraulic head increasing. In turn, this leads to most surface melt ponds equilibrating at sea level. Under-ice melt pond development continues as does the formation of false bottoms. During stage 3, internal melting, desalination, and decay of the ice cover continue. Surface ponds widen and deepen with associated decreases in large-scale albedo. Clearly, in this phase the thermal processes affecting the ice cover can no longer be considered to be one-dimensional. Stage 4 marks the end of the melt season. At this time, roughly 40% of the surface meltwater generated at the SHEBA site remained associated with the ice cover in one of three reservoirs: surface ponds, under-ice melt ponds + false bottoms, and meltwater within the ice matrix. This latter term constituted the largest single melt pool, averaging 15% of the net meltwater retention. This study is important, not only because of its detail and high quality, but also because related long-term summer experiments are unlikely in the near future as the result of increasingly unstable ice at reasonably accessible locations.

8.4 Theories

8.4.1 Simulated Profiles

There now have been several attempts to develop models that result in detailed predictions of the salinity profile at any time based on information of the growth history of the sea ice. The most ambitious of these for FY ice has been developed by Cox and Weeks (1988a), in that it calculates both FY ice thickness and the salinity profile as a function of time during the ice growth season. Its starting point is the thin sea ice growth model developed by Maykut (1978, 1986), although any growth model could be used provided that it gives values of the ice thickness and temperature profile as a function of time. The factors that were considered were initial entrapment, brine expulsion, and gravity drainage. In the case of brine expulsion the amount of brine expelled from each layer of the model was assumed to be rejected directly out of the ice sheet. This probably overestimates brine loss by this process. In estimating gravity drainage empirical equations were used based on the NaCl ice experiments of Cox and Weeks (1975). However, in these experiments the ice produced was frozen to the sides of the cylindrical container and was therefore not free-floating. This reduces the level of the upper surface of the brine in the brine drainage system and lessens the effectiveness of gravity drainage. Even considering these rather ad hoc assumptions, the model predictions were quite similar to profile properties observed on real sea ice. For instance, using representative air temperatures for the Arctic Basin the average salinity vs. thickness plots showed an initial rapid drop in mean salinity, followed, after a thickness of ~30 cm was reached, by a more gradual falloff. These results are similar to the field observations summarized by Kovacs (1996). In addition, the individual salinity profiles show the characteristic C-shapes that are frequently observed in FY sea ice. My current assessment of this approach is that it appears to be promising. If I were to try to develop it further, I would start by trying to add an improved gravity drainage model based on the work of Eide and Martin (1975), as well as a more realistic treatment of the fate of expulsed brine. I also note that the Cox and Weeks approach has been applied successfully by Eicken (1992, 1998) in his studies of antarctic salinity profiles, and by Maksym and Jeffries (2000) in their studies of flooding. By expanding the model to treat the effect of surface melting on the salinity profile, one should be able to simulate profile property changes from FY ice to second-year ice and ultimately to MY ice. If such a model could then be verified against field observations, its predictive capabilities would be useful.

8.4.2 Mushy Layer Theory

In the previous sections I have primarily discussed approaches to the salinity distribution problem that proceed by attempting to isolate the different mechanisms involved in incorporating and modifying the sea ice salinity profile. Here the hope is that once these mechanisms are properly understood, they can be combined into a useful predictive model. The Cox and Weeks model has been such an attempt. Mushy layer theory takes a very different tack in that it looks at values averaged over the microscale for the layer of interest, in the specific case here, sea ice. The theory was initially developed to deal with situations occurring in metals in which a solidifying ingot may contain both solid crystals of the metal as well as amounts of remaining metallic melt. Another obvious application is in the solidification of magmas where more mafic crystals may form a matrix containing residual melt in the interstices. Fields where mushy layer theory has found applicability include crystal growth, electrical engineering, geology, geophysics, metallurgy, and oceanography (Davis et al. 1992; Huppert and Worster 1985). Clearly in some of these areas the isolation of individual contributing mechanisms may be difficult. What has been needed is a theory that is more robust and gets results without having to know too much about the mechanistic details.

That said, let me add that I am not fond of applying the term *mushy layer* to sea ice, as it is misleading. In that a so-called mushy layer is taken to be "the two-phase region comprising essentially pure crystals of one of the components of the melt bathed in liquid enriched in the rejected components" (Worster and Wettlaufer 1997), this means that if any brine is present in a layer, it is by definition mushy. In turn this means that all sea ice is mushy in that it always contains some brine. Now, I know what the word *mushy* means: a thick, soft, pulpy, porridgelike mass. One could hardly complain if the word *mushy* were applied to grease ice or to a brine-saturated snow layer on the upper ice surface prior to its refreezing, or perhaps even to the thin, weak dendritic skeletal layer that exists on the underside of a sheet of actively growing sea ice. However, clearly the great percentage of sea ice is definitely nonmushy in the strict sense of the word. However, enough of this grousing as it seems that we are stuck with the mush.

One argument in favor of a more general, averaged theory over one that is more focused on mechanistic details is the fact that once brine channels form, they appear to have little if any regard for the structure of the preexisting sea ice. As best as is known at present, brine channel formation proceeds much the same independently of whether the ice matrix is frazil or highly aligned congelation (Worster and Wettlaufer 1997). The efficiency of the convective circulation that utilizes the brine channel network, plus the network's apparent disregard of preexisting structural details, is probably the reason that the sequence of salinity profiles in frazil and in congelation appear similar. One would not expect this to be the case if the primary factor controlling brine drainage were the permeability of the ice matrix.

In the theory the dependent variables are taken to be the local mean temperature T, the local mean concentration of the interstitial liquid C (in our case brine), the local mean solid fraction ϕ (ice + solid salts) and the local mean velocity u. These parameters are related by three equations expressing the conservation of heat

$$\rho C_p \left(\frac{\partial T}{\partial t} + u \bullet \nabla T \right) = \nabla (k \nabla T) + L \frac{\partial \phi}{\partial t} , \tag{8.24}$$

the conservation of solute

$$(1-\phi)\left(\frac{\partial C}{\partial t} + u \bullet \nabla C \right) = (C - C_s) \frac{\partial \phi}{\partial t} , \tag{8.25}$$

and a momentum equation

$$u = \Pi(-\nabla p + \rho g). \tag{8.26}$$

In equation 8.24 C_p is the heat capacity and L is the latent heat of fusion. Equation 8.26 is Darcy's equation, which is used to describe the flow in a porous media with permeability Π which, in turn, is a function of ϕ. Next it is necessary to express ρ in terms of the local temperature and brine concentration and to remember that ϕ is determined by the constraint that the sea ice is in local thermodynamic equilibrium, that T and C are coupled by the Nernst relation. Then, writing the above relations in dimensionless form, it is possible to identify three important dimensionless groups (for details see Wettlaufer et al. 1997a, 1997b ; Worster 1997; or Worster and Wettlaufer 1997). These groups are a Stefan number that is the ratio of the latent heat of solidification to the sensible heat necessary to cool the brine

$$S = \frac{L}{C_p \Delta T} , \qquad (8.27)$$

a compositional ratio which gives the difference in composition between the solid and liquid phases relative to the variations in composition of the liquid

$$C^* = \frac{C_s - C}{\Delta C} , \qquad (8.28)$$

and finally, a Rayleigh number which gives the buoyancy force relative to the viscous dissipation in the porous medium

$$Ra = g\beta\Delta C\Pi(\phi)h / \kappa\nu . \qquad (8.29)$$

Here g is the acceleration due to gravity, $\beta\Delta C = \beta(C_o - C_b)$ is the difference in the density of the brine across the ice layer where C_o is the initial concentration of the liquid and C_b is the liquidus composition corresponding to the surface temperature of the ice, assuming that the Nernst condition holds. The other terms are the thickness of the ice (h), the thermal diffusivity (κ), the kinematic viscosity of the liquid (ν), and the permeability (Π), which is a function of ϕ, the fraction of the layer that is solid (i.e., ice). It is this parameter that primarily determines the onset of convection. As they point out, if convection does not occur, the density changes in the ice resulting from the salinity profile would be purely controlled by diffusion with convection occurring only in a narrow compositional boundary layer existing in the liquid at the sea ice–seawater interface. They term this convective type the *boundary-layer mode* and note that although it has a small length scale and does not penetrate far into the ice (perhaps only into the skeletal layer?), it is the first to become unstable. The other convective mode, which they term the *mushy-layer mode*, occurs on a much larger scale and is driven by buoyancy internal to the ice. It therefore penetrates the whole mushy layer and results in a much larger salt flux.

At present the theory has been applied to thin sheets of laboratory-grown NaCl ice and also to the formation of ice in leads. Their experiments indicate that when an ice sheet initially starts to form, the included brine does not have sufficient negative buoyancy to overcome the resistance provided by the ice crystals and remains trapped within the ice. That this could occur was noted earlier by Farhadieh and Tankin (1972), who observed a lag time of ~80 seconds in laboratory experiments before brine was seen draining from the ice. Although Wettlaufer and his coworkers observed weak convection of brine starting at an ice thickness of 2.4 cm, they did not observe significant drainage until an ice thickness of ~7 cm was reached, corresponding to a growth time

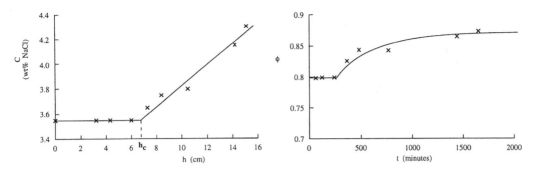

Figure 8.30. The experimental results of Wettlaufer et al. (1997a) obtained by freezing a NaCl–water solution having an initial composition of $C_0 = 35.5‰$ NaCl by weight. Part (a) shows the composition of the underlying solution as a function of the thickness of the ice layer. Part (b) presents the calculated values of the solid fraction of the layer as a function of time.

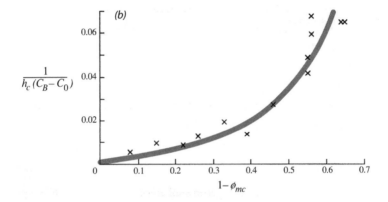

Figure 8.31. The crosses represent conditions when a significant brine flux began in each freezing experiment shown in Figure 8.30. In that $(h_c\Delta C)^{-1}$ is proportional to the permeability, the trend observed when these values are plotted against the mean liquid fraction $(1-\varphi_{mc})$ is consistent with the expected form of the permeability function, i.e., that permeability increases with liquid fraction and has a positive curvature, increasing rapidly at large values of the liquid fraction (Wettlaufer et al. 1997a).

of several hours. Figure 8.30 summarizes their results obtained with a salt solution of 35.5‰. As can be seen in part (a), until the ice thickness reached a critical value h_c the composition of the underlying solution remained constant, signifying that the brine rejected by the growing ice remained entrapped within the ice. A visual examination of the ice after the critical value of h_c had been exceeded showed that brine drainage channels extending the full thickness of the ice sheet were now developed and that strong convective plumes of brine were exiting the ice via these channels, corresponding to a fully developed mushy-layer convective mode.

Then, assuming that $(h_c\Delta C)^{-1} \propto \Pi(\phi_c)$, they plotted $(h_c\Delta C)^{-1}$ versus the liquid fraction at the critical thickness $(1-\phi_c)$ for all their test runs. The results, which are shown in Figure 8.31, indicate a good separation between conditions that result in strong brine drainage (mushy-layer mode) versus weak to negligible brine drainage (boundary-layer mode). The authors also note that the general form of the separating function is physically reasonable in that the permeability both increases with increases in the liquid fraction and rises rapidly as the liquid fraction approaches unity. More recently Feltham et al. (2006) have shown that the fundamental equations of mushy layer theory reduce to equations similar to those underlying the Maykut/Untersteiner thick ice growth model, provided that similar approximations are made in both cases. Although these results are impressive, it is my view that mushy layer theory, as applied to sea ice, is still very much a work in progress. More will be said concerning these results in the chapter on the development of ice in leads.

A related study has recently been published by Nagashima and Furukawa (1997), who froze a 30‰ water–KCl solution while observing the compositional profile in front of the solid–liquid interface via the use of a Mach–Zender interferometer. They found that the solid–liquid interface remained planar for the first 100 seconds. During this period the solution composition at the interface rapidly increased from 30.1 to 50.4‰, at which time the interface geometry changed from planar to cellular. Once this had occurred the interface composition, as measured at the tips of the cellular branches, dropped to a steady state value of ~3.7‰ as the result of solute being captured in the intercellular grooves characteristic of the skeletal layer.

Finally, in closing this section one should mention the recent work by Oertling and Watts (2004). This study applies the general theory for momentum, heat, and species transport during the solidification of binary systems developed by Bennon and Incropera (1987a, 1987b) to the freezing of $NaCl$–H_2O solutions as a model for the growth of young sea ice. Although in a general sense B & I theory is similar to mushy layer theory, there are significant differences between the two approaches. Mushy layer theory suggests that a constitutive equation relating permeability to local liquid fraction is not necessary in that only asymptotic states are examined. It also neglects the transient portion of the momentum equation. Oertling and Watts include such relations and use enthalpy rather than temperature as the independent variable in the heat equation. Comparisons with related experimental results such as those of Cox and Weeks (1975) are quite impressive. However, the model does not support the observed time delay in the initiation of brine drainage mentioned earlier. More impressive is the fact that the details of the simulations appear to be similar to the structural aspects of brine drainage as observed in real sea ice (Figure 8.32). In that such approaches are computationally intensive, it would be currently difficult to utilize them in large-scale sea ice growth, drift, and deforma-

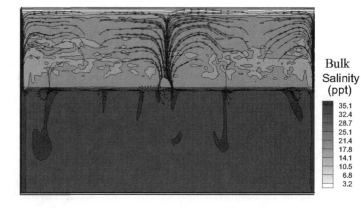

Figure 8.32. Results of a numerical simulation by Oertling and Watts (2004) of sea ice growth in a two-dimensional rectangular enclosure with a surface temperature of −10°C. Shown are black streamlines and shaded bulk salinity contours indicating brine drainage into the underlying reservoir. The ice thickness was 3.36 cm at 180 min.

Bulk
Salinity
(ppt)

35.1
32.4
28.7
25.1
21.4
17.8
14.1
10.5
6.8
3.2

tion models. Even so, they should be highly useful in verifying and extending brine drainage correlations as observed in the field.

8.4.3 Percolation Theory

Finally, Golden and a variety of coinvestigators have recently applied a percolation theory model to sea ice (Golden 1986, 1995, 1997a, 1997b, 1997c, 2001, 2003; Golden et al. 1998). This theory was initially developed to analyze property variations in materials where the connectedness of a specific component determines the bulk behavior. This general approach has received considerable attention within the mathematics and physics community and has been used to study a wide variety of processes such as flow through porous and fractured media such as rocks, soil, and firn, and also the property variations of various types of disordered conductors and absorbing composites. One reason for the general interest in this class of models is both their generality and the fact that they are possibly the simplest purely probabilistic models that exhibit a type of phase transition. In the case of sea ice the specific component of interest is, of course, brine and its connectivity. One of the questions of interest in this chapter, as discussed earlier, is the specification of conditions under which sea ice becomes impermeable (i.e., the Rule of Fives; Golden et al. 1998).

To gain an impression of this approach consider as a simple percolation model the d-dimensional integer lattice $\mathbf{Z}^d$ and the square (or cubic) network of bonds joining nearest neighbor lattice sites (Golden 2003). To each bond, with the probability p, where $0 \le p \le 1$, assign a value of 1 if the bond is open, and with probability $1-p$ assign a value of 0 if the bond is closed. Groups of connected open bonds are called open clusters, with the size of the cluster being the number of open bonds that it contains. In such models there is a critical probability p_c where $0 \le p_c \le 1$, referred to as the percolation threshold, at which the cluster size diverges and an infinite cluster appears, allowing the open bonds to percolate. In two dimensions $p_c = 0.5$ and in three dimensions $p_c \approx 0.25$. Figure 8.33 shows typical open cluster patterns in $d = 2$ for $p = 1/3$ and $p = 2/3$. For $p \ge p_c$ the infinite cluster density $P_\infty(p)$ is defined as the probability that any point is contained in the infinite cluster. At the percolation threshold, the infinite cluster has a self-similar fractal structure, where $d_f \le d$ is the fractal dimension. A graph of $P_\infty(p)$ for $d = 2$ is also shown in Figure 8.33. Although percolation models strictly only deal with the geometrical aspects of con-

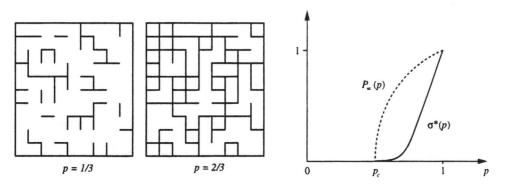

Figure 8.33. Representative configurations of the two-dimensional lattice in a bond percolation model showing below ($p = 1/3$) and above ($p = 2/3$) the percolation threshold $p_c = 1/2$, and graphs of the infinite cluster density $P_\infty(p)$ and the effective conductivity $\sigma^*(p)$ (Golden 2003).

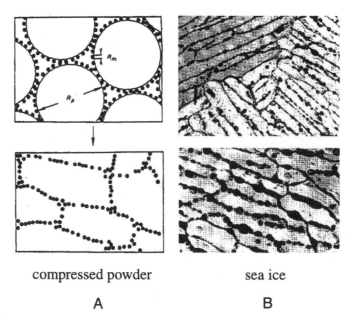

compressed powder   sea ice

A   B

Figure 8.34. A comparison of the microstructure of (a) a compressed powder of large polymer particles of radius R_p and small metal particles of radius R_m (Malliaris and Turner 1971), and (b) sea ice (Arcone et al. 1986) (from Golden et al. 1998).

nectedness, they can also be utilized in the study of transport properties such as permeability. In our case, consider the bonds as a pipe network with effective fluid permeability $\kappa^*(p)$ exhibiting behavior $\kappa^*(p) \sim (p - p_c)^e$ where e is the permeability critical exponent. Critical exponents such as e are believed to exhibit universality in that they generally depend only on dimension and not on the type of lattice.

If the above model is applied to sea ice with open bonds representing brine and closed bonds representing ice, then p_c in $d = 3$ would be ~25%, a value that is clearly much larger than the 5% suggested by the Rule of Fives. Even continuum models such as ellipsoidal brine inclusions randomly distributed in ice, which have been used in the analysis of sea ice data, show critical volume fractions in the range of 20–40% (DeBondt et al. 1992). Assuming that the Rule of Fives is correct, the problem then becomes determining whether a microstructure that is reasonably similar to that of sea ice reduces p_c to values near 5%. Golden et al. (1998) have suggested that composites produced by compressing a powder of large polymer particles of radius R_p that contains smaller metal particles of radius R_m of a conducting material provide such an example. One reason this comparison is attractive is that the resulting structure of the composite roughly matches the structure of real sea ice, as shown in Figure 8.34. Another is the similarity between plots of the variations in the electrical conductivity of compressed metal–polyethylene mixtures versus metal volume, on the one hand, and those of the liquid permeability of sea ice versus temperature, on the other; both suggest critical percolation thresholds where conductivity and permeability drastically decrease (Figure 8.35). In studies of compressed powders it has been shown that the ratio $\xi = \dfrac{R_p}{R_m}$ of the radii of the larger polymer particles to the smaller metal particles is a key factor in determining the conduction threshold. It has also been shown that the critical volume fraction of the small metallic spheres necessary

A

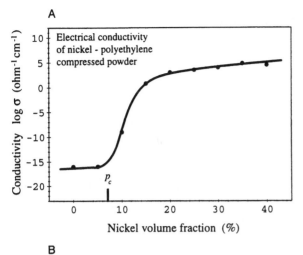

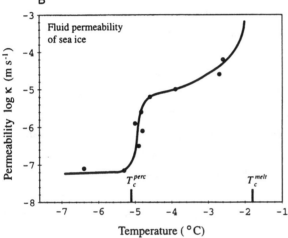

Figure 8.35. A comparison of (a) the electrical conductivity of a compressed powder of large polymer particles of radius R_p and small nickel particles of radius R_m (data from Malliaris and Turner 1971) and of (b) the fluid permeability $\kappa(T)$ of thin young sea ice as a function of surface temperature (data from Ono and Kasai 1985). Note that the transport properties of both materials exhibit critical behavior characteristic of a percolation transition. Also indicated for $\kappa(T)$ is a second transition at the melting point where log $\kappa(T)$ must increase rapidly (Golden 2001).

for percolation is approximately given by $p_c = \left(1 + \frac{\xi\phi}{4x_c}\right)^{-1}$ where ϕ is a reciprocal packing factor and x_c is the critical surface area fraction of the larger particles that must be covered by the smaller particles for percolation to occur (Kusy 1977). Microstructural measurements combined with conductivity measurements provide estimates of x_c and ϕ of 0.42 and 1.27, which Golden also suggests are reasonable for sea ice. Related measurements on sea ice provide an estimated value for $\xi \approx 24$, which suggests a p_c value for columnar sea ice of roughly 5%. It also should be noted that the exact value of p_c would be expected to vary with changes in the crystal structure of the ice. This is in agreement with the field observations of Ono and Kasai (1985), who, as noted earlier, found that frazil ice had slightly higher values of p_c than columnar ice, a difference presumably caused by a more random distribution of brine inclusions.

Further details of this approach are beyond the scope of this book. Readers wishing to explore this topic further will find that the Golden (2001, 2003) papers not only review the general subject and its application to sea ice, they also provide extensive reference lists of related papers dealing with other materials.

8.5 Inclusion Geometry

The reader quite possibly has noticed that in the above material attention has primarily been focused on the volumes of brine present in the ice (a discussion of the volumes of included gases can be found in Chapter 10 as part of the discussion of density variations). Although considerable attention has been given to the shapes of large features such as brine drainage tubes, little has been said about the changes in the geometry of the brine pockets that occur in the relatively large volumes of ice that occur between brine drainage tubes. In a way this is surprising in that, as will be seen in the chapter on sea ice property variations, one theory on the variation in the tensile strength of sea ice assumes that as brine volume changes, brine pockets change shape in systematic ways (Assur 1958). However, the reasons for increased interest in this subject are more indirect than direct, being driven by a need to better understand the optical properties of sea ice and their variations during the course of annual temperature and salinity cycles (Grenfell 1983, 1991; Light et al. 1998). Careful measurements of the geometries of both brine and gaseous inclusions in the ice are fairly recent. Other related missing observations are adequate descriptions of the solid salts in the ice. To put it simply, "What does this stuff look like?" Fortunately the status has started change. To conclude this chapter, I will summarize some aspects of this recent work.

One of the earliest attempts to examine the structural details of brine pockets and included salt crystals in more detail can be found in the paper by Sinha (1977), who was interested in expanding available techniques for examining the internal structure of sea ice. Although he presented photomicrographs made at magnifications of 30× and 100× that showed complex brine pocket patterns as well as precipitated salts within the pockets, the most interesting aspect of his paper was the use of a scanning electron microscope to examine brine pockets both directly and via the examination of polyvinyl formvar replicas of the ice. Observations were made at both –10°C and at –30°C (Figures 8.36a and 8.36b). Although positive identifications of the solid salts were not made, it is reasonable, based on knowledge of the phase diagram, that the solid salt crystals present at –10°C were

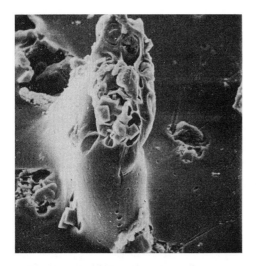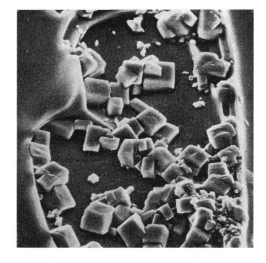

Figure 8.36. Two views of replicas of brine pockets made with a scanning electron microscope: (a) –10°C, in which the crystals are believed to be $Na_2SO_4 \cdot 10\ H_2O$ and occur near the bottom of the brine pocket; and (b) –30°C, in which the crystals are believed to be $NaCl \cdot 2\ H_2O$ (magnification of 500×) (Sinha 1977).

$Na_2SO_4 \cdot 10\ H_2O$, whereas at $-30°C$ the salt crystals were primarily $NaCl \cdot 2\ H_2O$. Magnifications as high as 500× were used. Note that in Figure 8.36a the $Na_2SO_4 \cdot 10\ H_2O$ crystals are concentrated at the bottom of the brine pocket. Figure 8.36b together with two other photos taken at $-30°C$ using lower magnification are the only currently available photos of brine pockets showing $NaCl \cdot 2\ H_2O$ crystals. Note that in contrast to the $Na_2SO_4 \cdot 10\ H_2O$ crystals, the $NaCl \cdot 2H_2O$ crystals appear to be reasonably uniformly distributed within the brine pocket. Sinha did not attempt to obtain a systematic set of measurements of brine pocket and salt crystal sizes.

Starting with the work of Arcone et al. (1986), and continuing with studies by Perovich and Gow (1991, 1996), Eicken (1991, 1993), Light et al. (2003a), Cole and Shapiro (1998), and Eicken et al. (2000), our knowledge of brine pocket geometry has gradually expanded. Examination of both horizontal and vertical thin sections has revealed that although there is considerable small-scale variability, the majority of visible inclusions are of brine and are frequently elongated in the vertical (growth) direction. For instance, Cole and Shapiro observed brine inclusion shapes ranging from spherical to vertical elongations exceeding the horizontal diameter by 15×. The most detailed examination of this subject published to date is by Light et al. (2003a), whose analysis also utilizes much of the previous work. This is useful in that most studies have been restricted to a limited size range. Figures 8.37, 8.38, and 8.39 show some of the results. Note that in all cases the best fit was provided by a power law. Figure 8.37 combines measurements by Light et al. with a curve obtained by Perovich and Gow. On the average 24 inclusions per mm^3 were observed by Light et al. as compared to 1.6 per mm^3 by Perovich and Gow. Figure 8.38 shows the observed size distribution for brine inclusions as a function of the length of the inclusions. Finally, Figure 8.39 shows the vertical-to-horizontal aspect ratio presented as a function of the inclusion length. The Light et al. paper also includes the calculation of the distribution of equivalent spheres based on the brine inclusion data. Such information has proven to be useful in scattering calculations. Figure 8.40 shows a log-log plot of gas bubble

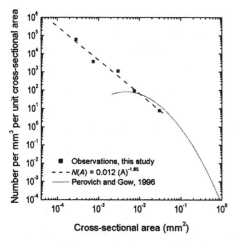

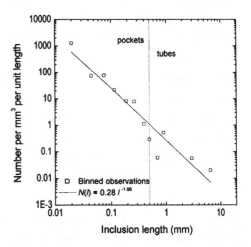

Figure 8.37. A comparison of the number density of brine inclusions vs. the horizontal cross-sectional area of the inclusions. Data from Light et al. (2003a) and Perovich and Gow (1996). The power law fit gave an r^2 value of 0.98 (Light et al. 2003a).

Figure 8.38. The observed size distribution for brine inclusions at $-15°C$ plotted as a function of the length of the inclusion. The power law fit gives an r^2 value of 0.92. The dashed line divides pockets from tubes (Light et al. 2003a).

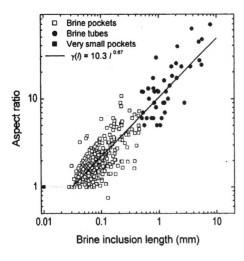

Figure 8.39. The vertical-to-horizontal aspect ratio as a function of inclusion length for brine inclusions observed at −15°C. The power law fit gives an r^2 value of 0.77. The solid black square represents numerous small inclusions that were at the resolution limit (Light et al. 2003a)

Figure 8.40. Gas bubble distributions observed in FY ice. Data are from Gavrilo and Gaitskhoki (1971); Grenfell (1983); Light et al. (2003a); and the SHEBA Field experiment.

observations collected from several different sources. Again a power law appears to work well. Gas bubbles were generally spherical and were primarily found to be contained within the brine tubes and channels as opposed to within the ice matrix. Bubble radii ranged from 0.004 to 0.07 for observations made at −15°C. A discussion of the amount and composition of the gas in sea ice can be found in Chapter 10.

Finally, Figure 8.41 shows brine pockets containing mirabilite crystals in photographs taken at −15°C. Note that as in Figure 8.36a the crystals occur either at the bottoms of the brine pockets or in constrictions. The crystals tend to have rounded edges and irregular shapes. The largest crystal visible in Figure 8.41b has a diameter of 0.14 mm. Light et al. (2003a) suggest that some coarsening may have occurred since initial precipitation. This is certainly a possibility, but at present it is still not possible to decide whether Ostwald ripening is a significant process in sea ice. The authors also note that mirabilite crystals are rarely observed in the smaller brine pockets. Their explanation is that there is insufficient Na_2SO_4 present in a small brine pocket to result in a crystal large enough to be observed with a system having their magnification capabilities. They suggest that in their study this critical size is $\sim\ell \geq 0.06$ mm. They also note that it is still not clear whether salt precipitation occurs quickly (they held their samples at a constant temperature for 24 hours before they took photographs). In the early 1960s I did some studies on ice grown from Na_2SO_4–$NaCl$–H_2O solutions. Although I cannot place a hard number on salt precipitation and solution rates, it was my impression that both occurred fairly rapidly (i.e., within a couple of hours). Certainly their use of 24 hours appears to be conservative. However, it would be nice to have some quantitative checks on these matters.

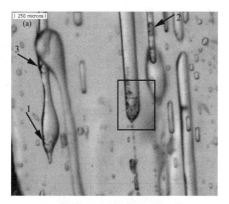

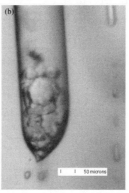

Figure 8.41. (a) Photograph of a thin section showing individual Na2SO4 10H2O (mirabilite) crystals. (b) An enlargement of the box in (a) (Light et al. 2003a).

8.6 Conclusions

In concluding this chapter I would like to note the major advances that have been made during the last half-century in our understanding of the salinity profile of sea ice and its changes with initial growth conditions and time. In 1950 all that we had were a few profiles and one possible mechanism (brine pocket migration) that, ironically, has since been discredited as an important process. We now know that there are a variety of interrelated processes that affect the salinity profile. We also have several models that can serve to guide future studies. One thing is clear: just obtaining a few additional salinity measurements on an opportunistic basis will not get us far. What are needed are carefully controlled experiments that look at different aspects of the problem in detail. Such studies can be carried out either in the field or in the laboratory. I would guess that the most effective approach would be to push laboratory studies and when appropriate, follow them up with focused field tests. A particularly interesting recent advance is the development of the ability to directly measure both the salinity and solid fraction in a growing sheet of sea ice in situ (Notz et al. 2005). This impedance technique is nondestructive, capable of very high temporal and spatial resolution, and can be automated as a stand-alone, battery-powered system. However, it has one major drawback in that it cannot be applied to preexisting sea ice; it must be in place before the ice starts to form and be allowed to freeze in as the ice sheet develops. Recent field results obtained applying this impedance technique to young sea ice forming in a Svalbard fjord show that there does not appear to be a sharp salinity discontinuity between the underlying seawater and the lowest level of the skeletal layer at the bottom of the ice (Notz and Worster 2008). In exploring this topic further, additional information on the changes in the geometry of the platelet tips as a function of growth velocity would be very useful, with more dendritic tips favoring a gradual compositional transition and sharp tips favoring a more discrete change.

Current results have clearly verified a problem long suspected by fieldworkers: that during the collection of samples by coring there is an appreciable loss of brine from the lower portion of the core. This is not a problem in the collection of samples for structural or property studies, as in these cases it is not the in-situ salinity that is important but the salinity of the sample at the time of testing. Where it is a major problem is in studies where the in-situ salinity profile of the ice sheet is needed. Perhaps future comparisons between cored profiles and in-situ profiles as determined by the impedance technique will prove to be sufficiently systematic to allow one to accurately estimate in-situ salinity profiles from cored samples. The next few years should offer opportunities for significant advances.

9 Sea Ice Growth: The Details

Scientific reasoning:
Inductive,
Deductive,
Seductive.
Unknown

9.1 Introduction

Although there was a brief earlier discussion of sea ice growth in Chapter 4, this is clearly a subject that merits a more detailed examination. As noted in Chapter 2, scientific studies of sea ice growth started with the analysis of the snow-free sea ice growth problem by Stefan (1889, 1891). In fact, his analysis was really primarily applicable to freshwater ice. As field observations continued to accumulate over the years, it gradually became clear that the growth and decay of sea ice were affected by many factors that Stefan did not consider. In fact, as will be seen there is a large and still developing literature on this general subject. As the reader will undoubtedly conclude at the end of the present chapter, there remain aspects of this subject that are, at best, only partially understood.

Before confronting the details, it is useful to consider just what aspects of sea ice growth deserve the most attention. A sense of the answer to this query can be achieved by examining Figure 9.1. This figure shows a representative histogram of the observed probabilities of the occurrence of ice with different drafts in a portion of the Arctic Ocean ice pack. As the draft of a piece of sea ice is an excellent proxy for its thickness, it can be seen that in pack ice there are several different classes of ice thickness. First, there are the draft measurements between 0 and 1.7 m. This is FY ice that has formed during the present winter in leads. As leads are always opening and closing, there is invariably some very thin FY ice present that has just formed. In the fall all the FY ice will be thin, as only the thickest classes of FY ice will survive the summer melt period to become second-year ice. In the present example there is a low peak at ice drafts in the range of 0.8 to 1.0 m. This ice thickness range represents FY ice that started to form during the early part of the ice growth season when large ice-free areas within the ice pack are frozen over and have now reached this thickness, perhaps in January. Next, note the large amount of ice with drafts in the 2.0 to 3.5 m range. These drafts are representative of second-year and MY ice. As has typically been the case in the central part of the Arctic Basin, this was the ice with the greatest areal extent. Finally, there is the long exponential falloff in observed ice drafts starting at ~3.5 m and continuing up to 10 m. This tail is not the result of ice growth but is produced by deformation within the pack, although growth processes subsequent to deformation contribute to the solidification of the deformed masses. These thicknesses represent the pack ice equivalent of mountain building. It is clear that small mountains are very frequent whereas high mountains are rare. The processes involved will be discussed in detail in Chapter 12. Therefore, there are only two thickness types

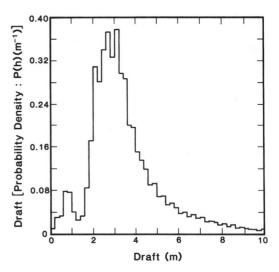

Figure 9.1. A representative histogram for the Beaufort Sea of the different probabilities of occurrence of sea ice of different drafts based on observations reported by Wadhams and Horne (1980).

of interest in the present chapter: first-year and old. However, as will be seen, the two ice growth models in common usage do not correspond directly to these two thickness types. Instead the thick ice model applies to both MY ice and to thicker FY ice, whereas the thin ice model behaves exactly as advertised due to its limitation to ice that is largely less than 50 cm thick. As noted in Chapter 4, there are large variations in the range of applicability of the thin ice model resulting from possible variations in the thickness of the snow cover, which, in turn, affects the difference between the air temperature and the temperature of the upper ice surface.

In the following I will deviate a bit from the historical sequence of ice growth model development and utilize what seems to me to be a more natural sequence, by first considering thin first-year ice growth, then multiyear ice growth, and finally decay, melting, and melt pond formation. The advantage here is that several simplifications that can be made in the analysis of the growth of thin FY ice cannot be made when thicker FY ice and MY ice are considered. For instance, as noted in Chapter 4, the temperature profiles in thin sea ice are linear. Thin ice has also not been in existence long enough to allow either an appreciable snow cover to develop or a significant heat flux from the ocean. Historically, Maykut and Untersteiner (1969, 1971) initially considered the more complex thick FY/MY situation and then Maykut (1978, 1986) simplified matters to focus on newly formed thin sea ice.

9.2 A Thin FY Ice Model

By the early 1960s it was clear that empirical degree-day curves such as those utilized by Lebedev (1938), Zubov (1945), Anderson (1961), and Bilello (1961) could achieve excellent after-the-fact fits of ice growth observations. Nevertheless, the fitted constants in their equations clearly contained implicit assumptions concerning temporal variations in the properties of the ice, snow thickness, amount of incoming radiation, and ocean heat flux. Unfortunately, for the empirical curves to be used for forecasting it was necessary to assume that these terms were essentially constant. Considering typical year-to-year variations in weather, it was clear that large errors could occur in forecast ice thicknesses even for identical field locations. When it was necessary to forecast for areas where such parameters were probably quite different, the predicted ice growth curves became increasingly uncertain. During the same general time period Russian investigators such as Kolesnikov (1958) and Doronin (1969) explored the possibilities of utilizing more rigorous, closed-form solutions in the analysis of sea ice growth. During time periods when computational capabilities were generally limited such approaches were attractive in that they could minimize the number of calculations.

Unfortunately this possible advantage came with an attached price, a loss of flexibility. A summary of this approach can be found in Doronin and Kheisin (1977) and in Doronin (1997).

What was needed was a flexible, physically realistic treatment that explicitly identified all the important factors affecting sea ice growth. Such an approach has the distinct advantage of, at the least, clearly identifying areas of uncertainty and, at best, serving to focus research on improving the parameterization of these more uncertain terms. Just such a thin ice model, as developed by Maykut (1978, 1986), has served as the foundation for much of the more current work on this subject. This model, in its simplest form, considered a thin layer of young sea ice of thickness H that has no snow cover. In such thin ice the temperature profile is to a good approximation linear. Then the value of the conductive heat flux through the ice (F_c) can be calculated from

$$F_c = k_i \left(\frac{T_f - T_o}{H} \right) \tag{9.1}$$

if the thermal conductivity of the ice k_i and the temperature of the ice surface T_o are known, in that during ice growth the seawater at the lower surface of the ice is fixed at its freezing point ($T_f = \sim -1.8°C$). If one wishes to consider the more complex situation of a thin layer of snow existing on top of the ice, the conductive heat flux within the combined snow and ice layer can be calculated from

$$F_c = \frac{k_i k_s}{k_i h_s + k_s H} (T_f - T_o) = \gamma (T_f - T_o). \tag{9.2}$$

Here k_s and h_s are the thermal conductivity and thickness of the snow cover, γ is the combined thermal conductance of the ice and snow layers, T_f remains the freezing temperature at the ice–ocean interface ($\sim -1.8°C$) and T_o becomes the surface temperature of the snow instead of the ice. If it is assumed that the value of T_o is that which results in a balance between the heat lost and the heat gained at the surface, one can determine its value by calculating the surface heat balance. Terms that Maykut considered were as follows:

F_r incoming shortwave radiation

aF_r reflected shortwave radiation where a is the ice albedo (i.e., the ratio of the reflected to the incident shortwave radiation)

I_o net influx of radiative energy which passes into the interior of the ice

$F_{L\downarrow}$ incoming longwave radiation

$F_{L\uparrow}$ emitted longwave radiation

F_s sensible heat flux

F_e latent heat flux

F_c conductive heat flux

If a flux toward the ice or snow surface is taken as positive and a flux away from the surface as negative, and if the ice surface temperature is taken to be always below the freezing point in that we are considering thin ice growth, the energy balance at the ice surface can be written as

$$(1-\alpha)F_r - I_o + F_{L\downarrow} - F_{L\uparrow} + F_s + F_e + F_c = 0. \tag{9.3}$$

The problem now becomes one of assessing the values of each of the terms in equation 9.3 and solving for the surface temperature of the ice T_o necessary to balance the relation. To get some sense of the difficulty of carrying this out, we can introduce some preliminary numbers into the different terms of equation 9.3. In doing this I will start on the left-hand side of equation 9.3 and gradually work toward the right-hand side.

9.2.1 Shortwave Radiation

One should recall that all matter radiates energy according to the relation

$$F = \varepsilon \sigma T^4 \tag{9.4}$$

where F is the emitted radiation, ε is the emissivity (typically a material constant), σ is the Stefan–Boltzman constant $5.67 \times 10^{-8}\,\mathrm{W\,m^{-2}\,°K^{-4}}$, and T is the radiative temperature of the material. If $\varepsilon = 1$, the emitting body in question is referred to as a *blackbody*. Incidentally, the Stefan of the Stefan–Boltzman constant is the same Stefan who also concerned himself with the growth of sea ice. The majority of the emitted radiation that concerns us lies in two distinct symmetrical spectral bands, as shown in Figure 9.2. The lower-wavelength band, which shows a peak intensity at $\lambda \sim 0.5\mu m$, originates at the sun with a surface temperature of 6000°K, whereas the higher-wavelength band, which shows a peak at $\lambda \sim 12\mu m$ originates from the Earth and its atmosphere at a temperature of ~250°K. As the result of the quite clean wavelength separation at ~4μm, the radiation associated with 250°K has commonly been referred to as longwave (F_L) and that associated with 6000°K as shortwave (F_r). Another important relation, known as Wien's Law,

$$\lambda_{\max} = \frac{2897}{T}, \tag{9.5}$$

gives the wavelength of the maximum emission (μm) as a function of the temperature $(°K)$ and shows that these parameters are inversely proportional to one another.

If, as is frequently the case, measurements of F_r are not available for a site of interest, a number of different relations have been used to provide estimates. Some of these relations are even specifically tuned to represent polar surface conditions. The question then becomes what relation or relations gives the best agreement with field observations. This particular query has been explored in detail by Key, Silcox, et al. (1996), who compared estimates made using both

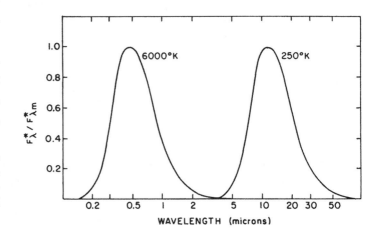

Figure 9.2. Normalized blackbody radiation per unit wavelength calculated for temperatures of 6000 and 250°K (Fleagle and Businger 1980).

shortwave and longwave parameterizations with field observations made at Resolute Bay, Canada, and at Barrow, Alaska. For clear-sky conditions the best agreement was provided by a relation initially suggested by Zillman (1972) for observations made in the Indian Ocean, and later modified by Shine (1984), by changing the coefficients so that it gave a better fit to arctic fluxes as calculated by a radiative transfer model. The relation is

$$F_{clr} = \frac{S_o \cos^2 Z}{\cos Z + (1 + \cos Z) e_a \times 10^3 + 0.046}. \tag{9.6}$$

Here F_{clr} is the downwelling shortwave radiation at the surface under clear-sky conditions, S_o is the solar constant, e_a is the partial pressure of water vapor, and Z is the solar zenith angle, which in turn is a function of latitude, day, and hour. For cloudy conditions Shine suggested

$$F_{cld} = \frac{(53.5 + 1274.5 \cos Z) \cos^{0.5} Z}{[1 + 0.139(1 - 0.9345\alpha)\tau]}. \tag{9.7}$$

Here the parameters a and τ are the albedo of the surface and the optical depth of the clouds. These factors attempt to include in the estimates the effects of cloud thickness and multiple reflections between the surface and the cloud base. The difficulty is that many times estimates of a and τ are not available. Combining (9.6) and (9.7) produces Shine's recommended relation for estimating F_r for all conditions, or

$$F_{r(all)} = [(1 - C) F_{clr} + C F_{cld}] \tag{9.8}$$

where C is the cloud fraction. Several other simpler relations are also available that include the effects of the fractional cloud cover C to the second or third power. For details and numerous references refer to Key, Silcox, et al. (1996).

During much of the year (autumn to spring) only a small percentage of the downwelling shortwave actually affects the ice, as the albedos of snow-covered sea ice are quite high, with values ranging from 0.81 to 0.87, and either sun angles are very low or the sun is continuously below the horizon. As will be discussed later, exact values for a are dependent on the exact frequency (λ) of the incoming radiation (i.e., $a\lambda$). For typical ice growth problems the wavelength-integrated or total albedo a_t is frequently used, where

$$\alpha_t = \frac{\int \alpha(\lambda) F_r(0, \lambda) d\lambda}{\int F_r(0, \lambda) d\lambda}. \tag{9.9}$$

A sense of the range of a_t values for materials of interest to this book can be obtained from Figure 9.3 (Perovich 1996, 1998). Several obvious trends can be seen. For example, snow has the highest a_t, starting with newly fallen snow (0.87) and decreasing to 0.77 when the snow starts to melt. Next is frozen white ice (0.70), followed by melting white ice (0.68–0.56). Although Figure 9.3 only gives one value (0.52) associated with bare (snow-free) FY ice, studies by Weller (1972) show gradually increasing a values as such ice thickens and loses brine (Figure 9.4). Other measurements of a_t on cold, thick FY ice by Grenfell and Maykut (1977) suggest slightly higher a_t values up to 0.6. For cold ice thicker than ~0.8 m, albedo shows little change with thickness.

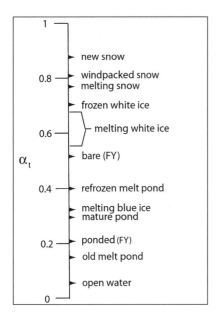

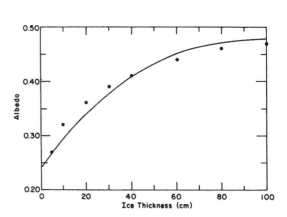

Figure 9.3. A sampling of observed total albedo values for different sea ice types as reported by Perovich (1996).

Figure 9.4. Surface albedo of bare sea ice as a function of thickness (Weller 1972). Parameters for the fitted polynomial can be found in Cox and Weeks (1988b).

All the features with lower albedos are associated with the presence of meltwater in one form or another, starting with a refrozen melt pond (0.40) and continuing with melting blue ice, where the blue indicates either the presence of considerable water within the ice or the absence of bubbles. This is followed by a sequence of melt ponds of increasing maturity (0.29–0.15). The listing concludes with a value for open water of 0.06 to 0.07. Because of the very large differences between the albedos of even melting snow and features such as melt ponds, it is clear that the effect of a given amount of incoming shortwave is greatly different in summer than in winter. More will be said about this later.

As sea ice is a translucent material, part of the net shortwave that reaches the ice surface is transmitted through the surface layer and does not immediately affect temperature and mass changes at the surface. As a result, as noted in equation 9.3, the amount of shortwave that actually enters the ice, I_o, is usually taken to be a fairly small percentage i_o of the net shortwave radiation, or

$$I_o = i_o (1-\alpha) F_r. \qquad (9.10)$$

Therefore the first two terms of equation 9.3 become $(1-a)(1-i_o)F_r$. Values that have been used for i_o by Ebert and Curry (1993) range from 17 to 35%. Untersteiner (1961) suggested a value of ca. 30%.

Sediment and other particulate matter are frequently entrained in sea ice. This is particularly the case for ice formed over the shallow shelves in the Arctic. Individuals needing to take such matters into account should examine Light et al. (1998), a study that combines multispectral observations with structural–optical modeling. It was found that particles with effective radii of > 30 μm have little effect on the bulk optical properties of sea ice. However, even apparently clean ice can contain trace amounts of particles (5–10 gm^{-3}) with effective radii of ~9 μm; these amounts can reduce albedo values by as much as 5–10% in the visible part of the spectrum. The results indicate that particulates principally affect radiative transfer in the visible range whereas the structure of the ice affects radiative transfer in the near-infrared.

It is this combination of low values of incoming shortwave and high albedos, coupled with the fact that fall and winter are also the seasons when most new thin ice growth occurs, that allows one to neglect shortwave terms in many thin ice growth simulations.

9.2.2 Longwave Radiation

The most important term dominating the surface heat exchange over sea ice is the longwave balance, the value of which, F_L, is determined by the difference between the incoming and outgoing longwave terms $F_L \equiv F_{L\downarrow} - F_{L\uparrow}$. Here the incoming or downwelling longwave $F_{L\downarrow}$ from the atmosphere is

$$F_{L\downarrow} = \varepsilon^a \sigma T_a^4 \qquad (9.11a)$$

where ε^a is an effective emissivity for the atmosphere (a value determined by the vertical structure of temperature and humidity in the troposphere), σ is the Stefan–Boltzman constant, and T_a is the radiative temperature of the atmosphere usually taken to be the air temperature at some specified height. Annual total values for $F_{L\downarrow}$ are typically more than twice the value of the incoming shortwave radiation F_r. Only during the summer melt period, when F_r values are large, are such values a few percentage points greater than $F_{L\downarrow}$.

Similarily the outgoing longwave term is

$$F_{L\uparrow} = \varepsilon^i \sigma T_i^4 \qquad (9.11b)$$

where ε^i and T_i are the emissivity and temperature of the snow or ice surface exposed to the atmosphere. In spite of the magnitude of $F_{L\downarrow}$ relative to F_r, the net longwave term (F_L) is typically negative, as a result of the radiative temperature of the ice or snow surface usually being colder than the atmosphere. As might be expected, net longwave losses are least during the summer, when $T_a \approx T_i$ and the water vapor content of the air is high. During the winter large longwave losses result from the decreasing cloud cover and from the fact that as the result of the very cold temperatures, the absolute amount of water vapor in the air near the surface is very low. Unfortunately, good time series of F_L values for polar locations are rare, primarily as the result of difficulties in keeping the glass domes of sensors free from frost and condensation. Therefore, ε^a values are frequently estimated using empirical relations.

Key et al. (1996) have also examined the effectiveness of the seven varied relations that have, to date, been used to estimate incoming longwave values under arctic conditions. The relation that represented the Barrow and Resolute Bay clear sky data with the least error was that of Efimova (1961). In this relation

$$F_{L\downarrow(clr)} = \sigma T^4 (0.746 + 0.0066e) \qquad (9.12)$$

e is the near-surface partial vapor pressure of water. Efimova's relation was developed from a study of year-round Russian land station data. It is similar in form to a relation suggested earlier by

Brunt (1932), who used e directly instead of $\sqrt{e}$ to account for higher fluxes under conditions of low humidity. To include the effects of clouds a variety of different relations have been used. For example Marshunova (1961) suggested

$$F_{L\downarrow(all)} = F_{L\downarrow(clr)}(1+xC) \tag{9.13}$$

where C is the cloud fraction. The value of x was determined by a comparison with year-round arctic observations and varied seasonably ranging from 0.16 in the summer to 0.31 in the winter. The most successful relation in the Key et al. study was suggested by Jacobs (1978), who found that assuming $x = 0.26$ in equation 9.13 gave a best fit to meteorological observations collected in Baffinland. Combining the Efimova and the Jacobs relations gives

$$F_{L\downarrow(all)} = \sigma T^4 (0.746 + 0.0066e)(1 + 0.26C) \tag{9.14}$$

as the recommended relation. A slightly simpler relation that gives generally comparable results is that of Maykut and Church (1973)

$$F_{L\downarrow(all)} = F_{L\downarrow(clr)}\left(1 + 0.22C^{2.75}\right) \tag{9.15}$$

based on data taken over a year's time at Barrow, Alaska.

9.2.3 Turbulent Heat Exchange

Here again there are two terms to consider: the sensible and latent heat fluxes F_s and F_e. Measurements of these terms are also not common and they are frequently estimated using the bulk parameterizations

$$F_s = \rho_a c_{p(a)} C_s u(T_a - T_o) \tag{9.16}$$

and

$$F_e = \rho_a L C_e u(q_a - q_o). \tag{9.17}$$

Here ρ_a and $c_{p(a)}$ are the density and the specific heat of the air, T_a and T_o are the air temperature (at some reference height) and the surface temperature, u is the wind speed (also at the reference height), L is the latent heat of vaporization, q_a and q_o are the specific humidities (at the reference height and at the surface), and C_s and C_e are bulk transfer coefficients, whose values are taken to be 0.00175. Values for L can be obtained from

$$L = \left[2.5 \times 10^6 - 2.274 \times 10^3 (T_a - 273.15)\right]. \tag{9.18}$$

Here L is in joules per kilogram and T_a is in degrees Kelvin. Values for q can be expressed in terms of the partial pressure (e) of water vapor and the total atmospheric pressure (P) as

$$q = \frac{0.622e}{P - 0.378e} \approx \frac{0.622e}{P} \tag{9.19}$$

in that $P >> e$. Here e and P are in millibars (mbar). If we then express the variation in e with T as a fifth-order polynomial, the difference in specific humidity can be obtained from

$$(q_a - q_o) = \frac{0.622}{P_o} \left[a \left(f T_a^4 - T_o^4 \right) + b \left(f T_a^3 - T_o^3 \right) + \cdots + e(f-1) \right]. \tag{9.20}$$

In this equation P_o is the atmospheric pressure at the surface (1013 mbar) and f is the relative humidity at the reference level, and it is assumed that at the ice or snow surface the air is saturated ($f = 1$). Values of the fitted constants a, b, c, d, and e can be found in Maykut (1978) or in Cox and Weeks (1988b). In his modeling for the central Arctic during the growth season Maykut assumed that f was equal to 0.90 for most of the winter period. A review of observations from both the Antarctic and the Arctic and including an extensive data set obtained on the SHEBA project shows that near-surface water vapor over sea ice is always near to saturation ($f = 1$) and sometimes slightly supersaturated (Andreas et al. 2002). The reasons for this will be discussed in the chapter on leads. Field observations using profile techniques over MY ice (Leavitt et al. 1978) suggest that F_e is negligible during the winter and during the summer has values similar to that of F_s. Values obtained by Doronin (1963) under apparently similar circumstances are appreciably higher than Leavitt's by a factor of 2 to 3. Maykut (1978) suggests that some of these differences may be the result of uncertainty in the selection of suitable turbulent transfer coefficients. One thing appears clear: transfer coefficients over thicker ice types where the atmosphere may be stable are undoubtedly significantly different from those over open water and thin lead ice, where atmospheric conditions are clearly unstable (Andreas 1980; Lindsay 1976).

9.2.4 Conductive Heat Flux

As noted earlier, if it is assumed that the temperature profile is linear, then the conductive heat flux through the ice can be calculated from equations 9.1 or 9.2, depending upon whether the ice is snow-free or not. To carry this out, the input terms needed in addition to the environmental terms are the thermal conductivity of the sea ice (k_i) and the snow (k_s). There have been a number of different attempts to produce a simple procedure for estimating k_i. One equation frequently used because of its simplicity was suggested by Untersteiner (1961), who related k_i to brine volume, ice temperature, and the conductivity of pure ice as follows:

$$k_{si} = k_i + \beta S_{si} / T. \tag{9.21}$$

Here k_i is the thermal conductivity of pure ice, S_{si} is the salinity of the sea ice in ‰, T is the temperature in °C, and $\beta = 0.13 \times \text{W m}^{-1}$. The value for the conductivity of pure ice k_i (W/m°K) can be obtained from

$$k_i = 9.828 \exp(-0.0057\,T). \tag{9.22}$$

Note that at temperatures near 273°K, the value of the thermal conductivity of ice is about 2.2 W/m °K, a value about four times that of water. An alternative approach to estimating k_{si} has been suggested by Ono (1968), who used the relation

$$k_{si} = k_i (1 - v_b) + k_b v_b \qquad (9.23)$$

where the thermal conductivity of pure brine k_b (W/m °K) was taken as

$$k_b = 4.186 \times 10^4 \left[1.25 \times 10^{-3} + 3.0 \times 10^{-5} (T - 273.15) + 1.4 \times 10^{-7} (T - 273.15)^2 \right]. \qquad (9.24)$$

In equations 9.22 and 9.23 T_i and v_b are the temperature and the brine volume in the upper level of the ice sheet where v_b can be calculated using equations 6.2 and 6.3 (Frankenstein and Garner 1967). Estimating values for k_{si} will be discussed further in Chapter 15.

The other input parameter that is needed is the thermal conductivity of the snow resting on the upper surface of the ice. Although the characteristics and properties of the snow cover on sea ice will be discussed in some detail in Chapter 15, here let us follow Maykut (1986) by noting that the thermal conductivity of snow is appreciably less than that of ice by a factor of ~5, and taking its value to be $k_s = 0.31 \text{ W m}^{-1} °K$. A more flexible approach would be to estimate the density of the snow from its thermal history and then calculate its thermal conductivity from

$$k_s = 0.0688 \exp\left(0.0088 T_s + 4.6682 \rho_s\right) \qquad (9.25)$$

where T_s (°C) and ρ_s (Mg/m³) are the temperature and density of the snow (Yen 1981).

As I am now going to move on to some results of the thin ice model, the reader will probably realize that there are at least two additional terms that could be added to equation 9.3. For instance, what about the possibility of ice melting, and of heat from the ocean? Justifications for ignoring these terms in the thin ice model are usually as follows. As thin ice growth primarily occurs during the winter, melting is not a significant factor. There are several reasons for ignoring the ocean heat flux values. First, measurements are commonly lacking. In addition, many sites of interest are near shore, where the heat capacity of the shallow water column is small. Finally, over the short periods of time required for appreciable thin ice growth, oceanic terms are typically small compared to surface heat losses.

9.2.5 Thin Ice Results

By now the reader should realize that ice growth calculations that take all the appropriate parameters into account require considerable "bookkeeping." In his thin ice paper Maykut has used polynomial fits to the time series of energy fluxes estimated by Maykut and Untersteiner (1971) for the first day of the month for the surface of MY ice in the central Arctic Basin. Although he did not present the polynomial coefficients, presumably equivalent estimates of these coefficients can be found in Cox and Weeks (1988b, Appendix A). One should remember that these values were based on field observations collected on drifting stations primarily in the 1950s and 1960s.

As there is good evidence for significant climatic changes in this geographic region, flux estimates based on current measurements would be different.

There is now enough information to solve the upper boundary condition (equation 9.3) for T_o for a given snow and ice thickness. This is done by substituting equations 9.2, 9.4, 9.10, 9.16, 9.17, and 9.20 into equation 9.3. The result is

$$(1-\alpha)(1-i_o)F_r+F_{L\downarrow}-\varepsilon^i\sigma T_o^4+\rho_aC_pC_su(T_a-T_o)+0.622\rho_aLC_eu(fe_s-e_{so})+\gamma(T_f-T_o)=0. \qquad (9.26)$$

As has been shown, the input consists of snow, ice, and air properties and a variety of external parameters related to environmental conditions. Clearly a critical external parameter varying with the environmental history is the snow depth h_s.

The lower boundary condition at the growing sea ice–seawater interface can be obtained by combining equation 4.5 (setting $F_w = 0$)

$$F_c=-\rho_iL\left(\frac{dH}{dT}\right) \qquad (9.27)$$

with equation 9.2, producing

$$-\rho_iL\left(\frac{dH}{dT}\right)=\frac{k_ik_s}{k_ih_s+k_sH}(T_f-T_o). \qquad (9.28)$$

Now if T_o is known, one can use equation 9.28 to calculate growth rates for any specified values of H and h_s.

Figures 9.5, 9.6, and 9.7 show examples of the use of the above model. The first figure, Figure 9.5, clearly shows that the change in T_o with H is strongly dependent on the season of the year, with T_o decreasing rapidly with increasing ice thickness when the air is cold (March). In the spring and fall this dependence is significantly less in that the ice surface temperature is near the melting point. The second figure, Figure 9.6, shows the seasonal variation in T_o for several different ice thicknesses. Here the 3-m MY curve was obtained from the thick ice model of Maykut and Untersteiner (1971) and includes the effect of a seasonally varying snow cover. The final figure, Figure 9.7, shows the striking effect that variations in the thickness of snow have on the surface temperature of 0.1-m-thick sea ice during different times of the year. Looking at this figure gives one an appreciation of the difficulties one encoun-

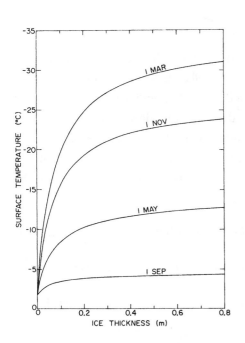

Figure 9.5. Predicted changes in the surface temperature of young sea ice with thickness and season (Maykut 1978).

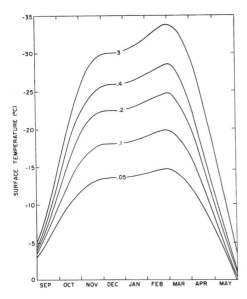

Figure 9.6. Seasonal variations in surface temperatures over different thicknesses of sea ice in the central Arctic Basin (Maykut 1978). The lines of constant thickness are in meters. The 3-m thickness calculation assumes a seasonally varying snow cover and uses the thick ice model of Maykut and Untersteiner (1971).

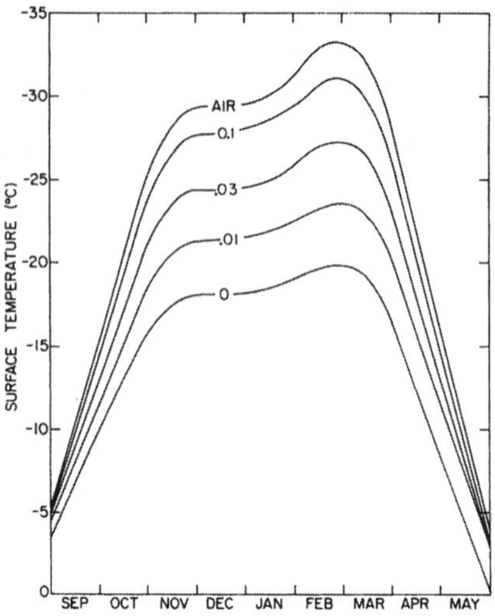

Figure 9.7. Seasonal variations in surface temperature over 0.1-m-thick sea ice calculated for a variety of different snow depths (in meters). The assumed air temperature values are also shown (Maykut 1978).

ters when trying to estimate thin ice thicknesses from airborne radiometer measurements that determine T_o. A number of additional results from the model can be found in Maykut (1978). More will be said about these matters in the chapter on polynyas and leads, locations where thin, rapidly growing sea ice is a prominent feature.

9.3 A Thick Ice Model

Fortunately, we will not have to replow most of the ground that we have covered in the thin ice model, as the considerations discussed there apply equally well to thick sea ice. However, some of the flux terms need to be considered in more detail and at least two additional terms need to be added. This additional effort is necessitated by the fact that thin ice models rarely look at the summer period, as thin ice invariably melts. Thick ice models, on the other hand, are invariably used to look at ice growth and decay through the complete year and in many cases through multiyear cycles. Therefore, terms that deal with evaporation and melting have to be considered, as do the effects of melt pond formation. As a result equation 9.3 becomes

$$(1-a)F_r - I_o + F_{L\downarrow} - F_{L\uparrow} + F_s + F_e + F_c + F_m = 0 \tag{9.29}$$

where F_m is the heat loss resulting from the melting of ice and snow, which can be expressed as

$$F_m = \left[\rho L \frac{d(H+h_s)}{dt} \right]_0 . \tag{9.30}$$

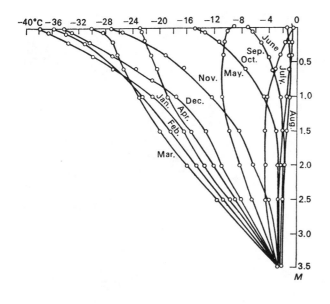

Figure 9.8. Mean monthly temperature profiles from multiyear sea ice as observed on Russian North Pole stations (Yakovlev 1963).

When T_o is at the melting temperature, any surplus energy flux arriving at the surface will result in melting and a change in $H+h_s$. During summers when F_r becomes a large term there is also penetration of shortwave radiation into the ice, resulting in internal melting. Therefore, the optical properties of the ice and melt ponds become of importance. Also we can no longer kid ourselves into believing that the oceanic heat flux $F_w=0$ and will have to attempt to take this term into account. As will be seen, even if F_w is small, over time it can exert a large effect on ice thickness. Finally as the ice becomes thicker than ~0.5 m, the assumption that the temperature gradient across the ice and snow is linear is no longer a reasonable approximation. That this is the case can be clearly seen in Figure 9.8, which presents a series of temperature profiles obtained on 3.5-m-thick MY ice in the early 1950s by Yakovlev (1963). Now let us look at some of these changes in some detail.

9.3.1 Additional Shortwave Considerations

During the spring and summer, incoming shortwave radiation clearly cannot be neglected. However, even during the peak of the summer melt period, values for F_r are only slightly larger than incoming longwave radiation $F_{L\downarrow}$. The reasons here are low sun angles and persistent cloudiness. However, these factors are offset by the very long days, resulting in monthly totals for the central Arctic during this time period being equal to or exceeding values measured at such midlatitude locales as New York or Chicago (Maykut 1986). Although F_r values peak at the summer solstice, the values are not symmetrical around the solstice, with April and May values being larger than those of July and August. This asymmetry is believed to be the result of an increase in cloudiness during the summer and a decrease in the diffuse radiation resulting from a decrease in the surface albedo. Although at midlatitudes clouds can reduce the values of F_r as measured at the surface by as much as 80 to 90%, in the central Arctic, because of the low water content of the clouds and their relative thinness, decreases in excess of 50% are rare.

During the start of the melt period, the albedo of the snow decreases from values of 0.80–0.87 to values in the range of 0.70–0.75. Once the snow has melted away, the albedo of the bare ice is strongly dependent on the age of the ice. As was noted earlier, thick ($H > 0.8$ m), cold FY sea ice has albedos in the range of 0.5 to 0.6. However, once surface melting starts a values become quite variable ($a = 0.3$–0.5), depending on the specific wavelength being considered and the exact state of the upper ice surface. As pointed out by Maykut (1986), if the ice is well drained, resulting in a whiteish appearance, albedos in the higher end of this range are observed, presumably because

Figure 9.9. Views of two different melt ponds showing their complex shapes and differing degrees of perforation through to the underlying ocean.

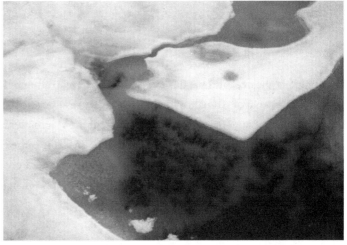

of the increased scattering from the large volume of included gas in the form of bubbles and irregular drained channels. If the ice is water-saturated, a values can be as low as 0.3.

A major factor controlling the effective albedo of sea ice during the summer is the nature of the surface melt ponds. In the Arctic such features invariably occur and at times can cover as much as 60% of the total ice area, although maximum values of 30% are more likely. Good, high-quality measurements of melt pond areas and characteristics have been rare in spite of the importance of this information (Fetterer and Untersteiner 1998). Some sense of the difficulties encountered in describing melt ponds can be seen in Figures 9.9a and 9.9b, which show melt ponds characterized by variable depths and different degrees of perforation through to the underlying ocean. An interesting field study of the hydraulic linkages between albedo, surface morphology, melt pond distribution, and ice properties (Eicken et al. 2004) shows that albedo is critically dependent on the melt pond hydrology—a parameter which, in turn, is controlled by melt rate, surface topography, and the permeability of the ice sheet.

In the Antarctic, melt ponds appear to be less ubiquitous. An explanation for this difference advanced by Andreas and Ackley (1981) posits that the reason is not that melting does not occur in the Antarctic but that evaporation and sublimation are much larger factors than simple melting. Considering the windiness of the Southern Ocean, their explanation appears to be reasonable.

As determined by Grenfell and Maykut (1977) and shown in Figure 9.10, the values of the spectral albedos ($a\lambda$) of snow are not strongly dependent on wavelength, as are the values of bare ice and melt ponds. This dependence is weakest in the visible portion of the spectrum (400–700 nm), with a strong decrease occurring at red wavelengths. The decrease in $a\lambda$ values at $a\lambda$ >700 nm is caused by the large amount of liquid water present in the ice. For shorter wavelengths it is the properties of the

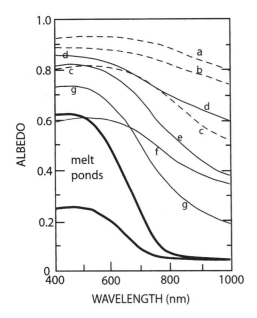

Figure 9.10. Spectral albedos as measured over varied ice surfaces by Grenfell and Maykut (1977): (a) dry snow; (b) wet new snow; (c) melting old snow; (d) frozen MY ice; (e) melting MY ice; (f) melting FY white ice; (g) melt pond with a 3-cm-thick ice cover. Within the area marked as melt ponds (without a surface ice cover as in g), Grenfell and Maykut report several sets of measurements, all with the same general shapes as the two boundary curves. In general, newly formed melt ponds have higher albedos than old, well-developed ponds.

near-surface layer of the ice that are important, as light penetration is small at these wavelengths. As wavelength increases, the properties of ice deeper in the sheet become important. If melt ponds do form, they are a major factor in controlling the surface heat balance, resulting in a values of 0.4 or less, for shallow ponds, to values as low as 0.15 for deep ponds, where the ice cover may be totally perforated. In such cases the melt pond albedo approaches that of the open ocean as the pond widens.

In that scattering and absorption attenuate the light as it passes through the ice and snow, it is important to consider this factor in any study that focuses on the late spring, summer, and early fall period. This can be done by considering the spectral extinction coefficient

$$\kappa_\lambda = \frac{1}{-F(z,\lambda)}\frac{dF}{dz} \tag{9.31}$$

where $F(z, \lambda)$ is the downward shortwave flux at depth z and wavelength λ. Values of κ_λ for different ice types are shown in Figure 9.11. Note that the largest attenuation occurs in the snow pack and that once radiation penetrates below the surface layer of the ice, κ_λ values are reasonably constant. There is also a lower extinction rate in the blue end of the spectrum than at the red end. This difference accounts for the tendency of ice to appear blue when viewed using light that is transmitted through the ice. If one thinks in terms of the so-called e-folding distance (the distance required to reduce the intensity by $1/e = 37\%$), one finds that for typical sea ice these distances vary from 24 m at 470 nm (blue) to 8 m at 600 nm (red) to 2 m at 700 nm (near infrared; Perovich 1998). The radiative flux at any depth below the ice surface can be found from Beer's law

$$F(z,\lambda) = F(0,\lambda)\exp^{-\kappa_\lambda z} \tag{9.32}$$

obtained by integrating equation 9.31. Here $F(0, \lambda)$ is the net total shortwave radiation of wavelength λ at the surface of the ice. More useful terms are F_z, the total flux in the ice given by

$$F_z = \int_0^\infty F(z,\lambda)d\lambda. \tag{9.33}$$

The limits on this integral are usually given as either 300 to 3000 nm or 250 to 2500 nm.

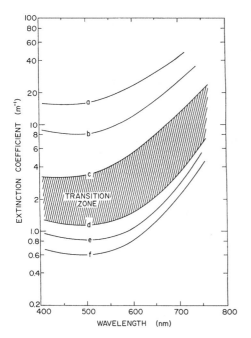

Figure 9.11. Spectral extinction coefficients for different types of sea ice and snow (Maykut 1986, after Grenfell and Maykut 1977): (a) dry compact snow; (b) melting snow; (c) surface layer on melting MY ice; (d) interior of MY ice; (e) interior of FY ice; and (f) ice beneath an old melt pond.

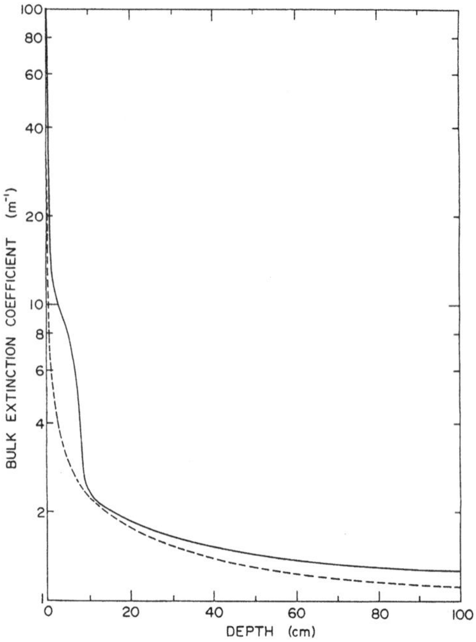

Figure 9.12. Bulk extinction coefficients in FY blue ice (dashed curve) and in MY ice (solid curve) (Maykut 1986, after Grenfell and Maykut 1977).

The bulk extinction coefficient κ_z is defined as the ratio of the total energy absorbed at depth z to the total flux at the same depth or

$$-\kappa_z = \frac{\int \kappa_\lambda F(0,\lambda)e^{-\kappa_\lambda z}\,d\lambda}{\int F(0,\lambda)e^{-\kappa_\lambda z}\,d\lambda} \quad . \tag{9.34}$$

Equation 9.32 then gives

$$F_z = F_o e^{-\int \kappa_z\,dz} \tag{9.35}$$

where

$$F_o = \int_o^\infty (1-\alpha)F(0,\lambda)\,d\lambda \tag{9.36}$$

is the net total shortwave radiation at the surface. Values of these bulk coefficients for both FY and MY ice have been calculated by Grenfell and Maykut (1977), who found that that they are very strongly depth dependent in the upper 10–20 cm of the ice and only weakly dependent at greater depths (Figure 9.12). The most recent work in this general area is by Light et al. (2008) and is particularly focused on bare and ponded FY ice observed during the melt season as part of the SHEBA program. They report integrated irradiance extinction coefficients for the interior of the ice that range from 0.65 to 0.98 m^{-1} at 600 nm. These values are appreciably smaller than the previously reported values of ~1.5 m^{-1} (Grenfell and Maykut 1977), indicating that more light penetrates deeper into the ice and into the ocean than previously thought. The results also support the earlier conclusion of Perovich (2005) that ponded ice transmits three to five times the amount of solar radiation to the ocean than does an equivalent thickness of nonponded ice.

The most fully developed model study to date of the effect of sea ice on the solar energy budget is that of Jin et al. (1994), who divided the solar radiation spectrum into 24 spectral bands and considered the effects of ice property changes such as salinity, density, and snow cover variations on the solar energy distribution in the entire system. Although this approach is computationally intensive and has not been directly coupled with an ice growth model, the results of the study are of interest in that they reinforce portions of the previous discussion. The results show that for bare ice it is the scattering, determined by the air bubbles and the brine pockets in the very top layer of the ice, that is most important in the solar energy absorption and partitioning within the whole system. This is particularly the case for MY ice, which characteristically has a well-drained surface layer containing a large volume fraction of air bubbles. Because air bubbles are more effective scatterers than brine pockets, radiative absorption is more sensitive to air volume variations than to brine volume variations. It was also found that ice thickness has a significant effect on the radiative balance in the atmosphere–sea ice–ocean system while the ice is thin. As thickness increases in the range 0 to 70 cm there is an increase in the radiative absorption in the ice and a decrease in the ocean and in the entire system. However, once ice thickness exceeds 70 cm, the total absorption in the atmosphere and in the entire system is no longer sensitive to ice thickness change. When a layer of water simulating a melt pond is present on top of the ice, the total absorption of energy by the whole system increases as a result of the drop in the surface albedo. In addition radiation penetration into the ocean increases whereas the change in the atmosphere is slight, a result supported by the work of Light et al. (2008) mentioned above. Again stressing the importance of melt ponds in the summer heat balance, it was found that a melt pond depth of only 5 cm can absorb about half of the total energy absorbed by the whole system. The presence of clouds was found to moderate all the sensitivities of the absorptive amounts in each layer to variations in ice properties and thickness.

9.3.2 Oceanic Heat Flux

This term is the "loose cannon" in sea ice growth calculations, in that only rarely are detailed observations of ablation at the underside of the ice available that allow estimates to be made of F_w. Early field observations by Badgley (1966) and theoretical considerations by Untersteiner (1964) and Maykut and Untersteiner (1971) suggested that in the MY ice region of the Arctic Basin the average value of F_w was roughly 2 W m^{-2}. This value has frequently been used because of the lack of site-specific data, as it generally matches estimates of the rate of heat loss from the underlying layer of Atlantic water (2.0–2.6 W m^{-2}) made by Panov and Shpaikher (1964). However, there is no general agreement that this is the primary mechanism fixing F_w, as mixing with colder shelf water and the absorption of shortwave energy in open leads have also been suggested as possible contributing mechanisms. Model calculations by Ebert and Curry (1993) suggest that the annual cycle for the heat flux at the base of the ice is roughly a bell-shaped curve with a peak of slightly under 7 W m^{-2} on day 220 and values of effectively zero during the winter, between day 300 and day 80 (Figure 9.13). When

Figure 9.13. The annual cycle of the heat flux at the base of the ice F_w as estimated by Ebert and Curry (1993).

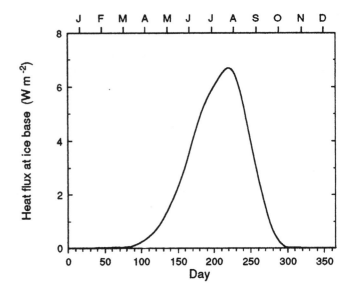

their curve is summarized as an annual average, the predicted value is 1.8 W m^{-2}, in reasonable agreement with the 2 W m^{-2} value mentioned earlier. Other annual estimates have been appreciably higher. For instance Fleming and Semtner (1991) suggest a yearly average value of 5 W m^{-2}.

Examination of F_w values determined from actual field observations support the idea that there are definitely seasonal and regional variations. For instance, Maykut and McPhee (1995), who examined data collected on MY ice in the Beaufort Sea in 1975 during AIDJEX, found a strong seasonal dependence with an annual average of 3 to 5 W m^{-2}, and peak values during the summer as high as 40 W m^{-2}. In the same general area in 1993–1994 Perovich et al. (1997) obtained an annual value of 4 W m^{-2} and a summer average of 9 W m^{-2}. Additional data, also from the Beaufort Sea, were obtained during the yearlong (October 1997–October 1998) deployment of the SHEBA program (Perovich and Elder 2002). A strong seasonal cycle was observed: values ranged from a few W m^{-2} in the October–May period, with a steady increase through June and July, to a peak value of 35 W m^{-2} occurring in late July during a period of appreciable ice motion. The correlation between drift speed and heat flux is physically reasonable as speed can, in a simplistic sense, be taken as a surrogate for turbulent mixing. Annual averages for MY ice ranged from 7.5 W m^{-2} for an undeformed site, to 10.4 W m^{-2} for a melt pond site, to 12.4 W m^{-2} for a site on an old ridge. Although regional variations are to be expected, their patterns are not as well known as might be desired. Maykut (1986) has suggested that much larger F_w values should be expected in the vicinity of Fram Strait as the result of the proximity of the warm West Spitzbergen Current. Working in this general region, Wettlaufer (1991) found that even on a scale as small as a few hundred meters there was considerable spatial variability in F_w values and that changes in ocean water mass characteristics strongly affected F_w values.

The most detailed regional look at this problem, to date, is the study by Krishfield and Perovich (2005), who combined seawater temperature and salinity observations obtained on a variety of earlier drift stations with relations suggested by McPhee (1992) in estimating the annual temperature cycle in the mixed layer beneath the ice. These values were then combined with drift rates to estimate F_w values. They found that even in winter F_w values were not negligible. This was particularly true in the Transpolar Drift Stream, a situation that suggests positive contributions to F_w values from sources other than solar heating such as synoptic storms (Steele and Morison 1993; Yang et al. 2001). Although counseling the reader that the interannual variations in their resulting ΔT_f values are, to some degree, specified by the nature of the parameterization, Krishfield and Perovich suggest that there is an overall positive trend in the F_w values of 0.2 W m^{-2} per decade, with the largest variations occurring in the southern Beaufort Gyre.

Data from the Antarctic are even less plentiful. As discussed by Andreas and Ackley (1981) the general rarity of surface melt ponds on antarctic pack ice can be taken to indicate that the seasonal disappearance of the pack is primarily the result of heat in the upper water column. There are two possible sources here: the absorption of shortwave radiation at the ocean surface and the advection of heat from the deeper ocean. Unfortunately, the relative contributions of these very different sources are usually not well established. Model studies to date have proven to be somewhat schizophrenic on this subject. For instance Parkinson and Washington (1979), using a large-scale model, obtained the best comparisons with observations when constant F_w values of 25 W m^{-2} were used. Somewhat less high values of 16 W m^{-2} were obtained by Gordon and Huber (1990). Gordon has argued that a heat flux of as much as 30 W m^{-2} is required to explain the rapid dissipation of the ice cover in the spring. On the other hand, Hibler and Ackley (1982, 1983), via the use of a dynamic/thermodynamic model, were able to simulate the annual ice cycle in the Weddell Sea by using a heat flux as low as 2 W m^{-2}. Studies by Robertson et al. (1995) and by Lytle and Ackley (1996) have also suggested that for the western Weddell Sea F_w values were low (in the range of 2–7 W m^{-2}). A long-term study (12 years) by Heil et al. (1996) of the role of the oceanic heat flux in affecting the growth of antarctic fast ice used a multilayer thermodynamic model driven by meteorological observations to determine the F_w values by difference. They found that at their field site in the Prydz Bay region near Mawson Station, F_w values averaged 7.9 W m^{-2}, with the yearly means varying between 5 and 12 W m^{-2}. Seasonal values varied from 0 to 18 W m^{-2}. The pattern was as follows: during the early growth period F_w values were generally < 6 W m^{-2} whereas during the late winter F_w values as high as 18 W m^{-2} were more representative. Furthermore, there appeared to be a decadal trend in the data, with values dropping from an average of about 10 W m^{-2} to 6–8 W m^{-2}. It is suggested that during the period of rapid ice growth, the thermohaline convection resulting from the growth is sufficient to cool deeper, warmer waters, causing the water column to become isothermal and lowering F_w to near zero. In the austral spring, when ice growth has decreased to near zero and thermohaline circulation has ceased, warmer Circumpolar Deep Water can increasingly move into the Prydz Bay area, resulting in a gradual increase in F_w values.

What is one to make of all this? Obviously, estimating the oceanic heat flux in the absence of direct determinations is not simple. In the Antarctic it is clear that there appear to be large regional and seasonal variations associated with varying oceanographic conditions. In the Arctic matters appear to be simpler, but this may be deceiving. Individuals wishing to explore these matters in more detail should start by studying the 2008 book by McPhee, an individual who has contributed much of the original work on this subject.

9.3.3 The Model

As was noted earlier, once sea ice reaches a thickness of > 80 cm the assumption of a linear temperature gradient across the thickness of ice and snow is no longer realistic. There also can be an appreciable temporal lag between surface temperature changes and the growth response of the ice. This can be seen in Figure 9.8 and also in Figure 9.14, which summarizes temperature and

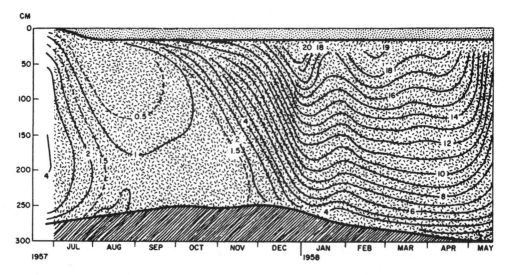

Figure 9.14. Temperature and thickness variations of an MY ice floe studied during the drift of Ice Station Alpha by Untersteiner (1961).

thickness variations in an MY ice floe studied during the drift of Ice Station Alpha (Untersteiner 1961). For instance, note that although the ice surface starts cooling in September this is not felt at the bottom of the ice until mid-December. Also of interest is the fact that the surface warming that occurred in January damps out before it can reach the bottom of the ice sheet and affect the growth rate.

The elements of the model are as follows. Note the differences from the thin ice model.

1. The surface heat balance equation is identical except for the addition of the final term that considers surface melting,

$$(1-\alpha)F_r - I_o + F_{L\downarrow} - F_{L\uparrow} + F_s + F_e + F_c + \left[\rho L \frac{d}{dt}(h+H)\right]_o = 0 \,.$$

2. The temperature distribution within the ice is calculated from a modified version of the heat conduction equation

$$(\rho c)_i \frac{\partial T_i}{\partial t} = k_i \frac{\partial^2 T_i}{\partial z^2} + \kappa_i I_o \exp(-\kappa_i z) \,. \tag{9.37}$$

Here the last term describes the depth-dependent internal heating that results from the absorption of shortwave radiation, as discussed earlier. The terms are ρ (density), c (specific heat), k_i (thermal conductivity of the ice) and κ_i (extinction coefficient of the ice).

3. The temperature distribution in the snow is calculated from

$$(\rho c)_s \frac{\partial T_s}{\partial t} = k_s \frac{\partial^2 T_s}{\partial z^2} \,. \tag{9.38}$$

Here the parameters are identical to those in equation 9.37, except for the fact that the subscript s indicates that snow is being considered.

At first thought, one might expect that the equation for calculating the temperature distribution within the snow would be similar to that for ice, with the only changes occurring in the values of the appropriate constants. This is certainly possible, if detailed observations of κ_s values are available. However, it was decided that because κ_s values for snow are both varied and high, with values for cold snow in the range of 1.3 to 1.7 cm^{-1} (Mellor 1964a), it would be easier to simplify the relation by dropping the last term in equation 9.37. The justification for this is that because of the high κ_s values, essentially all of the incoming radiation is absorbed in the first few centimeters of the snow cover. Considering the roughness of real polar snow covers, this is essentially equivalent to assuming that all the radiation is absorbed at the surface. Another aspect of this assumption is that if the snow thickness is more than a few centimeters, the amount of shortwave radiation I_o entering into the sea ice layer is essentially zero (i.e., the last term in equation 9.37 only becomes of importance after the snow pack is melted).

4. Finally, there are two balances of fluxes that need to be considered. The first occurs at the snow–sea ice interface, where

$$k_s \left(\frac{\partial T_s}{\partial z} \right)_h = k_i \left(\frac{\partial T_i}{\partial z} \right)_h . \qquad (9.39)$$

The second occurs at sea ice–ocean interface, where

$$k_i \left(\frac{\partial T_i}{\partial z} \right)_{h+H} - F_w = \left[\rho L \frac{d}{dt} (h+H) \right]_{h+H} . \qquad (9.40)$$

As pointed out by Maykut (1986), one of the primary differences between the thin and the thick ice models is that in the thick ice model the conductive heat fluxes at the boundaries depend on the local temperature gradients. The thick ice model also has the advantages of being both physically realistic and, when driven by reasonable estimates of the environmental conditions in the Arctic Basin, resulting in temperature and mass changes that are in general agreement with field observations.

Unfortunately the above series of equations cannot be solved exactly and requires the use of a finite difference procedure to obtain a solution. Maykut and Untersteiner (1969, 1971) used a technique initially suggested by Sauliev (1957a, 1957b). Later authors have used other procedures (Cox and Weeks 1988b; Ebert and Curry 1993; Maksym and Jeffries 2000). Whatever the procedure, the finer the vertical resolution demanded, the longer the required computational time. Because of these time constraints, the full Maykut and Untersteiner model has typically not been used in large-scale multiyear simulations of pack ice behavior. Instead, a streamlined procedure developed by Semtner (1976) has been used that agrees with the more involved model to within 25 cm under most conditions and is much faster. More will be said about these matters in the chapter on ice dynamics. This said, because of the incredible increase in computer capabilities between the late 1960s and the present, almost every working scientist has computing

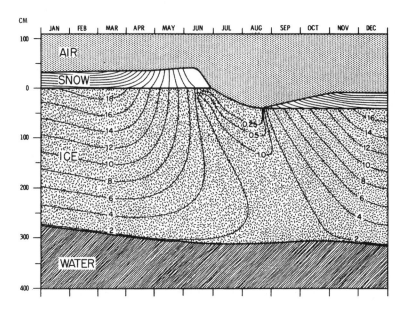

Figure 9.15. Predicted values of the equilibrium temporal temperature and thickness variations in an MY floe in the central Arctic as calculated from the Maykut and Untersteiner model (1969), using the values for the heat budget of the central Arctic as summarized by Fletcher (1965).

capabilities sitting on his or her desk that with a little programming effort would allow them to use the full Maykut and Untersteiner model in a wide variety of process studies.

9.3.4 Model Results

One of the more important results of the thick ice model has already been presented in Figure 4.5. This figure examines the scenario in which FY ice survives the summer although losing a percentage of its thickness. In subsequent years, assuming that there are no appreciable climatic trends and that the losses of ice during the summers are constant, winter ice growth gradually declines as a result of the insulating effect of the gradually thickening ice that survives the subsequent summers and serves as an insulating layer. Ultimately a steady state is reached in which the summer ablation exactly equals the winter's ice growth. Figure 9.15 shows the equilibrium temperature and thickness changes in an MY floe during a year based on the heat budget suggested by Fletcher (1965) for the Arctic Basin. Note that the agreement between the calculated estimates and the observed values as determined by Untersteiner (1961) on a floe studied during the drift of Ice Station Alpha (Figure 9.14) is generally very good.

If the ice thickness has reached an equilibrium value and then the climate regime shifts so that less ice ablates during the summer, the thicker ice surviving the summer results in less growth during the winter, which will ultimately result in a new, somewhat thicker, equilibrium thickness. If the opposite occurs, with an increase in the amount of surface ablation, the thinner ice in the fall results in increased winter growth, which ultimately balances the mass loss at the surface and achieves a new thinner equilibrium. Maykut and Untersteiner have carried out a number of simulations that examine the effects of such variations. In one, if an ice thickness of 340 cm was assumed as the initial value, the steady-state ice thickness averaged 288 cm and varied between 270 and 314 cm depending upon the time of year. The time required to reach the new equilibrium

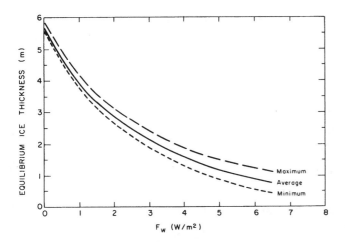

Figure 9.16. The equilibrium thickness of arctic sea ice as a function of the annual value of the oceanic heat flux (Maykut and Untersteiner 1969, 1971). Also included are curves showing the absolute maximum and minimum.

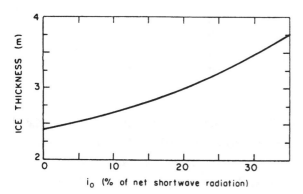

Figure 9.17. The average equilibrium thickness of arctic sea ice as a function of i_o, the percentage of the incoming shortwave radiation that actually enters the ice (Maykut and Untersteiner 1969, 1971).

was 38 years. In general, the time required to reach equilibrium is a few years if the equilibrium thickness is small, say 1 m. On the other hand, if the equilibrium thickness is large, say 6 m, the time to equilibrium becomes tens of years.

Consider the effect of variations in the value of the oceanic heat flux on the equilibrium thickness, assuming that no other parameters are changed. Figure 9.16 shows a graph of the annual value of H_{eq} versus F_w. Also included are curves representing the absolute annual maximum and minimum values. Again the heat budget used is Fletchers. If the average value of F_w exceeds ~7 W m⁻² the ice melts completely during the summer. The figure also shows that if $F_w = 0$ the maximum thickness to which sikussak could grow in the climate of the Arctic Basin as of ~1960 is ~6 m.

One of the more interesting series of simulations carried out by Maykut and Untersteiner examines the effect of variations in i_o, the percentage of incoming shortwave radiation that penetrates into the ice (see equation 9.10), on the resulting equilibrium thickness. Their results are shown in Figure 9.17 and are somewhat counterintuitive in that as the amount of shortwave entering the ice increases, the equilibrium thickness also increases. The explanation is that as i_o increases, the amount of energy available for surface melting decreases. Admittedly, the shortwave that enters the ice causes internal heating and a resulting increase in the amount of liquid within the ice. However, over an annual cycle this has little effect on the mass balance. Maykut (1986) has compared this situation to the formation of melt ponds on the ice surface, which, provided they do not drain, also have little effect on the annual mass balance.

Figure 9.18. The average equilibrium thickness of arctic sea ice as a function of maximum annual snow depth (Maykut and Untersteiner 1969, 1971).

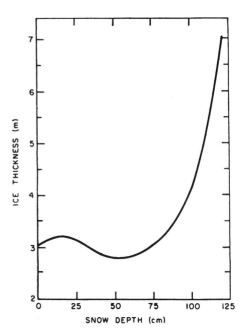

The final simulation that we will consider examines the relation between maximum snow depth on the sea ice at the end of the winter and the equilibrium thickness. The results are shown in Figure 9.18. Again the results are somewhat surprising in that one might anticipate that the more snow, the less loss of ice during the summer and the thicker the equilibrium thickness. Although this is true for snow depths in excess of 75 cm, for depths less than this value the equilibrium thickness oscillates at around 3 m. This appears to be caused by two competing effects. If there is less snow during the winter, the ice is colder and more sea ice growth occurs. In the summer however, in that the snow is thinner, the ice is exposed sooner, and with the associated increase in albedo, more ice melts. In the thickness range between 0 and 75 cm these two effects appear to roughly balance each other. However, at snow thicknesses in excess of 75 cm the decrease in the amount of summer ablation becomes the overriding factor and equilibrium thickness rapidly increases with snow depth.

In the above discussion the reader will note that sometimes I refer to Maykut and Untersteiner (1969) and other times to Maykut and Untersteiner (1971), or to both. Although these two references are fundamentally the same paper, the 1969 version is the "long form" with matters spelled out in considerably more detail and many more scenarios examined. It is definitely worth the trouble to locate a copy of the 1969 RAND report as the two papers are sufficiently different that it is useful to read both. In applying these simulations to current activities in the Arctic Basin, it is important to recall that there appears to have been an appreciable climatic shift since Fletcher made his heat balance estimates. I do not know of a reference that provides estimates of these changes in a form useful for model calculations.

9.4 Further Efforts

Although the reader may be exhausted, the subject of the growth of sea ice is not. In closing this chapter I will try to give the reader some sense of a few of the current activities in this area.

9.4.1 Scattering and the Phase Relations

The reader may have noticed that in the discussions of the albedo of sea ice there was no mention of any relations between the optical properties of the ice and the structure and chemistry of the ice as discussed in Chapters 6, 7, and 8. I am pleased to note that this obvious gap has been filled by a study by Light et al. (2004). This work, which builds on an earlier theoretical study by

Grenfell (1983), utilized FY sea ice collected at Barrow and studied in a freezer in Seattle. A two-dimensional Monte Carlo radiative transfer model was used to derive inherent optical properties from the optical measurements (Light et al. 2003b). Three temperature regimes were identified with the respective temperature ranges of $T > -8°C$, $-8 > T > -23°C$, and $-23°C > T$. These regimes, of course, exactly correspond to the temperature ranges in which no solid salts are present in the ice (except perhaps a tiny amount of $CaCO_3•6\,H_2O$), where $Na_2SO_4•10\,H_2O$ is present and where both $Na_2SO_4•10\,H_2O$ and $NaCl•2\,H_2O$ are present. Within each regime there was a distinct relationship between the ice microstructure and the associated optical properties via the phase relations. In cold ice, volume scattering is dominated by the size and number distribution of the precipitated hydrohalite crystals; at intermediate temperatures it is controlled by changes in the distribution of the brine inclusions, gas bubbles, and mirabilite crystals; and at warm temperatures it appears controlled by temperature-dependent changes in the real refractive index of the brine and by the escape of gas bubbles from the ice. Scattering coefficients can exceed 3000 m^{-1} for cold ice and average ~450 m^{-1} in the intermediate regime, reaching a minimum of ~340 m^{-1} at $-8°C$. In all of the three regimes the scattering is strongly forward peaked.

9.4.2 Further One-Dimensional Thermodynamic Models for Pack Ice

Most approaches to the sea ice growth problem are one-dimensional in that they examine the vertical heat and mass changes in a laterally uniform column of seawater–sea ice–atmosphere. However, in attempts to simulate the behavior of pack ice, a given computational node may represent an area of many square kilometers. Given the typical lateral variability of pack ice, this area commonly contains a variety of ice thicknesses and ages as well as areas of open water in the form of leads or polynyas. Is it possible to translate information on these different ice types and their areal extents in such a way as to produce a fictional one-dimensional model whose behavior nevertheless provides an accurate averaged description of the area represented by the node?

This is the problem explored by Ebert and Curry (1993), who focus on aspects of the ice cover such as surface albedo and leads that are important in interactions with the atmosphere. Factors considered include melt pond development and the amount of meltwater that ultimately runs off the floe, the absorption of solar radiation in and below the leads and the lateral accretion, and ablation on the edges of floes. The effect of a prescribed sea ice divergence rate was also included. A wide variety of sensitivity tests was carried out to study the effects of changes in the major external forcing variables. It was found that the equilibrium sea ice thickness was extremely sensitive to changes in the downward shortwave and longwave fluxes and to atmospheric temperature and humidity, moderately sensitive to changes in the ocean heat flux, and relatively insensitive to small changes in wind speed and snowfall.

Particular attention was paid to four positive feedbacks and two negative ones. These are as follows:

1. *Surface albedo feedback.* A decrease in the surface albedo increases the absorption of radiation at the surface, leading to increased melting and a further reduction in albedo. A consequence can be a reduction in the thickness and/or extent of the ice.

2. *Conduction feedback.* Thinner ice covers experience more conduction of heat from the ocean through the ice, leading to further surface warming and earlier melting.

3. *Lead solar flux feedback.* A greater proportion of shortwave radiation enters the ocean if the ice is thinner. This warms the mixed layer resulting in greater basal ablation.

4. *Lead fraction feedback.* Thinner ice covers undergo greater lateral ice ablation and accretion, resulting in a larger lead fraction in the summer and a smaller one in the winter. As a result more solar radiation is absorbed in the summer and less heat is lost to the atmosphere in the winter, resulting in a warmer mixed layer and greater basal ablation.

5. *Outgoing longwave flux feedback.* The increased upward flux from a warmer surface decreases the surface net flux, causing cooling.

6. *Turbulent flux feedback.* A warmer surface decreases the downward sensible and latent heat fluxes (or increases the upward fluxes). This decreases the net surface flux, resulting in surface cooling.

The Ebert and Curry paper is a useful source of information even if one only intends to utilize a "simple" one-dimensional model such as that of Maykut and Untersteiner, in that several alternate parameterizations with the appropriate constants are suggested for some of the processes. Unfortunately the details of much of their paper are beyond the scope of this book. I would like to conclude my discussion of this approach by showing a wonderful "wiring" diagram from this paper (Figure 9.19). This figure provides a schematic of the interactions between the external forcing (**bold**) and parameter values (*italics*), and internal variables (ovals) and fluxes (boxes). The solid arrows represent positive interactions, the short-dashed arrows represent negative interactions, and the long-dashed arrows represent interactions that may be either positive or negative depending upon the season. The thick borders around net flux and ice thickness are to emphasize their importance. I could explain this diagram but I will leave that to Ebert and Curry. Besides, I am certain that anyone who has read the present chapter is now so endowed with sea ice growth insights that the diagram will be intuitively obvious. Frankly I have always found complex wiring diagrams to be an excellent alternative to counting sheep as a counter to insomnia. Once one has mastered Figure 9.19 consideration should be given to graduate work in wire-ology by examining Fig. 3 (pp. 24–25) in *Earth System Science: Overview* (National Aeronautics and Space Administration 1986). This figure, known informally as the Bretherton Diagram in honor of the chairman of the committee that prepared the report, purports to show how all fluid and biological earth processes relate to each other. Note that sea ice occupies a small corner of one of the many boxes in the diagram. Now mentally scrunch all the wired connections relative to sea ice shown in Figure 9.19 into Fig. 3. See how simple everything becomes.

9.4.3 Micrometeorological Models

The Maykut and Untersteiner model has typically been driven by monthly averaged meteorological fields that have been extrapolated down to time steps of one day. It also utilizes a fairly simple parameterization of the snow cover. The Ebert and Curry model both includes more interactions within an ice pack and uses a slightly more complex parameterization of snow characteristics. However, there are other one-dimensional models that examine some aspects of sea ice growth in some

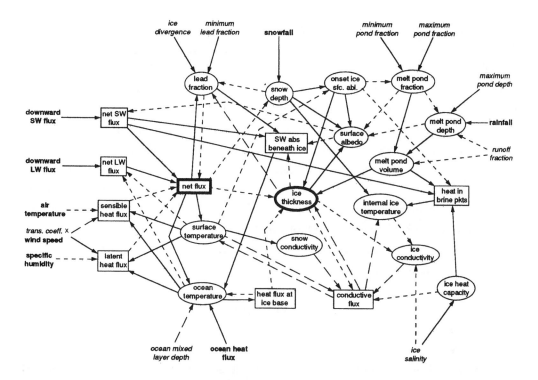

Figure 9.19. A schematic of the interactions between external forcing (**bold**) and parameter values (*italics*), and internal variables (ovals) and fluxes (boxes). The solid arrows represent positive interactions, the short-dashed arrows represent negative interactions, and the long-dashed arrows represent interactions that may be either positive or negative depending upon the season. The thick borders around net flux and ice thickness are to emphasize their importance (Ebert and Curry 1993).

detail via the use of the so-called SNTHERM model (Jordan 1991; Rowe et al. 1995). This approach models the snow cover as a one-dimensional, layered mixture of dry air, ice, liquid water, and water vapor. The model accumulates snow at 2-cm intervals and estimates initial densities using an empirical relation developed at the drift station. Horizontal transport and advection are considered, as are variable snow thickness, wind packing, and grain growth with time. The surface exchange with the atmosphere considers the usual radiation sources and computes the bulk transfer coefficients from estimates of more typical micrometeorological parameters such as the roughness lengths for momentum, heat, and moisture. For details refer to the original paper. Even more important than the details of the analysis is the fact that the study uses meteorological data collected during the drift of the Russian research station North Pole-4 (April 1956–April 1957). The field observations were collected on a 3-hourly basis and then interpolated to hourly time steps. The data are now available through a NSIDC CD-ROM (1996). Although this data set was initially analyzed by Nazintsev (1963, 1964a, 1964b) and the results were included in Russian atlases, it is good to see a reanalysis using more modern techniques. It is particularly encouraging to see data from the Russian NP stations becoming more generally available. The model was able to simulate temperature profiles in both snow and sea ice over the complete annual cycle. The temperature traces show the correct damping of amplitude fluctuations with depth and are in good agreement with the trends over time in both second-year and multiyear sea ice. Particular attention is paid to the mass, thermal, and radiative exchanges during the summer melt season. Both the simulated and the observed snow temperature profiles show a subsurface maximum resulting from absorbed solar radiation. Also, both the timing and the rate of modeled snow ablation agree with observations. In addition, the total amount of snow melting, sea ice melting, and basal sea ice growth are in the range reported by Nazintsev (1963). However, the authors feel that the model overestimates the amount of sea ice growth, a suggestion that is not particularly surprising in that the sea ice property simulation was rather rudimentary. During the summer, shortwave radiation is the main term in the surface heat budget. In the winter, although the net longwave balance was the main heat loss term, the flux of sensible heat from the air to the surface mitigated these losses in that it was nearly a mirror image of the emitted longwave flux. The Jordan et al. (1999) paper has an extensive reference list and also serves to introduce the reader to the second and third authors, Ed Andreas and Aleksandr Makshtas, both of whom have played important roles in the development of a more micrometeorological approach to sea ice studies. See, for instance, their papers Andreas (1996, 1998) and Makshtas (1991) for additional references.

9.5 Problems

When one contemplates the literature on the growth and decay of sea ice, a number of problems are apparent. Perhaps the most obvious is related to the fact that as a result of the large variability in almost all of the environmental parameters that affect ice growth and decay, unless data input is constrained by observational reality it is possible to obtain considerable undesirable variability in the resulting thickness estimates. For instance, consider the problem of estimating the amount of snow on the ice. Current approaches to this, although reasonable, are nevertheless simplistic in that averages are used and most of the snow is assumed to accumulate during the early winter, when appreciable amounts of open water still exist. One possibility for improving such estimates for specific years would be to use estimates of the amount of open water obtained via satellite to help with snowfall estimates, along the lines explored by Overland et al. (2000) in estimating regional sensible and radiative heat fluxes during the winter period. The problem here is that the presence of open water typically results in the presence of clouds, which in turn limit the useful sensor systems to passive microwave or radar because visible or IR systems do not penetrate clouds. Also I know of no work on the redistribution of existing snow during storms, although this is clearly a major effect, with snow scoured from regions of level ice and deposited where there is an appreciable surface roughness such as that created by pressure ridging. Of course it is the level regions that are losing snow where ice growth calculations are primarily applicable. Then there is the whole subject of the feedbacks between growth rates, the salinity and gas amounts included in the ice, and the resulting thermal properties of the ice; a subject that will be discussed in Chapter 10. Although the literature on sea ice thickness variations resulting from thermodynamic processes is large, there is still some way to go before everything influencing ice growth is suitably categorized and understood.

10 Properties

The trouble with sea ice properties is that there are so many.

N. Untersteiner

As noted above, at first glance the list of sea ice properties seems endless. To add insult to injury, each separate property invariably does not have a constant value but shows considerable variation. Admittedly, most materials do exhibit some small variations in their property values. However, in sea ice the variations can be extreme. These changes are, of course, the result of the fact that sea ice in its natural state normally exists at temperatures that are within a few degrees ($< 40°C$) of the melting point of the primary phase comprising the material, that is, ice I(h). Therefore, small changes in either the specimen temperature or the bulk composition of sea ice can result in significant property changes. It is also frequently the case that the property values obtained depend on the growth history of the specific ice sheet under study, as well as on the nature of the measurement procedure.

To treat the subject of property variations in sea ice in detail would require far more space than is available here. For instance, there is a very large literature on the strength of sea ice as the result of the importance of this parameter in a number of applied problems. What I hope to achieve here is to provide the reader with a sense of the natural variation observed in a few of the more important properties, as well as of why these variations occur. Also included will be a discussion of the models that have been used to correlate and occasionally to extrapolate such data to conditions where testing is either not possible or simply has not, as yet, been carried out. In the past I have written several long discussions touching on different aspects of this general subject. In that they include considerable material not covered here, the reader may want to examine them for treatments of specific subjects (Schwarz and Weeks 1977; Weeks 1998b; Weeks and Ackley 1986; Weeks and Assur 1968, 1972; Weeks and Cox 1984; Weeks and Mellor 1984).

10.1 Density

Density is one of the properties of sea ice that proves to be important in a wide variety of both fundamental and applied problems. As a result, density values have frequently been determined during field studies, starting in 1899 during the experimental cruises of the *Yermak* (Makarov 1901). The focus then was on icebreaker design. A few years later, during 1922–1925, Malmgren (1927) made a number of density measurements during the drift of the *Maud*. In his case the study focused on geophysical parameters related to ice growth. Malmgren's results clearly showed that the density of sea ice was far from constant (0.857 to 0.924 Mg m^{-3}) and that MY ice typically had a lower density than FY ice.

10.1.1 Measurement Techniques

Although in principle the measurement of density in the field would appear to be straightforward, in practice this is far from the case. As discussed by Timco and Frederking (1996), there are several different approaches that have been used and each has its advantages and disadvantages. These are as follows:

1. Mass/volume techniques

Here a sample is either taken from the sheet in the form of a well-defined shape such as a cylindrical core, which is then cut into segments, or a rough block is removed from the sheet and trimmed into a well-defined shape. The dimensions of the final specimen are then determined and used to estimate the sample volume (V). The mass of the sample (M) is then measured by weighing and used to calculate the bulk density (ρ_i) using the simple relation

$$\rho_i = \frac{M}{V}. \tag{10.1}$$

The difficulty here is in precisely determining the dimensions that one uses to calculate the sample volume, as sea ice samples are rarely as uniform as one would desire and perfectly planar cuts in sea ice are rare. Also, slush frequently freezes to sample surfaces changing their shapes. In principal all this is easy to correct, but such corrections require equipment such as milling machines or lathes (rarely available in the field and heavy) and time. During field operations time is a commodity frequently in short supply. More importantly, unless the ambient temperature is very cold ($< -23°C$), the longer the time between removing a sample from the ice sheet and weighing the prepared specimen, the more likely brine will drain from the samples and be replaced by air, resulting in erroneously low densities. Also, spinning a sample on a lathe is hardly a procedure designed to retain brine in the sample. Nevertheless, as noted by Timco and Frederking, this is the most common method of determining sea ice density.

2. Displacement techniques

In this case the load (P) necessary to submerge a sea ice sample of known volume in a liquid is determined and the density is calculated using

$$\rho_i = \rho_w - \frac{P}{V} \tag{10.2}$$

where ρ_w is the density of the liquid. This procedure has similar problems to the previous technique and is also appreciably slower. More importantly, if the sea ice is porous, the submersion liquid can enter the pores by displacing the air resulting in an erroneously high density. As a result this technique has rarely been used.

3. Specific gravity techniques

Here the sea ice sample is first allowed to reach thermal equilibrium at the ambient temperature and then weighed in air. Then it is reweighed while suspended in some suitable liquid. Malmgren (1927), who appears to have been the first to use this procedure, used "petroleum" as a liquid.

More recently, Nakawo (1983) has used 2,2,4-trimethylpentane. Assuming that the specific gravity of the immersion liquid is known at the test temperature, the specific gravity of the sea ice sample p_{si} at that temperature is given by

$$p_{si} = \frac{W}{W - w} \times p_l. \tag{10.3}$$

Here W is the weight of the sample in air, w is its weight when submerged in the liquid, and p_l is the specific gravity of the liquid. As was the case with the previous technique, if the submersion liquid enters the pores of the sample an erroneously high density will result. Malmgren tested to see if this appeared to be a problem by noting whether the measured weight of the submerged sample changed with time. No appreciable change was noted. Needless to say, this is essentially a time-consuming laboratory procedure.

In a few cases where these procedures have been used, the volume of gas present in the ice was also measured independently (Nakawo 1983; Urabe and Inoue 1986). The purpose here was not specifically to determine the density of the sea ice but to test to see whether the gas content of the ice could adequately be determined by comparing the observed density of the sample with the theoretical density of gas-free sea ice of the same salinity and at the same temperature. Agreement has proven to be quite satisfactory.

4. Freeboard and ice thickness techniques

This approach provides a measure of the overall density of the complete ice sheet. Therefore, the resulting value is not applicable to studies that utilize samples from different vertical levels in the ice sheet. Also the technique implicitly assumes that the ice sheet is in free-floating isostatic equilibrium. As will be discussed in the chapter on pressure ridges, this assumption does not hold in the near vicinity of ridges. This procedure also requires accurate estimates of f_b, the average height of the surface of the floe above the sea surface (i.e., the freeboard), as well as of the total thickness (h_i) of the area of the ice sheet under study, the thickness (h_s) and density (ρ_s) of the snow cover, and the density of the underlying seawater (ρ_w). Incidentally it was this procedure that was used by Makarov (1901) during his studies from the icebreaker *Ermak*. For a more recent study using this general approach see Ackley et al. (1976).

10.1.2 Results

Timco and Frederking (1996) have tabulated the field results obtained by 21 different investigators. The FY ice densities ranged from 0.75 to 0.96 whereas the MY values varied from 0.72 to 0.94 $\mathrm{Mg\,m^{-3}}$. The results indicate that ice samples obtained from above the waterline frequently show lower densities than ice samples obtained below the waterline. These changes undoubtedly are caused by the brine that drains from the above-water portion of the sea ice being replaced by air.

Considering that there are only a few phases present in sea ice, including ice I(h), brine, solid salts, and gas, it would appear reasonable to attempt to calculate its bulk density from the densities and amounts of the separate phases. Such calculations have been carried out by several individuals starting with Zubov (1945), who developed a table of density values for gas-free sea ice at temperatures of –23°C and warmer. He did not specifically consider the presence of solid

Figure 10.1. The calculated density of gas-free sea ice as a function of temperature and salinity (Anderson 1960).

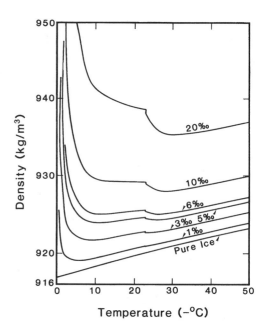

salts in his calculations, even though values of their precipitation temperatures had been published earlier by Gitterman (1937). The next set of calculations for gas-free sea ice was carried out by Anderson (1960), building on the studies of the chemistry of sea ice brine by Nelson (1953) and Nelson and Thompson (1954) and the phase relations as worked out by Assur (1958). Anderson's curves are shown in Figure 10.1. In that the details of these calculations were not published, it is not clear if he specifically considered the presence of $Na_2SO_4 \cdot 10\, H_2O$. However, he definitely considered the presence of $NaCl \cdot 2$ H_2O, as can be seen from the break in the density curves at ~ –23°C. The figure clearly shows several prominent density trends that have been supported by later work. At temperatures near freezing, density is strongly temperature dependent. However, as temperature decreases, this dependence decreases. Another interesting variation occurs in more saline ice (> 10‰), which at temperatures warmer than –25°C shows a continuous decrease in density with decreasing temperature. Anderson's results have been supported by the results of both Cox and Weeks (1983) and Nazintsev (1997). The fact that the Anderson and the Cox and Weeks results are in very good agreement, with a modal difference of only +0.0001 Mg/m³, should not be too surprising in that both sets of calculations are built on Nelson's experiments concerning the phase relations. Nazintsev's results (his Table 1.2.7) are also generally in excellent agreement, with significant variations (> 0.004 Mg/m³) only appearing at very low temperatures (–30°C) and high salinities (> 10‰). It is not currently known which of these two calculated data sets best approximates reality. In that sea ice with salinities > 10‰ and temperatures colder than –30°C should be rare, this problem is rather academic.

As pointed out by Arnold-Aliab'ev (1929a, 1934), the fact that the ratio of the volume of gas in the ice to the total volume of the sample is given by

$$\varepsilon = \frac{n}{1-n} = \frac{\rho_o - \rho}{\rho}, \qquad (10.5)$$

where n is the gas porosity of the sample, ρ_o is the density of gas-free sea ice of the same salinity and temperature, and ρ is the sample density, gives one an easy method of determining gas volumes. Even more encouraging is the observation that in the few cases when calculated gas-free sea ice densities have been combined with bulk sample densities in order to calculate the amount of gas in the ice, and the results compared with direct measurements of gas amounts, the agreement has been quite satisfactory (Figure 10.2). This agreement also supports the conclusion that

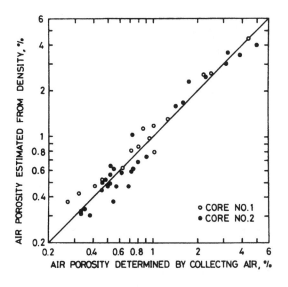

Figure 10.2. A comparison of the air porosity as determined from density measurements with values determined by direct measurement of air content (Nakawo 1983).

the gaseous inclusions in sea ice are at approximately one atmosphere pressure; a conclusion also supported by the observations of Gosink and Kelley (1982) during their work on air–sea ice–ocean gas exchange. Clearly the direct determination of gas amounts in large numbers of samples would be a time-consuming task, a suggestion supported by the fact that such measurements are rare. A description of Russian techniques for the direct determination of gas in sea ice has been published by Savel'yev (1954) and translated by E. R. Hope. The only problem remaining with the mass/volume technique is that as the amount of gas included in FY ice is frequently very small, dimensional errors in sample preparation can be large enough to occasionally result in gas volume estimates that are negative, a physical impossibility.

As far as average densities representative of the complete ice sheet are concerned, detailed ice profiles collected on the 1971–1972 AIDJEX stations give average MY ice densities between 910 and 915 kg m^{-3} (Ackley et al. 1974). They also found that the higher the freeboard, the lower the average density as given by the empirical relation

$$\rho = -194\,f + 974 \tag{10.6}$$

where ρ, the bulk density, is in kg m^{-3} and f, the freeboard, is in meters. For most purposes, unless precise density values for specific ice samples are required, 910 kg m^{-3} should serve as a reasonable estimate.

10.2 Gas Content and Composition

Now that we have established that it is possible to estimate gas volumes in sea ice using indirect methods, what is known about the amounts and composition of this material? First let us examine the different ways that gas can become included in sea ice. Tsurikov (1979) lists nine and discusses this subject in considerable detail, expanding on an earlier discussion by Savel'yev (1963). In my view, the more important of these are:

1. The release and entrapment of gas occurring during the formation of either frazil or congelation sea ice.
2. The inclusion within the sea ice of gas either formed on the seafloor or within the water column below the ice. Such inclusions can sometimes be distinguished by their lenticular forms.

3. The replacement of brine draining from the ice with air.

4. The release of gas from entrapped brine as the result of further freezing.

5. The formation of vapor-filled inclusions as the result of internal ice melting.

6. The formation of infiltration ice by the flooding of the overlying snow cover and its subsequent refreezing.

7. The formation of gaseous inclusions from oxygen produced by photosynthesis by algae living within the ice.

Although Tsurikov's discussion of these mechanisms is primarily theoretical, it provides insight concerning how little is really known about the effectiveness of the above mechanisms in the real world. His attempt to estimate gas amounts resulting from mechanisms 1, 4, and 5 results in volumes of 15 to 20%, much higher values than are typical of normal FY sea ice. Process 3 would be expected to be most effective in the above-sea-level portion of MY ice, in that this ice has lost most of its salinity by drainage and flushing. At times this can be quite extreme. I have occasionally seen surface ice on old floes and in the upper parts of pressure ridges that had a porosity of at least 20% and in a few cases of 40 to 50%. The result was essentially a very open latticework of pure ice. In that process 2 occurs external to the formation of the ice, its gas input could presumably vary over a wide range and might be expected to be largest over shallow-shelf seas where gas production from biological activity could be significant. Process 6 could also be quite significant. Fortunately it usually is clear, from the examination of vertical sections of the upper part of the ice sheet, if infiltration ice has formed. In that such flooding starts at the ice surface and subsequently moves upward, there is also a possibility that most of the air initially included in the snow will be swept up and out of the ice. I know of no papers bearing directly on process 7. Clearly, the process would be seasonally restricted as it would require sunlight. Therefore, it could make no contribution to gas incorporation in FY ice until the spring. Looking at the above mechanisms one can also surmise that the gas entrapped by mechanisms 3 and 6 would undoubtedly be air. On the other hand, the gas produced by mechanism 2 and 7 might be expected to be compositionally quite different from air.

So what do field observations tell us? First, as might be expected from the earlier discussion on sea ice densities, clearly FY ice contains less gas than does MY ice. This fact is obvious even considering the problems associated with working on cores in the field. Figure 10.3 shows the direct determinations of Nakawo and also indirect measurements of Richter-Menge et al. (1986). In this latter study the samples were carefully machined, allowing for improved estimates of sample volume. With rare exception gas volumes (V_a) are less than 30‰, with the most frequent values occurring in the range between 2 and 20‰. The highest values were primarily from samples obtained from the upper part of the ice sheet, a fact that led Nakawo to suggest that there is a correlation between growth rate and the volume of included gas. He also observed a direct correlation between sample salinity and the amount of included gas. In that a direct relation has clearly been established between growth velocity and initial salinity, this would tend to support this suggestion. The fact that the Richter-Menge et al. data set does not show such trends is not significant in that all of their samples were taken from ice sheet depths greater than 60 cm.

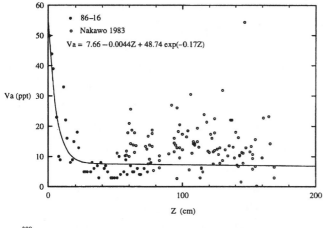

Figure 10.3. Gas volumes in FY ice from Eclipse Sound, NWT, and the Beaufort Sea north of Prudhoe Bay, Alaska, as determined by Nakawo (1983) and Richter-Menge et al. (1986). Z represents the location of the midpoint of the sample.

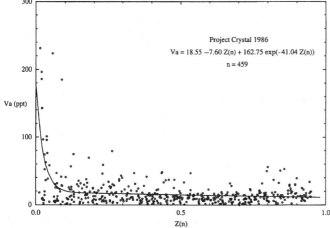

Figure 10.4 Gas volumes obtained from MY ice sampled on Ice Station Crystal (located between 84 and 85°N north of Nord, Greenland) plotted vs. normalized ice thickness Z(n).

Figure 10.4 presents V_a values determined (Weeks, unpublished field notes) from undeformed MY ice sampled at Station Crystal located at 84–85°N to the north of NE Greenland. Again most values are quite small (10 to 20‰), although values in excess of 40‰ are more frequent than in FY ice. The highest values were in excess of 200‰ and, as expected, were obtained from near-surface ice. High values were also obtained from samples located from near the bottom of the sheet at normalized depths of > 0.95. However, as these values were believed to be the result of brine drainage during sampling, they were not plotted. V_a values for a significant number of machined samples collected from MY pressure ridges located in the Beaufort Sea can be found in papers by Cox, Richter, Weeks, and Mellor 1984, 1985; and Cox, Richter, Weeks, Mellor, and Bosworth 1984. Considering the typical internal structure of the ice found in pressure ridges, it is not surprising that the highest gas volumes should occur in such ice and, of these values, that the largest occur in the upper parts of ridges. The ice in the 1984 report contained a particularly large amount of gas, with a mean value of roughly 300‰ and a maximum of 660‰.

Studies of the chemical composition of the gas included in natural sea ice have been limited. The earliest appears to have been by Bruns (1937) on ice collected in the Barents Sea. His results, as well as the analyses carried out by Matsuo and Miyake (1966) on samples collected from both the Antarctic and the Okhotsk Sea, are tabulated in the paper by Tsurikov (1979). The O_2, CO_2, and O_2/N_2 values are reproduced in Table 10.1. The fact that in the atmosphere the ratio

$O_2/N_2 = 0.27$, and that observed values are both appreciably greater than and less than this value, definitely supports the convention of calling this material gas instead of air. Tsurikov also notes that the oxygen content of the included gas never reaches the value that would correspond to saturation in seawater at its freezing point $(O_2/N_2 = 0.54)$. He suggests that this may be caused by the oxidation of organic matter and by the depletion of oxygen as the result of respiration by marine organisms living beneath the ice. He rejects Savel'yev's (1963) earlier suggestion that the observed higher CO_2 and lower O_2 values are the result of activity by organisms living within the ice. In my view the evidence for this conclusion is not particularly convincing.

The most careful study of gas content and composition in FY "sea ice" to date is by Tison et al. (2002), with the ice in question experimentally grown in the Arctic Environmental Test Basin of HSVA in Hamburg. During the freezing runs, the basin was segmented into two sections. In the first section a homogeneous current of 0.065–0.085 m s^{-1} was induced, and the "seawater" consisted of Hamburg tap water with NaCl added to achieve a salinity of 35‰. In the second section a current was not induced and "seawater" was approximated by a combination of tap water and "Instant Ocean Salt," again with a final salinity of 35‰. The purpose of the experiment was to obtain reference levels of total gas content and concentrations of atmospheric gases (O_2, N_2, CO_2) incorporated in the ice in the absence of appreciable biological activity. The range of total incorporated gas was between 3.5–18 mL STP kg^{-1} (STP=Standard Temperature and Pressure), values similar to the field results of Matsuo and Miyake discussed earlier. Bulk salinities were similar in both of the tank sections although at a given depth the salinity was higher in the quiet zone as compared with the current zone. These differences are discussed in terms of the solute boundary layer at the growth interface with currents favoring a thinner boundary layer, less buildup of gas at the interface, and more limited nucleation of bubbles. Surprisingly high CO_2 concentrations were observed in all samples (18–57%), values that were almost exclusively above values observed in natural sea ice. These values are believed to result from two factors:

Table 10.1 Results of the analyses of the chemical composition of gas included in natural sea ice. Composition is given in % by volume.

Sample #	O_2	CO_2	O_2/N_2	Source
Antarctic				
1	21.3	0.98	0.27	Matsuo and Miyake (1966)
2	20.6	24.3	0.38	
3	29.0	0.53	0.42	
4	26.5	1.0	0.37	
Sea of Okhotsk				
5	29.0	0.60	0.42	
6	31.0	0.41	0.46	
7	23.8	0.84	0.32	
8	25.4	0.93	0.35	
Barents Sea				
1	18.3	—	0.22	Bruns (1937)
2	15.9	0.6	0.19	
3	16.5	0.5	0.19	
4	18.5	0.4	0.22	

(1) the use of $CaCO_3$-rich tap water resulting in the precipitation of $CaCO_3$ and associated CO_2 degassing, and (2) uncontrolled bacterial activity.

The final piece of information that one would like to have concerning the gaseous inclusions in sea ice concerns their shapes and size distributions. To date, this has received little attention with most attention focused on the size and shapes of brine pockets. Available studies include those of Grenfell (1983), Perovich and Gow (1991), and Light et al. (2003a). Note that the observed size distributions in FY ice can reasonably well be described by a power law (see Figures 8.38 and 8.39). Bubble shapes are generally spherical, particularly in contrast to brine pocket shapes, which generally are not spherical.

10.3 Thermal Properties

To make realistic calculations of the growth of sea ice via the equations discussed in Chapter 9, a knowledge of the temperature dependence of the thermal properties of pure ice, snow, seawater brine, and seawater salts are a necessary starting point. Then, if these pure component properties can be fed into a model that simulates the structural characteristics of sea ice, it should be possible to make reasonable estimates of the thermal properties of real sea ice that then can be tested against field measurements. In the following I will try to provide the reader with a sense of the present status of this activity. From what has already been said about the structure and compositional variation in real sea ice, it is clear that its thermal properties will not be constant. If fact, they are not only highly variable, some are directionally dependent. As there is quite a large literature on this subject, a useful starting point is the review paper by Yen (1981), which summarizes the subject up through 1973. The following discussion is based on his review. However, only a small portion of the material that he covers is presented.

10.3.1 Ice
The heat capacity of ice (c_i) is defined as the heat required to raise the temperature of one unit of mass of the material by one unit of temperature at either constant pressure (c_p) or constant volume (c_V). Formally this is expressed as

$$c_p = \left(\frac{\partial H}{\partial T} \right)_P \qquad \text{or} \qquad c_V = \left(\frac{\partial U}{\partial T} \right)_V \qquad (10.7)$$

where H and U are the enthalpy and internal energy per unit mass, respectively. In a pure material such as Ice I(h) the unit of mass is frequently considered to be 1 mole. In composite materials such as sea ice, the gram or kilogram is more commonly taken to be the mass unit. As the heat required to warm up the air and water vapor in the interstices is quite small, the heat capacity of dry snow and ice are essentially the same. The relation between c_V and c_p is

$$c_p - c_V = \frac{\gamma_c^2 V T}{\omega_T} \qquad (10.8)$$

where V is the volume of the ice, γ_c is the coefficient of volumetric expansion, and ω is the compressibility. The value of c_V is ~3% less than c_p at the melting point, and the difference between the two values decreases with decreasing temperature. Ashworth (1972) has grouped the available data into several different temperature ranges and in the temperature range of interest here (150 < $T \leq 273°K$) recommends the following linear relation for c_p:

$$c_p = 2.7442 + 0.1282T. \tag{10.9}$$

Here the units of c_p are J/mol K and the correlation coefficient is 0.9737. Near the melting point, the value of c_p is about half the heat capacity of liquid water.

The latent heat of fusion of ice L_f is defined as the change in enthalpy when a unit mass of ice is converted isothermally and reversibly into liquid water. At one atmosphere pressure, the value of L_f at 0°C is 335 kJ/kg. In the temperature range down to –10°C, L_f decreases linearly to 274 kJ/kg. At lower temperatures, the decrease rate is somewhat less. Yen notes the interesting fact that at the ice–liquid–vapor triple point, the value of L_f is only about 12% of the value of the latent heat of sublimation (2838 kJ/kg) L_p, a fact suggesting that only about 12% of the hydrogen bonds break when ice melts.

At temperatures near 273°K the thermal conductivity of ice ξ_i is about 2.2 W m^{-1} K^{-1}, or about four times larger than the thermal conductivity of water at the same temperature. Considering the scatter between the values of ξ_i obtained by different authors, Yen suggests that equation 10.10 be used to compute values of ξ_i

$$\xi_i = 9.828 \exp\left(-0.0057\,T\right) \tag{10.10}$$

where ξ_i is in units of W m^{-1} K^{-1} and T is in K. The temperature range covered is ~40 to ~273°K. A more recent analysis by Slack (1980) gives the following estimates of ξ_i (0°C, 2.14 W m^{-1} K^{-1}; –23°C, 2.4 W m^{-1} K^{-1}; –73°C, 3.2 W m^{-1} K^{-1}). Based on these results, Pringle et al. (2007) have used

$$\xi_i = 2.14 - 0.011T \tag{10.10a}$$

where T is in °C. Considering that most sea ice exists at temperatures warmer than –30°C, the uncertainty of these values is estimated to be ~5%. As Pringle et al. (2007) have pointed out, much of the uncertainty in using effective-medium models to estimate the thermal conductivity of snow and of sea ice is due to the uncertainty in the value of ξ_i. As the large variations in the thermal properties of snow and sea ice are not the result of variations in ξ_i but are the result of variations in the amount of trapped gas and brine, it appears to me that, at present, in calculations of sea ice properties, one can safely assume that $\xi_i = 2.3$ W m^{-1} K^{-1} and ignore its variation with temperature.

There appears to be no significant difference between the ξ_i values for laboratory-grown single crystals of ice, single crystals from glaciers, and polycrystalline commercial ice. However, limited measurements of ξ_i by Landauer and Plumb (1956) suggest that ξ_i values measured parallel to the c-axis may be 5% higher than values measured perpendicular to this axis. A more recent discussion of this topic in light of current solid-state theory concludes that although the suggested orientation difference is reasonable, the difference is probably less than 5% in that the oxygen–oxygen distances of the four different bonds involved are all nearly equal in length (Slack

1980). In addition, data on both the elastic constants and on the thermal expansion coefficient show only a small anisotropy. As Landauer and Plumb's results were not particularly definitive, it is surprising that there appear to be no additional experimental studies on this topic.

10.3.2 Snow

The thermal conductivity of snow (ξ_{sn}) is appreciably more complicated than that of pure ice. In snow heat is transferred by conduction through the interconnected ice grains, by conduction, convection, and radiation across the air space of the included bubbles, as well as by the movement of vapor by sublimation and condensation. Because of this mix of processes, measured values of the thermal conductivity of snow are commonly referred to as effective values ($\xi_{sn(e)}$). Data available at the time of Yen's survey suffer from several problems in that both snow type and vapor diffusion were usually not adequately assessed. Also, measurement techniques varied from author to author. Nevertheless, based on a combined data set from seven different authors, Yen found that $\xi_{sn(e)}$ values could be simply expressed as a function of the bulk density of the snow (ρ_{sn}) by

$$\xi_{sn(e)} = 2.22362 \left(\rho_{sn} \right)^{1.885}, \tag{10.11a}$$

a relation not that dissimilar to

$$\xi_{sn(e)} = 2.846 \left(\rho_{sn} \right)^{2} \tag{10.11b}$$

suggested over 90 years earlier by Abel (1893). For equation 10.11a the correlation coefficient was 0.86. As Yen points out, this relation gives a reasonable estimate of $\xi_{sn(e)}$ when extrapolated to the density of pure ice, but gives a lower estimate of ξ_{a} when $\rho_{sn} \to \rho_{a}$. Yen has also attempted to extract the effect of temperature on $\xi_{sn(e)}$ from the same basic data set. His suggested relation is

$$\xi_{sn(e)} = 0.0688 \ \exp \left(0.0088 \ T + 4.6682 \ \rho_{sn} \right). \tag{10.12}$$

Here and in equation 10.11a, the units of $\xi_{sn(e)}$, T, and ρ_{sn} are respectively W m^{-1} K^{-1}, K, and g cm^{3}. When $\rho_{sn} \to \rho_{i} = 0.917$ and $T = 0°C$, the estimated value of $\xi_{sn(e)} \to \xi_{i}$ 4.98 W m^{-1} K^{-1} is ~2.3 times the estimated value for pure ice of 2.14 W/m K at 0°C. For $\rho_{sn} \to \rho_{a}$ and 0°C, $\xi_{sn(e)} \to \xi_{a} = 0.0688$ W m^{-1} K^{-1}, a value ~2.8 times the value for air at 0°C ($\xi_{a} = 0.0247$ W m^{-1} K^{-1}).

This general subject has been reexamined by Sturm et al. (1997) using 488 ξ_{a} measurements on natural snow covers. The data set also combines conductivity and temperature measurements with descriptions of snow type. The conductivity measurements were obtained via the use of a needle probe system especially adapted for use in snow. The probe was 1.5 mm in diameter, 20 cm long, and only heated over the distal 12 cm to minimize end effects. Snow temperature and texture were also determined. These results are shown in Figure 10.5. The predictive relations suggested were

$$\xi_{sn(e)} = 0.138 - 1.01 \rho_{sn} + 3.233 \rho_{sn}^{2} \ \{0.156 \le \rho_{s} \le 0.6\} \tag{10.13}$$

$$\xi_{sn(e)} = 0.023 + 0.234 \rho_{sn} \ \{\rho_{sn} < 0.156\} \tag{10.14}$$

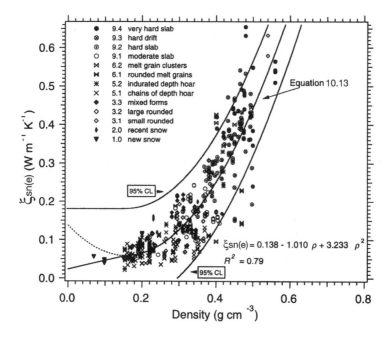

Figure 10.5. The thermal conductivity of snow as determined by use of a probe measurement system as a function of density and snow type (Sturm et al. 1997). Equation 10.13 has been fitted using all the data (n = 488). Below a density of 0.156 g cm^{-3}, a linear regression with a y intercept of 0.023 W m^{-1}K^{-1} was used. A dashed line shows the continuation of equation 10.13. The 95% confidence limits are also shown.

where again $\xi_{sn(e)}$ is given in W m^{-1}K^{-1} and ρ_{sn} is in g cm^{-3}. When equation 10.13 is extrapolated to the density of pure ice, its prediction for $\xi_{sn(e)}$ is close to the correct value ($\sim$2.2 W m^{-1}K^{-1}). As seen in the figure, equation 10.13 predicts a minimum value of the conductivity at a density of 0.156 g cm^{-3}. As Sturm et al. had no experimental data, they chose to use a linear extrapolation to the conductivity of dry air at zero density (0.023 at $-15°$C). Other possible regression relations are also discussed. The question then becomes which of these data sets and relations is most successful in dealing with snow covers on sea ice. More will be said about this in Chapter 15.

A theoretical effective-medium model for the variation in $\xi_{sn(e)}$ is also available (Schwerdtfeger 1963b). This model is based on Maxwell's (1891) much earlier effective-medium theory for the variation in the electrical conductivity of heterogeneous media. The equation is

$$\xi_{sn(i)} = \frac{2\xi_i + \xi_a - 2\eta\left(\xi_i - \xi_a\right)}{2\xi_i + \xi_a - \eta\left(\xi_i - \xi_a\right)}\xi_i. \qquad (10.15)$$

Here $\xi_{sn(i)}$ is taken to be the thermal conductivity of dense snow, a material that can be approximated as ice containing isolated gas bubbles, and η is the porosity of the snow. The term η is related to ρ_i and ρ_{sn} by the equation

$$\eta = 1 - \left(\rho_i / \rho_{sn}\right). \qquad (10.16)$$

In that $\xi_a \ll \xi_i$ and using equation 10.16, equation 10.15 becomes

$$\xi_{sn(i)} = \frac{2\rho_{sn}}{3\rho_i - \rho_{sn}} \xi_i . \tag{10.17}$$

In the lower density range (down to 0.15 g/cm³)

$$\xi_{sn(a)} = \frac{(2+s)s}{(1+s)^2} \xi_i . \tag{10.18}$$

Here s is a constant related to porosity by $\eta = 1/(1+s)^3$. For example, if $\rho_{sn} = 0.4$ and $\eta = 0.536$, then $s = 0.212$ and from equation 10.18, $\xi_{sn(a)} = 0.7023$ W m⁻¹ K⁻¹ at 0°C. Additional models are available for very low-density snow, a commodity infrequently found in the polar regions. The important thing to remember is that such models are static, they do not per se consider the transport of heat and mass by forced convection within the snow. As will be seen in Chapter 15, the polar regions are windy places and these could be significant effects. Refer to Yen's (1981) review or to Schwerdt-feger's (1963b) and Sturm et al (1997) papers and Chapter 15 for additional details.

10.3.3 Sea Ice

The thermal properties of sea ice are appreciably more complex than those of either pure ice or snow. Only at very low temperatures can sea ice be approximated as a static mixture of ice, salt crystals, and gas. At more typical temperatures, one must consider the specific properties of the above phases plus those of brine. In addition, one must also consider the thermal effects of the temperature and compositionally dependent phase changes. For instance, consider the specific heat of sea ice, which is the total amount of heat required to raise the temperature of the sea ice constituents by one unit of temperature plus the heat associated with the phase change as the brine becomes less saline by melting pure ice at the brine–ice interfaces. It is the thermal effect of these phase changes that causes sea ice to have an unusually large specific heat. It should be noted that both the specific heat and the related latent heat of fusion do not depend on how the phases are arranged.

The first field measurements of the specific heat of sea ice were completed at temperatures down to –23°C during the 1922–1925 drift of the *Maud* by Malmgren (1927). Some years later Untersteiner (1961) reexamined these measurements and suggested that they could be represented by

$$c_{si} = 2.092 + 1.72 \times 10^{-2} \left(\frac{S_{si}}{T^2}\right) \tag{10.19a}$$

where c_{si} is the specific heat in J kg⁻¹ °K⁻¹, the salinity of the ice S_i is in ‰, and the temperature T is in °C. This relation has frequently been utilized in large-scale model calculations because of its computational simplicity. Additional justification for this relation can be found in Ono (1967). In that different solid salts precipitate (or dissolve) at specific temperatures, analytical models of sea ice thermal properties are usually considered to only apply in specific temperature ranges.

The simplest temperature interval is between 0.0°C and –8.2°C where, if the possible effects of CaCO₃•6 H₂O are ignored, the included brine in the ice can be considered to have the same chemistry as the seawater from which the ice formed. More importantly, no significant amount of solid salts are precipitated or dissolved in this temperature range. In addition, the thermal capacity of salt in 4‰ sea ice is only 3.3472 J/kg K and can be considered to be negligible. Similarly the effects of the heat of crystallization (or dilution) can also be taken to be negligibly small in this temperature range. Therefore the specific heat of sea ice can, within this interval, be written as (Pounder 1965; Schwerdtfeger 1963a)

$$c_{si(0\,to-8.7)} = -\alpha\, L_f\, \frac{S_{si}}{s^2} + \frac{S_{si}}{s}\left(c_w - c_i\right) + c_i. \tag{10.19b}$$

Here a is the coefficient of the relation between the fractional salt content of the brine s (in grams of salt per gram of water) and the temperature T, S_{si} is the salinity of the sea ice (in grams of salt per gram of sea ice), and c_i and c_w are the specific heats of pure ice and of water. In that the relation between s and T is linear $(s = aT)$, and as a is the slope of s versus T, it is a negative quantity. Then replacing s in (10.19b) results in

$$c_{si(0\,to-8.7)} = -\frac{S_{si}}{\alpha\, T^2}\, L_f + \frac{S_{si}}{\alpha\, T}\left(c_w - c_i\right) + c_i \tag{10.20}$$

plus an ac_i term that is small and usually omitted. Here L_f is the latent heat of fusion for pure ice. Calculations for the temperature interval –8.6 to –23°C are more involved in that the thermal effects associated with the precipitation (or dissolution) of $Na_2SO_4 \cdot 10\,H_2O$ must be considered. Procedures for carrying out the appropriate calculations can be found in Schwerdtfeger (1963a) and in Pounder (1965).

A bit later Ono (1967, 1968) reexamined this general problem using a slightly different analytical approach, which, substituting numerical values for parameters and component properties, gives (Pringle et al. 2006)

$$c_{si} = 2.113 + 0.0075T - 0.0034S_i + 8.4 \times 10^{-5}\left(S_i T_i\right) + 18.04\left(\frac{S_i}{T^2}\right) \tag{10.21}$$

Here the first term $(2.113\ \text{J g}^{-1}\,°\text{C}^{-1})$ represents the specific heat of pure ice, S_i is in ‰, and T is in °C. Figure 10.6 is based on these results. Note that at temperatures colder than ~ –4°C, c_{si} is relatively independent of temperature, whereas at temperatures warmer than –3°C, c_{si} becomes strongly temperature dependent.

A related equation for the latent heat of fusion for sea ice developed by Ono (1967) is

$$L_{si} = 333.40 - 2.113T_i - 0.114S_i + 18.04\frac{S_i}{T_i}. \tag{10.22}$$

Here L_{si} is in J g^{-1}, and again T_i is the ice temperature in °C and S_i is the salinity of the ice in ‰. A plot of these results is shown in Figure 10.7. As was the case in Figure 10.6, at near-melting temperatures L_{si} is strongly temperature dependent, whereas at temperatures colder than –6°C, L_{si} is relatively independent of temperature. Comparisons of L_{si} values obtained from equation 10.22 with calorimetric laboratory and field measurements show excellent agreement (Johnson 1989; Trodahl et al. 2000).

For sea ice a particularly interesting thermal parameter is the thermal conductivity (ξ_{si}), in that its value is dependent on the structural arrangement of the ice and brine as well as the

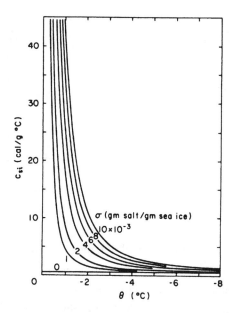

Figure 10.6. Specific heat of sea ice as a function of temperature and salinity (Ono 1967 as modified by Yen 1981). 1 cal/(g°C) = 4190 J/(kg °K).

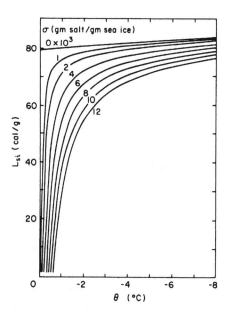

Figure 10.7. Latent heat of fusion of sea ice as a function of temperature and salinity (Ono 1967 as modified by Yen 1981). 1 cal/g = 4190 J/kg.

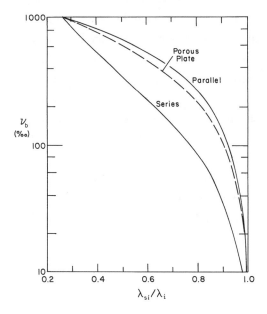

Figure 10.8. The ratio of the computed thermal conductivity values of sea ice relative to that of pure ice vs. brine volume for three different structural arrangements of the brine in the ice (Anderson 1958).

temperature and salinity of the ice. That this should be the case was first suggested by Anderson (1958a, 1960), who used the effective-medium relations suggested by Maxwell (1891) in his property simulations. His results are shown in Figure 10.8. Here in both the series and parallel simulations the brine is assumed to be in uniform continuous layers. As can be seen, as long as the brine volume is fairly high, the thermal conductivity measured perpendicular to the c-axis (the parallel case) is appreciably larger than when measured parallel to the c-axis (the series case). Possible effects of this difference on selective growth rates were discussed in Chapter 7. One should note, however, that these conductivity differences have never been experimentally verified on real sea ice although it would be easy to collect suitable samples of aligned ice in the fast ice at locations like Barrow, Alaska. I note that this difference is primarily of academic interest in that sea ice crystals oriented with their c-axes aligned parallel to the direction of heat flow (vertical) are rare, primarily occurring in initial ice skims.

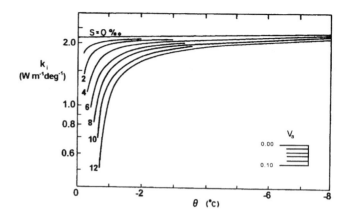

Figure 10.9. Thermal conductivity of sea ice as a function of salinity and gas content (Ono 1967, 1968).

Several other investigations, including those of Schwerdtfeger (1963a), Nazintsev (1964b), and Ono (1968), have examined the variation in the thermal conductivity via the use of related model calculations. The differences in their results are largely associated with assumed differences in the geometry of the inclusions. As Ono's results have frequently been cited as being the easiest to use, they are shown in Figure 10.9. The figure shows that the thermal conductivity of sea ice is greatly reduced relative to that of pure ice at near-melting temperatures when the brine volume is high. At lower temperatures where the brine volume is low, the thermal conductivity of sea ice asymptotically approaches the value for pure ice, provided there is no air in the ice. Figure 10.9 suggests that warm, saline ice forming in the early winter should grow more slowly than either cold ice formed later in the year or low-salinity ice (Wadhams 2000). How much this effect is countered by the fact that rapidly growing ice traps more salt, and therefore contains more brine, has not been investigated. One also needs to remember that while the thermal conductivity of brine is approximately 25% that of pure ice, the conductivity of air is less than 1% of that of pure ice. Therefore, even the small amounts of gas present in FY ice can have an appreciable effect on the thermal conductivity. In MY ice, which characteristically contains more gas, its effect on the thermal conductivity can be appreciable. The small inset scale in Figure 10.9 indicates the amount the gas bubble-free value should be reduced by different gas bubble fractions.

As was the case for the specific heat, the first field measurements of the thermal conductivity were by Malmgren. Later these values were also used by Untersteiner (1964) as the basis for a formula for estimating the thermal conductivity ξ_{si} as a function of the temperature and salinity of the sea ice:

$$\xi_{si} = 2.03 - 0.117\left(\frac{S_{si}}{T}\right). \tag{10.23}$$

Here 2.03 is the thermal conductivity of pure ice in W m^{-1} K^{-1} at temperature $T(°C)$, and S_i is the salinity of the ice in ‰. Although Malmgren's measurements appear to be largely on MY ice, this relation has frequently been used in model studies of FY ice areas.

Since the drift of the *Maud*, several other investigators (Lewis 1967; Nazintsev 1964a, 1964b; Pringle et al. 2007) have completed direct thermal conductivity measurements on either artificial or natural sea ice. The most recent work on ξ_{si} by Pringle et al. (2007) includes a thorough analysis of in-situ temperature profiles in landfast FY ice in both the Arctic (Barrow) and the Antarctic (McMurdo Sound) as well as a review of earlier data. The thermal conductivity values that were obtained were found to be in good agreement with the effective-medium predictions obtained using the bubbly ice model suggested by Schwerdtfeger (1963a) over the density, salinity, and temperature ranges $840 < \rho < 940$ kg m^{-3}, $0 < S_{si} < 10‰$ and $-30°C < T < -1.8°C$.

For these conditions the full model output was adequately approximated by

$$\xi_{si} = \frac{\rho_{si}}{\rho_i}\left(2.11 - 0.011T + 0.09\frac{S_{si}}{T} - \frac{(\rho_{si} - \rho_i)}{1000}\right) \tag{10.24a}$$

Here $\rho_i = 917$ kg m^{-3} was taken as the density of pure ice, and the thermal conductivity values of the brine, pure ice, and air were estimated from $\xi_b = 0.523 + 0.013T$, equation 10.10a, and $k_a = 0.025$. The authors also explored the use of a bubbly brine model specifically applicable to FY ice in the density range $\rho = 890$ to 930 kg m^{-3}. In this case the air volume reduces the conductivity of the lower conductivity brine rather than the ice. As a result, the bubbly brine model results in slightly higher estimates of the thermal conductivity; a difference that increases as the density decreases. In that the last term in equation 10.24a is small if $\rho s_i \approx \rho_i (\rho \geq 890$ kg m$^{-3})$, the authors recommend the use of

$$\xi_{si} = \frac{\rho}{\rho_i}\left(2.11 - 0.011T + 0.09\frac{S}{T}\right) \tag{10.24b}$$

as it fits both models to within ±2%. The estimates provided by equation 10.24b are suggested as applicable to both FY and MY ice. A figure showing the results of their field measurements as well as their and earlier authors' measurements and model estimates can be found in Pringle et al. (2007, Fig. 7). As was anticipated, their estimates as well as their field results are between 10 and 15% higher than the estimates provided by the Untersteiner (1964) relation (equation 10.23), which, as noted, was primarily developed for MY ice. Malmgren's (1927) results were also low in that they came from the near-surface layers of MY ice where higher gas volumes would be expected. There was good agreement with the results of Nazintsev (1964b), obtained from artificial laboratory-grown sea ice. However, the results of Lewis (1967) were consistently higher than those of Pringle et al. This is somewhat surprising in that they would appear to be the most directly comparable, The reason for this difference is not known, but Lewis did assume that all heat was transported with no dispersion by the dominant frequency component of the surface temperature variation, and also that the oceanic heat flux was constant (Pringle et al. 2007). Moreover, accurate in-situ measurements of this kind are difficult owing to experimental concerns (disturbance of both the ice and the heat flow regime by the emplacement of the measurement system) and by the applicability of the heat flow analysis to the measured temperatures.

Where I have always had problems with effective-medium models is not with the cold near-surface ice but with the warm, high-brine-volume ice at the base of the ice cover. My reasons are as follows. All the thermal property models are static in that nothing moves. Yet we know that in real sea ice brine drainage tubes are alternately pumping cold brine into the ocean and inhaling warmer seawater. It was always hard for me to believe that brine drainage tubes and other possible convective processes would not have an appreciable effect on effective thermal conductivity values. Fortunately, these questions have now been at least partially resolved by Pringle et al. (2006, 2007), as well as by related studies in McMurdo Sound, Antarctic, by McGuinness et al. (1998) and Trodahl et al. (2001). Results that are even more striking have been obtained by Lytle and

Ackley (1996) based on observations on appreciably warmer second-year sea ice. According to the effective-medium models, the ξ_{si} values from molecular–diffusive conduction should continue to fall at temperature values warmer than $-10°C$, where the rising brine volume is predicted to decrease the thermal conductivity. On the contrary, the effective conductivity increases, an effect that presumably is associated with increased convective flow of the brine in the ice. Pringle et al. (2007) report that in their 1957 days of in-ice thermistor string measurements from both the Arctic as well as the Antarctic, only 22 distinct convective events could be recognized. The events appear to be of two different types: those occurring near the base of the ice during the winter (~lowest 25 cm) and those occurring late in the season, which can extend further into the interior of the ice. Some of the above events immediately followed the first sustained surface warming in the spring (Backstrom and Eicken 2006). A possible explanation in this latter case is that the surface warming resulted in an increased volume of brine in the upper part of the ice. The associated connectivity and permeability increases, in turn, enabled the discharge of the head of dense, cold brine formerly trapped in the upper ice through the already well-connected brine network below. Good estimates of the magnitude of the heat transfer associated with these convective events are not yet available.

The Pringle et al. (2007) paper also examines the possible contribution of brine drainage channels to the heat flow through a sheet of sea ice. Their calculations indicate that a brine channel spacing of 15 cm will, during the growth season, result in a heat flow of approximately 3.5 MJ m^{-2}, a value corresponding to about 3 cm of new basal ice growth or two days of conductive heat flow up into the ice assuming a temperature gradient of 10°K m^{-1}. This suggests that during the winter–spring time period, the heat flow associated with brine drainage tubes is only a few percent of the total heat flux.

Perhaps the most useful of the thermal parameters is the thermal diffusivity ϑ, whose value is determined by the ratio of the thermal conductivity to the product of the density and the specific heat ($\vartheta \equiv \xi/\rho c$). The reason for its importance is that it is directly related to the rate of temperature change observed in the ice (i.e., $\rho c \partial T/\partial t = \nabla \cdot \xi \nabla T$). In that all three of the parameters commonly vary with vertical position in a sea ice sheet, that ϑ should also vary is not surprising. Figure 10.10 presents the results of Ono's (1968) calculations of the variation of ϑ as a function of temperature and salinity. In that as temperature rises, ξ decreases while both terms in the denominator (ρc) increase, ϑ is a stronger function of temperature than are its constituent parameters. Therefore, during the sea ice growth season ϑ would be ex-

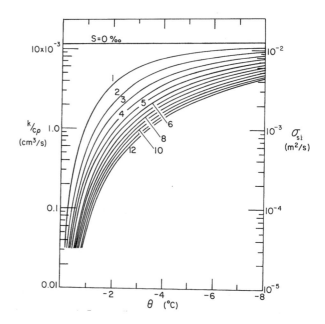

Figure 10.10. Thermal diffusivity of sea ice as a function of salinity and temperature (Ono 1968).

pected to show a maximum value in the cold upper-sea ice layers and a minimum value in the warm lower layers. In addition, ϑ values in young sea ice would be expected to be much lower than in MY ice. It should also be noted that the values of ϑ are not appreciably affected by changes in the gas content of the sea ice. This insensitivity can clearly be seen by examining Table VI in Schwerdtfeger (1963a). A similar insensitivity to changes in gas content occurs in the thermal diffusivity of freshwater ice (Bergdahl 1977:70–71).

Pringle has suggested to me that, in his opinion, improved methods of observing and addressing the thermal effects of convective processes in sea ice are definitely needed. I certainly would agree. A better understanding of how this all fits together with mushy layer theory and brine drainage tubes would also be very useful.

10.3.4 Thermal Expansion Coefficients

Because of its importance in a variety of sea ice geophysics and ice engineering problems, such as thermal-crack-induced noise and thermally induced ice stresses on offshore structures, some of the earliest experimental work on sea ice focused on determining thermal expansion coefficient values (Malmgren 1927; Pettersson 1883). This work concluded that the range of such values is large compared with freshwater ice and that the volumetric expansion coefficients can be either positive or negative. Later Cox (1983) developed an analytical description of thermal expansion, concluding that trapped gas and brine in sea ice have no appreciable effect on thermal expansion and that the coefficients for sea ice must be the same as for freshwater ice during both warming and cooling. Although his analysis was very reasonable, no observational data were presented in support of his conclusion. More recently Johnson and Metzner (1990) revisited this subject, starting with a careful reanalysis of earlier work. They concluded that the Pettersson and Malmgren experiments, although carefully done, were flawed as the result of changes in the relative amounts of the different phases present in sea ice, which resulted in brine expulsion when air-free sea ice cooled and an increase in internal porosity when sea ice warmed. As a result the samples tested by Pettersson and Malmgren could not be considered to be closed systems. In addition, Johnson and Metzner carried out a series of experiments to determine the linear coefficients of expansion on sea ice samples with salinities of 2 and 4‰. During the initial warming of both sets of samples from −15°C to −4°C, the thermal expansion coefficients were found to be the same within the experimental error as freshwater ice, as reported by Butkovich (1959) and Doronin and Kheisin (1977). Subsequent cooling resulted in coefficients that, although somewhat lower than for freshwater ice at the higher test temperatures, approached the freshwater values as the ice cooled. Therefore, the results of the initial cycling agree with the theoretical conclusions of Cox. On a second thermal cycling there was a small decrease in the experimentally determined coefficient, and a small hysteresis loop appeared in the results. The causes of the hysteresis are to date unknown. Butkovich (1959) also reported similar results for freshwater ice. In addition, measurements were made on samples oriented so that the expansion and contraction were either parallel or perpendicular to the direction of growth (i.e., either parallel to the a-axis directions of the crystals or in the direction where random c-axis orientations are to be expected). No clear difference was observed. This means that for sea ice the coefficient of volume expansion

β_v is simply $\beta_v = 3 \times \beta_\ell$ where β_ℓ is the linear coefficient. Measurements on higher-salinity sea ice do not appear to be available.

10.4 Mechanical Properties

10.4.1 Some Background

To analyze the behavior of sea ice as a component in a wide variety of applied problems, an understanding of the various factors affecting its mechanical properties is essential. Some obvious examples of applied problems include bearing capacity, pressure ridge mechanics, icebreaker design, and ice forces on offshore mobile and fixed structures. In the following I do not intend to provide a detailed discussion of these different applications. What I do hope to provide is an overview of property variations in light of the environmental history of the ice in question. This should allow the reader to judiciously select property values appropriate to the particular application that is under consideration. As my treatment of sea ice mechanics will be brief in comparison to the size of the literature on the subject, the reader would be wise to scan Mellor (1986), who also provides a review of basic mechanics, and Gavrilo (1997), who views the subject based on the Russian experience.

The reader will recall from Chapter 2 that the first documented field measurements of sea ice properties were its change in thickness with time (Weyprecht 1879). These observations, which may have contributed to Stefan's interest in the subject and the development of Stefan's equation (Stefan 1891), allowed scientists to predict ice thicknesses as a function of time. The obvious next step needed was a knowledge of the mechanical properties of sea ice. This process was started roughly 20 years later as part of a research effort led by Admiral Makarov based on the icebreaker *Yermak* (Krylov 1901). The applied problem of interest here was the development of an analytical approach to improved icebreaker design. During the 1930s and 1940s related studies were carried out by Russian investigators, but the level of activity was low (Arnol'd-Aliab'ev 1939; Weinberg 1940). The problem up until this time was that there was no framework for treating the large strength variations observed in the field. In his book, it is clear that Zubov (1945) understood that knowledge of both ice temperature and salinity would be essential to understanding strength variations. As was noted in Chapter 7 of the present book, the missing ingredient was an understanding of the structure of the ice. That this was the case was first realized by Tsurikov (1940, 1947a, 1947b), who attempted to analyze variations in sea ice strength in terms of structural models, initially considering only air content and later introducing brine volume based on Malmgren's calculations. Although his models contained errors and were based on an incorrect structural model, his approach was clearly on the right track. Unfortunately, other Russian workers did not appear to be receptive to Tsurikov's approach. An expanded discussion of Tsurikov's contributions can be found in Weeks and Assur (1968).

The development of a more successful theory followed in the 1950s as the result of field operations at Hopedale, Labrador, and Thule, Greenland. The great advantage here was that the same project contained studies both of the internal structure of the sea ice (Weeks 1958b) and of its strength (Assur 1958; Weeks and Anderson 1958b). Model development proceeded rapidly (Anderson 1958a, 1960; Anderson and Weeks 1958; Assur 1958). In the following I will briefly sketch how this occurred. The purpose of the project was to develop an improved basis for estimating the bearing capacity of sea ice in order to extend the operational season when sea ice runways could be safely utilized. In that a bearing capacity theory existed courtesy of Heinrich Hertz (1884; plate on an elastic foundation), it would appear that all that had to be accomplished was to determine the strengths and elastic parameters of sea ice, stick them in the appropriate equations, and crank out the answers. As usually happens, matters were hardly that simple. One value that was definitely needed was the tensile strength of sea ice. Unfortunately there was, at that time, no suitable technique for carrying out a pure tension test on sea ice, even under laboratory conditions. Therefore a variety of different strength tests were performed (ring tensile, flexure, up and down in-situ cantilevers, unconfined compression, shear; Butkovich 1956, 1959; Weeks and Anderson 1958b). An examination of this data set clearly shows several interesting characteristics. Different test types gave quite different values and the scatter was large. However, using the same test, warm ice was weaker than cold ice and, at least in some tests, there was an appreciable decrease in strength with increasing brine volume. One thing appeared clear: because of the scatter a large number of tests would be required. What would have been preferred would be to have the results of a large number of in-situ cantilever or beam tests. Unfortunately, unless one is working with very thin ice, performing in-situ cantilever or beam tests is a very time-intensive process. Therefore, in general this was not a realistic option. One test type appeared to be very attractive: the ring tensile test (Assur 1958; Ripperger and Davids 1947). Sample collection was easy in that this test used cylindrical samples taken by a corer. Also, as the stress levels required to cause failure were not large, a small portable press suitable for field operations could be used.

At the same time, surface rubbings of small beam samples forced to fail in flexure showed that the resulting fracture surfaces preferentially ran parallel to the (0001) planes of the sea ice crystals (Figure 10.11; Anderson and Weeks 1958; Tabata 1960). In addition, an examination of thin sections of sea ice showed that both gas bubbles and brine pockets were concentrated along the same (0001) planes. Furthermore, one could see that when the volumes of brine and gas were large, the percentages of the failure planes that were ice were reduced

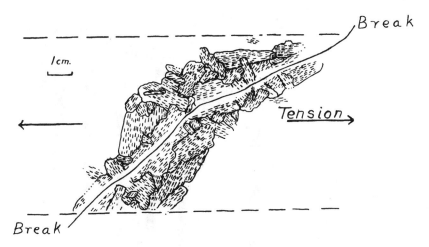

Figure 10.11. Rubbing of the horizontal surface of a simple beam of sea ice that has failed as the result of tensile stresses. The tic marks indicate the orientations of the (0001) planes. All c-axes are essentially horizontal.

(less ice-to-ice contacts as the result of the brine and gas inclusions). This allowed failure to occur at lower stress levels. Conversely, when the volumes of brine and gas were small, the failure planes were essentially all ice and the strength should approach that of pure ice with a sea ice substructure.

What was needed was a model that expressed the failure strength of the sea ice (σ_{si}) in terms of the volume and geometry of the inclusions. Different versions of such a model can be found in Assur (1958), Anderson (1958a, 1960), and Anderson and Weeks (1958). In the following I will use the approach utilized by Assur, who suggested that the strength variation observed in sea ice could be expressed as

$$\frac{\sigma_t}{\sigma_o} = 1 - \psi .$$

(10.25)

Here σ_o is the basic strength of sea ice (i.e., the strength of an imaginary material that contains no brine or gas, but still possesses the sea ice substructure and fails as the result of the same mechanism[s] that cause failure in natural sea ice), and ψ is the "plane porosity" or relative reduction in the area of the failure plane caused by the presence of the brine and gas inclusions. The critical value of ψ in the failure plane is

$$\psi = f(v) = f(v_a + v_b)$$

(10.26)

where v is the void volume or porosity, and v_a and v_b are, respectively, the volume of gas and brine in the sea ice. In the following consider FY ice, where $v_a \gg v_b$ so that v_a can be neglected. Figure 10.12 shows Assur's (1958) simplified model of the brine pocket geometry in sea ice. Here the relative brine volume is

$$v_b = \frac{F_g}{a_o b_o g_o}$$

(10.27)

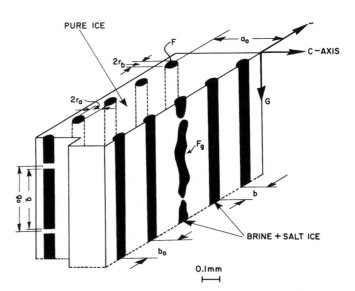

Figure 10.12. Diagram of an idealized portion of a crystal of NaCl or of sea ice (Assur 1958). The parameters used to describe the geometry of the brine inclusions are indicated.

where F_g is the average area of a brine pocket in the BG plane. Note that a_o in Figure 10.12 is the plate width λ discussed in chapter 7. Then defining $\beta_o \equiv \frac{b_o}{a_o}$ and $\gamma \equiv \frac{g}{g_o}$, the reduction in the cross-sectional area because of the presence of the brine pockets is

$$\psi = \frac{2 r_b g}{b_o g_o} = \frac{2 r_b \gamma}{a_o \beta_o}. \tag{10.28}$$

The question now becomes, how do the geometric parameters r_b and γ vary with v_b? Although a variety of different assumptions can be made, only three are typically considered. The simplest of these is to assume that geometric similarity is preserved along the B axis (Figure 10.12) and that the width as well as the relative length (γ) of the brine inclusions remains constant. In this situation r_b must change proportionally to v_b and an equation of the following form results:

$$\frac{\sigma_f}{\sigma_o} = 1 - c\ v_b. \tag{10.30a}$$

However, if the average length and spacing of the brine cylinders remain constant, then changes in r_b will be reflected only in the BC cross section. If geometric similarity is preserved in this cross section, then the form of the equation becomes

$$\frac{\sigma_f}{\sigma_o} = 1 - c\ v_b^{1/2}. \tag{10.30b}$$

Finally, if all brine pockets remain of a similar shape during changes in r_b (geometric similarity in space), then all linear dimensions will change proportionally to $\sqrt[3]{v_b}$. The resulting equation is of the following form:

$$\frac{\sigma_f}{\sigma_o} = 1 - c\ v_b^{2/3}. \tag{10.30c}$$

These three models can be represented as straight lines in σ_o, v_o^k space where $k = 1$, $1/2$, and $2/3$, respectively. Here the σ_f axis intercept is σ_o and

$$c = v_o^{-k}$$

where v_o is the volume of brine required for the ice to have zero strength.

Specific models that have been used in structural studies include the constant-width model ($k = 1$), in which case

$$F_g = 4 r_a r_b \tag{10.31}$$

and

$$v_o = \frac{2 r_a}{a_o} = \frac{d_o}{a_o}. \tag{10.32}$$

Here d_o is the minimum width of a parallel brine layer before it splits as the result of interfacial tension to form individual brine pockets. An estimate of $d_o = 7 \times 10^{-3}$ cm has been obtained by Anderson and Weeks (1958) and a value of 11×10^{-3} has been obtained on NaCl ice by indirect calculation (Weeks and Assur 1963). Regardless of what this number should be, when a brine layer becomes thinner than this value it necks, causing ice-to-ice bonds to form. When the first of such bonds forms it establishes the top of the so-called skeletal layer, as ice platelets below this level are not laterally connected and have zero strength. The thickness of the skeletal layer should be neglected in bearing capacity calculations. For thick ice this correction is not very important; for thin ice it can be.

In the elliptical cylinder model $(k=2)$, let

$$\varepsilon = \frac{r_b}{r_a} \tag{10.33}$$

and

$$F = \pi r_a r_b = \frac{\pi r_b^2}{\varepsilon} \tag{10.34}$$

causing equation 10.30b to become

$$\frac{\sigma_f}{\sigma_o} = 1 - 2\left[\sqrt{\frac{\varepsilon \gamma}{\pi \beta_o}}\right]\sqrt{v_b} \; . \tag{10.35}$$

If one wishes to consider circular cylinders such as described by Langleben (1959), the interruption of the brine pockets in the vertical direction can be neglected by setting $\varepsilon = 1$. Then equation 10.35 becomes

$$\frac{\sigma_f}{\sigma_o} = 1 - \left[\frac{2}{\sqrt{\pi \beta_o}}\right]\sqrt{v_b} \; . \tag{10.36}$$

Further expansion of these ideas can be found in Weeks and Assur (1972).

The question then becomes which of these models, if any, most closely describes real sea ice. One way to resolve this would be to observe the shape changes in real brine pockets during known changes in temperature. This is far more difficult than it would seem and has never been done. One of many problems is that in making the thin section sufficiently thin to allow decent viewing, one frequently slices through the interesting brine pockets. This is particularly the case if the brine pockets are roughly vertically elongated cylinders, as frequently appears to be the case.

10.4.2 Ring-Tensile Strength

An alternative approach is to perform a large number of tensile tests and then examine the resulting data set to see which plot results in the best linearization. As field data on natural sea ice showed a great deal of scatter, it was decided to try this approach using NaCl ice, a material that was structurally similar to sea ice and, if studied at temperatures warmer than –21.2°C, was comprised simply of ice and brine (Weeks 1962). The test used in this study was the so-called ring-tensile test. Such tests use sections of cylindrical ice cores in which an axial hole has been

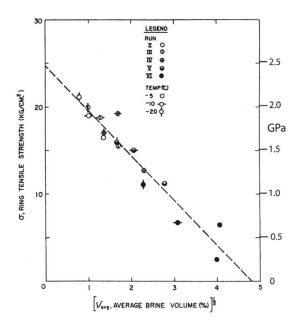

Figure 10.13. Average ring-tensile strength values for NaCl ice vs. average $v_b^{1/2}$. All tests occurred at temperatures warmer than –21.2°C (Weeks 1962).

drilled along the centerline of the core segment. The test then consists of subjecting the specimen to a compressive stress normal to the axis of the cylindrical sample. The sample fails as the result of the buildup of tensile stresses in the vertical plane. The maximum stress occurs at the inner hole and is oriented perpendicular to the applied load.

Even though the NaCl ice studied was grown under controlled conditions in a cold room, the strength results did not permit a clear choice to be made between the different brine volume exponents as the correlation coefficients proved to be similar. Ultimately, it was decided to use the term $v_b^{1/2}$ for two reasons: it facilitated comparisons with previous work and, more importantly, when the least-squares relation was extrapolated to $v_b^{1/2}=0$, the resulting estimate for v_o (2.42 MPa) was roughly similar to the strength of freshwater ice as determined using the same test procedure. In spite of the above rather tenuous foundations, plotting sea ice strength values versus $v_b^{1/2}$ has become a rather common practice, particularly in studies of tensile strength variations. As will be seen, at times it is very successful in linearizing the data. Figure 10.13 shows the results of the tests using NaCl ice.

Other interesting results based on this data set are as follows. The ring-tensile strength of ice containing NaCl•2 H$_2$O is independent of both temperature and the volume of solid salt present in the temperature range studied (–25 to –35°C). In addition, its strength is similar to the strength of freshwater ice as determined using the same procedures. This suggests that the failures are occurring in the ice matrix. Such observations were counter to the suggestion, current at this time (Assur 1958; Peyton 1966, 1968), that the precipitation of NaCl•2 H$_2$O served to reinforce the ice, resulting in a higher overall strength. Also the effects of thermal cycling appear to be largely the result of brine drainage. Finally, systematic changes in the growth-velocity-dependent plate width (a_o or λ) result in systematic changes in the v_o intercept values (Weeks and Assur 1963). Although these latter correlations look good, I have never been certain that this result is correct in that there are other factors such as grain size that vary vertically in directionally frozen ice sheets that could possibly produce such changes in v_o values. Further theoretical considerations based on these results can be found in Assur and Weeks (1963).

At roughly the same time other investigators carried out additional ring-tensile tests on a variety of sea ice samples of different ages and from different locations, including the Antarctic (Dykins 1963; Graystone and Langleben 1962; Hendrickson and Rowland 1965; Langleben and Pounder 1964). Their results were generally similar to the NaCl ice results, although the σ_o values were

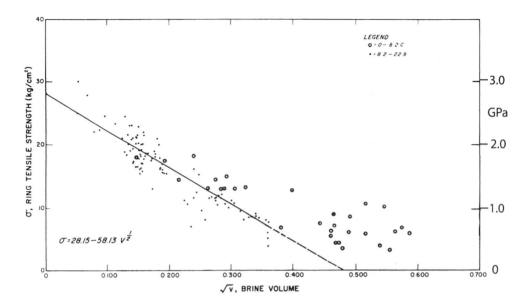

Figure 10.14. Ring-tensile strength of sea ice vs. the square root of the brine volume (Frankenstein 1969). The open circles represent tests performed in the temperature range 0 to –8.2°C; the closed circles, –8.3 to –22.9°C.

slightly higher. There was no clear evidence that the precipitation of $Na_2SO_4 \cdot 10\,H_2O$ had an effect on the ring-tensile strength. A detailed discussion of these results can be found in Weeks and Assur (1968). Finally, in 1969 Frankenstein published a major study of the ring-tensile strength of sea ice, reporting the results of over 1400 tests. In Figure 10.14 every data point represents roughly nine tests. This figure clearly shows that nature was not behaving as it was supposed to behave. At $v_b^{1/2}$ values of less than 0.40 (cold ice with low brine volumes), sea ice and NaCl ice behave in a similar manner in that σ_f versus $v_b^{1/2}$ plots prove to be linear. However at $v_b^{1/2}$ values greater than 0.40, σ_f values do not continue the systematic decrease toward zero but instead stay roughly constant at an average value of 0.66 MPa. (Note that the sea ice and NaCl ice data sets are not in disagreement in that the NaCl ice data does not contain tests on warm, high salinity ice, i.e., $v_b^{1/2}$ is always less than approximately 0.4 using ‰.) Perhaps this is the result of the actual stress concentration factor in warm ice being less than the theoretical stress concentration factor calculated assuming perfectly elastic behavior. As best I know, this subject has never been investigated.

Now let us summarize the good and bad features of the ring-tensile test as applied to the study of sea ice.

GOOD

1. It is clearly a test that is sensitive to changes in the small-scale internal structure of the material.
2. Sample preparation is comparatively easy and the time necessary to perform a test is small, allowing the possibility of large sample sizes.
3. Ring-tensile tests are easy to perform in the field.
4. The σ_o values estimated from the linearized plots of σ_o versus $v_b^{1/2}$ are in general agreement with the strength of freshwater ice as measured using the same test.

BAD

1. The stress concentration factor, which should be a constant determined only by the geometry of the test, appears to be temperature sensitive at near-freezing temperatures.
2. Although the estimated σ_o value for sea ice and the σ_f value for freshwater ice are similar, they are both appreciably higher than results obtained by true uniaxial tension tests. A detailed discussion of the problems of interpreting the results of ring-tensile tests can be found in Mellor and Hawkes (1971).

As the primary reason for performing tensile tests on sea ice was not to study the effect of variations in the structure of the material on strength but to obtain numbers that could be used in applied sea ice problems such as bearing capacity and ice forces on structures, the ring-tensile test has fallen into disuse. The same can be said for the Brazil test, which is similar to the ring but without the central hole. In sea ice the Brazil test has been found to give results that are lower than values obtained from uniaxial tests. They also show a larger scatter than ring tests.

Considering that the ring-tensile test is no longer in favor, one might ask why I have devoted space to discussing its results. The answer is simple. Much of the current thinking concerning how sea ice's state influences its mechanical properties is based on the results of ring-tensile tests. Therefore I believe that understanding this bit of history is useful.

10.4.3 Pure Tensile Strength

During what might be called the ring-tensile period, work continued on developing techniques that would allow pure uniaxial tension tests to be carried out on sea ice samples. There were several problems. For one, the sample needs to be perfectly aligned. To achieve this, the sample needs to be carefully machined. In addition, the machining should be carried out at a fairly low temperature ($< -20°C$) to minimize brine drainage. Third, some procedure needed to be devised to maintain contact between the sample and the plattens of the test machine. Otherwise the tensile failure will occur between the ice and the platten instead of within the ice sample. As might be expected, this is a particular problem at near-melting temperatures when large brine volumes occur. Fortunately, procedures for mounting end-caps on tensile samples are now available (Cole et al. 1985; Lee 1986; Richter-Menge et al. 1993), although they are time consuming. Even so, pure tensile tests are essentially a laboratory procedure. To date, the amount of data available on the tensile strength of sea ice is limited to five sets ranging from that of Dykins (1967), using sea ice grown in a tank in a cold room in California, to that of Kuehn et al. (1990), who used both natural and artificially grown sea ice. The most recent study appears to be by Richter-Menge and Jones (1993), who utilized lead ice obtained in the Beaufort Sea and transported to New Hampshire for testing. In all cases the tensile direction was in the horizontal plane of the ice sheet. Strain rates varied from 10^{-3} to 10^{-7}. Mean values for these different data sets are shown in Figure 10.15. Note that as would be anticipated there is a general decrease in the tensile strength with increasing temperature (Figure 10.15a), and also that the dropoff in strength with increasing porosity (brine + gas) is clearly not linear (Figure 10.15b). As can also be seen in Figure 10.15a, at each test temperature the higher the strain rate, the higher the strength. At all temperatures, it was observed that the stress versus strain curves were essentially linear (elastic) at strain rates of 10^{-3},

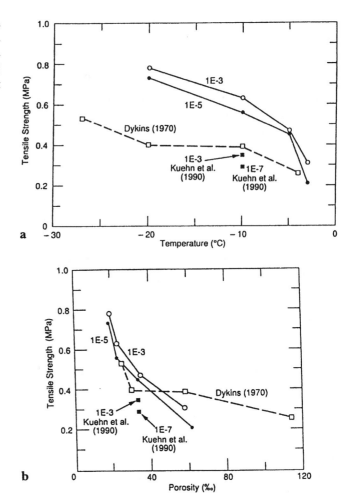

Figure 10.15. Maximum tensile stress a function of (a) temperature and (b) porosity (Richter-Menge and Jones 1993). Plotted are the mean values at each test condition.

whereas at values of 10^{-5} they were clearly nonlinear. Figure 10.16a shows the Richter-Menge and Jones data presented in $\sqrt{\sigma_t}$ versus $\sqrt{v}$ plots, where v represents the porosity expressed in ‰. This representation was suggested earlier as an alternative model by Weeks and Assur (1972). However, they never utilized it in fitting data. Here the basic relation is assumed to be

$$\frac{\sigma_t}{\sigma_o} = \left[1 - \left(\frac{v_b}{v_o}\right)^{\frac{1}{2}}\right]^2 \qquad (10.39)$$

which when expanded becomes

$$\frac{\sigma_t}{\sigma_o} = 1 - 2\left(\frac{v_b}{v_o}\right)^{\frac{1}{2}} + \frac{v_b}{v_o}. \qquad (10.40)$$

Although Richter-Menge and Jones found that $\sqrt{\sigma_t}$ versus $\sqrt{v_b}$ plots resulted in slightly higher correlation coefficients than did σ_t versus $\sqrt{v_b}$ plots (0.75 vs. 0.60), these differences did not prove to be statistically significant. Ultimately they decided to utilize the $\sqrt{\sigma_t}$ versus $\sqrt{v_b}$ representation in that it resulted in estimates of v_o that they believed to be more in line with petrographic observations of the apparent porosity of the skeleton layer. An alternate way to analyze

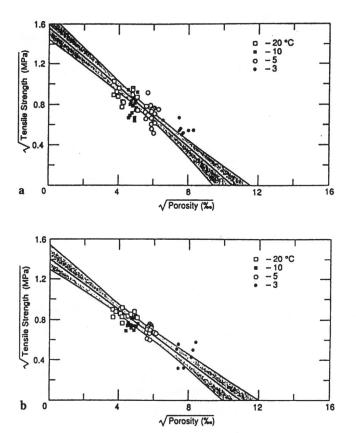

Figure 10.16. Least-squares regression analysis of uniaxial tension data using a $(\sigma_t)^{\frac{1}{2}} - (v_b)^{\frac{1}{2}}$ model at strain rates of (a) $10^{-3}s^{-1}$ and (b) $10^{-5}s^{-1}$ (Richter-Menge and Jones 1993). The shaded area represents the 95% confidence interval.

this data set would be to neglect the small effect of strain rate variations. In this case, a fit to the data yields $\sigma_t = 4.278v_t^{-0.6455}$ with an r^2 value of 0.72 (Timco and Weeks 2009). Here σ_t is in MPa and v_t is in ‰. Although this relation is clearly convenient, it obviously becomes unrealistic at low porosities.

The set of tensile tests obtained by Peyton (1966) is of particular interest as he was able to systematically vary the orientation of the axis of his horizontal samples relative to the direction of the preferred c-axis alignment observed in the lower portion of the ice at Barrow, Alaska. As a result his tests clearly show the relation between both tensile and compressive strength and crystal orientation (Figure 10.17). The notation is as follows: the first number gives the angle between the axis of the test specimen and the vertical, and the second number gives the angle between the sample and the mean c-axis direction of the ice crystals. One should also note that the ratio of the strength obtained from vertical cores is roughly three times that obtained from horizontal cores. Similar results have been obtained by several other authors.

Other results that should be mentioned here are the tensile values obtained by Cox and Richter-Menge (1985b) on ice collected from MY pressure ridges. It should surprise no one that there were large variations in their observed tensile strengths that correlated with variations in the structure of their samples. They noted that in brecciated ice, failure commonly occurred in the finer-grained portions of the samples when the columnar portions of the sample were oriented in the hard-fail direction. Also, when there was a large range in the grain sizes present in a sample, there did not appear to be a significant correlation between grain size and tensile strength.

Figure 10.17. The average failure strength as measured in direct tension (open circles) and in compression (solid circles) plotted vs. sample orientation (Peyton 1966). The orientation notation can be found in the present text.

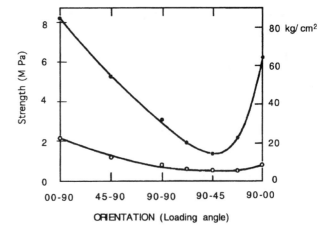

This is somewhat surprising in that tests on freshwater ice have suggested that strength decreases as grain size increases (Currier and Schulson 1982), although not everyone feels that this correlation is warranted at near-melting temperatures (Sinha 1983).

Uniaxial tensile strength measurements on undeformed MY sea ice have been obtained by Sammonds et al. (1998) using ice collected from a ~7-m-thick floe located near Graham Island in the Canadian Archipelago. In contrast to the ice collected from MY pressure ridges mentioned above, this ice was quite massive and showed no obvious blocky structure. Salinities were uniformly quite low (< 3‰) and porosities were relatively high. The temperature and strain rate ranges used were from –3.5 to –40°C and 10^{-2} to $10^{-7}s^{-1}$. Strength values ranged between 0.47 and 1.02 MPa for horizontal loading and 1.39 to 1.44 MPa for vertical loading. Over the temperature and strain rate range utilized there was no significant correlation between tensile strength and either brine volume or void volume.

10.4.4 Flexural Strength

Because of the difficulties in obtaining good measures of the direct tensile strength of sea ice via field tests, it was quickly recognized that there was a need for proxy field tests that would provide related information. This brings us to flexural strength, a measure that, similar to the ring-tensile test, possesses its own set of problems. Flexural strength measurements go back to the earliest days of sea ice research (Makarov 1901; Arnol'd-Aliab'ev 1929b), were restarted immediately after World War II (Butkovich 1956, 1959; Petrov 1954–1955; Smirnov 1961), and have continued to the time of the writing of this book (Shapiro and Weeks 1993, 1995). Even though it is known that the flexural strength is more of an index value than a fundamental material property, I will be very surprised if we do not see more of this type of testing. What are the problems? Most importantly, the equations that are used to calculate flexural strength assume that the ice in the beam is both homogeneous and perfectly elastic. If this were the case, then the flexural strength would be approximately equal to the tensile strength. Although this may be true for some materials, it is certainly not the case for sea ice. Considering this problem, why do people continue to use this test? There are several reasons. Many real sea ice failures occur in flexure. Therefore the test is a reasonable approximation to reality in many applications. Also, the test can be carried out in the field using portable equipment as failures typically occur under comparatively small loads. From a research viewpoint the test is very flexible, allowing the investigator to study the effects on strength of changes in sample orientation in the horizontal plane as well as variations in the vertical profile. It also allows one to vary, within limits, the volume of ice subjected to stress. In addition, the nature of the samples makes it easy to examine correlations between the observed failure surface and visible flaws in the ice. Finally, by increasing the size of the sample until it equals the total ice thickness, one ends up with an in-situ beam. In this case one measurement provides a strength value representative of the complete ice sheet, provided that the in-situ temperature profile across the sheet is maintained. For such tests the assumption that the sample is vertically homogeneous is clearly far less true than it is for small samples. Although there are techniques for dealing with the effects of vertical variations in the elastic modulus in such tests (Assur 1967; Kerr and Palmer 1972), they have only rarely been utilized (Frankenstein 1970; Frederking and Häusler 1978). In addition, as the sheet being tested gets thicker and thicker, the equipment necessary to load an in-situ beam to failure becomes heavy. For simple beams this is not the case, but assuming that the thickness of the beams being tested remains constant, the number of tests necessary to obtain a complete profile of a thick ice sheet can become large. In both cases, the amount of work involved can be significant.

What have the results of such flexural tests shown us? They clearly show that both in frazil ice and in ice that has a random c-axis orientation in the horizontal plane, the orientation of the sample in the horizontal plane has no effect on the observed strength. In addition, when a strong c-axis alignment exists, there are appreciable differences between horizontal beams cut in the hard-fail direction (the long axis of the beam cut perpendicular to the c-axis) and beams cut in the easy-fail direction (Figure 10.18). In addition, as the stressed volume of the sample increases, both the scatter and the flexural strength decrease. Finally the location of the failure surface appears to be less controlled by the subgrain structure and more by the presence of obvious flaws such as the presence of large brine drainage channels and thermal cracks.

To complete our survey of this test type, one also needs to mention in-situ cantilever beam tests. The only advantage associated with cantilevers is that sample preparation requires slightly less cutting and that the equipment necessary to perform a test is simpler. Such tests have all the difficulties associated with full ice sheet simple beams plus the fact that the shape of the test specimen produces a possible stress riser located at the butt of the beam. For instance, in tests run on

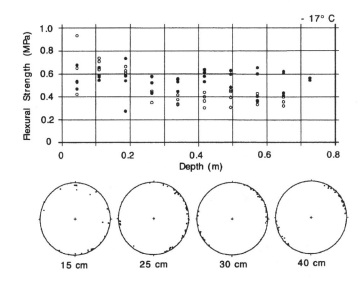

Figure 10.18. Normalized flexural strength vs. depth of ice collected at Barrow, Alaska (Shapiro and Weeks 1993). The open circles indicate beams tested in the easy-fail direction (c-axis parallel to the long axis of the beam) and the solid circles indicate beams tested in the hard-fail direction (c-axis perpendicular to the long axis of the beam). Also shown are fabric diagrams obtained at four different levels in the ice sheet.

freshwater ice, it has been found that the flexural strengths of simple in-situ beams were as much as twice the values obtained from similar cantilever tests (Gow et al. 1978; Timco and O'Brien 1994). In freshwater ice, procedures for relieving these butt stresses have been developed (Frederking and Svec 1985; Svec and Frederking 1981; Svec et al. 1985). For sea ice it is generally believed that there are enough natural flaws present in field samples that such procedures are not necessary.

Studies by different investigators on the effect of variations in the loading rate $\dot{\sigma}$ on the flexural strength σ_f of sea ice appear to show contradictory trends. For instance, Tabata (1960, 1966, 1967) and Tabata et al. (1975) have found that at stress rates of up to 0.3 MPa s^{-1}, there is a linear increase in σ_f with ln$\dot{\sigma}$. However, results of Enqvist (1972) suggest that if corrections are made to correct the inertial forces associated with the displacement of water during such tests, the increases disappear and σ_f becomes essentially independent of $\dot{\sigma}$. This latter result is appealing inasmuch as tensile strength has been found to be essentially independent of $\dot{\varepsilon}$.

Recently Timco and O'Brien (1994) surveyed all the available flexural strength data on both lake and sea ice, including both free beams and cantilevers. There clearly is a large increase in strength with decreasing temperature. As similar tests on freshwater ice do not show a significant temperature effect, it is reasonable to suggest that this temperature variation is produced by variations in the volume of brine and gas in the ice. Also, both the large beams (failure planes > 100 cm^2) and the small beams (failure planes < 100 cm^2) gave similar results, but with the smaller beams showing a slightly reduced scatter. Figure 10.19 shows the results of all these sea ice tests plotted versus $\sqrt{v_b}$. Again, note that for $\sqrt{v_b}$ values of > 0.400 flexural values become essentially constant, as was the case for the ring tensile strength values. The y-axis intercept value (1.76 MPa) is a mean value obtained from simple beam tests on freshwater ice. Although the form of the equation used to fit the data now contains a negative exponential, the empirical equation continues to contain the $\sqrt{v_b}$ term. Surprisingly, there do not appear to be any flexural strength observations available on either SY or MY ice.

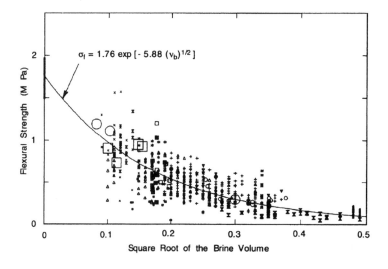

Figure 10.19. Flexural strength vs. $\sqrt{v_b}$ for all available tests on sea ice (n = 939). Although there is appreciable scatter, the correlation between the tests made by 14 different authors is quite good. For details on the data groupings and data sources see Timco and O'Brien (1994).

10.4.5 Uniaxial Compressive Strength

Studies of the uniaxial compressive strength of sea ice (σ_c) started in the 1950s (Butkovich 1956, 1959). Early studies of interest include the work of Peyton (1966, 1968), who showed that changes in c-axis alignments result in systematic changes in both tensile and unconfined compressive strengths (see Figure 10.17). His results also showed a pronounced decrease in the compressive strength as the square root of the brine volume increased. However, the exact form of the decrease is controversial (see the discussion in Weeks and Assur 1968). Unfortunately, Peyton's observations were not plotted directly but instead took the form of a series of corrections that, in retrospect, have made his results difficult to interpret. Even considering these reservations, I believe that the trends shown by Peyton's data are probably correct.

The magnitudes of the σ_c values measured by the early investigators are all of somewhat dubious quality as the test machines that they used were commonly soft (Frederking and Timco 1984a; Sinha and Frederking 1979). As a result, reported failure strengths could be appreciably lower than values obtained using stiff machines. Recently this problem has been resolved by the use of large, heavy, and very stiff presses. As such equipment is not readily transportable, the samples tested either have to be transported from the field to the laboratory or grown in the laboratory. Here I will focus on these latter tests (Frederking and Timco 1984a; Kuehn and Schulson 1994; Sinha 1984; Wang 1979, 1981). Testing was usually carried out at several different temperatures, strain rates, and sample orientations (Figure 10.20). Some of the results are as follows:

1. For every temperature and orientation studied to date, a transition from ductile to brittle behavior occurs if the strain rate exceeds some critical value.
2. The above transition occurs at strain rates that are roughly an order of magnitude lower for samples of unaligned columnar-grained sea ice when the compressive axis is parallel to the growth direction than for samples oriented at 90° to this direction (Figure 10.21).

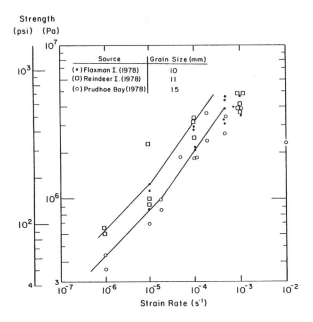

Figure 10.20. The compressive strength of unoriented columnar sea ice at −10°C as a function of strain rate. Also shown are the effects of changes in grain size and strain rate (Wang 1979, 1981).

Figure 10.21. The ductile to brittle transition strain rate of vertically and horizontally oriented laboratory-grown sea ice vs. test temperature (Kuehn and Schulson 1994).

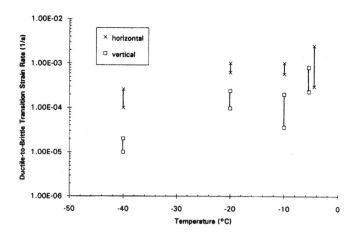

3. For both orientations the ductile and the brittle failure strengths generally increase as temperature decreases.

4. Only the ductile strength increases with increasing strain rate.

5. The ratio of the vertical to horizontal strength is ~3.6 in the ductile regime and ~2.0 in the brittle regime. A number of investigators have reported similar values.

6. As the amounts of brine and gas in the ice increase, the strength decreases. The exact functionality of this decrease does not appear to have been well documented to date.

A particularly useful data set has been assembled by Timco and Frederking (1990, 1991) who coupled 283 small-sample compressive strength tests on FY sea ice with associated observations on strain rate $\dot{\varepsilon}$, test temperature, sample salinity, density, ice type, crystal type, and orientation. For horizontally loaded columnar ice, the uniaxial strength could be described by

$$\sigma_c = 37\dot{\varepsilon}^{0.22}\left[1-\sqrt{\frac{v_T}{270}}\right] \ . \tag{10.41}$$

Simarily, for vertically loaded columnar ice the test results could be summarized by

$$\sigma_c = 160\dot{\varepsilon}^{0.22}\left[1-\sqrt{\frac{v_T}{200}}\right] \tag{10.42}$$

and finally, for granular ice by

$$\sigma_c = 49\dot{\varepsilon}^{0.22}\left[1-\sqrt{\frac{v_T}{280}}\right] \ . \tag{10.43}$$

In these relations the compressive strength is in MPa, v_T is in ‰, and the range of strain rate considered was between $10^{-7}s^{-1}$ and $10^{-4}s^{-1}$.

The most recent study of the unconfined compressive strength of sea ice has been by Moslet (2007), who utilized a specially designed field-portable testing system to study Svalbard fast ice. There were two particular foci in this multiyear study (~540 samples): to minimize the time between the removal of the sample from the ice sheet and testing, and to provide additional information on the strength of warm, decaying sea ice (0 to −11°C). As is obvious, this clearly is a

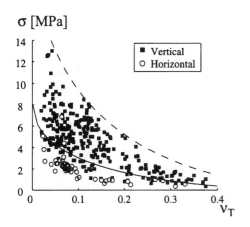

Figure 10.22. Maximum strength (unconfined compression of in-situ vertical and horizontal samples) as a function of total porosity. The lines are fit to maximum values for the vertical (dashed line) and horizontal (solid line) strength (Moslet 2007).

temperature range in which brine drainage upon sample removal from the ice sheet is a particular problem. Although trends were generally similar to the results of earlier work, there were differences. These were as follows:

1. Sea ice retains substantial strength at higher porosities than has been suggested by earlier studies (e.g., Timco and Frederking 1990).
2. In the temperature range (0 to −11°C) and for the strain rate (10^{-3} s^{-1}) utilized, temperature alone does not indicate whether the failure would be ductile or brittle. However, for a given temperature, higher strength values were obtained when the failures were brittle.
3. Brittle failures only occurred if the volume of air was less than 7%.
4. At a given stress rate, there was very little difference between the results of brittle and ductile failures.
5. The ratio between strengths measured in the vertical and the horizontal directions does not appear to be constant. Instead it varies from unity for cold ice to as high as 4–5 for warm ice. The change in the ratio appears to be roughly continuous (Figure 10.22).
6. Strength and behavior are not unique functions of state and therefore cannot be predicted by ice properties alone. At least in some cases, path may also be of importance.

I think that this work clearly shows that additional work will be necessary before the variations in the unconfined compressive strength of sea ice are completely understood. I also believe that ice sheet and ice sample history (path) have an effect in determining sea ice strength. Just how important it is remains to be determined. Clearly it is less important than brine and void volume variations. I also note that the Moslet (2007) paper includes a quite thorough listing of earlier studies using similar testing procedures.

The purpose of much of the above testing has been to provide design data pertinent to the problem of ice forces on offshore structures. In such calculations, the desired information is the compressive strength of thick FY ice. As tests involving the full thickness of ice sheets can be both difficult and expensive, only a very limited number of such tests have been completed (Chen and Lee 1986; Lee et al. 1986). Fortunately, models have been developed that give reasonable estimates of full thickness compressive strengths based on small-scale tests (Kovacs 1996; Timco and Frederking 1990, 1991). For instance, the relation suggested by Kovacs is

$$\sigma_c = B_2 \dot{\varepsilon}^{\frac{1}{n}} \varphi_B^m \qquad (10.44)$$

where the values of the parameters B_2, n, and m are approximately 2.7×10^3, 3, and -1, respectively, and $\dot{\varepsilon}$ and ϕ_B are the ice strain rate and its bulk porosity (‰). The units of the horizontal compressive strength σ_c are in MPa. The Timco and Frederking approach would appear to be more conventional, basing its estimates on a combination of small-scale tests, ice thickness, and air temperature. To the best of my knowledge, a critical study comparing these two approaches has not been undertaken. Although such approaches could clearly be convenient, at present it would be prudent to develop additional comparative data sets prior to using these estimated values in design calculations.

There is also a limited amount of data available for undeformed second-year and MY ice (Sinha 1984, 1985), with the columnar ice showing strengths that are similar to values obtained from FY columnar ice. However, frazil ice proved to be appreciably stronger than columnar ice. In regions where MY ice exists, peak stresses are believed to occur when offshore structures are impacted by MY pressure ridges. As a result there have been a significant number of unconfined compressive strength tests completed on ice collected from this type of ridge (Cox, Richter, Weeks, and Mellor 1984; Cox, Richter-Menge, et al. 1985). The results are consistent with test data on columnar FY ice oriented in the hard-fail direction, as well as with granular sea ice results reported by Wang (1979). When the variation in ice strength within and between different ridges was examined, it was found that in all cases the main factor contributing to the observed variance was the differences within cores (Weeks 1985). When the effect of internal sea ice structure was examined, highest strengths were observed when the loading occurred parallel to the crystal elongation. When the elongation direction approached the plane of maximum shear, there was an appreciable decrease in strength. A similar decrease in strength occurred when the samples contained a mix of granular and columnar ice (Richter-Menge and Cox 1985).

If quick and easy field estimates of the unconfined compressive strength of the sea are needed, one could also explore the possibility of using the axial double-ball test described by Kovacs (1993). His results suggest that the results of this simple rapid small-sample test are generally comparable to the significantly more demanding results from unconfined compressive tests. Finally, a useful review of data on the uniaxial compressive strength of freshwater ice as a function of strain rate and temperature has recently been published by Jones (2007). A comparison of these data with similar tests on iceberg ice shows that the strengths are similar only at very low strain rates ($< 10^{-6}$ s^{-1}). At higher strain rates, freshwater ice is consistently stronger. It is suggested that this difference results from the presence of preexisting, partially healed cracks in the iceberg samples.

10.4.6 Confined Compressive Strength

There also has been considerable interest in the compressive strength of sea ice as determined under different types of confinement conditions. Here the ultimate purpose has been to determine the complete failure envelope for the material. True triaxial tests ($\sigma_1 > \sigma_2 > \sigma_3$) are rare and require specialized equipment (Gratz and Schulson 1994; Hausler 1981). Even biaxial tests ($\sigma_1 > \sigma_2 = \sigma_3$), frequently referred to as conventional triaxial tests, can require involved procedures (Cox and Richter-Menge 1985a).

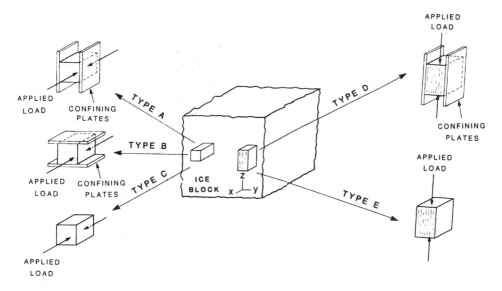

Figure 10.23. Geometry for confined compression tests showing the five confinement arrangements used (Timco and Frederking 1983, 1986). Note that when columnar ice is being tested, the z-axis is not only physically aligned in the vertical, it is also parallel to the direction of crystal elongation.

Both the failure stress and the failure mode are sensitive to confinement, with the failure mode changing from axial splitting to shear faulting in the loading plane (Smith and Schulson 1994). The transition from brittle to ductile has been found to occur when the strain rate falls below a critical level, with the transition strain rate first increasing and then decreasing with increasing across column confining stress (Schulson and Nickolayev 1995). In addition, the confined compressive strength of columnar ice is substantially different depending on the direction of the confinement relative to the long axes of the columnar crystals. When columnar freshwater ice is constrained from moving in the plane of the ice sheet, as shown in the type A loading condition in Figure 10.23, the strength was increased by a factor of four at a strain rate of 10^{-7} s^{-1} (to about 2.4 MPa) and by a factor of two at 10^{-4} s^{-1} (to about 20 MPa; Frederking 1977). When the crystals were confined in a direction normal to the axis of elongation, as shown in the B loading condition, little or no change was observed. When similar tests were carried out on granular sea ice (Timco and Frederking 1983), the confined compressive results were about 20% higher than for the unconfined specimens. In contrast, when columnar specimens were tested, there was up to a fourfold increase in the uniaxial strength when the loading conditions were of Type A. No significant change was observed when the conditions were of type B.

Timco and Frederking (1986) later expanded their series of tests, obtaining results for all five test configurations shown in Figure 10.23. Their purpose was to develop the failure envelope for the material. Here the term *failure envelope* refers to a relation that describes when the ice yields for any combination of compressive or tensile stress states. In this particular study the ice was loaded in one direction while confined in a second direction, thereby allowing deformation in only one direction. By measuring both the applied and the confining loads as well as knowing the orientation of the confined ice, it is then possible to determine the failure stresses for various combinations of stress states. Figure 10.24 shows a number of the test points plotted on the x–y plane of the failure envelope.

Figure 10.24. Test results showing points on the failure envelope for columnar sea ice on the *x–y* plane at a strain rate of $2 \times 10^{-4}\text{s}^{-1}$ and $T = -2°C$. For a detailed description of the testing procedures and these types of plots refer to Timco and Frederking (1986).

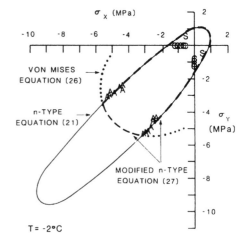

Figure 10.25. The brittle failure envelopes at –10°C for both FY sea ice (Richter-Menge and Jones 1993; Schulson et al. 2006) and freshwater ice (Iliescu and Schulson 2004). All the ice had a similar S2-type structure with the *c*-axes oriented randomly in the horizontal plane.

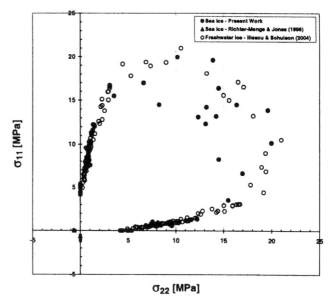

Further work on this subject by Schulson et al. (2006) has combined measurements of the brittle compressive strength of FY S2 sea ice (S2 = *c*-axes randomly oriented within the horizontal plane) with the earlier measurements of the tensile strength of similar sea ice by Richter-Menge and Jones (1993), thus allowing the complete brittle failure envelope to be constructed. The envelope was found to be symmetric about the loading path $R = (\sigma_2/\sigma_1) = 1$ as the result of the isotropic character of the material within the horizontal plane. However, the envelope is asymmetric with respect to the compressive/compressive and the tensile/tensile quadrants as the result of the relative weakness of sea ice under tension. Figure 10.25 shows a comparison of the brittle failure envelopes at –10°C for FY sea ice with that of freshwater ice having a similar structure (from Iliescu and Schulson 2004). Three different failure modes were observed. When the ice was unconfined $(R = 0)$, axial splitting occurred along the loading direction, with the splits running parallel to the long axis of the columnar grains. There appeared to be no preference for fracture along the grain boundaries. Under moderate confinement $(0.2 < R \leq 1)$, failure occurred by coulombic shear faulting. Finally, when the ice was loaded under high confinement $(R > 0.2)$, failure occurred in a raftlike manner accompanied by splitting across the columns. Comparable failure modes have been observed in earlier studies of saline ice (Schulson and Nickolayev 1995). The fact that the failure envelopes appear to be similar in both fresh and sea ice suggests that the processes controlling fracture are also similar. The authors suggest that these small-scale mechanisms also play an important role in fracture and deformation within the pack ice. The apparent similarity of the shapes of the failure envelopes on the experimental scale as well as on the geophysical scale is one of several factors supporting this contention. These matters are discussed further in Chapter 16.

The study of MY ice by Sammonds et al. (1998) discussed in section 10.4.3 also included a number of triaxial compression tests. Four main types of behavior were observed under uniaxial compression:

1. At high strain rates (10^{-3} to 10^{-2} s^{-1}), brittle fracture characterized by multiple axial splitting occurred.
2. If even a small confining pressure was present splitting was inhibited, with fracture occurring via the formation of a narrow shear fault inclined at 45 ± 3 degrees.
3. As confining pressure increased, plastic deformation was accompanied by substantial cracking activity.
4. At the highest confining pressures, cracking was completely inhibited and the deformation became completely plastic.

The unusual 45° orientation of the observed shear fractures is believed to be the result of the fact that low-stress slip and cleavage occurs in the basal planes of ice crystals. Clearly, additional data on the multiaxial strength of old ice types would be desirable.

A related in-situ strength measure favored by Canadian investigators can be obtained via the use of a borehole jack. In such tests the jack is placed in a precut cylindrical hole. Pressure is then applied hydraulically to the front and back load plates located inside the body of the jack and activated by a pump located on the ice surface. Oil pressure and plate displacement are then measured and used to determine the failure stress of the ice. The advantage of this test is that it can be used during both the summer and the winter as it does not require samples to be removed from the ice sheet. It can also comparatively easily obtain values from different levels in thick ice. The problem with the test is that the stress conditions and stress state are complex, making it difficult to compare its results with more conventional procedures.

Test results for FY ice have been published by a variety of authors (Johnston 2006; Masterson 1996, 2009; Masterson et al. 1997). In general the borehole strength of cold FY ice during the winter is in the range of 20 to 30 MPa. However, during the spring the strength decreases very rapidly, dropping to values as low as 10% of the midwinter values (Johnston 2006; Johnston et al. 2001, 2002, 2003a, 2003b). A number of tests are also available for old ice types (Johnston et al. 2003b; Sinha 1986, 1991).

10.4.7 Shear Strength

There have been several different types of shear tests applied to sea ice, including the use of torsion, direct shear, and punching, with each giving somewhat different results. The problem appears to be associated with the difficulty of generating a stress condition that corresponds to the condition assumed in analyzing the results. Therefore, care should be exercised in combining the results of different test series unless one is certain that the testing procedures are essentially identical and the ices tested are comparable.

The earliest studies (Butkovich 1956) loaded the central section of a core at right angles to the long axis of the specimen, resulting in failure along two circular surfaces that were oriented normal to the direction of grain growth. For sea ice with a salinity of ~6‰, shear strengths were roughly 1.6 MPa at temperatures between −5 and −6° C. At temperatures between −10 and −13°C, the shear strength increased to 2.3 MPa. Next Pounder and Little (1959) performed direct shear tests on sea ice at a variety of temperatures, obtaining strength values ranging from 0.02 to 1.0 MPa. Later Paige and Lee (1967) and Dykins (1971) used apparently identical shear systems that resulted in failures normal to the axis of the core being tested. In the Page and Lee antarctic field study the cores were vertical, resulting in failures that were normal to the long axis of the columnar grains (i.e., transverse to the structure of the sea ice crystals). Strengths were in the range of 0.4 to 1.3 MPa. In the slightly later Dykins laboratory study, cores were oriented both vertically and in two perpendicular horizontal directions. Surprisingly, these orientation changes did not appear to significantly affect the associated shear strength. Test values were appreciably lower, in the range of 0.14 to 0.30 MPa.

In all of the above test procedures there is the possible problem of uncertain normal stresses on the failure plane. As a result, more recent studies utilized an asymmetric 4-point bending system to produce shear failures in sea ice specimens (Frederking and Timco 1984b, 1986) as this procedure was felt to lead to less uncertainty in the generated stress field that produces the failure. It was found that when temperatures were higher shear strengths were lower, with total porosity providing a better correlation to strength changes than brine volume alone. In general, horizontal shear strength values were higher than values obtained when the shear failure was oriented vertically (i.e., when the failure plane was oriented parallel to the crystal elongation). There was one exception to this in that at −2°C the vertical shear strength was higher than for other orientations. Frederking and Timco do not have an explanation for this apparent anomalous behavior; neither do I. It also should be mentioned that similar anomalous values were noted by Dykins in his work. Figure 10.26 shows the results of the studies mentioned above. The curve presented uses the Paige and Lee values combined with the values determined by Frederking and Timco. The equation for the curve shown in the figure is

$$\sigma_s = 1500\left(1 - \sqrt{\frac{v_t}{390}}\right)$$

(10.45)

where v_t is the total porosity in ‰ and σ_s is the shear strength in kPa. Shear strength determinations on old types of sea ice appear to be nonexistent. Clearly more testing is needed.

10.4.8 Creep

The amount of effort devoted to the study of the long-term creep deformation of sea ice appears to be limited. There are several reasons for this. For one, it is very difficult to perform long-term tests on natural sea ice without the occurrence of significant brine drainage during the testing period. At best this would mean that although the total void volume might remain essentially constant during

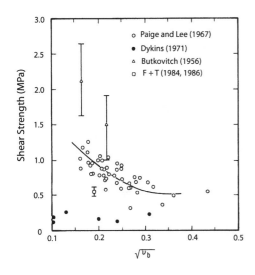

Figure 10.26. Shear strength as a function of the square root of the brine volume (Frederking and Timco 1986). Note that the data from the literature are for columnar ice whereas the results from the Frederking and Timco study are for granular ice.

a test, the ratio of the volume of air to the volume of brine in the sample would be continuously increasing. Only if the testing times were short would such changes not prove to be a problem. Although there are procedures for dealing with such problems, such as encasing the sample in a rubber membrane, to the best of my knowledge they have not been used on sea ice. Another factor here is that when one examines sea ice failures on the scale of engineering structures, the failures appear to be brittle. Features associated with long-term creep such as observed in glaciers are either absent or rare. Fortunately, there are a limited number of papers on the creep of columnar-grained freshwater ice (Gold 1965, 1972, 1983). One would expect that sea ice might behave in a similar manner but with higher deformation rates at similar stress levels. Direct evidence for this difference is available from studies of the effect of a liquid phase on the creep of pseudo-glacier ice. Studies of that subject have shown that when grain boundaries are sufficiently wetted, creep is enhanced. Of specific interest here is the work of de LaChapelle et al. (1995), who produced isotropic polycrystalline ice with NaCl brine volumes of 30 and 70‰ and then subjected it to compressive creep tests at temperatures of −13 and −24°C. The liquid content was obtained from the phase diagram for the tests at −13°C. In the tests at −24°C no liquid phase was present as the NaCl occurred as the solid dihydrate NaCl•2 H_2O. Figure 10.27 shows the results. Also shown for comparison are results for isotropic pure polycrystalline ice and a dashed line representative of basal glide on single crystals of pure ice. As can be seen, the strain rate of ice containing 30‰ brine is appreciably higher than the strain rate of pure ice. The strain rate of the ice containing 70‰ brine is higher still (over an order of magnitude greater than observed in the pure ice). It is also interesting to note that both saline ice types (30 and 70‰) show changes in the slope of the creep curves that are similar to the changes observed in pure ice as axial stress increases; $n \approx 1.8$ at lower stresses, changing to $n \approx 3$ at higher stresses. The results of the creep tests on the NaCl•2 H_2O ice are shown in the inset box. The dihydrate appears to be softer than pure ice at −24°C in that its presence clearly enhances the creep rate. The reader should not assume that because the presence of NaCl•2 H_2O softens ice, the same is necessarily true for other included solid phases. For instance, ice containing a fine uniform dispersion of particles of amorphous silica shows creep rates 10 to 30 times less than that of pure ice (Nayar et al. 1971).

10.4.9 Elastic Modulus

Of all the mechanical properties of ice, the elastic or so-called Young's Modulus (E) of sea ice (i.e., the proportionality constant between stress σ and strain ε or $\sigma = E\varepsilon$) has probably been the most studied. At least it appears to have been of interest over the longest period of time. For

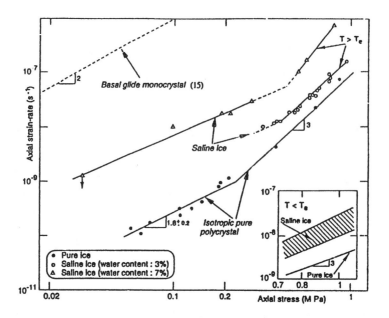

Figure 10.27. Minimum creep rate as a function of the applied stress for both pure and saline polycrystalline ice at −13°C. The dashed curve is representative of data for basal glide in single ice crystals and is given for comparison. The point with an arrow indicates that it is an upper limit for the strain rate. The inset gives the strain rate as a function of stress at −24°C for saline and pure ice; data within the hatched zone correspond to samples with a salinity between 5 and 15‰. T_e is the eutectic temperature of NaCl·2 H_2O (de LaChapelle et al. 1995).

instance, experimental results for pure ice were available as early as the 1820s and have been summarized in tabular form by Weinberg et al. (1940) and Voitkovskii (1960). Not surprisingly, this early work showed considerable scatter. In the following, measurements of E will be considered in two subsets: dynamic and static.

10.4.9.1 Dynamic Measurements   Dynamic measurements of E are typically determined by either measuring the rate of wave propagation in the ice or by exciting the natural resonant frequencies of different vibration modes. As the induced displacements are very small, anelastic effects are also typically small. As a result, dynamic measurements tend to be more reproducible than static values. The values are also invariably larger and show less scatter. It should be noted that this does not mean that dynamic values are necessarily better. The "best" values depend upon the type of application that is being considered. For instance, one would not want to apply dynamic measurements to an application where measurable strains are occurring.

In-situ seismic determinations of E have been reviewed by Weeks and Assur (1968). These values vary from 1.7 to 5.7 GPa when measured by flexural waves and from 1.7 to 9.1 GPa when determined by body-wave velocities. This difference is reasonable in that the flexural wave velocity is controlled by the overall properties of an ice sheet whereas the body wave velocity is controlled by the high-velocity channel in the commonly colder, less saline, and stronger upper section of the ice. That there is a pronounced increase in the value of E with decreasing temperature was clearly shown by Listov based on seismic observations on sea ice properties in the Laptev Sea (see Peschanskii 1960). At roughly the same time the results of Anderson (1958b) showed that there was a pronounced decrease in E as brine volume v_b increases. As a result, one would expect to observe

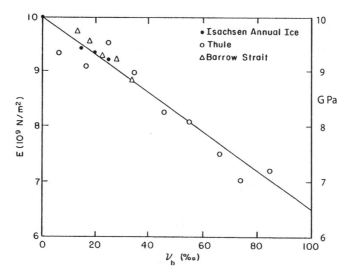

Figure 10.28. The elastic modulus of cold, arctic sea ice as determined from small sample tests plotted as a linear function of brine volume (Langleben and Pounder 1963).

significant seasonal changes in elastic parameters such as the longitudinal plate wave velocity. That this is the case has clearly been shown by the data collected by Hunkins (1960), who documented systematic velocity variations during the year on Arctic Ocean pack ice ranging from a low of ~2.3 km s^{-1} in August to a high of 3.2 km s^{-1} in late February.

Most dynamic measurements are not in situ but are, instead, determined from small, presumably homogeneous, samples that have been removed from the ice sheet. A typical series of tests is shown in Figure 10.28 (Langleben and Pounder 1963). Note that the E values at $v_b \approx 0$ are characteristically in the range of 9 to 10 GPa and are similar to measurements obtained from freshwater ice. Note also that E appears to decrease as a linear function of v_b. The results shown in the figure are averages based on over 300 measurements. Measurements on second-year and MY ice give values that are roughly 4 to 5% lower but show similar trends. At v_b values greater than 0.15, there is some evidence that E becomes a very weak function of v_b (Slesarenko and Frolov 1974). Still, dynamic determinations of E are almost completely from FY columnar sea ice in which the c-axis orientation in the horizontal plane is unknown. It would be useful to have measurements on other FY ice types such as aligned ice as well as on different MY ice types. In that this would appear to be a simple test to perform, it is surprising that so few values are available.

A theory has been developed (Bergdahl 1977) that deals with the linear variations in E with changes in brine volume in a manner somewhat similar to the theory suggested earlier by Assur for related changes in ice strength. It starts with several simplifications: only stress in the vertical direction is considered (the G direction, see Figure 10.13); the volume of gas in the ice is considered to be negligible ($v_a = 0$); the vertical interruptions between brine cylinders are neglected; the deformation of the brine inclusions and the ice matrix are assumed to be the same; and the loading is taken to be dynamic. Then

$$\varepsilon = pK = \sigma_1/E_1 = \sigma/E \qquad (10.46)$$

where ε is strain, p is pressure in the brine pockets, σ_1 is stress in the ice, σ is stress averaged over cross section A, K is the compressibility of the brine, and E and E_1 are the elasticity of pure ice and the bulk elasticity of sea ice, respectively. From the above

$$\sigma = E\varepsilon \quad p = \varepsilon/K \quad \sigma_1 = \varepsilon E_1$$

and the average stress is

$$\sigma = \sigma_1 (1 - A_b) + pA_b \qquad (10.47)$$

where A_b is the relative cross-sectional area of the brine inclusions. These equations give

$$E = E_1 + (1/K - E_1)A_b \qquad (10.48)$$

or, if $A_B = v_b$,

$$E = E_1 + (1/K - E_1)v_b \qquad (10.49)$$

suggesting that E should be a linear function of the brine volume, as has been shown experimentally. Further details can be found in the original paper or in Weeks and Ackley (1986), a more easily obtained reference.

10.4.9.2 Static Measurements   Static measurements of E are more variable and difficult to interpret than are dynamic measurements because sea ice invariably behaves viscoelastically when subjected to appreciable stresses for finite periods of time. Nevertheless, it is these effective E values (E') that are applicable to most engineering problems such as bearing capacity or ice forces on structures. The most extensive work on the static modulus of sea ice is by Dykins (1971), who tested small beams in bending. His stress–strain curves, obtained at stress rates of 0.25 MPa s^{-1}, were quite linear. His plots of E' versus temperature suggest discontinuities at −8.2 and −22.9°C, where $Na_2SO_4 \cdot 10\ H_2O$ and $NaCl \cdot 2\ H_2O$ precipitate. Unfortunately the

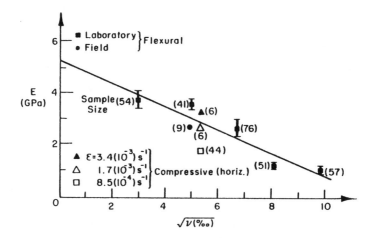

Figure 10.29. The effective elastic modulus of sea ice vs. the square root of the brine volume (Vaudrey 1977).

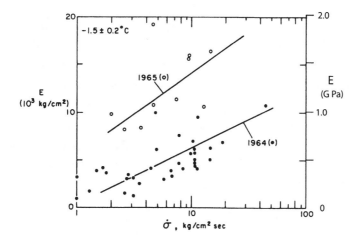

Figure 10.30. Relations between the effective elastic moduli E' and the stress rate $\dot\sigma$ at −1.5°C as determined by in-situ beam tests (Tabata et al. 1967).

testing was not sufficiently detailed to definitely establish these effects. Figure 10.29 shows E' plotted against v_b. Note that these static values are in general agreement with the seismic values of Anderson discussed earlier. Finally, Vaudrey (1977) has utilized data from both large beams in the field and small beams in the laboratory in determining the variation of E' with v_b. In this case the E' values appear to be linear when plotted versus $\sqrt{v_b}$. Obviously more comparable values would be useful.

The best studies of the time dependence of E' in sea ice are by Tabata and his associates. These studies can be separated into two groups: those using small sample beams (Tabata 1960, 1967) and those using in-situ cantilevers (Tabata 1966; Tabata and Fujino 1964, 1965; Tabata et al. 1967). In the beam tests, log E' appears to increase with log $\dot\sigma$, approaching the dynamic value at large values of $\dot\sigma$. The results of the cantilever tests are shown in Figure 10.30. Again, a pronounced change in E' with $\dot\sigma$ is observed. The difference between the two field seasons is believed to be the result of changes in the amounts of infiltrated snow ice present (more in 1965). It is possible that this large difference is explained by the fact that Tabata's tests were commonly performed at temperatures very near the freezing point.

Subsequent tests by Lainey and Tinawi (1981) demonstrate how E' values for laboratory saline ice can change with temperature and loading rate: i.e., E' increases with increasing loading rate and decreasing temperature, with the high $\dot\sigma$ tests giving values that are in general agreement with those reported by Vaudrey (1977). As was the case for the dynamic values of E', available static observations are primarily from FY ice. Cox, Richter, Weeks, Mellor, and Bosworth (1984) and Cox, Richter, Weeks, and Mellor (1984) provide modulus values obtained using samples from MY pressure ridges. Their results are comparable in magnitude to values reported by Tratteberg et al. (1975) for freshwater ice. Again the effective modulus was observed to increase with increasing strain rate and decreasing temperature. The considerable scatter in the data was attributed to the large variations in the internal structure of the samples (Richter-Menge and Cox 1985). Considering how pressure ridges form, such variations are hardly surprising.

10.4.10 Poisson's Ratio   As with the effective elastic modulus, it is the effective Poisson's ratio μ' that is of interest in most ice engineering problems. The only available direct measurements of μ' appear to be by Murat and Lainey (1982), who determined the longitudinal and transverse strains on simply supported beams loaded in flexure. The tests were performed at different temperatures and loading rates on columnar ice of the S2 variety that had an average salinity of 5‰. It was found that μ' decreased with increasing stress rate $\dot{\sigma}$ and decreasing temperature. At very low values of $\dot{\sigma}$, μ' tended toward the expected limit of 0.5, and at high values of $\dot{\sigma}$, μ' tended toward 0.33. A general expression was also obtained, giving μ' in terms of the stress rate and the dynamic Poisson's ratio at the temperature of interest:

$$\mu' = 0.24 \left(\dot{\sigma} / \dot{\sigma}_1 \right)^{-0.29} + \mu_D. \tag{10.50}$$

Here the dynamic Poisson's ratio is estimated from

$$\mu_D = 0.333 + \left(6.105 \times 10^{-2} \right) \exp \left(T/5.48 \right) \tag{10.51}$$

where T is the ice temperature in °C (Weeks and Assur 1968). Murat and Lainey also assumed that the strain rate could be approximated by

$$\dot{\varepsilon} = \dot{\sigma} / E' \tag{10.52}$$

where E' is the effective elastic modulus, obtaining

$$\mu' = \left(2.4 \times 10^{-3} \right) \left(\dot{\varepsilon} / \dot{\varepsilon}_1 \right)^{-0.30} + \mu_D \tag{10.53}$$

where $\dot{\varepsilon}_1$ is a unit strain rate (1 s^{-1}). When compression tests were performed on sea ice having aligned c-axes, it was found that μ' values ranged from 0.8 to 1.2 in the horizontal direction normal to the columnar crystals, and from 0 to 0.2 in the vertical direction parallel to the columns.

A theoretical examination of the effects of vertical variations in μ' in floating ice sheets on the mechanical response of the sheet can be found in Hutter (1975). He suggests that, in many cases, it is not necessary to consider such variations.

There appears to be no determinations of μ' from old ice types. Considering the fact that μ' appears to be influenced by parameters such as loading rate and direction, temperature, grain size, and structure, additional measurements would clearly be useful.

10.4.11 Scale Effects
The fact that scale effects occur in lake ice has been known for some time. For example, using data collected by Butiagin (1966a, 1966b) on beams of river and lake ice, Weeks and Assur (1972) were able to show that there was a pronounced linear increase in the ratio σ_s / σ_L (where σ_s is the flexural strength of a control beam with a cross section of 70 cm² and σ_L is the flexural strength of identical ice tested under identical conditions) when plotted against L, where L is the square root of the area of the failure surface. For sea ice the experimental support for such changes is a

bit more mixed. For instance, Parsons et al. (1992) have analyzed the results of a series of three-point bending tests carried out on simply supported beams of sea ice, freshwater ice, and iceberg ice samples of a similar size. For the sea ice samples it was found that as expected, the failure strength dropped as the sample size increased. It was also found that the results could be well fitted by using a three-parameter Weibull distribution (a function that has commonly been used to analyze this type of data). The decrease was found to scale with $V^{-1/12}$ where V is the volume of the specimen. A similar dependence has been found in coal and granite as well as in ice failing under compressive stresses. The usual basic assumption here is that the sample fails when the stress at the weakest flaw (generally assumed to be the largest flaw) exceeds some critical value. As the volume under stress increases, the chance of a large flaw existing in that volume also increases, resulting in the observed decrease in strength with increasing sample size. Unfortunately the observed $V^{-1/12}$ dependence is not what would be expected if a weakest-link model were to hold.

As was pointed out in Chapter 7, sea ice has many potential types of flaws. For instance, on the smallest scale there are lattice defects within the pure ice comprising the platelets. As the size of the scale increases, other defects include brine pockets, gas bubbles, and grain size variations, all of which are pervasive features distributed throughout sea ice at submillimeter to centimeter spacings. Then there are brine drainage channels and networks, which although larger are also less frequent, with spacings of tens of centimeters to meters. Unfortunately the rules, if any, that govern the spacings of brine drainage networks are currently not well understood. These features are particularly interesting in that they frequently extend completely through ice sheets. During recent field tests at Barrow (Cole et al. 1995), crack propagation through large full-thickness sections of highly aligned sea ice was investigated. In several of the tests an attempt was made to force the crack to run in the hard-fail direction. In some of these cases the crack would run in the hard-fail direction for a distance of several meters and then make a right-angle turn, running across the rest of the sample in the easy-fail direction. Excavation at the site of the right-angle bend in the track of the crack revealed the presence of a large brine drainage tube. Brine drainage tubes are of a particularly interesting scale relative to the size of laboratory test specimens that have lateral dimensions of ~20 cm. If a large drainage tube is spotted in a test specimen, the specimen is commonly discarded as flawed. On the other hand when truly large specimens, say with a size of 50 cm or larger, are tested, it is usually impossible to identify brine drainage features as thick sea ice samples are opaque. Therefore the flaw is not identified, possibly resulting in a reduced strength measurement.

A discussion of recent measurements of crack propagation velocities in both freshwater and sea ice can be found in Petrenko and Gluschenkov (1996). The differences between ice types are striking. For instance, in lake- and laboratory-grown freshwater ice velocities varied from a few hundred to 1320 m s^{-1}. In sea ice, on the other hand, crack velocities were very low, in the range of 0.85 to 18 m s^{-1}, provided that brine was present in the ice. As this difference in crack velocity disappears when the ice is colder than −30°C (i.e., when essentially all the brine is frozen), the velocity difference is believed to be the result of the dynamic resistance of the liquid inclusions. Preliminary calculations have been able to provide estimates of the velocity differences that are in reasonable agreement with the experimental observations. Additional discussion of the results of large-scale field programs focused on crack propagation in floating ice sheets can be found in Kennedy et al. (1994) and Dempsey et al. (1999).

It might be well to comment here on a point picked up by a reviewer. In that there have been a number of fracture toughness tests performed on sea ice, why are these measurements not discussed in detail? (Here fracture toughness is a material property that describes the stress required to make a crack of known size propagate.) There are several reasons. For one, there does not appear to be agreement on how best to perform such tests. As a result there are questions as to how best to combine and analyze existing data sets. Also it is not clear, at least to me, as to how these values should be utilized in treating ice engineering problems. What can be said is that the size-independent fracture toughness of FY sea ice is on the order of 250 kPa m$^{0.5}$ (Dempsey 1989) and that there are no measurements on old ice.

Perhaps thermal cracks are the defining features of the next larger scale of flaws. There certainly are individuals who think that this is the case (Lewis 1993, 1995; Lewis et al. 1994). An important feature of thermal cracks is that their presence in large samples is not necessarily obvious. It is well established that thermal cracking is a frequently occurring process as one can hear its acoustic signal associated with the passing of cold fronts (Milne 1972). We also know some aspects of the mechanics of the phenomenon (Bazant 1992, 2001; Evans 1971; Evans and Rothrock 1975; Evans and Untersteiner 1971; Lewis 1993). It is interesting to note that Bazant's analysis suggests that the critical temperature difference necessary for thermal cracking decreases in proportion to the thickness of the ice sheet to the $-3/8$ power. He notes that this may be the reason that large fractures appear to form more often in thick ice than in thin ice. Unfortunately, field observations either supporting or rejecting these suggestions do not appear to exist. We also do not have realistic distributions of "active" thermal cracks at specific instances of time or understand the spacial distribution of these features or their life cycles as effective flaws. Let me explain. When such cracks form they can extend either partially or completely through the sheet. A crack that extends only partway can thus be either dry or wet depending upon the exact location of the crack. Dry cracks presumably weaken the ice for appreciable periods of time (perhaps weeks to months). On the other hand, if when a crack forms it is filled with seawater, it will immediately start to heal as the walls of the crack are below the freezing point of the seawater. How long the refreezing takes will, of course, depend on the initial width of the crack and the vertical temperature profile in the ice sheet. Furthermore, the ice that forms in narrow cracks frequently has its c-axis aligned both in the horizontal plane and parallel to the axis of the crack (Petrich et al. 2007). This is the orientation that gives the highest possible tensile strength in the direction normal to the crack, causing the crack to be stronger than the surrounding ice. Therefore, based on field observations of sequentially formed cracks occurring between the headlands of islands, new cracks do not follow old cracks but form as separate entities. This means that the refrozen crack is no longer serving as a flaw even though the thickness of the ice in the crack will always be less than the thickness of the surrounding ice sheet (when a crack floods, the waterline in the crack is at sea level, not at the ice surface). What does all this mean? For one, it is doubtful that you can assess the importance of thermal cracks as effective flaws in ice floes by simply counting them via acoustic methods or as seen on the ice surface where they hide beneath the snow cover. Probably the development of new measurement techniques will be required. It should also be noted that there are other features besides cracks that could serve to localize failures. This is particularly true in MY ice, where thickness variations and subsurface irregularities occur as part of the normal growth cycle even if we neglect thickness variations resulting from deformation (Hallam et al. 1987).

Presumably in our hierarchy of flaws we have now moved to a scale where cracks leading to features such as leads are sufficiently large to be obvious. As samples containing large, obvious cracks are typically not selected for strength testing, the effect of such features on the overall strength of regions of sea ice cannot readily be ascertained from sequences of laboratory tests. We have now entered scales that include floe–floe interactions such as rafting and ridging.

Why is there so much interest in the question of flaws and size effects on strength? Consider Figure 10.31, in which indentation pressure is plotted against the contact area (Sanderson 1988). Note that as the contact area increases over 10 orders of magnitude, the indentation pressure decreases by 4 orders of magnitude. Note also that contact pressure is a more complex measure of ice failure than measures such as tensile strength. The data groups included in the figure are: (a) laboratory tests on lake and sea ice; (b) medium-scale tests that typically involve the complete thickness of the ice sheet; (c) field data from full-scale, man-made island–ice sheet interaction programs and from experiments where measurements were taken while MY floes were split during collisions with towerlike natural islands (Masterson 2009); and (d) values of the "gross compressive strength" used in mesoscale computer simulations, such as those of Hibler (1979, 1980b) and Pritchard (1977, 1980a), that typically utilize grid scales of 40 to 125 km. The problem of interest here is the design and construction of oil production platforms capable of existing on a semipermanent basis at inhospitable locations on the continental shelves of the Arctic. Such platforms have to take the ice as it comes and confidently designing them is far from simple. If one uses ice strength values that are too high, large amounts of money are spent for no purpose. If values are too low, you save money during construction but risk the possibility of structural failure during operations and perhaps even an oil

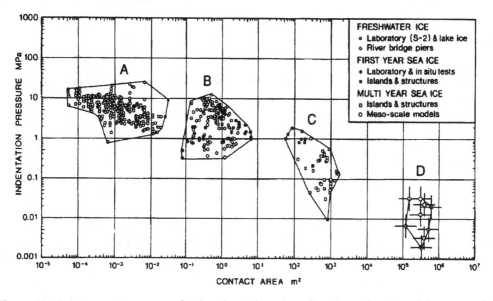

Figure 10.31. A pressure–area curve for the edge indentation of ice. An explanation of the nature of the different data groups can be found in the text (Sanderson 1988).

spill with all its expenses and negative impacts. Consider the political fallout of a photograph of an oil-covered polar bear. In fact, you would undoubtedly use not one but several different strength values in a design scheme. For instance, the values assumed to apply during point loads on the panels on the side of an offshore structure would undoubtedly be larger than the values used in estimating the total maximum load acting on the structure. The reason is that when loads on very wide structures are considered, the domains of specific failures are small relative to the width of the structure and these domains do not all fail at the same time. As a result there is a spatial and temporal averaging that results in a reduced average load (Kry 1980a, 1980b). For additional details relative to the problem of designing offshore platforms capable of surviving in heavy pack ice, the reader should refer to Sanderson (1988), to Masterson (2009), and to Timco et al. (2000), paying particular attention to the references of papers by Croasdale.

Recent thoughts on scaling have been summarized by Overland et al. (1995), who approached the problem from the point of view of hierarchy theory. This approach offers a way to view the interaction in complex multiscaled systems such as sea ice. In general it has been found that in such systems, state variables that exist at level $n + 1$ vary more smoothly over larger spatial scales and evolve more slowly than those at level n, with the variables on the larger scale commonly appearing as constraints, drivers, or boundary conditions affecting the next-lower level. Also, processes occurring at higher levels commonly remain unaffected by the details of processes occurring at the lower levels. Typically the effect of a lower level on the next-higher level comes through the development of an emergent property that is an aggregate of one or more of the lower processes. For instance, when an ice floe transmits stress to a neighboring floe or to a fixed structure, it does it through a series of localized failures that vary in both space and time. The stress as seen by the other floe only exists as a spatial and temporal average and cannot be studied by only investigating individual localized failures. On the other hand, it may be possible to parameterize the averaged processes so that they can be transferred to the next higher scale. In general it has been found that phenomena occurring on level $n + 2$ are too large and too slow to have much direct impact on level n. The same may be said of level $n - 2$, where processes are generally too small and too fast, appearing as a sort of background noise.

What does this tell us relative to designing research in sea ice mechanics? It appears to me that this says that it is important to know exactly what the scale controlling processes are for a specific problem. For instance, are thermal cracks important in controlling the failure of ice floes, or do we jump straight from brine drainage tubes to floe boundaries, or is there some intermediate-scale process that has so far gone unnoticed? It also says that, aside from the pure intellectual pleasure of understanding how everything hooks together, one primarily has to examine the mechanics of scales $n + 1$ and $n - 1$ to understand what goes on at scale n. The hierachical approach suggests that individuals interested in the large-scale behavior of pack ice need only concern themselves with parameterizing the aggregate result of floe–floe interactions and do not need to become involved in the mechanistic details of processes occurring within floes. It strikes me that a technique useful in bridging some of these scale gaps would be computer simulations using particle models in which the particles can be either ice floes or crystals, depending upon the problem. If done correctly such models should allow one to realistically simulate both the $n - 1$ processes within floes as they affect the interactions between floes at the n level, and also the resulting emergent behavior of aggregations of floes at the $n + 1$ level. Examples of papers exploring such possibilities include Hopkins (1996), Hopkins et al. (2004), and Hopkins and Thorndike (2006).

It should be noted that not everyone agrees that hierarchy theory is an essential component in an explanation of all aspects of scale effects in sea ice. For instance, in the next chapter it will be argued that identical mechanisms operate on both the hand specimen scale and on the grand scale of large-scale floe–floe interactions (Schulson 2001a; Schulson and Duval 2009). Discussion of this subject is far from over. An appreciation of the diversity of opinions and approaches currently active in the study of scaling laws as applicable to both problems in ice mechanics and ice dynamics modeling can be gained from a study of the varied papers in a related IUTAM Symposium (Dempsey and Shen 2001).

10.4.12 Conclusions

The reason that Professor Untersteiner was so correct in the quote that opened this chapter was not the fact that sea ice has many mechanical properties. A similar list can be made for almost any solid. It was the fact that every one of these properties shows large variations within the natural temperature range in which sea ice exists. Estimates of the seasonal variations in the elastic modulus and the flexural strength of sea ice in the Arctic Basin are shown in Figures 10.32 and 10.33 (Gavrilo 1997; Gavrilo et al. 1991). As might be expected, values are highest in January and February and lowest in August and early September. In selecting values for specific applications, I believe that it is important to examine the testing details very carefully so that there is the best possible match between how the parameter of interest has been determined and the application

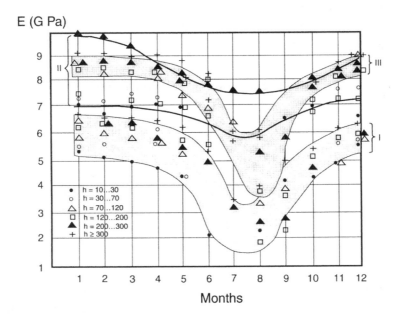

Figure 10.32. The seasonal variability of estimated values for the elastic modulus E for sea ice of different thicknesses: (a) static values from Gold's (1977) formula $E = E_o \left(1 - \sqrt{v_b}\right)^4$, where v_b is the brine volume in relative units; (b) dynamic values calculated from average longitudinal wave velocities; (c) dynamic values calculated from data on the temperature and salinity of the ice and the salinity of the included brine using the formula of Berdennikov (1948) (Gavrilo 1997).

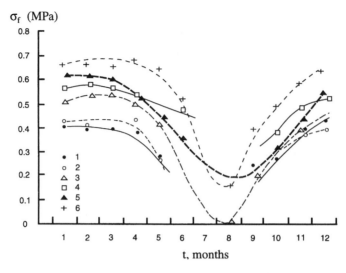

Figure 10.33. Seasonal variability of the flexural strength for ice from the Arctic Basin. The thickness ranges are (1) 10–30; (2) 30–70; (3) 70–120; (4) 120–200; (5) 200–300; and (6) above 300 cm (Gavrilo 1997).

in which it will be used. Perhaps for some materials one can safely pick property values out of a handbook. For sea ice this could prove to be rather risky.

One might notice that I have not mentioned snow in the above sections on mechanical properties. The reason is that in my view, snow does not play a significant role in determining the mechanical behavior of sea ice. A possible exception is the contribution of entrained snow to the properties of the piles of ice blocks that comprise pressure ridges or brash ice. However, even here the ice is fractured before the snow is entrained.

A discussion of the material covered in this section but with more emphasis on the engineering aspects of the subject can be found in Timco and Weeks (2009).

10.5 Electromagnetic Properties

10.5.1 Introduction

Although the electromagnetic properties of sea ice are as variable as its mechanical properties, its treatment will be appreciably briefer. My reasons are that my experience with electromagnetic properties is appreciably less than my experience with mechanical properties. However, it is my hope that the following discussion will, at least, provide the reader with a basis for further literature searches on this varied subject. To date the individuals most interested in the electromagnetic properties of sea ice are those from the remote sensing community, as understanding the electromagnetic properties of the material being sensed is frequently essential to the interpretation of remote sensing observations. A starting point here is the monograph titled *Microwave Remote Sensing of Sea Ice* edited by Carsey (1992), and in particular the included paper on the physical basis of sea ice remote sensing by Hallikainen and Winebrenner (1992). Other useful general references include Schanda (1986) and the monumental three-volume effort by Ulaby, Moore, and Fung (1982–1986).

As has been demonstrated in the earlier chapters of this book, sea ice essentially consists of four phases: ice I(h), brine, gas, and solid salts. In that each of these phases has different dielectric properties, ideally one would like to develop a formulation that expresses the dielectric properties of sea ice as a function of the relative proportions of the participating phases. In reality, the dielectric properties of sea ice are dominated either by the relative amounts of ice and brine or, when the amount of brine is small, the amounts of ice and air. Fortunately, the solid salts in sea ice do not have a strong effect, as their dielectric properties at frequencies between 0.1 and 40 GHz are sufficiently close to those of ice that their presence in small amounts is not generally distinguishable. As a result, dielectric formulations for sea ice are generally two-phase formulations accounting for the variations of either ice and brine or of ice and gas (usually taken to be air). As will be seen, the dielectric properties of both brine and air are sufficiently different from those of ice to provide an appreciable contribution to the dielectric properties of sea ice. Other factors are the exact shapes of the brine and gas inclusions, and their spatial arrangements. Finally, as the dielectric properties of the component phases are dependent on the frequencies utilized by the measurement systems, it is essential that this also be considered. In that the frequencies of primary interest in respect to sea ice are in the microwave range (1 to 60 GHz), and as the electromagnetic frequency (f) and the free-space wavelength (λ_o) are related by

$$\lambda_o = c/f, \tag{10.54}$$

where c is the speed of light ($3 \times 10^8 \, \text{m s}^{-1}$), this means that we will be primarily interested in wavelengths in the range of 30 cm to 5 mm where all-weather observations are possible.

Now for a brief review of some basics (Hallikainen and Winebrenner 1992). We will be concerned with the relative permittivity

$$\varepsilon = \varepsilon' - j\varepsilon' \tag{10.55}$$

($j = \sqrt{-1}$) where ε is a complex number that characterizes the electrical properties of materials. Here the real part of the relative permittivity ε' gives the dielectric contrast with respect to free space ($\varepsilon'_{air}=1$), whereas the imaginary part of the permittivity ε'' gives the electromagnetic loss of the material. These terms ε' and ε'' will be referred to respectively as the dielectric constant and the dielectric loss factor. Furthermore, if κ_a is the power absorption coefficient and if it is assumed that its value is not dependent on depth in the media and that there is no scattering from particles within the media, then

$$\delta_p = \frac{1}{\kappa_a} = \frac{\sqrt{\varepsilon'}}{k_o \varepsilon'} \quad \text{provided} \ \varepsilon' << \varepsilon'. \tag{10.56}$$

Here δ_p is the penetration depth and k_o is the wave number in free space ($k_o = 2\pi\lambda_o^{-1}$). The penetration depth is typically taken as the maximum depth in the medium that contributes to remote sensing parameters such as the brightness temperature and the backscattering coefficient.

10.5.2 Ice

There is a very large literature on the dielectric characteristics of pure ice, with references going back into the 1800s. For our purposes useful references include Evans (1965), Hobbs (1974), Hallikainen and Winebrenner (1992), and Petrenko and Whitworth (1999). Some sense of the size of the literature on this general subject is provided by the fact that in Evans' 1965 review of dielectric studies pertinent to the natural occurrence of ice, an annotated bibliography of 72 references is provided. Fortunately, it is not necessary here to review the bearing of variations in the dielectric properties of ice on its molecular characteristics, a matter of considerable interest in the physics literature. One must also remember that there are several different versions of freshwater ice. First, there is frozen distilled water that, if bubble free, can be considered an idealization of natural bubble-free ice. Even here there can be differences owing to variations in grain sizes, shapes, and orientations. Natural occurring freshwater ice can, in addition to similar structural variations, contain small amounts of chemical impurities within its atomic structure, a condition referred to as a solid solution and known to result in measurable changes in ice's dielectric properties. Fortunately, the substances (e.g., NH_4F, HF, $NaOH$) that can substitute appreciably in the ice lattice do not typically occur in significant amounts in naturally occurring ice or in seawater. The ice phase in sea ice is also believed to contain similar amounts of impurities. However, as these amounts are very small and their exact variation in sea ice has not been studied, the dielectric properties of the ice phase in sea ice are typically assumed to correspond closely to those of distilled-water ice.

Experimentally it has been found that for pure ice at low frequencies (< 1 MHz), the value of the permittivity is dominated by the reorientation of water molecules in the superimposed electric field. If ε' and ε'' are measured as a function of frequency, the permittivity decreases as the frequency increases, exactly as expected for a Debye relaxation process (Debye 1929) having a single relaxation time (Auty and Cole 1952). The classic way of representing the Debye theory is to plot the real (ε') versus the imaginary (ε'') parts of the permittivity. If the Debye theory holds, the resulting so-called Cole–Cole plot should be a semicircle. As can be seen in Figure 10.34, pure ice is an excellent example of a Debye relaxation process. This figure also shows that the permittivity values are different when measured parallel or perpendicular to the c-axis.

At higher frequencies (between 10 MHz and 1000 GHz) the dielectric constant of ice is approximately constant, with a value of $\varepsilon'_i = 3.17$. The dielectric loss factor ε'', which is of the order of 10^{-2} to 10^{-4}, shows a minimum centered at 1 to 5 GHz (Cumming 1952; Mätzler and Wegmüller 1987). The considerable scatter observed in the loss data appears to be caused by a combination of impurities in the natural samples plus inaccuracies in the measurements, at least partially resulting from the fact that the value of ε'' is so small. As will be seen, in comparison to the loss values for sea ice, loss values for pure ice are relatively insignificant. The air in sea ice is also relatively lossless and has a ε'_a value of 1.

10.5.3 Brine

Of the four components in sea ice, it is the brine that shows the greatest variability in its dielectric properties in the frequency range of primary interest (0.1 to 40 GHz). For starters, the complex permittivity for pure water is unusually high as a result of its polar molecular structure and the fact that in a liquid the molecules are free to rotate. Its dielectric behavior follows the Debye equa-

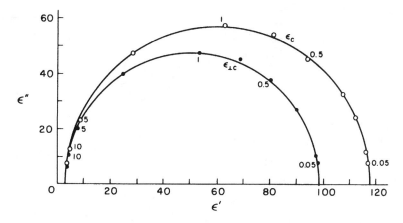

Figure 10.34. A Cole–Cole plot of the real and imaginary parts of the permittivity of ice at –20.5°C (Auty and Cole 1952). The frequencies in kHz are marked against the plotted points. The different curves are for measurements made parallel to (ε_c) and perpendicular to ($\varepsilon_{\perp c}$) the c-axis. (see Glen 1974). Note that the plots are nearly perfect semicircles.

tion with both the real and the imaginary parts of ε_w varying with frequency and temperature. Numerical relations for these variations are available in the literature (Klein and Swift 1977; Lane and Saxton 1953; Stogryn 1971). In addition, as dissolved salts increase the ionic conductivity, the dielectric loss factor for seawater (32–34‰) is higher than that of pure water at frequencies below 10 GHz. The appropriate modified Debye equation is

$$\varepsilon_{sw} = \varepsilon_{sw\infty} + \frac{\varepsilon_{sw0} - \varepsilon_{sw\infty}}{1 + j2\pi f\tau_{sw}} - j\frac{\sigma_{sw}}{\varepsilon_0}. \tag{10.57}$$

Here ε_{sw}, ε_{sw0}, and $\varepsilon_{sw\infty}$ are, respectively, the complex permittivity, the static permittivity, and the high-frequency limit of the permittivity of seawater, τ_{sw} is the relaxation time for seawater, and f is the frequency. Numerical values for the terms of this equation can be found in the literature (Klein and Swift 1977; Stogryn 1971; Weyl 1964). As the salt concentrations in the brine in sea ice are appreciably higher than those for seawater, measurements of the dc conductivity and the dielectric properties of seawater brine in equilibrium with ice between –2.8 and –25.0°C have also been determined, as have estimates of ε_{b0}, $\varepsilon_{b\infty}$, and τ_b (Stogryn and Desargeant 1985). Figure 10.35 shows the dielectric properties of brine based on these results. Both ε'_b and ε''_b typically decrease with decreasing temperature. The rapid increase in ε''_b with decreasing frequency below 3 GHz is caused by the increase in the ionic conductivity (the third term in equation 10.57). The dielectric constants at any given frequency are dependent on the concentration of salts in the brine. In turn, this concentration is specified by the temperature, via the phase relations, as discussed in Chapter 6. Therefore the dielectric constants of the brine should be a unique function of temperature unless it can be shown that there are significant changes in the ratios of the chemical components as the ice ages. The sample salinity only tells one how much brine is present.

10.5.4 Sea Ice

As sea ice is essentially a mixture of ice, brine, and gas, and as we know the dielectric properties of each of these component phases, we can now anticipate certain aspects of its composite

Figure 10.35. The dielectric (a) constant and (b) loss factor of brine as a function of frequency with temperature as a parameter (Stogryn and Desargant 1985). The figure is from Hallikainen and Winebrenner (1992).

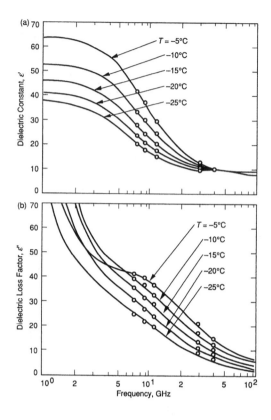

dielectric properties. For example, because ε'' for brine (or water) is several orders of magnitude higher than that for pure ice (e.g. 10 to 100 at 1 GHz as compared to 10^{-4}), one would expect ε'' for sea ice to be largely accounted for by its presence even though the amount of brine in sea ice is frequently small. On the other hand, although the real part of the dielectric constant ε' is only roughly one order of magnitude higher than that of brine, this is offset by the fact that the relative volume of ice in sea ice is much larger than the volume of brine. As a result it is the properties of the ice phase that are primary in controlling variations in ε'.

There have been a number of experimental studies of the dielectric properties of natural sea ice as well as of approximations to the natural material produced in the laboratory using seawater, artificial seawater, and NaCl solutions. A detailed summary of many of these results can be found in Hallikainen and Winebrenner (1992). Here I will simply touch on a few important points. Much of the available data are for frequencies of 1 GHz, 4 to 5 GHz, and 10 to 16 GHz, as these frequency bands have been used by microwave remote sensing systems in the study of sea ice. Higher-frequency data, although available, are less common.

Particularly detailed studies using both natural and laboratory-grown sea ice as well as NaCl ice, have been carried out by Arcone et al. (1986), Vant (1976), and Vant et al. (1974, 1978). The Vant et al. (1978) reference is one of the few that reports results from MY ice. Figure 10.36 is one of the several plots presented by Hallikainen and Winebrenner based on data from Vant (1976) and Hallikainen (1983). Although all these curves were based on measurements at 1 GHz, the trends are similar to those obtained at higher frequencies. For instance, the higher the salinity of the ice, the higher the permittivity values. Note that the low-salinity ice from the Gulf of Finland has properties similar to MY ice even though the low-salinity values are the result of very different processes. (The low salinity of the ice from the Gulf of Finland is caused by the local seawater having a low salinity. The low salinity of MY ice, on the other hand, is the result of brine drainage during the aging process.) As the temperature nears the freezing point, permittivity values increase rapidly, frequently in a nonlinear manner. These results are telling us quite clearly that the most important factor in controlling the dielectric properties of sea ice is the amount of liquid it contains, i.e., the brine volume.

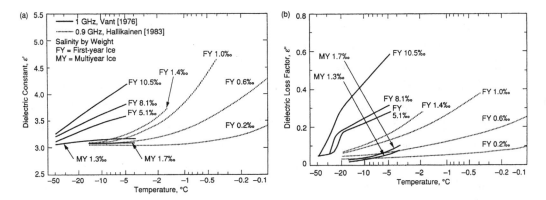

Figure 10.36. Values of ε′ and ε″ as obtained at 1 GHz by Vant (1976) and Hallikainen (1983) for sea ice. The figure is from Hallikainen and Winebrenner (1992).

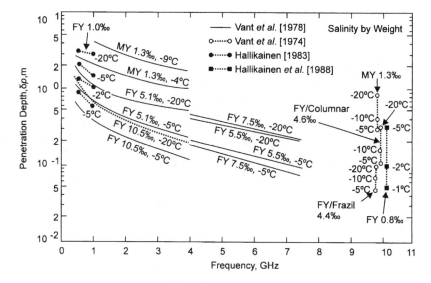

Figure 10.37. Calculated penetration depths for different types of sea ice based on experimental determinations of the power absorption coefficients as measured by various investigators (Hallikainen and Winebrenner 1992).

Actually, more important than the permittivity per se is a value that it specifies, the penetration depth δ_p, which determines the maximum depth in the ice, which contributes to the backscattering coefficient and the brightness temperature—important remote sensing measures. As can be seen in Figure 10.37, in the 1-to-10-GHz range δ_p values fall off as frequency increases (or as wavelength decreases). For instance, for FY ice δ_p values vary from 5 to 100 cm. For MY ice equivalent values are 30 to 500 cm. Actually, these values are a bit misleading, as such estimates assume that the ice sheet has a constant temperature and salinity. The only time that this would be almost true for a real sea ice sheet is during the summer, when the ice becomes essentially isothermal. Even then, the salinity profile complicates matters in that the lower part of the ice sheet has higher salinities than the upper part (see Chapter 8) and is therefore more lossy. Nevertheless, the general trends indicated are quite important. In FY sea ice, penetration of electromagnetic radiation into the ice is very limited and, as a result, scattering from brine and gas inclusions within the ice sheet is restricted. On the other hand, in low-salinity MY ice, penetration depths are larger and scattering within the ice becomes important.

There have been several attempts to model the dielectric characteristics of sea ice using a variety of different mixing relations. For instance, Addison (1970) used the so-called Maxwell–Wagner–Sillars (Maxwell 1891, 1913; Sillars 1937) equations in fitting his data on artificial sea ice obtained in the MHz frequency region. He concluded that although the model seemed reasonable, the data suggested that only roughly one-quarter of the brine was in inclusions that could be described as ellipsoidal, as assumed by the model. There were also reservations concerning whether or not the model would be successful at frequencies other than the MHz range. Vant et al. (1974) also based a model on the experimental fact that ε'_m, the dielectric constant of the mixture (i.e., sea ice), appeared to be linearly related to

$$\varepsilon'_m = \varepsilon'_i / (1 - 3v_b) \tag{10.58}$$

where ε'_i is the dielectric constant of pure ice and v_b is the brine volume. This linear correlation was first suggested by Hoekstra and Cappillino (1971) in their study of saline ice produced by very fast freezing. Interestingly, the highest linear correlation coefficient obtained by Vant was not for frazil ice ($r = 0.812$), where the liquid inclusions would be expected to be spherical, but for columnar sea ice ($r = 0.897$), where elongated shapes appear to be more common. They suggested that this difference might be explained by the fact that in their columnar ice measurements the brine pockets were oriented perpendicular to the electric field, minimizing the full effect of their elongated shapes. Later Vant et al. (1978) had reasonable success in the 0.1-to-40-GHz range using a confocal model initially sugested by Tinga et al. (1973). Here the brine inclusions were assumed to be symmetrical prolate spheroids with an axial ratio of 20. In addition, the best agreement was obtained by assuming that the brine inclusions were oriented at ~40° to the vertical.

General agreement between theory and experiment has been obtained by Stogryn (1987), who applied what was termed a "bilocal approximation to strong fluctuation theory" to data obtained over the frequency range 0.1 to 40 GHz and the temperature range −2 to −32°C. As seen in Figure 10.38, the theory gives a reasonable fit to ε'' data for frazil ice. For ε'', although the shapes of the theoretical curves correspond to the experimental trends, measured values are consistently somewhat higher than the experimental. An alternate approach, in which the emissivity values are estimated via a stochastic representation of the sea ice structure in terms of a two-dimensional autocorrelation function, can be found in Lin et al. (1988). Stogryn suggests that for such modeling efforts to advance further improved measurements will be required, including data in which the polarization of the electric field is parallel as well as perpendicular to the growth direction. Such measurements will allow the testing of models that treat anisotropic dielectric properties. He also notes the importance of measurements on specimens whose density, salinity, and temperature, as well as microstructural properties, are well described. Only when such results are available can theoretical developments be adequately tested and rise above the criticism of simply being fancy exercises in curve fitting. Saying this is one thing but doing it is another, as the amount of work involved is large. A good example

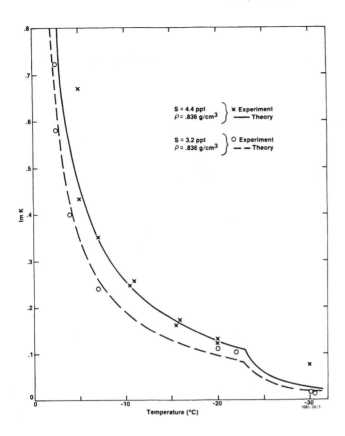

Figure 10.38. Comparison of the theoretical and experimental values of the dielectric loss factor of frazil sea ice at 10 GHz (Stogryn 1987).

of a study proceeding along these lines is that of Arcone et al. (1986). This paper also provides data sets that clearly show that although ε', ε'', and the transmission attenuation are strong functions of brine volume, the relations are really not linear, tailing off at higher brine volumes. Theoretical considerations as well as data supporting the speculation that preferred c-axis alignments should have a significant effect on the electromagnetic properties of sea ice can be found in Golden and Ackley (1981) and Morey et al. (1984). As might be expected, the magnitude of this effect typically increases with depth in the ice as the strength of the c-axis alignment increases. The latter paper also presents data supporting the contention that ε', although strongly dependent on v_b, is less dependent on the orientation of the brine inclusions. In contrast, ε'' is strongly dependent on inclusion orientation and less dependent on v_b. The study concludes that because the physical state of the ice is continually changing, only trends in the relationship between the electromagnetic properties of sea ice and its brine volume and inclusion geometry can be established. This is essentially saying that the EM properties of sea ice are nonthermodynamic and instead are a function of path, i.e., a function of the history of the ice. I think that there is little doubt that this is true. This conclusion also clearly applies to the mechanical properties in that plate width is a function of growth rate. Nevertheless, the question still remains as to just how significant these effects are. If they are small, life becomes much simpler.

Questions that one might have concerning whether temperatures where the precipitation of different solid salts starts will appear as break points on permittivity–temperature plots can be at least partially answered by examining Figs. 7, 8, and 9 in Fujino (1967). Even if one allows Fujino a bit of artistic license in drawing in the curves, it is clear that distinct break points do occur in the vicinity of –22 and –50°C, temperatures where $NaCl\cdot2\ H_2O$ precipitation starts and where the last of the brine presumably solidifies. There is no indication of a break point in the vicinity of –8.7°C, where the precipitation of $Na_2SO_4\cdot10\ H_2O$ occurs. I find this to be surprising.

10.5.5 Snow

In closing this section some mention should be made of the electromagnetic properties of snow. Additional details concerning the characteristics of typical snow packs occurring on arctic sea ice covers can be found in Chapter 15. Snow on sea ice can be separated into two classes, dry snow and wet snow, with the former being clearly the simpler case in that it is comprised of ice and air. As stated earlier, in the frequency range of most interest here (10 MHz to 1000 GHz), the value of ε' is essentially constant at 3.17 and is not appreciably dependent on temperature. As a result the ε' value for dry snow (ε'_{ds}) is only a function of the density of the snow. Between 3 and 37 GHz, Hallikainen et al. (1986) found that experimental measurements on dry snow could be well described by two linear relations:

$$\varepsilon'_{ds} = 1 + 1.9\rho_{ds} \text{ for } \rho_{ds} \leq 0.5 \text{ g/cm}^3 \qquad (10.59)$$

and

$$\varepsilon'_{ds} = 0.51 + 2.88\rho_{ds} \text{ for } \rho_{ds} \geq 0.5 \text{ g/cm}^3. \qquad (10.60)$$

Here the units of ρ_{ds} are g/cm^3.

Two-phase mixing formulae can also be useful in fitting and extending the observed variations in the dielectric properties of dry snow. For example, Ulaby et al. (1986, vol. 3) have used the model for spherical ice inclusions initially suggested by Tinga et al. (1973), obtaining

$$\varepsilon''_{ds} = \frac{0.34 V_i \varepsilon''_i}{\left(1 - 0.417 V_i\right)^2}. \qquad (10.61)$$

Here V_i is the volume fraction of ice, ε''_i is the loss factor for pure ice, and V_i is related to the snow density by

$$V_i = \frac{\rho_{ds}}{0.916} \qquad (10.62)$$

where 0.916 is taken to be the density of pure ice in g/cm^3. These relations imply that typical snow covers found on sea ice with densities in the 0.30–0.40 g/cm^3 range will have dielectric losses of roughly 15 to 22% of that of pure ice. Several other more complex relations can be found in the literature that express ε''_{ds} in terms of frequency, temperature, grain size, and density. References to this work can be found in Hallikainen and Winebrenner (1992).

The electromagnetic behavior of wet snow is more complex than that of dry snow, in that wet snow is comprised of a mixture of ice crystals, liquid water (sometimes saline), and air. As a result, the permittivity of wet snow is sensitive to changes in frequency, temperature, snow bulk density, volumetric water content, and the shapes of both the ice grains and the water inclusions. As the permittivity of water is appreciably higher than that of either ice or air, the dielectric behavior of wet snow is primarily a function of the volume fraction of water. The appearance of plots of variations in ε'_{ds} and ε''_{ws} are essentially compressed versions of those of water at 0°C. When mixing formulae are used, best agreement occurs when it is assumed that inclusion shapes

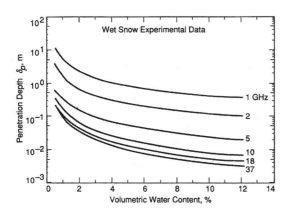

Figure 10.39. Penetration depths for wet snow presented as a function of volumetric water content with frequency as a parameter (Hallikainen and Winebrenner 1992).

are asymmetric and also depend on the water content. Figure 10.39, based on a modified Debye-like model, shows electromagnetic penetration depths as a function of frequency for several different liquid water contents. The important thing to note is that a liquid water content of even a few percent results in a pronounced decrease in the penetration depth. This means that if wet snow is present as the upper layer on a sheet of sea ice, microwave and radiometer observations will only provide information on the nature of the snowpack. The underlying sea ice will be screened from observation.

I recommend that individuals interested in exploring this subject in more detail start by reading Hallikainen and Winebrenner (1992), the reference from which much of the above material was extracted. At the same time it would be useful to begin to acquaint oneself with how variations in the electromagnetic properties of snow and sea ice affect how these materials are perceived via the use of different remote sensing techniques (Carsey 1992; Martin 2004).

10.6 Concluding Remarks

Reflecting a bit on the above chapter, there are a few important points that should stay with the reader. Sea ice property values all show variations. In situations where ice temperatures are only slightly below freezing, these variations can be quite large. Remember that even though the top of a sea ice sheet may have a temperature of –40°C, the bottom of the sheet, invariably in direct contact with seawater, is at the freezing point of ~ –1.8°C. Fortunately, the temperature profile between these two points is frequently nearly linear. There are of course exceptions, such as blocks of ice located in the upper parts of pressure ridge sails, which will be nearly isothermal at the local ambient temperature as they are no longer in contact with seawater. However, these are volumetrically rare in comparison to the amount of ice that has either one or two surfaces, usually the bottom and a side, in direct contact with seawater.

As a result, when one deals with real-world problems, it will invariably be necessary to take ice property variations into consideration. Fortunately, in addition to the ice temperature, sea ice properties are reasonably well specified by a very few parameters, namely the relative volumes of brine and gas in the ice. As the amount of gas in most FY ice is small, in such material it is the brine volume that is important. Conversely, as all but the lowest layer of MY ice is highly desalinated, in such ice the gas volume becomes the more important. In many cases the relations between the ice property of interest and the controlling parameters are relatively simple. For instance, strength commonly scales with $\sqrt{v_b}$ and the elastic modulus with v_b, the relative volume of brine in the ice. In the case of gas, the scaling is usually directly with v_g, the volume of gas present in the ice.

As far as the actual numbers are concerned, there is no reason to remember them. It is where to find them that is important. If this chapter is successful, it will provide the reader with a general feel for how and why property variations occur, and more importantly it can be used as a guide to more fundamental sources. One final point: when examining original sources pay particular attention to exactly how the property was determined to be certain that the test is appropriate for your intended use. For example, using ring-tensile strengths in bearing capacity calculations could dampen one's spirits.

11 Polynyas and Leads

Nature does nothing uselessly.

Aristotle)

11.1 Introduction

Polynyas and leads share certain common features. In particular, when air temperatures are below freezing they share the presence of areas of open water for short periods of time and areas of thinner ice classes for longer periods. Both situations result in direct interactions between the atmosphere and the ocean, under climatic conditions where thick layers of sea ice would be expected to isolate these entities from one another. Here I will start with polynyas as many, if not most, of the causes of polynyas are simple. This does not mean that the processes occurring in the polynyas are simple.

11.2 Polynyas

There is some confusion in the literature concerning the exact meaning of the term *polynya*. The word itself is Russian and, similarly to many other Russian sea ice terms, comes from the Pomor people of the White Sea region. In a recent review of Russian sea ice terminology (Borodachev et al. 2000), a polynya is defined as a stable area > 5 × 5 km in size composed of open water and primary ice types of any compactness, or a zone of rarified ice of any age gradation occurring within a sea ice area that has a compactness of 6/10 or greater, or between such an ice area and the coast. Here the critical word is *stable*, which implies both that the feature commonly occurs at some specific location and that, when it occurs, it exists for some appreciable period.

There have been other definitions. For example, Armstrong et al. (1966) have described a polynya as any enclosed seawater area in pack ice, other than a lead, that is not large enough to be called open water. A number of modifiers are then added such as unstable, stable, and recurring. Lastly, WMO terminology simply defines a polynya as any nonlinear-shaped opening enclosed in ice. In my view, all this is slightly confusing.

In the present book I will follow the Russian usage such that the term *polynya* will be used to refer to an area *within* a region of extensive pack ice that

1. is of significant size, having lateral dimensions of at least 5 km and more frequently of the order of 100 km,

2. is comprised of open water and of primary ice types that are appreciably thinner than would be expected considering the regional climatology and the thicknesses of the ice in the areas surrounding the polynya,

3. when present persists for a significant period of time (several days to months).

Note that using this definition a polynya does not have to be nonlinear, as in the WMO definition. The reason for this change is that many coastal polynyas have one dimension that is very much larger than the other. Note also that in the above definition, polynya implies recurrence in the same general geographic region. This agrees with the current usage of the term in the technical literature. Admittedly the distinction between a polynya and a flaw lead, which by definition is located along a coast between the fast ice and the pack, can be a bit fuzzy. In my view, if a flaw lead that develops during the winter is large with offshore dimensions in terms of at least five kilometers, and if it also occurs frequently and persistently, then polynya is the better term.

Figures 11.1 and 11.2 show the general location of some of the more prominent polynyas located in the Arctic and Antarctic, respectively (Martin 2001). Several things are immediately obvious. Polynyas are not rare features and they commonly occur along coastlines. In the Arctic all of the major polynyas are located along coastlines with one exception; the Kashevarov Bank polynya in the Okhotsk Sea (K in Figure 11.1). Note also that there are no known polynyas occurring in the main basin of the Arctic Ocean. In the Antarctic the same general pattern is repeated, although the number of noncoastal polynyas has now increased by three (the Weddell [W], Maud Rise [M], and Cosmonaut Sea [C] polynyas), with the Weddell being quite large (when it occurs). Not surprisingly, polynyas are frequently classified into coastal and open-ocean types.

11.2.1 Coastal Polynyas

Coastal polynyas occur when recently formed ice is continuously removed from its formation site and replaced by open water or by thinner ice types.

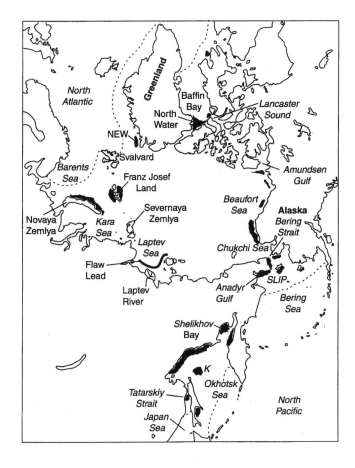

Figure 11.1. Polynya locations in the Arctic (Martin 2001).

Because the heat loss from coastal polynyas goes into ice growth, such polynyas are frequently refered to as latent heat polynyas (Smith et al. 1990). One can think of this type of polynya as being the result of large differences in advection rates; i.e., such polynyas form when ice moves out of a region much faster than it is replaced by ice moving into the region. This advective difference produces open water and, if freezing is occurring, thin ice types. Most coastal polynyas are in some way related to the geometry of the coast or of the fast ice or ice shelves in the near vicinity of the polynya that, associated with specific wind and current directions, allows ice to be removed but prevents other ice from moving into the region to replace it. Typical long-shore lengths for coastal polynyas are 100 to 500 km. Offshore lengths are characteristically less, in the range of 10 to 100 km.

In the Arctic a prominent polynya known as Whalers Bay frequently occurs to the north of Svalbard, and another (the Storfjorden polynya) occurs in a south-facing fjord located within the island complex. Other polynyas occur in the Barents Sea on the west side of Novaya Zemlya and around the complete circumference of Franz Josef Land, as well as at coastal sites on the mainland. Polynyas also occur during the early winter at near-coastal sites in the Kara Sea. In the Laptev Sea a long flaw lead, sometimes referred to as a polynya, frequently forms between the fast ice and the pack ice in the area to the north of the Laptev Delta. Polynyas also occur off the coast of Severnaya Zemlya. In the Chukchi Sea a polynya frequently occurs to the northwest of the Alaskan coast. In the Beaufort Sea a polynya referred to as Westwater typically forms in early winter in Amundsen Gulf to the southwest of the small Inuit community of Sach's Harbour. In the Bering Sea polynyas occur along the coasts of both the Alaskan Seward Peninsula and off Siberia in the region of the Anadyr Gulf. A particularly well-studied polynya occurs off the south coast of St. Lawrence Island (SLIP). Polynyas also occur further south along the coast of the Okhotsk Sea in Zaliv Shelikova as well as somewhat to the west just south of Magadan. Even further to the south a polynya occurs in the southern portion of Tatarskiy Strait and off the southeast coast of Sakhalin Island. A particularly large polynya known as Northwater occurs in the area between Canada and Greenland at the north end of Baffin Bay. A large contributing factor here is clearly the develop-

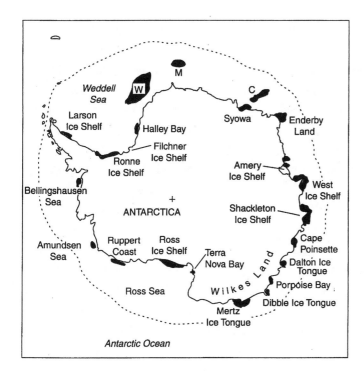

Figure 11.2. Polynya locations in the Antarctic (Martin 2001).

Figure 11.3. An aerial photograph that shows an ice arch developed across the north central portion of the photograph. The wind is blowing from the southeast toward the northwest as shown by the orientation of the cloud streets. Photograph location is unknown.

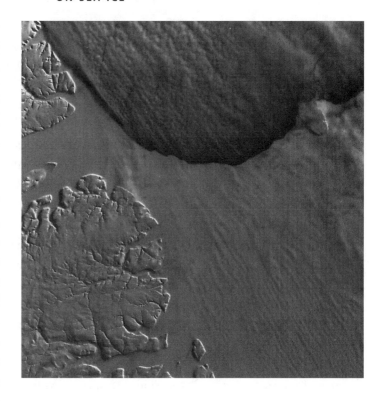

ment during the winter of a fast-ice arch between Greenland and Ellesmere Island that terminates the flow of pack ice from the north into the region of the polynya. Figure 11.3 shows a small example of such an arch that has formed between the coastline in the northwest part of the photograph and the island in the northeast portion of the photograph. The cloud streets indicate a strong wind blowing off the fast ice from the southeast to the northwest. No pack ice is in sight, although it undoubtedly exists further out to sea.

In that the unexpected breakup of an area of fast ice can present one with problems related to operational safety, one might think that there would be numerous studies of fast ice stability and breakup. This does not appear to be the case. However, there are some exceptions, among them the recent investigation of factors contributing both to the stability and to the breakup of the fast ice cover along the northern coast of Alaska and northwest Canada, particularly in the vicinity of Barrow (Mahoney, Eicken, Graves Gaylord, et al. 2007; Mahoney, Eicken, and Shapiro 2007). The first of these studies utilized land-based radar that provided essentially continuous observations, and the second study was based on observations every two to three days. It was found that ice edge stabilization is strongly related to the advance of the ice edge to a water depth of ca. 18–20 m and that an isobar map of this water depth provides a good estimation of the ice edge location during midwinter. Comparisons of current results with earlier work by Stringer (1974, 1978), Barry (1979), and Barry et al. (1979) show that between 1996 and 2004 fast ice formed one month later along the Chukchi coast than it did in the 1970s. There was no appreciable timing difference along the Beaufort coast. In that the onset of fast ice correlated well with the incursion of pack ice into nearshore waters, it is suggested that these later dates result from the recent northern migration of the location of the edge of the pack during the late summer (see the discussion in Chapter 18). The timing of breakup correlated well with the onset of above-melting temperatures, an event that is also occurring earlier in the year resulting in a lessening of the time when a stable fast-ice cover exists. One surprising result was that there was an appreciable variation in the strength of the radar return (flickering) from targets near the ice edge just prior to the ice detaching and becoming a part of the pack, an effect first noted in the 1980s by Shapiro (1987). It is believed that these variations in signal strength represent small-scale motions or vibrations in the ice, presumably due to decreases in the effectiveness of the grounded ridges that stabilize the outer fast ice edge. Preliminary analysis suggests that the anchoring strength supplied by such ridges is two to three orders of magnitude larger than the decoupling stresses associated with wind and water stresses. If these estimates are generally true, the authors suggest that there must be additional decoupling processes at work; suggested possibilities include sea level surges and thermal erosion of grounded keels. Another surprising result is that during much of the winter, the correlation between atmospheric events and fast ice response appeared to be limited. That noted, I still recommend against working near the ice edge during periods of strong offshore winds. I also recommend that particular attention be paid to cracks that form in the fast ice parallel to the ice edge. One indication that something may be about to happen is when these cracks start working, i.e., the ice on one side of the crack starts moving up and down relative to the ice on the other side.

Classic coastal polynyas are common along the coast of Antarctica at locations where strong katabatic winds force the pack to move away from the coast, and the presence of the antarctic landmass and its adjoining fast ice prevents other ice from moving into the region. Examples are to be found along the edges of the Amery, the Ronne, and the Ross Ice Shelves, as well as at a variety of other coastal locations. In that katabatic winds rapidly subside once they move out onto the flat sea ice, the offshore dimensions of such features are usually limited to a few tens of kilometers. Polynyas also occur in the lee of ice tongues such as the Mertz, Dibble, and Dalton that can extend as far as 160 km off the coast. The Terra Nova Bay polynya is also of this type (Bromwich and Kurtz 1984; Bromwich, Parish, and Zorman 1990). Also contributing to the formation of some of the antarctic polynyas is the effect of dominant easterly winds. These combined effects can result in polynyas that are very large. For example, the area of the Mertz Ice Tongue Polynya is roughly 20,000 km². A quite thorough listing of polynya locations and contributing processes can be found in Morales-Maqueda et al. (2004).

The reader will note that most attention, to date, has been focused on polynyas that are obvious during the winter, in that their impact on ocean circulation and ocean–atmosphere heat exchange are greatest at this time of the year. Recently, however, there has been increased interest in large areas of open water occurring within the sea ice zone during the spring and summer. Perhaps the best example is the Ross Sea polynya, with a total area of ~400,000 km², considered to be the most biologically productive polynya in the world (Reddy et al. 2007). In such polynyas, although wind forcing remains an important factor, the effect of extreme katabatic winds appears to have lessened whereas the importance of air temperature has increased (Arrigo et al. 1998). In addition, shortwave radiation must be considered, a factor that is lacking during the winter.

11.2.2 Open-Ocean Polynyas

Open-ocean polynyas are less common. They are produced when oceanographic conditions cause warm water located at depths in the water column to move upward to near-surface levels, where above-freezing surface water results that is capable of preventing new ice formation and melting previously formed ice that drifts into the affected area. Such polynyas are referred to as sensible heat polynyas, in that the heat lost to the atmosphere goes into cooling the water col-umn. As noted, the only known pure open-ocean polynya in the Arctic occurs in the Okhotsk Sea at a location where the Kashevarov Bank rises from a depth of over 700 m to within nearly 100 m of the surface. There it is believed that a strong tidal resonance over the bank generates the increased heat flux that produces the polynya, whose typical diameter is ~100 km (Polyakov and Martin 2000). Rather special cases of sensible heat polynyas occur at several locations within the Canadian Arctic Islands such as Fury and Hecla Strait and Bellot Strait, sites that in the winter are surrounded by extensive areas of thick fast ice. At these locations the channel geometry results in extremely strong tidal currents through very narrow passages. The result is what is sometimes referred to as a velocity system, where the large volume of water passing through a restriction is sufficient to prevent ice formation even though water temperatures are very near the freezing point (Den Hartog et al. 1983; Topham et al. 1983).

The largest open-ocean polynyas occur in the Antarctic, specifically, the very large Weddell Sea polynya (W), with a maximum area of ~2 to 3.5×10^5 km^2, and the relatively smaller Maud Rise (M) and Cosmonaut Sea (C) polynyas, with areas of ~10^5 km^2. Although the two smaller polynyas are persistent features, the appearance history of the Weddell Sea polynya is rather perplexing. The feature was first noted in studies of passive microwave imagery of the Southern Ocean obtained during 1973–1976. It does not appear to have occurred since. It has been sug-gested that the Weddell polynya originated as an eddy located over the Maud Rise that after sepa-ration from the rise migrated into the Weddell Sea, where it was noticed. If this is the case, then the number of Southern Ocean polynyas should probably be reduced to two. Both the Maud and the Cosmonaut polynyas are sited in regions where large amounts of comparatively warm water occur beneath a weak pycnocline. As a result any upwelling is capable of bringing warm water to the surface. Both these polynyas also appear to be self-maintaining during the winter in that the denser water resulting from the subsequent surface cooling then sinks, driving further convec-tion that brings additional warm water to the surface. However, during the summer this process can be turned off either by a lack of sufficient surface cooling or by the introduction of a low-salinity (i.e., low-density) freshwater cap resulting either from advection or the local melting of sea or glacier ice.

11.2.3 Polynya Processes

In the Arctic the existence of polynyas has been known for very long periods of time. In fact the locations of many Inuit communities were undoubtedly chosen because the nearby polynyas meant good hunting during the winter (Schledermann 1980). However, knowing that a polynya exists and understanding it scientifically are two different matters. Even today only a few arctic polynyas have been studied in detail. In the Antarctic even less work has been done. Fortunately, because of our gradually improving understanding of the importance of these features, this is changing. As the reader probably can anticipate, working in and around most polynyas can be both difficult and dangerous. To understand a polynya one should study it when the processes involved are working full tilt. In most cases this means mounting a major winter expedition such as the recent ANARE program in the polynya near the Mertz Ice Tongue. Because of the highly variable open-water and thin ice conditions, working in a polynya during the winter preferably requires ship or, at the very least, helicopter support. Ideally the ship should be an icebreaker. Unfortunately such major allocations of logistic support are rarely available. As a result, much of what we know about polynyas comes from satellite remote sensing and by piecing together information gathered at more benign locations such as leads. With that said, there still is a lot of information available concerning the different processes that occur in polynyas.

First consider coastal polynyas, which, because of their relative accessibility, are better understood than open-ocean polynyas. The necessary ingredient at most such locations is a strong and persistent cold wind blowing in an offshore direction. As this wind blows thicker, earlier-formed ice offshore, an area of open water develops near the coast. As the fetch increases, a wave field develops in which both wave amplitude and wavelength increase with increasing distance from the upwind edge of the polynya. The polynya edge may be located either right at the coastline or at the edge of the fast ice, if such ice exists. In larger polynyas, if the wind speed is greater than 5 to 10 m/s the interaction between the waves and the wind stress creates a Langmuir circulation within the water column. This circulation consists of a series of rotating vortices with their horizontal axes aligned roughly parallel to the wind direction, as diagramed in Figure 11.4. Adjacent vortices rotate in opposite directions, with the diameter of the vortices extending either to the seafloor, in shallower, well-mixed water, or to the top of the halocline, in deeper water. The extremely effective vertical mixing resulting from the above process rapidly cools the upper portion of the water column to the freezing point. Once this has occurred, in shallower waters slightly supercooled seawater can be swept down to the seabed, where it can feed the growth of anchor ice crystals that have nucleated on material on the seafloor. In addition, frazil ice crystals form both at the surface and within the water column. Possible observations of this process using upward-looking sonar (ULS) deployed in the SLIP polynya have been reported by Drucker et al. (2003), who observed dispersed scatters, presumably frazil crystals, in the water column at depths of 5 to 20 m when seawater temperatures were within 0.01°C of the freezing point or perhaps slightly supercooled. Recent studies

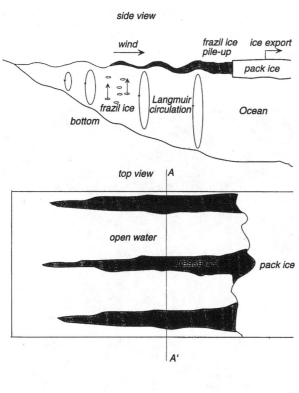

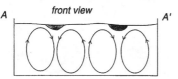

Figure 11.4. Schematic showing the circulation pattern in a well-developed Langmuir circulation (Martin 2001).

of Langmuir circulation in ice-free shallow seas (Gargett et al. 2004) have shown that when the vertical scale of the Langmuir cells becomes equal to the water depth, major episodes of sediment resuspension can occur as the result of the intense turbulence. This could lead to the appreciable entrapment of suspended sedimentary material in the ice that is forming. I am not aware of any specific studies of such incorporation in active polynyas, although this is a subject of some interest (Pfirman et al. 1995; Weeks 1994). Clearly, all the necessary elements for such incorporation occur in coastal polynyas: shallow water, a well-mixed water column, high and sustained winds, a well-developed Langmuir circulation, and the formation of both frazil and anchor ice. Observations by divers have documented that after severe autumn storms anchor ice occurs on the sea floor and sediment-laden frazil ice occurs on the surface (Reimnitz et al. 1987). In addition, there are several observations of pack ice containing large amounts of sediment down the drift path from polynya regions (Pfirman et al. 1989). A paper that uses the relative percentages of the different Fe-oxide mineral grains incorporated within different sea ice samples to determine the provenance of the samples has recently been published by Darby (2003).

As these frazil crystals grow they reject brine, resulting in a downward salt flux. As the ice crystals increase in size, they float to the surface where the Langmuir circulation herds them into long plumes of grease ice located on the convergence zones that are oriented parallel to the wind direction. Another associated phenomenon that assists with the herding of the ice into elongated streamers is the fact that, once the plumes of grease ice form, the drag between the wave fronts and the edges of the grease results in a slight curvature of the wave fronts toward the plumes. Figures 7.2 and 11.5 are aerial photographs showing long streamers of grease ice that have formed under such conditions. In both figures it is possible to tell the wind direction by the fact that the heads of the "tadpoles" are located downwind (in Figure 7.2 the wind is blowing from the lower right to the upper left part of the photograph, and in Figure 11.5 it is blowing from the upper left to the lower right). As the grease thickens, the formation of pancakes within the plumes is also frequently observed. This is particularly true in polynyas that result from katabatic drainage winds,

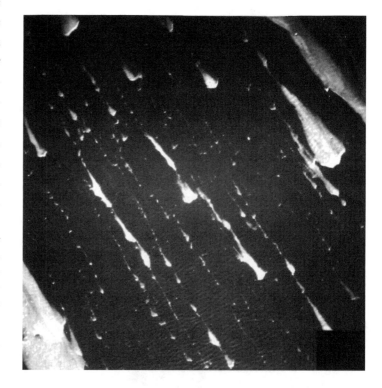

Figure 11.5. Streamers of grease ice presumably comprised of crystals of frazil ice as seen in an aerial photograph. The wind direction is aligned with the long axis on the streamers and is blowing from the upper left to the lower right of the photograph.

in that as such winds decrease in intensity as distance from the coast increases, the frazil stream-ers and the bands resulting from their merger usually catch up with the pack ice, resulting in the formation of larger composite floes. A detailed description of the ice formation processes in the Mertz Glacier polynya can be found in Lytle et al. (2001). The geometry of the Mertz polynya is particularly complex, being influenced by the blocking aspects of the ~160-km-long Mertz Gla-cier Tongue, by seasonal changes in the geometry of the fast ice cover, and by the exact locations of grounded icebergs (Massom, Hill, et al. 2001).

In the case of open-ocean polynyas, two critical questions must be answered in a positive sense if a polynya is to form. First, does the water column below the surface layer contain suf-ficient heat to produce a polynya? Second, are the dynamic conditions sufficient to force this warm water to the surface? One reason why open-ocean polynyas are so rare in the Arctic is that in most places the "lid" placed on the water column by the large influx of low-salinity, low-density water from the large Russian rivers is sufficiently stable to prevent warmer, more saline water located lower in the water column from ever reaching the surface. Obviously the Kashevarov Bank polynya in the Okhotsk Sea, located between 144.8°–146.7°E and 55.4°–56.0°N, can hardly be considered to be in the high Arctic. Although direct oceanographic observations have not, to date, been made here during the winter when the polynya is present, summer observations plus remote sensing plus modeling are sufficiently clear as to indicate what is occurring. For in-stance, examination of a variety of satellite images shows that the areas of low ice concentration defining the polynya are roughly bounded by the 200-m contour of the bank while extending 75 to 125 km south of the bank and 25 km to the north of the bank (Alfultis and Martin 1987; Polyakov and Martin 2000). Furthermore, these observations indicate that over the last 20 years the polynya has changed from a feature that was persistent over time periods of several months to a more transient feature with a shorter lifetime. Apparently the suggestion that the reduced ice concentration over the bank was the result of tidal forcing was first made by Kuz'mina and Sklyarov (1984). Second, Kowalik and Polyakov (1999) were able to demonstrate through a series of model calculations that the vertically homogeneous water structure observed over the bank during the summer was the result of tidal action. Finally, Polyakov and Martin (2000) were able, by combining satellite and meteorological observations with modeling, to make a strong case for tidal forcing of the polynya. In their 1987 study Alfultis and Martin estimated the mean upward oceanic heat flux for February to be ~50–100 W m^{-2}. In the more recent Polyakov and Martin study, a slightly higher average upward heat flux value of 110 W m^{-2} was suggested.

It should be mentioned that the Whaler's Bay polynya located off the north coast of Svalbard is believed to be at least partially the result of the inflow of warm (+3°C to +4°C) water from the West Spitzbergen Current (WSC). Other polynyas where warm ocean waters have in the past been sug-gested to have an appreciable influence include North Water, located at the north end of Baffin Bay, and Northeast Water (the NEW polynya), located off the northeast tip of Greenland (Smith et al. 1990). However, recent studies suggest that such input, if present, is small (Morales-Maqueda et al. 2004; Schneider and Budeus 1995). Another polynya that one might expect to show the effects of the upwelling of warmer water from the WSC would be the Storfjorden polynya located in a south-facing fjord within the Svalbard island group. Again recent studies show that this does not appear to be the case, with the primary factor controlling polynya formation being strong northerly winds.

In defining a specific cause to a given polynya one should always remember that in an advective polynya driven by strong wind forcing, a side effect would be expected to be upwelling of water from deeper in the column to replace the water being stripped away by the wind at the top of the water column. If this water proves to be above freezing when it reaches the surface, then the associated heat flux will contribute to the growth of the polynya. The question is whether there would be a polynya if the wind forcing did not occur. In the case of the Kashevarov Bank polynya there is little doubt that the polynya would occur. In the cases of both North Water and Northeast Water this conclusion appears less certain.

The offshore antarctic polynyas are particularly interesting in that their occurrence appears to be the result of a combination of topographic and oceanographic factors. First consider the Maud Rise polynya. The Maud Rise is a seamount that has a horizontal diameter of ~100 km and rises from a 5000-m-deep abyssal plain to an elevation within 1600 m of the sea surface. In addition, the vertical density gradients in the Eastern Weddell Sea are small in comparison to other portions of the World Ocean, meaning that if warm water exists at depth it should be comparatively easy to cause it to move toward the surface where it could affect the existence of sea ice. As the result of this geometry, an isolated column of water referred to as a Taylor column (Ou 1991) is believed to form above the seamount. When a steady current impinges on an isolated feature such as the Maud Rise, the fluid column is compressed vertically on the upstream side and stretched on the downstream side. This results in anticyclonic (upstream) and cyclonic (downstream) vortices that corotate about the seamount. If the current is sufficiently strong, a steady current will flow over the seamount and sweep both vortices downstream. If the current is weaker, the cyclonic vortex will be swept downstream but the anticyclonic vortex will remain trapped on top of the seamount as a Taylor column. This trapped water, which is colder, less saline, and denser than the surrounding water, blocks the horizontal flow of water over the rise. This causes an acceleration of the regional southwest flow around the column and results in a ring-shaped closed circulation around the Rise, sometimes referred to as a "halo." As the regional flow continues toward the southwest, it contributes to a downstream pool of water that has a slightly elevated temperature and is associated with the Taylor column over the rise (Muench et al. 2001). The combined effects of these features strongly influence the local heat fluxes in the upper ocean (McPhee et al. 1996) and extend the influence of the Maud Rise bottom topography on the upper-ocean heat flux to an area that is larger by a factor of >2 than the horizontal area of the Rise itself. Muench et al. (2001) suggest average upward heat fluxes of 25 W m^{-2} and localized values in excess of 100 W m^{-2}. These are clearly values that would be expected to keep an area ice free. Another interesting phenomenon that occurs above the Maud Rise is a halo of ice deformation that is coincident with the low ice concentrations that are produced by the Taylor column circulation (Lindsay, Kwok, et al. 2008).

As mentioned earlier, it has been suggested that the seldom-seen Weddell polynya is the result of an eddy of warm water breaking free of the Maud Rise and migrating toward the west. This sounds dubious to me in that it would require a significant increase in the current speed in the vicinity of the Rise. Other possibilities include the spinoff of a large quantity of warm water from the Antarctic Circumpolar Current into the Weddell Sea. Although not impossible, what is different about the three years (1973–1976) to produce such changes?

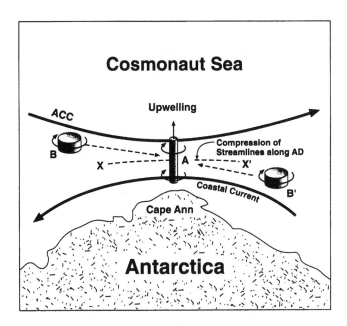

Figure 11.6. A schematic used by Comiso and Gordon (1996) to illustrate their vorticity stretching concept as applied to the formation of the Cosmonaut polynya, Antarctica. The relative vorticity of a water column as it moves from either B or B′ to A becomes more negative, resulting in a clockwise rotation as the water is carried into a region of increased shear as the Coastal Current and the Antarctic Circumpolar Current are pressed together along XX′ at the Antarctic Divergence (AD) by the configuration of the antarctic coastline.

The Cosmonaut polynya, which occurs further to the east (~52°E, 65°S) and to the north of Cape Ann on the coast of Enderby Land, is the third Southern Ocean polynya. Although not always present, it is a reasonably persistent feature always located at about the same location, with an average size of ~7×10^4 km². There appear to be two primary modes of formation (Comiso and Gordon 1987, 1996). The first of these occurs early in the winter and appears to be initiated during a storm when an embayment occurs in the ice edge. The second occurs during midwinter and is associated with the formation of a polynya near Cape Ann. Because of the northern projection of Cape Ann and the eastward flow of the Antarctic Circumpolar Current off the coast, the water in the westward-flowing Antarctic Coastal Current is stretched as it moves by the Cape as the result of a vorticity-conserving reaction. This stretching enhances upwelling and moves warm water located deep in the water column into the surface layer (Figure 11.6). As a result new ice formation is prevented and existing ice is melted. Additional studies of the atmospheric conditions in the near vicinity of the polynya by Arbetter et al. (2004) and Bailey et al. (2004) add further support to this explanation. However, they note that their reconstructions of meteorological data combined with the satellite microwave observations suggest that there appears to be a one- to two-day lag between the passage of atmospheric low pressure systems and decreases in the area covered by sea ice in the region of the polynya. Furthermore, their model simulations suggest that polynyas such as the Cosmonaut, although atmospherically initiated, are subsequently maintained by oceanic heat fluxes. A natural extension of this idea would be a wind-driven sensible-heat polynya. Clearly more field studies and modeling will be required to sort out the relative contributions of these different factors.

11.2.4 Models

In that many latent-heat polynyas can be approximated as being either one-dimensional or two-dimensional in a vertical sense, polynya formation is a phenomenon that lends itself to model development. The first attempt to do so was by Lebedev (1968), who suggested that if meteorological conditions were sufficiently frigid to produce ice in a polynya, a limiting maximum size would result. It was not until almost 20 years later that this idea was explored further by Pease (1987), leading to a frequently referenced model that estimates polynya widths. Her model has several advantages. It is both simple and physically realistic, and provides additional insights into the polynya-forming process.

Let X_p be the width of the polynya at time t, V_i the advection rate of the solidified ice from the shore, and F_i the rate of frazil ice production in the polynya given as the ice thickness added per unit time. Then allow this ice to be swept downwind and allowed to pile up as a uniform slick of thickness H_i against the thicker ice that marks the seaward edge of the polynya. If the thicker ice is also moving downwind under a wind stress with a velocity V_i, the change in the polynya width with time is given by

$$\frac{dX_p}{dt} = V_i - \frac{F_i X_p}{H_i}. \tag{11.1}$$

If there is no appreciable ice production $(F_i = 0)$, then equation 11.1 shows that the width of the polynya will increase at a constant rate $(V_i t)$ and an equilibrium will never be reached. If V_i is constant, and assuming that the value of H_i does not change, then equation 11.1 is a simple differential equation with the solution

$$X_p = \frac{V_i H_i \left[1 - \exp\left(-t F_i / H_i\right) \right]}{F_i} \tag{11.2}$$

provided that one starts with a closed polynya $[X_o = 0$ at $t = 0]$. Equation 11.2 implies that a maximum polynya width of

$$X_{max} = \frac{V_i H_i}{F_i} \tag{11.3}$$

will occur, but unfortunately only after an infinite period of time. However, if the polynya is taken to be fully developed when it reaches 95% of its maximum possible width, then the time to achieve this width can be determined by setting

$$X_p = \frac{0.95 V_i H_i}{F_i} \tag{11.4}$$

and solving for t or

$$t_{95\%} = \left(\frac{H_i}{F_i}\right) \times \ln 20 = 3.0 \left(\frac{H_i}{F_i}\right). \tag{11.5}$$

This implies that the time required for a polynya to develop depends only on the freezing rate scaled by the collection thickness H_i. One might think that this implies that the rate of polynya growth is independent of wind speed. This is not the case, as both H_i and the freezing rate are functions of wind speed. However, it does suggest that in determining the rate of polynya growth, air temperature is more important than wind speed.

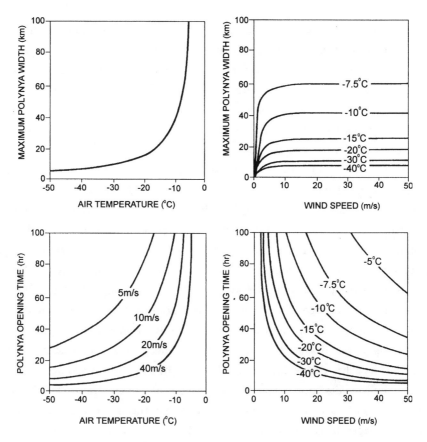

Figure 11.7. Variation of the maximum polynya width X_p and opening time t_{95} with air temperature and offshore wind speed. Note that coastal polynyas reach a stable size within typical synoptic timescales for low temperatures but not for temperatures approaching freezing. The breakover point is a function of both wind speed and air temperature (Pease 1987).

Pease then estimated the freezing rate by using a bulk parameterization for the surface heat flux and introducing environmental parameters representative of the winter climate of the Bering Sea (she was particularly interested in the polynya on the south side of St. Lawrence Island). It was also assumed that the value of the advection rate of the solidified ice from the shore (V_i) is 3% of the wind speed. The results are shown in Figure 11.7. Wadhams (2000) has summarized Pease's results as follows:

1. At very low temperatures, polynyas reach a maximum width within the persistence time of typical weather systems (a few tens of hours). At temperatures near freezing most polynyas are unable to reach equilibrium before atmospheric conditions change.
2. Polynya opening time decreases as wind speed increases.
3. Maximum polynya width is very dependent on air temperature, with the width increasing as the temperature increases.
4. For a given air temperature, maximum polynya width increases with wind speed. However this holds only up to a wind speed of ~5 m s^{-1}. At higher wind speeds, the width remains fairly constant.

I am not aware that this latter prediction has ever been verified by field observations. I would agree with Wadhams (2000) that this prediction is both interesting and surprising.

One should remember that the exact results of the model are dependent on what values are assumed for the effective frazil collection thickness. As noted by Pease, the linearity of the response as shown in her paper is the result of the simple form of the frazil ice growth model. As shown by a variety of photographs in her paper that illustrate the numerous holes (presumably associated with strong convective motions in the water column) that are commonly observed in accumulating frazil sheets, constant thicknesses are not a characteristic of ice sheets formed in the open ocean. It is easy to point out these "missing links"; it is not easy to obtain the field data required to resolve them.

Because of the difficulties in obtaining direct in-situ observations in active polynyas, several investigators have based their studies on a combination of remote sensing to keep track of changes in the overall size of the polynya and the amount of open water as well as the different ice types within it. These observations are then compared with polynya size estimates made from either the model developed by Pease or by using related models such as those developed by Ou (1988) and Chao (1999). The difference between these models is that Pease's model is kinematic in that it assumes that the frazil forming in the open water is instantaneously collected at the edge of the retreating older and thicker ice that defines the initial polynya boundary, whereas the later models consider the fact that the frazil has a finite drift speed. Also, in many polynyas the geometry of the coastline in relation to the wind direction is a factor that must be considered (Darby et al. 1995). How the complex geometry of a site can affect the nature of two of the antarctic coastal polynyas is well described in the papers by Grosfeld et al. (2001) and Massom et al. (2003). Factors considered by these authors include coastline and ice shelf geometry, the location of glacier tongues that protrude out to sea as much as 160 km, and, finally, the locations of large grounded tabular icebergs.

Useful papers, in that they provide considerable detail concerning how the heat balance estimates required to drive the estimates of the growth rates of the different ice types (typically frazil and congelation) observed within the polynya were made, are as follows: Cavalieri and Martin (1994), Haarpaintner et al. (2001), Martin and Cavalieri (1989), Martin et al. (2004), and Van Woert (1999). The Martin et al. reference provides a useful compact review of how different authors have utilized passive microwave imagery in these studies.

The difficulties encountered in these investigations are considerable. For instance, it is rare to have meteorological observations from a site actually located within the polynya. This can be a particular problem in investigating polynyas driven by katabatic winds in that such winds are strongly topographically forced, causing small lateral changes in station position to possibly result in significant changes in wind speed. In addition, katabatic winds commonly exhibit pronounced decreases in wind speed as the distance from the coast increases. In addition the air temperature rises as the distance the air has transited over the polynya increases. See, for instance, the attempt to consider these factors in the Van Woert (1999) study of the Terra Nova Bay polynya in the Antarctic. Presumably if the wind speed is controlled by the movement of a regional weather pattern over the area, a decrease in the velocity over the polynya would be less of a factor. However, an increase in the air temperature would still be observed. A mesoscale (2-km-resolution) study of surface fluxes and boundary-layer clouds based on model calculations and aircraft observations of a Beaufort Sea polynya can be found in Mailhot et al. (2002).

One poorly known factor that is essential to the realistic implementation of all of the above models for estimating the maximum width of a coastal latent heat polynya is the appropriate thickness H_i of the solid-ice equivalent of the frazil ice generated per unit area. As Drucker et al. (2003) point out, in that this is not a true thickness this value is difficult to determine. Moreover, field observations of frazil ice thicknesses accumulated at the edges of leads give highly varied numbers. Assumed values vary considerably in the range between 2 and 48 cm, although most values lie between 5 and 20 cm. Some authors simply assume whatever value is appropriate to give model estimates that generally agree with satellite observations of the size of the polynya under study. Based both on field observations and flume experiments, one would expect that appropriate H_i values would be a function of the characteristics of the wave field that develops within the polynya, which, in turn, should be a function of wind speed and fetch (Martin and Kauffman 1981).

11.2.5 Importance

Considering the relatively small total area occupied by polynyas compared to the overall area affected by pack ice, one might well ask what the big deal is. The answer is that polynyas are extremely important from several different points of view. First, although polynyas typically contain open water and thin ice, they serve as ice and, more importantly, brine factories. By this I mean that if one considers a given geographic coordinate within a polynya, the amount of ice that forms there and the amount of brine rejected into the underlying water column is far in excess of those amounts in an identical climate under conditions where a polynya does not form. This, in turn, has significant oceanographic consequences that are believed to operate on a global scale. Finally, polynyas, in that they provide a location where marine mammals such as belugas, narwhals, walrus, and seals can surface and obtain air, also offer predators a hunting location (Stirling 1980, 1997). Although most people do not know where the nearest polynya is located, the same cannot be said for polar bears. At times the bear concentration can be sufficiently high to make working conditions very interesting for scientists.

There have been several studies focused on estimating the relative amounts of ice formed and of the brine rejected in the different polynya systems of the Arctic and Antarctic. As in the model studies discussed earlier, there are obviously uncertainties in these estimates. However, as will be seen, the numbers resulting from such studies are so large and consistent that there can be little doubt about the oceanographic importance of polynyas. The realization that during the winter the shelf regions of the polar oceans must be serving as sources of cold saline water comes from more conventional oceanographic studies instead of studies of processes within the polynyas themselves. Speculations on the formation of dense water to the west of Novaya Zemlya date back to Nansen in the early 1900s. The possible importance of such processes was more recently discussed by Aagaard et al. (1981), who noted that there must be a flux of water with a salinity of 34.75‰ and a temperature of ca. −2°C if the Arctic Ocean Intermediate Water, which occurs within the halocline, was to be maintained. Two possible mechanisms were suggested, one of which postulated that the cooling and salinization occurred as the result of the formation of sea ice on the continental shelf. Moreover, their data suggested that the source region occurred from Svalbard to the Laptev Sea. At about the same time Swift et al. (1983) noted that the Barents Shelf was contributing to the deep water of the Eurasian Basin and that this water was slightly saltier than the deep waters of both the Greenland

and Norwegian Seas. They suggested that possible sources of such water would be the Barents and Kara Seas, with the Barents Shelf being the most likely.

To explore this possibility further Martin and Cavalieri (1989) carried out a study that was to serve as a model for a number of studies that were to follow. The elements of the study were as follows. First they obtained satellite imagery, specifically passive microwave imagery, for the region of interest for a several-year period (1978–1982). Although the interpretation of passive microwave imagery as applied to the identification of thin ice types can be a bit problematic, passive microwave has the great advantage of being independent of weather and light conditions, an essential characteristic for observations of polynyas during the arctic winter, a time of continuous darkness and frequent cloudiness. Then, by obtaining time series of the size of the polynya from the imagery and coupling this with calculations of heat loss from the open water and from the different thin ice types (frazil is always considered and sometimes frazil and congelation are considered separately), the rate of ice production in the polynya could be estimated. Then, based on the initial salinity of the water and a knowledge of representative salinities of different thickness of thin ice generated, the amount of salt rejected back into the water could also be estimated. For the case of frazil ice, the laboratory results of Martin and Kauffman (1981) were used, which indicate that on the average the salinity of frazil ice S_f can be related to the salinity of the water S_w as follows:

$$S_f = 0.31\, S_w. \qquad\qquad (11.6)$$

As might be expected the details of these calculations vary slightly from author to author. However, these differences would be expected to be small relative to the possible uncertainties in the input parameters. The Haarpaintner et al. (2001) paper describes in some detail the estimation procedures that his group used and also considers the formation of both frazil and congelation ice. A useful additional discussion of uncertainties can be found in the paper by Van Woert (1999).

Estimates of the volumes and effective thicknesses of sea ice that grows during the winter in different arctic polynya areas can be found in Martin and Cavalieri (1989) and Cavalieri and Martin (1994). For the different Russian polynyas the range of the total ice growth estimates are as follows: 11 to 16.5 m (Franz Josef Land), 7.2 to 13.6 m (Novaya Zemlya), and 13.6 to 19.0 m (Severnaya Zemyla). These numbers are very different from the ~2 m ice thicknesses that would be expected if the polynya areas had been covered with FY fast ice.

Although the above numbers are impressive, the few similar estimates available for antarctic polynyas are even larger. For example, the Mertz polynya is estimated to have a surface area of 20,000 km², with an effective daily ice production rate of 4 to 8 cm d^{-1} over the entire polynya (Bindoff et al. 2001; Lytle et al. 2001). In the more extreme areas daily ice thicknesses reach values of up to 25 cm d^{-1}, corresponding to annual equivalent thickness of ~50 m (Roberts et al. 2001). A ranking of 13 different antarctic coastal polynyas in terms of estimated total annual ice production can be found in Tamura et al. (2008). The largest producer is the Ross, with an estimated production of 390±59 km³, followed by the Darnley (181±19 km³) and the Mertz (120±km³). The total estimated annual ice production for all 13 of the identified coastal polynyas is 1410±75 km³. This amounts to roughly 10% of the sea ice produced in the Southern Ocean.

11.2.6 Oceanic Consequences

A number of recent papers have focused on exploring the oceanic consequences of polynya formation. Examples include Chapman and Gawarkiewicz (1997); Chao (1999); Comiso and Gordon (1996); Grigg and Holbrook (2001); Hunke and Ackley (2001); Signorini and Cavalieri (2002); and Winsor and Björk (2000). Exploring the details of this subject is beyond the scope of the present book. However, I will try to briefly summarize the general conclusions of this work. The salty brine that is the product of the polynya "ice factories" results in the formation of more saline and colder water in the polynya regions. This is particularly true in the lower part of the water column. Once formed, this dense water can exit the shelf by a variety of mechanisms, which have been summarized by Baines and Condie (1998) as follows:

1. If the source of dense water is both large and extensive, the fluid may descend the slope as a deep broad sheet governed largely by geostrophy and mixing.
2. If the source is weaker, a process similar to (1) may occur, but only partway down the slope until Ekman drainage takes over the downslope transport.
3. If the source is weak, Ekman drainage may take over from the top of the slope.
4. Local topographic features such as submarine canyons may channel the flow rapidly down the slope.
5. Eddies containing bodies of dense water may form on the slope and aid in the downslope transport.

As these authors note, the details for a particular site will be very topographically dependent, and different mechanisms may become dominant at different points during the descent. As the formation of the dense water that drives these processes is intermittent, one should not expect these processes to occur all the time or all at the same time.

In some sense this process can be thought of as an oceanic equivalent of the katabatic winds that flow off of the large ice sheets. During this process the density of the salt-rich layer is gradually modified by changes in its salinity and temperature resulting from mixing with other local water masses. As might be expected, the details of these changes vary from site to site. Those who might have doubts about the drainage of the dense, saline water off of the shelves should examine Figure 11.8, which shows potential temperature (left) and salinity (right) at a profile located in the western Weddell Sea (from Baines and Condie 1998, after Muench and Gordon 1995). The active downslope flow is obvious.

Once the density of the mixed fluid on the slope becomes equal to the ambient density of the ocean at a given depth, the downward-flowing water can leave the slope and feed a specific layer in the ocean. In the Arctic the total amount of dense water that is formed in the coastal polynyas is believed to be in the range of 0.7–1.2 Sv (Cavalieri and Martin 1994). These numbers are in range of estimated fluxes (1–1.5 Sv) required to feed the cold halocline layer that thermally insulates the surface cold, fresh arctic layer from the deeper warm, saline Atlantic water. However, in that some unknown fraction of the Barents, Kara, and Laptev brine must also feed the Eurasian Basin deep water, the coastal polynyas must not be the only source of water renewing the halocline layer. The Chukchi polynyas are also believed to be a possible source for deep water

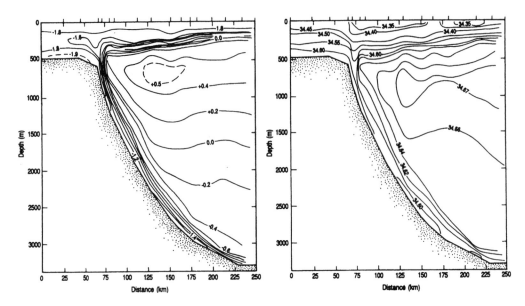

Figure 11.8. Potential temperature (left) and salinity (right) profiles from an oceanographic section across the antarctic continental slope located in the western portion of the Weddell Sea (Baines and Condie 1998, after Muench and Gordon 1995). Downslope flow was clearly occurring at the time of these observations.

formation although this appears to be a less active process (Winsor and Björk 2000). Other brine generation mechanisms that have been suggested include overall freezing on the shelves and the response of the ice to infrequent storms.

In the Antarctic the water produced in the polynyas is believed to contribute to the formation of Antarctic Intermediate Water. However, more importantly, it is also believed to be the primary source of Antarctic Bottom Water, the cold saline lower layer of the world ocean. It is my impression that there is still much to be learned about the details of the interactions between the different polynya regions and the offshore water masses in the World Ocean. In both the Arctic and the Antarctic, the shelves appear to have a dual oceanic presence in that they act as positive estuaries during the summers, with a net outflow of low-density water at the surface, whereas in the winters they act as negative estuaries producing net outflows of high-density saline water at depth.

During the last 20 years interest in the polynya regions of the polar oceans has increased significantly. I expect this high interest level to continue for some time, with particular emphasis on better understanding the details of the processes involved and the resulting water mass transformations. Individuals wishing to explore this subject further should start by consulting the extensive reference list provided by Morales-Maqueda et al. (2004).

11.3 Leads

11.3.1 Definitions and Contrasts

The definitions of a lead found in the literature vary a bit. For instance, WMO defines a lead as being any fracture or passage through sea ice that is navigable by surface vessels. Armstrong et al.'s (1966) view of the term is similar in that they also consider a lead to be a navigable passage through floating ice. Whether the vessel doing the navigating is a canoe or an ocean liner is not specified. The Russians take a more restrictive view and consider a lead to be any fracture or passageway of considerable lateral extent through sea ice that is wider than 1 m (Borodachev et al. 2000). They also note that leads form via the initiation and widening of cracks. This definition is close to the thinking of people who have worked on the ice, who consider that one can jump across a crack but not across a lead. In the Russian definition the phrase *considerable lateral extent* is very important. In the present book I will follow the Russian definition and focus on the classic long leads that form during the winter when the ice coverage of the sea is nearly 10 tenths and the air temperatures are well below the freezing temperature of seawater. This means that once a lead forms, open water will only exist for a very short time before an ice cover starts to form converting open leads into refrozen leads.

Before we become immersed in the details, it is useful to reflect for a moment on some obvious differences between leads and polynyas. Although there are different types of polynyas, all polynyas share the following characteristic. They only occur at locations where the topography of the land–seawater interface has assumed a geometry that in some way disrupts the flow regime of either the seawater or the pack ice. In coastal polynyas the disruptive influence frequently is the presence of a coastline oriented so that the wind is consistently blowing offshore. Other possibilities include the blocking effects resulting from sea ice arching across straits, tongues of glacier ice projecting well offshore, and locales where grounded icebergs stabilize the existence of blocking regions of fast ice. The same may be said for offshore polynyas. Although the region of their occurrence may generally be covered with deep water, the polynya is in some way connected to the presence of topographic features such as seamounts.

The above restrictions do not apply to leads, which, although commonly the result of a divergent flow in either the atmosphere or in the ocean producing a divergent flow in the pack ice, are capable of occurring anywhere on an ice-covered ocean. They typically are transient features that start to refreeze soon after they form. In that the driving force behind lead formation is frequently the wind, the duration of lead-forming events would be expected to be related to the time of passage of synoptic weather systems, typically a few days. In addition, although the dilation of a given lead may last a few days, only a small portion of the lead will actually be open water at a given time, with the rest of the lead being covered with rapidly thickening thin ice. Relative to thicker FY and MY ice, leads introduce significant amounts of heat and water vapor into the atmosphere and salt into the upper ocean. However, they are not nearly as effective in these matters as polynyas, which can maintain large areas of open water and very thin ice for long periods of time. Another difference between leads and polynyas is that the fetch over a typical lead is much less than over a polynya. This suggests that fully developed Langmuir circulation is far less common in leads than in polynyas, and also that the amount of suspension freezing and the incorporation of bottom sediments in the ice will be less in leads. Of course, a major factor here is that many leads occur over deep water, meaning that they do not interact directly with the seafloor. Instead, the convective processes generated by lead formation only reach the halocline at the base of the well-mixed surface layer.

11.3.2 Lead Distributions

It has been known for some time that the spatial distribution of leads shows patterns that can change with region and season (Marko and Thompson 1975, 1977). For instance, leads may be more common in one region than another. At certain locations and times of the year, they may also tend to exhibit preferred orientations determined by the strain patterns within the ice pack (Erlingsson 1988). Since the 1950s, much of our knowledge concerning leads has come from the upward-looking sonar (ULS) systems deployed on U.S. and British submarines making transits of the Arctic Ocean. As these cruises were not primarily scientific, the resulting data came with several drawbacks. For one, the data were not generally available. Also, the sampling was along line tracks and was spatially and temporally discontinuous. Even so, it was found that lead spacings fitted a negative exponential at moderate spacings (400–1500 m), with the caveat that at both small and at very large spacings there was an excess of leads (Wadhams 2000). Lead widths, on the other hand, were found to be well fitted by a power law of the form

$$P(w) = Kw^{-n}. \tag{11.7}$$

Here $P(w)$ is the probability of the occurrence of a lead having width w per unit increment. It was found that the best fit was obtained when n equaled 1.45 for leads that were less than 100 m wide and that equaled 2.50 for wider leads (Wadhams 1981b, 1992; Wadhams et al. 1985).

Although the above information was useful, it was not what was really needed, which was a series of sequential analyses of area-wide lead patterns. Once data from operational satellites became available, such studies became possible. However, as will be seen, these data sets were still hardly ideal. Two such studies have recently examined portions of the Arctic Ocean. To date, there appear to be no similar studies in the Antarctic. The first of these studies (Lindsay and Rothrock 1995) used images obtained by the Advanced Very High Resolution Radiometer (AVHRR) during 1989. This system, which has been flown on a number of NOAA satellites, provides data on several visible and infrared channels, with a nominal resolution of 1.1 km. This said, because of the very high thermal and brightness contrasts between the warm, dark leads and the cold, bright ice floes, it is possible to easily see many subpixel-sized leads on cloud-free AVHRR images. In using such images, some limitations must be kept in mind; both visible and IR imagery of the ocean surface are masked by the presence of a cloud cover whereas visible imagery also is limited to periods of daylight. There are also problems related to calibration and to exactly how a lead is defined because, as many leads are subpixel in scale, there are many pixels that have mixed brightness values. The sampling units were a series of 270 cells that are 200 km square and were located in both the central Arctic and in its peripheral seas. Cloud-covered areas were treated as missing data. Once a sampling unit was selected, the lead and floe width distributions were obtained via a series of 200 random transects. It is suggested that such a random sampling technique is more directly comparable to aircraft and submarine observations that transect leads at random angles. One of the first steps in such studies is the development of an operational definition that allows one to extract useful information from the imagery. In this particular case, Lindsay and Rothrock decided to define leads in terms of their observed thermal and brightness levels. This was accomplished by defining the parameter potential open water δ as "the fraction of a pixel that must be open water for it to have the observed temperature or albedo if the pixel is composed of some mixture of open water and thick ice." Therefore, δ ranges from zero, for a pixel composed only of cold or bright thick ice, to one, for a pixel displaying the temperature and brightness of open water. For example, the potential open water based on temperature is

$$\delta_T = \frac{T_{sfc} - T_b}{T_{ow} - T_b}, \quad T_{sfc} > T_b \tag{11.8}$$

$$\delta_{T=0,} T_{sfc} \leq Tb \tag{11.9}$$

Here, T_{sfc} is the surface temperature as estimated for each pixel, based on the use of a multichannel algorithm suggested by Key and Haefliger (1992); T_b is the background temperature of the thick ice; and T_{ow} is the temperature of open water assumed to be fixed at −1.8°C. Recent comparisons between T_{sfc}, as estimated from AVHRR thermal channels, and near-surface temperatures, as measured from in-situ data buoys, indicate very good agreement ($r^2 = 0.92$; Yu et al. 1995), although winter air temperatures were consistently 1.4°C warmer than snow surface temperatures. A similar relation can be developed to calculate δ_A based on albedo. Figure 11.9 shows monthly mean δ_T values for the central Arctic and the peripheral seas for all months except June and August, when albedo was used as the result of low temperature contrasts. The results, although not surprising, are useful in that they provide an independent validation of impressions gained over the years via field observations. For instance, the values for the central Arctic are less than 3% during the winter and spring and increase to 5 to 7% during the summer. Also, δ values from the peripheral seas are consistently larger than in the central Arctic, manifesting the fact that the ice there is both more mobile and variable. Lead and floe width distributions were also determined via the use of 200 randomly placed transects across each sampling cell. In general it was found that there is a sharp decline in the number of both floes and leads as size increases. Figure 11.10 shows log-log plots of both the lead and floe number densities along a track, expressed as the number of leads or floes per kilometer of track per kilometer of width increment. The letters a and b indicate two binary images shown in the original paper. Figure 11.11 shows the monthly averages of the power law exponents for leads (top) and floes (bottom). The monthly average lead width for the central Arctic ranges from 2 to 4 km during the winter, rising to 7 km in September. Here it is important

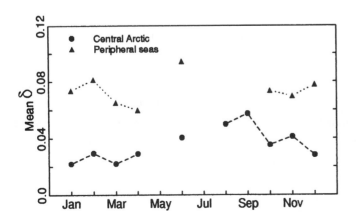

Figure 11.9. Monthly mean open water based on temperature for all months, except June and August, when albedo was used. There was an average of 22 cells for each month for the central Arctic and 17 cells for the peripheral seas (Lindsay and Rothrock 1995).

Figure 11.10. Lead (left) and floe (right) number densities along a track (expressed as the number of leads or floes per kilometer of track per kilometer of width increment) for two binary images (cells *a* and *b*) shown as log-log plots. The width increment (bin size) is 1 km (Lindsay and Rothrock 1995).

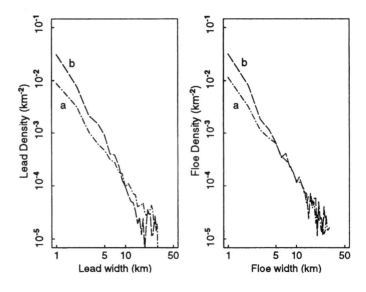

Figure 11.11. Monthly averages of the power law exponents for leads (top) and floes (bottom). There was an average of 11 cells each month for the central Arctic and 14 for the peripheral seas (Lindsay and Rothrock 1995).

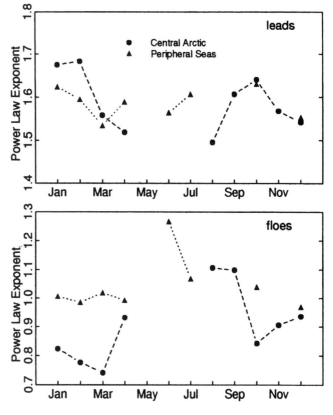

to remember that these numbers represent both open water plus thinner ice classes. One should also recall that based on field experience, there are invariably numerous smaller leads that are below the resolution of the satellite imagery. A comparison of the average exponent value for leads ($b = 1.6$) with those of Wadhams, as determined from ULS profiles, reveals that the submarine data show a steeper drop in the number of large leads, with $b = 2.0$ based on data collected on a traverse between Fram Strait and the North Pole and $b = 2.29$ in Davis Strait. The important thing here is that both the AVHRR and the ULS data appear to be well described by power laws.

The second study is by Miles and Barry (1998), who developed a five-year winter lead climatology for the western Arctic. The time period was 1979 to 1985 and in this case the data used were obtained by the visible and thermal-band systems flown by the Defense Meteorological Satellite Program (DMSP). The nominal ground resolution was 0.62 km. Although this resolution was higher than the 1.1 km available in the Lindsay and Rothrock study, it was only available for the western Arctic (150°E to 90°W, between the west side of the New Siberian Islands and the west coast of Ellesmere Island). As in all such imagery collected by polar orbiting satellites, there is a region that is not imaged near the Pole (~ >83°N). The parameter that they chose to examine was termed *lead density*, defined as the total length of leads in each sampling cell divided by the area of the cell. For instance, assume that there are two 200-km-long leads in a 200×200 km cell. The lead density (ρ_L) is therefore 400 km/40,000 km^2, or 0.01 km^{-1}. Note that the widths of the leads are not considered. As the authors note, ρ_L units are meaningful only in a relative sense, in that many small leads cannot be detected and the mapping and interpretation are somewhat subjective. Nevertheless, the parameter is still useful in examining the spatial and temporal variability of leads. Figure 11.12 shows an example of a lead map made manually from a DMSP thermal image. Note the areas affected by the presence of cloud cover. Sample-to-sample comparisons of lead densities from the different years show that although there is considerable interannual variation, 50% of the pairings had similar means and variability, whereas 50% were significantly different. The mean lead densities for two 3-month seasons, referred to as early winter (November–December–January, or NDJ) and late winter (February–March–April, or FMA), are shown in Figure 11.13. During both seasons the same pattern appears, with the lowest values in the East Siberian Sea and the highest values in the Canadian Basin. In addition, both the density gradient and the absolute value are higher in NDJ than in FMA. The authors interpret this to mean simply that for the same region, when the ice is thinner (NDJ), there are more leads.

A particular interesting aspect of the Miles and Barry study is their treatment of lead orientation. After noting that the leads in their study area characteristically showed diamond-shaped or rectilinear patterns similar to those that have been observed by a variety of earlier workers on satellite imagery, and that the characteristic crossing angles were ~30°, they decided to utilize the mean angle. For instance, if there is only one lead, its orientation is noted. If there are two crossing leads, as would occur in a rectilinear pattern, then the average angle was used. For example,

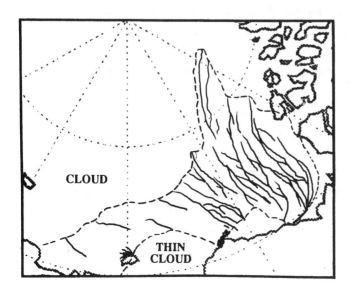

Figure 11.12. An example of a lead map made manually from thermal imagery obtained on 21 March 1984 by a DMSP satellite (Miles and Barry 1998).

Figure 11.13. The mean lead densities in the Beaufort and Chukchi Seas for November–December–January (NDJ) and February–March–April (FMA) for the years 1979–1985 (except for 1980–1981). Lead density units are 10^{-3} km^{-1} (Miles and Barry 1998).

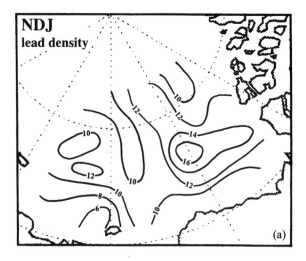

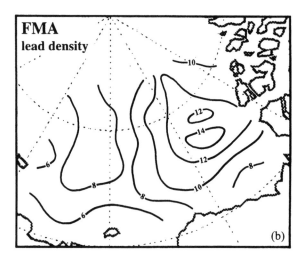

if two leads cross at angles of about 105° and 135°, the mean orientation of 120° is used. Although this procedure results in the losses of the two dominant orientations of a rectilinear lead pattern, in principle they can be recovered from the mean by assuming that the theoretical relation between the internal angle of friction and the angle of intersection holds, which predicts fractures at ±16° to the principal axis of compression (Erlingsson 1991). Figure 11.14 shows mean lead orientations observed during NDJ and FMA during 1979–1985 (except for 1980–1981). Note the predominant north–south orientation in the Beaufort Sea area, in contrast to the east–west orientation in the East Siberian Sea. Note also that although these patterns are similar in both early and late winter, the strength of the orientations are stronger in the early winter when the ice is thinner and less resistant to deformation. When comparisons were made between these observed orientations and several parameters believed to affect pack ice movement, several interesting but not totally unexpected correlations were found. The parameters investigated, based on data collected by the Arctic Ocean Buoy Program, include atmospheric pressure, ice velocity, ice speed, and the velocity gradients (divergence, vorticity, and shear). The positive correlation between divergence and lead density was surprisingly low ($r^2 = 0.46$), considering that divergence is a direct measure of the opening and closing of the pack. As the authors note there are several possible reasons for this lack of a strong correlation. First, the parameter ρ_L does not take the widths of the leads into account. Second, the divergences used are the values determined at the time of the observation. Leads, on the other hand, may be apparent well after their time of formation. A third point is that leads may result from local divergences that are not reflected in the large-scale divergence estimates provided by the buoy motions. Finally, leads may form in the absence of large-scale divergences as the result of the fact that shear may

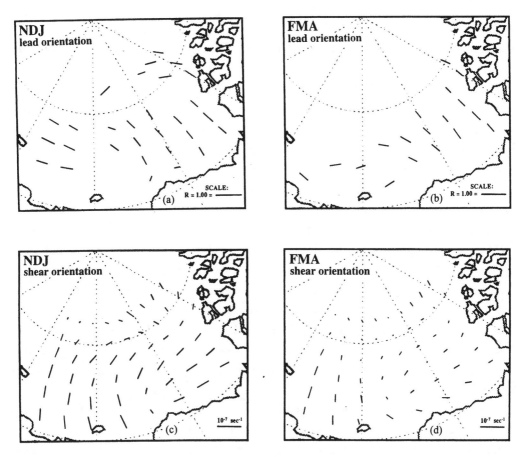

Figure 11.14. The mean orientation of leads observed during November–December–January (NDJ) and February–March–April (FMA) and the mean ice shear fields observed during the same time periods. The vector lengths indicate the strength of the preferred orientation and the intensity of the shear (Miles and Barry 1998).

be expressed locally by some fractures opening and others closing. I would guess that the first two of these four factors are the most important.

Figure 11.14 also shows the shear magnitudes and orientations determined during the same time periods as the lead orientations. The r^2 value is now +0.65, indicating that 65% of the observed lead orientation variations can be explained by variations in the shear values. Note also that the mean orientations of the leads are almost exactly orthogonal (90.7°) to the shear orientation. This is to be expected based on results from ice fracture theory (Erlingsson 1991; Mellor 1986) that indicate, based on small sample testing, that fractures generally form at 90° to the principal stress. Although not the specific focus of this study, the authors point out that their observations indicate that the basic deformation pattern in pack ice is a rectilinear pattern with a crossing angle of ~30°. The importance of this angle will be discussed in more detail in the next section.

I hope that the above discussion will convince the reader that even if it is possible to obtain the sequential imagery required to study sea ice deformation, it is far from simple to extract numerical measures from the imagery that adequately describe the physical processes under study.

11.3.3 Lead Formation

It is clear that leads are initiated by the buildup of stresses in the ice until a critical value is reached that causes a crack to nucleate and propagate laterally. Although I have witnessed a considerable amount of deformation while working at pack ice stations, I can only report a near miss when it comes to being present at the "birth" of a lead. During one of the AIDJEX pilot programs, I was working in a hut located some distance from the main camp and its continuous drone of generators. Suddenly the quiet was broken by a loud sound similar to that produced by firing a rifle. I immediately went outside to see if someone was trying to drive a polar bear away from the hut. However, everything appeared normal and there was no one near. "How odd," I thought, and went back to work. About an hour later, when I started back to camp I was startled to find that a 3-m-wide lead was located close to the hut and that it was gradually widening. Clearly the sound that I heard was associated with the formation of the initial crack, which then opened sufficiently to become classified as a lead. Later aerial reconnaissance showed that this particular crack/lead went from horizon to horizon. Another interesting feature of such cracks is that they do not appear to show a strong tendency to utilize weak preexisting features in the pack, although they presumably initiate at some weakness. Although such cracks are hardly straight, they appear to transit thick MY floes and thinner FY floes equally well. I would guess that the following statement would not usually apply to ice islands, which have been used extensively by the Russians as sites for their NP stations.

Also, just because a crack forms does not mean that it will develop into a lead. While I was on another drift station, the oceanographers cut a hydrohole through the MY floe on which the camp was sited in order to utilize a very heavy winch/cable system to undertake oceanographic profiling through the complete water column (3000–4000 m). Late one evening, while oceanographic sampling was proceeding, a crack appeared, intersected the hydrohole, and gradually started to widen. This resulted in considerable panic as the winch system was difficult to move and if a lead developed there was an appreciable chance of losing the winch into the lead. The crack widened a bit, sat there, and refroze. Nothing more happened and the hydrohole continued to be used throughout the lifetime of the station. Figure 11.15 is a photograph of this particular crack. As in the previous lead, observations the next morning showed that the hydrohole crack was part of an extensive system that extended a very long distance (at least tens of kilometers) from camp in both directions. I personally think that the presence of the hydrohole had nothing to do with the initiation of this crack.

It has only been recently that crack initiation in pack ice has received the attention that the subject clearly warrants. There are two main reasons for this lack of interest. Foremost, until the satellite era there was no reliable method for observing the life history of cracks and leads in the polar oceans. Admittedly, one could easily see leads from locations on the ice. However, a person on the ice was much too close to discern either the extent or the patterns of these features considering that lead systems frequently are tens of kilometers in length. Aerial photography could be a possible observational technique. However, in that photography is limited by the presence of clouds, one could never be certain that it would be possible to obtain the desired sequence of photographs. Also, the considerable costs of operating the aircraft would have to be borne by the research project. Projects with such large budgets are invariably difficult to arrange. Once polar-orbiting satellites became

Figure 11.15. Photograph of a crack that had recently formed in the vicinity of a camp occupied during the AIDJEX pilot study in 1971. This particular crack did not open far enough to become classified as a lead.

available, many of the above problems were solved. Satellites are far, far more expensive than aircraft; however, these funds do not come out of your budget and frequently agencies such as NASA and ESA fund projects just to stress the fact that the data from their satellites are useful. Satellites also facilitate the collection of data at regular repeat cycles such as three days. Nevertheless the problem with clouds obscuring leads remain if systems operating at either visible or infrared frequencies are used. It was only in 1991 that this final obstacle to obtaining reliable large-scale all-weather observations of the ice pack was removed with the launch of a satellites carrying a SAR (synthetic aperture radar) system. (Readers unfamiliar with the general capabilities of current satellite-borne remote-sensing systems as applied to sea ice research should scan Appendix F, which briefly describes the positive and negative aspects of a few of these systems.)

To date there have been two main approaches to studying the lead initiation problem. In the first of these, the crack and lead patterns as seen on satellite imagery have been examined and comparisons drawn with crack formation and propagation results as primarily obtained from failure tests on small samples. One of the primary results here is the suggestion that the failure mechanisms involved are independent of scale. The second approach also utilizes the natural failure patterns but stresses the concept that there appears to be a spatial hierarchy in the deformation of pack ice, with different mechanisms working at different scales. The question appears to be whether one of these concepts is correct and the other is wrong, or whether they both could be correct (or wrong).

The first of these approaches is primarily the result of the work of Schulson and Hibler. The reader will recall, from the discussion of ice mechanics in the preceding chapter, that Schulson's primary focus has been on the strength of both freshwater and sea ice and the interrelations between the stress conditions, the structures of the samples, and the nature of the failures. Hibler, on the other hand, has primarily focused on the development of model simulations of pack ice behavior. In recent studies using both Landsat and SAR imagery they have examined the shapes of long leads in the Arctic Ocean (Schulson 2004; Schulson and Hibler 1991, 2004). For instance, note in Figure 11.16 two different cracks, the most recent fracture running NW to SE, and an earlier, now refrozen, fracture running SW to NE. Note that the recent fracture, although proceeding in one general direction, is not straight. Instead it is comprised of a series of generally long, straight line segments separated by a series of shorter offsets that are oriented at angles of up to ~45° off the general trend of the crack. Note also that after a crack has formed differential strike–slip motion along the initial crack will open a series of leads spaced along the original crack. These openings are generally rhombohedral in shape, as clearly shown along the SW–NE trending refrozen fracture. An idealized diagram of two such cracks, assumed to have formed as a result of a far-field compressive stress, is shown in Figure 11.17. Similarly shaped cracks occur during small-scale compressive testing of samples of both freshwater and sea ice, and are known as wing cracks. Their development is believed to be as follows (Ashby and Hallam 1986; Brace and Bombolakis 1963; Horii and Nemat-Nasser 1985). As stress increases an initial short crack forms that is inclined at ~45° to the loading direction (Figure 11.18a). As the stress increases further, extensions (the wings) form at about one-third of the failure stress at both ends of the initial crack (Figure 11.18b). Although the orientations of the wings are parallel to the direction of the compressive load, the cracks develop to relieve the local tensile stresses that build at the crack tips as the result of differential sliding along the surface of the initial crack. In the laboratory situation the assumption that the far-field stress is compressive is clear from the nature of the test used (commonly unconfined compression). In the field, this assumption is supported by field

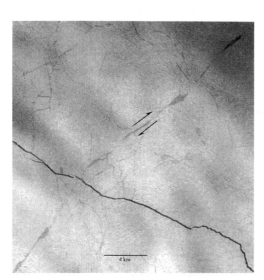

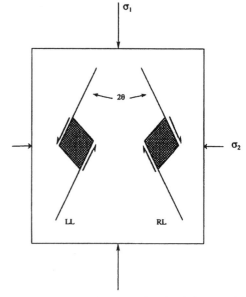

Figure 11.16. A Landsat image of an area of the Beaufort Sea showing two intersecting leads. The lead running from SE to NW formed recently and appears to contain either open water or very thin sea ice. The second lead, which trends SW to NE and is marked by two arrows indicating the direction of the movement along the crack, is clearly older and refrozen (Schulson 2004).

Figure 11.17. A schematic sketch of rhomboidal openings, such as are observed along the SW to NE trending lead shown in Figure 11.16, showing right lateral (RL) and left lateral (LL) displacements along the attendant conjugate lineaments. The lineaments intersect at an acute angle 2θ and are oriented as shown to the axes of principal stress (Schulson 2004).

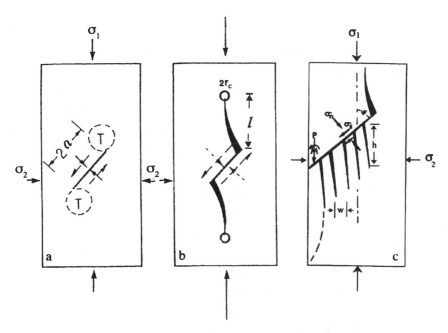

Figure 11.18. A schematic showing the formation of secondary cracks resulting from sliding along a parent flaw under an applied compressive stress: (a) buildup of localized tension at the tips of the parent flaw; (b) wing cracks growing from the tensile zones shown in (a); (c) comb cracks growing from one side of the parent flaw (Schulson 2004; Schulson and Hibler 1991, 2004).

measurements of in-situ stress that indicate that typically stresses are compressive during crack formation (Richter-Menge and Elder 1998). In the laboratory the defects serving to localize the initial cracks are believed to be irregularities in the internal grain structure that serve as stress risers. In the field the initiating defects are uncertain. Likely possibilities include thermal cracks that appear to have the appropriate scaled length (> 60 m). It is not too difficult to think of other possible stress risers in pack ice.

Other aspects of wing cracks are short, secondary "comb" cracks that stem from one side of the parent crack (Figure 11.18c). These secondaries are also oriented approximately parallel to the direction of the maximum compressive stress and form a set of slender columns that are fixed at one end and free at the other, somewhat like the teeth of a comb. Images of pack ice examples of what are interpreted to be cracks of these types can be found in Schulson (2004). Based on these observations of visual similarity between the laboratory and the field, Schulson then makes a mental scale jump of roughly nine orders of magnitude (10^{-4} m to 10^5 m) and suggests that the basic failure mechanisms resulting in lead formation in sea ice are scale invariant (i.e., that the same mechanisms operate on a wide variety of scales).

Here, I cannot present the varied arguments in support of this hypothesis in adequate detail and refer the reader to the Schulson and the Schulson and Hibler papers referenced above, as well as to the very recent discussion of the subject in Schulson and Duval (2009). Are they right? I don't know. To be cynical, with a little careful selection one can presumably find crack and lead shapes that could support a variety of hypotheses. However, the information that they present in support of their hypothesis is impressive. What I really like about their hypothesis is that it leads one to questions and potential answers that if further verified would greatly increase our understanding of the mechanics of pack ice. For instance, why does the sea ice cover behave in such an apparently brittle manner considering that average ice sheet temperatures are typically within 20 to 30°C of the melting temperature of ice? Clearly, this is a temperature range where plastic deformation might be expected to be dominant. A possible explanation lies in the typical rates of loading. If the rates are low, creep within the tensile zones will allow the stresses to relax, with ductile behavior ensuing. However, if rates are high, wing and comb cracks are initiated. Tentative calculations support this hypothesis as applied to pack ice, provided that the flaws initiating the brittle behavior have a length in excess of ~60 m (Schulson and Hibler 2004). The transition strain rates required for brittle behavior appear to be quite low as ice exhibits a relatively high creep resistance, a characteristic that appears to result from the fact that dislocation glide within the lattice of ice I(h) is sluggish (Shearwood and Whitworth 1991), resulting from the unique requirements for protonic rearrangement (Glen 1968). The fact that the fracture toughness of ice is very low compared to that of other materials is another factor favoring a low transition strain rate into the brittle regime. I might add an observational aside here. I have looked at a lot of sea ice. I have seen very few examples of deformation that obviously occurred in the ductile mode, and in all cases the ice involved was very thin (< 10 cm) and therefore very warm, containing a large brine volume at the time of the deformation. Almost invariably the ductile behavior was noted in newly formed ice that had grown in recently opened leads. Clearly, the features exhibiting the ductile behavior had nothing to do with the formation of the leads. These matters will be discussed in more detail in Chapter 12.

The scale-independent hypothesis also provides a possible explanation for the very long intersecting fractures or zones of fractures that are frequently observed in satellite imagery of pack ice. First, it is suggested that very long fractures form via the linking of *en echelon* arrays of wing cracks as shown in Figure 11.19. Figure 11.20 shows a lead in the Beaufort Sea that appears to have formed by such a mechanism. A related sketch showing how a long fracture zone could be formed via the linking of *en echelon* arrays of comb cracks can be found in Schulson (2004). Such lineaments recur year after year and appear to be comprised of wide diamond-shaped bands of damage that often intersect at acute angles of $2\theta \sim 20°$ to $40°$.

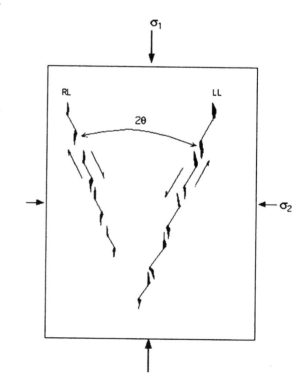

Figure 11.19. A schematic suggesting a mechanism for brittle compressive shear faulting such as lead formation based on the linking of *en echelon* arrays of wing cracks (Schulson 2004).

Figure 11.20. An oblique aerial view of a lead that had recently formed in the Beaufort Sea north of Alaska. The shape of the lead resembles what might be expected if it formed by a wing crack mechanism.

These features are believed to be similar to the large strike–slip faults such as the San Andreas that occur in the Earth's crust. When conjugate pairs of such features form they are located on either side of the direction of the maximum compressive stress. These have been termed *coulombic faults* by Schulson (2002), in contrast to *plastic faults*, which occur under higher degrees of lateral confinement. C-faults are believed to typically be inclined at 28 to 30° to the direction of shortening whereas P-faults are inclined at 45°. If Schulson is correct in surmising that the typical faults observed in lead formation are C-faults, it follows that

$$\frac{\sigma_1}{\sigma_2} < \frac{\left(1+\mu^2\right)^{0.5} - \mu}{\left(1+\mu^2\right)^{0.5} + \mu}. \tag{11.10}$$

Here σ_1 and σ_2 are the maximum and the minimum principal stresses and μ is the coefficient of internal friction. See Schulson (2002) for details. To date there have been no suggestions as to why C-faults do not always develop as conjugate pairs, as is frequently the case in small-sample testing. If I were to hazard a guess, I would suggest asymmetrical boundary conditions, as might be expected near coasts, as a possibility.

An additional factor taken to support the argument for the importance of a similar mechanism in both laboratory studies of ice mechanics and in the large-scale deformation of pack ice is the fact that the failure envelope as determined in the laboratory (Schulson et al. 2006) has a very similar shape to failure criteria that have proven to be useful in geophysical scale simulations.

As interesting as the overarching wing cracks hypothesis is, not everyone is convinced that it is true. Specifically, Overland et al. (1995) suggest that different mechanisms may well be operating at different scales, citing the well-documented observations from ice mechanics discussed in Chapter 10 to the effect that failure stresses determined on small samples are a factor of $\sim 10^3$ larger than values determined during the failure of complete ice sheets. The general thinking regarding the reason for this change is roughly as follows. There clearly are a number of different types of flaws that occur naturally in sheets of sea ice. These range from tiny brine pockets with dimensions of a fraction of a millimeter, to wing cracks, to brine drainage tubes, to thermal cracks with dimensions of tens of meters or larger. Each of these flaw types is capable of initiating failure provided that the local stress exceeds some critical value associated with the flaw type. When one studies the strength of small samples of sea ice, samples that contain obvious large flaws such as large cracks or voids are invariably discarded as unsuitable for testing. On the other hand, when testing involves the total thickness of the ice sheet and sample sizes increase to a meter or more, samples still may be rejected if an obvious flaw such as a thermal crack transects the sample. However, the more numerous smaller-sized flaws that exist within the sample now do not result in rejection as they are unseen. Once the scale of the test becomes sufficiently large to include entities such as floes, a variety of floe–floe interactions associated with rafting and ridging become possible that were experimentally excluded in smaller-scale tests. Therefore the decrease in the values of parameters such as whole-sheet indentation pressure as sample size increases is hardly unexpected (Figure 10.31).

The Overland et al. (1995) argument is that the degree of disconnectedness between scales accounts for the organization of the system. Classically, state variables at level $n + 1$ vary smoothly over larger spatial scales and evolve more slowly than those at level n. More importantly, the $n + 1$ level remains relatively unaffected by level n processes, although level n is constrained by the dynamics set at the higher level. Conversely, phenomena at levels higher than $n + 1$ are too large and too slow to have a direct impact on level n. They then stress the idea, taken from the hierarchy theory of complexity (O'Neil et al. 1986), that in systems where scale discontinuities occur, emergent properties can occur in which "the whole is greater than the sum of the parts." That is, it may not be adequate to know all the values of the state variables at level n to know the state of the system at level $n + 1$. As an example they cite regional ice mechanics where understanding processes that occur on a single floe does not, in itself, allow one to understand floe–floe interactions. In essence, floe–floe interactions produce an emergent property.

The original hierarchical structure suggested goes from sample scale (millimeters to tens of centimeters) to floe scale (1 to 100 m) to small regional scales (0.1 to 10 km) to large regional scales (> 10 km). These authors have suggested that based on studies of satellite imagery, a better breakdown might be similar to floe scale (< 1 km), multifloe scale (2–10 km), and aggregate scale (10–75 km). Independent of the scale hierarchy versus wing crack arguments, the Overland et al. (1995) paper contains some very interesting observations on the regional lead patterns in the Arctic Ocean, which for some regions and some times appear to be related to coastal orientations even at distances of several hundred kilometers.

As can be seen from the dates on many of the papers referenced in the above paragraphs, the debate concerning lead formation is ongoing and unresolved. Each of the arguments has its strong and weak points. My personal take on the two approaches is as follows. The "wing crack" hypothesis is clearly the more highly developed and potentially useful. Because of this, it should be easier to verify or to reject. As far as the "hierarchy" hypothesis is concerned, although many elements of it are clearly true, I find it to be too vague to currently be very useful. The most recent work bearing on these different approaches is a study of the dispersion of pairs of drifting IABP buoys by Rampal et al. (2008). The observed scaling properties were somewhat similar to observations made on turbulent fluids, with the deformation being both very heterogeneous and intermittent at all scales. The authors conclude that these observations suggest a deformation accommodated by a multiscale fracturing/faulting process. Stay tuned. Also note the related discussion on scale effects in Chapter 10 (section 10.4.11).

11.3.4 Ice Formation in Leads

During the ice growth season, once a crack forms and begins to open, new ice starts to form within a very short time. There is nothing unusual about this ice, and its structure and composition have been discussed in detail in earlier chapters. However, it is still useful to review the general trends shown by ice in leads in contrast to ice formed in polynyas and at the edge of an expanding ice pack. Essentially, by definition, leads are relatively narrow, elongated reaches of open water or very thin ice scattered through extensive areas of thicker ice. This means that waves are less of a significant factor in many leads as these are damped by the presence of the surrounding ice. This also means that as the effective fetch in most leads is small, features such as Langmuir circulation have less time to become well developed. In addition, the fact that a lead is opening as the result of wind stress does not necessarily mean that it is locally windy, as it is now well known that stress can be transmitted long distances through the pack. This fact, requiring the development of the concept of the so-called internal ice stress, will be discussed in the chapter on dynamics. Freezing ice in a lead is frequently like freezing ice in a swimming pool, with a typical result being classic congelation ice that develops from the initial formation of an ice skim. That said, it is not at all rare to find the buildup of frazil ice along the downwind sides of leads. This is particularly true in windier regions such as the Southern Ocean. Figure 11.21 shows a representative example of this process. If the ice at the lead edge is thin and the buildup of frazil is thick, a significant portion of the frazil can be swept underneath the floe, resulting in an ice floe that is a complex interlayered stack of alternating congelation and frazil ice. Such internal floe structures have frequently been reported in structural studies of antarctic pack ice. The leads that were studied in

Figure 11.21. Frazil ice accumulating along the downwind edge of a lead in the Bellingshausen Sea, Antarctica.

Figure 11.22. An aerial photograph of a large lead, Beaufort Sea, showing a significant amount of rafting.

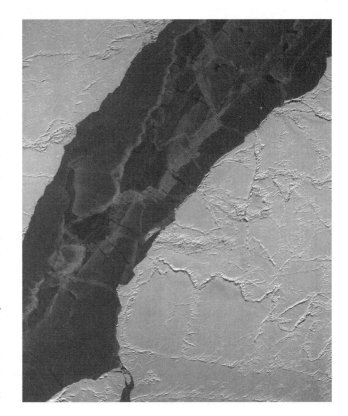

the Arctic Ocean during SHEBA varied in width from 3 to 400 m (Pinto et al. 2003). When wind speeds were high, frazil formed in the larger leads where it commonly became organized into linear bands oriented parallel to the wind via the Langmuir circulation. When the irregularly shaped frazil mats came together at the downwind edge of the lead, gaps between the mats formed 5-to-10-m-wide pools of open water that persisted for several hours. I would guess that a significant portion of this frazil originates as snow blown off the floes located on the upwind side of the lead.

One common feature of lead ice, and a warning that the ice may still be very thin, is the extensive development of frost flowers on some leads. As noted in Chapter 7, the observation that frost flowers characteristically do not form when wind speeds are in excess of 3 to 5 m s⁻¹, tells one that the ice under the frost flowers is probably classic congelation. Another feature frequently observed in refrozen leads are rectilinear outlines indicating that finger rafting and simple rafting have occurred. Figure 11.22 is an aerial photograph of a large wide lead in the Beaufort Sea. Note the large amount of rafting occurring in the lead, as shown by a decrease in the grayscale where the ice thickness has been doubled. As will be discussed in Chapter 12, such deformation features primarily form when the ice is very thin.

11.3.5 Thermodynamic Importance

The thermodynamic importance of leads in pack ice has been clear since Badgley (1966) first estimated that in the Arctic Ocean, the total heat flux from open leads was roughly equal in magnitude to the flux through the remainder of the pack, even though leads only account for < 10% of the surface area. Later Maykut (1978) expanded this idea, providing a more detailed analysis that suggested that open leads, thin ice (< 1 m), and thick ice (> 1 m) contribute equally to the total heat flux in areas of pack ice within the Arctic Basin. It is not difficult to convince anyone who has spent time living in the pack ice during the winter that leads have an appreciable affect on local atmospheric conditions. I offer as evidence Figure 11.23, a photograph taken across a newly open lead in the Beaufort Sea. The amount of water vapor (and

Figure 11.23. A photograph showing water vapor streaming into the atmosphere above a recently opened lead in the Beaufort Sea.

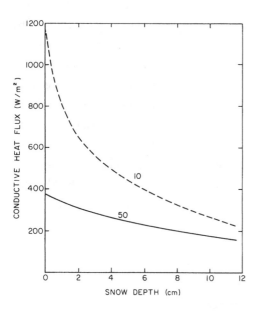

Figure 11.24. The effect of snow depth on heat conduction during the midwinter (Ta = −34°C). The dashed curve shows values for 10-cm-thick sea ice and the solid curve assumes that the ice is 50 cm thick (Maykut 1986).

heat) transferred from the ocean to the atmosphere appears impressive. In this particular case the temperature difference between the air and the water was approximately 40°C. The calculations that these extremely large temperature and moisture gradients do indeed lead to large heat and moisture fluxes into the arctic boundary layer were verified by Andreas et al. (1979), who measured sensible and latent heat fluxes above open leads exceeding 400 W m^{-2} and 100 W m^{-2}, respectively. When flying over the ice pack one frequently sees long, low, "linear" clouds that clearly are directly associated with specific leads. Even when leads become ice-covered, the turbulent fluxes, although appreciably reduced relative to open water, can still be very impressive. For instance, Alam and Curry (1998) estimated that turbulent fluxes over leads covered with sea ice 20 cm thick were still 10 times greater than over the surrounding, significantly thicker, ice pack.

A particularly important factor affecting the amount of heat released to the atmosphere by a lead is exactly when and how much snow is deposited on lead ice while the ice is still quite thin. As can be seen in Figure 11.24, if the conductive heat flux for 10-cm-thick snow-free sea ice is 1200 W m^{-2}, a 1-cm-thick layer of snow will reduce the flux by one-third to ~800 W m^{-2}. Considering typical variations in snow covers on new ice largely resulting from drifting snow, the effect on the thicknesses of ice developing in newly opened leads can be significant.

In the early 1970s I was involved in an experiment to measure local scale ice deformation in the near vicinity of the AIDJEX camp in the Beaufort Sea (Hibler, Weeks, et al. 1974). This was accomplished by using a laser to accurately measure the distances to a number of targets (corner reflectors) located at a variety of distances from the camp. A major unanticipated problem proved to be associated with the opening and closing of leads. When a lead opened between the laser and the target, we invariably lost the target. It was as if a white wall had instantly appeared to block the laser beam. Fortunately, within a few hours, the target invariably reappeared as the result of the lead either closing or, more probably, skimming over with a very thin crust of ice, thereby significantly decreasing the moisture flux into the atmosphere.

11.3.5.1 Atmospheric The atmospheric importance of leads has received considerable attention recently as the result of the LEADEX and SHEBA programs. One of the purposes of these programs was to develop knowledge that would allow better coupling between models focused on sea ice behavior and more general atmospheric models. To accomplish this, a better understanding was needed of exactly how the heat and moisture generated at the surface of a lead or by a system of leads distributes itself in the atmosphere. For instance, are the effects of leads purely local, or can they be of sufficient magnitude to affect regional climate? The conventional wisdom on these matters was that the influence of leads was limited to the lowest few hundred meters of the arctic atmosphere. The reasoning was that the strong surface inversions that characteristically occur over the pack in the winter would effectively cap the lead, limiting the upward motion of the warm, moist air formed over the lead. What were missing were direct observations to verify this speculation. Surprisingly, when such information began to become available via the use of an airborne infrared laser, it was found that the condensate from some leads reached heights up to 4 km (Schell et al. 1989), and that in one case a lead-induced cloud extended as much as 250 km downwind from the generating lead. Even though these may be extreme examples, they clearly are cases where a lead is producing an effect of some regional importance. The problem then becomes one of separating leads that produce regional effects from those that only affect the atmosphere in the near vicinity of the lead.

A variety of models of differing complexity have been used to explore this problem. Although discussing the details of the differences between these models is not appropriate for a book focused on sea ice itself, what I will do is discuss the results of the large eddy simulation (LES) model as developed by Glendening (1995) as a guide to some of the general trends shown by the model results. There are two parts to the LES model. First, it estimates the heat flux over the lead from the heat transfer coefficients as determined by Andreas (1980), Andreas and Murphy (1986), and Ruffieux et al. (1995). This part is, in some form, common to all such models. Then it attempts to parameterize how this heat will be vertically distributed. In the LES case, the individual three-dimensional thermal elements that comprise the plume are resolved. The model lead investigated was assumed to be linear, with an infinite length and a width of 200 m. The thermal plume was created as the net effect of the individual thermal eddies created over the lead. Figure 11.25 depicts the growth of such a plume, normalized by the average flux over the lead for two different wind directions $\phi = 90°$ and $\phi = 5°$, where $\phi = 90°$ indicates that the wind direction

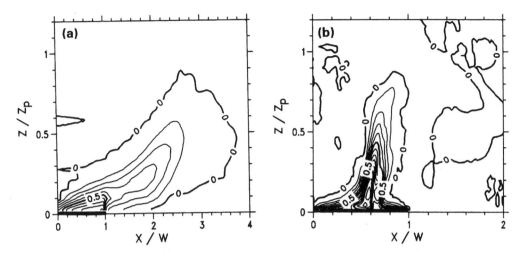

Figure 11.25. Vertical turbulent heat flux (sum of resolved and subgrid fluxes) normalized by the average surface heat flux over the lead for a 200-m-wide lead as calculated using a large-eddy simulation assuming wind angles across the lead of (a) $\varphi = 90°$ (perpendicular to the axis of the lead) and (b) $\varphi = 5°$ (Glendening 1995).

is perpendicular to the axis of the lead. The vertical axis shown in Figure 11.25 is scaled relative to the predicted plume height Z_p as calculated by

$$Z_p = \left[\frac{W^2 Q_s}{\rho c_p \Gamma U} \right]^{1/3}. \tag{11.11}$$

Here W is the width of the lead, Q_s is the surface heat flux, ρ and c_p are the density and heat capacity of the air, respectively, $\Gamma \equiv \partial\theta/\partial z = 10°K \ km^{-1}$ is a measure of the base-state stratification (assumed to be constant), and U is the wind velocity component normal to the lead. Considering the fact that $U = |V|\sin\phi$, where $|V|$ is the geostrophic wind speed assumed to be 2.5 m s^{-1}, and that the surface heat flux is given by $Q_s = \rho c_p C_h |V| \Delta T_s$, where C_h is the heat transfer coefficient and ΔT_s is the ice–water temperature difference, here taken to be 27°K, equation 11.11 transforms into

$$Z_p = \left[\frac{W^2 C_h \Delta T_s}{\Gamma \sin\phi} \right]^{1/3}. \tag{11.12}$$

This relation is clearly not applicable when airflow is nearly parallel to the lead ($\phi = 0°$) or when the vertical stratification is neutral ($\Gamma = 0$). Limited data presented by Glendening (1995, his Fig. 2) suggest that equation 11.12, which estimates the highest level at which the vertical heat flux is still positive, gives a good estimate of the maximum plume height. Note that in Figure 11.25, the horizontal scale is normalized by dividing by the width of the lead in order to emphasize the relation between the surface heating area and the structure of the plume. Another important horizontal scale is that of the atmospheric response given by $UP/2\pi$ where P is the Brunt–Väisällä or buoyancy frequency (the frequency at which a vertically displaced parcel of fluid will oscillate within a statically stable environment).

As pointed out by Glendening (1994a), plume development depends on the strength of the cross-lead flow, with the maximum vertical turbulence typically occurring at a distance of $UP/2$ from the center of the lead. The cross-flow component for which maximum vertical development occurs significantly downwind of the lead, rather than above it, is given by $|V| > S/P$, where

$|V| \equiv U/\sin\phi$ is the wind speed and $S = W/\sin\phi$ is the slant distance (fetch) across the lead. When there is either a narrow lead or a strong cross-flow, the individual thermals do not have enough time to achieve a maximum updraft velocity while over the lead. As a result, the largest vertical motions occur over the ice downwind of the lead, as seen in Figure 11.25a. If the mean vertical motions are weak, they will be responsible for only a small percentage of the total vertical heat transport. On the other hand, if the cross-flow is weak, it does not remove heat rapidly and a relatively strong mean updraft can develop. As Figure 11.25b shows, the maximum vertical turbulence then occurs over the lead itself, only slightly offset from its center. For more details on the LES approach refer to Glendening (1994a, 1994b) and Glendening and Burk (1992). The LES model gives lower plume height estimates than the earlier Serreze et al. (1992) model, which does not consider the lateral entrainment of the colder air surrounding the plume. Although neither of these models consider the effect of water vapor on the plume height, it is argued that its inclusion will not appreciably change the plume height estimates. The correct estimation of the height and general shape of the plume is important for several reasons. It is required to estimate the effect of leads on large-scale climate models. It is also needed to improve estimates of the amount of heat released by leads that is recaptured by the ice downwind from the lead. For instance, the LES model indicates an appreciable warming of the boundary layer, which in turn affects the snow–ice surface temperature up to 5 lead widths downwind under typical arctic conditions. Again, this is a process that is dependent on the angle that the wind crosses the lead.

More complete models have also been utilized that focus on the processes within the plume, including the formation of clouds and ice crystals. For instance, Pinto et al. (1995) and Pinto and Curry (1995) have used one-dimensional models to study thermodynamic, microphysical, and radiative processes that result from the presence of lead-induced clouds and of the formation of ice crystals in lead plumes. Their results indicate that cloud formation over leads significantly affects the longwave heating rate profile, resulting in strong cooling at the top of the cloud and reduced longwave cooling at the surface. They also found that when radiative transfer, condensation, and precipitation process are neglected, there is an underestimation of some 10 to 20% of the depth of the region affected by the plume and that the stability of the atmosphere above the surface temperature inversion becomes important in controlling the thickness of the surface boundary layer. In addition, a cloud-resolving model used by Zulauf and Krueger (2003), which included treatments of cloud microphysics and radiative transfer, increased plume depth by 25% over estimates provided by dry simulations. Further recent exploration of the application of a LES model to leads can be found in Lüpkes et al. (2008).

What are still rare are model calculations that can be compared to direct observations on the lead that is being simulated. An exception here is the study by Pinto et al. (2003) carried out during the SHEBA experiment that considers the following information for each lead studied: location, orientation, maximum width, hours open with water present, ice types (congelation/frazil), frost flower occurrence, albedos, and average temperatures over the lead. Lead widths varied from 3 to 400 m. Separate exchange coefficients were used for open water and ice. The model used also considered that the fetch of the lead could change with time. It was found that with sufficiently detailed input, the model was able to accurately simulate the thermodynamic effects of the individual leads. Not surprisingly, it was also found that horizontal variations in the surface conditions on a lead have a significant effect on the heat and moisture fluxes into the atmosphere. Important observations include the fact that unless the presence of frost flowers is included, real daytime surface temperatures will be warmer than calculated. In addition, best results for the turbulent heat fluxes required the use of lower values for the surface roughness length for nilas than had previously been estimated by Guest and Davidson (1991). As suggested by simulations discussed above, measurements of the sensible heat fluxes taken 70 m downwind from a lead proved to be a function of across-lead fetch, upwind atmospheric stability, and open-water fraction. Furthermore, these particular flux values remained above background for two days in spite of 11.5 cm of ice growth on the lead. Similar results were obtained on other leads. The atmospheric influence of a 400-m-wide lead was found to extend 2.5 km downwind when the wind direction was normal to the lead. After the lead opened, the sensible heat flux associated with the lead averaged about 80 W m^{-2} above background for the first 12 hours. A 15% increase in specific humidity was also observed.

Clearly, in the last 40 years since Badgely's original paper pointed out the atmospheric importance of leads, a great deal has been learned. Let us assume that in the next decade or so progress continues until we are able to predict the atmospheric effects of a specific lead with a great deal of confidence. Even with this capability we will, in my view, still only be part of the way toward the goal of coupling sea ice models that treat dynamics and thermodynamics with models that simulate large-scale meteorological conditions. To accomplish this there is a need for approaches that treat ensembles of leads. Fortunately, some thought has been given to how this might be accomplished. As Pinto et al. (2003) showed, accurate simulation of the leads that they studied required detailed knowledge of the rapidly changing ice conditions in the lead. Even with greatly improved remote sensing capabilities for determining the status of factors such as the presence or absence of features such as frost flowers, it appears to me to be unrealistic to believe that it will ever be possible to deal with regional effects through the consideration of individual leads. The computational and data collection requirements would be staggering. A more "simple-minded" approach that deals with either averages or, better yet, the distribution of factors such as lead widths will be required. This subject has been explored by Maslanik and Key (1995) in terms of bulk transfer coefficients that either use fixed transfer coefficients or consider only atmospheric stability. As the rate of turbulent heat transfer from leads is a function of the fetch encountered by the air as it moves across the lead, and as the fetch both varies from lead to lead and along the same lead, they assumed that the fetch distribution could be well described by a negative exponential

$$f(X)=\frac{1}{\lambda}e^{-X/\lambda} \qquad (11.13)$$

where $f(X)$ is the distribution of lead widths X with a mean value of λ (Key and Peckham 1991). The values used were based on upward-looking submarine sonar data supplied by McLaren (1989). In calculating the sensitivity of the turbulent fluxes to changes in the fetch, the wind speed, the air temperature, and the surface temperature, the equations suggested by Andreas and Murphy (1986) and by Andreas (1987) were used in combination with an energy balance model, meteorological observations, and ice thickness data. Figure 11.26 shows the estimated changes in the vertical sensible heat flux transfer coefficient C_{H10} at a reference height of 10 m as a function of fetch and surface wind speed for surface air temperatures of (a) –10°C and (b) –20°C. Note that higher wind speeds extend the range of fetch over which C_{H10} is affected. The fluxes from an open lead were found to decrease by 34% as the fetch increases from 10 to 100 m. As the lead refreezes this effect decreases appreciably and becomes negligible when the ice becomes thicker than 0.3 m. It was found that if open or newly refrozen leads make up 2% of the region under study, then an increase in the mean fetch from 10 to 100 m will result in a decrease of approximately 2 W m^{-2} in the area-averaged flux from the pack. Importantly, a study was carried out in which fluxes calculated based on the actual observed distribution of lead widths were compared with both the flux estimated from the negative exponential distribution constructed from the observed mean width, and the flux estimated using only the mean width. Related experiments were completed using the log-normal distribution. The differences in the flux estimates were extremely small—negligible if the measurement uncertainties in determining lead widths are considered. Although individual lead widths must still be measured in order to determine a mean, the ability to use a single representative flux calculated from that mean would be an important simplification.

Other related studies of importance include the study by Lindsay and Rothrock (1994), in which they utilized high-resolution AVHRR infrared imagery to compute regional surface sensible heat fluxes. This approach, although limited to cloud-free periods, shows considerable promise in that snow surface temperatures estimated by this technique are consistently 1.4°C colder than nearby air temperatures (Yu et al. 1995). A related paper by Walter et al. (1995) then compares regional heat fluxes calculated using the Lindsay and Rothrock approach with heat fluxes measured by an aircraft-based gust-probe system. Although the agreement was not perfect, I feel that the results are very encouraging. Follow-up comparative studies could be very profitable and lead to the development of tested seasonally dependent heat transfer coefficients useful on a large scale. Once near-surface heat transfer under clear-sky conditions can be confidently estimated, it may be possible to extend these estimates to cloudy condi-

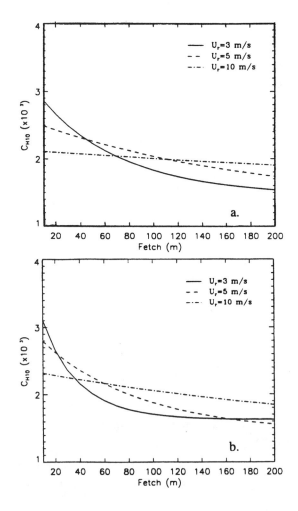

Figure 11.26. Sensible heat flux transfer coefficient calculated as a function of fetch and surface wind speed (U_r) for surface air temperatures of (a) –10°C and (b) –20°C. (Maslanik and Key 1995).

tions based on the results of model experiments such as those described earlier. One current result of considerable interest is that even when it is quite cold (–15°C) and windy, regional sensible heat fluxes over sea ice, including the effects of leads, are generally small (< 10 W m^{-2}). The reason presumably is that even though the heat flux over truly open leads can be very large, the time is very brief before a given lead is covered by a skim of ice thick enough to appreciably decrease the heat flux into the atmosphere.

11.3.5.2 Oceanographic The first attempt to develop an appropriate numerical model of the oceanographic consequences of lead formation appears to have been by Schaus and Galt (1973), who considered that transfer within the lead was primarily the result of forced convection. Ten years later, Kozo (1983) produced a more realistic model that also considered the circulation within a lead, which is driven by brine generation caused by ice formation at the surface of the lead. He also examined the effects of advective processes (i.e., that the ice and upper level of the ocean are frequently moving relative to one another). As Kozo is a meteorologist, his ocean model was similar to his atmospheric model, with appropriate scaling changes made in the thickness of the oceanic boundary layer. The model was successful in predicting the local circulation that results from the formation of a lead: that brine release resulting from ice formation increases the density of the surface water, leading to downward convection and the outward flow of dense water away from the center of the lead along the top of the pycnocline. Volume continuity then causes near-surface water to flow inward from the sides of the lead to the center of the lead. Specifically, for the case of free convection (zero ambient current), the model correctly predicted both the direction and velocity (~5 cm s^{-1}) of jets associated with this circulation. Figure 11.27 shows the velocity distribution in such a lead with an assumed ice drift velocity of zero. As can be seen, the general velocity distribution is in agreement with the suggestions of Kozo. When a 5 cm s^{-1} current along the lead or a 2.5 cm s^{-1} across the lead was added (or equivalently a similar ice drift rate relative to the water at the base of the mixed layer was considered) either a reduction occurred in the convective pattern or it was moved entirely to the down-

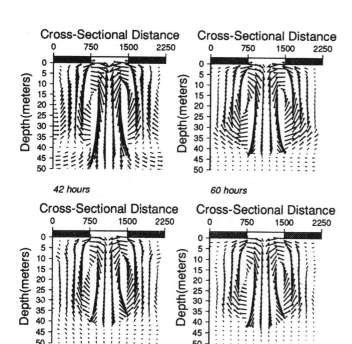

42 hours

60 hours

Figure 11.27. The velocity distribution as modeled, at the indicated time intervals after the lead opened, by Kantha (1995) for a lead studied in the field by Smith and Morison (1993). The ice drift velocity is zero. Note the deep impinging plume at the lead center and the convective rolls at the ice edges.

Figure 11.28. The salinity increase in ‰ above the initial mixed layer value as modeled at intervals of six hours by Kantha (1995) for the advective situation initially considered by Kozo (1983). The relative fluid motion is from the left to the right at 5 cm s⁻¹.

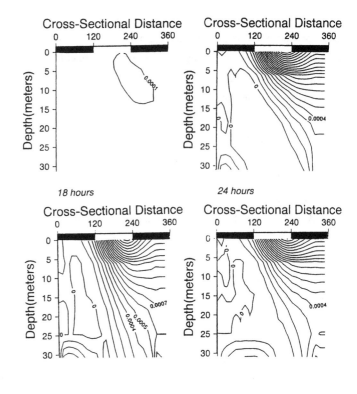

stream side of the lead, as shown in Figure 11.28. These results were later supported by modeling results obtained by Smith and Morison (1993), whose work additionally explored the development of criteria that would allow one to predict, based on a knowledge of the environmental conditions, whether the flow regime within a lead would be primarily by free, mixed, or forced convection. In scaling the governing equations they suggested that a parameter that they termed the lead number L_0, where

$$L_0 = \frac{q\,d}{U\,u_*^2},\qquad(11.14)$$

would be useful in characterizing the flow within a lead. Here q is the buoyancy flux resulting from the salt flux caused by freezing, d is the depth of the mixed layer, U is the ice velocity relative to the surface water, and u_* is the friction velocity, where $u_* = C_D^{1/2}U$ with C_D, the drag coefficient. Therefore L_0 is the ratio of the pressure gradient term to the turbulent stress term in the lateral momentum equation. When $L_0 < 1$, free convection should be the dominant process, with salt being mixed downward in sinking plumes of dense water. On the other hand, when $L_0 > 1$ forced convection should dominate, with complete mixing of the salt rejected during the growth of ice occurring near the surface as the result of stress between the water and the ice. When $L_0 \approx 1$, both processes should be of roughly equal importance. Additional discussion of alternative criteria related to L_0 can be found in Kantha (1995). The reader should note that the simulations shown in Figures 11.27 and 11.28 are not those of either Kozo or Smith and Morison, but instead Kantha (1995), who based his work on the published descriptions of the leads studied by the earlier authors. I have decided to use Kantha's results instead of the original results because he considered the time-dependent decrease in the ice growth rate and the associated decrease in the salt flux, whereas Kozo as well as Smith and Morison considered these values to be constant. A more detailed discussion of the differences between these sets of simulations can be found in Kantha (1995).

The reader may well think that although these simulations are fine, it would be desirable to see some real data that directly support the idea of vertical convection in leads. This can be found in Fig-

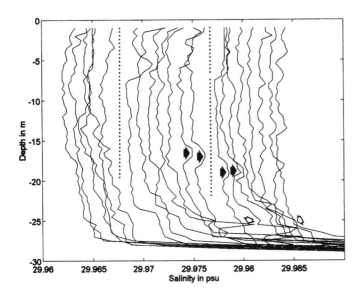

Figure 11.29. Time sequence of individual vertical distributions of salinity derived from CTD cast data at lead site 14 (LEADEX, Beaufort Sea). These are presented in three groupings, separated by vertical dashed lines. Within each grouping the profiles are separated by 5-minute intervals; between the groups the intervals were approximately 30 minutes long. The solid arrows indicate examples of sinking, brine-enriched water parcels, and the open arrows indicate saline, tonguelike features consistent with the spreading of brine-enriched water outward from the lead along the pycnocline (Muench et al. 1995).

ure 11.29, which presents the time sequence of individual vertical salinity profiles based on observations made on a 80-m-wide lead with an ice cover of approximately 10 cm studied in the Beaufort Sea on 12 April 1992 during LEADEX (Muench et al. 1995). As noted in the caption, the figure presents three groups of data separated by vertical dashed lines. In each group the profiles were obtained at 5-minute intervals, and the time intervals between the groups were roughly 30 minutes. The solid arrows indicate sinking brine-enriched water parcels. The sinking rates are estimated to be approximately 0.2–0.3 cm s^{-1}, values that are consistent with the mean downward vertical currents observed at the site. The vertical scales for these parcels appear to be on the order of 1–2 m. The open arrows near the bottom of the profiles indicate saline tongues of water that are consistent with the spreading of brine-enriched water outward from the lead along the top of the pycnocline.

When one considers interactions between leads and the atmosphere, there clearly are significant differences between winter, on which investigators have focused most of their attention, and summer. During summer leads freeze over slowly if at all, as differences between air temperatures and ocean surface temperatures are very small. As a result, the temperature and moisture gradients that control the fluxes into the atmosphere are drastically lessened. Shortwave radiation also becomes a significant factor. These are all trends that lessen the atmospheric importance of leads. The oceanographic situation, on the other hand, is quite different in that the processes that drive the downward flux of brine and the circulation within the lead can turn off completely, with leads changing from locations of instability to locations of stability. A major factor here is the meltwater that flows into leads from the surface of the surrounding ice floes, producing a stable surface layer of water that is both less saline and warmer than the underlying water. Furthermore, the increased incoming shortwave radiation heats this upper water layer, contributing to the effect. As a result, the warmer water in the lead contributes to the melting of the sides and the lower surfaces of the adjoining floes, expanding the lead and thinning the surrounding floes.

The fact that this is an interesting problem amenable to analysis has caught the attention of several scientists, starting with no less a luminary than N. N. Zubov (1945), who derived a simple equation describing the process by assuming that all the solar energy absorbed by the water was used in lateral melting and that ice thickness changes are negligible:

$$Q_w dt = A_w \left(1 - \alpha_w\right) F_r dt = \rho_i L_f H dA_w. \tag{11.15}$$

Here a_w is the albedo of the water, ρ_i is the density of the ice, L_f is the latent heat of fusion of the ice, F_r is the shortwave solar energy flux, H is the ice thickness, and t is time. If one lets $\xi_w = (1 - a_w) F_r / \rho_i L_f$, then the solution to (11.15) can be written as

$$A_w = A_{w0} e^{\xi_w t / H} \tag{11.16}$$

where A_{w0} is the area of open water at time $t = 0$. In that one would think that equation 11.16 is clearly an oversimplification, the problem was examined further by Langleben (1972) in a study of ice decay along the eastern coast of Canada. He modified Zubov's relation by taking the thinning of the ice surrounding the lead into account. However, in doing this he introduced another simplification—that any decrease in the thickness of this ice was only the result of melting at the upper ice surface as determined by the magnitude of the net shortwave radiation flux, or

$$A_i \left(1 - \alpha_i\right) F_r dt = -\rho_i L_{fi} A \, dH. \tag{11.17}$$

Here the terms are similar to those in equation 11.15 with the subscript i indicating ice. Integration of this relation gives the change in ice thickness with time:

$$H = H_0 - \xi_i t \tag{11.18}$$

where H_0 is ice thickness at time $t = 0$ and $\xi_w = (1 - a_i) F_r / \rho_i L_f$. If equation 11.18 is substituted into equation 11.15, then

$$A_w = A_{w0} \left(1 - \frac{\xi_i t}{H}\right)^{-\mu} \tag{11.19}$$

where $\mu = \xi_w / \xi_i = (1 - a_w)/(1 - a_i)$. Using values similar to those used by Zubov, Langleben's relations predicted the disappearance of ice in half the time predicted by Zubov's relations. More importantly, they were in better agreement with his field observations.

A much more detailed look at this general problem followed 15 years later (Maykut and Perovich 1987), and included an examination of the fact that not all the solar energy absorbed by the water goes toward lateral melting of the floe edge. Instead, if an appreciable amount of heated water is swept beneath the neighboring floes, it will result in the melting of the underside of the floe further thinning the ice. The question is, what are the factors that determine the partitioning of the absorbed energy between melting the floe edge and melting its underside? Note that the melting of the upper surface of the floe proceeds independently of these processes and is not affected by them. On the other hand, the amount of surface melt definitely controls the amount of relatively fresh water entering the lead. Maykut and Perovich examined the problem by first considering the heat and mass changes occurring in a single lead under steady-state conditions, and then attempted to generalize the results to a consideration of lead systems. A schematic showing the energy fluxes in the lead and adjoining ice is presented in Figure 11.30. The different terms considered are listed in the figure caption. I will not discuss the development of the applicable equations in detail. However, individuals who have studied the material presented in Chapter 9 should recognize much of this material. Maykut and Perovich considered several different cases, the first of which they termed *instantaneous heat transfer* (IHT). This assumed that the energy transfer to the ice edge was so rapid that the water temperature (T_w) remains at the freezing point. As they note, this is similar to the Zubov–Langleben analysis except for the fact that the amount of shortwave radiation transmitted through the lead and absorbed at the bottom of the surrounding ice does not contribute to the change in the width of the lead with time dW/dt. A consequence of the above assumption is a linear increase in lateral melt rate with increasing lead width. This result is unreasonable in that as W increases it becomes increasingly unlikely that all the heat absorbed in the lead can be effectively transmitted to the lead walls.

They also attempted a boundary layer parameterization (BLP) using the form of a relation suggested by Josberger (1979) to calculate vertically averaged lateral melt rates observed in laboratory studies. A major problem here is that when the laboratory results were compared to field

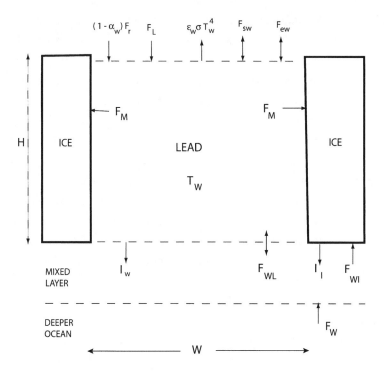

Figure 11.30. A schematic illustration of energy fluxes in both a lead and the surrounding water as considered by Maykut and Perovich (1987). The terms are as follows: T_w = water temperature in the lead; F_r = incoming shortwave radiation; F_L = incoming longwave radiation; $\varepsilon\sigma T_w^4$ = longwave radiation emitted from the surface of the lead; F_{sw} = sensible heat flux; F_{ew} = latent heat flux; F_m = average horizontal heat flux at the lead wall; I_w = amount of shortwave radiation transmitted through the lead and absorbed below the bottom of the surrounding ice; F_{wl} = flux of heat associated with the water exchange between the lead and the underlying ocean; I_i = flux of shortwave radiation through the ice to the ocean; F_{wi} = oceanic heat flux to the bottom of the ice; and F_w = heat exchange with the deeper ocean.

Figure 11.31. A comparison of the steady-state lateral melt rates predicted for the central Arctic by the laboratory BLP (curve 1), the field BLP (curve 2), the IHT assumption (curve 3), and the Zubov–Langleben equations (curve 4). The ice floe serving as the lead boundary was assumed to have an initial thickness of 3 m (Maykut and Perovich 1987).

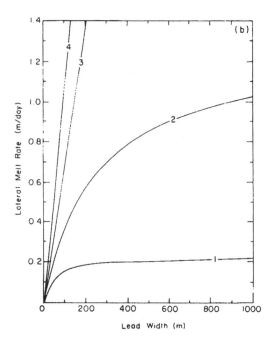

measurements (Perovich 1983), it was found that the Josberger relation underestimated observed melt rates by a factor of more than 5. As Maykut and Perovich point out, this clearly suggests that in the polar regions other factors not present in the laboratory significantly affect melting rates. Obvious possibilities include currents, wind-induced mixing, and mechanical erosion. Figure 11.31 shows some of the results of these calculations based on conditions in the central Arctic. As can be seen, the Zubov–Langleben (Z/L) formulation shows a rapid linear increase in melt rate as the lead widens (curve 4). The IHT assumption gives similar linear results as shown in curve 3, with the differences between Z/L and IHT being less than might be anticipated. The use of the BLP results in significantly reduced melt rates at large lead widths. This is particularly true if the BLP based on the laboratory results is used (curve 1). If the BLP is fitted to field observations (curve 2), the dropoff is less dramatic but still pronounced. Unlike the IHT case, values for T_w are generally greater than T_a for the BLP cases. Therefore the turbulent fluxes result in a net heat loss instead of a gain.

Maykut and Perovich have also considered the melting of the horizontal boundaries in the ice pack and their effect on ice decay. Although it is possible to suggest appropriate equations, confident estimates of the values of some of the terms are not readily available. As a result it is necessary to make reasonable assumptions, such as assuming that the energy input at the bottom of the ice is proportional to the total amount of energy present in the underlying water and to the ice concentration (G). In considering the effects of variations in lead widths on dG_0/dt, it was found, as expected, that dG_0/dt values were largest when leads were narrow and that these values dropped off rapidly as leads widened. Complex lead geometries were also explored, leading to the conclusion that in general, the lead geometry that has the greatest lead perimeter per unit area will experience the greatest amount of lateral melting.

In an attempt to obtain additional field observations on summer processes, Perovich and Maykut (1990) undertook a series of field observations during the peak of the melt season on an isolated transverse lead located in the fast ice near Mould Bay, NWT. Initially, the water was vertically uniform and at the salinity-determined freezing point to a depth of 20 m. Currents were weak and largely tidal. By the end of the field observations, the water column had developed three distinct layers: a well-mixed, nearly fresh surface layer comprised primarily of meltwater; a very stable half-meter-thick halocline centered a bit below the bottom of the ice; and a thermally stratified constant-salinity layer extending at least to 25 m. Of particular interest was the fact that the halocline developed a pronounced temperature maximum that was about 2°C warmer than the surrounding water. This temperature maximum is believed to result from the trapping of shortwave energy in the lower portion of the pycnocline, thereby preventing it from contributing to melting the lower surface of the overlying ice cover. Analysis suggests that only one-quarter of the total shortwave energy deposited in the lead was actually utilized in melting the lower surface of the ice. To the best of my knowledge, this was the first time that such a thermohaline structure had been described. Clearly, the fact that the general situation could be characterized as static was very important. The closest analogy to these observations appears to occur in ice-covered freshwater lakes. When similar observations were made in the ice pack just a few kilometers to the north of Prince Patrick Island, the temperature maximum in the pycnocline was not observed. The difference between the two sites appears to be the result of the presence or absence of mechanical mixing.

To summarize the above observations, it is clear that in the summer, significant amounts of solar energy enter the polar oceans, both through the ice and through leads. In bays and fjords and regions where the conditions are relatively static and oceanic circulation is limited, freshwater input from the melting of ice floes and from the nearby land can result in a stratified upper ocean, in effect a solar pond, that traps most of the shortwave radiation absorbed below the pycnocline. If these conditions continue through the summer, this stored heat will be released in the fall as the result of brine rejection associated with the growth of new ice destroying the vertical stability. Consequently, the initial thickening of both new ice and the floes surviving the summer will be somewhat slower than would be expected if the water column is initially at the pressure freezing point. In most situations, however, both ice motion and oceanic turbulence will be sufficiently intense to prevent such a stable situation from forming. There the introduced heat will contribute directly to the melting of both the side walls of the leads and the undersides of the floes, with the melting rates depending with the relative ice–water velocity, the temperature difference, and the characteristics of the laminar sublayer at the ice–ocean interface (McPhee et al. 1987). In pack ice another factor would be the occasional passage of a pressure ridge keel which would presumably act as a paddle moving through the water, further destroying any tendency for a stable vertical density distribution to form. Locations where stable vertical density distributions would be very unlikely to form would include the complete Southern Ocean ice pack, where winds, waves, swell, and rapid ice motions are business as usual.

What is needed in this area of sea ice research? The theory appears to be reasonably well in hand, or at least capable of modification and expansion so that it describes most realistic situations. Still, there does not appear to be an adequate assessment of the partitioning of energy between the melting of floe edges and floe undersides. Clearly, there is more melting on the floe edges than on the undersides. For instance, associated with the observations on the changes in the thermohaline structure of one of the leads (Sarah's Lake) that formed in the vicinity of the SHEBA camp, it was estimated that the net average heat flux expended in melting the floe edge was ~820 W m^{-2}, values that translate into lateral melt rates in the range of 0.14 to 0.23 m d^{-1}. Furthermore, Perovich et al. (2003) have estimated that for at least part of that summer, the lateral melt was about 5% of the total mass loss (surface, bottom, lateral). During early August, at the peak of the melt period, lateral melting was estimated to amount to as much as 30% of the total ice loss. However, as the relative areas involved are vastly different, this does not necessarily mean that during most of the summer the floe edges are the most important heat sink. Additional careful fieldwork will definitely be required. As Perovich and Maykut (1990) point out, time- and depth-dependent data on the current velocity, optical properties of the water, and horizontal distribution of salinity and temperature at varied locations within the leads, as well as under and within the adjoining ice floes are needed. Furthermore, these observations should be maintained throughout the complete melt season. Easy to say, difficult to accomplish.

As might be expected, one effect of such new data sets is frequently the stimulation of expanded modeling efforts. For example, Skyllingstad et al. (2005) have used an oceanic version of the large-eddy-simulation (LES) model discussed early during the discussion of atmospheric plumes associated with the opening of leads in the winter. Because of the intensive computational requirements of the model the simulations were unfortunately short in duration (≤ 4 hours). Therefore, it has not yet been possible to run simulations long enough to reach a steady state or that would extend through the complete summer melt period. Even so, the results are very informative, particularly because the discussion in the paper effectively combines model results with field observations. In the following I will only touch briefly on a few of these results.

As anticipated, the larger the lead the greater the amount of edge melting. An important factor here is the fact that identical wind stresses are able to generate stronger currents in larger leads and these currents move warm water from the center of the lead to the lead edge. In addition, over time stronger winds are more effectively able to transport the cooler water initially located next to a wall away from the ice, ensuring that cold water resulting from the melt process cannot collect at the ice edge, which thereby isolates the ice from the warm water in the center of the lead. If the winds are weak and there is no appreciable surface melt, the local lead heat balance is dominated by entrainment mixing at the base of the surface layer, with the average lateral heat loss being insignificant at ~ 15 W m^{-2}. Further, if the surface melt is zero, only calm conditions can lead to appreciable warming of the lead water. During SHEBA an analysis of maximum surface melt values (~ 2.5 cm d^{-1}) and lead fraction (0.05) suggests that only about 1% of the available meltwater ended up in leads, with the remainder either ponding on the ice surface or pooling beneath the ice cover. For strong stratification to develop in summer leads, surface meltwater is clearly an essential factor. If it is sufficiently windy, low-salinity water does not collect in the lead. Instead it will be forced along the edge of the lead, as well as under the ice. The results for low-wind situations are quite different in that the meltwater flowing into the lead does not have enough momentum to overcome the stratification at the lead edge and is trapped. Clearly optimum conditions for developing a stable freshwater layer in a lead occur when the winds are light and the freshwater flux off the floe is large. If the winds are weak, heat storage within the lead is roughly double the value observed when the winds are strong. Furthermore, both entrainment and wall flux have larger values with a strong wind than with a weak wind. Although one might think that melting would be maximized on the downwind edges of leads, the simulations suggest that locations where current velocities at the ice–water interface are largest have the highest melt rates. The downwind edges of leads are dominated by convergence and consequently have weaker currents. However, one should note that the simulations in Skyllingstad et al. (2005) do not consider melting resulting from the generation of waves.

Although the model results support the conclusion that lateral melting is greatly enhanced when leads contain a fresh layer, it is not at all clear that the change in lead size resulting from this effect is large enough to cause appreciable additional melting through the so-called ice–albedo feedback effect. In fact, Perovich et al. (2002) showed that during a period of strong warming and low winds during SHEBA there was little change in the lead fraction. It was only later during August when strong winds occurred that the lead fraction appreciably increased. As Skyllingstad et al. (2005) note in the last line of their paper, given the combined results of SHEBA plus the simulations, it appears that the improved representation of surface ponds in coupled ocean/ice models may be of greater importance than leads. Unfortunately ponding is a phenomenon that has received little attention until very recently (Eicken et al. 2002; Fetterer and Untersteiner 1998; Flocco and Feltham 2007; Lüthje et al. 2006; Skyllingstad and Paulson 2007). Clearly interest is picking up.

I would like to close this chapter by stressing that working around leads and polynyas, and in fact in any region where large areas of thin ice are present, can be very dangerous. The problem is simple: small ice movements can result in open water or thin ice areas that are difficult to see under visibility conditions that are frequently encountered in the polar regions (low light, blowing snow, frost flowers). Just because the ice appears to be unchanged from yesterday does not mean that ice conditions are necessarily exactly the same.

12 Deformation

But 'ice floes' have flaw
God hems and haws
As the curtain He draws
O'er His physical laws
It may be a lost cause.
Modified from an ode by Isadore Singer
as quoted by Roman Jackiw, *Physics Today* (1996)

12.1 Introduction

It was only in the early 1500s, as the result of the voyages of men such as Frobisher and Davis, that more accurate descriptions of arctic ice conditions became available (Zukriegel 1935). Through the succeeding centuries, it became generally known that deformation in a variety of forms was an integral part of the drift of pack ice. The attention given to this subject is hardly surprising inasmuch as the early explorers were primarily concerned with sea ice as an impediment to the movement of shipping. The fact that the most difficult areas of ice to transit were those that had been deformed left a lasting impression. For instance in many early books there are drawings of ships in pack ice where pressure ridges are depicted as towering, menacing shapes quite out of proportion to their more typical dimensions. Many times these illustrations were not drawn in the field by individuals who were on the expedition, but were prepared later by illustrators based on verbal descriptions. It was clear that the crew members were impressed by ridges. The first major scientific work limited strictly to the study of sea ice, *Die Metamorphosen des Polareises* by Karl Weyprecht (1879), devoted the second of its eight chapters to "Hummocks of the sea ice." The early scientific investigators working in the polar oceans such as Nansen (1897, 1900–1906), Makarov (1901), and Arctowski (1908) attempted to document all aspects of sea ice, including deformation features. Too little was known about sea ice to allow them to focus on specific problems. However, by the time of the *Maud* expedition, scientists such as Malmgren (1927) had begun to concentrate on specific problem areas. It is hardly surprising that investigators did not find the jumbled piles of ice that comprise ridges and hummocks to be particularly appealing. Why study "chaos" when you could study equally interesting problems in undeformed ice that appeared to be orderly? One can hardly argue with such decisions. For instance, although an excellent discussion of the current knowledge of ridges as of ~1940 can be found in Zubov (1945), it is clear that it was pieced together from subsidiary observations made by investigators while they were primarily involved with other matters. An exception is the book by Burke (1940), which discusses pressure ridges from the point of view of an icebreaker captain.

With the increased interest in sea ice since the 1950s, the importance of ice deformation gradually began to surface. As expected, the earliest projects that specifically focused on deformation were in relation to perceived applications. As mentioned in Chapter 2, the first of these (mid-1960s) was the Arctic Surface Effect Vehicle (ASEV) project, which required information about both the large-scale and the small-scale surface roughness of the Arctic Ocean's ice cover so that giant hovercraft could be designed that would race over the frozen surface of the Arctic Ocean at 200 knots. Of course, this goal was quite unrealistic. Although such a vehicle was never built, much of the data collected during the project later proved to be of considerable use for quite different purposes. What the project did accomplish was to allow investigators to spend time examining ridges in an attempt to understand how they formed.

A second project followed soon after (1970s), and was definitely realistic. It was concerned with ridging, and in particular with large multiyear ridges impacting tanker ships and offshore production oil facilities. Here one must remember that the discovery well for the Prudhoe Bay super giant oil field was almost located in the ocean on the coast of the Beaufort Sea. At that time, serious consideration was being given to transporting the oil out of Prudhoe Bay by marine tankers. A reading of the literature of polar exploration as well as experience gained during the two experimental cruises of the S.S. *Manhattan* drove home the fact that when pack ice went into compression, a ship could frequently be "nipped" by the ice. At best, its speed would be drastically slowed or more frequently reduced to zero. At worst, the ship could receive structural damage and even sink. Figure 12.1 was taken on the second *Manhattan* cruise and shows first-year

ice failing against the ship during a compressional event. The ice first failed against the side of the ship, and then the broken blocks were forced to pass completely beneath the ship. Considering that the *Manhattan* had a draft of ~20 m, this was quite an impressive event. One might ask how it was possible to verify that the ice was passing completely beneath the ship. This information came from the engine room staff, who reported that when they placed their hands on the external plates the external flexing and pounding was obvious. They described it as "Maniacs with sledgehammers beating on the hull of the ship." Essentially the ship had become part of the ridge.

Figure 12.1. Ice deforming against the side of the S.S. *Manhattan* during a compressional event, Northwest Passage, Canadian Arctic Archipelago.

Another aspect of ridging that interested the oil industry was the gouging of the seafloor by ridge keels as they are pushed along by the movement of the nearshore ice pack. This interest is easy to understand, as peak gouge depths of 6 to 8 m were known to exist along the Beaufort Coast. A more detailed description of this phenomenon can be found in the next chapter. In addition, the failure of sea ice against a fixed object such as a production site in the offshore region results in an ice pileup that has many similarities with naturally occurring ridges. When I describe the above studies as a project, I do not mean to imply that these studies were funded by any one entity or even that they occurred at the same time. I only mean that all these studies had an overall purpose: the safe extraction and transportation of oil in the environment of the Arctic Offshore.

During roughly the same time period, the Arctic Ice Dynamics Joint Experiment occurred (1969–1976; Pritchard 1980b). Although AIDJEX was primarily focused on ice motion and its prediction, it served as a platform leading to the development of a more coherent view of sea ice behavior. In this view, which is essentially the current view, the different aspects of the deformation of sea ice are regarded as an integral component of pack ice behavior. In the pack the thermodynamics and the dynamics are interlinked in a complex dance. Understanding this dance is essential to understanding the behavior of the world's sea ice covers.

Many of the more recent programs focused on ridging can, in a general sense, be considered to be extensions of the earlier work on ridging started by the discovery of oil at Prudhoe Bay, Alaska, in that the focus is on the applied effects of ridging. Groups involved include Finnish and Swedish investigators concerned with winter shipping in the Bay of Bothnia, and Canadian investigators concerned with the role of ridges in determining peak ice forces. A paper of particular interest in relation to the current chapter is the recent review of FY ridging by Timco et al. (2000). The first 62 pages of this report present a summary of the morphology and properties of FY ridges, and the final 95 pages discuss the interactions between deformed ice and offshore structures.

12.2 Terminology

The descriptive terms used to designate the different types of deformation features occurring in sea ice can be somewhat confusing, particularly when one compares papers written in English with those written in Russian. In this book I will sometimes use the word *ridging* to refer to sea ice deformation in general. Included under this rubric could be pressure ridges, shear ridges, hummocks, rubble fields, and rafts. In Russian the term *hummocking* is frequently used in this same general manner. Otherwise, I will generally use terms as defined in the WMO Sea Ice Nomenclature (see Appendix C).

12.3 Field Observations

12.3.1 Arctic

The initial step in forming any type of ridge is the formation of a lead or, at the least, a crack. No lead or crack, no ridge. This means that if you set up your tent in the middle of a uniform floe of a thickness greater than, say, 10 to 20 cm, you do not have to worry about the ice buckling and incorporating you into a ridge. Now a ridge might start to form at the floe edge and work its way over to where you are sleeping, but this takes time as ridging is commonly not a very rapid process. Insofar as ridging in ice of sufficient thickness to serve as a tent site is also accompanied by a wide variety of vibrations and noises, ranging from squeaks to squeals to low woofs, I can assure you that you would be wide awake well before you were in any danger. A more probable risk, although still very small, is that a crack could form and open beneath your tent, resulting in your taking a refreshing midnight dip in the Arctic Ocean. I have heard that this actually happened to two members of the staff of one of the Russian NP stations. Apparently they were not injured. However, I bet that they had trouble going back to sleep. Figure 12.2 shows a lead widening beneath a building during the AIDJEX program. Although the building was saved, the camp was split into two segments that became separated by an inconvenient distance.

Does the above discussion mean that buckling never occurs in sea ice? No, not at all; see, for instance, Figure 12.3. However, buckling that is not associated with deformation at the edge of a floe appears to be rare, primarily occurring in very thin ice (< 10 cm, frequently < 5 cm). Here I refer to out-of-plane flexures occurring *within* a uniform sheet of sea ice away from the edge of the floe. Besides, anyone crazy enough to spend time sleeping on sea ice that is less than 10 cm thick deserves to get wet. As will be seen later, buckling is just one of the failure types that can occur when two or more floes mechanically interact, resulting in the formation of sea ice deformation features.

Now that I have assured you that ridges do not jump out of nowhere and grab you, let us consider how frequently ridges form. This is not an easy question to answer, for several reasons. For one, lead formation and deformation are believed to be locality dependent. In addition, the fact that a lead occurs does not necessarily mean that a ridge will form. One might suggest simply counting the number of ridges within a specified area as an index of ridging activity. Although this is informative for FY ice covers, it is often difficult to know what period of time the ridges represent if old ice is present, as the age of such ice is frequently not well known. Some help here

Figure 12.2. A crack in the process of widening into a lead, Big Bear Ice Camp. AIDJEX Field Program, Beaufort Sea.

Figure 12.3. Folds that have developed in a very thin ice cover, Beaufort Sea.

can be obtained by examining Gordienko's (1962) summary of Russian experience from 1937 through 1961 on the North Pole Stations. During 14 annual drift cycles, there was fracturing and ridge building 95 different times within the camps and 353 times outside of the camps, although still in the general camp area. As the result of this activity, it was necessary to reestablish the camps 57 different times. Also, I note, based on newspaper reports, that the most recent Russian drift station (~2003) was being prepared for closure when it had to be abandoned as the result of breakup. The experience on Western drift stations, although less extensive, is similar. Both drift stations Alpha and Charlie were abandoned because of severe deformation near the camps. On the AIDJEX stations, active leads and ridges were frequently located within 300 m or less of the camps, and several times fractures passed either through (Figure 12.2) or very near the camps (Figure 11.15), even though the actual camp areas were small. In evaluating this information one must remember that when all of the above stations were initially sited, they were placed on the thickest ice available, where deformation was not considered to be likely. Clearly, whenever one occupies an on-ice field site in the Arctic Basin, one can expect to see some type of active ice deformation occurring in the near vicinity within a few days time. Features related to ice deformation are both spatially and temporally common within the Arctic pack.

A second generalization regarding ridging is that if there are a variety of different ice thicknesses in a region, when the ice comes under compression, it is invariably in the thinner ice classes where failures occur and ridges develop. Many times the thinner ice classes are the only ice types that undergo deformation. Figure 12.4 shows an example of this general tendency. In the middle of the photograph, there is a rounded old floe. This is clear from its undulating surface topography. All the deformation is in the thinner FY ice surrounding this older, thicker, obviously stronger floe. Observations that strongly support this generalization were made by Koerner (1973) during the crossing of the Arctic Ocean by the British Transarctic Expedition (89°N to 81°N along 30°E long). He found that although the ice traversed was primarily old ice (75%), only 13% of the pressure ridges contained old ice and 81% of the ridges were composed of ice that was

Figure 12.4. A large undeformed MY floe surrounded by highly deformed FY sea ice, Northwest Passage, Canadian Arctic Archipelago.

less than 50 cm thick. The most frequent ice type in the ridges was young ice (10–30 cm, 57%). Ridges of thin ice were at least 27 times more frequent than they would be if ridge occurrences were independent of ice thickness. Pressure ridging is, therefore, an extremely selective process. In any given area of the pack, when the ice converges it is the thin ice that is deformed first, followed by the next-thicker ice class as deformation continues, and so on until ridging finally occurs in the thickest ice. The fact that MY ice is frequently not deformed in the ridging process, even though it is the most common ice type, at least in the central part of the Arctic Ocean, suggests that the forces involved are commonly not sufficiently large to cause it to deform. North of Alaska, it was generally considered wise to site field camps planning long-term deployments at least 200 km north of the coast. The general idea here was to avoid regions that contained large areas of FY ice, even considering the fact that camps are invariably sited on thicker and presumably safer MY floes. It is also my impression, based on Russian experience, that in the central Arctic Basin, problems with breakup are frequently experienced in the region where the Beaufort Gyre separates from the Transpolar Drift Stream. It should be noted, however, that although sequential deformation by increasing ice thickness classes is a strong tendency, it is not perfect. It is not unusual to see fragments of MY ice incorporated into a ridge that is primarily composed of thin ice types, whereas nearby ice areas of intermediate thickness remain undeformed. When this occurs it is frequently the result of geometrical considerations. For example, if the closing of a refrozen lead causes a large pileup of thin ice blocks on the edge of the MY floe that forms the edge of the lead, asymmetric loading may result in the edge of the MY floe failing in flexure. It is these sorts of tendencies that make selecting the location for a field camp difficult. First you attempt to find a thick MY floe whose surface is smooth enough to serve as a landing strip. Realizing that these requirements are essentially mutually exclusive, insofar as thick old floes rarely have smooth upper surfaces, you frequently have to compromise and attempt to find an old floe that is in the near-vicinity of a refrozen lead that contains undeformed FY ice of an adequate thickness to serve as a landing strip. Then the question becomes, how long will the landing strip survive? Probably long enough to establish the camp. But will it still be operational when you want to leave? It is definitely a good idea to select locations with more than one possible runway. Easy to say, not always easy to do.

I will now briefly describe the different types of deformation features that one sees in the Arctic.

12.3.1.1 Rafts   Immediately after the formation of nilas in a new lead, several types of deformation features can develop that only occur in thin ice. Such ice, in addition to its thinness, has certain distinctive characteristics. For one, it has a very high salinity as the result of the fact that it is growing rapidly. Because the ice is very thin, it is also very warm. Also, the temperature of the upper surface of such ice is usually within a few degrees of the freezing temperature of the underlying seawater, even if the air temperature is appreciably colder. This is quite important as the amount of brine present in the ice will be large, resulting in the ice being both weak and flexible. For example, a slab of dark nilas with a thickness of < 5 cm will frequently not be sufficiently strong to support its own weight when it is lifted out of the water. When thicker ice comprising either side of a lead moves slightly, the nilas in the lead breaks and deformation features referred to as finger rafts frequently develop as the two sheets of nilas interact. Figures 12.5 and 12.6 show different views of these features. The first figure shows classic finger rafts as viewed from an icebreaker. At times, the regularity of the spacing of these features is striking. A diagrammatic sketch of the geometry of these features is shown in Figure 12.7a and b. Descriptions of these and related features can be found in Fukutomi and Kusunoki (1951), Holmes and Worthington (1953), Weeks and Anderson (1958a), and Dunbar (1960, 1962). One way to help visualize the geometry of finger rafts is with your fingers. Place both your hands with the palms down and the fingertips touching each other. Now place the first

Figure 12.5. Finger rafts as viewed from the bridge of an icebreaker (U.S.S. *Nathaniel B. Palmer*) operating in the Amundsen Sea off the coast of the Antarctic.

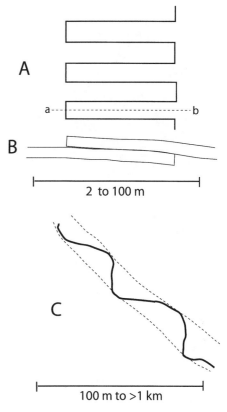

Figure 12.6. Aerial view of finger rafting occurring in the thin ice of a recently refrozen lead, Beaufort Sea.

Figure 12.7. Diagram illustrating the geometric relations typical of rafting. (a) presents a plan view of a series of finger rafts. (b) gives a cross section of the raft along section a–b. (c) gives a plan view of a large sinuous ridge typical of the Arctic Ocean. The heavy line gives the surface trace of the ridge. The area between the two dashed lines is where the two sheets probably overlap.

finger of your right hand on top of the first finger of your left hand. Then place the second finger of your right hand under the second finger of your left hand, Then, in a similar manner, do the same for your third and fourth fingers. Finally, push your hands together as far as possible, and you have a finger raft. If you will then compare the geometry of your fingers with the sketch of finger rafting presented by Weeks and Anderson (1958a), you will find that they do not agree. Believe your fingers; the sketch is incorrect. Credit for publishing the first diagrams that correctly show the geometry of finger rafting between two sheets of the same thickness goes to Holmes and Worthington (1953) and Volkov and Voronov (1967).

The corner angles in finger rafts may be as low as 70 degrees, although many angles are close to 90 degrees for thin ice. In fact, the consistent appearance of ~90-degree angles, plus the absence of rubble at the front end of the raft, can be taken to indicate that the rafting in question occurred when the ice was very thin (< 10 cm). In many cases the sides of the thrusts are extremely straight, although some are curved. The latter appear to be the result of floe rotation during

Figure 12.8. Aerial view of a thin ice area (gray nilas) that has undergone extensive finger rafting and simple rafting, Arctic Ocean (AARI Files). Note the pronounced change in the gray scale at locations where the ice thickness has doubled.

thrusting. The dimensions of these thrusts are quite varied. They typically are a few tens of meters measured in the direction of the thrusting although thrusts of up to several hundred meters have been reported (Weeks and Anderson 1958a). In some cases a thrust can develop a large number of layers; ten is the maximum that I have seen documented. In very thin ice the change in gray scale caused by doubling the ice thickness is particularly striking (Figure 12.8). See also Figs. 3, 4, and 5 of Dunbar (1960) and the covers of the 1962–1963 copies of the *Polar Record*.

There are two aspects of the finger rafting problem that need to be considered. First, how can sea ice that is so thin (< 10 cm) and weak (unable to support its own weight when lifted out of the water) be thrust over another ice sheet for distances occasionally in excess of 100 m (a thrust length-to-thickness ratio of 1000/1) while frequently undergoing little or no deformation? This is possible because of the fact that as the upper ice sheet slides over the lower sheet, there is a rapid drainage of brine from the overriding sheet onto the surface of the lower sheet. This liquid layer drastically reduces the friction between the two interacting sheets. Incidentally, this mechanism is not speculation. When one examines newly formed finger rafts closely, the layer produced by the drained brine is quite visible. Another factor is that the upper surface of dark nilas is typically covered with a saline "scum" that is slippery. This means that when thin ice rafts, the lower sheet is prelubricated, to which the overriding sheet adds even more lubrication. Incidentally, when examining recently formed thrusts, it is wise to limit your walking to the overthrust zone, as when you step out of this zone the ice thickness may decrease by a factor of two, with a commensurate decrease in bearing capacity. I note, however, that Green (1970) has pointed out that brine drainage is not absolutely necessary for finger rafting to occur, inasmuch as thin (0.2 to 1.5 cm) sheets of lake ice can develop finger rafts. However, he notes that because of the thinness of the interacting ice sheets, water was observed leaking on to the surface just ahead of the advancing fingers as the overriding plate deflects the underlying sheet. For lake and river ice, interacting sheets thicker than 2 cm do not appear to form finger rafts, but instead break irregularly.

In a number of cases, open folding has been observed on the upper sheets of thrusts. When this occurs the upper sheet typically separates from the lower under the anticlines. The fold axes are frequently perpendicular to the length of the thrust sheet, indicating that the movement is parallel to the length of the thrust sheets. A similar conclusion can be reached by studying the movement "tracks" of the thrusts over the snow or frost flowers on the upper surfaces of the underthrust sheets.

This brings us to the second perplexing problem associated with the formation of finger rafts. Why does this often very regular over-under-over-under interfingered pattern occur? Two possibilities have been suggested, both of which depend upon the conditions at the time of formation. The first, initially suggested by Weeks and Anderson (1958a) and expanded by Dunbar (1962), assumes that when two thin ice sheets with slightly irregular edges interact, the foremost part of the advancing sheet will be deposited on the top of the edge of the other sheet by the crest of an advancing wave. When the wave crest passes on, the adjoining ice will tear and slide under the interacting ice sheet with the following wave trough. Once a tear is established, the ice will continue to fracture along the same line because of the pronounced stress concentration at the tip of the tear. Dunbar (1962), who observed thrusts forming in association with the bow wave of an icebreaker, described the process as follows. "First, a series of small folds form in the ice on the divergent crests of the bow wave. The ice then ruptures along the crest lines and the two pieces effectively instantaneously fall back with an interlocking overlap." I have watched this process numerous times and I have always found it difficult to see exactly what is happening. Obviously the flexing of the ice sheets by the bow wave is part of the process. As Dunbar notes, the thrusts seem to appear instantaneously. First, they are not there. Then like magic, there they are.

It is also known that finger rafts form under conditions when wave action is unlikely (calm conditions at locations well within the ice pack). An explanation for finger raft formation under such conditions has been advanced by Fukutomi and Kousunoki (1951), who also note that when two sheets of ice come together, there are invariably projections on their boundaries. Depending on the exact geometric relations, the first projection to contact the adjoining ice sheet will ride either over or under it. This causes edge loading of the adjoining sheet, producing a series of oscillating positive and negative deflections as one moves away from the point load. Their calculations suggested that the half wavelengths of the oscillating positive and negative deflections are of the same order of magnitude as the widths of the thrusts. Also, as their field observations show that the overthrusting sheet is never the thicker of the two sheets, their calculations show that significant deflections can only occur when the interacting sheets are very thin.

Recently Vella and Wettlaufer (2007) have explored this explanation more thoroughly through a series of tank experiments using thin sheets of sealing wax floating on water. When two sheets of the same thickness interacted, finger rafts formed. Consider Figure 12.9: here, if a small portion of a floe A overrides floe B at point C, floe A will be slightly lifted from its free-floating position while floe B will be slightly depressed. The response, as observed along the edges of the sheets, will be a series of oscillatory deflections modulated by an exponential decay. In that the initial perturbations of the sheets are of opposite sign (floe A is deflected upward while

Figure 12.9. Plan view of two floes colliding (Vella and Wettlaufer 2007). A small protrusion in floe A leads to a small area of overlap C. This overlap produces oscillations that decay away from C in the vertical position of the floes. The sign of these displacements is indicated by the +/– symbols in the figure. Note that the oscillations along the free edge are exactly out of phase, causing the two floes to alternately ride over and under one another during compression (arrows).

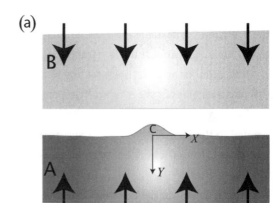

floe B is deflected downward), and as the thickness and properties of both sheets are the same, the deflections will remain out of phase as one moves along the free edge (Figure 12.9b). If further compression occurs the floes will then alternately ride over and under one another, resulting in a series of interlocking fingers. The authors stress that although the finger-rafting phenomenon has primarily been observed occurring in sea ice, it should occur in any two interacting sheets whose deflection responses approximate

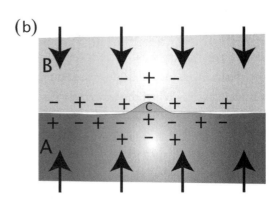

those of a plate on an elastic foundation. They then develop an analysis of this behavior in terms of the elastic properties of the plates. Of particular interest here is the position of the zero points in the deflection profile in that these determine where the floes can most easily ride over (or under) one another. They found that the smallest distance providing zero deflection was ≈4.507 m, with the next root occurring at ≈7.827 m. They then assume that the distance between these two roots determined the finger width, with the position of the subsequent fingers being determined once the initial fingers are in place. As the wavelength of the pattern is twice the finger width, in dimensional terms this results in

$$\lambda = 6.64\,\ell_*. \tag{12.1}$$

where λ is the wavelength and the characteristic length ℓ_* is given by

$$\ell_* = \left(\frac{\beta}{\rho g}\right)^{1/4}. \tag{12.2}$$

In equation 12.2, β the bending rigidity of the floes is a function of the elastic modulus E, the plate thickness h, and Poisson's ratio μ

$$\beta = \frac{Eh^3}{12\left(1-\mu^2\right)}. \tag{12.3}$$

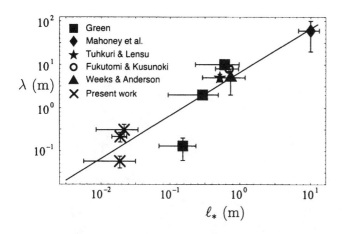

Figure 12.10. Wavelength λ, of the fingering pattern for floating sheets with different characteristic length scales (Vella and Wettlaufer 2007). The data shown are a combination of observations on ice obtained in the field and experiments using thin sheets of wax floating on water (labeled X Present Work). The solid line shows the theoretical prediction based on equation 12.1.

To test this prediction Vella and Wettlaufer combined their wax results with the available field observations (Figure 12.10). Considering the uncertainties in knowing the exact elastic values and thicknesses to insert into the above equations, I personally find these results to be quite encouraging. They also note that as $\ell_* \sim h^{3/4}$, there should be a strong correlation between the wavelength of the rafting and the thickness of the ice in which it occurs. They envision rafting as propagating along the ice edge as a zipper; once rafting occurs at one location the displacements there are sufficient to cause the next raft to form, etc. Assuming that the rafting travels at the speed of gravity waves in water covered by an elastic sheet, they estimate the wave speed for sea ice 10 cm thick to be ~5 m s^{-1}. This is a speed sufficiently high to make visual observations difficult. Finally, they suggest that finger rafting frequently occurs after a uniform thin ice sheet, such as would form in a newly refrozen lead, fails in buckling. Such an event results in two sheets, each with a ragged edge and exactly the same thickness, that can evolve via the zipper scenario into a series of finger rafts. Another corollary of this explanation is that unless the interacting ice sheets are essentially of the same thickness and have similar elastic properties, finger rafting is not possible as the spacings of the positive and negative deflections will no longer match.

When you see a finger raft pattern on the surface of a floe, this tells you that you are looking at FY ice in that the rectangular pattern of the rafts would commonly be destroyed during the summer melt period. It also tells you that the ice was usually very thin when the pattern formed. What it does not tell you is how thick the ice is at present.

Although finger rafting is the poster child of ice deformation because of its unusual surface pattern, simple rafting appears to be more common than finger rafting. Here simple rafting refers to a deformation process by which one piece of ice overrides another but without the regular over-under sequences characteristic of finger rafting. When viewed from above, simple rafting in thin ice is indicated by the lighter color produced by the doubled ice thickness and the topographic step produced by the edge of the upper ice sheet. These features are obscured by both subsequent ice growth and by snow. In addition, when rafting occurs in ice that is more than ~20 cm thick, the upper ice sheet has become sufficiently opaque to mask any color change resulting from the doubled thickness. Although both finger rafting and simple rafting occur in nilas, as ice thicknesses increase beyond ~8 to 10 cm regularly spaced finger rafts become rare (Vella and Wettlaufer 2008). Figure 12.11 provides an example of simple rafting. Note that flooding still occurs, lubricating the sliding.

Figure 12.11. Simple rafting, Beaufort Sea. Note the pronounced flooding of the upper surface of the underthrust portions of the interacting sheets.

Note also that we are observing an over-under sequence with the ice on the right in the lower half of the photograph sliding over the ice on the left, whereas in the upper part of the photograph it is the ice on the left that is riding over the ice on the right. The over-under patterns still occur but the repeat intervals are now larger, very irregular, and more difficult to recognize. Additional examples can be found in section 12.3.1.3.

12.3.1.2 Folds and Buckles The other deformational features that occasionally occur in very thin ice are folds. As might be expected, when there is a compressive motion normal to the axis of the lead, the fold axes will be parallel to the lead axis. If there is a shearing motion along the re-frozen lead, the fold axes will be inclined to the trace of the lead and the folds will form *en echelon*. Two examples of such folding can be found in Gakkel' (1959, Figs. 3 and 4). When such fold-

ing occurs, it can readily be used, as diagrammed in Figure 12.12, to determine the sense of the relative motion across the lead (Volkov and Voronov 1967). This is particularly true if the folding of the new ice in the lead is also

Figure 12.12. Diagram showing the general orientation of en-echelon folds and fractures in relation to the relative movements of the sides of a lead (after Volkov and Voronov 1967).

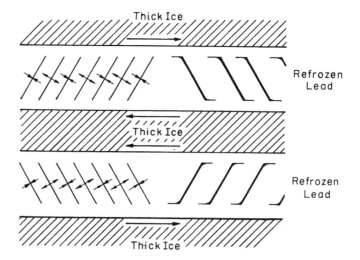

accompanied by crack formation. If the compressive or shearing motion of the surrounding floes continues, the fold in the thin ice in the refrozen lead will ultimately fail, with small-scale ridges and rafts resulting. If the pressure relaxes before failure occurs, one might think that the folds would disappear if the ice were to behave elastically. This is usually not the case as the ice is frequently too warm and the brine volume is too large. As a result, the deformation is primarily plastic. Also a small amount of additional growth while the ice is in a flexed position can lock the fold into place after the forces causing the deformation have ceased, even if the ice were to behave in an elastic manner.

My impression is that folding is more common than it appears. In the few striking examples of folding that I have seen, the troughs of the folds were "outlined" in white by small amounts of drifting snow. Without the snow, the folds, which are invariably small, would have been close to invisible. On the other hand, once a finger raft has formed, it becomes a nearly permanent feature of the ice surface unless destroyed by further deformation or surface melt.

I stand by my earlier statement that buckling and folding at locations away from floe edges, although possible in very thin ice, appear to be rare in pack ice that is thicker than 20 cm. Nevertheless there are locations where thick sea ice shows what seems to be buckling deformation. However, their settings are hardly representative. For instance, the 2.5-m-thick FY fast ice in front of the McMurdo Ice Shelf has been buckled into a series of folds by the forces exerted on it by the ice shelf as it moves forward by ~23 cm/day during the winter. The folds have a wavelength of ≈25 m and an amplitude of 2 m. Buckling also occurs in the thicker shelf ice, with a wavelength of 50 m and amplitudes of up to 3 m. Excellent photographs of this area are shown in MacDonald and Hatherton (1961, pp. 865–66, Figs. 1, 4, and 5). Hochstein (1967) has estimated the axial stresses necessary for both the sea and the shelf ice to buckle and has concluded that the deformation must be quasi plastic.

12.3.1.3 First-Year Ridges   According to the WMO nomenclature, a (pressure) ridge is a line or wall of broken ice forced up by pressure. In addition, the submerged volume of the broken ice under a ridge forced downward by pressure is termed a keel, whereas the above-sea-level portion of the pileup is called a *sail*. Figure 12.13 shows an example of a typical pressure ridge sail as seen in the Arctic Ocean. Note that the ice when the ridge formed was fairly thin (~0.3 m). Note also that there has been at least one storm since the ridge formed, as a snowdrift has built up against the left side of the ridge. The ridge is not very wide but it clearly is quite long. Also, the track of the ridge is not straight. Instead, its pattern might be described as an irregular snake dance. How do these features form? We can gain some idea by examining the following photographs. Figure 12.14 shows the very early stages of ridge formation. Note that the thrusting interaction between the two sheets results in the end of the overriding sheet breaking into a number of blocks. There are several reasons for this change from what occurs in nilas. As the ice is thicker, it is colder and less flexible, and instead of yielding, as happens in a simple thrust, it fractures. Also, although there clearly is water involved helping to lubricate matters, the saline surface layer characteristic of nilas is no longer present, a change that increases the friction between the interacting sheets and favors fracturing. Although not present in this particular example, in many cases appreciable snow will be present. Now not only does the overriding sheet have to serve as a snowplow, but any water involved freezes rapidly,

Figure 12.13. A typical sinuous ridge observed in the Beaufort Sea. Sail heights vary from 1 to 2 m. Note that a small snowdrift has started to form in the lee of the ridge.

Figure 12.14. The early stages of ridge formation, Beaufort Sea. Note the cusp-shaped cracks in the downthrust sheet and that the edge of the overthrust sheet is being fractured into a number of small blocks. Ice thickness is estimated at 10 cm.

further increasing the friction. Note also the cusplike shapes of the cracks that have developed in the underthrust sheet. Such cusps are frequently observed on the edges of nonuniformly loaded ice sheets. Figure 12.15 shows the continuation of this process on somewhat thicker ice. The cusps are now larger. After the crack forms, the ice within the cusp gradually rotates as it is broken and incorporated into the ridge. Figure 12.16 shows a similar ridge developing in even thicker ice. Here it is clear that although compression is necessary to cause the two ice sheets to overlap, the breaking is not the result of the compression. It is the result of asymmetric loading in which the pileup in the overriding sheet deflects the underriding sheet, ultimately causing it to break in flexure. In short, the failures are tensile failures, not compressional failures. Again, it is the weakest link that fails insofar as the tensile strength of sea ice is significantly less than its compressive strength. Figure 12.17 is

Figure 12.15. Continuation of ridge formation in thicker ice. Again, note the presence of cusp-shaped cracks on the underthrust sheet. Ice thickness estimated at 20–25 cm.

Figure 12.16. Ridge formation in even thicker ice, Beaufort Sea. Note the rotation of the underthrust block. Ice thickness estimated at 35–50 cm.

interesting as at location 1 the sheet on the right is clearly overriding the sheet on the left. At locations 2 and possibly also at 3, the sheet on the left appears to be overriding the right, although at 3 it is difficult to be absolutely certain. This points out another trend: the bigger the ridge the more difficult it is to ascertain exactly what is occurring. This is particularly true if one is standing on the ice, where one tends to get lost in the clutter of ice blocks. If one has the opportunity to view a ridge from a low- flying aircraft it is invariably much easier to see the overall geometry of the feature.

Figure 12.7c shows my understanding of the geometry of the long sinuous ridges that are frequently seen in the Arctic Ocean. When Figure 12.7c is compared with Figure 12.7a, after noting that the horizontal scales of (a) and (b) are different, one can see that the general patterns of large sinuous ridges are quite similar to finger rafting except that the regularity of the over-and-under patterns has significantly decreased. Even so, by carefully examining the surface pattern on large ridges that show sinuous surface traces, one can frequently estimate the extent of the over-

Figure 12.17. Ridge building in Viscount Melville Sound, Northwest Passage. Note that in the vicinity of number 1, the sheet on the right is the overriding sheet while in the vicinity of numbers 2 and 3, the overriding sheet appears to be on the left.

lap. I have found this to be frequently true even in FY ice that is nearly 2 m thick. Unfortunately this is not always the case. Figure 12.18 shows an aerial photo of a large compressional ridge in the Beaufort Sea. Although one rotated block is clearly visible located to the right of A, the over-under pattern of this ridge is not clear (much of the trace is relatively straight). Perhaps the scale of the deformation is sufficiently large that the initial structure has been destroyed via incorporation into the ridge. In the smaller ridge located near B, the nature of the overlapping structure is clear. I should also note that one time I examined finger rafting that had occurred in ice that was over 1 m thick at the time of the rafting. Clearly, this is an exception that proves the rule. The extent of the overlap was in excess of 50 m. The feature was spotted from the air where the telltale finger-rafting pattern was obvious. On landing I attempted to examine the hinges but I was unable to see what had happened because of broken blocks and drifted snow. I am also certain that if I had not initially seen these features from the air, I would have had considerable difficulty recognizing them as finger rafts on the ground. As the area where

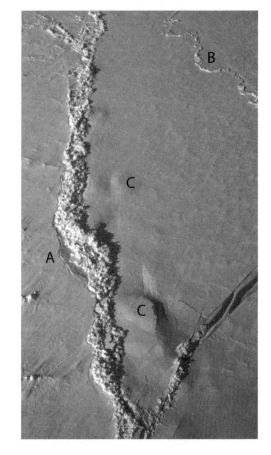

Figure 12.18. An aerial photograph of a large compressional ridge in the Beaufort Sea. In the smaller ridge located near B, the nature of the overlapping structure is clear. On the larger ridge near A, a rotating block can be seen. The surface bulges that can be seen near the letter C are believed to be indicate of the presence of podsov.

these rafts formed had originally been continuous fast ice that fractured and then came under pressure, the interacting sheets were of the same thickness. Vella and Wettlaufer (2008) have recently argued, based on plate theory, that the existence of rubble is essential for the formation of rafting, much less finger rafting in ice of up to 1 m in thickness. My field observations noted above suggest that matters may not be this simple.

A classic description of somewhat similar simple rafting in thick ice can be found in Zubov (1945), as described by Gordeev (1940). During this event the icebreaker *Sedov* found itself included in the overthrust portion of 2-m-thick ice. The total overlap was on the order of 30 to 40 m. When the thrusting was completed, divers were able to examine conditions beneath the ship where they found that the ice thickness exceeded 10 m. The *Sedov* case is similar to the experience of the *Manhattan* (see Figure 12.1), except that the ice underthrusting the *Sedov* was appreciably thicker.

What you are usually able to examine in the field is, of course, the sail of the ridge. Occasionally these can be quite impressive (Figure 12.19). As best I can tell from looking at this photograph, this ridge is largly comprised of thick FY ice. Figure 12.20 shows ridges that are partially comprised of MY ice. This was a very interesting site located in highly deformed ice in the Beaufort Sea. The relatively flat ice on which the helicopter has landed is obviously the remaining portion of a thick MY floe, i.e., the portion that has not been incorporated into the surrounding ridges. Note that although most of the deformed ice is FY, there are also blocks of thicker MY ice in the ridges. For instance, look just to the right of the helicopter. Also note the crack located in the MY ice in front of the helicopter. Clearly the MY ice was under considerable stress and, if the surrounding ridges had continued to grow, the MY floe would have gradually been incorporated into the surrounding ridges. If you had had the misfortune to have been camped on this floe while the deformation was underway, you would have had difficulty escaping, as the surrounding ice is extremely rough. We were only able to access the site via helicopter. To add to the difficulty,

Figure 12.19. A large pressure ridge sail as seen in the Beaufort Sea. Note that at least some of the ice blocks incorporated into the ridge are fairly thick and that the high point (~10 m) of the ridge is of a limited lateral extent.

Figure 12.20. Intense deformation, Beaufort Sea. Note that although most of the ridges are composed of thinner ice types, there are also blocks of MY ice in the ridges. The undeformed ice on which the helicopter is sitting is definitely MY.

the ice in the ridges would have been moving. It would have been like running a gauntlet of large moving gears where entrapment would not have improved your health.

The highest free-floating sail that I am aware of was located in the Beaufort Sea and had a free-board of 13 m (Kovacs et al. 1973). Although it was part of a larger ridge system, the high portion was quite localized: a hill, much like the pileup shown in Figure 12.19. As will be discussed later, really high sails primarily appear to occur in the nearshore area, where grounding is possible. Note the apparent bulges in the relatively undeformed ice near the letter C in Figure 12.18. They are probably the result of significant amounts of ice being thrust beneath the sheet at these locations.

This brings us to an important point that must be remembered in looking at any ridge. It is easy to think that the keel of a ridge is simply a scaled mirror of the sail. This is not necessarily so, as the two environments are very different. On the surface everything conspires to minimize the lateral dimensions of the sail: cold snow to push out of the way, freezing seawater acting like cement, and the weight of the blocks. Definitely high friction. In the keel the situation is completely different. Nothing is directly exposed to the arctic atmosphere. The only short-term cooling source is the residual cold in the ice being incorporated into the ridge. The amount of lubricating seawater is essentially limitless. Also, the force pressing a submerged block up against the underside of the overlying ice is proportional not to its weight but to its buoyancy force, which is equal to the weight of liquid displaced by the block. Inserting some reasonable numbers into these relations, it is easy to show that the force exerted by a submerged block against the underside of the ice is only ~13% of the force that would be exerted by the same block if it were placed on the upper surface of the ice. An easy way to convince yourself of this difference is to cut out two small blocks of sea ice of identical size. Pick one up and set it on the ice upper surface. Push the other one down and sideways so that it comes up on the underside of the surrounding ice. The difference in the degree of effort required is striking. Incidentally, pushing the cut blocks down and out of the way is commonly how access holes are prepared. One only expends the extra energy required to haul a block onto the surface if samples are needed.

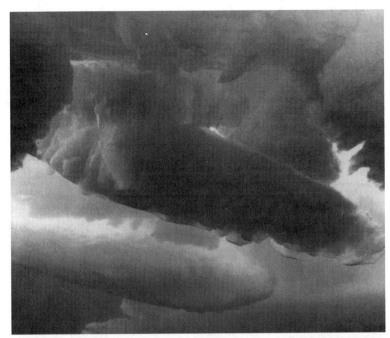

Figure 12.21. Pieces and plates of ice presumed to be in the keel of a ridge (Arctic Submarine Laboratory Files). The Russian term for such ice is podsov.

Figure 12.22. Podsov (i.e., pieces and plates of ice pushed under an ice sheet as the result of rafting and hummocking (AARI Files, courtesy of V. D. Gristschenko). Note the pronounced growth of frazil crystals in the edges and sides of these submerged plates.

Figure 12.21 is a view of a number of blocks in the underside of a ridge. The general look is quite different from that of the ice in the sail. In general, the lateral dimensions of the blocks appear larger and there appears to be less crushing than in sails. The Russians have a word for such ice, *podsov*. This term refers to the pieces and plates of ice that are pushed under an ice sheet as the result of rafting and hummocking (Borodachev et al. 2000). There is no equivalent phrase in English. Note in Figure 12.21 that the sides and edges of the blocks do not appear to be clean fracture surfaces, but instead look somewhat soft and fuzzy. This is probably the result of additional ice growth resulting from the residual cold in these downthrust blocks. This phenomenon is quite evident in Figure 12.22, where numerous frazil crystals can be seen forming on the downthrust blocks. Again, the podsov is not comprised of blocks of ice but of slabs whose lateral dimensions are much larger than their thickness.

12.3.1.4 Shear Ridges and Stamukhi   The deformation features that have been described above are all believed to be the result of ice movements that are primarily compressive (i.e., the direction of closure is roughly normal to the direction of the initial crack or lead). There are, however, ridges that have formed during deformation in which the relative motions are primarily parallel to the initial crack or lead. I will refer to these features as *shear ridges*. In the WMO no-menclature such ridges are simply grouped as a type of pressure ridge. However, it is my impres-sion that shear ridges are sufficiently different to warrant being considered separately. They are the sea ice equivalents of strike–slip faulting as seen, for instance, on the San Andreas fault and they have quite distinctive features. For instance, they can be quite long (tens of kilometers) and are either very straight or very slightly arcuate. They do not display the sinuous patterns charac-teristic of compressional ridges. Their cross-sectional shapes are also distinctive. Their landward side is frequently an irregular ramp leading up to the ridge crest. The seaward side is different: a vertical wall. Whether or not the verticality continues all the way to the bottom of the ridge is apparently unknown. Figures 12.23 and 12.24 are oblique aerial photographs of shear ridges occurring along the Beaufort Coast. Figure 12.23 shows the result of the intense deformation that frequently can be observed in this area. Figure 12.24 provides a closer view of the essentially vertical seaward side of this ridge type. Figure 12.25 is a photograph of the seaward side of a shear ridge presumed to be grounded in ~20 m of water in the Chukchi Sea offshore of Barrow, Alaska. This figure also shows another frequent feature of this class of ridges: the presence of slickensides on the wall indicating that the direction of relative motion was primarily horizontal. These fea-tures appear to correspond exactly to slickensides as seen in faulting in geological materials. The surface of the slickensides was grooved and glassy with a polished appearance. Whether or not shear ridges only occur when all or a portion of the ridge is grounded is unknown. Clearly many of them are at least partially grounded. In my experience, they do not appear to be common in the far offshore regions.

Figure 12.23. An aerial view of a large shear ridge, Beaufort Sea.

Figure 12.24. A closer view of the essentially vertical seaward side of a shear ridge, Beaufort Sea.

Figure 12.25. The seaward side of a large shear ridge located along the Chukchi coast near Barrow, Alaska (Weeks and Weller 1980; photograph courtesy of Buster Points).

Along coastlines with wide, comparatively shallow continental shelves, one frequently sees another quite spectacular ice deformation feature: so-called *stamukhi* (singular: *stamukha*). This Russian term, which has come into common usage in the non-Russian literature, refers to a large pileup of deformed sea ice that is grounded in water depths of up to 20 m or more. Heights (free-boards) of these features frequently exceed 10 m. Along the coast of the Canadian Archipelago, stamukha heights of up to 30 m have been reported (Taylor 1978). Sverdrup, Peary, and Stefansson have also reported ridge sails of 30 m from areas north of Greenland (see Zukriegel 1935). Stamukhi are truly impressive features: islands of deformed ice as can be seen in the oblique aerial photograph shown as Figure 12.26. In this figure the ridges comprising the stamukha generally appear to be oriented roughly perpendicular to its long axis. For descriptive purposes taking the top of the photograph to be north, two shear ridges can be tentatively seen running to the northwest from the southern tip of the stamukha. Note also the arcuate ridges running northwest from the northern tip of the stamukha. Two other linear deformation features bound portions of the stamukha on both its eastern and western sides. It appears to me that this stamukha developed in several stages. Clearly the deformation involved was major.

Figure 12.26. Stamukha along a line of shoals located to the WSW of Cross Island located north of Prudhoe Bay, Beaufort Sea. The longitudinal length of the stamukha is ~5 km. Note also the presence of arcuate portions of a shear ridge.

A very interesting example of a stamukha is the grounded sea ice feature located ~160 km to the west of Barrow, Alaska, in the Chukchi Sea (Kovacs, Gow, et al. 1976). This feature was apparently first noticed in 1971 by photo interpreters at the U.S. Navy Fleet Weather Facility during an examination of imagery from the NOAA-1 satellite. The clue that something unusual was occurring at this location was the existence of a polynya that appeared essentially motionless in a field of moving pack ice. Examination of the location by icebreaker and aircraft revealed the presence of a large area (> 30 km²) of grounded sea ice. Subsequent analysis of earlier satellite imagery indicated that this "island" of ice developed every winter over a shoal that rises to within 20 to 25 m of the sea surface and that the feature has been present since at least 1960. In the past this feature has occasionally survived the summer, although I would be surprised if this is true at present. The one time I was able to fly over this feature, although it was clear that the majority of the ice in the stamukha was normal sea ice, several ice island fragments were also included and well may have initiated the pileup by providing the initial grounding. Similar, but probably not identical, ice island fragments were noted by Dehn when he flew over the island earlier in 1972. Two of my close associates, Austin Kovacs and Tony Gow, had a very interesting visit to this site. They were flown in, planning to spend several nights investigating different aspects of the island. However, after one night they requested evacuation back to Barrow. There were several nontrivial problems. For one, the island was so rough that it was extremely difficult to move from one location to another. The only comparatively smooth ice was a narrow border around the edges of the island. Unfortunately, this ice was constantly breaking away and moving to the west with the Chukchi Sea ice pack. Taking such an excursion might have been interesting but it would not have been pleasant. Finally, it was very difficult to get any sleep because of the numerous other occupants of the island, polar bears. As you might expect from the discussion in the previous chapter, the fact that there was invariably open water and thin ice associated with this grounded feature gave seals easy access to the surface, thereby providing the bears with dinner. Clearly in the world of polar bears, the presence of the stamukha and the associated polynya was not a secret. A few days after Kovacs and Gow left the island, a bear with a neck diameter of 137 cm was tagged there. It was estimated that he weighed ~700 kg. Dining must have been excellent.

The above discussion brings up a question of sea ice jargon. Kovacs, Gow, et al. (1976) have referred to the feature as an island of grounded sea ice, while I have used the Russian term *stamukha*. In addition, Stringer and Barrett (1976) have applied the term *floeberg* to the feature. I agree that the feature is indeed an island of grounded sea ice. I also point out that by definition a stamukha is a large pileup of deformed sea ice that is grounded in water depths of up to 20 m or more. I therefore conclude that, as a stamukha is also an island of grounded sea ice, the two expressions are equivalent. So, take your pick. Do you want to use five English words or one Russian? I prefer the Russian term because it clearly has historical precedence. I also note that the U.S. Geological Survey group has used stamukhi in describing features of this type along the Beaufort coast. What Kovacs and I agree on is that the feature is not a floeberg. Kovacs, Gow, et al. (1976) have gone to the trouble of tracking down the origins of the term *floeberg*. It appears to have been first been used on the Nares expedition of 1875–1876. In 1887 Sir Clements R. Markham, who was a member of that expedition, defined floeberg as follows in the 1887 edition of the *Encylopedia Britannica*: "In the palaeocrystic…sea, there are floes from 80 to 100 feet thick—and the smaller pieces broken from them have been very appropriately named floebergs." WMO's more current definition is "a massive piece of sea ice composed of a hummock or a group of hummocks, frozen together and separated from any ice surroundings." As Kovacs et al. note, this implies that a floeberg is a piece of something even larger that is also presumably composed of a hummock or a group of hummocks. I take this to mean that although both large and impressive, floebergs are portions of even larger and presumably more impressive stamukhi.

As a result of the efforts of investigators from the U.S. Geological Survey, considerable information is currently available on the stamukhi located along the Beaufort Sea coast of Alaska between Prudhoe Bay and Barrow. Individuals interested in exploring this subject further should examine Reimnitz, Barnes, et al. (1972), Reimnitz, Toimil, and Barnes (1978), Reimnitz and Barnes (1974), and Reimnitz and Kempema (1984). Information concerning the internal structure of the large ridges and stamukhi occurring in the Pechora Sea, the Sea of Okhotsk, the Caspian Sea, the Sea of Azov, and the Russian coast of the Arctic Basin can be found in Kharitonov (2008). This paper also serves as a guide to a number of other Russian papers on related subjects.

12.3.1.5 Rubble Fields   Here I will use the expression *rubble field* to designate areas where essentially all the sea ice has been disturbed and the surface consists of irregularly broken slabs and blocks of sea ice. A photograph of a rubble field is shown in Figure 12.27. There appears to be no particular pattern to the deformation except that effectively everything is broken, creating an ice surface that, although not comprised of high ridges, is nevertheless extremely rough and very difficult to transit. The term *rubble field* contrasts with the expression *hummock field* that is sometimes used to describe this type of deformation in that rubble field does not imply that the surface is comprised of a series of separated hillocks. Frequently, when a surface pattern is evident, it more closely approximates plowed furrows than hummocks. In my experience, it is frequently difficult to perceive any surface pattern in these features. The extent and degree of development of rubble fields appear to vary greatly. Many rubble fields appear to have formed when

Figure 12.27. View of a rubble field located off the coast of the Beaufort Sea.

Figure 12.28. A rubble field located in the Chukchi Sea north of Barrow, Alaska.

the ice was quite thin (< 0.3 m). By looking closely at Figure 12.28, one can distinguish at least two different thicknesses in the rubble. Rubble fields composed entirely of MY ice do not appear to be common. If they exist, they certainly would be impressive. Occasionally during the formation of a rubble field, an individual slab will be pushed into a vertical position (Figure 12.29). The resulting feature is referred to using the Russian term *ropak*. Again, there is no English equivalent. At dusk with very low sun angles, a rubble field with an occasional ropak sticking up in the air can provide a rather unworldly appearance. Trying to cross a rubble field, particularly when its surface is covered with just enough new snow to obscure the underlying ice and not enough snow to support your weight is, at best, very time-consuming and at worst can easily result in a trip to a medical facility. A discussion of these difficulties plus a number of photographs of rubble can be found in Barker et al. (2006).

Figure 12.29. The vertical slab is referred to as a *ropak* in Russian (AARI files).

12.3.1.6 MY Ridges Although many deformation features do not survive the following summer's melt season, some do. These features gradually undergo a gradual metamorphism that is of interest for several different reasons. For one, surface melting gradually rounds off the edges of the blocks that comprise ridges. With time, the protruding portions of the blocks completely melt away and the surface of the ridge changes from rough to smooth. Experienced ice observers who fly repeatedly over a particular sea ice area are frequently able to utilize the degree of smoothness of the upper portions of ridges to differentiate between features that are FY, SY, and old. Usually after a ridge has survived more than three summers, a terminal smoothness is approached and such observations are less useful. For instance, in the ridge in Figure 12.30 one can still see remnants of its original blocky structure, whereas in Figure 12.31 this has all been erased. However, in both of these cases the elongation of these features indicates that they were originally ridges as opposed to localized pileups. Contrast these figures with Figure 12.32. There is no obvious pattern to the spatial distribution of these hummocks. There are two possible explanations. One is that the hummocks originated as blocks or piles of blocks in a rubble field and that they were later rounded to their present shapes. The other is that this topography is purely the result of localized differences in melting with no deformation involved. I definitely favor the first of these alternatives. Additional clues would be the internal structure of the hummocks and the degree of integration of the surface drainage. For such a surface to develop purely by differential melting,

Figure 12.30. An old pressure ridge that is at least two years old. Note that the blocky structure of the ridge can still be discerned.

Figure 12.31. An old ridge that is clearly older than the ridge shown in Figure 12.30. Note that all indications of the initial structure of the initial blocky structure of the ridge have been erased.

Figure 12.32. The hummocky surface of an old floe, Viscount Melville Sound, Northwest Passage. I think that the surface topography here originated as a rubble field and then was modified to its present state by surface melt and drainage development.

the surface drainage would have to be well integrated, allowing meltwater to drain off the floe. Otherwise, a rubble field would tend to smooth with time.

Although it is common to find blocks of MY ice in pressure ridges, this is usually the result of an FY pileup building on the edge of an MY floe. Ultimately, the asymmetric loading causes a crack to form that runs parallel to the edge of the MY floe. As the pileup continues, the broken MY segment is rotated and gradually incorporated into the ridge, at which time the process starts all over again. There are few descriptions of ridge formation in very thick ice. One of the best is by Yakovlev (1955), which will serve to end this descriptive section on ice deformation features. He states that "On 14 February 1951 we beheld a grandiose spectacle—the formation of pressure ridges in an old field. This was preceded by intense compression, as a result of which the old sealed cracks burst. After this the first pressure ridge arose at the eastern boundary of the field; it began to move quickly toward the center of the floe, grinding before it ice 3–4 m thick. After this a second ridge formed in front of the first, then a third, etc. The ridges moved gradually grinding whole sectors of ice between them. Huge blocks were broken off the field easily, they rose up under the ridges and piled one upon the other. A network of cracks appeared before the ridges.

The ridges reached a height of 7 m. Because of this powerful hummocking, the relief of the comparatively level old floe changed abruptly in several hours. In place of the level ice, a 'mountain country' arose with rows of ridges, between which was ground and hummocked ice."

Individuals who would like to examine further examples of ice deformation should refer to a number of papers by Kovacs and his coinvestigators. During the 1970s and early 1980s, as part of OCSEAP (Outer Continental Shelf Environmental Assessment Program), Kovacs was able to spend a considerable amount of time investigating the wide variety of both FY and MY deformation features that occur along the coasts of the Beaufort and Chukchi Seas. References of particular interest are as follows: Kovacs (1971, 1976, 1983b, 1984); Kovacs and Gow (1976); Kovacs and Mellor (1974); Kovacs and Sodhi (1980, 1988); Kovacs, Weeks, et al. (1973); and Kovacs, Gow, et al. (1976). There are also a number of interesting papers focused on FY ridging in the Baltic. For instance, see Kankaanpää (1991), Kraus (1930), Lensu et al. (1998), Leppäranta and Hakala (1992), and Palosuo (1974). Although the Kraus reference is a bit of a bibliographic obscurity, it is well worth examining as one of the earliest studies of impressive coastal ridges.

12.3.2 Antarctic

Although it is reasonable to assume that somewhere in the Southern Ocean, all of the ice deformation features discussed above occur, in general in the Antarctic things are different. If I were to guess where one might find arctic-type deformation features, it would be in the western part of the Weddell Sea. There very heavy ice consistently occurs and appreciable deformation might be anticipated as the result of the blocking effect of the Antarctic Peninsula. Around the remainder of the continent, the main factor affecting ice deformation is the presence of the Southern Ocean to the north. As stated previously, to a good approximation the antarctic sea ice area can be considered a marginal ice zone. Because of the vertical motions associated with the frequent penetration of heavy swell deep into the antarctic ice pack, rafting appears to be particularly common, with several similar ice thicknesses frequently being stacked on top of each other. Large floes appear to be rather rare in that they are frequently converted to small floes by flexing resulting from the passage of swell. These smaller floes then bang together, resulting in broken edges and rotated blocks. The result is a topography that might be described as cluttered and irregular. Long sinuous ridges appear to be rare. Also, as will be shown later, the heights of the ice pileups are appreciably lower. In contrasting the two regions one might say that in the Antarctic both the amount as well as the intensity of the deformation is less. A major factor here is the fact that, as antarctic sea ice moves to the north, the amount of space available to it in the Southern Ocean is continuously increasing. As a result the ice is generally diverging, a trend not conducive to either very frequent or very large amounts of deformation.

12.4 Properties of Individual Ridges

12.4.1 Arctic

Since the discovery of oil at Prudhoe Bay, Alaska, considerable effort has been expended on determining the properties of individual ridges, as a knowledge of these properties is essential to the design of offshore oil production structures for the shelf seas of the Arctic Ocean. Here under properties I include a variety of ridge characteristics related to their geometry, composition, and strength. Let me start by saying something about how these data were collected. As might be expected, a variety of techniques have been used.

For ridge geometry, simple drilling and side-looking sonar (SLS) have been the most frequently used methods. The trouble with drilling is simple: if the ridge is thick and the desired sampling interval is small, the total thickness of ice that must be penetrated becomes large. In some cases when a ridge is thick and a large number of holes are desired, hot water drills have been used. The problem here is that such specialized drilling systems are heavy to transport and time-consuming to set up. Attempting to obtain cores from large ridges is even more difficult. For starters, the coring position is frequently in a location that is difficult to access (e.g., on top of a tilted block near the top of the ridge sail). In addition, the collection of cores invariably results in the person performing the coring becoming soaked with seawater, which freezes, coating one's parka and gloves with ice. Finally, because in many FY ridges the blocks are poorly bonded together, one frequently finds that when one brings the core barrel back to the surface, it does not contain a core. Even if one is only attempting to profile the surface of a sail, floundering around trying to find a stable site on which to stand while avoiding dropping one's leg into one of the numerous voids between the blocks can be a significant challenge. I once saw Austin Kovacs completely disappear from sight as the result of unexpectedly dropping into a large void between some blocks in a snow-covered FY ridge. I won't tell you what he had to say about this event. (For more details on drilling and coring see Appendix D.)

The advantage of SLS is that it is usually less labor-intensive in the field. First, an access hole is cut in the undeformed and hopefully thin ice near a ridge, and an acoustic device (usually fixed on a metal pole) is lowered a variety of different distances below sea level. The shape of the ridge is then determined from the differences in the acoustic travel times between the "pinger" and the side of the ridge.

Although obtaining data in the field on ridge geometry, as well as finding where it hides in the literature, can be challenging, some recent publications have greatly simplified matters. Here I refer to the papers of Burden and Timco (1995) and Timco and Burden (1997). The first of these papers, entitled *A Catalogue of Sea Ice Ridges*, tabulates observations from 22 different sources on the characteristics of 176 ridges, of which 112 were FY and 64 were MY. The FY ridges were further separated into those studied and presumably formed in the Beaufort Sea (46) and those formed in more temperate regions (66), including Labrador Sea, Northumberland Strait, Bering Sea, Gulf and Bay of Bothnia, and Gulf of Finland. Particularly useful is the fact that these reports include a number of data sets that have not been published or widely circulated, and which were obtained from contract studies for oil companies. There is considerable scatter in the data. Even so, a number of trends are quite clear. Figure 12.33 gives a schematic of the shape of an

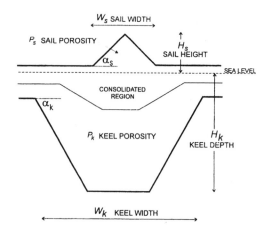

Figure 12.33. A schematic of the cross section of an FY ridge showing the definition of the terms used in the Timco and Burden (1997) analysis.

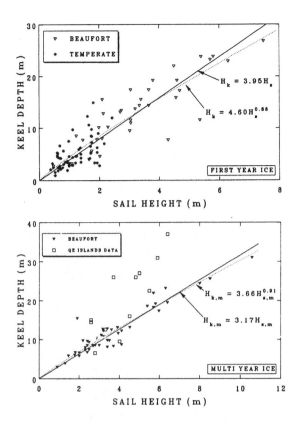

Figure 12.34. Keel depth versus sail height for FY ridges (Timco and Burden 1997).

Figure 12.35. Keel depth versus sail height for MY ridges (Timco and Burden 1997).

FY ridge showing the definitions of the terms used in the analysis. Figure 12.34 shows a plot of keel depth versus sail height for FY ridges. As can be seen, ridges with deep keels generally have high sails. Moreover, there appears to be no significant difference between ridges from the Beaufort Sea and those from more temperate regions except that, not surprisingly, all the very large ridges studied were from the Beaufort. An analysis of the variability of the keel-to-sail ratio for this same group of ridges shows that the distribution is clearly not normal, showing a pronounced positive skew. In that keel depth proved to be both related to the sail height and to the keel width, the keel width should also be related to the sail height. The data indicate that this is also true ($W_k = 14.85\,H_s$). Correlation coefficients were surprisingly high ($r^2 = 0.7$ to 0.9), considering the natural variability of the ridges studied. Equally important were the parameter pairs where no significant correlation was found: sail width and sail height, sail height (or keel depth) and level ice thickness, average keel angle and keel depth, and average sail angle and sail height.

Less information is available on MY ridges (64), of which most (53) were studied in the Beaufort Sea, with an additional 11 being studied in the Canadian Queen Elizabeth (QE) Islands. As with FY ridges, there are a wide range of shapes and sizes. In general, MY ridges tended to be broader and more rectangular than the more triangular shapes of FY ridges. Figure 12.35 shows a keel depth versus sail height plot for the MY data. Several observations are of interest. For one, by comparing Figure 12.34 with 12.35 one sees that for a given sail height, FY keels are ~8% deeper than MY keels. Assuming that when the MY ridges were less than one year old, they were similar to the ridges in Figure 12.34, this difference could be the result of relatively more keel ablation than sail ablation during summers. A factor here could be the ablative losses associated with the

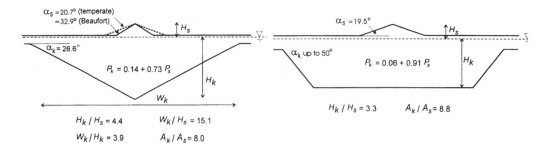

Figure 12.36. Diagrams showing the average geometric features of FY and MY ridges as summarized by Timco and Burden (1997).

"ice pump" process proposed by Lewis and Perkin (1986). It was also found that MY ridges from the Queen Elizabeth Islands area show deeper keels for given sail heights than do ridges from the nearby Beaufort Sea. Timco and Burden (1997) do not suggest an explanation. I find this difference rather surprising unless these ridges are grounded. However, this does not appear to be the case. Figure 12.36 summarizes the shape differences between FY and MY ridges.

Another interesting aspect of FY ridges is the relation between the sail height and the thickness of the ice being incorporated into the ridge. This relation has been studied by Tucker and Govoni (1981) and by Tucker, Sodhi, et al. (1984) in ridges located off the Alaskan coast of the Beaufort Sea. The authors reasoned that if the analysis of ridge mechanics by Parmerter and Coon (1972) proved to be generally correct, then it should be possible to document a relation between block thickness and ridge height from field observations. The Parmerter and Coon analysis suggested that the height of a compressional ridge depends on the strength of the host ice sheet. As the host is loaded with more and more blocks as the ridging continues, it ultimately fails. When this occurs the ridge begins to build laterally instead of vertically, with the blocks broken from the parent ice sheet becoming a part of the ridge. In that the critical bending stress depends on the ice thickness, there should be a relation between the thickness of the blocks in the ridge and the height of the ridge. Figure 12.37 shows the field results. Although there is considerable scatter, it is clear that, as anticipated, sail height is a function of the thickness of ice forming the ridge. The powers on the different fitted curves represent different assumed failure mechanisms ($t^{0.24}$ = flexural failure and $t^{0.5}$ = buckling failure). In fitting the relation $h = 5.24\ t^{0.5}$, only data points that exceeded the best fit relation ($h = 3.71\ t^{1/4}$) by more than one standard deviation were used (see the original paper for details). Although ice conditions differed appreciably between 1980 and 1981, when these studies were made, there does not appear to be a year-to-year difference in the relations between sail height and thickness. In their 1981 paper Tucker and Govoni based the justification for using a $h \propto at^{1/2}$ relation on a geometric argument. In the 1984 study they examined the functional relationships via the use of an energetic analysis that suggested that if the failures were the result of buckling, a $t^{1/2}$ relation would be expected. This was in contrast to flexure failures, where a $t^{1/4}$ relation should occur. They then take the observation that $h \propto at^{1/2}$ gives a better fit to the overall data set to support the contention that buckling may be an important mechanism in pressure ridge formation. Kovacs and Sodhi (1980) came to a similar conclusion in their study of grounded ridges and coastal ice thrusting. Incidentally their paper contains a thorough literature review of this type of phenomenon as well as many informative photographs.

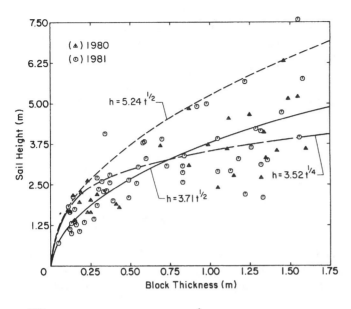

Figure 12.37. Sail height as a function of block thickness (Tucker, Sodhi, et al. 1984). For details concerning the different fitted curves see the current text.

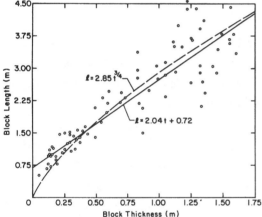

Figure 12.38. Block length versus block thickness observed in the ridges studied in the Beaufort Sea by Tucker, Sodhi, et al. (1984).

Although the Tucker et al. results can not be termed definitive, I see no reason to rule out buckling as a possible ridging mechanism. I also see no reason to change my earlier comment in this chapter that a person sleeping in the interior of a floe does not need to be concerned with becoming incorporated into a ridge unless a crack forms, thereby converting their tent site from the floe interior to the floe edge. In the jumble that comprises an active ridge, it is certainly possible to conceive of more than one failure mechanism operating simultaneously.

Another interesting result of the Tucker, Sodhi, et al. (1984) study can be seen in Figure 12.38, where the lengths (longest axes) of the blocks in the ridges that they studied are plotted as a function of block thickness. Clearly, the thicker blocks are longer. The best-fit linear relation gave a correlation coefficient of 0.91. Their analysis, using reasonable estimates for the mechanical properties of the ice, suggests that the observed block lengths are less than the lengths predicted for both flexural failures and buckling failures. They suggest that this result should not be taken as evidence that flexural and buckling failures do not occur. Instead, they believe that it is more likely that other failure modes such as shearing and crushing are also present. They also note that secondary failures occurring in the ridge would reduce the block sizes regardless of the initial failure mechanisms.

Several other ridge characteristics are of interest, specifically, their salinity, temperature, and porosity profiles. In regard to these matters FY and MY ridges are very different. In that FY ridges

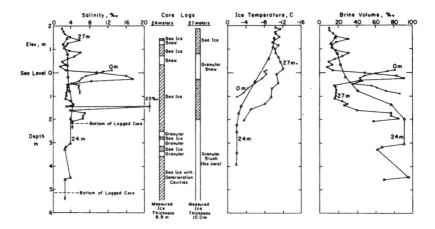

Figure 12.39. Salinity, temperature, and brine volume profiles plus core logs from an FY ridge located offshore of Barrow, Alaska (Weeks et al. 1971). Note that at both locations cored it was not possible to obtain samples from the lower portions of the ridge, a fact attesting to the difficulty of obtaining good core samples from recently formed ridges.

are commonly composed of blocks of comparatively thin FY sea ice piled together with some snow and seawater in the interstices, one would expect FY ridge salinity profiles to show considerable local variability. Soon after formation, the salinity profile is rather flat in that the ridging process would be expected to tend to average out the normal vertical trends found in undeformed FY ice. There do not appear to be many examples of profiles through FY ridges. One example can be found in Figure 12.39. Note the two salinity spikes. Note also that the lower portion of the ridge is essentially at the freezing temperature of seawater. This means that the blocks in the lower portion of the ridge are at best weakly frozen together, accounting for the difficulty in core recovery. As the ice in the ridge sail is located above sea level, brine there can readily drain downward. Therefore, salinity values in the upper portions of ridges would be expected to decrease more rapidly than comparable values in the above-sea-level portions of nearby undeformed ice that is similar to the ice incorporated into the ridge. A recent detailed study of the structure and composition of several FY ridges located in the northern portion of the Gulf of St. Lawrence is definitely to be recommended (Johnston and Barker 2000).

There appear to be only a few studies of the changes in the characteristics of a FY ridge as a function of time (Leppäranta, Lensu, et al. 1995; Høyland 2002a, 2002b). In that this clearly is a useful thing to do, one might wonder why such studies are so rare. Two obvious reasons are logistical difficulties and problems resulting in the loss of the study site, an event that can occur for a number of reasons. Even in the northern Baltic, where it is comparatively easy to implement field operations, carrying out such a study takes considerable planning, dedication, and a bit of luck. The results of the study are quite interesting. First one should note that winter seawater salinities in the study region are ~3.5‰. This means that the initial salinities of the ice in the ridge were probably ~1‰ (see Gow et al. 1992). Such ice can hardly be considered to be typical sea ice, which would be expected to have salinity values more in the range of 5–8‰. Nevertheless, the study provides a very general insight into the compositional changes that would be expected in more saline arctic ridges. For example, considerable brine drainage occurred in the upper portion of the sail, resulting in very low salinities. Even at a depth of 0.5 m, salinities were only in the range of 0.2–0.3‰. The ridge formed on 22 January (1991) and broke up on 8 May. During this period, it was possible to visit the site three different times and to document the considerable structural evolution that occurred. For instance, as the ridge aged its external geometry became smoother. In addition, the thickness of the consolidated layer within the ridge, where the interblock sea-water had frozen, increased from 0.52 to 0.93 to 1.02 m. When these values are compared to the thickness of the surrounding level ice determined at the same time (0.31 to 0.55 to 0.58 m), ridging results in a consolidated core that is somewhat thicker than the nearby undisturbed level ice. It was possible to obtain good agreement between the observed consolidated layer growth and estimated values obtained using a one-dimensional heat flow model based on observed air or ice surface temperatures and the growth of nearby undeformed sea ice. At the same time there was a decrease in the overall bulk porosity of the ridge (0.281 to 0.203 to 0.175). Contributing factors here are believed to be melting and gradual consolidation in the upper portion of the ridge and mechanical erosion and improved packing in the lower portion. Using data collected by Leppäranta and Hakala (1992) in the Baltic and also by Beketsky et al. (1996) in Sakhalin, Timco and Burden (1997) found that the relation between the bulk porosity of FY ridge sails P_s and that of the associated keels P_k could be expressed by the simple relation $P_k = 0.14 + 0.73 P_s$. In that the sample size is small (10) and there is considerable difference in the average values for the two regions studied, I would suggest that this relation be used with caution.

The profile characteristics of multiyear ridges have received considerable attention. As just noted, the exceptions to this occur at locations such as the Baltic, where MY ridges do not exist as the ice there does not survive the summer. The primary reason for this heightened interest is practical. Although large FY ridges can be formidable obstacles that must be considered in the design of polar shipping and offshore oil production platforms, the fact that the ice blocks in the lower portions of such ridges are poorly frozen together makes such ridges less formidable design objects than the massive, well-bonded, low-salinity ice typically found in old ridges. For instance, look at Figure 12.40. Salinity values in the above-sea-level portions of the ridge are very low and the general profile shape is similar to profiles observed in thick undeformed MY ice (see the discussion concerning MY salinity profiles in Chapter 8). Note that the salinity of the below-sea-level portion of the ridge is ~2‰, a value that is even lower than typically found in the lower parts of undeformed MY ice. Why this occurs is as follows. During the summer, very low-salinity meltwater from the above-sea-level portions of the ridge percolates downward into the inter-block interstices. Presumably a freshwater–seawater interface forms that gradually reaches lower and lower within the ice as the surface melt continues. Clearly this is an unstable situation in that the low-salinity interstitial water has a freezing point of ~0°C and is located at sites where the temperature is dominated by the freezing point of normal seawater (i.e. –1.8°C). Ultimately, the interstices become filled with essentially freshwater ice. The end result is a ridge that has a core of very low-salinity ice. Furthermore, all the interblock voids that existed when the ridge initially formed are filled with ice, thereby welding together the blocky structure of the ridge. The fact that MY ridges typically have massive, low-salinity cores was first noted by Russian icebreaker captains who found that breaking through them was very difficult, usually involving extensive backing and ramming (Burke 1940). I note, however, that Burke's explanation of the origin of

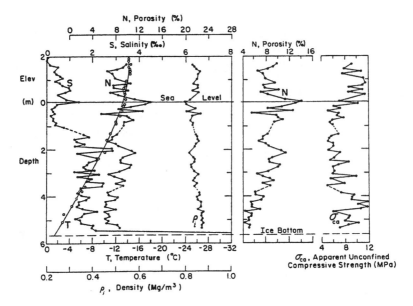

Figure 12.40. Salinity, porosity, temperature, and brine-free density profiles of the ice in an SY ridge located along the Beaufort Sea coast of Alaska (Kovacs 1983a). Also shown are the apparent unconfined compressive strength and the porosity of the ice at –10°C.

such ridge cores appears to be incorrect in that he suggested that such ridge cores were the result of intense compression during the initial ridge formation process. Although there is definitely some crushing during ridge formation, it is not a major failure mechanism. Furthermore, it is now well established from drilling that newly formed compression ridges have appreciable interblock void volumes. When one travels across sea ice areas where MY floes are common, it is frequently possible to examine MY ridges that have been split in a direction transverse to the main ridge axis. During OCSEAP, Austin Kovacs and I were able to examine a number of such ridge cross sections. Invariably one could see the original blocky structure resulting from the ridging process as well as establish that all the original intrablock voids were now filled with ice. Figure 12.41 provides an example of such a cross section. I have been told that a ridge must pass through two melt seasons before the voids in ridge sails become completely filled with ice (Alan Gill, pers. comm. 1972). I would guess that the voids in the below-sea-level portions of a ridge fill more rapidly. However, I know of no direct observations concerning differences in these rates. Additional MY profile data can be found in Kovacs et al. (1973) and Wright et al. (1978). Recent thermodynamic model development relating to both the consolidation and the melting of pressure ridges can be found in Marchenko (2008), who varied his input parameters in order to simulate variations in natural conditions.

12.4.2 Antarctic

Although numerous authors have noted in passing that FY ridging in the Antarctic appears different than in the Arctic, this impression has not been supported by in-situ fieldwork until the recent study by Tin and Jeffries (2003). The following comments are based on an examination of 42 deformed sea ice features observed in almost 19 km of profile data obtained by in-situ drilling. This is in contrast to the study of Arctic ridges by Timco and Burden (1997), which was based

Figure 12.41. Photograph of the side of a stamukha fragment (i.e., a floeberg) located in the Beaufort Sea. Note the blocky structure associated with the individual ice blocks comprising the stamukhi and also that large-scale voids appear to be absent (Kovacs and Gow 1976).

on an examination of the geometric properties of 112 FY ridges. It is important to note that these two data sets were collected for quite different reasons. In the Arctic the focus was on large ridges as entities that could cause peak forces on ships and offshore structures. Investigators went in the field, looked around, and selected several well-defined, impressive ridges for study. In the Antarctic, the purpose of the study was not specifically to study ridges but to gain information on variations in sea ice thickness. Later the ice thickness profiles were examined and found to contain 42 features that could be considered ridges. In doing this Tin and Jeffries found that the commonly used Rayleigh ridge selection criterion was not adequate for their purposes. (The Rayleigh criterion takes individual ridges to be features that have a sail higher than some defined cutoff value and contain a crest enclosed by troughs on both sides which descend at least halfway towards the local level ice horizon.) The problem was that it identified narrow peaks within extensive deformation features as individual ridges. Similar problems with the Rayleigh criterion have been noted by Melling and Riedel (1995) in their study of sea ice draft via the use of moored subsea upward-looking sonar. In fact, they suggest that the Rayleigh criterion is not an effective ridge discriminator for keels with drafts of less than 5 m. In that the mean draft of the Tin and Jeffries data set was only 0.82 m, a new identification scheme was needed. The scheme used was based on the assumption that all FY ice that is not level is the result of deformation, assuming that small-scale surface features not associated with ridging have been eliminated by the use of a suitable cutoff value. As will be seen in the following section, the cutoff value that they chose (0.3 m) was appreciably smaller than values that have been used by other authors in studies using lasers and sonar profilers. The important point here is that it is important to be certain that the differences in the results of different studies are not simply caused by how different authors define a ridge.

They note that in the Antarctic, ridges appear to be more point-type roughness features like hillocks. They also frequently appear as short linear features that define the edges of floes and are composed of a small number of fairly thick blocks. The long, sinuous ridges frequently seen in the Arctic appear to be singularly lacking. When the Tin and Jeffries results are compared with the Timco and Burden data set, it is found that although the antarctic ridges are smaller than those observed in the Beaufort Sea, they are similar to ridges observed in the subarctic. In addition, antarctic ridges tend to be flatter, with slope angles of nearly half the values observed in the Arctic. The ratio between keel area and sail area also appears to be nearly four times higher in the Antarctic than in the Arctic. Finally, when the relation between the keel depth and the thickness of the level ice that is forming the ridge is examined for antarctic sites, Tin and Jeffries found the following striking differences from results obtained at other localities. For example, Tucker, Sodhi, et al. (1984) and Melling and Riedel (1996) found that their data sets relating the thickness of the level ice (L) forming a ridge and its keel depth could be summarized as $H_k = aL^{0.5}$ where the constant a had values of 20 and 16 respectively. For a similar set of ridges studied in the Baltic by Kankaanpää (1991), a equaled 15. Results in the Antarctic gave an a of 5.

Tin and Jeffries offer three possible explanations for these differences. There are differences in sampling, in parent ice characteristics such as strength, and in the duration and magnitude of the driving forces. Although I agree that these are all factors, I suggest that there is an additional factor that should be added thanks to the differences in the geographic arrangement of the land masses in the two polar regions. The Arctic Ocean is essentially surrounded by land. It is also clogged by thicker ice as it is farther north ($\sim > 75°N$) than the Southern Ocean is south ($\sim < 70°S$). When storms occur in the north, the ice has nowhere to go and intense deformation occurs. In the Southern Ocean the ice is largely unrestricted in that it can always move to the north. Additional space is always available, large stresses occur less frequently, and the amount of deformation is accordingly less. The final point to be made here is that the current information strongly supports the conclusions reached by the earliest explorers: that sea ice there is less deformed than in the Arctic and that the style of the deformation is also different. Additional evidence in support of these conclusions will be found in the next section.

12.5 Remote Profiling

Before 1966, although there was some discussion in the literature concerning the distribution of values obtained from the measurement of the physical characteristics of sea ice such as its salinity and strength, there were very few observations relating to the thicknesses of undeformed ice in the pack, much less as to the distribution of ice thicknesses in deformed areas. The problem was, of course, not a lack of interest but a lack of methods for collecting suitable long-track data. Nevertheless it was possible from the early literature to piece together what might be expected when suitable sampling techniques became available. For instance, Nansen (1897) found that in the central Arctic Basin, although there were numerous 5-to-7-m-high ridges, ridges with heights of 8 m were rare and ridges with heights of 10 m were not observed. In addition the highest ridge surveyed from the *Sedov* was only 6.1 m, whereas the highest ridge sighted by Gakkel' and Khmyznikov (1938) in the Chukchi Sea was 7.2 m. This general range of maximum ridge heights was also supported by the observational data collected by the Birdseye flights over the Arctic Ocean (Wittmann and Schule 1966). Considering that the number of smaller ridges appeared to be very large, one could anticipate that the probability distribution function (PDF) would show a strong positive skew and quite possibly would prove to be some form of a negative exponential. In short, it was reasonable to assume that the PDF would prove to be nonnormal. What was needed were methods for rapidly profiling both the upper and lower surfaces of the pack so that data sets could be obtained against which different PDFs could be tested as representative. Such methods are now available.

12.5.1 Laser Profilometry

Suitable methods for rapidly profiling the upper surface of sea ice became available in the 1960s with the installation of laser distance-measuring systems on low-flying aircraft. These systems accurately provide a record of travel time versus the distance between the aircraft and the upper surface of the ice. The initial laser profilometry was carried out with a Spectra-Physics Geodolite 3A laser profiler operated by NAVOCEANO from the Birdseye aircraft (Ketchum 1971). The instrument produces a continuous-wave, coherent light source that transmits an amplitude-modulated laser beam. A measurement of the phase delay between the transmitted and the reflected light is used to determine the distance to the target. As the width of the transmitted light beam is 10^{-4} radians, this corresponds to a 3-cm-diameter spot on the ice surface if the aircraft is flying at an altitude of 300 m. The response time of the instrument is 10 millisec. Although the distance accuracy of the laser when operated at a height of 330 m has been reported to be 3 cm, tests reported by Mock et al. (1973) suggest that 10 cm is probably a realistic estimate. This accuracy is comparable to that possible via the use of high-quality photogrammetric techniques.

The problems with the results of laser profilometer flights are as follows. Although the results appear to be a straight-line track across the ice, this is not the case as the result of the pitch and yaw of the aircraft. In fact, the exact location of the sampled "line" is frequently unknown even if photographs of the track are available. In addition, the range output signal varies from zero to full-scale as the measured distance from the aircraft increases by one range step. In that this range step is frequently set at a value of a few meters in order to obtain high-resolution measurements of small-scale features of the surface roughness, this means that if the distance between the aircraft and the ice surface increases beyond one step, the instrument output rapidly steps back to zero before continuing its increase. It is, of course, necessary to remove these low-frequency steps from the record as they are the result of gradual changes in the elevation of the aircraft and have nothing to do with actual surface changes on the ice.

In that the aircraft motion is generally low-frequency whereas the ice roughness is high-frequency, one might think that the aircraft motion problem could easily be handled by using a variety of high-pass filtering techniques. Indeed, for some purposes such as spectral analysis, this has proved to be adequate (Hibler and Le Schack 1972). However, it has not proved to be adequate when one wishes to study the distribution of ridge heights, as there is an overlap between the surface roughness spectrum, which has a finite variance at very low frequencies, and the aircraft motion spectrum. As a result a high-pass filter removes a part of the roughness profile and in particular underestimates the heights of the higher ridges. A way around this problem has been developed by Hibler (1972), who used the fact that the ice surface profile is basically a one-sided noise trace in that the roughness always rises from approximately sea level. As a result, if a high-pass filter is applied to the initial profile, the resulting profile can be processed to obtain a series of minimum points. Although these minimum points will probably not have exactly the same elevation (the sea surface), their elevations will differ far less than will the ice surface elevations.

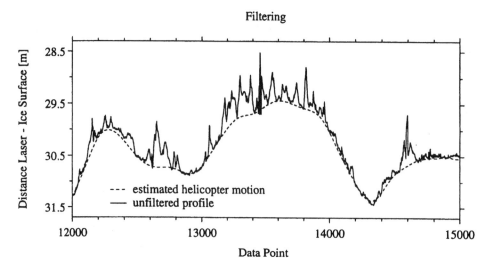

Figure 12.42. An example of unprocessed laser profiler data (the solid line) from the Weddell Sea, Antarctica (Dierking 1995), as well as the helicopter motion (the dashed line) estimated using procedures initially suggested by Hibler (1972a). The corrected ice surface profile is obtained by subtracting the two curves.

Therefore, a profile drawn through them would contain significantly less low-frequency variance to be lost in a high-pass filtering operation. The curve of minimum points is then low-pass filtered to obtain an estimate of aircraft motion that is, in turn, subtracted from the original profile to obtain the corrected profile. This process appears to work particularly well in ridged areas as the computer normally selects minimums that are located on either sides of pressure ridges. In addition, field observations show that such locations are commonly depressed below sea level with the occurrence of subsequent flooding. Figure 12.42 shows the results of applying Hibler's three-step filtering process to a series of laser measurements made in the Weddell Sea using a helicopter as a platform (Dierking 1995). The solid line (a) is the unfiltered profile and the dashed line, curve (b), is the estimated aircraft motion. The corrected ice surface profile is then obtained by subtracting curve (b) from curve (a).

The first problem that arises once laser profiles of the upper surface of the ice are ready for analysis is that one must decide how a ridge should be defined. There have been two different approaches. The first, most common, approach uses the so-called Rayleigh criterion discussed in the previous section, which defines a ridge as a local maximum that is at least twice as high as the neighboring minima. As such it ensures that a broad deformed area that shows multiple peaks will be counted as one independent peak instead of as several. However, not all authors have found the Rayleigh criterion to be adequate for their purposes (Tin and Jeffries 2003). In addition, a cutoff value is usually defined whose purpose is to exclude small roughness features such as sastrugi that should not be considered as ridges in that they are not the result of deformation. Typical cutoffs for laser profiles have values that range between 0.5 and 1.0 m.

12.5.2 Laser Results

There is not as much laser data as one might expect, given that it clearly is more cost-effective to operate a laser from an airplane than to operate sonar from a submarine. In addition, aircraft can

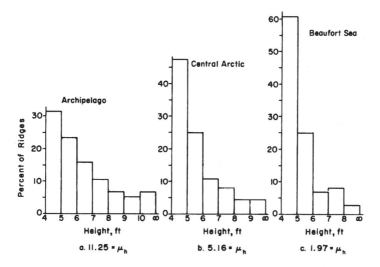

Figure 12.43. Ridge height distributions observed in February 1973 (Hibler 1975). Each distribution was taken from a laser track that was ca. 40 km long. The locations are approximately (a) 83°N, 85°W; (b) 87°N, 162°W, and (c) 70°N, 139°W. The cutoff value used was 4 feet (1.22 m).

operate in areas near coasts where shallow water limits submarine operation. Even so, there are sufficient laser data available to permit one to examine differences between the Arctic and the Antarctic, as well as seasonal and regional differences in the Arctic.

First, what do laser results look like? Figure 12.43 shows ridge height distributions for three different areas of the Arctic (Hibler 1975). As can be seen, the falloff with increasing height is clearly exponential in nature. There are also significant differences in the distributions from the different regions, with the region located slightly north of the Canadian Archipelago being the most highly deformed (i.e., largest number of ridges per km [11.25] having heights above 1.22 m plus comparatively more high ridges). In binning laser data, several factors must be considered. If the sampled length is too short, the number of ridges measured will be too small to provide adequate statistical estimates of the parent population. If it is too long, interesting spatial details in the amount of ridging will be lost. In studies along the north coast of Alaska, where appreciable differences in the degree of ridging occur, 20 km was found to be a useful sampling length for each bin. Note that in Figure 12.43, 40 km was used. One must next decide what operational definition should be used in defining a ridge. As noted earlier, the most commonly used criterion is the Rayleigh. Other possibilities are the so-called two-feet criterion (a peak must be at least two feet above nearby minimum points; Hibler 1975), and the "I know a ridge when I see one" criterion. The important point here is that if one wishes to make comparisons with other data sets, it is important to use the same criterion.

Next, if one wishes to summarize the characteristics of the data in each bin, one has to decide what distribution function should be taken as representing the data. There have been two main possibilities here. The first was suggested by Hibler et al. (1972), who derived the function via a variational calculation based on two assumptions regarding the nature of ridging: first, that all ridge height arrangements yielding the same net deformation are equally likely and second, that all ridge cross sections are similar in a geometric sense. Furthermore, it was assumed that the cross-sectional areas of all ridges were equal to the square of the ridge height times a constant proportionality factor. The resulting relation gives the following probability density function (PDF) p for the ridge height h:

$$p\left(h; h_0, \lambda_1\right) = 2\sqrt{\frac{\lambda_1}{\pi}}\; \frac{\exp\left(-\lambda_1 h^2\right)}{\mathit{erfc}\left(h_0\sqrt{\lambda_1}\right)}, \quad h > h_0 \tag{12.4}$$

where h_0 is the minimum ridge height considered (i.e., the cutoff height), λ_1 is the distribution shape parameter, and erfc is the complementary error function. The distribution shape parameter λ_1 is related to the average ridge height μ_h by

$$\mu_h = \frac{\exp\left(-\lambda_1 h_0^2\right)}{\sqrt{\lambda_1 \pi}\; \mathit{erfc}\left(h_0\sqrt{\lambda_1}\right)}. \tag{12.5}$$

When tested against both sonar and laser data, this model gave reasonable fits and was used by Hibler, Mock, and Tucker (1974) for several years.

An alternate approach was also suggested by Wadhams (1976a, 1978a). Here the distribution, chosen empirically, was of the form

$$p\left(h; h_0, \lambda_2\right) = \lambda_2 \exp\left[-\lambda_2\left(h - h_0\right)\right], \; h > h_0. \tag{12.6}$$

As above, h_0 is the cutoff height. In equation 12.6 the distribution shape parameter λ_2 is the inverse of the ridge height standard deviation σ_h and the average ridge height is

$$\bar{\mu}_h = h_0 + \lambda_2^{-1}. \tag{12.7}$$

This distribution has two important advantages. First, it is easier to use in that no iterative solution is required, and second, it appears to fit the field observations on the high ends of the distributions better than the Hibler model. As a result this is the approach that has been most frequently used. Its only drawback is that it is purely an empirical choice. One should also note that equation 12.7 is a special case of the Hibler model if the ridges are assumed to be rectangular in cross section.

Dierking (1995) has discussed the pros and cons of the Hibler and Wadhams distributions in some detail, pointing out that it has been suspected for some time that the model function that gives the best fit to a ridge height distribution is to some degree dependent on the choice of the ridge identification criterion (Lowry and Wadhams 1979). A first guess here (Wadhams 1980) was that if a ridge is defined as a local maximum exceeding some cutoff value, then a distribution of the Hibler type $[\exp(-\lambda_1 h^2)]$ would be favored. On the other hand, if a Rayleigh criterion is used to select the ridges, then the simpler exponential distribution $[\exp(-\lambda_2 h)]$ would be preferred. Dierking points out that things do not appear to be that simple in that both his and the Lytle and Ackley (1991) antarctic data sets found that the Hibler distribution gave the better fit, even though the Rayleigh criterion was used as the selection criterion. He suggests that a factor may be whether the ridges studied are primarily single ridges whose heights are determined by a combination of ice thickness and ice strength, as contrasted with multiple ridges that have reached a limiting height and are now building laterally in contrast to vertically, a process that would result in several closely spaced crests. If such lower crests are ignored by the Rayleigh criterion, smaller ridge heights will be counted less frequently than would be the case if a simple threshold algorithm were used. In such a case, both the frequency and structure of multiple ridges will play an important role in determining whether the Wadhams or the Hibler distribution gives the best fit. Dierking suggests that both of the above distributions should be considered as working approximations that hopefully will lead to a more realistic ridge height distribution that also considers both the ice properties and the ridge formation processes.

More recently a quite different approach has been used to explain the roughly exponential nature of the tail of the sea ice thickness distribution. After noting that the sea ice thickness statistics are the result of an interplay between two different processes, the thermal and the mechanical, Thorndike (2000) considers the evolution of a large number of ice "particles." During each time step, ice thickness changes both as the result of the thermodynamic factors and deformation. When deformation occurs, the new thickness becomes the sum of the thickness of the two interacting particles. This essentially assumes that the primary deformation mechanism is by rafting. Which ice particles participate in the deformation process is not a function of thickness, as is usually assumed; it is only proportional to their areal abundance. An essential feature of the model is that there are many combinations of thicknesses that can result in a ridge of a given total thickness. Qualitatively the tail of the thickness distribution can be understood as thin ice piling on thicker ice in a sequence of events, each of which has a very low probability. To get very thick ice, such as is observed at the exponential tail of the thickness distribution, requires a sequence of several improbable events to occur within the thermal relaxation time. The thicker the ice, the longer the sequence of events and the lower the probability. This view of the thickness distribution combined with sequential ULS profiles obtained during submarine cruises lends further support to the observational conclusion stressed in the present book, that rafting plays an extremely important role in the naturally occurring deformation within the ice pack in the Arctic (Babko et al. 2002).

The one other distribution of importance here is the distribution of ridge spacings. Again, there are two possibilities. As pointed out by Hibler et al. (1972), if ridges are randomly distributed along the sampling track, then the distribution of ridge spacings would be expected to be

$$p(s; h_0, \lambda_3) = \lambda_3 \exp(-\lambda_3 s) \quad h > h_0 \tag{12.8}$$

where h_0 is again the cutoff height and λ_3 is the distribution shape parameter. For such a distribution the average spacing is $\mu_s = \lambda_3^{-1}$. However, Wadhams and Davy (1986) and Lewis et al. (1993) have found that for the arctic data sets that they have examined, a log-normal distribution gives a significantly improved distribution. In such a case, the log-normal PDF is

$$p(s; h_0, \theta, \mu_{\ln}) = \frac{\exp\{-[\ln(s-\theta) - \mu_{\ln}]/2\sigma_{\ln}^2\}}{\sqrt{2\pi}(s-\theta)\sigma_{\ln}} \quad s > \theta, h > h_0. \tag{12.9}$$

Here θ is the cutoff and $\mu_{\ln}$ and $\sigma_{\ln}$ are the mean and standard deviations of $\ln(s-\theta)$. The average ridge spacing is $\theta + \exp[\mu_{\ln} + \sigma_{\ln}^2/2]$. Equation 12.9 is applicable if $\ln(s-\theta)$ is normally distributed. Dierking (1995) has also examined his Weddell Sea data set with respect to these matters and found that the log-normal relation definitely gave a better fit. As noted by Cramer (1946), a log-normal distribution is to be expected if the effect of an event that causes a decrease in the distance between two ridges is directly proportional to the distance between them. An alternate way to think about these matters is to assume that if there were no lead formation, the distribution of ridge spacings would be random and the simple exponential would be observed. Once divergence occurs associated with the formation of leads, this then extends the tail of the distribution in that the larger the ridge, the larger the average distance between ridges and the more likely this distance would include a large number of leads, presumably favoring a log-normal distribution.

In examining laser data on lateral variations in ridging, a number of different parameters have been used, in addition to the obvious ones such as $\bar{\mu}_h$ the mean ridge height, $\bar{\mu}_\#$ the mean number of ridges per km, and the latter's inverse $\bar{\mu}_s$ the mean ridge spacing. These other parameters include different versions of the so-called ridging intensity γ, specifically,

$$\gamma_1 = \mu_h / \mu_s \qquad (12.10)$$

a term that is proportional to the aerodynamic form drag produced by the ridges, and

$$\gamma_2 = \mu_{h^2} / \mu_s \qquad (12.11)$$

which is related to the average thickness of ridged ice (Lewis et al. 1993). A related parameter γ_3 that has been used replaces μs in equation 12.11 with λ_1, the distribution shape parameter from the Hibler distribution (see equation 12.4). Finally, a somewhat simpler parameter that is not conceptually tied to the Hibler distribution, although also initially suggested by Hibler, Mock, et al. (1974), is

$$\gamma_4 = \bar{\mu}_\# \left(\bar{\mu}_h\right)^2 \cot\theta \qquad (12.12)$$

where θ is the assumed ridge slope angle and γ_4 is the area of deformed ice under the laser path, which is proportional to the topside volume of deformed ice along the track. Because laser profiles cross ridges at a variety of angles, values used for θ are usually appreciably less than slope angles measured normal to ridge axes during in-situ studies. If the crossing angles are random, then the volume of deformed ice per unit area above sea level is $(\pi/2)\gamma_4$ and an estimate of the total volume of deformed ice per unit area in ridges is $10(\pi/2)\gamma_4$, a value that assumes a 9 to 1 ratio of the amount of ice in keels to that in sails.

12.5.2.1 Arctic   One might think that over the years the Arctic would have become a nearly continuous crisscross of laser tracks. This is definitely not the case. Most of the data are from nearshore areas. Typically, areas sampled have been those where either ship traffic or offshore oil exploration are likely. Longline profiles in the central portion of the Arctic Basin are comparatively rare, although some data are available that were collected during the U.S. Navy Birdseye flights. Fortunately there is enough data available to allow one to make some general statements relative to ridging patterns. What are not available are sufficient repeat tracks over time to allow one to examine trends.

Perhaps the most intensely sampled area is off the north coast of Alaska (Tucker et al. 1979) and offshore of the Mackenzie Delta in the western Canadian Arctic (Wadhams 1976a). In the

Alaska study, replicate 200-km tracks were flown oriented approximately normal to the coast off six locations varying from Point Lay in the west to Barter Island (Kaktovik) near the Canadian border. Flights were made in sequence in February, April, August, and December. However, because laser sampling is cloud- and fog-limited, not all flights were possible with a total of 17 track lines completed. In the data analysis, ridges were discriminated using the Rayleigh criterion and a cutoff value of 0.9 m. A systematic seasonal variation in $\bar{\mu}_h$ was observed, with values being lowest in the summer and early winter (1.1 to 1.2 m), increasing to values as high as 1.8 m by late winter. However, at any given time $\bar{\mu}_h$ did not appear to show a strong systematic spatial variation. In general, $\bar{\mu}_\#$ values were higher in the Beaufort Sea than in the Chukchi Sea. The largest $\bar{\mu}_\#$ values typically occurred 20 to 60 km off the coast. The patterns in the variation in the values of the ridging intensity γ_4 (equation 12.12) were found to be quite similar to those observed of $\bar{\mu}_\#$. This is simply a statement of the fact that at a given time when the ridging intensity or the amount of deformed ice increases, it is not because the ridges are appreciably higher; it is because they are more numerous.

The Wadhams (1976a) study concentrated on the offshore region from Barter Island on the west to the near-coastal area off the west coast of Banks Island on the east. As the cutoff value was 0.9 m, it is possible to compare his results directly with those of Tucker et al. (1979), although it is important to note that the data sets were from different years. Most of Wadhams' data were collected during the summer. This is unfortunate in that in the Tucker study, the only summer profile is from Barrow. As a result these two data sets do not overlap either spatially or temporally. The Barrow mean height values of ~1.25 m are slightly lower than in most of the data collected by Wadhams, whose values ranged from slightly less than 1.2 m to 1.63 m. Interestingly, the majority of the highest mean height values occurred in the nearshore areas to the north of the Tuktoyaktuk Peninsula. When presented on a plot of $\bar{\mu}_h$ versus $\bar{\mu}_\#$ (Figure 12.44), these values appear to be anomalously high. It would be interesting to know how many of these ridges were grounded. This figure also shows that if these anomalously high values are excluded, the remainder of the data suggest a simple linear increase in mean ridge height as the number of ridges per kilometer increase. Similar trends have been observed by Gonin (1960), based on stereo air photography, and by Hibler et al. (1972), using ULS data. Clearly, the highest ridges appear to occur where the ice is most highly deformed. Wadhams was also able to obtain a limited amount of data during the late winter of April 1975, with the sampling line running north from the Mackenzie Delta along the 135°W longitude

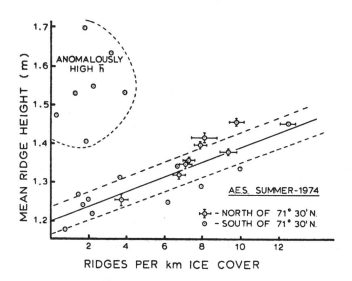

Figure 12.44. Mean ridge height values plotted against ridge frequency (ridges per km of ice cover) based on laser profilometer data collected during the summer of 1974 in the Canadian portion of the Beaufort Sea (Wadhams 1976a). The cutoff value used was 0.91 m.

line from ~70° to 76°N. There is a gradual increase in the number of ridges per kilometer as one moves to the north. However, there is no clearly defined highly deformed zone near the coast, as was frequently the case in the Tucker et al. data set.

The only regional view of ridging based on laser data is to be found in Hibler et al. (1974b). Some of these results are shown in Figure 12.45. The parameter plotted here is the ridging intensity parameter γ_3 (see the discussion between equations 12.11 and 12.12), which utilizes the shape parameter λ_1 from the Hibler distribution. As stated earlier, this parameter is believed to be directly proportional to the volume of ice occurring in ridges. The γ_3 patterns are quite interesting in that they show that the ridging intensity varies significantly with season. They also show that for a given season and location, γ_3 values can vary appreciably from year to year. In addition, the ice located just to the north of Greenland and the Canadian Arctic Archipelago clearly appears to be more highly deformed than the ice in the central Arctic Basin, which, in turn, is more deformed than the ice in either the Beaufort or the Chukchi Sea. The existence of this highly deformed zone of thick ice north of Greenland and the Canadian Arctic Archipelago has been noted in both exploration narratives and aerial observations, and appears to be a fairly consistent feature both

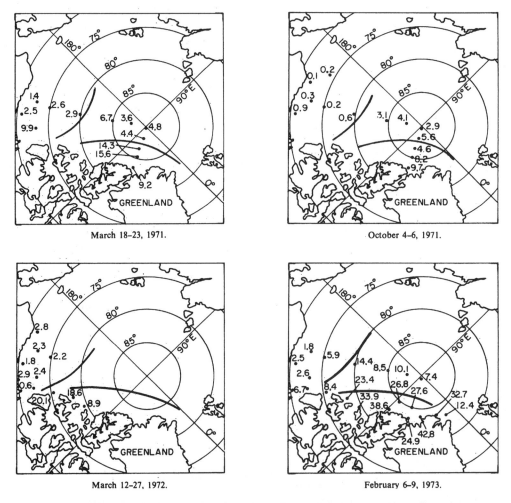

Figure 12.45. Distribution of ridging intensity γ_3 values observed in the Arctic Ocean during 1971–1973 for ridges higher than 1.22 m (Hibler 1975). Units are m^2km^{-1}.

historically and at present. Recently its existence has been noted in comparative laser studies in conjunction with the satellite ICESat (Forsberg and Skourup 2005; Hvidegaard and Forsberg 2002). The probable occurrence of such a highly deformed coastal zone is also supported by a modeling study by Rothrock (1975a, 1975b) that predicts decreasing ice stress (which presumably inversely correlates with deformation as one proceeds from Svalbard to Alaska), plus a zone of maximum shear to the north of the Archipelago. As Hibler et al. (1974b) note, the fact that high ridging occurs just north of the Archipelago is hardly surprising as this is the location where the ice in the Pacific Gyre encounters the fact that the Archipelago is comparatively motionless, moving a few cm/yr, as contrasted with the ice, which moves a few km/day.

There are no observations on the Russian side of the basin. This is an artifact of the Cold War in that U.S. and Canadian planes were not able to land in Siberia to refuel. To the best of my knowledge, this still is a data void. Perhaps when this void is filled it will be possible to relate the ridging intensity more closely to the circulation patterns within the Basin. More will be said about these matters in the discussion of ULS data and in the chapter on modeling.

12.5.2.2 Antarctic   Profiling data from the Antarctic, although limited in both extent and in temporal coverage, are nevertheless sufficient to allow some regional comparisons to be made. At present, there are four data sources, three of which are derived from laser profilers (Weeks, Ackley, et al. 1989 [Ross Sea], and Dierking 1995 and Granberg and Leppäranta 1999 [Weddell Sea]). The fourth study (Lytle and Ackley 1991), which also was located in the Weddell Sea, utilized a shipboard acoustic sounder. The Rayleigh criterion was used as a ridge selector in all four studies. I will start with the Ross Sea study as the ice there is presumably more generally representative of antarctic pack ice. The platform was a C-130 aircraft, the total line track imaged was 2696 km, the cutoff value was 0.91 m, and the observations were made during the late winter (4–15 November 1980). As was observed in the Arctic, the frequency distribution of individual ridge heights was well described by a negative exponential (Figure 12.46a). There appeared to be no systematic regional variation in ridge heights. This was also the case for ridge frequency, with the exception that the track nearest the east coast of Victoria Land showed appreciably higher frequencies than the other tracks. This presumably was the result of the blocking effect of the coast. The distribution of ridge frequency values was strongly positively skewed, although tests for log normality were not made. The differences between these results and similar measurements made in the Arctic are pronounced. For instance, in the Beaufort Sea, a 3-m-high ridge is encountered on the average every ~55 ridges; in the Ross Sea there is one every ~960 ridges. Furthermore, the ridge probabilities for the Ross Sea in the late winter, when ridges would be expected to be most intense, are lower than those observed in the Chukchi Sea in the summer (Figure 12.46b).

The Lytle and Ackley (1991) and Granberg and Leppäranta (1999) Weddell Sea observations were both made in its eastern portions, well away from the heavy ice of the western Weddell. In fact, the Lytle and Ackley location, at between 2 and 8°E to the north of Queen Maud Land, is a bit to the east of the Weddell proper, whereas the Granberg and Leppäranta location at 27°W is more centrally located. The timings of these two measurement sets are, however, quite different, in that they were made in the late winter (August 1986) and early summer (December/January

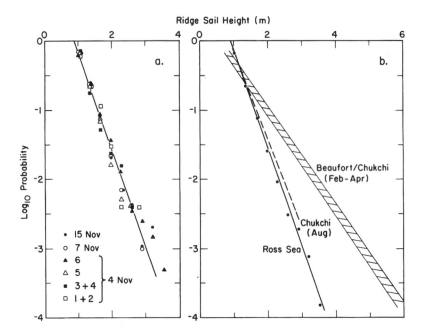

Figure 12.46. Log–linear plots of the probability density function of ridge heights for (a) the different Ross Sea sampling runs, and (b) the Ross Sea data set treated as a whole (Weeks, Ackley, et al. 1989). Flights were made in the late winter. Also shown in (b) are the results from similar data sets obtained in the Beaufort and Chukchi Seas in late winter and in the Chukchi Sea in the summer (Tucker et al. 1979).

1989), respectively. The cutoff values were 0.75 and 0.91 m, meaning that only the Granberg and Leppäranta values are directly comparable to the Ross Sea values. In the central Weddell both the ridge height and frequency values were somewhat larger than the values observed in the Ross Sea. Although frequencies were similar to values reported in the Arctic, the ridge heights were much smaller. In the far eastern Weddell, height and frequency values were similar to those from the Ross Sea. In all cases, the height frequency data were found to be a negative exponential. The log-normal distribution for fitting the ridge frequency data worked well in the central Weddell but was less adequate in the Lytle and Ackley study. In all cases, the correlation between the mean height of the ridges and the frequency of ridge occurrence was limited.

The Dierking (1995) study, which occurred during the winter of 1992, cuts across the Weddell, ranging in latitude from 60 to 72°S and in longitude from 50°W to 0°W. Again the heavy ice area of the western Weddell was avoided. The cutoff at 0.8 m was between the values used on the other studies. The height values were well described by a negative exponential and the frequency distributions by a log-normal relation. Again heavy ridging was most common near coastlines. Although Dierking found an increase in the mean height of the ridges as the frequency of ridge occurrence increases, there was considerable scatter in the data. Mean height values ranged from 1.09 m to 1.38 m. The number of ridges per kilometer showed considerable variation (3.2 to 52.6). Figure 12.47 is a frequency–height plot prepared by Granberg and Leppäranta that summarizes the results of the different antarctic laser studies. Note that the value termed Dierking-rough was based on values determined within a few kilometers of the coast. Also included are comparable data from the central Arctic, offshore Greenland, and the Baltic.

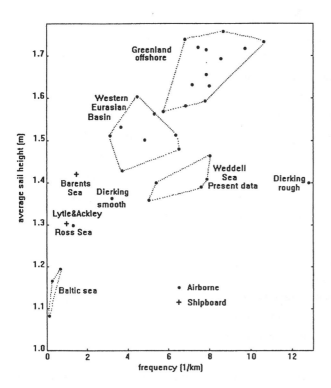

Figure 12.47. Frequency–height diagram of ridging (Granberg and Leppäranta 1999), including values from the central Arctic (Wadhams 1981a), Baltic Sea (Lewis et al. 1993), Barents Sea (Overgaard et al. 1983), Ross Sea (Weeks, Ackley, et al. 1989), and the Weddell Sea (Dierking 1995; Granberg and Leppäranta 1999; Lytle and Ackley 1991) labeled as present data). The cutoff is 1 m.

Although we will be able to add to our understanding of ridges in the Arctic by examining ULS data, the Antarctic offers us no such advantage as there have been no under-ice submarine cruises in the Southern Ocean. Nevertheless the laser data allow us to come to several conclusions concerning the similarities and differences between ridging in the Arctic and the Antarctic. These are as follows:

Similarities

- In all cases, ridge sail height distributions are negative exponentials in type.
- Ridge sail spacings appear to follow a log-normal distribution, although there are some exceptions.
- Mean ridge heights are seasonally dependent, with the lowest values occurring in the late summer and early fall and the highest values occurring during the late winter.
- At a given time, the most highly deformed ice commonly occurs near coastlines.
- There appear to be consistent regional differences in ridging intensity in both the Arctic and the Antarctic.

Differences

- Both the mean ridge height and the ridge frequency are generally higher in the Arctic than in the Antarctic.
- The positive correlation between ridge frequency and ridge height appears to be less strong in the Antarctic than in the Arctic. This difference could be the result of the fact that rafting appears to be appreciably more common in the Antarctic than in the Arctic.

12.5.3 Upward-Looking Sonar (ULS)

ULS profiles of the undersurface of sea ice suffer from many of the same problems discussed in the laser profiling section, as submarines also pitch and roll as they cruise beneath the ice. Fortunately the under-ice environment is more stable than the atmospheric. Submarines are also much larger and heavier than aircraft such as the Twin Otters that have frequently been used as laser platforms. In addition, submarines characteristically operate less far removed from the ice surface that is being studied. Therefore, irregular and large platform motions are less of a problem in ULS observations. The principal problem with ULS data is caused by the larger cone diameter of the sonar pulse. This was particularly true of data from early submarine cruises in that they used sonar systems operating in the kHz range with large cone angles (~20°). As a result, the sonar "illuminated" an area of 70 m² at sea level, when the echo sounder was at a depth of 45 m, and 390 m² when the echo sounder was at 105 m. The problem with this is that the ice producing the first echo is not necessarily directly overhead and that narrow stretches of thin ice will possibly not show at all in the records. The overall effects of such large beam widths are (Wadhams 1984):

1. overestimation of the mean ice draft;
2. underestimation of the number of pressure ridges;
3. underestimation of pressure ridge slopes and distortion of their shapes; and
4. loss of information on fine-scale spatial roughness.

Fortunately the absolute draft of a ridge is correct, providing that the ridge does not become lost by being merged with another ridge.

In addition, the early under-ice cruises recorded the ULS data on paper charts. This meant that further analysis required costly and time-consuming conversion of the analog records to a digital format. A particularly difficult problem occurs when one wishes to study temporal changes in ridging and sea ice thickness distributions by comparing the results of recent cruises that utilize sonar systems, which have cone angles of 2 to 3 degrees and thereby minimize the cone angle effects, with earlier results obtained in the 1950s and 1960s. In this case, not only must the older records be digitized, but the records must be resampled to create records that are comparable. A discussion of this problem and a procedure for creating comparable records can be found in Wensnahan and Rothrock (2005). Until recently the largest problem with the earlier U.S. cruises had been gaining access to the data, in that only a limited amount of data from these cruises had been released. Fortunately, the Royal Navy decided to promptly release the ULS data collected on their arctic cruises. As a result, much of what has been known about the ice thickness in the Arctic has been based on their comparatively few cruises. Credit for analyzing these data goes to Peter Wadhams and his associates at Cambridge University. Individuals particularly interested in ridging should study his discussion of this subject in his book (Wadhams 2000), as well as more detailed treatments in his research papers, several of which will be referenced in the following discussion. The Wensnahan and Rothrock paper notes that it will soon be possible to begin to make realistic comparisons between the results of the recent SCICEX cruises made in the early 1990s and at least some of the older data. In that the results of recent studies (Rothrock et al. 1999) clearly suggest that appreciable thinning of the arctic ice pack is under way, comparisons with older ULS data are extremely valuable. This material will be discussed in more detail in Chapter 18.

12.5.3.1 ULS Results The procedures used to analyze ULS data are generally similar to those already described in the discussion of laser data. The ridge selection criterion used is commonly the Rayleigh criterion. The maximum points that are selected as a first step in removing the vertical motion of the submarine are either vertical distances to open water or to the lower surface of thin FY ice that occurs in recent leads. In that maximum keel depths are appreciably larger than sail heights by a factor of ~4.5 (Figure 12.39), cutoff values in ULS studies have commonly been larger (~5 m) than values used in laser studies. While laser studies have largely concentrated on ridges, ULS studies have taken a more holistic approach and attempted to look at the ice thickness distribution as a whole. Figure 12.48 shows some representative probability density functions of ice draft obtained at locations near the North Pole (85–90°N) along longitude 70°W (Wadhams 1981a, 2000). Here I will start by looking at the tail of the distribution, the portion that is the result of ridging.

As can be seen in Figure 12.48, after peaking at an ice thickness in the range of 3 to 6 m, values representative of the thickness of undeformed MY ice, the occurrence probability of the thicker, presumably deformed ice falls off in an exponential manner. As with ridge sails, this portion of the distribution has been found to be well described by a simple negative exponential function (equation 12.6; McLaren et al. 1984; Wadhams 1992; Wadhams et al. 1985; Wadhams and Horne 1980). Again as with ridge sails, the spacings of ridge keels have been found to follow a log-normal distribution (equation 12.9; Key and McLaren 1989; Wadhams and Davy 1986). Figure 12.49 shows a fit of wide-beam keel spacing data from the Eurasian Basin using log-normal functions with different thresholds (0 and 15 m). In general, good fits have been obtained using either a 2-parameter or a 3-parameter log-normal distribution with the threshold set to either twice the spacing of successive data points or to the surface beam diameter, whichever is the greatest (Wadhams 2000).

The great majority of ULS studies are similar to laser profiles in that they both are simply line tracks across the lower or the upper topography of the sea ice. This makes the interpretation of parameters such as ridge slopes and widths difficult without making additional assumptions. For instance, in a 1978(a) study, Wadhams found that it was necessary to assume that ridge orientations were horizontally random in order to make corrections. Fortunately, one particularly interesting study made using data collected during a 1987 cruise (Davis and Wadhams 1995) has been able to avoid this

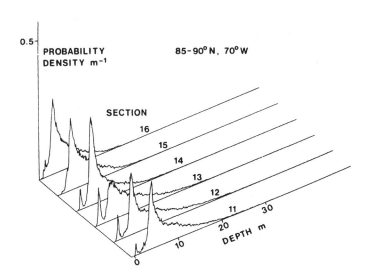

Figure 12.48. Representative probability density functions of ice drafts from the Arctic Basin (Wadhams 1981a, 2000). Each function is derived from 100 km of submarine upward-looking sonar profiles.

Figure 12.49. An example of wideband keel spacing data from the Eurasian Basin fitted by a log-normal distribution. The open circles have a threshold of zero whereas the closed circles have a threshold of 15 m (Wadhams and Davy 1986, Wadhams 2000).

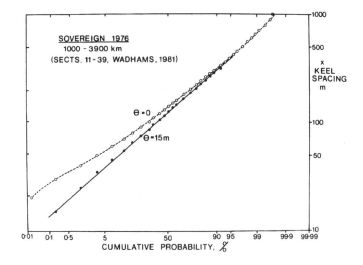

problem by utilizing a combined analysis of ULS and side-scan sonar (S-SS). This appears to have been the first cruise where S-SS was utilized (Wadhams and Martin 1990). In the study, a large number of ridges (729) were identified that were observed on both systems. It was then possible to convert observed keel slopes to real keel slopes. As noted above, both keel width and the keel slope values were well described by log-normal distributions with correlation coefficients greater than 0.95 and 0.99, respectively.

Useful trends based on ULS observations are as follows:

- keel widths and slopes fit a log-normal distribution;
- keels are wider than sails;
- ridge slopes are variable;
- lead widths can be fit to a power law using an exponent of 1.45 for leads less than 100 m wide and 2.50 for wider leads; and
- lead spacings can be fit by a simple negative exponential at moderate spacings (400–1500 m); however, at larger and smaller spacings there is an excess of lead pairs (Wadhams 2000).

The most recent study of the SCICEX ULS data set is by Percival et al. (2008), who examined variations in the mean draft of the ice $(\bar{H})$ as obtained via a variety of different sampling patterns on the length (L) of the profiled segments. They found that a nonstandard statistical model was necessary in that the variance (σ_L^2) of $\bar{H}$ falls off more slowly than L^{-1}, the autocorrelation sequence does not fall rapidly to zero, and the spectrum does not flatten with increasing wave number. As a result ice draft exhibits so-called long-range dependence; a condition that can be modeled by assuming a fractionally differenced process where σ_L^2 is proportional to $L^{-1+\delta}$ where δ is an observationally determined constant. They report that, based on the SCICEX data set, for L values of 50 km, $\delta = 0.29$ independent of the mean draft. Values for a variety of sampling patterns are also provided.

One might note that there is no section on submarine-based ULS results from the Southern Ocean. The reason for this absence is political as opposed to technical, in that the Antarctic is a nuclear-free zone. In addition, the presence of frequent deep-draft shelf icebergs would definitely complicate data collection along longline sampling tracks.

One final difference between laser and ULS data should be noted. Analysis of laser data has been focused on determining the number of pressure ridges and their characteristics. Although this has also been an important aspect of ULS studies, ULS data have also contributed significantly to the current understanding of changes in the ice thickness distribution in the Arctic, another subject discussed in Chapter 18.

12.6 Models

There have been two general approaches to modeling deformation in sea ice. First there have been the mathematical models, starting with the kinematic models of ridging and rafting developed by Parmerter and Coon (1972) and Parmerter (1975). These were followed in the 1990s by the development of dynamic models of ridging and rafting by Hopkins et al. (1991) and Hopkins (1994, 1998). This, in turn, led to attempts to test the models by comparisons with both field observations (Lensu et al. 1998) and ice tank tests (Hopkins et al. 1999; Tuhkuri et al. 1998). In the following I will try to provide the reader with a sense of these efforts. I will not attempt to describe the formal developments in detail. I will start by discussing rafting, although sequentially the kinematic ridging model was developed first. My reasons are that this appears to be a more natural sequence and also that the rafting development appears to be less arbitrary.

12.6.1 Rafting Models

As Parmerter (1975) points out, rafting starts when two ice sheets of equal thickness t are forced into contact along their edges, assumed to be linear, by an external force F. Although the edges of each sheet are assumed to be smooth and straight, they do not necessarily have to be vertical. Let ϕ and ϕ' be the angles between the two edges and the vertical as shown in Figure 12.50. If the lead opens and closes without shearing, then $\phi' = -\phi$. In such a case, the sheet in which $\phi > 0$ will become the overriding sheet. Parmerter's discussion also applies to the more general case where shearing has occurred along the lead before it closes, in which case $-\phi \leq \phi' \leq \phi$. This is the case diagrammed in Figure 12.53 where the sheet on the left will override the sheet on the right. Here the interaction forces per unit length are the axial force F, the vertical forces $P_1 = P_2 = F\tan(\phi - \theta_2)$ and the moment $M_1 = M_2 = Ft/2$. Sign conventions are indicated in the figure. The ice sheets are then treated as plates on elastic foundations provided that they are not lifted completely out of the water or completely submerged. In that the height of the ice out of the water is ξ_h, where h is the ice thickness, $\xi = (\rho_w - \rho_i)/\rho_w \approx 0.1$, and ρ_i and ρ_w are the bulk densities of the ice and seawater, respectively, the beam on the left will not be lifted completely out of the water whereas the beam on the right will be partially submerged. As noted, the submerged portion of the sheet (indicated in Figure 12.50 by the encircled 1) can no longer be treated as a beam on an elastic foundation, but must be considered as a beam under a combination of axial loading F, transverse loading due to buoyancy q, end shear P_1, and end moment M_1. As can be seen in the original paper, these computations can become complex.

The resulting analysis suggests a number of quite interesting conclusions. These are as follows:

Figure 12.50. The initial stages of simple rafting. The upper figure represents the initial contact of the two sheets. The middle figure represents an intermediate stage in the process and the lower figure represents the incipient stage of actual rafting (Parmerter 1975). In the figure F is the axial force, δ and θ are the deflections and rotations at the ends of the beams, and M_1, M_2, and M_3 are moments.

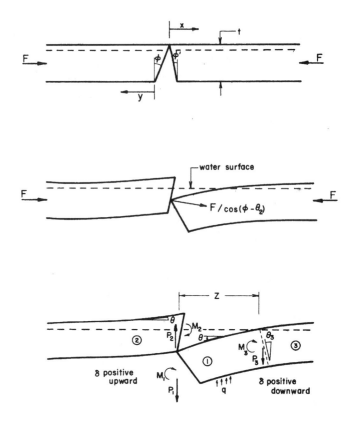

1. The force required to raft ice is only about 40% of the force required to buckle a semi-infinite plate on an elastic foundation. This is in excellent agreement with the field observations that rafting is very common in thin ice whereas buckling is comparatively rare. (This does not mean that buckling is impossible under all situations. Where it can occur is at locations where higher stresses are possible, for instance, in cases where thinner ice is deforming against MY ice or ice is deforming against fixed objects such as rocky islands or offshore structures; Kovacs and Sodhi [1980, 1988].)

2. The maximum stress always occurs in the submerged portion of the ice sheet and increases as the thickness of the sheet increases.

3. Simple rafting can only occur provided that the stresses resulting from the process do not exceed the fracture levels. Insofar as fracturing leads to ridging instead of rafting, it is therefore possible to estimate the maximum thickness under which rafting can occur as greater thicknesses will lead to stresses that will fracture the ice. The suggested relation is

$$h_{max} = 14.2 \frac{\left(1 - \mu^2\right)}{\rho_w g} \cdot \frac{\sigma_t^2}{E} . \qquad (12.13)$$

Here g is the gravitational acceleration and μ, σ_t, and E are, respectively, Poisson's ratio, the tensile strength, and the elastic modulus of the ice. By introducing reasonable values for these parameters into the above equation, Parmerter obtained an estimate of ~17 cm for the maximum ice thickness in which simple rafting can occur provided both interacting sheets are of the same thickness. This value is in good agreement with the earlier field observations of Weeks and Kovacs (1970), who noted that the rafting-to-ridging transition frequently occurred at ice thicknesses of roughly 20 cm. No explanation is offered for how rafting might occasionally occur in thicker ice. An exploration of the possibility that the presence of rubble can ease the development of rafting in thicker ice can be found in Vella and Wettlaufer (2008).

Recently Tuhkuri et al. (1998) have described a series of scale model tests that explore different aspects of rafting and ridging. Scale model testing is a technique frequently used in engineering studies of sea ice mechanics to deal with complex problems to which formal solutions are not, as yet, available. Froude scaling is typically used. In this approach similitude between the ratios of inertial and gravitational forces in the model and full scale are required. For instance, if the geometric scale factor is λ, the forces should scale as λ^3, velocities as $\sqrt{\lambda}$ and the flexural strength and the elastic modulus as λ. The problem is producing a modeled ice with the suitable scaled characteristics. A variety of different recipes have been utilized in the past, some with proprietary ingredients admixed with a bit of voodoo. In the present paper both the procedures for preparing the model ice and its scaled properties are well described. The authors suggest that the results can be interpreted as model tests in which $\lambda \approx 10$.

It was found that when both interacting ice sheets were of the same thickness, only finger rafting occurred. This was the case even though the experiments were designed to simulate simple rafting. Although the ice sheets did fracture during the lifting/submerging process, the fractured floes always maintained their geometric arrangement and ridging did not occur. This was true even when thicknesses were in the ridging range as suggested by Parmerter's criteria. I would think that important factors here should be whether or not the upper ice sheet fractured and, if it did, the condition of the upper surface of the underthrust ice sheet. For instance, if in the model the upper surface of the underthrust sheet was slippery, the overthrust sheet might hold together even if it was fractured. Under similar conditions in the real world, the presence of cold snow on the top of the underthrust sheet would commonly provide a more highly frictional environment resulting in the rotation of blocks and the formation of ridges. The other major result of the above scale model studies was that when the thicknesses of the interacting ice sheets were significantly different, ridging invariably occurred. These results will be discussed in more detail after the discussion of numerical ridging models.

12.6.2 Ridging Models

The initial ridging model developed by Parmerter and Coon (1972) was not the first model of a pressure ridge. In fact, the initial ridge model appears to have been suggested by Makarov (1901). Others were contributed by Burke (1940) and by Wittmann and Schule (1966). However, all these attempts are simply static descriptions of a conceptual representative ridge shape and, as such, are similar to the models developed by Timco and Burden (1997) that are shown in Figure 12.36. The Parmerter and Coon model was quite different in that it is kinematic. By this I mean that it considers the different processes that move ice blocks vertically and laterally in conjunction with a force balance and breaking stress calculations. As a result it provides a step-by-step description of the growth of a ridge. More importantly, it produces ridge profiles and predicts ridge heights that can be compared with field observations. It also provides estimates of the necessary driving forces that are within the range of the driving forces believed to occur in real sea

Figure 12.51. The flow diagram for the ridging model of Parmerter and Coon (1972). The symbols are as follows: P and M are the stations at the edge of the ice sheet and at the center line of the rubble filled lead, respectively; θ and φ are the angles of repose above and below the ice; D is the distance between stations; h is the ice thickness; a is the fraction of rubble placed under the sheet; β and y are the fractions of ice transferred laterally below and above the ice; σ and σ_{cr} are the stress and fracture stress of the sea ice; ρ_w and ρ_i are the densities of seawater and ice; E is Young's modulus; v is Poisson's ratio; X and Z are measures of the rubble thicknesses both above and below the ice sheet; and Y is the deflection of the ice sheet.

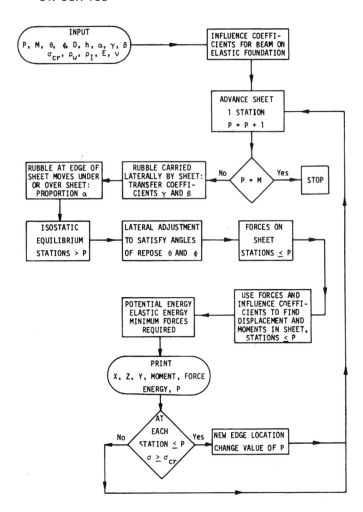

ice. The model implicitly assumes that ridges are formed from blocks broken from the parent sheet by bending. This bending is, in turn, the result of the unbalanced vertical forces that occur when blocks are thrust upon the surface of the sheet. The process is assumed to begin when two ice sheets of equal thickness move toward each other, closing a lead that has straight parallel sides. Within the lead, the ice consists of broken pieces of ice, i.e., rubble, that is of the same thickness as the ice on either sides of the lead. At $t = 0$ it is assumed that the edges of the lead have moved together sufficiently so that the accumulating rubble has reached the same thickness h as the surrounding floes. Subsequently, the closing proceeds in a series of equal steps in both space and time. As can be seen in Figure 12.51, which shows a flow diagram, the model requires a number of input parameters in addition to those that describe the response of the sheet as a beam on an elastic foundation. These include, for instance, the proportion a of the rubble that moves over (and under) the sheet at the lead edge, and the coefficients that describe both the lateral transfer of rubble by the sheet and the angles of repose of the rubble in the sail and in the keel. During each step, the forces that the displaced rubble exert on the sheet are calculated and, if the bending stresses exceed the strength of the ice, the ice is assumed to fail at the location where the stress is exceeded. When this occurs all stations to the right of the break are taken as being full of rubble. In that this model is two-dimensional (i.e., conditions along the lead are taken as invariate), the

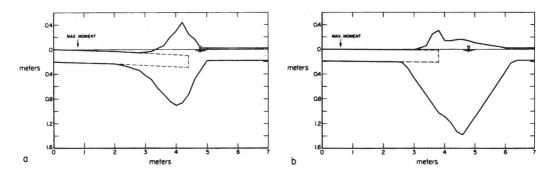

Figure 12.52. Computed profiles for ridges composed of sea ice 20 cm thick (Parmerter and Coon 1972).

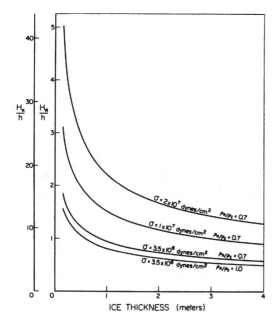

Figure 12.53. Limiting heights of sails (H_R) and keels (H_K) normalized against thickness vs. ice sheet thickness.

plate solutions are identical to the solutions for a beam on an elastic foundation (Hetenyi 1946), with the plate rigidity replacing the beam stiffness.

Although at first glance this model may appear to be a bit contrived, its results offer considerable insight into the ridging process. The forces necessary to drive the model are within the range of forces expected in nature. Also, the model produces an apparently endless variety of ridge shapes that are similar to those observed in nature. For instance, Figure 12.52 shows two profiles computed for an ice thickness of 20 cm. Note the misalignments of the sails and keels. Note also that the ridges are obviously not in isostatic equilibrium. Both misalignments and nonisostatic adjustment are frequently observed characteristics of real ridges. It was found that although the value of α used in the majority of the simulations was 0.5, consistent with finger rafting observations where overriding and underriding are roughly 50:50, realistic profiles were obtained with α varying within the range of 0.4 to 0.8.

Finally, the model has proven to be useful in examining what was referred to as a limit cycle, where this term refers to cases where the central rubble pile reaches a certain height and then grows laterally into a rubble field. Results of such limit calculations show that the limiting thicknesses of both sails and keels gradually increase as the thickness of the participating ice sheet increases. This result is in excellent agreement with the field observations of Tucker, Sodhi, and Govoni (1984). Furthermore, if these limiting heights are normalized against thickness, it can be seen that the ratio of ridge height to ice thickness is much greater for thin ice than it is for thick ice (Figure 12.53). For instance, although ridges developed in ice 15 to 25 cm thick may be 5 to 6 times the thickness of the participating sheet, this is definitely not the case for ice thicker than

1 m. Individuals interested in the details of the Parmerter and Coon model should examine both the 1972 paper and a later 1973 paper published in the *AIDJEX Bulletin*, as this later paper fills in a number of details that are not discussed in the earlier paper.

The next significant advance in the modeling of ridging and rafting came almost 20 years later, as the result of interactions between Hopkins, Hibler, and Flato while all three were at Dartmouth College. Hibler and Flato were interested in the development of improved approaches for incorporating the ridging process into large-scale sea ice dynamics simulations (Flato and Hibler 1995). In attacking such a problem, Hopkins possessed specialized capabilities. By this I mean that he had experience with two-dimensional computer simulations of systems containing large numbers of polygonal blocks (Hopkins 1992). The result was the first dynamic model of the development of rafts and ridges (Hopkins and Hibler 1991a, 1991b; Hopkins et al. 1991). The initial model produced a two-dimensional simulation of the compression of a rubble-filled lead between two MY floes. As in the earlier Parmerter and Coon ridging model, the question of the origin of the rubble was not addressed. This ultimately led to an improved dynamic model in which a lead containing an intact ice sheet was deformed between two appreciably thicker MY floes (Hopkins 1994). An example of such a simulation is shown in Figure 12.54. I find these models to be quite impressive in that the forces on and the positions of every block are analyzed during every step of the ridging process. In the model the ice sheet is considered to be a dynamic linear viscous elastic material. Specific processes considered include flexural failures as well as buckling; the calculation of realistic block lengths by locating each break at the location in the parent sheet where the tensile stress exceeds the strength; the secondary flexural breakage of rubble blocks; inelastic contacts between rubble blocks; frictional sliding between blocks; and separate friction coefficients for submerged and above-water contacts, buoyancy factors, and water drag. Detailed descriptions can be found in Hopkins (1992, 1994). The model allows one to examine the evolution of both the sail and keel profiles, the forces involved, and the energetics during each step of ridge development. For instance, the fact that both the frictional and inelastic processes are considered separately definitely provides an improved picture of the overall process (Figure 12.55). It should be noted that all of these models assume, as did the Parmerter and Coon model, that the ice sheet breaks in flexure (including buckling). The results would not apply to situations where the primary failure mechanism is crushing. Although crushing can definitely occur during ridging, it is my feeling based on field observations that it is a relatively small component of the overall process.

In the following I will consider first the results from simulations where thin ice in a lead is deformed against appreciably thicker surrounding floes. In these cases, the thin lead ice characteristically initially overrides the thicker floe. The large degree of curvature involved in this process commonly results in buckling, which acts to terminate any tendency for rafting. As a result the ridging is invariably localized, with the interaction between the sheets usually limited to distances of 10 to 20 m even though 100 m of ice is always pushed into the ridges in the simulations. A part of the keel in front of the edge of the thicker floe serves as a platform supporting the downward component of the sail-building force. Sail growth continues as long as the sheet is capable of transmitting the force and the platform is able to support the downward component. Buckling occurs when the sheet is no longer able to transmit the force. When the platform is un-

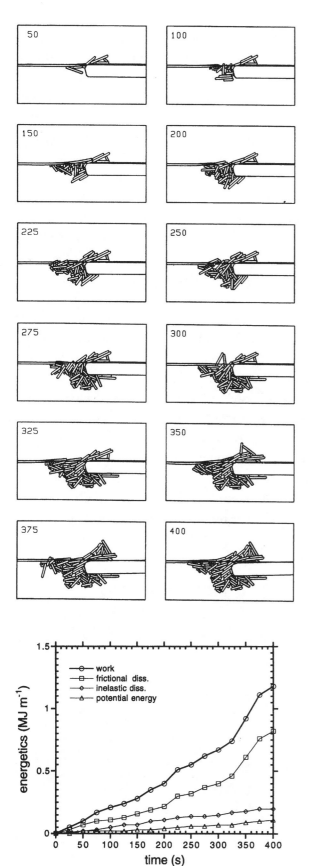

Figure 12.54. Sequential snapshots from a numerical ridging experiment (Hopkins 1994). The thickness of the ice being deformed is 30 cm; the dry and wet friction coefficients were 1.0 and 0.6, respectively. Each frame is 12 m × 20 m.

Figure 12.55. Energy budget for the numerical experiment shown in Figure 12.57 (Hopkins 1994).

able to support the force, it collapses. These patterns can be seen in Figure 12.54. Rubble is added to the keel at the surface whenever the progress of the sheet over the platform is obstructed, resulting in breaks due to buckling and flexure. It is also possible for the edge of the sail to become unstable and collapse onto the platform, adding to the growth of the keel. Occasionally the sheet deflects downward, pushing rubble underneath the thick floe. Force versus time plots from these simulations show low-level sustained forces with numerous large sharp spikes. Hopkins suggests that the sharp spikes that occur during periods of keel building are associated with buckling failures. By far the greatest energy sink in the process is friction, which proves to be eight times larger than the potential energy remaining in the completed ridge. The final shapes of the ridges are particularly sensitive to the values assumed for dry friction in the sail (μ). If μ is large, smaller, more compact sails and larger keels result. If $\mu < 0.4$, sails become extremely wide. The amount of force required to enlarge a sail increases with the size of the sail and with μ, and is limited by the ability of the sheet to resist buckling. In that the supply of broken ice is taken to be constant, a reduction in sail volume is accompanied by an increase in keel volume.

Later Hopkins et al. (1999) applied similar numerical simulations plus physical modeling to the rafting problem in which both sheets have identical thicknesses. In the numerical simulations where the thicknesses of the interacting sheets were approximately equal, rafting always occurred. When the thicknesses of the interacting sheets were appreciably different, ridging always occurred. Mixed ridging and rafting was possible between these two extremes. These trends also proved to hold in the physical model studies (Hopkins et al. 1999). Changes in the rate of convergence of the interacting ice sheets did not result in any noticeable systematic effects. Figure 12.56 shows values of the nondimensional force versus the nondimensional displacement for a series of simulations in which the major thickness h_1 is held constant while the thickness ratio varies from 2/8 to 8/8. Clearly the force increases with increases in the value of the thickness ratio. The reasons for this are 1) that the probability of rafting increases as the thickness ratio approaches a value of one, coupled with the fact that 2) rafting forces are higher than ridging forces. The smooth increase in the work versus displacement graphs with increasing thickness ratio suggests that the transition from ridging to rafting is quite seamless. It is concluded that, other than the importance of the thickness ratio, the conditions during the initial contacts between

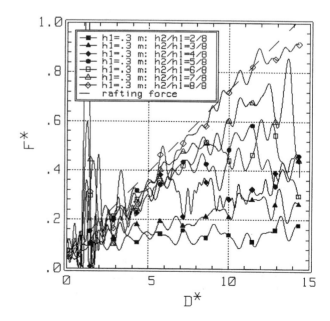

Figure 12.56. The nondimensional force versus the nondimensional displacement based on simulations in which the major thickness h_1 is held constant while the thickness ratio is varied from 2/8 to 8/8 (Hopkins et al. 1999).

the sheets do not appear to have a significant effect on whether the interacting sheets will raft or ridge. Interacting sheets that are thicker and more unhomogeneous in thickness will tend to ridge. Although thin homogeneous sheets will always raft, thick homogeneous sheets can also raft, as field observations have shown. The reader might ask, "If this is so, why are rafts in thick ice apparently less common than rafts in thin ice?" One possible answer to this query is that rafting in thick ice requires stress levels that are rarely attained in the pack because there is commonly thin ice in the near vicinity that deforms and lowers the stress.

In all of the above simulations the total amount of ice that was allowed to deform was held constant at 100 m in order to examine the different factors affecting the deformation and to limit the amount of computational time involved. However, Hopkins (1998) has also performed a series of simulations in which the amount of ice being deformed was relatively unlimited in order to better explore the complete deformation cycle. On the basis of these runs, he suggests that there are four stages in the development of pressure ridges. The first stage starts when the edges of the interacting sheets first touch, and ends when the sail reaches its maximum height. During this stage the maximum sail height appears to be reached at about 200 m of lead ice extent and appears to be independent of the thickness of the ice involved. During the second stage the keel broadens and thickens. This stage ends when the keel has reached its maximum thickness. The third stage ends when the lead ice is consumed. During this stage the ridging force remains constant at ~1/3 of the buckling force. Ridge growth proceeds leadward, resulting in a rubble field of ~ uniform thickness. The thickness of the rubble field is again specified by the buckling strength of the sheet ice as this limits the force available to compress the floating rubble. During the fourth stage the rubble between the thicker floes is further compressed. The maximum force available for compression is limited by the buckling force of the floes themselves, whereas the force resisting compression is proportional to the square of the thickness of the rubble.

During both of these two initial growth periods, the sail height as well as the keel draft appear to increase, with the square root of the extent of the lead ice deformed into the ridge. Figure 12.57 shows the maximum average sail heights and keel drafts based on Hopkins simulations, compared with the limiting sail height envelope obtained by Tucker, Sodhi, and Govoni (1984) and the limiting keel draft envelope obtained by Melling and Riedel (1995). In both cases the agreements between the numerical simulations and the curves fitted to field observations are impressive. As Figure 12.57 shows, the maximum sail height is proportional to $\sqrt{h}$ where h is the lead ice thickness. As Hopkins points out, this can be rationalized as follows. The two-dimensional volume of lead ice in a ridge is the product $h \times L$ where L is the extent of the lead ice pushed into the ridge. Then, if the value of L required for a sail to reach a maximum height is independent of h, it follows that the maximum sail height is proportional to $\sqrt{h}$ as sail height is proportional to the square root of the ice volume. The effects of variations in the thickness of the floe ice were also investigated. It was found that within the range investigated (2.0–3.5 m) there was no observable effect.

As Hopkins points out, although the agreement between field observations and numerical simulations is encouraging, this may be a bit misleading. The reason here is the value (100 MPa) used for the elastic modulus E in the simulations. This value is appreciably smaller than the values usually cited for typical lead ice (1–2 GPa; see Chapter 10). It was necessary to use this low value to keep the ice sheet from failing in a brittle manner via buckling during the two initial stages of ridge

Figure 12.57. Maximum average sail heights and keel drafts based on numerical simulations (Hopkins 1998) compared with the limiting sail height envelope obtained by Tucker, Sodhi, et al. (1984) and the limiting keel draft envelope obtained by Melling and Riedel (1996).

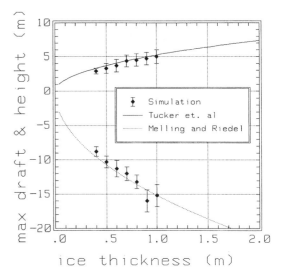

development. Such a failure event would limit the ability of the oncoming lead ice to push a train of blocks up the leadward slope of the sail, thereby limiting the maximum sail height that could be achieved. It is also possible that the values measured by Tucker, Sodhi, and Govoni (1984) and Melling and Riedel (1995) do not actually represent maximum heights and drafts. Another possibility is that true ridging forces are much lower than the simulations would imply. If this is the case, then one would have to discard the assumption that the ridging force is limited by the buckling strength. A final possibility is that the limiting values were not reached because the MY floes were unable to support the pileups.

12.7 Conclusions

I think that the amount of material in this chapter attests to the appreciable progress that has been made since 1960 in understanding the natural history of deformation within the pack. It certainly would be useful to have more detailed descriptions of ridge development while the deformation is in process. Of course this is easy to say and not so easy to achieve in that, although one frequently can guess where ridging may be probable, one rarely knows when, or even if, it will actually happen. Although the reader may think that I have covered all known papers on pressure ridging, this is far from the case. There is an appreciable literature on pressure ridge and rubble field development and its role in determining peak ice forces on offshore structures. A good place to start exploring this literature would be Timco et al. (2000).

13 Sea Ice–Seafloor Interactions

One of the great obstacles to progress is not ignorance,
but the illusion of knowledge.
Daniel Boorstin, *The Discoverers*

13.1 Introduction

Inuit hunters have known for a long time that both sea ice and icebergs could interact with the underlying seafloor, as seafloor sediments could occasionally be seen attached to these icy objects. As early as 1855 no less a scientific luminary than Charles Darwin speculated that gouging icebergs could traverse isobaths, concluding "that an iceberg could be driven over great inequalities of [a seafloor] surface easier than could a glacier." However, it was not until 1924 that similar notice of the possibility of sea ice interacting with the seafloor began to appear in the scientific literature (Kindle 1924). The subject then remained a seldom-studied curiosity (Carsola 1954; Rex 1955) until the 1970s, when serious studies of the phenomenon were initiated (Barnes and Reimnitz 1974; Kovacs 1972; Kovacs and Mellor 1974; Reimnitz and Barnes 1974; Shearer and Blasco 1975; Shearer et al. 1971). There were several reasons for this hiatus. For one, the technology for examining what was occurring on the seafloor was not adequately developed. However, probably foremost was the fact that whatever was occurring between the sea ice and the seafloor was not causing sufficient trouble to have arrived on anyone's list of problems that needed to be investigated. This was particularly true considering the fact that investigating underwater phenomena is invariably expensive.

This all changed with the discovery of oil at Prudhoe Bay, Alaska. Because the discovery well, Prudhoe #1, was almost in the water, the possibility of major offshore oil fields occurring to the north of Alaska and Canada immediately came to mind. It was also apparent that if major oil resources occurred to the north of Alaska and Canada, similar resources might be found on the world's largest continental shelf, located to the north of the Russian mainland. It is now known that major oil fields and gouging occur in all three of these regions. For instance, Figure 13.1 is a sketch based on divers' observations describing typical conditions during gouging at Badarskaya Bay, located on the Russian coast of the Arctic Ocean. Offshore oil fields are typically tied together by subsea pipelines that transport the oil from the individual wells to a central collection point where tanker pickup is possible, or to the coast where it can be fed into a pipeline transportation system. As subsea pipelines and ice-induced gouging of the seafloor do not sound particularly compatible, it was immediately obvious that if offshore development was to occur over the arctic continental shelves, the gouging phenomena would have to be investigated and understood. One would need to know the widths and depths of the gouges, the frequency of gouging events

Figure 13.1. A sketch of active gouging based on observations by Russian divers at Badarskaya Bay, Southern Kara Sea (courtesy of Dr. V. Ryabinin).

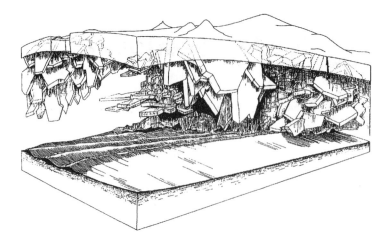

Figure 13.2. A fathogram of an ice-gouged seafloor. Water depth is 36 m. The record was taken in the Beaufort Sea NE of Cape Halkett, Alaska, by Barnes and Reimnitz.

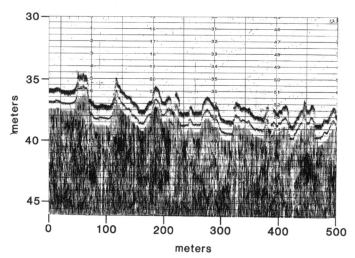

in time and space, the water depth range in which gouging occurs, and the effective lifetime of a gouge after its initial formation. Also of importance is the type of subsea soil and the nature and extent of the soil deformation below the gouges, as this information is essential in calculating safe burial depths for subsea structures such pipelines and well heads.

In the following I will try to summarize some of the results of these investigations. I note that there is no commonly agreed upon terminology for this phenomenon, which has been referred to as scouring, scoring, plowing, and gouging. Here I will use the term *gouging* as a matter of personal preference as I feel that it more accurately describes the process.

13.2 Observational Methods

A variety of techniques have been used to study gouging. Typically a fathometer is used to resolve the seafloor relief directly beneath the ship (Figure 13.2). At the same time, a side-scan or more recently a multibeam sonar system provides a map of the seafloor on either side of the ship (Figure 13.3). Total sonar swath widths have been typically 200 to 250 m, not including a narrow area directly beneath the ship that is not imaged. The simultaneous use of these two different types of records allows one to both measure the depth and width of the gouge (fathometer) at the

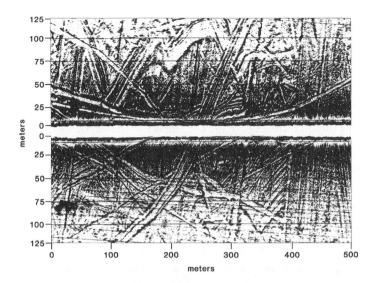

Figure 13.3. Sonograph of an ice-gouged seafloor. Water depth is 20 m. The record was obtained in the Beaufort Sea 20 km NE of Cape Halkett, Alaska, by Barnes and Reimnitz.

Figure 13.4. A three-dimensional sonar image of a highly gouged area of the floor of the Beaufort Sea. The approximate location is 70 03 04.5168N and –133 46 24.2057W. The water depth is 28 m. The large gouge is ~4 m deep. The image was processed from data obtained by Dr. Stephen Blasco and is shown here courtesy of the Geological Survey of Canada.

point where it is crossed by the ship track, and also to observe the general orientation and geometry of the gouge track (sonar). Along the Alaska coast the ships used have been small, allowing them to operate in shallow water. In addition, the requirement for precise measurements means that such operations are typically restricted to reasonably calm seas and relatively ice-free conditions. Figure 13.4 is a three-dimensional image of an intensely gouged area of the floor of the Canadian Beaufort Sea obtained by Steve Blasco of the Geological Survey of Canada. The main gouge is approximately 4 m deep. The procedures used to make Figure 13.4 overemphasize the steepness of the sides of the gouges, making them appear to be vertical when, in fact, they are not. The maximum depth is, however, correct.

In a few cases divers have been used to examine the gouges. To date, submersibles do not appear to have been used in the Arctic, as water depths affected by the ice are relatively shallow.

However, they have been utilized at a few deeper-water sites off the Canadian east coast to obtain direct observations on gouges produced by icebergs.

13.3 Results

As might be expected, the most common features gouging the shelves of the Arctic Ocean are the keels of pressure ridges, although gouges resulting from grounded ice island fragments also occur. As noted in the section on pressure ridges the distribution of keel depths of pressure ridges is approximately a negative exponential: there are many shallow draft ridge keels whereas deep-draft keels are rare. Ice deformation resulting in extensive ridging is also particularly intense in some nearshore regions. Such a location is the U.S.–Canadian shelf of the Beaufort Sea, where the pack grinds along the coast moving from the west to the east. I have occasionally seen > 100-km-wide bands of severely deformed ice that parallel that coast. The old rule of thumb that states "If you wish to establish an ice camp that would survive for a significant period of time in the Beaufort Sea, you should site it at least 150 km north of the Alaskan coast and the further off-shore the better" definitely considered these conditions. I note that one of the deepest pressure-ridge keels reported to date was located in this same region (47 m, McClure Strait; Lyon 1991) as were very high-pressure ridge sails (grounded, > 20 m). I have also been told that a 57- and a 49-m keel have been observed in ULS records (Kovacs, pers. comm.) although, to the best of my knowledge, these observations have not as yet been published. Large linear (in plan) shear ridges that extend for tens of kilometers roughly parallel to the coast are also common. Such ridges are frequently anchored at shoal areas where large accumulations of highly deformed sea ice can build up. Such grounded ice features, commonly referred to using the Russian term *stamukhi*, can have freeboards in excess of 10 m and lateral extents of 10 to 15 km. A thorough study of one such stamukha located along the Beaufort coast can be found in Reimnitz and Kempema (1984). Impressive photographs of ice features caused by intense deformation along the coast of the Beaufort Sea can be found in Kovacs (1976), as well as in Kovacs and Gow (1976) and Kovacs et al. (1976).

As the offshore ice pack moves against and along the coast, it exerts force on the sides of any grounded ice feature, causing it to scrape and plough its way along the seafloor. Considering that surficial sediments along many areas of the Beaufort Sea Coast of Canada and Alaska are fine-grained silts, it is hardly surprising that over a period of time such a process can cause extensive gouging of the seafloor sediments. Figure 13.5 is a photograph of active gouging occurring on the Alaskan Beaufort Coast. The relatively undeformed first-year sea ice in the foreground is moving to the west, away from the viewer, and pushing against a piece of grounded multiyear sea ice (indicated by its rounded upper surface that formed during the previous summer's melt period). As a result the grounded MY fragment is pushed along the coast. The interaction between the first-year and the multiyear ice has resulted in a pileup of broken first-year, which when the photograph was taken was higher than the upper surface of the multiyear ice. Also evident is the track cut in the first-year ice as it moves past the multiyear ice. The fact that both the multiyear ice and the pileup of first-year ice are interacting with the seafloor is indicated by the presence of bottom sediment in the deformed first-year ice and on the far side of the multiyear ice. This situation, in which the grounded ice is

Figure 13.5. A photograph of active ice gouging occurring along the coast of the Beaufort Sea (photo by G. F. N. Cox).

pushed along by the surrounding ice pack, appears to be common. This view is supported by a study of the relative forces involved in gouging by Kovacs and Mellor (1974), who found that when the sizes and inertias of typical individual ice masses involved directly in gouging are considered, only shallow gouges (< 1 m) are possible. It appears that only when these masses ground and are then pushed along by the moving ice pack are deep gouges possible.

The maximum water depth in which contemporary sea ice gouging is believed to occur is on the order of 50 to 60 m, in keeping with the ~57 m value for the deepest pressure-ridge keel observed to date. I also note that gouges occurring in water > 50 m deep on the Chukchi Shelf have been reported to have the appearance of being "recent" by Barnes and Reimnitz (1974). Here "recent" refers to a gouge that has at least one of the following characteristics: (1) sharply defined relief features; (2) steep side slopes lying at the angle of repose for the sediment type involved; (3) absence of soft sediment or soupy ooze on the gouged floor; (4) absence of burrow marks, trails, and tracks within the gouge; (5) unfilled tear fissures inside the gouge and the flanking ridges; and (6) traverse ripple marks confined to the inner gouge flanks. Although there appears to be no reason why somewhat deeper keels are impossible, keels capable of gouging in such deep water are obviously very rare. Wadhams (2000) estimates that at any given moment the deepest keel present in the Arctic Ocean has a draft in the range of 55 to 58 m.

Sonar data have indicated the presence of gouges in water up to 65 m deep off the north coast of Alaska and Lewis (1977a) has observed gouges in water up to 80 m deep north of the Mackenzie Delta. However these deeper gouges are believed to be historical relicts that formed when sea level was lower than at the present. The deepest gouges known have recently been discovered at water depths of from 450 m to at least 850 m on the Yermak Plateau located to the northwest of Svalbard in the Arctic Ocean proper (Vogt et al. 1994). They are believed to be caused by icebergs and are presumed to have formed during the Pleistocene when sea level was much lower than at present.

Repetitive plowing of the seabed has a somewhat similar effect to a farmer's repetitive plowing of his land. In both cases the soil is loosened to a depth at, or just below, the depth of the plow or keel. On the Beaufort Shelf this vertical zone has been referred to by a variety of names, including the ice gouge zone, the ice remoulded zone, the ice keel turbate zone, and the paleogouge zone. Below this zone the soil is undisturbed and stiffer, as revealed by cone penetrometer tests. On the Beaufort Shelf the depth of the turbate zone varies with water depth, sediment type, and location. The particular significance of this zone is that it is a useful index of the deepest depth affected by ice gouging. In water depths greater than ~15 m on the Beaufort Shelf, the turbate zone appears to be less than 4 m thick.

As pressure-ridge keel depths are well described by a negative exponential distribution, it is reasonable to surmise that in a local area the distributions of gouge depths into the seafloor might be described by a similar function. That this is indeed the case was first shown by Lewis (1977a, 1977b), and has been further documented by a variety of other investigators. Figure 13.6 shows a semilog plot of the number of gouges observed versus the depth of the gouges into the seafloor for several areas along the coast of the Beaufort Sea. Note that in all four of the areas studied, a simple negative exponential function fits the data quite well. Figure 13.7 shows the relative frequency of occurrence of gouges of differing depths grouped into classes according to water depth. Here only data from offshore areas unprotected by barrier islands were used. Note that in the water depth range studied, the parameter λ in the probability density function

$$f_X(x) = \lambda \ e^{-\lambda x} \quad x \geq 0 \tag{13.1}$$

undergoes a pronounced decrease as water depth increases, indicating that deep gouges are far less likely to occur in shallow water than in deeper water. For instance, along the Beaufort Coast in water 5 m deep, a 1-m gouge has an exceedance probability of approximately 10^{-4}; that is, 1 gouge in 10,000 will on the average be expected to have a depth equal to or greater than 1 m. In water 30 m deep, a 3.4-m gouge has the same probability of occurrence (Weeks et al. 1983, 1984). Studies have shown that 3-m gouges in the seafloor are not rare, and 8-m gouges have been reported in the vicinity of the Mackenzie Delta. Although Reimnitz and Barnes (1974) have noted the occurrence of a 10-m gouge in the Mackenzie Delta offshore, the details of its origin remain speculative. The maximum likelihood estimate of the free parameter λ of the exponential distribution is the reciprocal of the sample mean x:

$$\hat{\lambda} = 1/\overline{x}. \tag{13.2}$$

This distribution has a number of advantages. Most importantly it fits the data quite well, as demonstrated in Figures 13.6 and 13.7. Also, the probability that a random variable will assume a value in the interval (x_1, x_2) is

$$P[x_1 \leq X \leq x_2] = \int_{x_1}^{x_2} f_X(x)dx = \lambda \int_{x_1}^{x_2} e^{-\lambda x}dx. \tag{13.3}$$

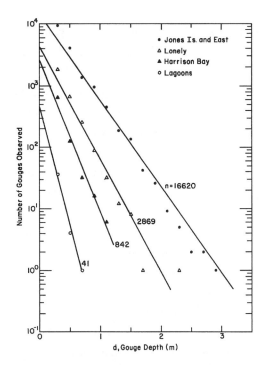

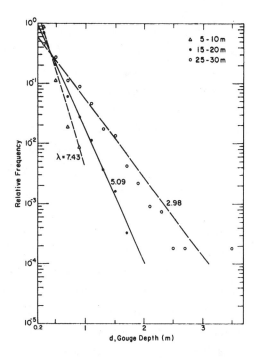

Figure 13.6. A semilogarithmic plot of the number of gouges observed versus the depth of the gouges into the seafloor for four regions along the Alaska coast of the Beaufort Sea (Weeks et al. 1984). The regions are as follows listed in order of decreasing mean water depth: • Jones Island and east; Δ Lonely; ▲ Harrison Bay; and O Lagoons.

Figure 13.7. The relative frequency of occurrence in three different water depth classes of gouges of differing depths in the seafloor based on data collected from offshore areas unprotected by barrier islands along the Alaska coast of the Beaufort Sea (Weeks et al. 1984).

The cumulative distribution function (CDF) can be found by integration

$$F_X(x) = P[X \le x] = \int_0^x f_X(u)du = 1 - e^{-\lambda x} \qquad (13.4)$$

where $x \ge 0$. Finally, because there is considerable interest in gouges having depths equal to or greater than some specified value, this can be obtained from the complementary distribution function $G_X(x)$, where

$$G_X(x) = P[X \ge x] = 1 - F_X(x) = e^{-\lambda x}. \qquad (13.5)$$

$G_X(x)$ is a simple function to graph as it is a straight line on semilog paper and has a value of 1 at $x = 0$. Therefore the simple relation

$$P[D \ge d] = \frac{n[D \ge d]}{N} = e^{-\lambda d} \qquad (13.6)$$

can be used to estimate $n[D \ge d]$ the expected number of gouges with depths greater than or equal to d, given that N gouges have occurred. In fitting such data it is common to apply a cutoff by only considering gouges with depths equal to or greater than some specified value such as

0.2 m. The reason for using such a cutoff value is that it is difficult to unambiguously identify small gouges from similar features formed by other mechanisms. A discussion of the problems associated with dealing with such terminated data sets can be found in Weeks et al. (1983).

That the maximum number of gouges occurs at some intermediate water depth is clearly shown in Figure 13.8a, which presents a plot of gouge density versus water depth. The maximum gouge density occurs at a water depth of 22–25 m. Figure 13.8b shows the gouge intensity, which is a parameter that estimates the total volume of the sediments involved in the gouging by plotting the product of the maximum depth × the maximum width × the gouge density. Note that the peak of the intensity plot occurs at ca. 37 m (i.e., in somewhat deeper water than the maximum gouge density).

Considering that the keel depths of the incoming ice follow a negative exponential, it also follows that in deep water there are only a few keels capable of hitting the bottom. As the ice moves into shallower and shallower water, the number of keels with sufficient drafts to hit the seafloor increases exponentially. At first glance one might think that the amount of gouging would continue to increase as the coast becomes nearer and nearer. However there are several other factors at work here. For one, once a keel grounds in deep water it is no longer available to cause gouging in shallower water because, at least along the northern coast of Alaska, ice motions tend to parallel the coast with only a slight onshore component developing after the initial grounding occurs. Also the total mass of the ice in a given ridge is important, as only large keels have sufficient mass necessary to produce the downward force required to allow them to cut deep gouges in the seafloor. Because small keels lack this mass, even if there are large numbers of such keels, they are incapable of producing deep gouges. The most deeply disturbed seafloor occurs in the water depth range between 20 and 35 m as the result of the fact that many gouges occur in

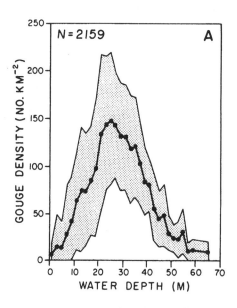
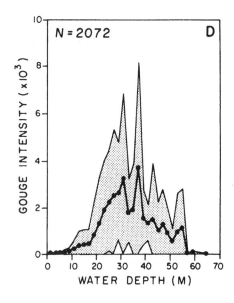

Figure 13.8. The (a) gouge density (number/km²) and (b) the gouge intensity (× 10³) plotted vs. water depth, Alaska coast of the Beaufort Sea (Barnes, Rearic, et al. 1984). The parameter gouge intensity estimates the total volume of sediment affected by the gouging.

this water depth range and deep gouges are still possible. In theory, the largest number of gouges should occur in shallow water, as ridges with small keels are the most frequent. However, in shallow water wave effects and currents can erase any gouges that form. Another consideration is that as more keels ground they impede the motion of the ice pack along the coast. This also reduces the amount of possible gouging in the nearshore region.

As might be expected, the nature of gouging varies with changes in seafloor soil types. Along the Beaufort coast where there are two distinct soil types, the gouges in stiff sandy clayey silts are typically more frequent and slightly deeper than those found in more sandy soils. Presumably the gouges in the more sandy material are more easily obliterated by wave and current action. It is also reasonable to assume that the slightly deeper gouges in the silts provide a better picture of the original incision depths. An extreme example of the effect of bottom sediment type on gouging can be found between Sakhalin Island and the eastern Russian mainland. Here the seafloor is very sandy and the currents are quite strong. As a result, even though winter observations have conclusively shown that seafloor gouging is common, summer observations show that the gouges have largely been erased by infilling.

Gouge tracks in the Beaufort Sea north of the Alaskan and western Canadian coast generally run roughly parallel to the coast, with some unprotected regions having an excess of 200 gouges per kilometer. This is clearly shown in Figure 13.6, in that the degree of exposure to the moving pack ice decreases in the sequence Jones Islands–Lonely–Harrison Bay–Lagoons. Histograms of the frequency distribution of distances between gouges are also well described by a negative exponential, a result suggesting that spatial gouge occurrences may be described as a Poisson process. Spatial occurrence is hardly uniform, with significantly higher concentrations of gouges occurring on the seaward sides of shoal areas and barrier islands and fewer gouges on the more protected lee sides (Reimnitz and Kempema 1984). In any comparative study of the number of gouges in a given area, care must be taken to ascertain that the counting conventions used in the different studies are identical, as some authors count every individual gouge while others group multiple gouges if it appears that they were produced by the same ice mass. I personally feel that considering each identifiable individual gouge as a separate entity is the most straightforward approach, particularly if one of the purposes of the data collection is risk assessment for bottom-founded structures.

The second type of feature producing gouges on the surface of the outer continental shelf of the Arctic Ocean is the ice island. As noted earlier, these features are, in fact, an unusual type of tabular iceberg formed by the gradual breakup of the ice shelves located along the north coast of Ellesmere Island, the northernmost of the Canadian Arctic Islands. Once formed, an ice island can circulate in the Arctic Ocean for many years. For instance, the ice island T-3 drifted in the Arctic Ocean between 1952 and 1979, ultimately completing three circuits of the Beaufort Gyre (the large clockwise oceanic circulation centered in the offshore Beaufort Sea) before exiting the Arctic Ocean through Fram Strait. The lateral dimensions of ice islands can vary considerably, from a few tens of meters to over ten kilometers. Thicknesses, although variable, typically are in the 40- to 50-m range; ice islands possess freeboards in the same range as exhibited by larger sea ice pressure ridges. In the study of fathometer and sonar data on gouge distributions and patterns no attempt is usually made to separate gouges made by pressure ridges from gouges made by ice islands, as the ice features that made the gouges are commonly no longer present. However, it is reasonable to assume that many of the very wide, uniform gouges are the result of ice islands interacting with the seafloor. Ice island fragments were quite common along the Alaska coast in the 1970s, where they frequently formed the core of large ice pileups such as the one at Katie's Floeberg, a virtual island of deformed ice located on a 20-m-deep shoal in the Chukchi Sea roughly 100 km to the west of Barrow, Alaska. Several photos of these features can be found in Kovacs and Mellor (1974) and in Kovacs and Gow (1976). In more recent years, ice island fragments appear to have been rare in this same general area. A useful compilation of information concerning ice islands can be found in Jeffries (1992).

13.4 Applications

It is relatively easy to characterize the state of the gouging existing on the seafloor at any given time. It is another matter to answer the question of how deep one must bury a pipeline, a cable, or some other type of fixed structure beneath the seafloor to reduce to some acceptable level the chances of it being struck by moving ice. To answer this question one needs to know the rates of occurrence of new gouges. Unfortunately such values are poorly known at most locations, as they require replicate measurements over a period of many years so that new gouges can be counted and the rate of infilling of existing gouges can be estimated. The best available data on gouging occurrence along the Beaufort Coast have been collected by the Canadian and U.S. Geological Surveys and indicate that gouging frequently occurs in rather large-scale regional events related to periodic storms that drive the pack ice inshore—about once every four to five years. Because there is no current technique for absolutely dating the age of existing gouges, there is no current method for determining whether a particular gouge that one observes on the seafloor formed during the last year or sometime during the last 6,000 years (after the Beaufort Shelf was submerged as the result of rising sea levels at the end of the Pleistocene). It is generally believed that gouges observed in water less than ~20 m deep formed quite recently. For instance, during the late summer of 1977, when the Beaufort Coast was comparatively ice free, strong wave action during late summer storms obliterated gouges in water less than 13 m deep and caused pronounced significant infilling of gouges in somewhat deeper water (Barnes and Reimnitz 1979). It is generally estimated that such summer storms have an average recurrence interval of no more than 25 years (Reimnitz and Maurer 1978). On the other hand, gouges in water deeper than ~60 m are believed to be relatively old as they are presumed to be below the depth of currently active gouging. The lengths of time represented by the gouges observed in water depths between 20 and 50 m is less well known. Unfortunately this is the water depth range in which the largest gouges are found. Stochastic models of gouging occurrence that incorporate approximate simulations of subsea sediment transport suggest that if seafloor currents are sufficiently strong to exceed the threshold for sediment movement, gouge infilling will occur within a few years (Weeks et al. 1985). I believe that it would be useful to further develop such simulations using more recent sediment transport models. At the least this would help focus the collection of pertinent sedimentological and oceanographic parameters, many of which are poorly known for the polar shelves.

To date several different procedures have been used to estimate the depth of gouges with specified recurrence intervals in problems that consider the safe burial of offshore structures such as pipelines. One possibility is to use the scour budget approach suggested by Lewis (1977a, 1977b). This assumes that the seafloor as seen at a given time represents a steady-state condition, with the number of new gouges per unit time equaling the number of gouges infilled by sedimentation plus the number of new gouges superimposed on existing gouges. The essential assumption here is statistical time invariance. I believe that this approach, although possibly useful in deeper water, is of doubtful applicability in water depths of < 20 m. Another approach assumes that an increasingly large proportion of the bottom is regouged before the entire bottom is gouged (Barnes et al. 1978). In this approach, if it is assumed that 10% of the seafloor is gouged each year, then in the second year only 19% has been gouged, as 1% of the gouges occur in areas already gouged. This can be expressed as a polynomial

$$G_t = 1 - (1 - K)^T \tag{13.7}$$

where G_t is the fraction of the bottom gouged since time T_0, K is the fraction of the bottom gouged each year, and T is the time measured relative to T_0.

Several possible approaches have also been suggested as to how to combine information on pressure ridge keels, pack ice drift, and observed distributions of gouge depths to estimate required burial depths (Wadhams 1983a, 2000). The first combines the observed negative exponential distribution of keel drafts with the distance that the ice drifts over the site during a year to calculate the number of gouging events per year and the expected return period. A second approach uses the set of spacings between the crossings of a given depth horizon by the profile of the lower surface of the ice to calculate a return period for gouging. A third possibility is to use extreme value statistics generated by separating the profile of the lower surface of the sea ice into a number of equal (spatial) sampling intervals and determining the deepest ridge keel in each interval. A ranking is then prepared in order of decreasing depth and the resulting values are plotted on exponential extreme value probability paper. As the result is typically close to a straight line it can be extrapolated to keel depths not present in the record. This approach gives return periods for the southern Beaufort Sea of 9 months for an impact by a 30-m keel, 24 years for a 40-m keel, and 1010 years for a 50-m keel. A related study by Wang (1991) is focused on the numerical simulation of extreme gouge depths. Although these approaches are interesting, they have one major drawback. To date, ice thickness profiles and ice movement information are limited for water depths where ice gouging is occurring. As will become apparent in Appendix F on remote sensing, techniques operating from above the ice have been unable to obtain this information to date owing to the undesirable electromagnetic properties of sea ice. The successful application of under-ice techniques such as upward-looking sonar (ULS) also appears to be difficult. Manned submarines are certainly out; they do not operate under ice covers in water depths of less than 60 m. One might think that profiles obtained further offshore in deeper water could be substituted for unavailable nearshore profiles. However, as available submarine profile data show that ridging characteristics change as the Beaufort Coast nears and ridging becomes more intense (Wadhams and Horne 1980), such substitutions could result in misleading information being used. One possibility might be to combine data from bottom-founded ULS systems with ice movement data obtained by ice-surface-based Global Positioning Systems (GPS) or satellite-based systems such as Synthetic Aperture Radar (SAR). Producing such a data set would not be easy considering that placing bottom-founded ULS systems at locations where gouging is probable is hardly a scheme designed to maximize the life of the ULS system.

To date the preferred approach to calculating gouging probabilities for structures such as pipelines has been to obtain site-specific replicate side-looking and depth-sounding sonar observations of the seafloor in the vicinity of the proposed line. This provides estimates of gouging depths and rates. Ideally, such information should be collected over a several-year period. Limited available information indicates that the PDFs of new gouges are similar but with somewhat smaller λ values than PDFs obtained from complete seafloor gouge sets in the same region (Weeks et al. 1984). This trend appears reasonable because deeper gouges receive more fill per year than do shallow gouges (Fredsoe 1979). Using such information the total number of gouges that will occur during the proposed lifetime of the pipeline N is

$$N = \bar{g}\ TL\sin\theta \tag{13.8}$$

where $\bar{g}$ is the average number of gouges km^{-1} yr^{-1} occurring along the pipeline route, T is the proposed lifetime of the pipeline (years), L is the length of the line (km), and q is the angle between the route and the trend of the gouges. If one only considers 1 contact in T, $N[D \geq d]$ in equation 13.6 equals 1 resulting in

$$e^{-\lambda(x-a)} = \frac{1}{\bar{g}TL\sin\theta} \tag{13.9a}$$

or, rearranging,

$$x = a - \frac{1}{\lambda}\ln\left[\frac{1}{\bar{g}TL\sin\theta}\right]. \tag{13.9b}$$

Here a is the cutoff value and x is the depth of the gouging relative to the local seafloor.

Needless to say, considerably more observational and theoretical work is needed to provide a more solid foundation for making such estimates. Comparisons between the burial depth estimates necessary to allow one hit in 1,000, 100, and 10 years as calculated by Pilkington and Marcellus (1981), Wadhams (1983a) and Weeks et al. (1984) can be found in the latter reference. There are considerable differences in the estimated burial values. The most obvious missing information concerns rates of gouging, as well as differences between new gouge sets and the complete gouging history as seen on the sonar records. Multiyear replicate observations that would provide this information have only rarely been made either by private industry or by government agencies (Rearic 1982, 1986).

Finally, it should be noted that even if one were completely confident in one's ability to estimate burial depths that would reduce impact probabilities to a specified level, one would have only solved a portion of the problem. The question here is how much additional burial is required beyond the value of the design gouge in order to reduce the soil compaction caused by the gouge to a level that will not result in damage to the pipeline. A consideration of these additional aspects of the burial safety problem can be found in Palmer (1997), Palmer et al. (1990), and Konuk et al. (2005a, 2005b). A comprehensive bibliography of the literature on ice gouging by both sea ice and icebergs up through the early 1980s can be found in Goodwin et al. (1985). Detailed documentation of Canadian mapping of gouging in the Beaufort Sea can be found in Meyers et al. (1996). Finally, one should realize that gouging is not the only sea ice process that "digs" significant excavations in the shelves of the polar oceans. Other culprits include ice wallows and strudel scours (Reimnitz and Kempema 1982). Fortunately the damages done by these processes are neither as intense nor as extensive as that resulting from gouging. In the case of wallowing by large ice masses the problem becomes as much one of side-thrusting as of top-loading.

Many of the earlier studies on this subject have been focused on the shores of the Beaufort and Chukchi Seas as the result of the major oil strikes at Prudhoe Bay, Alaska, and in the vicinity of Tuktoyaktuk, N.W.T., Canada. In recent years the possible areas of interest have broadened considerably to include the Russian Shelf of the Arctic Ocean, offshore Sakhalin, Svalbard, the Barents Sea, and portions of the Bay of Bothnia. Also, there has been a shift in the focus of the studies from the general geophysics of the problem to investigations more closely related to offshore design. A number of recent references can be found in Liferov and Høyland (2004), which describes a study in which a FY pressure ridge was produced artificially by cutting blocks from level ice, allowing them to weld together for a four-week period, and then pulling the "ridge" onto a beach and measuring the pulling force, displacements, keel failures, and gouges. Although sea ice gouges obviously occur in nearshore waters off the Antarctic, this subject remains unstudied for two main reasons. First, in the Antarctic sea ice gouging is a minor problem compared to the damage inflicted on the seafloor by icebergs, and second, to date there has been no interest in exploring for oil off the antarctic coast. Such activities would clearly be technically difficult, expensive, and politically involved.

14 Marginal Ice Zone

Researchers who insist on clarity at all costs
rarely make important discoveries.
H. Gintis

14.1 General Characteristics

The marginal ice zone, or MIZ, is any portion of the polar sea ice cover sufficiently near to the ice-free "open" ocean such that interactions with the open sea result in the modification of the properties of the ice so that they are different from properties deeper within the pack. Another possible definition would be to consider the MIZ as the zone at the edge of the ice pack whose width is the lateral distance over which the penetration of waves can fracture the ice, thereby changing the morphology of the floes. A thorough examination of processes within the MIZ involves not only ice mechanics and thermodynamics but also fluid mechanical aspects of both the atmosphere and ocean. That this, in all its parts, is a very large subject can readily be demonstrated by considering the size of some of the more broadly based review papers (Wadhams 1986, 166 pp.; Squire 1998, 65 pp.; Squire 2007, 26 pp.; and Squire et al. 1995, 54 pp.), as well as symposia concerned with this subject (Muench 1983, 259 pp.; Muench et al. 1987, 509 pp.). Even in Wadhams's (2000) more introductory recent book, the MIZ warrants almost 20% of the 238 pages devoted to sea ice. This tells us that not only is the MIZ important, but working scientists and, more importantly, funding agencies think so.

I am not certain exactly where the term MIZ originated (pronounced *Ms.*). It appears to be fairly recent, as it is not mentioned in earlier papers by the U.S. Navy (1952), or by Armstrong et al. (1966), or even by WMO (1989). As it definitely sounds like a U.S. Navy or NASA acronym, the credit possibly goes to the U.S. Navy Arctic Submarine Laboratory. Whatever the source, the expression has definitely been accepted by the research community as useful and is firmly entrenched in the literature. This is in contrast with most specialized sea ice terms coined by the submarine community (Roberts and Armstrong 1964), which rarely surface in the scientific literature.

Consider two possible MIZ scenarios. In the first, the wind is blowing normal to the pack–ocean boundary and toward the ice. When this occurs the ice is compressed and the boundary between the ice-free and the ice-covered portions of the sea can be very sharp: essentially a line that separates ice-free ocean from pack ice that has a concentration of ~100%. If there is a significant swell running, the motion of the water flexes and breaks the ice with the greatest amount of breakage and, as a result, the smallest floes occurring at or very near the ice edge. Frequently the ice at the actual edge is closer to a mixture of brash and slush than to a series of well-defined floes. As the wave field moves further into the ice, the interactions between the waves and the ice result in the higher-frequency waves being damped. As a result, there is a systematic increase in the floe size as one moves into the pack until at the inner MIZ boundary the presence of the open ocean is no longer discernable. This distance is, of course, not a constant but is dependent on the nature of the wave field and the thickness distribution and mechanical properties of the ice. In that the atmosphere over the open ocean is warmer than that over the ice, such on-ice winds are frequently associated with stratus cloud formation and atmospheric boundary layer modifications that extend for distances of ~100 km into the ice (Glendening 1994b; Kantha and Mellor 1989). If the ice edge is retreating, the MIZ is frequently a zone of both pronounced surface melting and floe breakup. If the ice edge is expanding, the initial ice formation will be frazil, with larger and larger composite floes developing as part of the pancake cycle as discussed in Chapter 7 (see Figures 7.43, 7.44, and 7.45). This is particularly the case in the Southern Ocean during the winter.

If the wind is still blowing normal to the pack–ocean boundary but away from the ice, the MIZ frequently becomes increasingly diffuse (Figure 14.1). An alternate possibility is that a series of irregular ice edge bands form that are oriented roughly perpendicular to the wind direction. These bands consist of compact ice separated by regions of essentially open water (leads). Band development appears to be a particularly frequent phenomenon in the Southern Ocean and will be discussed later in this chapter. Other complexities arise if the wind is blowing parallel to the ice edge (Guest et al. 1944). In the northern hemisphere, if the ice lies on the left when looking downwind (parallel-left), the Ekman transport produced in the ocean boundary layer can result in an upwelling that introduces additional heat from below into the area of the ice edge. If the situation is parallel-right, then there is

Figure 14.1. Scattered small floes just inside the ice edge, Fram Strait.

an interaction with the local ice breeze that is the result of the strong temperature gradient between the cold air over the ice and the warm air to seaward. A possible result is a strong surface front located outside the ice edge. For example, Shapiro et al. (1989) observed a front located ~100 km outside of the Barents Sea ice edge that exhibited a temperature change of 6°C in 25 km. Tables summarizing these types of effects can be found in Guest et al. (1994) and in Wadhams (2000). Here I mention these effects to stress the fact that the atmosphere–ice–ocean interactions occurring in the MIZ are not simple.

14.2 Regional Characteristics

More detailed descriptions of regional MIZ characteristics can be found in Wadhams (2000) and Squire (1998). A particularly detailed, but now somewhat dated, description is also available in Wadhams (1986). My goal in the following is to present a more compact version of this information.

An important point made by Wadhams (2000) is that a true MIZ has its features permanently determined by its contact with an ocean in which long-period, large-amplitude waves are frequently present. Although MIZ characteristics are possible in any region where either fast or pack ice abut on the open sea, there are a number of locations, particularly in the Arctic, where geometric constraints sufficiently limit the fetch so that large-amplitude waves are rarely achieved. In such areas the width of the MIZ may be small to undiscernable. Examples include the summer ice edges north of Russia, Alaska, and Canada, as well as the channels of the Canadian Archipelago and the Gulf of Bothnia. Regions where the MIZ is well developed include the Bering, Greenland, Barents, and Labrador Seas, as well as Baffin Bay. In the southern hemisphere, the great majority of the ice pack can be considered to be part of the MIZ, with the main exceptions being portions of the western Weddell Sea and the Ross Sea. One should also note here that the above statements apply to pack ice conditions unmodified by global warming. A case in point is the Chukchi Sea, where recent ice edge recessions have greatly increased the fetch length during the late summer and fall and the associated probability of large-amplitude waves.

14.2.1 Eastern Arctic

Here the dominant features are the strong ocean currents that import warm water from the south. For instance, the West Spitzbergen Current has a temperature of +3 to +6°C at a depth of 150 m and a salinity of > 34.9‰. Its flow paths are both complex and topographically controlled, with a portion of the current turning to the west resulting in a recirculation referred to as the Return Atlantic Current (RAC). This ultimately becomes part of the East Greenland Current (EGC), which transports ice (the East Greenland Drift Stream) and polar water southward out of the Arctic Ocean. Water from the RAC is characteristically associated with the East Greenland Polar Front, which separates the cold, low-salinity water exiting the Arctic Basin from the warmer, more saline water of the Greenland Sea. In the winter the ice has, in the past, extended as far south as Kap Farvel (60°N) at the southern tip of Greenland. In recent summers the ice edge has retreated as far north as Fram Strait (80°N), although historical retreats to 74°N are more representative. There are large year-to-year variations. The ice edge here, in conjunction with the ice edge in the Barents Sea, not only presents the longest stretch of MIZ in the Arctic, but faces the Greenland

and Norwegian Seas, regions where major storms are common. I certainly can vouch for this. I have crossed the Denmark Strait between Iceland and East Greenland four different times during September–October. Although all these crossings were rough, one was particularly memorable. Our ship, which normally was able to cruise at 14 knots, was only able to make 2 knots. If the weather had deteriorated further, we would have been advancing to the rear.

The ice edge in the Labrador Sea has, in the past, occasionally been an extension of the East Greenland Drift Stream. By this I mean that during extreme winters sea ice as well as icebergs have traveled around Kap Farvel and then proceeded northward up the west coast of Greenland as part of what is there referred to as the West Greenland Current. Although a few icebergs may still possibly be able to make this transit, it is doubtful that under present climatic conditions even initially thick MY floes are robust enough for this trip. What sea ice survives initially drifts up the west coast of Greenland to the northern part of Baffin Bay. There it joins locally formed sea ice plus, during the summer, sea ice that initially formed in the Arctic Ocean and has moved to Baffin Bay by transiting south out of the Lincoln Sea, through the Robeson Channel and Nares Strait. Ultimately this highly varied ice pack joins the south-flowing cold boundary current that moves first along the coast of Baffin Island, then Labrador, ultimately ending when it encounters the Gulf Stream just off the Grand Banks. During winters there is a considerable amount of FY ice that forms during this journey. During the portion of this trip through the Labrador Sea, this ice can be exposed to very heavy wave action with the development of a prominent MIZ. Further to the north in Davis Strait and Baffin Bay, the ice is more protected in that only a limited range of wind directions are effective in building large-amplitude waves. During summers the southern extent of this ice retreats to off northern Baffin Island.

14.2.2 Western Arctic

The primary area of interest in the western Arctic is the Bering Sea and to a lesser extent the Sea of Okhotsk. Although small, deteriorated MY floes have, in the past, been sighted just south of the Bering Strait, these are rare occurrences. At the end of the summer the ice edge is located north of the Bering Strait in the Chukchi Sea. Once ice development starts in October the ice edge gradually progresses to the south, passing back through the Strait into the Bering. Ultimately the ice edge moves some 1000 km to the south, reaching its maximum extent in March near the shelf edge, where the water depth increases from the ~200 m characteristic of the shelf to the ~4000 m characteristic of the North Pacific Ocean. Here the major factor limiting further ice advance is increased oceanic heat. In fact, at maximum extent the ice edge corresponds to the edge of an upper-ocean frontal system maintained by the input of low-density water along the ice margin. Ice blown across this front melts rapidly. Pease (1980) has compared sea ice development in the Bering Sea to that of a conveyor belt and has estimated that during a typical winter the Bering Sea MIZ replaces itself as many as eight times.

There are some fundamental differences between the situation in the Western Arctic and that in the Eastern Arctic. Essentially 100% of the ice in the Bering Sea is FY ice, less than a year old. In addition, because of the constriction at the Strait most of the ice in the Bering Sea forms there. Furthermore, as most of the ice growth occurs in the northern part of this sea, there is a significant flux of salt into the upper part of the water column as the result of this growth.

Particularly effective examples of this process occur in several polynyas located in the northern Bering (Figure 11.1). However, melting invariably takes place further to the south. In short, sea ice production in the northern portions of the Bering Sea acts as an effective salt generator. In contrast, the East Greenland Drift Stream is primarily composed of old ice. Although this ice is appreciably more desalinated than the ice in the Bering, the desalination has occurred a year or more earlier somewhere to the north in the Arctic Ocean. However, on melting this ice serves as a source of freshwater, just the opposite of the western Arctic.

14.2.3 Antarctic

With the exception of the ice in the Weddell Sea and in portions of the Ross Sea, the circumpolar antarctic ice pack is clearly the widest and the longest MIZ in the world. The Southern Ocean is legendary for its intense storms and the fact that the longest-period waves in the World Ocean are generated there. In most cases the complete width of the antarctic pack can be considered to be part of the MIZ in that these waves are sufficient to break ice floes hundreds of kilometers from the outer ice edge. For example satellite altimetry over the time period 1985–1988 indicated average wave heights in the Southern Ocean of more than 4 m throughout the year, with average values of more than 6 m in the Pacific sector (Campbell et al. 1994). I have made several crossings of the Southern Ocean, none of which would be called smooth. One of them was definitely memorable. It was during the austral summer at the completion of a tourist trip down the west side of the Antarctic Peninsula. The ship was the *Kapitan Dranitsyn*, a Russian icebreaker. When we started north the weather looked OK and I figured that we ought to be able to sneak across without encountering any significant weather. That was not to be, as in the middle of the Drake Passage we sailed into a tightly wound-up polar low that was not on the weather map. Peak wind speeds were 70 knots; the maximum recorded roll was 47° with 55° being the roll of no return. My wife was actually thrown out of her bunk onto the floor and slid completely across the room before I was able to stop her. In the same storm another ship had its grand piano break loose in the lounge. That certainly must have been interesting. Glad I wasn't playing. I also note that these events were during the summer when the weather is generally better.

In many ways the antarctic MIZ is similar to the MIZ in the Bering Sea, albeit on a much grander scale. Both are primarily composed of FY sea ice and the conveyor belt analogy appears to be equally applicable. In the Bering the drift of the ice is generally toward the south. In the Antarctic, the drift of the ice is generally to the north-northeast. In the winter the actual ice edge is fixed by locations where the rapid formation of frazil ice is possible. This is the process that initiates the pancake cycle that has been discussed in Chapter 7. This cycle is important in that, as it is not conduction-limited, it permits an ice cover of a reasonable thickness to form in areas where a large oceanic heat flux is possible. One should remember here that FY ice in the Antarctic rarely has thicknesses in excess of 1 m. The retreat of the ice in the spring appears to occur by two different processes, depending on location. At most locations, the retreat of the pack occurs as the result of melt at its northern edge outpacing the general northward drift of the pack. An important point noted by Wadhams (2000) is that although the pancake cycle proceeds through a number of stages from small to large (i.e., frazil → small pancakes → large pancakes → small pancake floes → large composite floes → etc.), melting is not the reverse of this sequence in that the floes do not revert to pancakes. At locations where ice survives the summer such as the Weddell Sea, melting occurs essentially in situ as the result of a combination of above-freezing air temperatures, insolation, and oceanic heat.

14.3 Wave–Ice Interactions

14.3.1 Field Observations

The name of the person who first noticed that sea ice is extremely effective in damping out high-frequency ocean waves appears to be lost in the mists of time. In that this is such a striking effect, this "eureka" moment probably occurred long ago. The Inuit certainly knew of the effect, as did the early explorers and whalers. What is known is the name of the person who first took this on as an interesting scientific problem, Gordon Robin. Robin, an Australian geophysicist who later in his career became the director of the Scott Polar Research Institute, presumably first noticed this damping during his Southern Ocean transits as part of the Norwegian–British–Swedish Maudheim Expedition of 1949–1952. This very successful scientific operation has the distinction of being the first truly international antarctic expedition. In addition to Robin, its staff included such well-known polar specialists as G. Liljequist, E. F. Roots, V. Schytt, and C. Swithinbank. Later, in 1959–1960, Robin (1963) started systematic study of this effect in the Weddell Sea by using a shipboard wave recorder to determine variations in wave amplitude and period while in the pack. He was fortunate in that during his observations the sea state north of the ice edge remained relatively constant while the ice conditions (floe size and thickness) changed as the ship's distance from the ice edge changed. Figure 14.2 is a plot of the wave spectra collected on the northbound leg of the cruise. The caption gives the nature of the ice

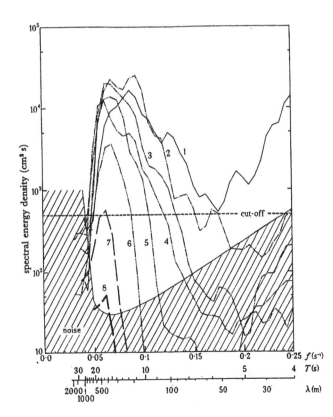

Figure 14.2. Wave spectra collected by Robin (1963) during the outward voyage of the R.R.S. *John Biscoe*, Weddell Sea (Squire 1998). The sea or ice conditions during the different collection periods are as follows: (1) open ocean (64.2°S); (2) 10-m-diameter floes (61.9°S); (3) 20-m-diameter floes (66.5°S); (4) 30-m-diameter floes (65.8.5°S); (5) 40-m-diameter floes (67.1°S); (6) 50-m-diameter floes (67.7°S); (7) 3-km-diameter floes, 1–1.5 m thick (68.8); (8) 5-km-diameter floes, 2–3 m thick (70.3°S).

conditions associated with each spectra. Clearly the ice is functioning as a low-pass filter removing the high-frequency (short-period) waves near the ice edge and then systematically attenuating all but the low-frequency (long-period) waves as the distance into the pack increases.
Robin also demonstrated that when long-period swell (> 10 s) moved through large-diameter
floes, the floes reacted to the wave passage by bending. On the other hand, if the waves were
short and the floes were small (< 40 m × 1.5 m thick), the floes acted as simple rigid rafts. The
most energy passed through when the floes were ≤ 1/6 wavelength in diameter. When the floes
were ≥ 1/2 wavelength in diameter, no appreciable penetration occured.

Later in the 1970s this problem was explored further by Wadhams (1975) as part of his
doctoral thesis at Cambridge by using an airborne laser profilometer (similar to that utilized by
Hibler in his studies of pressure ridge statistics—see Chapter 12—to investigate wave penetration into the Labrador Sea ice pack). Later Wadhams (1978b) was also able to use an upward-
looking sonar (ULS) system mounted on a hovering submarine to explore wave penetration into
the Greenland Sea MIZ. He found that the wave energy attenuated exponentially as the distance
into the MIZ increased, and that the attenuation rate increased with the square of the wave frequency. Finally, by adjusting the observations from the more open Labrador Sea MIZ to the
conditions encountered on the more concentrated Greenland Sea MIZ, he was able to show that
they were comparable, as seen in Figure 14.3: a line with a slope of –2 adequately fits each data
set even though the ice conditions are quite different.

As time has passed, measurements of wave passage into the MIZ have gradually become
sophisticated, with the use of wave buoys outside the ice edge plus floes instrumented with ac-
celerometers, tilt meters, and strain gauges (Squire and Moore 1980). Because of the frequently
less than ideal working conditions in the MIZ (see Wadhams 2000, p. 193), recent studies have
also used synthetic aperture radar (SAR) to study wave penetration (Liu, Holt, et al. 1991; Liu
and Mollo-Christensen 1988; Liu and Peng
1992; Liu, Vachon, et al 1991; Liu et al. 1992;
Wadhams and Holt 1991). Although the use
of SAR requires extensive data processing,
this is a small price to pay as SAR has many
advantages in that it is all-weather, works day
or night, and, most importantly, provides an
essentially instantaneous wide-swath "picture" of wave conditions both outside and
inside the MIZ.

Squire (1998) has summarized the results of both the surface and the satellite field
observations as follows:

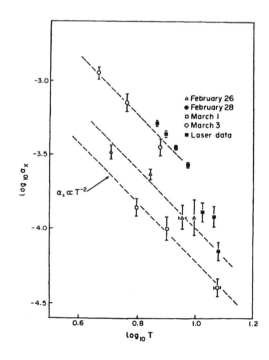

Figure 14.3. Log–log plots of the amplitude attenuation coefficient vs. wave period for the submarine and laser profilometer data sets collected
by Wadhams (1978b).

1. Wave attenuation vs. distance into the MIZ is a negative exponential, with the attenuation coefficient decreasing with increasing wave period over most of the spectral range. In heavy compact ice the coefficient typically varies from 2×10^{-4} for the longest swell to 8×10^{-4} for 8–9s waves, corresponding to e-folding distances of 5–1.2 km.

2. There is some evidence of a rollover for short-period waves (< 6–8 s), where the decay rate diminishes as the wave period shortens.

3. Inside the ice field the spread of the directional spectrum gradually increases, becoming essentially isotropic within a few kilometers of the ice edge.

4. In most cases when the ice edge is compact, only a small percentage of the wave energy is reflected from the outer edge of the ice field. (However, in some ice conditions SAR data indicate that total reflection is possible.)

14.3.2 Theory

Not being a theoretician I do not intend to provide the reader with a blow-by-blow account of the development of the different theoretical approaches to waves in sea ice. What I hope to accomplish is to provide the reader with a sense of the different approaches that have been used, their fundamental assumptions, how well they agree (or disagree) with observations, and, finally, where to find the mathematical details.

There have been two fundamentally different approaches. In the first the sea-keeping and flexural responses of each floe are treated separately in order to assess how that floe interacts with the incoming wave train. Then these results are combined over the entire MIZ. In the second approach, the ice is modeled as a continuum that behaves according to some assumed constitutive relationship and then the model parameters are "tuned" to achieve a best fit. Both approaches have their strengths and weaknesses.

14.3.2.1 Solitary Floe (Scattering) Theories   Here a solitary ice floe is assumed to be floating on an infinitely deep ocean and its response to an incoming wave train analyzed. Included in the analysis is the flexural behavior of the floe as well as its six rigid-body "shiplike" responses (heave, surge, sway, pitch, roll, and yaw). To calculate the floe deflections, a thin-plate model is typically used with the properties of the floe entering as density, thickness, Poisson's ratio, and effective elastic modulus. An approximate treatment of this problem was developed in the 1970s and a detailed discussion of this work can be found in Wadhams (1986). In the 1990s this earlier work was superseded by the work of Meylan and Squire (1993a, 1993b, 1993c), who were able to obtain a complete solution for the matching of velocity potentials across the ice floe edge. These authors have also provided a finite depth version of their development (Meylan and Squire 1994), focused on the breakup of fast ice. Figure 14.4 shows the roll response multiplied by the floe's length for floes of different lengths for a thickness of 1 m (Meylan and Squire 1994). The 1-m floe length is shown to demonstrate rigid-body resonance, which occurs when the thickness and length of a body are similar. Both the approximate and the more developed model predict that the ice in a MIZ will act as a low-pass filter. In its simplest form, using only surface fitting and considering only a single floe with a specified diameter d, the model results in an amplitude attenuation coefficient a_x, which generally decreases with increasing wave period except for a number of

Figure 14.4. The roll response multiplied by the ice floe's length for floes of different lengths (Meylan and Squire 1994). The ice has a constant thickness of 1 m.

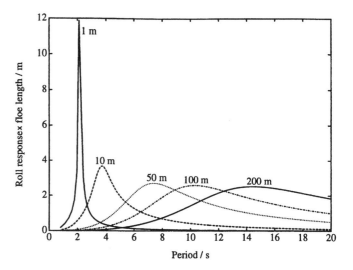

resonant periods when a_x goes to values near zero. In that these resonant periods are dependent on d, in real ice fields where there is a distribution of floe diameters resonances are smeared out by averaging. The result is an overall decrease in a_x as wave period increases. Although single-floe models predict a "rollover" in attenuation at wave periods of less than 5 s, this frequently is not observed in field data. Again, this is presumed to be the result of averaging the effects of floes of different sizes.

In that they do not treat floe–floe interactions, the application of these models is also limited to MIZ of low concentration. However, when ice–ice interactions are added, Meylan and Squire (1994) find that the serial application of the solitary floe results still provides an excellent approximation to reality unless the floes are very close together. In such cases, a variety of dissipative mechanisms result in poor agreement between theory and observation. These models also do not predict the observed broadening of the directional wave spectra as the distance into the MIZ increases. In fact, the models suggest just the opposite, that the spectra will become narrower as the waves move into the MIZ. As Squire (1998) notes, this result follows from the two-dimensional nature of the theory. For a more adequate theory a full three-dimensional model will be required in which the wave-making possibilities of the individual ice floes in the MIZ are considered. In that calculated wave-induced strains frequently exceed strains considered necessary for failure, the theory is in agreement with observations that in real MIZ, failures are common and can be highly regular.

14.3.2.2 Continuum Theories   The three different interrelated approaches that have been tried here have been referred to as the mass-loading model, the flexible sheet model, and the viscous model. The earliest of these approaches was the mass-loading model (Keller 1997; Keller and Weitz 1953; Peters 1950; Shapiro and Simpson 1953; Weitz and Keller 1950). Here the ice was assumed to be a continuum comprised of noninteracting mass points having no coherence or rheological properties. Although this approach was initially posited as applying to all types of ice, it has been found to be generally unsuccessful except when describing wave propagation in either frazil or pancake ice. This approach predicts that as a wave enters the ice the wavelength will be reduced, resulting in the waves requiring greater amplitude and steepness. The waves will also be refracted toward the normal to the ice edge. Both of these effects have been observed in frazil and pancakes in the Chukchi Sea by Wadhams and Holt (1991). In that there is no interaction assumed between the mass points in the ice cover, there is also no progressive wave decay. Squire (1993) has used a version of this model to consider a pack ice area in which the concentration varies spatially. He has found that in general the predicted attenuation coefficients introduced to produce damping increase too rapidly with frequency.

There are two versions of the most commonly used continuum model. Both assume that the ice behaves as a thin (or thick) continuous plate. In the first variation, both an elastic constitutive law and damping are applied a posteriori. In the second, a viscoelastic relationship is introduced from the start. In these theories wave refraction also occurs as waves enter the ice. In addition, refraction can occur either toward or away from the normal, depending upon the wavelength under consideration and the flexural rigidity L of the ice cover, a term which, in turn, depends on its thickness, elastic modulus, and Poisson's ratio. In the case of fast ice, the meaning of L is clear-cut and all but the very long waves bend away from the normal as they enter the ice. As Squire (1998) points out, this is not so clear for a normal open-ocean MIZ, where L is a parameterization whose value can vary from 0, where the continuum model reverts either to the mass loading model discussed earlier, or to its engineering value, as specified by the physical and mechanical properties of the ice. In short, for real MIZ with nonzero compactness, L is determined by fitting the theory to field observations. Figure 14.5 shows the critical-angle curves for several different thicknesses of solid sheet ice resting on deep water (Fox and Squire 1994). For conditions to the right of each curve the incoming wave generates a transmitted wave that is propagated into the ice sheet. However, if conditions are either on or to the left of the curve, incoming waves will be perfectly reflected at the ice edge and no propagating mode is produced within the ice. As can be seen, the thicker the ice, the more likely it is that total reflection will occur. Squire (1998) notes that the SAR observations of Liu, Holt et al. (1991) of the MIZ in the Labrador Sea MIZ showing 12s waves being totally reflected by 1-m-thick consolidated sea ice would appear to be possible only if the sea ice continuum has appreciable rigidity. The most recent work exploring different aspects of the application of an elastic-plate model to wave attenuation and floe breaking in the MIZ is by Kohout and Meylan (2008). Their model predicts an exponential decay of energy in agreement with observations. A comparison of the theoretical attenuation coefficients to field data suggests that their model is applicable to large floes for short- to medium-wave periods. Their floe-breaking model based on the wave attenuation model suggests an underprediction of the attenuation coefficients at long periods.

Several individuals have explored the usefulness of assuming that sea ice in the MIZ behaves according to some

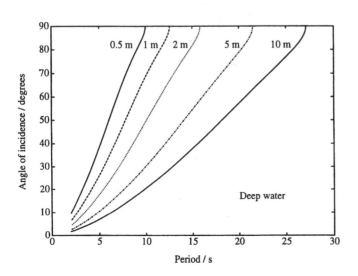

Figure 14.5. Critical angle curves for deepwater waves entering an ice cover (after Fox and Squire 1990).

type of viscoelastic relationship. For example, Wadhams (1973) found that a type of power law with a coefficient of 3, similar to that used by Glen (1955b) in his studies of the creep of polycrystalline freshwater (glacier) ice, could be used to describe the attenuation of ocean waves in pack ice as observed by Robin (1963). Other related approaches include that of Weber (1987) and of Liu and Mollo-Christensen (1988). In the first of these, the ice cover was assumed to be a thin, highly viscous Newtonian fluid. In the second it was assumed that the observed attenuation resulted from the presence of a hypothesized viscous boundary layer beneath the ice.

Such models do predict the observed rollover at short wave periods, and by picking the "right" free parameter values can be made to agree with the results of field observations. However, like the previous models, they also have their weaknesses. For example, to obtain reasonable fits to field observations extremely large variations are required in the free viscosity parameter (Liu et al. 1992). Squire (1998) notes that in Liu's approach a number of different types of energy dissipation are all assumed to be viscous. As a result, a large eddy viscosity is required to provide sufficient loss to agree with field observations.

14.4 Ice-Edge Bands

As noted earlier, when the wind direction is essentially off the ice, bands of fairly compact ice alternating with stretches of nearly open water frequently occur near the outer edge of the MIZ. Band widths are characteristically of the order of 1 km (parallel to the wind) whereas lengths are appreciably longer (on the order of 10 km). As the bands are located at or near the ice edge and are moving seaward, the ice involved is frequently melting. Another characteristic feature is that the outer (seaward) edges of the bands are frequently very clean and uniform, whereas the inner edges are more irregular. Bauer and Martin (1980) have been able to obtain careful observations of the movement of such bands in the Bering Sea. They observed that over a 14-hour period a band moved at an average speed of 0.4 m s^{-1} at an angle of 25° to the right of a 10 m s^{-1} wind. The long axis of the band remained normal to the wind direction. During the same period the remainder of the pack, although moving in the same direction, only drifted at 0.15 m s^{-1}. As a result the band readily separated from the pack and moved out into open water, where it rapidly deteriorated.

One possible explanation of band formation that has been advanced by Wadhams (1983c) and is in general agreement with field observations is as follows. As the result of waves battering the ice edge during periods of on-ice waves and swell, the outer strip of the MIZ is characteristically more broken and both aerodynamically and hydrodynamically rougher than ice further inside the pack. As a result, when an off-ice wind starts to blow, this edge zone is blown seaward at a slightly higher rate than ice deeper inside the pack. This small difference is believed to be enough to initiate leads near the outer edge of the pack. Shortly after this has occurred, the fetch in the wider leads becomes large enough to allow an off-ice wind to produce an appreciable number of fetch-limited short-period waves in these leads. Once this has occurred, so-called wave radiation pressure also becomes a factor. This is a force F_{wr}, which occurs when a wave collides with a floating body. For deep-water waves, it can be estimated by

$$F_{wr} = \rho_w g (a^2 + a'^2 - b^2)/4. \qquad (14.1)$$

Here F_{wr} is the forward force exerted by the wave per unit width of the wave front, and a, a', and b are the amplitudes of the incident, reflected, and transmitted waves, respectively (Longuet-Higgins 1977). As a result of F_{wr}, any floating object will be pushed downwind. Floes that are shielded from the short-period waves in the leads will continue to drift at a rate specified only by the wind stress. However, more exposed floes are driven by both the wind stress plus the wave radiation pressure and therefore will drift faster. Once these floes encounter a less-favored floe they tend to attach themselves, collecting stray floes as they go. These clusters of floes ultimately will tend to form a band oriented normal to the wind, in that the forcing is downwind and the collection length in that direction is self-limited: only the floes on the upwind side of the band directly benefit from the wave radiation pressure. On the other hand, there is no limit to lateral expansion. Once a band is formed it becomes a stable entity in that the wave force on the upwind side exerts a compressive stress through the band that is balanced by an opposing stress resulting from the longer waves and swell on the downwind side of the band. Considerably expanded discussions of this mechanism of band formation can be found in Wadhams (1983c, 1986, or 2000).

The reader may not be too surprised to hear that the above is not the only possible explanation for band formation that has been suggested. For instance, different internal water wave and wind wave theories have been advanced by Muench et al. (1983) and Martin et al. (1983). In addition, McPhee (1981, 1982, 1986) has posited that the rapid melting commonly associated with the outer bands in the MIZ may modify the under-ice boundary layer sufficiently to reduce the water drag on this ice near the outer edge of the MIZ, leading this ice to break away. More purely meteorological explanations have also been advanced by Häkkinen (1986) and by Chu (1987). It is important to stress here that there is no reason why only one of these explanations should be correct and the others, wrong. It is quite possible that more than one process contributes to ice band formation. I think that it is safe to say that the definitive paper on ice band formation has yet to be written.

14.5 Conclusions

All that a brief chapter like this can do is to attempt to provide the reader with an impression of the highly variable ice conditions observed in the MIZ and of our current understanding of how and why these conditions develop. Some of these features are sufficiently persistent to have been named (e.g., the so-called Odden that develops off the east coast of Greenland). Individuals interested in exploring these matters in more detail should examine the more lengthy reviews and symposia that are mentioned in the first paragraph of this chapter. The symposia are useful in that they provide one with a glimpse of the oceanographic consequences of the MIZ such as eddies, jets, and meanders—subjects not discussed in the present chapter. For views of a range of such features, see Johannessen et al. (1994) and Wadhams (2000). An example of such an interesting, only recently reported ice-edge feature is the tongue of sea ice that develops in the Southern Ocean at ~85°E (Rintoul et al. 2008). The feature extends more than 800 km to the north of the general regional ice edge and covers an area greater than 200,000 km². The size of the feature appears to be largely related to wind variations, with northerly (southerly) anomalies inhibiting (promoting) the development of the feature. Current observations indicate that ice drift is generally strongly northward along the axis of the feature and that the feature is primarily the result of advection of ice from the south, as opposed to thermodynamically driven in-situ formation.

In conclusion, I would like to point out something that may already have occurred to the reader. This is the fact that over the past 50 years much of our knowledge of the MIZ has been contributed by three individuals initially associated with Cambridge University and SPRI: specifically Robin, Wadhams, and Squire. Without their efforts and insights, our current knowledge of the MIZ would only be a shadow of its current state. I would also note that studies of this particular aspect of sea ice currently remain very active. A demonstration of the level of this activity can be found in the bibliography of the recent review by Squire (2007), which specifically focuses on studies completed between 1995 and 2007 and which contains 166 references.

15 Snow

But where are the snows of yesteryear?
F. Villon

I used to be Snow White, but I drifted.
Mae West

One might think that there would be a very large literature concerning the snow that rests on top of the world's sea ice covers. This is not the case. There are several reasons that snow on sea ice has traditionally received far less attention than the underlying ice, even though it is clearly more readily accessible. First, snow was not perceived as the problem. It was the ice that could fail to support you and your equipment, beset and sink your shipping, gouge up your buried pipelines, destroy your offshore platforms, etc. Sure, snow was there, but it just went along for the ride.

This is not to imply that the importance of the insulative and reflective properties of snow in determining the growth and ablation rates of the underlying ice was not known; see, for example, the discussions in Malmgren (1927), Zubov (1945), Holtsmark (1955), and Maykut (1986). For example, Zubov references field observations by Ponomarev made in 1942 showing that ice in a snow-free roadbed proved to be almost 1 ½ times thicker than nearby ice covered by the natural snow cover. The presence of snow also smooths the upper surface of the ice, thereby modifying the ice–air drag coefficient (Andreas et al. 1993) as well as the bulk transfer coefficients for latent and sensible heat (Andreas 1987). Furthermore, its presence adds an additional layer of complexity to the interpretation of sea ice type and concentration via the analysis of data collected by satellite-borne microwave systems (Carsey 1992). Finally, through its effect on sea ice growth and ablation, snow affects the structure and dynamics of the upper ocean via the input of brine (Gordon and Huber 1990) and freshwater (Fahrbach et al. 1991).

It's just that until fairly recently, detailed observations of the characteristics of the snow on sea ice were either lacking or unavailable. Although every glaciologist has, at one time or another, measured a few snow thicknesses and densities and noted the presence of layers of different snow types, it is also fair to note that the study of snow is a speciality only practiced at a high level by a few individuals. As a result, detailed studies of the properties of the snow cover on sea ice are few. Even so, as I will try to show in this chapter, recent studies that have focused specifically on the properties of the snow cover on sea ice have, in concert with scattered earlier observations, allowed the development of a reasonably consistent picture of the nature of the snow covers existing on the sea ice in both the Arctic and the Antarctic.

15.1 Arctic

If you spend time during the winter walking around on sea ice in the Arctic Basin, you will soon reach the following conclusions:

1. There is not a great deal of snow on the ice. Average snow thicknesses are typically a few tens of centimeters, with thicknesses of over a meter only occurring rarely.

2. Snow distribution patterns are primarily the result of the wind-induced redistribution of snow during storms.

3. Snow thicknesses are quite variable, with the deepest snow occurring in drifts localized by the presence of rough ice that is the result of pressure ridging. This fact follows from conclusion (2).

4. Snowdrift features tend to be quite varied and complex. Although classic barchan dunes, with their crescentic wings pointing downwind, do occasionally occur on floes that have flat upper surfaces, they do not appear to be particularly common. When present they often form on new, undeformed ice in refrozen leads (N. Untersteiner, pers. obs.). This is in keeping with observations on sand barchans in more normal deserts that generally occur in flat areas that are devoid of vegetation and where the total amount of drifting sand is comparatively limited. I would guess that the wings of the barchans on sea ice are indicative of the prevailing wind direction as opposed to the storm wind direction, although I know of no studies on this subject. The most frequent dunes are longitudinal, with their long axes aligned in the direction of recent storm winds. Such dunes frequently start at pressure ridges and can extend downwind for considerable distances (tens of meters). I would guess that in many cases, the nose of the dune is located, not just at a ridge, but at a low point in the ridge that acts to channel the near-surface drift of snow during storms.

5. Sastrugi, the sharp irregular ridges that form on a snow surface as the result of wind erosion and deposition, are frequently observed (Figure 15.1). The WMO Glossary states that on drift ice the ridges of the sastrugi are parallel to the prevailing wind direction at the time of their formation. It should have added that insofar as sastrugi form during storms, it is the storm wind direction that is important.

6. At least in the Arctic Ocean, there appears to be a fairly consistent pattern in the variation in the snow thickness with the time of year. The basic controlling factors are simple:
   a. significant snowfall requires nearby sources of atmospheric moisture, and
   b. the moisture carrying capacity of very cold air is small compared to that of air at near-freezing temperatures.

When freeze-up starts in the fall, open leads are common and frequently the boundary between the pack and the open ocean is fairly close. In addition, the air then is capable of absorbing and transporting appreciable amounts of moisture. Therefore, after freeze-up snow accumulates rapidly for the first month or two. By this time the nearby leads are ice-covered and the open ocean is farther removed. In addition the air is now colder and less capable of transporting moisture. As a result, the amount of atmospheric moisture available to form snow has drastically decreased. Dur-

Figure 15.1. Sastrugi, Beaufort Sea. The dark object in the photograph foreground is a glove.

ing the remainder of the winter, the amount of additional accumulation is small although, as noted, the existing snow cover may undergo considerable redistribution during storms. In the spring, snow accumulation again slightly increases, with the maximum snow thickness occurring just prior to the onset of the melt season. Interestingly, in the Arctic Basin snowfalls are actually more frequent during the summer than during the winter. For example, observations during the SHEBA experiment indicated that ~40% of the annual snowfall occurred during the summer months (Sturm, Holmgren, et al. 2002). However, as this snow either falls into open water or onto the surfaces of melting floes, its lifetime is limited and it does not contribute to the mass balance of the floes. These observations are also consistent with the meteorological estimates that in the Arctic Basin maximum precipitation occurs in the late summer (July to September) and also with the rawinsonde-derived seasonal cycle of moisture flux convergence across 70°N (Serreze et al. 1995).

Some relevant information for the Arctic Basin is as follows. In the fall the formation of new ice in leads starts in August near the Pole and in late September to early October at more southerly sites, although dates may vary considerably from year to year. Many times the first new ice that forms is subsequently destroyed. Once a stable new ice cover exists, the buildup of snow can start. Although in the fall the buildup of a snow cover on MY ice can predate the formation of new ice on the leads, the time difference appears to be small and no appreciable consistent differences in snow thickness have been reported. Early snow observations include those of Yakovlev (1960), made on the Soviet Drift Stations 2, 3, and 4; of Untersteiner (1961), made on Drift Station Alpha; and of Hanson (1965), made on Alpha II and ARLIS II and later on AIDJEX (Hanson 1980). Based on these early field observations, similar trends were assumed by Maykut and Untersteiner (1969) in their initial ice growth simulations, e.g., that a linear increase in snow depth of ~30 cm occured during the fall (20 August to 30 October) followed by only 5 more centimeters of snow during the winter (1 November to 30 April), with the deposition of an additional 5 cm of snow in May.

Figure 15.2. Seasonal cycles of snow depth on MY arctic sea ice (Warren et al. 1999). The values shown are the means obtained on all NP stations where complete years (August–July) were sampled. Also included are the seasonal cycles for three other local areas in the Arctic Basin.

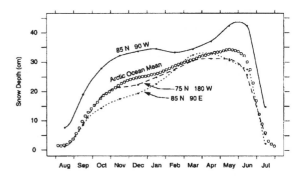

Considering the size of the Arctic Basin, these early observations would have to be considered sparse both spatially and temporally. It was only in 1999 that adequate studies of snow thicknesses and densities became available (Buzuyev et al. 1997; Warren et al. 1999). The critical difference here was the availability of the snow thickness and density observations obtained between 1954 and 1991 on Russian NP (North Pole) Drifting Stations 3 through 31 and related data obtained between 1959 and 1988 on the *Sever* Program of aircraft landings (see Chapter 2). Of these two data sets the NP data are the most important, consisting of snow thicknesses measured at 10-m intervals along lines starting at least 500 m from the NP stations (to eliminate drift effects associated with the presence of the stations) and extending outward another 500 to 1000 m. Two thirds of the lines were 500 m long, with the remaining third 1000 m in length. Densities were determined every 100 m along the lines. Although the ends of the lines were marked by stakes, there were no stakes spaced at 10 m along the lines. Measurements were performed once or twice a month at each station, resulting in a total of 499 line measurements during 1954–1991.

What does the NP data set tell us about the temporal and spatial variations in the snow cover? As can be seen in Figure 15.2, the seasonal thickness trend is similar to that assumed by Maykut and Untersteiner (1969): a sharp increase in snow thickness immediately after freeze-up, a more gradual thickness increase during the winter reaching a maximum thickness in the spring, and a rapid thickness dropoff in the spring–early summer. To study variations in the lateral distribution of the snow, the NP data for all 37 years were grouped according to the month that the data were collected, and plotted on maps and fitted with a two-dimensional quadratic function. Figure 15.3 shows the results for April and May, times when the snow pack can be expected to be at or near its maximum thicknesses. Similar plots for the other 10 months of the year can be found in Warren et al. (1999). The mean thicknesses were 33.7 cm for April and 34.4 cm for May, with maximum values of 40 and 42 cm estimated by the trend surfaces. Considering that the NP data sets include thickness values from snowdrifts associated with deformed ice, these results support the earlier statement that snow thicknesses on sea ice are typically in the range of a few tens of centimeters. Further support for this conclusion can be seen in Figure 15.4, a plot of the probability distribution function based on the 21,169 snow thickness measurements made during the one-year operation (1997–1998) of the SHEBA program in the Beaufort Sea. During that time period the camp drifted 2,800 km from 75°N, 142°W to 80°N, 162°W. Although the average snow depth was only 33.7 cm, the distribution has a pronounced positive skew, with a maximum thickness of 1.4 m presumably occurring in a snowdrift.

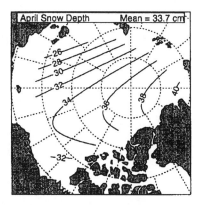
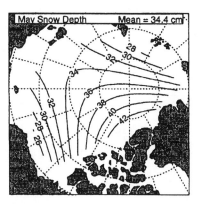

Figure 15.3. Two-dimensional quadratic functions based on data on mean snow depths measured on MY ice during April and May for the time period 1954–1991 (Warren et al. 1999).

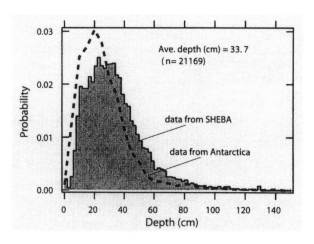

Ave. depth (cm) = 33. 7
(n= 21169)

data from SHEBA

data from Antarctica

Figure 15.4. The probability distribution function (PDF) for all SHEBA snow depths compared with a PDF for snow on ice in the Bellinghausen, Ross, and Amundsen Seas, Antarctica (Sturm, Holmgren, et al. 2002).

Warren et al. (1999) summarize their lateral distribution results as follows. Initially there is rapid accumulation of snow in the autumn north of Canada and Greenland as compared to the Alaskan and Siberian sectors. However, during the winter these two latter sectors catch up. As a result, in February there is a fairly uniform snow thickness of ~30 cm across the Arctic. In the spring, snow accumulation resumes north of Canada and Greenland. As a result the geographical pattern for May is similar to that of November. The regional differences are greatest in June, in that north of Siberia and Alaska the snow thickness declines rapidly whereas at higher latitudes north of Greenland accumulation is still occurring.

Warren et al. treated the less extensive *Sever* data set separately in that it is different in two important ways. Because of operational constraints, the *Sever* data were only obtained in the spring. Furthermore these sites were not restricted to MY locations as were the NP sites. As a result the *Sever* data include FY ice in coastal regions that were not sampled by the NP stations. The most important difference, however, is that the *Sever* data were grouped into different classes according to the microtopography of the ice surface at the site: snow on the prevailing ice of the landing area, sastrugi, drifts behind ridges, and the average of the windward and leeward sides of hummocks. Trend surfaces based on this data set can also be found in Warren et al. (1999). Some of the results appear to be reasonable: the depth of snow and the height of the sastrugi on the landing areas, as well as the thickness of snow on the sides of the hummocks, increase, as presumably

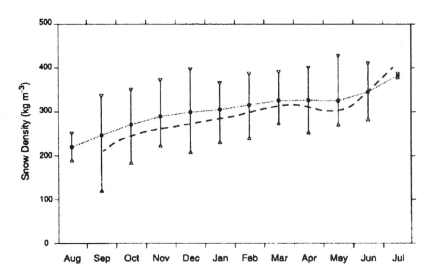

Figure 15.5. Mean snow density from the NP samples for each month (large solid dots; Warren et al. 1999). All available density measurements for each month are used, irrespective of year and geographic location. The error bars indicate one standard deviation. For comparison, the values of Loshchilov (1964), based on observations at stations NP-2 through NP-9, are shown as the dashed line.

does the age of the ice as one moves from offshore Siberia towards Fram Strait. Also snow depths in hummocked areas are thicker than in the flat areas where the aircraft have landed. On the other hand, the snow depth behind ridges appears to decrease as one moves toward Canada, while at the same time the height of snow around hummocks increases. Warren et al. find this to be peculiar, as do I. As a result they question whether the *Sever* data set can be taken as representative of the general trends. In that the climate of the Arctic Basin appears to be changing at present, one may have to wait for a long time for this question to be resolved.

The temporal variation in snow densities is also both interesting and fairly systematic, as can be seen in Figure 15.5. In September, values start at approximately 250 kg m^{-3} and by May have gradually increased to 320 kg m^{-3}. The scatter is considerable. At any specific time there appears to be little systematic geographic variation in the density values.

Studies that focus on changes over a season in the internal structure of the snow on sea ice appear to be rare. Fortunately there is one notable exception, completed as part of the SHEBA program between October 1998 and October 1999 (Sturm, Holmgren, et al. 2002). As this is both a very thorough study and appears to be representative, it can be used to provide additional insight into the nature of the snow covers of the Arctic Basin. For instance, at the end of the winter the average snow depth at SHEBA (33.7 cm) was nearly identical to the tabulated and mapped values obtained in the same sector of the Arctic over the time period between 1954 and 1991 (Radionov et al. 1997; Warren et al. 1999). The same can be said for the trends observed at SHEBA at other times of the year: the rapid accumulation of snow in the fall, the roughly constant thickness during all but the late winter when some additional accumulation occurs, and finally, the rapid decrease in snow thickness during the spring. As noted earlier in Figure 15.4, over the season the snow thickness values obtained during SHEBA varied from 0 to 150 cm and showed a positive skew. At any given time, thickness distributions were rather variable, a condition that persists even when snow depths are separated according to the type of ice acting as a platform (Sturm, Perovich, et al. 2002, Fig. 9). There were, however, some anticipated trends. For example, snow on thin ice was also thin and thickness variations were small, as the ice had not been present long enough for a thick snow cover to form. Also, snow on deformed ice was both thicker on the average and more variable.

The general stratigraphic relations observed in the snow pack at SHEBA in the spring are shown in Figure 15.6. The actual profiles as seen at the different sampling sites (195 snow pits) varied a bit, largely as the result of one or more layers being missing; usually these were layers located higher in the stratigraphic column. Variations were also noted in the thickness of the different layers and in their density, hardness, and degree of metamorphism. However, in general the stratigraphy was typical of snow that is deposited in a windy environment and subjected to pronounced vertical temperature gradients. Three basic snow types were present in the form of ~10 layers: depth hoar (layers b, c, d, and e), wind slab (layers f, g, and j) and recent (layers h and i). The lowest layer, which was deposited prior to the initiation of field observations, is presumed to have originated as ~10 cm of snow that ultimately converted to ice. The snow layers were the result of 10 fairly discrete weather events that usually lasted less than 24 hours. The recent (April–May) snow layers showed a wide range of grain sizes and shapes that were determined by the meteorological conditions during the snowfall (Magono and Lee 1966; Nakaya 1954). The two most prominent snow layers (wind slabs f and g) were the result of multiday storms that combined both wind and snow. However, as Sturm et al. note, even when these multiday events are included, the combined time for all the key weather events was only a small fraction of the entire winter (6%). As its name suggests, wind slab is the result of the tumbling action

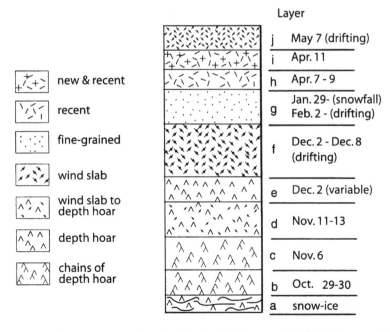

Figure 15.6. A generalized snow stratigraphy for the SHEBA area, 1997–1998 (Sturm, Holmgren, et al. 2002). The symbols follow the International Classification for Snow on the Ground (Colbeck et al. 1982). The dates indicate the approximate times when the layers were deposited. The snow ice was derived from snow that had fallen prior to the author's arrival in the field. Layers h and i were very similar and difficult to differentiate in the field.

to which drifting snowflakes are subjected during storms. This milling process produces small (0.3–0.8 mm) rounded grains that pack together tightly and sinter rapidly (within a few days), resulting in a strong dense layer. The snow involved is usually a combination of snow falling during the storm plus grains of preexisting snow scoured from the snow cover.

Superimposed on these two types of snow layers is the development of depth hoar, a low-density, brittle, highly permeable type of snow comprised of ornately faceted grains (5 to 15 mm) and having a low thermal conductivity. Layers *b* through *d* have been converted into classic depth hoar as the result of strong temperature gradients across the snow pack during the early part of the winter. In addition, during April and May diurnal temperature cycling produced small (0.5 mm), distinctively faceted depth hoar crystals within the surface snow (Birkeland 1998).

Now that we know the nature of the snow pack as described at SHEBA, what does this tell us about the nature of the snow packs at other sites in the Arctic Basin? Actually, quite a lot. The wind slab and new snow layers should have similar properties to the layers at SHEBA, although the sequence of layers and their thicknesses would vary as the result of differences in the site-specific meteorological conditions. Whatever the sequence of snow types and thicknesses, one would expect to find an appreciable portion of the snow thickness modified by depth hoar formation as the Arctic Basin provides ideal conditions for depth hoar formation. Everything necessary for the establishment of a pronounced temperature gradient across the snow pack is present: very cold surface air temperatures, comparatively warm temperatures at the base of the snow as the result of the approximately fixed temperature at the sea ice–ocean interface ($\sim -1.8°C$), and the thinness of the snow layer. The fact that two of the three types of snow found on the sea ice in the Arctic are wind slab and depth hoar explains the earlier observation that during the winter months arctic snow packs show very little densification. Work by Armstrong (1980) and Sturm and Benson (1997) has demonstrated that depth hoar is stiff in vertical compression and usually undergoes only limited compaction. In addition, the presence of depth hoar is a clear indication that a pronounced vertical vapor flux has occurred in the snow. This process, which over time removes ice from the affected layers within the snow pack, tends to counter densification by compaction. The result is a density that is nearly constant with time. Wind slab layers, on the other hand, are deposited at such high densities that, once in place, they show little tendency for further compaction. Finally, in the Arctic typical sea ice snow packs are too thin to provide the overburden pressure necessary for appreciable compaction in any snow type except recent snowfalls as their initial densities are very low.

Chapter 10 provided the reader with several empirical and theoretical equations (10.11, 10.13, 10.14, and 10.15) that have been used to predict the thermal conductivity of snow ξ_{sn} based on measured values of its density. However, ξ_{sn} is also known to be a function of snow temperature and texture (Mellor 1977; Sturm et al. 1997). Here texture is a factor that includes grain size, shape, and the degree of bonding. It is also known that texture can vary from one snow layer to another. Therefore one might ask how well the above relations predict the ξ_{sn} values of real Arctic Basin snow packs. The problem has been that although numerous density measurements of Arctic Basin snows are now available, only very recently have studies been completed that combined densities with textural descriptions plus thermal conductivity measurements (Sturm, Perovich, et al. 2002). At SHEBA measurements of ξ_{sn} were completed on approximately 10% of

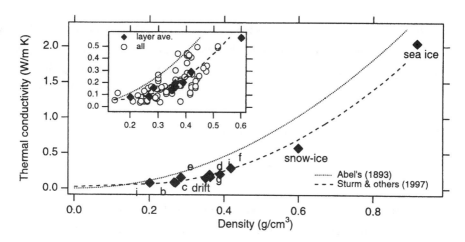

Figure 15.7. Average layer thermal conductivity (ξ_{sn}) values as a function of the average layer density (Sturm, Perovich, et al. 2002). The regressions shown are from Sturm et al. (1997) and Abel (1893). The inset shows the individual as well as the layer values. The letters refer to the layer designations of Figure 15.6.

the layers, on which density, hardness, and type were determined. The resulting values were as follows: averages varied from 0.078 W m^{-1} K^{-1}, with the highest for a layer of snow ice (layer a, Figure 15.6) to the lowest value for a layer of newly deposited snow (layers h and i combined). For nonicy snow the highest value (0.290 W m^{-1} K^{-1}) was for the hard wind slab, layer f. As might be expected, layer average values of ξ_{sn} increased smoothly with increased snow density. As can be seen in Figure 15.7, regression equations 10.13 and 10.14 of Sturm et al. (1997) provide ξ_{sn} estimates that are in good agreement with field observations at SHEBA. Earlier empirical relations such as Abel's predict higher conductivities than observed. This result should not be too surprising in that the measurement techniques used at SHEBA are exactly the same as those used to produce the earlier data set on which the empirical relations were based. These results, however, do suggest an important conclusion. There does not appear to be anything unusual about the snow on sea ice. This is not saying that the snow cover on the sea ice in the Beaufort Sea is exactly the same as the snow cover on the North Slope of Alaska or Siberia. What it says is that within a textural snow type, snows with similar densities appear to have similar thermal conductivities independent of location.

By using a combination of the measured ξ_{sn} values and the snow stratigraphy, Sturm et al. estimate the average bulk ξ_{sn} value of the full snow pack at SHEBA to be 0.14 W m^{-1} K^{-1}. They then estimated this same parameter by using observations on ice growth rates plus the temperature gradients in the snow. One might expect the estimates to be similar. They were not: the latter estimate of ξ_{sn} was 0.33 W m^{-1} K^{-1}, over twice as large as the value based on the probe measurements. The authors suggest that this mismatch is the result of the difference in the scale of the two measurement techniques. The probe measurements are clearly point values. Ice growth measurements, however, include considerable heterogeneity in both the snow and ice layers. If these two values had been in general agreement, it would have appeared to be saying that bulk values of ξ_{sn} obtained from point measurements can be used on a variety of spatial scales. Although this would be very convenient, it does not appear to be true. Certainly the lateral transfer of heat as the result of nonplanar geometric considerations is a possible explanation. In fact, Sturm et al. were able to show, via the use of a finite element model, that geometric considerations were definitely able to produce local "hot spots." However, when these were taken into account, the apparent conductivity only increased by a factor of 1.4, not enough to explain the observed factor of 2.3. It is also conceivable that the mismatch is the result of mechanisms that are not captured in a vertical "point source" profile. I refer the reader to Sturm, Perovich, et al. (2002) for a more detailed discussion of this problem. As the reader will discover, there are more possibilities than answers.

15.2 Antarctic

From what we already know about the environmental conditions in the antarctic sea ice zone, I would guess that it will not be too surprising to find that the snow cover on the sea ice there is a bit different than in the Arctic. First, the nature of the data sets in the Antarctic are quite different from those in the Arctic. In the Antarctic there are no equivalent long-term data sets concerning snow similar to those obtained on the Russian NP drift stations. There also have been no equivalent intensive drift station programs such as Alpha, T-3, AIDJEX, or SHEBA in the Antarctic. The closest would be the Joint Russian–U.S. Ice Station Weddell (ISW-1) that operated for four months during the austral fall (February) and early winter (June) of 1992. Although snow thicknesses and properties were determined in conjunction with thermistor-based vertical temperature profiles, to date I am not aware of these data being published as a separate study. At present, most of our detailed knowledge of the snow cover on the sea ice in the Southern Ocean comes from short-term ship-based observations, primarily those carried out during the last 10 to 15 years. However, as was the case for the Arctic, by piecing these data together and adding in scattered information collected at coastal stations, it is possible to construct a plausible description of the similarities and differences between the snow cover on the ice in the Southern Ocean and that on the ice in the Arctic Ocean (Massom, Eicken, et al. 2001).

In the following I list several known aspects of the environment of the antarctic sea ice zone and follow this by the associated snow characteristics.

1. As is the case in the Arctic, wind is the major factor determining how the snow is distributed on antarctic sea ice. As a result snow thickness is only roughly related to the frequency and duration of snowfall. In fact, at nearshore fast ice areas affected by katabatic winds blowing off the continent, there is little accumulation during the winter as all the snow is blown further out to sea. There, at locations 5–15 km offshore where the katabatic wind velocities drop off, snow thicknesses of as much as 1 m can be found on FY ice (Fedotov et al. 1998).

2. A summary of mean snow thickness observations combined with information on location and the age of the underlying ice can be found in Massom, Eicken, et al. (2001). Figure 15.8 shows frequency histograms of snow thickness values for five different sectors of the Southern Ocean. As can be seen, snow thicknesses vary widely both seasonally and regionally. With the exception of the snow-free zone near the coast scoured by katabatic winds, during the winter the thinnest snow occurs on the northernmost pack ice. This distribution is not an indication of less snowfall. Instead, it is primarily caused by most of

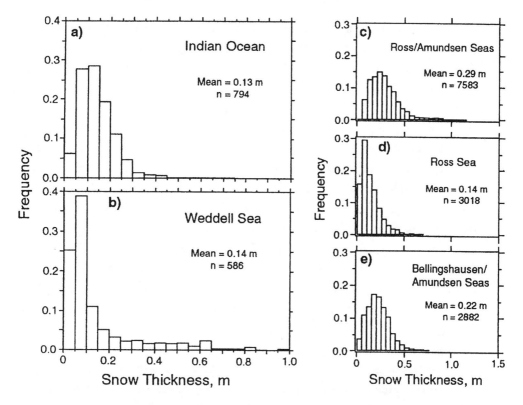

Figure 15.8. Frequency distributions of snow thicknesses as observed on the sea ice of five different sectors of the Southern Ocean (Massom, Eicken, et al. 2001). The seasons and sources of the observations are as follows: (a) Indian Ocean, winter 1995; (b) Weddell Sea, winter 1992; (c) Amundsen and Ross Seas, winter–spring 1994; (d) Ross Sea, autumn 1995; and (e) Bellingshausen and Ross Seas, winter 1995. Detailed source information can be found in Massom, Eicken, et al. (2001).

the ice there being thin FY. In that this ice has only formed recently, it has had little chance to accumulate a thick snow cover (Worby et al. 1996). In many cases, the presence of a unusually thick snow cover suggests that the floe being examined is more than one year old (Jeffries and Adolphs 1997). All the snow thickness distributions show pronounced positive skews, with individual values ranging from 0 to ~1.0 m. Mean values at any given time (0.13 to 0.22 m) are appreciably lower by one-half to one-third compared to similar values in the Arctic. Based on their combined experience, the 14 authors of the Massom, Eicken, et al. (2001) paper suggest that on FY ice, snow thicknesses greater than ~0.25 m represent values obtained from snowdrifts.

3. The positive tail of the snow thickness distributions generally appears to be less pronounced in the Antarctic than in the Arctic. Another way to say this is that the snowdrifts appear to be smaller in the Antarctic than in the Arctic. This is reasonable in that drifts are typically associated with ridges and antarctic ridges are known to be smaller and more irregularly distributed than arctic ridges. Although I believe this to be a true effect, it does not explain the distribution tails as shown in Figure 15.8 in that snow drifts around ridges were not included in the data used to prepare these histograms. Instead Massom, Eicken et al. (2001) suggest that the distribution tails are caused by the thick snow covers on very thick floes that are

presumably more than one year old. This tells me that if the snow thicknesses around ridges were included, the histograms would be even more strongly skewed.

4. The snow covers on the sea ice of both polar regions are similar in that they are both far from uniform slabs, a situation that should be expected considering that both locations are very windy. All arctic snow types appear to occur in the Antarctic, i.e., new snow, fine-grained snow, hard wind slabs, and depth hoar. However, some antarctic snow types noted to date do not appear to occur nearly as frequently in the Arctic; specifically soft and moderate slabs, icy layers, and saline slush. Hard slabs also appear rare in the Antarctic, being replaced by soft and moderate slabs. These differences are believed to be temperature related in that the warmer and even above-freezing temperatures frequently occurring in the Southern Ocean favor softer slabs, icy layers, and slush. Other characteristics of the snow on the pack of the Southern Ocean include the presence of hard, icy surface crusts, thicker ice layers within the snow cover, and the formation of superimposed ice produced by the percolation of meltwater down to the upper sea ice surface, where it refreezes. Many times this process results in an ice layer comprised of coarse, spherical grains. More will be said later about the formation of superimposed ice.

One reasonable question that one might ask concerns whether there are any indications of systematic changes in snow type as latitude decreases. The limited amount of published data bearing on this question is shown in Figure 15.9. As can be seen, although there are trends, r^2 values of +0.42 and −0.42 can hardly be called impressive. At least one of the trend directions is reasonably easy to rationalize. As one moves further north and warmer weather is encountered, including the possibility of rain on snow, the fact that icy and melt grain clusters occur slightly more frequently in the snow pack is not surprising. The fact that the development of depth hoar decreases with decreasing latitude at first glance also makes sense, as depth hoar development is driven by strong temperature gradients across the snow pack, a situation more frequently encountered at higher latitudes where air temperatures are lower. However, the more I think about this, the less clear the situation becomes. The missing link here is information on exactly where and when the ice and snow at each study site formed and the exact path taken to reach the sampling site. Perhaps what the negative slope is telling us is that when sampled these floes had either not drifted far from where they initially formed or, more likely, they had all drifted similar distances northward. What Figure 15.9 is definitely telling us is how few detailed descriptions exist of Southern Ocean snow packs, considering the vast expanses that they cover every austral winter.

5. Massom, Eicken, et al. (2001) summarize the observed variations in snow densities by relating the values to current weather conditions as follows: dry and cold; low density (200–300 kg m^{-3}); warm and windy: medium to high density (350–500 kg m^{-3}); and melt-refreeze conditions (400–700 kg m^{-3}). Local small-scale variability can be very pronounced. For instance, Massom et al. (1998) report variations of between 240 and 600 kg m^{-3} across a single floe based on 47 measurements taken at 1-m intervals. When these values are compared to the typical arctic density range of 250 to 320 kg m^{-3}, one can see that high densities appear to be more frequent in the Antarctic than in the Arctic. I would guess that this difference is more apparent than real and is at least partially due to the fact that most antarctic

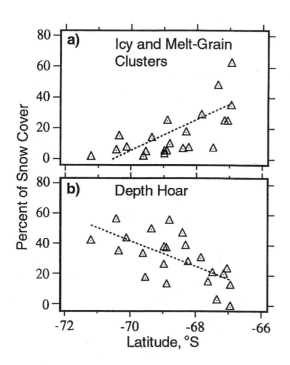

Figure 15.9. Variations in the percentage of (a) icy and melt-grain clusters, and (b) depth hoar vs. latitude (Sturm et al. 1998). The observations were made on a cruise in the Ross and Amundsen Seas between days 262 and 285, 1994. When similar plots of soft and moderate slabs and new and recent snow were made vs. latitude, the correlations were essentially zero.

measurements are made during the summer when icy layers are present and ship operations are comparatively easy. On the other hand, arctic measurements are commonly made in the winter when icy layers are rare and aircraft operations are favored.

6. Numerous investigators have reported the presence of wet to flooded snow at the base of the snow pack on the ice in the Southern Ocean. In the austral spring and summer such conditions are hardly surprising because of melting conditions. However, occasionally melting has been reported even during the winter months. A far more important cause of wet snow and flooding is the incursion of seawater onto the surface of the sea ice. This is a phenomenon that can occur any time of the year and is not at all rare (Eicken, Fischer, et al. 1995; Jeffries, Li, et al. 1998; Takizawa 1984; Wadhams et al. 1987). There are a variety of factors that contribute here. For instance, in the outer parts of the pack, wave–ice interactions can result in seawater being deposited on the upper ice surface. Here by *outer* I do not necessarily refer to ice within a few kilometers of the ice edge. During periods when an intense long-period swell has been running, fracturing of the ice sheet with associated flooding of the upper ice surface has been reported at distances of ~300 km south of the ice edge. Massom et al. (1999) report that one such event flooded an estimated 20% of the sea ice surface over a meridional band ~75 km wide. I observed such an event during the winter in the Bellingshausen/Amundsen Sea area. It was very impressive, converting large ice floes into fragments 2 to 3 m on a side in a matter of minutes.

Perhaps an even more important contributor to flooding is the fact that in many antarctic situations, the freeboard of the ice floes is very close to zero. In such cases an appreciable snowfall will depress the upper ice surface to below sea level, resulting in flooding (Jeffries, Hurst-Cushing, et al. 1998). Data tabulated by Massom, Eicken, et al. (2001) compiled from observations made on a number of different cruises indicate that slush occurred at the base of the snow pack at between 11 and 51% of the sites examined. It appears that the ease of flooding in the Antarctic is the result of repeated cycles of snow/ice formation. As a result the freeboard is always near zero and even a small amount of new snow is sufficient to depress the ice surface below sea level, initiating additional flooding and superimposed ice formation. However this works, it is clear that superimposed ice formation is much more common in the Antarctic than in the Arctic. This is just the opposite of what one might expect based on the fact that the snow cover on the sea ice is generally thicker in the Arctic than in the Antarctic. An additional important factor is the fact that the oceanic heat flux in the Antarctic appears to be at least twice that observed in the Arctic. As a result, for the same thickness of snow cover antarctic sea ice is typically warmer and thinner than arctic ice and as a result more porous and permeable. The development of superimposed ice is in itself a contributing factor. When flooding occurs, this raises the temperature of the snow–ice interface to the freezing point of seawater ($\sim -1.8°C$). This additionally lowers the growth rate of the sea ice while at the same time increasing the temperature gradient across the snow, a condition that favors the development of depth hoar. During the formation of superimposed ice, a thickness of snow that has a low thermal conductivity is replaced by a thinner layer of ice that has an appreciably higher thermal conductivity. The point that I am trying to get across here is that the sea ice in the Antarctic dances to a different drummer. At first glance it looks similar to its northern relative. However, this is a case where looks are deceiving. I would also suggest that this is an area where additional computer modeling would be very useful. All the ingredients exist: take one part Maykut and Untersteiner, throw in some mushy layer theory at the interface, add a recent snow code on top, and vary the ocean heat flux. Challenging, but definitely possible. It also has the advantage of being one-dimensional. In fact, a study along these lines (Jordan et al. 2002) using data collected on Ice Station Weddell has been able to achieve very good agreement with field observations. A description of the model used can be found in Jordan (1991).

7. As a result of the above processes, it is not surprising to learn that the snow covers on antarctic sea ice are very frequently saline. In fact, Massom reports that on some cruises, particularly in the Indian and West Pacific sectors, all the snow covers were saline. What is surprising to me is not the fact that these snow covers are salty, but that they are frequently extremely salty, with mean salinities for the total snow column often in excess of 8‰—a situation where the snow cover may actually be more saline than the underlying sea ice (Massom, Eicken, et al. 2001; Table 8). Extreme situations can occur when frost flower mats with salinities as high as 110‰ become the base of the snow cover (Figure 15.10). These salinity values, which are well in excess of that of normal seawater, are believed to be produced by wicking concentrated brine from the sea ice into the snow (Drinkwater and Crocker 1988). In some instances, upward brine expulsion may also be a contributing factor (Perovich and Richter-Menge 1994). Massom et al. (1998) also note that in the snow pack at heights of above 0.3 m background salinity values are generally < 1‰.

   Although it is clear that the snow on the ice in the Southern Ocean is consistently saltier than the snow cover of the Arctic Ocean, the difference may not be quite as large as it seems. One must remember that there are some fundamental differences in the sampling settings of these two regions. In the Antarctic most snow study sites have been located on

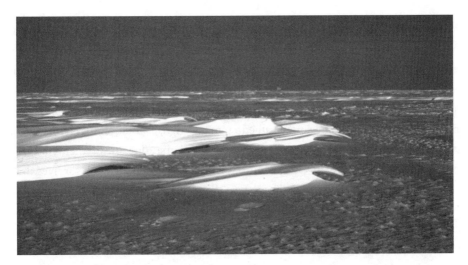

Figure 15.10. The flat "nubbly" surface in the foreground is an example of frost flowers that are gradually being infilled by drifting snow. The overhanging shapes of the undercut sastrugi demonstrate the fact that snow has an appreciable tensile strength (sand, on the other hand, does not).

FY ice that is thin (< 1 m) and relatively warm. In the Arctic most snow study sites have been located on MY ice that is thick (> 2 m), cold, and previously desalinated by passing through one or more summers. It's not that individual processes are different in these two parts of the world; rather, the oceanographic and atmospheric environments of these regions are sufficiently different to cause the dominant processes that affect the snow cover to change.

8. From what has already been said concerning the more varied snow types present on the ice cover of the Southern Ocean, one would expect the thermal conductivity of the snow there to be more varied than in the Arctic, although the total range of values would presumably be about the same. In attempting to estimate ξ_{sn} values for Southern Ocean snow packs, Massom et al. (1998) and Sturm et al. (1998) have grouped ξ_{sn} values according to snow type to obtain a mean and a range for each type. These values were then weighted according to the proportion of that snow type observed in the pit studies in a specified area or a particular cruise, to estimate the bulk ξ_{sn} value for the snow pack. Combining these results they obtained the following estimates: 0.112 (autumn–winter), 0.124 (spring), 0.138 (winter–spring), 0.141 (winter), and 0.164 W m^{-1} K^{-1} (winter). As was the case in the Arctic, these values are also appreciably lower than values that have typically been used in model studies ($\sim$0.31 W m^{-1} K^{-1}).

9. Based on what has already been said about antarctic FY sea ice, one might reasonably expect that antarctic MY sea ice would also prove to be different. This turns out to be the case. In the Arctic, by the end of the summer the surviving ice has essentially lost all its snow cover to melt. As a result, when the snow cover starts to accumulate it does so on a cooling, thickening sheet of desalinated ice that has an appreciable freeboard. As the upper layer of the sea ice is almost totally desalinated (see Figures 8.7, 8.8, and 8.9), the situation is somewhat similar to snow falling on lake ice or on the lower end of a glacier.

In the Antarctic, the situation at the end of the summer is totally different. Melt ponds commonly are either absent or comparatively rare in the south. This difference has been attributed to the relatively low humidity of the strong winds that drain off the continent (Andreas and Ackley 1981). The result is increased latent heat loss from the ice to the atmosphere, plus air temperatures that are generally below freezing. As a result the snow cover on the surviving sea ice generally remains intact and the surface albedo remains high, a situation that limits the input of shortwave solar radiation and its associated melting. Of course, one must remember that, as was shown in Table 1.1, only a relatively small proportion of the antarctic pack survives the austral summer. This ice, primarily located in the western Weddell, Amundsen, Bellingshausen, and eastern Ross Seas, has been able to minimize bottom melting by remaining near the continental edge. At the end of the summer, these floes, which usually have lost appreciable mass during the summer from bottom melting, are comprised of a thick snow cover resting on a thin layer of warm sea ice. The freeboard is either near zero or slightly negative. Once snow starts to accumulate in the fall, it adds to the negativity of the freeboard and furthers the formation of superimposed ice.

I first saw such floes in 1991 while in transit from McMurdo Station to Valparaiso, Chile, on the USCGC *Polar Sea*. Although we were unable to investigate them at the time, their appearance was strikingly different from the majority of the surrounding FY ice in that the snow cover was very thick (> 1 m), with the flat upper snow surface standing well above the other floes. The sides of the floes were frequently vertical walls of snow. Often it was impossible to determine the thickness of the actual sea ice as it appeared to be depressed below sea level. Fortunately in 1992 Jeffries, Veazey, et al. (1994) were able to sample several similar floes in the Ross and Amundsen Seas. All of the floes sampled had high flat tops and vertical walls. Snow thicknesses varied from 0.3 to 2.0 m, although a few floes (unsampled) had estimated snow thicknesses in the 3- to 4-m range. There was no indication of melting, either at the snow surface or within the snow cover. Individual snow densities varied from 91 to 543 kg m^{-3}, with mean values for individual pits varying from 240 ± 51 to 446 ± 41 kg m^{-3}. Although the snow densities were very high, there was no indication that the values were due to overburden. For example, a density of 400 kg m^{-3} was observed at a depth of only 0.2 m below the surface. Both the highest snow densities and the largest snow loads were observed in the Ross/Amundsen region.

The contribution of snow ice formation to the mass balance of the sea ice appears to be highly variable, with some authors suggesting that the contribution is relatively minor (Ackley et al. 1990; Eicken, Oerter, et al. 1994; Lange et al. 1990), and others suggesting a more significant role (Jeffries and Adolphs 1997; Jeffries et al. 1997, Jeffries, Li, et al. 1998). It currently appears that snow ice formation is particularly important in the Eastern Pacific sector. A contributing factor would be that this appears to be a region of enhanced snowfall.

15.3 Conclusions

If the reader only retains one concept from the above discussion, I would suggest the following: *Although the interactions between sea ice and its snow cover are reasonably straightforward in the Arctic, in the Antarctic the situation is much more complex.*

In the Arctic the snow gradually accumulates during the winter while the ice thickens. The thermodynamic situation is such that the snow–sea ice interface is always above sea level (i.e., the freeboard is always positive). In the summer the snow melts. The next fall, if the ice survives, the process starts all over again, producing even thicker sea ice and a similar snow cover. The only difference is that the snow can now start to accumulate on preexisting ice.

In the Antarctic the snow also gradually accumulates during the winter as the ice thickens. However, the thermodynamics have changed in that the air temperatures are higher and the oceanic heat flux is larger. As a result the snow–sea ice surface is frequently below sea level. This, coupled with a wave-dominated environment, results in the base of the snow pack frequently being flooded. This raises the temperature, causing the temperature gradient across the snow to increase and the temperature gradient across the ice to decrease. This in turn favors the formation of depth hoar in the snow cover and slows the growth of the underlying sea ice. The flooded area at the interface frequently freezes, thereby converting snow into ice. This process may reoccur several times during the winter. During the summer, if the snow either melts or is stripped off the sea ice layer, the sea ice is also destroyed. If the sea ice survives, its snow cover also survives. When snow accumulation starts in the fall, the freeboard again becomes negative and the cycle of flooding and superimposed ice formation starts all over again. Further "normal" thickening of the underlying sea ice layer is limited as the layer is well insulated. It is also undoubtedly very porous, facilitating flooding. However, other rather unpredictable factors such as the input of frazil and/or marine ice may possibly contribute to the thickening of the sea ice layer.

Individuals interested in exploring these antarctic matters further should first refer to Massom, Eicken, et al. (2001). Although I have relied rather heavily on this reference in preparing the present review, the original treats the subject in far more detail. It would also be worthwhile to obtain a copy of Jeffries (1998) as this volume contains several excellent papers that deal with different aspects of snow in the Antarctic. Although I have tried to avoid discussing remote sensing in the present volume, the reader does not have to be an expert to sense the added complexities that the remote sensor must consider when dealing with antarctic sea ice. Clearly using the same analysis procedures in both the Arctic and the Antarctic could prove to be misleading.

16 Ice Dynamics

Predictions are always difficult,
particularly when they involve the future.
Samuel Goldwyn

by W. D. Hibler III (with W. F. Weeks)

16.1 Some Background

As I remarked in "About This Book," this chapter is an update of a portion of the review titled "Modeling the Dynamic Response of Sea Ice" by Bill Hibler (2004). It also has been modified in an attempt to fill in some basics that were not discussed in the original. More importantly, several sections of the original have been omitted as it was felt that they offered more detail than was appropriate for the present book. It is hoped that this fact will provide the reader with an incentive to read the original paper. Although the present chapter assumes that the reader has some experience with large-scale computer simulations, I believe that it is profitable reading for individuals totally lacking such experience. What the reader will gain is an appreciation for the many different factors that a modeler must consider as well as the surprisingly complex interactions between the atmosphere, the sea ice, and the ocean. Individuals desirous of additional background material on sea ice drift and dynamics should also refer to the recent book by Matti Leppäranta titled *The Drift of Sea Ice* (2005), as well as to the many references cited in the following chapter.

Native peoples living in the polar regions have been interested in the stability of fast ice and the movement of pack ice for a very long time. As marine mammals are important food items whose availability often hinges on the hunter being able to operate successfully on fast ice and frequently even on pack ice, going onto the pack at the wrong time could, at best, result in a cold trip to an uncertain destination. At worst, it could result in one's permanent disappearance. For instance, the antarctic exploration literature contains several instances in which individuals attempted poorly timed fast ice crossings and presumably found themselves on northward-moving ice floes. At least, this is what is assumed to have happened in that the individuals involved simply disappeared. Because of the nature of the drift of the Southern Ocean pack such occurrences were invariably fatal. In the north, because of the more constrained nature of the drift, hunters who were skillful and lucky had a far better chance of survival under such circumstances. During the first AIDJEX Pilot Program, I had an opportunity to talk with an elderly hunter from Tuktoyaktuk, N.W.T. As a young man he had survived over a month on the pack before being able to work his way back to land. He remembered this event very clearly.

As noted in Chapter 2, scientific interest in the motion of pack ice goes back to Nansen and the drift of the *Fram* in the Arctic Ocean from 1893 to 1896. Nansen's belief that if he locked a ship into the ice pack off the coast of Siberia, it would exit the ice several years later off the east coast of Greenland was based on the fact that the *Jeannette*, a ship that had been abandoned near the New Siberian Islands in 1881, was discovered off east Greenland in 1884. Fortunately for Nansen, the *Jeannette*'s drift proved to be a representative forecast of the drift of the *Fram*. Subsequent analysis of the *Fram*'s movement across the Arctic Basin (Nansen 1902) showed that the wind-driven drift of the ship averaged 2% of the wind speed and also that the drift direction was 28° to the right of the wind direction. In analyzing this data set, Nansen considered the wind stress, the water stress (drag), and the Coriolis force. As his analytical solution was not very successful in predicting the observed drift, Nansen postulated the existence of a constant gradient current to explain the difference between theory and observation. The real usefulness of the *Fram* drift data was that it led Ekman (1905) to postulate the existence of a spiral in both the atmospheric and the oceanic boundary layers. The Ekman spiral has proven to be one of the fundamental building blocks of boundary-layer meteorology and oceanography.

Almost 20 years passed before the next paper appeared on this general subject (Vize 1923), although the Russians clearly had for some time been considering ways to improve their ice forecasting capabilities as the ability to successfully move shipping across the Northeast Passage was very important. Other individuals involved include Karelin, Lebedev, Nazarov, Shuleikin, Somov, Volkov, and Zubov. Much of this work focused on methods for estimating ice growth and decay. A discussion of these early efforts with appropriate references can be found in Armstrong (1955). There appears to have been considerable difference of opinion concerning the effectiveness of different approaches to ice forecasting. For example, Vize (1923) explored the possibility of developing useful correlations between climate and ice conditions for different regions of the Arctic. Zubov, on the other hand, stressed the importance of improving our understanding of processes occurring within the ice pack. One result of his efforts was the so-called Zubov's rule (Zubov 1945; Zubov and Somov 1940). This simple relation assumes that ice drift follows the geostrophic wind (i.e., that the wind direction is parallel to the isobars), and that the wind speed could be estimated from the pressure gradient. He then used Nansen's results by taking the difference between the wind direction and the ice drift direction to be 28° and the drift speed to be 2% of the wind speed. His final relation was $s = 13,000 \partial p / \partial x$ where s is the drift speed in km/month and $\partial p / \partial x$ is the pressure gradient in mb/km obtained from monthly pressure charts. Ultimately he compared the predictions of the rule with the observed drift during 1937–1940 of the beset ship *Sedov*. The results were encouraging, considering that the quality of the pressure field estimates for the Arctic Basin were generally poor at that time.

More recently this general approach has been advanced by Thorndike and Colony (1982), who used smoothed drifting buoy data and wind fields to develop a relation predicting ice motion on short time scales when ice drift is typically more dominated by wind than currents. Their starting relation was

$$\mathbf{u} = A\mathbf{G} + \bar{c} + \varepsilon \qquad\qquad (16.1)$$

where A is a complex constant and u, G, $\bar{c}$, and ε are vectors representing, respectively, the ice velocity, the geostrophic wind, the mean ocean current, and the part of the velocity that is neither constant nor a linear function of the geostrophic wind. The coefficient A involves both a scaling factor and a turning angle

$$A = |A| e^{-i\theta}. \qquad\qquad (16.1a)$$

The values of $|A|$ which relate the ice speed to the geostrophic wind speed proved to vary seasonally, with the largest values occurring in the summer when the ice is the most open. In addition, the turning angle measured clockwise from G to u is larger in the summer. Both these results imply that the state of the ice pack is important in affecting the drift of sea ice. It was also found that on a short time scale more than 70% of the variance of the ice velocity was explained by the geostrophic wind, with the remainder due to the mean ocean circulation. On longer time scales oceanographic effects became more important, with the percentage of the variance explained by the wind dropping to closer to 50%. In addition, the geostrophic wind was less successful in explaining ice motion at locations less than ~400 km from coasts. Clearly a number of different influences are at work here and would have to be considered in any effective sea ice model.

Other early work included the paper by Sverdrup (1928) who, in studying the drift of both the *Fram* and the *Maud*, considered the resistance or friction between the ice floes but did not add either a gradient current or water stress. In 1935 Rossby and Montgomery added a logarithmic boundary layer at the ice–water interface, as did Shuleikin in 1938. The next to the last of these initial, somewhat isolated, efforts was the study of Reed and Campbell (1960) who, in addition to including the Coriolis force, added logarithmic ice–ocean and ice–air boundary layers into the estimates of the wind and water stresses. Finally, Campbell (1965) added a gradient current and a term representing the internal ice stress. Additional detail on these papers and other early related efforts can be found in Campbell (1968).

This work set the stage for AIDJEX, the Arctic Ice Dynamics Joint Experiment, whose purpose was to collect a consistent set of ice drift observations over an extended time period and via the study of this information expand current modeling capabilities (Pritchard 1980b; Untersteiner et al. 2007). AIDJEX also contributed new concepts such as the ice thickness distribution as well as different rheological approaches for simulating ice–ice interactions within the pack. Perhaps even more important, AIDJEX was effective in introducing new scientific talent to the study of sea ice. Fortunately, the interest of these individuals in this field did not terminate with the end of the project, as many have remained active in this research area to this day. For instance, individuals associated with AIDJEX who have specifically contributed to model development and the study of ice dynamics include Ackley, Brown, Campbell, Colony, Coon, Hibler, Maykut, McPhee, Pritchard, Rothrock, Thomas, Thorndike, and Untersteiner. Considerably more will be said about these contributions in the rest of this chapter. It should also be noted that several of the fundamental concepts developed during AIDJEX are currently under attack as being inadequate (Coon et al. 2007).

16.2 Observations of Sea Ice Motion and Deformation

It is important to have an understanding of what observations tell us about the motion of pack ice. The general drift rates and some information on variability can be deduced from the mean monthly positions of several historically interesting drifting stations plotted in Figure 16.1. Two notable features in this image are that ice floes located north of a latitude of ~77° and west of a longitude of 180° drift over the pole and out the Greenland–Svalbard passage (Fram Strait). This route is frequently

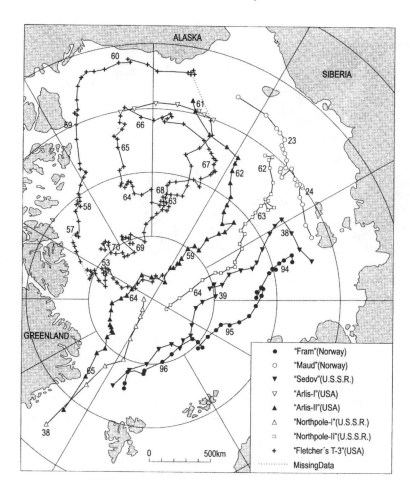

Figure 16.1. Tracks of several drifting stations in the Arctic Ocean. Monthly positions are plotted, with the year of each January 1 position indicated. (From Leppäranta 2005; redrawn from Hibler 2004; originally plotted by Hastings 1971).

referred to as the Transpolar Drift Stream. The second feature is the clockwise rotation north of the Alaska coast referred to as the Beaufort Gyre. Other features of interest are the presence of substantial drift (even in winter) parallel to the Alaska coast, and a relatively small component of motion perpendicular to the Canadian archipelago near the pole. A feature of drift tracks not shown in the figure is the meandering nature of the drift, with weeklong reversals in direction being common. Also, as drifting stations approach Fram Strait, the drift rates begin to increase and meandering decreases. This phenomenon continues after the ice exits from the basin, with drift rates being as large as 25 km per day in the East Greenland region (Vinje 1976). As a generality, the highest drift rates are invariably observed near the ice edge where floe–floe interactions are near zero. It's hard to sprint when you are in the middle of a crowd. In recent years, the presence of large numbers of remote buoys in the Arctic have made the ice-drift field the most precisely measured surface field of any world ocean.

Because ice drift generally follows the wind, the shifting of ice-drift patterns in response to large-scale oscillations in the wind field is currently of great interest. Recent work has emphasized patterns of wind fields coincident with the North Atlantic Oscillation (NAO) and the Arctic Oscillation (AO) indexes and their effect on ice drift (Polyakov et al. 1999; Proshutinsky and Johnson 1997; Wang and Ikeda 2000). The NAO and AO are measures of shifts in high-latitude wind fields in the northern hemisphere (Hurrell 1995). Under these classifications, there is a weakened anticyclonic flow together with a more concentrated Beaufort Gyre under so-called high-index conditions

(a)   (b)

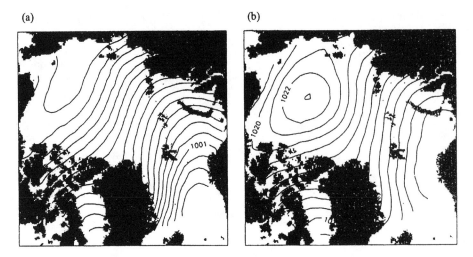

Figure 16.2. Mean sea level pressure for the months of December–March (1978–1996): (a) NAO index > +1; (b) NAO index < −1. (Contour interval −1 mbar.)

(Figure 16.2a). With a low index, the drift pattern tends to have stronger anticyclonic flow with a much larger and more pronounced Beaufort Gyre, together with greater recirculation of ice (Figure 16.2b). Differences in ice thicknesses deduced from sonar data in the Beaufort Sea between the early 1980s and the late 1990s can be related to these changes. In particular, Tucker et al. (2001) found significantly thicker ice in the Beaufort Gyre region (86°N, 146°W) in 1986–1987 during the low NAO index than in the 1990s, when a high index prevailed. They attributed much of this difference to longer residence ice time owing to the rotary ice circulation and hence greater accumulation of ice mass through ridging.

In the Arctic, a dominant term in the sea ice mass budget is the ice flux through Fram Strait. Probably the most detailed observations of outflow have been made by Kwok and Rothrock (1999) by the use of passive microwave satellite data. These results (Figure 16.3) show an areal outflow that varies somewhat with location in the Fram Strait. The outflow tends

Figure 16.3. Typical areal outflow estimates over the time period 1979–1996 (Kwok and Rothrock 1999). Illustrated here are the scattergrams showing observations as well as the correlation and regression line between monthly area flux and gradient in the monthly sea level pressure across the Fram Strait for October–May (dotted line) and December–March (dashed line).

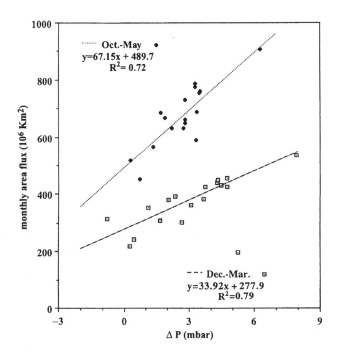

to correlate strongly with both the average geostrophic wind through Fram Strait and with the NAO. Moreover, the best-fit regression coefficient between the December–March area flux and geostrophic wind is one-half of the October–May regression coefficient. Also, the outflow magnitudes are significantly smaller during December–March for the same atmospheric pressure gradient across the Strait. This difference is most likely due to mechanical effects resulting from ice–ice interactions (see section 16.5.3). Although areal outflows are relatively precisely measured by this method, ice-mass outflow is more uncertain owing to lack of knowledge about the lateral thickness profile of the sea ice across the strait (Vinje et al. 1998). Typical long-term estimates (Aagaard and Greisman 1975; Kwok and Rothrock 1999; Vinje and Finnekasa 1986) of ice-area outflow and ice-mass outflow (based partially on moored upward-looking sonar observations) are ~0.1 Sv (3,154 km^3 per year) and ~900 km^2 per year for ice area. There is considerable inter-annual variability. Areal ice volume fluxes from interpolated buoy drift data (Colony 1990) are of a similar magnitude. Interestingly, as technology has presumably improved over the last 15 years, volume fluxes have decreased from highs of 4–5×10^3 km^3 per year (Vinje and Finnekasa 1986; Wadhams 1983b).

Concomitant with ice drift is ice deformation. This deformation fluctuates in a manner that is not directly related to the wind field, especially at high frequencies. As will be discussed, ice deformation is believed to be significantly modified by ice mechanics. Typical measurements of deformation are shown in Figure 16.4. These results were obtained from an experiment (Hibler, Weeks, et al. 1974) in the spring of 1972 in the Beaufort Sea as part of AIDJEX. In this experiment a detailed set of small-scale measurements of the temporal variability of the strain were obtained by using laser surveying techniques to measure the positions of a large number of reflectors mounted on towers on the ice. The smoothed deformation (Figure 16.4c) from this meso-scale array compares well with larger-scale observations. In addition to deformation occurring in response to large-scale wind fields, there is a definite tendency for deformation to oscillate at around 12-hour intervals (1 cycle/12 hrs = 0.083 cycle/hr; Figure 16.4d). In the polar regions this period coincides with the inertial period and also with the semidiurnal tidal period. This high-frequency character of ice deformation has been a general feature of deformation measurements almost everywhere, including the East Greenland marginal sea ice zone (Leppäranta and Hibler 1987) and the antarctic ice pack (Heil et al. 1998).

In terms of ice drift there also appears to be a seasonal character to the oscillations in drift (Thorndike and Colony 1980), presumably because of greater damping by increased ice interaction during the winter. The spatial structure of deformation, however, often tends to take place along linear-oriented patterns. This is effectively illustrated by spatial deformation patterns on a 5-km footprint obtained from processing synthetic aperture radar (SAR) imagery (Kwok 2001). Although such deformation patterns, termed *linear kinematic features* by Kwok (2001), have been noted for some time based on aerial and satellite imagery (e.g., Marko and Thompson 1977), the actual deformation taking place concomitant with the formation of leads has not normally been directly determined. Examination of SAR and visual imagery has also led to a consistency in lead intersection angles. In particular, Cunningham et al. (1994) demonstrated a remarkable tendency for leads to intersect at angles of about 40–50°, with the orientation of the leads tending to be symmetric about the principal axes of the strain rate tensor. There are also appreciable differences in ice

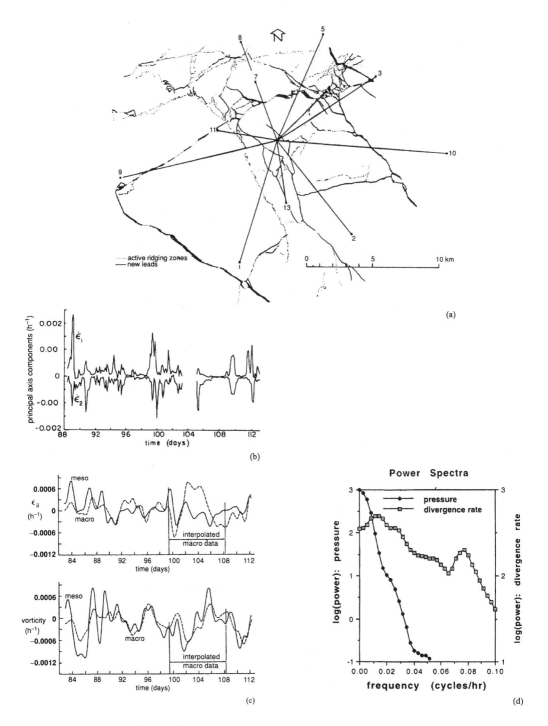

Figure 16.4. (a) Mesoscale strain array in the late winter/early spring of 1972 in the Beaufort Sea with an overlay of active leads and ridging zones on April 6, 1972.

(b) Ice deformation time series from the mesoscale strain array.

(c) Comparison of smoothed mesoscale and macroscale (strain triangle over ~100-km region) divergence rates and clockwise vorticity (Hibler, Ackley, et al. 1974).

(d) Spectra of atmospheric pressure and mesoscale ice divergence rate from the strain array shown in (a) (Hibler 1974).

deformation and production depending upon whether the region under study is located in FY ice characteristic of the seasonal sea ice zone (SSIZ) or deep in the basin in an area predominately containing thick, old ice types. These differences have recently been studied by Kwok (2006) through the use of four years of SAR data (1997–2000). Not surprisingly, a distinct seasonal cycle was observed in both zones, with both deformation and ice production being highest in the late fall and a minimum occurring in midwinter. Also, both ice deformation and production were highest in the SSIZ, where deformation-related ice production was 1.5 to 2.3 times greater than in regions of primarily old ice. When the data set was examined as a whole, it was found that seasonal ice growth in leads accounted for ~25 to 40% of the total ice production in the Arctic Ocean. Furthermore, the correlations between divergence in the ice and cyclonic motion in the atmosphere were found to be negligible in both zones during both the winter and the summer.

Important manifestations of deformation are large pressure ridges which, as discussed in Chapter 12, routinely reach depths greater than 20 m and have been observed to have keels as deep as ~50 m. In the central Arctic ~10% of the ridges were deeper than 20 m, whereas in the most heavily ridged region ~10% of the ridges were below 30 m. Observations of spatial and temporal variations of ridging in the western Arctic Basin (Hibler, Mock, et al. 1974) deduced from airborne laser profilometers generally show that the spatial variability of ridging correlates well with mean thickness estimates obtained from submarine sonar data (e.g., Bourke and Garrett 1987; Bourke and McLaren 1992). In addition to these basinwide variations, nearshore ridge studies (Tucker et al. 1979) off the north slope of Alaska and Canada show a buildup of ridging near the coast which decays several hundred kilometers offshore. Field studies show ice in this region to be highly deformed with substantial amounts of rubble (Kovacs 1976). The magnitude of ridging can vary significantly from year to year. Hibler, Mock, et al. (1974), for example, found the ridging intensity (a rough measure of the mean thickness of ridged ice) to be 50% larger off the Canadian archipelago in the winter of 1972 than in 1971, and to comprise (assuming triangular ridges) on the order of 2 m^3/m^2 of ice in 1972. Estimates of ridged ice-area fractions based on submarine transects are generally quite high: 60–80% by Williams et al. (1975), 44% by Wadhams and Horne (1980), and 38–43% by McLaren (1989). Observations of the physical and structural properties of ice in the Fram Strait region found that one-third of ice from cores drilled in multiyear ice (which comprised over 84% of the ice volume in the region) were identified as ridged; this was despite the fact that floes, which appeared from the surface to be ridged, were explicitly avoided (Tucker et al. 1987).

16.3 Stress Measurements

Fortunately, the development of the capability to make stress measurements efficiently (Cox and Johnson 1983) has provided a means to measure stress buildup in pack ice directly. Heretofore, estimates of stress have been largely based on comparisons of nonlinear sea ice models with buoy-based ice-drift observations. Representative sets of ice-stress measurements have now been made in both the Eastern Arctic (Tucker and Perovich 1992; Tucker et al. 1991) and the Western Arctic (Beaufort Sea) (Richter-Menge 1997; Richter-Menge and Elder 1998). Typical measurements of stress (Figure 16.5) in both thick and thin ice show a high degree of variability, with changes

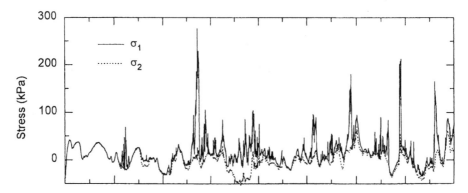

Figure 16.5. Principal stress components at a site ~300 m from the edge of a multiyear floe measured over a 160-day interval (Richter-Menge and Elder 1998).

appearing to be associated with the deformation field over a region. Records from sensors located near the edges of multiyear floes are suggestive of rapid stress buildup and release, implying loading and subsequent fracturing. In the centers of the floes, the stresses are smaller and have a somewhat reduced magnitude inasmuch as they represent the integrated stress around all the floe boundaries. In the Beaufort Sea the central floe stresses tend to have maxima in the range of 50 kPa. At edge sites where stresses are highly concentrated, the stresses may rise to 200 kPa.

A dominant component of measured stresses in pack ice is a diurnal cycle (Lewis and Richter-Menge 1998). This cycle tends to be suppressed in the absence of a daily temperature swing and is thought to be caused by thermal stresses in the ice floes owing to differential heating (Lewis 1998). Because of this strong diurnal variability, it is not clear whether semidiurnal variations (Tucker et al. 1991) are a separate phenomenon, or are simply harmonics of a daily thermal cycle.

An important characteristic feature of geophysical-scale ice mechanics is the stoppage of flow through narrow channels by "arching," as analyzed, for example, by Sodhi (1977) in the Bering Strait. This arching phenomenon forms a free surface downwind or downstream of a narrow passage, which can support enough stress to prevent ice flow through the channel. However, depending on the ice strength, increases in the wind or currents can break the arch, whose extent depends on the failure strength of the ice and the passage width. One of the most notable examples of recurring "arching" is in the narrow Nares Strait region above Baffin Bay. This phenomenon occurs every year after an initial period of ice flow through the Strait and plays a significant role in the formation of the North Water polynya by preventing the southward flow of ice into the North Water region. Relatively stationary "fast" ice is a mechanical/thermodynamic phenomenon prominent in shallow shelf regions of the Arctic. This phenomenon is somewhat related to the presence of arching in pack ice. Owing to the large shallow shelf seas in the Russian Arctic, fast ice is more extensive there and can extend far from the coast. Studies of the seasonal and interannual variability (Brigham 2000) of fast ice in the Russian Arctic show it to extend several hundred kilometers off the coast and to have a pronounced seasonal signal. Over the period 1975 to 1995, Brigham found the maximum extents to have a range of between 600,000 and 900,000 km^2 over the longitudinal band 58°E to 180°. Of this variability, only a small amount (~90,000 km^2) could be explained by a uniform linear reduction over two decades. Off the Alaska coast, fast ice often contains numerous ridging bands, indicating that it has been formed in incremental stages becoming gradually more extensive in character. Once formed, detailed measurements (Weeks et al. 1977) of the motion of the fast ice in the Prudhoe Bay area show it typically to move less than ~1 m horizontally. As might be expected, there are exceptions.

16.4 Modeling Sea Ice Drift and Deformation

16.4.1 Equations of Motion

Sea ice moves in response to wind and water currents and the internal stress in the ice. Considering sea ice to be a two-dimensional continuum, it obeys the normal Euler equations of motion which are generalizations of Newton's second law of motion (e.g., Fung 1997)

$$m \frac{D\mathbf{v}}{Dt} = \nabla \cdot \sigma + X \qquad (16.2)$$

where v is the ice velocity, m is the ice mass per unit area, σ is the second-order internal stress tensor in the sea ice due to ice interaction, and X is the total body force on the ice. The substantial derivative of the ice velocity is

$$\frac{Dv}{Dt} = \frac{\partial v}{\partial t} + (v \cdot \nabla)v. \qquad (16.3)$$

The interfacial stresses between the ice, atmosphere, and ocean are part of the body forces X in Euler's equation. Explicitly writing out these terms in the case that the interfacial stresses are known, the equation of motion becomes

$$m \frac{D\mathbf{v}}{Dt} = -mf\mathbf{k} \times \mathbf{v} + \tau_a + \tau_w + \nabla \cdot \sigma - mg\nabla H \qquad (16.4)$$

where, in addition to the symbols defined in equation 16.2, k is a unit vector normal to the surface, f is the Coriolis parameter, τ_a and τ_w are the body forces per unit area due to air and water stresses, and H is the height of the sea surface. In this formulation, τ_w is assumed to include frictional drag caused by the relative movement between the ice and the underlying ocean.

In the steady-state solution of the ice–ocean boundary layer (a good approximation at time scales longer than a few inertial periods), the air and water stresses are normally determined from idealized integral boundary layer theories, assuming constant turning angles (Brown 1980; Leavitt 1980; McPhee 1979):

$$\tau_a = c_a' \left(\mathbf{V}_g \cos\varphi_a + \mathbf{k} \times \mathbf{V}_g \sin\varphi_a \right) \qquad (16.5)$$

$$\tau_w = c_w' \left[\left(\mathbf{V}_w - \mathbf{v} \right) \cos\varphi_w + \mathbf{k} \times \left(\mathbf{V}_w - \mathbf{v} \right) \sin\varphi_w \right]. \qquad (16.6)$$

Here V_g is the geostrophic wind, V_w represents the geostrophic ocean currents near the ocean surface, c_a' and c_w' are air and water drag coefficients, and φ_a and φ_w are air and water turning angles. In practice, both the currents and the sea-surface tilt are estimated from geostrophic considerations by setting H equal to the dynamic height and computing currents by

$\mathbf{V}_w = (g/f)\mathbf{k} \times \nabla H$. In general, both c'_a and c'_w are nonlinear functions of the winds and currents. The two most commonly used formulations are (1) linear, where c'_a and c'_w are taken to be constant, and (2) quadratic, where

$$c'_a = \rho_a c_a \left| \mathbf{V}_g \right|, \tag{16.7}$$

$$c'_w = \rho_w c_w \left| \mathbf{V}_w - \mathbf{v} \right|, \tag{16.8}$$

with c_a and c_w being dimensionless drag coefficients [with typical values of ~0.0012 and ~0.0055, respectively (McPhee 1980)] and turning angles φ_a and φ_w of about 25° in the Arctic or −25° in the Antarctic. In linear models, a linear drag coefficient of $\rho(f\mathbf{K}_v)^{1/2}$ (where ρ is the density and $\mathbf{K}_v$ is the vertical eddy viscosity) is obtained from classical Ekman layer theory with a turning angle of 45°. In a real turbulent boundary layer the vertical eddy viscosity varies with depth and velocity (McPhee 1982). For this reason, quadratic drag laws with turning angles different from 45° give a better fit to observations (McPhee 1980) than the simple linear Ekman theory. However, it should be noted that the best fit to observations occurs with drag laws where the turning angle varies with the surface stress.

The dominant terms in the temporally averaged momentum balance in equations 16.4 to 16.8 are the air and water stresses and the ice interaction. The Coriolis term, tilt, and steady current terms are about an order of magnitude smaller. Also the inertia term in this time-averaged formulation is rather small, as it would take about one hour for ice to come to a steady-state drift state after the sudden imposition of a wind field. That the ice interaction term is the same order of magnitude as the wind and water drag has been verified by observation in enclosed areas, such as the Bay of Bothnia, where the ice sometimes hardly moves at all, even with significant winds (Leppäranta 1980; Omstead and Sahlberg 1977). However, as verified by a large number of numerical simulations, ice interaction typically opposes the wind stresses and hence does not prevent very high coherence between wind fluctuations and ice drift.

Although this steady-state formulation is used in almost all climate studies, typical observations of ice motion and deformation (section 16.2) show the motion to have considerable power at the inertial period, even in the absence of wind. An integral formulation (McPhee 1978) that retains this motion within the above framework considers the turbulence to be large, so that for all time scales longer than ~1 hr the boundary-layer transport and the ice motion are interdependent, with the integrated mass transport of the oceanic boundary layer $\left(\mathbf{M}_w = \rho_w \overline{h\mathbf{U}} \right)$ taken to be given by

$$\mathbf{M}_w = \left(\rho_w c_w / f \right) V \mathbf{V} e^{-i\beta} \tag{16.9}$$

where $\mathbf{V} = \mathbf{v} - \mathbf{V}_w$ is the ice velocity relative to the steady currents. In these expressions, $\mathbf{A} = A e^{i\delta}$ are two-dimensional vectors written in complex form, i.e., with δ the counterclockwise turning angle from the real axis. Substituting this expression into the equation of motion for the combined momentum $\left(\mathbf{M} = \mathbf{M}_w + m\mathbf{V} \right)$ for the ice and oceanic boundary layer results in

$$\frac{D\mathbf{M}}{Dt} = \tau_a + f\mathbf{k} \times \mathbf{M} + \nabla \cdot \sigma. \tag{16.10}$$

Expressing all momentum components in terms of the ice velocity $\mathbf{V}$, we obtain a more realistic equation of motion (in complex form) for the ice cover:

$$m\frac{D\mathbf{V}}{Dt} - \left(\rho_w c_w/f\right)\frac{D\left(VV\right)}{Dt}e^{-i\beta} + ifm\mathbf{V} + \rho_w c_w VVe^{-i(\pi-\beta)} = \tau_a + \nabla\cdot\sigma \qquad (16.11)$$

The main feature of equation 16.10 (or of equation 16.11) as compared with equations 16.4 and 16.6, is the presence of inertial oscillations largely driven by the oceanic boundary layer. These oscillations are undamped solutions of the homogeneous portion of equation 16.10 in the absence of ice interaction. They have periods of $(12/\sin\lambda)$, where λ is the latitude, and are a ubiquitous phenomenon in the upper ocean (Gill 1982). In the northern hemisphere, these oscillations describe a clockwise particle motion. In equations 16.4 and 16.6, these solutions are artificially damped out in less than an hour or so, which is unrealistic.

Note that in the steady-state limit, this more general expression (equation 16.11) reduces to the same quadratic drag result as expressed in equations 16.4 to 16.9. However, there is a major difference in the relative magnitude of the momentum balance components on time scales of less than a few days. The importance of the time-dependent character of the momentum equation can be demonstrated by comparing unsmoothed buoy drift characteristics with simulated values (Heil and Hibler 2002). Some of these characteristics are discussed in sections 16.5.3 and 16.6.6.

16.4.2 Deformation Scaling

To a large degree, the deformation or differential drift of ice may be considered to be a perturbation of the smoother ice drift, which, as noted above, tends to follow the wind field. This deformation is, however, critical to the dynamical evolution of the thickness of the sea ice cover as it affects open water creation and ridging. Because it is a "perturbation," however, the essential scaling can be quite different.

Aspects of deformation scaling may be examined (Hibler 1974) by using a linear version of the steady-state ice-drift equations coupled with a linear–viscous rheology, so the $\nabla\cdot\sigma$ term in the momentum balance is given by

$$\nabla\cdot\sigma = \eta\nabla^2\mathbf{v} + \zeta\nabla\left(\nabla\cdot\mathbf{v}\right), \qquad (16.12)$$

where $\mathbf{v}$ is the ice velocity, and η and ζ are constant shear and bulk viscosities, respectively. Expressing the geostrophic wind in terms of the atmospheric pressure by

$$\mathbf{U}_g = -\frac{1}{\rho f}\frac{\partial P}{\partial y} \qquad (16.13)$$

$$\mathbf{V}_g = \frac{1}{\rho f}\frac{\partial P}{\partial x}_g \qquad (16.14)$$

and taking the curl and divergence of the linearized momentum balance (equation 16.4) with c'_a and c'_w taken to be constant in equations 16.5 and 16.6 and $\nabla H = 0$, we obtain two linear equations for $\nabla\cdot\sigma$ and the vorticity $\omega\left(=\frac{1}{2}\left(\frac{\partial v_y}{\partial x} - \frac{\partial v_x}{\partial y}\right)\right)$:

$$\left[(\eta+\zeta)\nabla^2 - D\cos\theta\right]\nabla\cdot\mathbf{v} + \left[mf + D\sin\theta\right]2\omega = \frac{B\sin\varphi}{\rho f}\nabla^2 P \qquad (16.15)$$

$$-\left[mf + D\sin\theta\right]\nabla\cdot\mathbf{v} + \left[\eta\nabla^2 - D\cos\theta\right]2\omega = \frac{-B\cos\varphi}{\rho f}\nabla^2 P \qquad (16.16)$$

Similar equations for different components of the strain rate in a given coordinate system can also be obtained (Geiger et al. 1997).

Independent of the linear rheology assumption, these equations may be scaled with observed deformation data. Referring back to Figure 16.4 for deformation, we find typical values of divergence and vorticity to be $\sim 4\times 10^{-4}$ per hour and 8×10^{-4} per hour. Using the fact that the divergence of the surface wind field is ~ 0.14 per hour, the wind stress term in these deformation equations is about 10 times larger than the water stress terms in the equations of motion. Consequently, in this case the essential balance on scales of ~ 100 km is between the gradient of the surface stress and the gradient of the force due to the internal ice stress,

$$0 \approx \nabla(\tau_a) + \nabla(\nabla\cdot\sigma). \qquad (16.17)$$

Within the linear–viscous approximation in equations 16.15 and 16.16, the effect of high viscosity is to smooth the atmospheric pressure field spatially in order to obtain the local deformation. This linear procedure has also been applied directly to ice drift with best fits to observations yielding substantially higher viscosities in winter than in summer (Hibler and Tucker 1979). However, application of such linear models is limited to relatively homogeneous regions far from boundaries. As discussed in section 16.5, to explain ice mechanics successfully, highly nonlinear rheologies are generally required.

Limiting solutions of equations 16.15 and 16.16 for large viscosities in the infinite boundary limit (Hibler 1974) shows that the ice would be expected to converge if the surface winds were directed inward, which could occur owing to a low-pressure region. This is in contrast to the free drift solution, which applies as $\eta\zeta \to 0$, where $\nabla\cdot\mathbf{v} \sim \nabla^2 P$. The vorticity ω, however, has the same sign in either limit. Comparisons of these limiting cases with temporally smoothed observations (Figure 16.6) show the large viscosity limit to be in much better agreement with the observed deformation, indicating that the effects of ice interaction are likely to be very substantial. Moreover, least-squares regression fits of large viscosity solutions upon the divergence rate and vorticity yielded η and ζ values of $\sim 10^{12}$ kg/s with $\zeta \sim 2\eta$. The physical notion here is that the ice interaction prevents large deformation from occurring, and hence the divergence rate tends to follow the wind gradients, especially on shorter spatial scales. The viscous approximation may be justified for small deformation magnitudes if one considers sufficient spatial and temporal smoothing of the velocity field with plastic behavior on the local scale (Hibler 1977).

Although these linear solutions are only approximate, the main point is that ice mechanics alone can cause the ice deformation to differ substantially from the expectations based on simple linear drift rules such as that of Zubov (1945) or Thorndike and Colony (1982). Other relevant mechanisms can be repeated lead and ridging events (see section 16.5.3). This is not widely appreciated, as much interpretation of recent decreases in arctic atmospheric pressure (Walsh et al. 1996) has been based on anticyclonic wind forcing favoring convergent ice motion. This has also

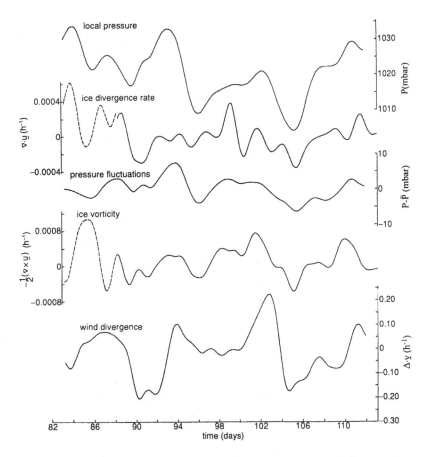

Figure 16.6. Comparison of experimental time series from AIDJEX 1972 deformation data and atmospheric observations of surface wind speed and pressure. The deformation data is taken from a densely sampled mesoscale array covering about a 20-km region, and the wind data are taken from surface observations of wind at three ice camps separated by ~100 km in a triangle. Pressure data from land stations were also used in the pressure plots (Hibler et al. 1974c).

been a common interpretation in the analysis of the effect of Arctic Oscillation wind patterns (Polyakov et al. 1999). Interestingly, recent detailed analysis of SAR data by Moritz and Stern (2001) yielded divergence rates having a zero correlation with the Laplacian of the atmospheric pressure field. This result emphasizes the necessity for using nonlinear sea ice mechanics in ice models to simulate deformation properly.

16.5 Sea Ice Mechanics

Typical nonlinear formulations of ice mechanics in dynamic thermodynamic models yield a stress gradient $\nabla \cdot \sigma$, generally in opposition to the wind stress (Hibler and Bryan 1987; Steele et al. 1997). When vectors are averaged over time periods of a few months or longer, the ice stress gradient tends to have a magnitude about 50% of the average wind stress (see Table 1 from Hibler and Bryan 1987). However, as examined below, owing to nonlinear ice interaction the daily stress gradient is a smaller fraction of the wind stress gradient. Because of these considerable effects, proper treatment of ice mechanics plays a key role in modeling the evolution of sea ice.

Successful descriptions of pack ice failure and flow have typically incorporated nonlinear plastic failure models with differing treatments of shear strength and flow rules. There are two general treatments of plastic failure that generally parallel the classic Von Mises and Tresca yield curves in solid mechanics (Malvern 1969). The essential idea in these plastic failure models is to have a rate-independent failure stress once plastic failure is initiated. Much of the initial development of plastic sea ice models was based more on aggregate energetic arguments (Coon 1974; Rothrock 1975a, 1975c; see also section 16.6 below) than on particular physical failure processes applicable to fractures or sliding friction between floes. Recently proposed rheologies (Hibler and Schulson 2000) have similarities to plastic and brittle failure constitutive laws developed in smaller-scale laboratory studies of ice (e.g., Schulson and Nickolayev 1995). These rheologies also have significant similarities to granular Coulombic failure models used in soils and rock mechanics (see, e.g., Dempsey et al. 2001; Schulson 2001a, 2001b).

16.5.1 Aggregate Isotropic Sea Ice Constitutive Laws

Aggregate isotropic descriptions of pack ice have typically sought to portray the collective behavior of pack ice with a large number of failing ridges and leads comprising the pack ice continuum. In these descriptions failure is considered to describe statistical aggregate behavior, with both the stress and strain comprising a reasonable statistical number of leads, fractures, and ridges. Depending on the constitutive law used to determine ice stress from deformation, this continuum description does not, in principle, prevent the formation of long-oriented failure features in simulations that are similar to those seen in SAR imagery. Considering sea ice to be a two-dimensional isotropic fluid, a general nonlinear constitutive law applicable to a relation between stress and deformation rate (Malvern 1969) is

$$\sigma_{ij} = 2\eta\,\dot{\varepsilon}_{ij} + \left[(\zeta - \eta)\dot{\varepsilon}_{kk} - P\right] \tag{16.18}$$

where repeated subscripts are summed over, σ_{ij} is the two-dimensional stress tensor, $\dot{\varepsilon}_{ij}$ is the two-dimensional strain rate tensor, and ζ, η, and P are functions of the two invariants of the strain rate tensor. In the general Reiner–Rivlin fluid there is also an additional term $\lambda\dot{\varepsilon}_{ik}\dot{\varepsilon}_{kj}$. However, since P can also be a nonlinear function of the strain rate invariants, this additional λ-term is usually omitted in sea ice rheologies. In constructing a law beginning with this general framework, one would like to take into account the following observational characteristics and/or intuitively reasonable assumptions: discontinuous slippage near shore; relatively coherent motion; possible lack of ice motion under considerable wind forcing; small tensile strength; and high compressive strengths. These last two features are also characteristic of the biaxial failure of laboratory ice (Schulson and Nickolayev 1995).

Although linear–viscous models are useful for mechanistic calculations, it became clear from theoretical and numerical modeling studies in the early 1970s that some of the basic phenomena of ice dynamics and ice deformation discussed above could not be explained by linear rheologies. For example, using a linear–viscous model employing both bulk and shear viscosities, it was found that best-fit viscosities of the order of 10^{11} to 10^{12} kg/s were needed to model deformation (Hibler 1974) and drift (Hibler and Tucker 1979) far from shore. Nearshore, however, investigations of satellite

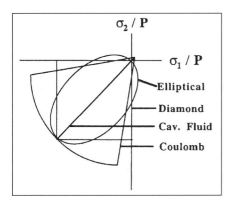

σ_2 / P

σ_1 / P

— Elliptical

— Diamond

— Cav. Fluid

— Coulomb

Figure 16.7. Typical "aggregate" isotropic yield curves proposed for sea ice.

imagery near coastal boundaries indicated viscosities as small as 10^8 kg/s (Hibler 1974), and model simulations of tidal-induced drift indicated that in very small channels nearshore viscosities might be as small as 10^5 kg/s (Sodhi and Hibler 1980). Such large variations indicated that some type of nonlinear behavior was necessary. The most recent work relative to this subject area (Rampal et al. 2008) has looked at the scaling of deformation within the ice pack by studying the motions of pairs of data buoys deployed by the International Arctic Buoy Programme (IABP). They find that sea ice deformation is both very heterogeneous and intermittent at all scales. As a result they feel that, as suggested in the previous chapter, sea ice deformation cannot be considered to be viscous and is much better simulated as a multiscale fracturing/faulting process. This result would appear to favor the Schulson–Hibler fracture approach discussed earlier.

One way to formulate a nonlinear rheology consistent with the above characteristics is to make use of plasticity theory (see, e.g., Pritchard 2001). This was first noted by Coon (1974), who based much of this proposition on the similarities between sea ice and granular media. The basic assumptions are plastic failure together with small or zero tensile stresses. For comparison, note that the two-dimensional Von Mises failure criterion (Malvern 1969) for a thin steel plate or shell is an ellipse centered at the origin. Some typical isotropic yield curves (Figure 16.7) proposed for sea ice along these lines are the elliptical yield curve (Hibler 1979) and the coulombic yield curve (Overland and Pease 1988; Smith 1983), which has a more direct explanation in terms of oriented failure along surfaces and will be discussed below. Initial plastic pack ice simulations utilized a circular yield curve for simplicity (Coon and Pritchard 1974). In addition, two-dimensional energetic arguments by Rothrock (1975a) lead to a proposed "teardrop" yield curve, and kinematic arguments by Bratchie (1984) lead to a "sine lens" yield curve, which also can be considered "scale invariant" (Hibler 2001). A "square" or diamond yield curve was also proposed by Pritchard (1975), based partially on frictional arguments. A particularly simple conceptual yield curve is the "cavitating fluid" (Flato and Hibler 1992), whereby ice resists compression but has no resistance to opening. This yield curve is useful where temporally averaged wind forcing is employed. With all such yield curves, to supply a complete description of the stress state in the system one also needs to have a way of obtaining stresses inside the yield curve. Moreover, if one wishes to model deformation and flow, one needs a flow rule to specify the stresses for different strain rates.

The specification of the stresses inside the yield curve has traditionally assumed elastic behavior (Pritchard 2001). With elastic behavior inside the yield curve and rate-independent stresses on the yield surface, energetic conditions typically lead to the so-called normal flow rule. This rule specifies that in coincident principal axes space, the ratio of the principal strain rate components is specified by a vector normal to the surface. Since energy may be stored and released by the elastic portion of the system, this constraint can be shown to necessitate a convex yield surface and a normal flow rule. If some other type of closure, such as rigid or viscous, is used, a normal flow rule is not required by this argument as there is never any storage of mechanical energy. A typical simple coulombic yield curve, for example, often assumes pure shear deformation for a variety of stress states. Numerical solution of the elastic–plastic formulation is typically by direct explicit methods (Pritchard 2001; Pritchard et al. 1977). Although elastic closure may be very valuable mathematically, since "plastic" pack ice rheologies seek to describe the collective behavior of pack ice, there appears to be very little, if any, physical basis for necessarily assuming elastic behavior at the large scale.

Another description of the stress states inside as well as on the plastic yield curve is the "viscous–plastic" (vp) rheology approach introduced by Hibler (1979, 1980a) for the purpose of modeling nonlinear flow for a wide variety of yield surfaces (see, e.g., Ip et al. 1991). Here, rigid flow is approximated by a state of very slow creep, a procedure that can be justified in part by considering a statistical aggregate of different deformation states undergoing rigid plastic flow in any control volume (Hibler 1977). In practice, this description may be thought of as essentially a highly nonlinear viscous fluid, as described in equation 16.18, with the viscosity adjusted to yield rate-independent stresses for large strain rates. This viscous–plastic formulation is widely used in a variety of ice–ocean and atmosphere–ice–ocean models. Particular rheology and solution procedures are given below. (The "viscous–plastic" formulation should not be confused with a "visco-plastic" or Bingham rheology. A visco-plastic rheology is similar to rigid plasticity except that at failure viscous, rather than plastic, flow occurs; see, e.g., Shames and Cozzarelli 1992.)

A third method, often used in soil mechanics, is to consider the system to be rigid inside the yield surface. With this procedure one needs some type of constraint to determine the interior stresses. A procedure for numerically solving for such interior stresses in the rigid approximation was proposed by Flato and Hibler (1992) for coulombic-type rheologies with no dilatation. Here one uses the constraint that there is no deformation for stresses inside the yield curve. This procedure has subsequently been extended to a more general coulombic yield curve, including dilatation, by Tremblay and Mysak (1997).

In the case of the elliptical yield curve with the viscous–plastic approximation and a normal flow rule, a simple closed set of equations (Hibler 1977, 1979) can be formulated relating the stress to the strain rate anywhere on or inside the yield surface. This yield curve and flow rule has been widely used in many numerical sea ice models (Lemke et al. 1990, 1997). Referring back to equation 16.18, for an elliptical yield curve with a normal flow rule, the nonlinear bulk and shear viscosities ζ and η can easily be shown (Hibler 1977, 1979) to be functions of the strain rate invariants according to

$$\zeta = \text{minimum} \left[\frac{P^*}{2\Delta}, \zeta_{\text{max}} \right] \tag{16.19}$$

$$\eta = \frac{\zeta}{e^2} \tag{16.20}$$

where

$$\Delta = \left[\left(\dot{\varepsilon}_{11}^2 + \dot{\varepsilon}_{22}^2 \right) \left(1 + \frac{1}{e^2} \right) + \frac{\dot{\varepsilon}_{12}^2}{e^2} + 2\dot{\varepsilon}_{11}\dot{\varepsilon}_{22} \left(1 - \frac{1}{e^2} \right) \right]^{1/2}. \tag{16.21}$$

Here P^* is the ice strength (equal to the pressure P for high strain rates), and e is a constant (typically $\sqrt{2}$). To approximate stress states inside the yield curve, ζ and η are capped at some large maximum values (ζ_{max} and $\eta_{max} = \zeta_{max}/e^2$) for small strain rates.

In order to ensure that there is no stress at zero strain rates, in recent formulations it has been insisted (Hibler and Ip 1995; Hibler and Schulson 2000; Ip et al. 1991) that the pressure P in equation 16.18 is given by

$$P = 2\Delta\zeta \tag{16.22}$$

where ζ is given by equation 16.19. When plastic flow is occurring, P will be a constant (P^*). However, for $\zeta = \zeta_{max}$, P will be less than or equal to P^*, and the stress state will lie on a smaller but geometrically similar yield curve, as shown in Figure 16.8, going through the origin. This condition, together with the above flow rule, ensures that the internal mechanical energy dissipation $\sum_{ij=1}^{2} \dot{\varepsilon}_{ij}\sigma_{ij}$ is positive definite. This result is in contrast to the earlier formulation of an elliptical viscous–plastic rheology by Hibler (1979), where P was taken to be a constant times P^*. As noted by Schulkes (1996), this energy condition is the main requirement for overall energetic stability of the viscous–plastic rheology, since it guarantees, via the continuum energy equations, that the kinetic energy will be bounded in the absence of body forces. Simulations (Hibler and Ip 1995; and see below) show the inclusion of equation 16.22 to substantially reduce excessive stoppage of ice motion.

The numerical procedure proposed by Hibler (1979) as part of a dynamic thermodynamic sea ice model represents probably the most direct method of solving numerically the viscous–plastic dynamical equations. In this direct implicit solution the coupled nonlinear momentum equations are linearized and solved implicitly by successive relaxation at each time step with a locally implicit solution of antisymmetric terms. Since the linear system of equations thus formed is positive definite (but not symmetric), other more efficient numerical solvers could probably be used. For example, similar relaxation procedures have been used by Harder and Fischer (1999). Because of this implicit solution there is no time step limitation on the momentum equation, although with time steps long compared with the forcing, the evolution of the nonlinear terms (Figure 16.9) will not keep up with the forcing changes unless repeated iterations at each time step are used. For example, such "updating" procedures were used at each time step by Harder and Fischer (1999), Hibler and Ip (1995), Hilmer et al. (1998), Ip (1993), and Kreyscher et al. (2000).

A more efficient "splitting" method was proposed by Zhang and Hibler (1997). This procedure solves only a portion of the momentum equations at each time step and hence is less dissipative than the Hibler (1979) method. Moreover, at large time steps the Zhang and Hibler (1997) method may have convergence problems. A more convergent "splitting

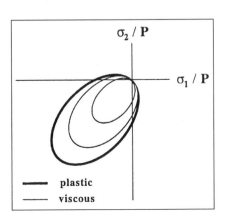

Figure 16.8. Stress states for small strain rates inside the yield curve configured in such a way as to always dissipate energy.

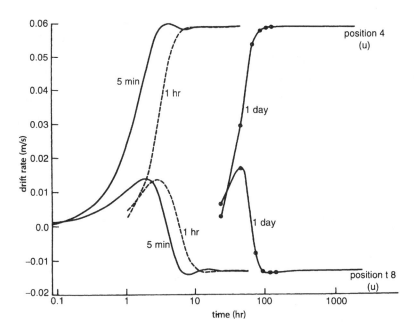

Figure 16.9. Approach to equilibrium of the ice velocity using different time steps in the integration of the viscous–plastic momentum equations. In all cases the ice was initially at rest and then a constant wind field was turned on. Position 4 is very near a boundary, whereas position 8 is near the center of the grid (Hibler (1979).

procedure" has subsequently been developed by J. K. Hutchings and colleagues. Such splitting techniques have been used in very high-resolution ice–ocean simulations (e.g., Maslowski et al. 2000). Hunke and Dukowicz (1997) and Hunke and Zhang (1999) developed an "elastic–viscous–plastic" (evp) solution procedure whereby an artificial elastic term is added so explicit time steps may be taken. This yields a more efficient solution on parallel computers. The Hibler (1979) procedure can also be concurrentized on parallel computers, but the initial setup time scales with the number of processors.

It should be emphasized that, given the same numerical discretization of the nonlinear viscous and pressure terms, all of the numerical solutions should converge to the same plastic solution for cases where analytic solutions exist (Leppäranta and Hibler 1985; see also section 16.4.1). However, different discretizations are typically used as well as time steps in comparisons (Arbetter et al. 1999; Hunke and Zhang 1999), so precise differences are at present poorly determined. Moreover, when the viscous–plastic (vp) solution was not updated at each time step, Hunke and Zhang (1999) obtained artificially long response times for the vp model as compared with the evp model. The original Hibler (1979) rectangular finite-difference discretization and solution procedure ensures that the viscous terms are locally and globally energy dissipative for each linear solution, and that the full rheology is dissipative if the equations are iterated to full plastic flow with the formulation of equation 16.22. Recent developments utilizing full curvilinear energy dissipative discretizations for spatially varying bulk and shear viscosities by Hunke and Dukowicz (2002), and a spherical coordinate dissipative finite-difference discretization by Hibler (2001), retain all metric terms. There are also a number of other numerical methods (e.g., finite-element methods) that can and have been applied to solving these equations, although the overall time-stepping procedures are usually similar to one of the methods described above.

16.5.2 Coulombic and Fracture-Based Isotropic Models

While they have many similarities to aggregate rheologies based largely on energy arguments, coulombic and fracture-based rheologies are based more on either failure across faults or sliding friction between discrete particles (floes in the case of sea ice; see the discussion in Chapter 12). In this sense, the classic plastic analog would be the Tresca failure criterion, where failure is considered to occur when shear stress across any surface reaches a critical value. The classic coulombic rheology (Overland and Pease 1988; Overland et al. 1998; Smith 1983; Tremblay and Mysak 1997) extends this concept to make the critical failure shear stress across any surface depend on the compressive stress on that surface. Because of this emphasis on failure surfaces and faults, this type of rheology provides a more natural formulation for treating oriented fractures and failure in sea ice and for building up a material description of sea ice based on such phenomena. In the case of fracturing it can be argued that similar rheological behavior should apply on both the large and small scales (Hibler 2001; Schulson and Hibler 1991). Such rheologies also provide a framework for describing likely mechanisms leading to oriented fractures and leads in sea ice (Hutchings and Hibler 2002; Schulson 2001a).

The simplest coulombic model may be conveniently described by considering an oriented failure surface in two dimensions. For the coulomb sliding friction model (Figure 16.10a), we imagine a flaw oriented at an angle θ relative to the most compressive stress σ_y applied to the local region. We consider this flaw and all other flaws in the region to have a local friction coefficient of μ. With the "coulomb model" assumption the flaw will fail when the shear stress τ along the flaw is related to the compressive stress σ across the flaw according to the frictional sliding approximation $\tau_c = \mu\,\sigma + b$, where b is a measure of the cohesive strength of the material and τ_c is the critical shear stress for failure. To simplify the mathematics, if we consider all stresses to have units of bars, this equation becomes $\tau_c = \mu\,\sigma + 1$. From the orientation of the flaw in Figure 16.10a, for a given confinement ratio r less than one, the flaw will fail for an external principal stress σ_y given by

$$\sigma_y \frac{(1-r)\sin(2\theta)}{2} = \mu\frac{(1+r)\sigma_y}{2} - \mu\frac{(1-r)\cos(2\theta)\sigma_y}{2} + 1. \qquad (16.23)$$

Equation 16.23 was obtained by expressing the failure equation of the flaw in terms of the external principal stresses σ_y and $r\sigma_y$. To find the angle of flaws most likely to fail, we can consider σ_y to be a function of θ and seek a minimum value of σ_y. Differentiating this equation with respect to θ and setting $d\sigma_y/d\theta = 0$, we obtain the classic result

$$\tan(2\theta_c) = \pm\frac{1}{\mu}. \qquad (16.24)$$

Rearranging equation 16.23, if flaws preferentially fail at the fixed angle θ_c, the relationship between the maximum external shear stress, σ_s and the compressive stress, σ_c is given by

$$\sigma_s = cos(2\theta_c)\sigma_c + sin(2\theta_c). \tag{16.25}$$

This is the principal axes space yield curve following from the local sliding friction model with the angle of internal friction given by $\beta = (\pi/2) - 2\theta_c = tan^{-1}(\mu)$.

Although useful, the classic coulombic rheology of Figure 16.10a has the drawback that only pure shear deformation is allowed. A useful physical model (Coon et al. 1998; Hibler and Schulson 2000) that can be used to extend the coulombic rheology to arbitrary dilatation is the "finite-width"-oriented flaw in Figure 16.10b. While loosely referred to as "finite width," the flaw may be made infinitesimally narrow without affecting the theory. In this limit it takes on features of a mathematical characteristic used in partial differential equations inasmuch as the strain and velocity derivative are still defined on the "strip," even though its width is arbitrarily small (Mendelson 1968). It can be shown (Hibler and Schulson 2000) that insisting on continuity of stresses at the interface of the flaw and minimizing the stresses as before leads to a preferred angle of failure that depends on the flow rule of the rheology of the thin ice, which is in general larger than the coulombic angle for a sliding surface. In the case of a more complex rheology applying to the thin ice, the preferred orientation can be solved numerically (see Hibler and Schulson 2000 for details). Since in numerical models the failure patterns take place along finite-width zones, these failure intersection angles are also the ones we expect from numerical simulations. As an example, a viscous–plastic formulation of a nonnormal flow rule coulombic yield curve (Figure 16.11) based on observed laboratory results (Schulson and Nickolayev 1995) was proposed by Hibler and Schulson (2000), and used in the "lead" in the finite-width model shown in Figure 16.10b. The predicted flaw angles tend to be much narrower than in the elliptical yield curve for diverging conditions. Confirmation for the theory comes from viscous–plastic boundary value simulations forced by stresses at the boundaries. In these simulations the "modified coulombic" rheology leads to shear zones of localized deformation (see Fig. 10 of Hibler and Schulson 2000) in good agreement with the theory. With an elliptical rheology, there are no clearly oriented failure results in this experiment.

The concept here is that this type of "coulombic" model should be much more successful (Aksenov and Hibler 2001) at simulating the oriented deformation patterns observed in the

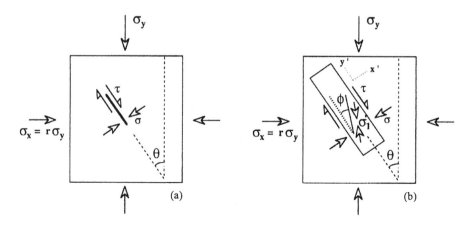

Figure 16.10. Schematic showing (a) a coulomb frictional sliding model and (b) a "finite-width" failure model allowing for dilatation.

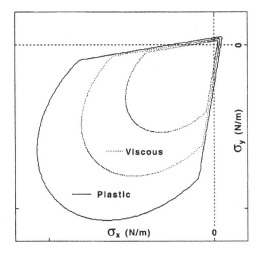

Figure 16.11. A coulombic yield curve used for a numerical calculation of composite failure of thick ice with an oriented thin-ice lead as shown in Figure 16.10. This yield curve was based on laboratory observations of the brittle failure of ice by Schulson and Nickolayev (1995). The dotted lines are the stress states for very small strain rates (Hibler and Schulson 2000).

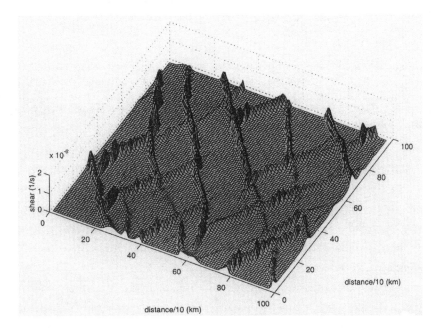

Figure 16.12. Shear strain rate simulated for a two-dimensional grid of ice with strength initially varying randomly and the wind forcing resulting in a uniform strain field. The ratio of principal strain rate axes is −1.4. The ice has been allowed to weaken in proportion to local divergence rate for 3.6 hours. When plastic equilibrium was achieved with no weakening, the diagonal pattern seen above was barely perceptible.

arctic ice pack. That this is indeed the case can be shown (Hutchings and Hibler 2002) beginning with an initial heteorogeneous ice strength field. With this initial condition, appropriate wind forcing causes the ice to weaken and fail along intersecting oriented zones (Figure 16.12) that appear very similar to observations (Kwok 2001). In this simulation a wind field resulting in a constant deformation rate with a −1.4 ratio of the principal strain rate components was utilized. With this wind field and initial condition, Hutchings and Hibler (2002) find intersecting failure features (Figure 16.13b) form with average spacing scaling with the square root of the average ice strength over the square root of the spatial wind stress gradient. This scaling may be physically argued on the basis of allowable rigid block motion dimensions with zero stress boundary

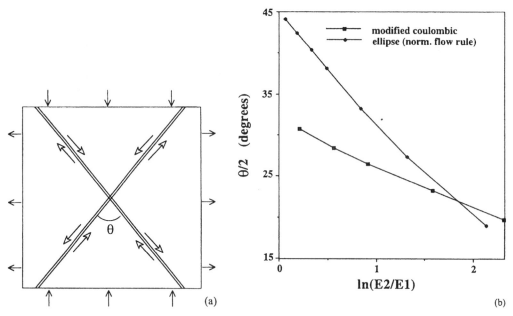

Figure 16.13. (a) Intersecting finite-width faults together with a specified divergent strain field used to estimate lead orientation for different rheologies. (b) Predicted lead orientation obtained by finding the minimum energy of the intersecting fault model for different rheologies vs. the natural logarithm of the ratio of the principal axis strain rate magnitudes. E2 is positive and E1 is the magnitude of the convergent principal axis value (Hutchings and Hibler 2002).

conditions at the "lead." Due to the strength heterogeneity, the initial plastic deformation field is rather chaotic without any clear oriented structure. However, with weakening due to divergence, local failure points nucleate the formation of intersecting failure lines, which form in only a few hours. Over this period the global averaged stress magnitude drops by about a factor of 5, and the stress in the oriented failure zones drops by several orders of magnitude. This general behavior is consistent with stress observations of leads (see, e.g., Figure 16.5).

By considering the steady-state equations of motion of an idealized intersecting set of leads, Hutchings and Hibler (2002) argue that the orientation of the leads should be predicted by an energy minimization calculation (for opening and hence weakening flaws) with respect to orientation of an intersecting set of "finite" failure zones. This energy minimization solution can differ from "characteristics" based on static solutions (e.g., Erlingsson 1991; Pritchard 1988b) and can be used for nonnormal flow rule rheologies. Applying the Hibler and Schulson (2000) procedure to an ice field in which the strength varies randomly yields the predicted angles shown in Figure 16.13b. The point here is that the ellipse (or any normal-flow-rule-based rheology) yields much larger (and hence less realistic) intersection angles than fracture-based rheologies. This is especially true in the limit of pure shear strain rates that are often encountered in typical sea ice deformation fields. The main physical reason for this is the cusp in the yield curve where different failure mechanisms take over (see, e.g., Schulson and Nickolayev 1995).

One other notable approach to modeling granular flow that is somewhat related to continuum coulombic models is the discrete element simulation of sea ice (e.g., Hopkins 1996). In this approach the motion of large numbers of discrete distributions of floes are solved individually with coulombic friction at contact points. Although not highly applicable to climate-scale simulations, such models are particularly useful for process simulations to estimate realistic thickness distribution ridging parameterizations in larger-scale models.

The main relevance of the "coulombic" rheologies to the evolution of the mass budget of sea ice lies in their deformation characteristics of shearing and dilating intersecting fractures, an interpretation which becomes more relevant in high-resolution models that can resolve local scales. With this interpretation, open water formation is dictated by the divergence rate except for certain deformation states (e.g., Hibler and Schulson 2000; Hutchings and Hibler 2002) requiring both opening and closing sets of leads. On the other hand, in the case of aggregate rheologies it is usually hypothesized that both divergence and convergence take place simultaneously over a wide range of deformation states (see section 16.6.2). As discussed in section 16.7.2, how this partition is made in numerical model simulations significantly affects ice-mass characteristics. It is also notable that certain coulombic rheologies are arguably scale-invariant (Hibler 2001) and hence would be expected to apply at very high resolution as well as at larger scales. This view is, however, in contrast to arguments by Overland et al. (1995), who argue for a scale break in rheology around ~10 km.

16.5.3 *Effect of Plastic Ice Interaction on Modeled Ice Drift*

Characteristics of plastic flow particularly relevant to the sea ice–mass balance are its tendency to yield large-scale ice drift similar to the wind forcing and the tendency for ice flowing through narrow channels to slow down and sometimes totally stop via the formation of static arches. Both these features occur because of the highly nonlinear character of plastic flow. Because of the first characteristic, large-scale ice motion fields will tend to correlate highly with wind field variations, even with high ice stresses. The second "arching" characteristic is important because of the critical control of relatively narrow outflow passages, most notably Fram Strait, on the ice-mass budget in the Arctic Basin. Some of these characteristics are examined below.

16.5.3.1 A Mechanistic One-Dimensional Plastic System   Some physical notions regarding plastic flow and sea ice drift may be illustrated by analyzing a special one-dimensional case of the momentum balance employing only a linear water drag term cu, an external constant wind stress τ, and a one-dimensional ice stress σ:

$$cu - \frac{\partial \sigma}{\partial x} = \tau \qquad (16.26)$$

Now consider a rigid plastic case with constant strength with rigid walls at $x = 0$ and $x = L$. For the plastic rheology, assume that $\sigma = -P$ for $\partial u / \partial x < 0$ and $\sigma = 0$ for $\partial u / \partial x > 0$. These assumptions define a rigid plastic rheology with no tensile strength. A solution for this case may be constructed by noticing (a) that for any convergent deformation, $\sigma = -P$, whereas for no deformation the stress is in the rigid state so that $0 \geq \sigma \geq -P$; and (b) that the maximum force would be expected to take place at the right-hand boundary since the wind stress has built up at this point. Under these assumptions, it is easy to show (see, e.g., Hibler 1985) that u takes on a constant value

$$u = \frac{\tau L - P}{cL} \qquad (16.27)$$

with the stress σ varying linearly over the whole length of the ice cover according to

$$\sigma(x) = -P + (cu - \tau)(x - L). \qquad (16.28)$$

This analysis is also easily extended to include nonlinear water drag terms. In this case, the steady-state momentum balance for this one-dimensional system is

$$\rho_a c_a u_g^2 = \rho_w c_w u^2 - \frac{\partial \sigma}{\partial x} \qquad (16.29)$$

where u_g is the geostrophic wind. By the same methods as employed in the linear case for a constant wind, the solution for this system is

$$u = \left(\frac{\rho_a c_a u_g^2 - P/L}{\rho_w c_w} \right)^{1/2} \qquad (16.30)$$

if $P < \left(\rho_a c_a u_g^2 \right) L$ and $u = 0$ otherwise. For a numerical comparison relevant to the Arctic Basin we take $L = 2.5 \times 10^3$ km, which is the scale of the basin, and $P = 8.25 \times 10^4$ N/m, a typical model-based ice strength for 3-m-thick ice in the Arctic Basin (Hibler and Walsh 1982). Keeping in mind that this ice strength may be a bit high, one obtains a comparison of free drift and ice interaction modified drift as shown in Figure 16.14. Note that for large wind speeds there is little difference between free drift and the rigid plastic result, whereas for smaller wind speeds the difference is very marked, with the rigid plastic system stopping totally.

16.5.3.2 Large-Scale Simulated Plastic Drift and Deformation Characteristics Comparisons of simulated and observed drift characteristics (Hibler and Ip 1995; Ip 1993) tend to support the basic elements of the mechanistic one-dimensional results described above: namely, the stoppage of ice drift under low wind speeds and the high correlation of simulated ice drift, albeit with substantial ice stresses. A number of these features are illustrated in Figures 16.16 and 16.17, taken from a series of multiyear simulations by Ip (1993) employing a hierarchy of plastic rheologies using a two-level dynamic thermodynamic model (Hibler 1979) with a full heat budget code (see Chapter 9). The main variables are the mean ice thickness per unit area, h, and the ice compactness, A, with the ice strength P related to h and A

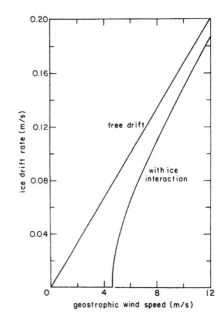

Figure 16.14. A comparison of free drift and ice-modified drift using typical one-dimensional plastic solution characteristics with constant strength and a fixed constant wind field and assuming quadratic boundary layer drag. Typical strength and length scales relevant to the Arctic Basin were assumed for ice interactions.

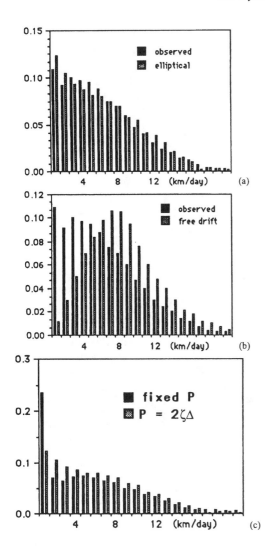

Figure 16.15. Comparison of observed and simulated buoy drift statistics over a seven-year period (1979–1985) for (a) an energy-dissipative elliptical rheology (see Figure 16.8), and (b) for free drift. In part (c) the effect of taking a constant pressure rather than energy-dissipative rheology (see equation 16.22) is shown (Ip 1993).

by $P = P^* h e^{-C(1-A)}$, where $P^* = 2.75 \times 10^3 \mathrm{N/m^2}$ (Hibler and Walsh 1982) and $C = 20$. Simulations were carried out between 1975 and 1985. The yield curves are basically as shown in Figure 16.7, except that the coulombic yield curve goes to the origin and has no tensile strength.

The characteristic stoppage of ice is illustrated in Figure 16.15, where the observed daily distribution of drift speeds is compared with values simulated with a plastic rheology in Figure 16.15a and with a "free drift" simulation with no ice interaction in Figure 16.15b. What is clear from these figures is that with a rheology including shear strength, the distribution of drift speeds demonstrates a higher fraction of low or zero ice motion, in general agreement with the simple one-dimensional model physics. However, the "free" drift model, or any direct linear model correlating ice drift with wind with fixed spatially invariant coefficients, does not reproduce the observations. It should be noted, however, that this realistic correlation of modeled ice drift requires an energy dissipative rheology. The use of a fixed pressure term, without the pressure being dependent on deformation (equation 16.22), results in excessive amounts of stoppage (Figure 16.15c).

Although this reduction of drift speeds can occur at low wind speeds, there is a very strong complex correlation (Figure 16.16a) between wind and simulated ice drift, as we would expect from physical considerations. It is also notable that there is a strong negative correlation between wind force and the ice force arising from the gradient of the ice stress tensor. This general behavior has also been noted by Steele et al. (1997) in a similar model calculation, and earlier by Hibler and Bryan (1987) in an ice–ocean model calculation. Depending on the season, the average magnitude of the stress gradient is about 10–30% of the average wind force (Figure 16.16b). If, on the other hand, vector force averages are taken (Hibler and Bryan 1987), then the ice force can account for a significantly higher fraction of the vector-averaged wind force.

Although drift characteristics are affected by ice mechanics, ice deformation is modified in a much more pronounced way. This is shown in Figure 16.16c, where the averaged deformation

Figure 16.16. (a) Complex correlation coefficient between simulated ice drift and observed ice drift, geostrophic wind velocity, and simulated ice stress gradient ($\nabla \cdot \sigma$). Three different rheologies and free drift (no ice interaction) were utilized. (b) Average daily magnitude of the wind stress and the ice stress gradient ($\nabla \cdot \sigma$) for four different rheologies as a function of time of year. (c) seasonally averaged divergence rates for three different rheologies and a free drift model. Statistics in (a)-(c) were compiled over a seven-year period (1979–1985) (Ip 1993).

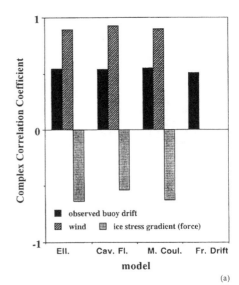

(a)

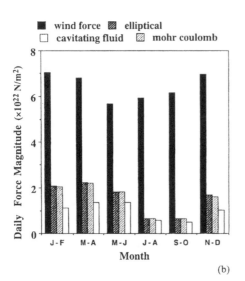

(b)

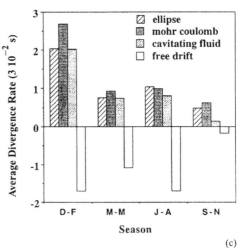

(c)

time series for the Beaufort Sea region is plotted for both a full rheology and the free drift case. Clearly, the effect of rheology, especially when combined with ice thickness evolution equations, is to cause a general divergence of the ice cover. This is in contrast to the free drift case where there is always a convergent condition because of the dominant anticyclonic condition over the Beaufort Sea. Although part of this condition is attributable to the reduced magnitude of the divergence rate, as argued in idealized linearized calculations above, it is also likely to be affected by the freezing of ice in leads formed by a divergent pack. This additional ice can then oppose convergence under subsequent changed wind fields.

16.5.3.3 Improvement of Simulations by "Inertial Imbedding"

As noted in section 16.4.2, most models utilize the basic equations of motion as specified by equations 16.4 to 16.6. However, these equations ignore the presence of considerable inertial power in the upper portions of the ocean, which significantly affect the higher-frequency motion of the ice drift. Simulations (Heil and Hibler 2002) including this inertial energy by utilizing equation 16.11 for the equation of

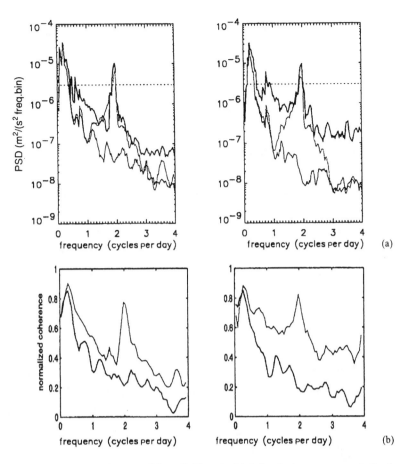

Figure 16.17. (a) Power spectral density of the x (left) and y (right) velocity components for the motion of a buoy (buoy A1; see Figure 16.24) over a several-month-long period in the central portion of the Arctic Basin. The heavy top curve represents the observations, the middle curve represents simulated values including "inertial imbedding" (equation 16.11), and the bottom curve is with no inertial imbedding (equations 16.4 to 16.6). (b) Normalized coherence of the x (left) and y (right) components of the observations and the two model results. The high coherence is for the model including "inertial imbedding" and the low coherence is without inertial imbedding (Heil and Hibler 2002).

motion improve simulations of buoy drift at all frequencies. This is graphically illustrated in Figure 16.17, which shows simulated and observed power spectra for the two different models. As in the previous section, the simulations were carried out in conjunction with "two-level" ice evolution equations described in section 16.6.4. The simulated power spectra now much more closely follow the observed drift of this central arctic buoy. Most important is the greatly enhanced coherence at all frequencies. This improved coherence is reflected in the overall correlation coefficients that are substantially greater for the inertial imbedded result ($R_x = 0.87$, $R_y = 0.84$) than for the nonimbedded model ($R_x = 0.59$, $R_y = 0.48$). These comparisons demonstrate both the importance of ice interaction in restraining the modeled ice velocity fluctuations to values close to the observed ice velocity, and the capability of nonlinear plastic rheologies (together with the ice evolution equations discussed below) to supply the appropriate amounts of high-frequency energy. An underlying reason for this improvement is the artificial damping of power by the passive water drag if the traditional momentum equations are used (see, for example, comparisons by Steele et al. 1997). In the imbedded equations 16.10 and 16.11, the only way the combined average integrated ice–boundary layer energy may be reduced is via ice interaction dissipation or lateral viscosity in the ocean unless some small artificial damping is used. (In reality, other ocean damping mechanisms exist, such as damping by internal waves.) These imbedded simulations also greatly improve the simulation of ice deformation, as discussed in section 16.6.6. To date little work has been carried out with this improved momentum formulation.

16.5.3.4 The Effect of Rheology on Outflow

An important aspect of plastic flow with a rheology with some cohesive strength is the ability to form some type of static arch when the material flows through narrow channels. Even if total stoppage does not occur, the flow can be significantly reduced by ice mechanics. This is particularly important for the Arctic Basin owing to the control that the ice export through the Fram Strait exerts on the mass budget, as well as affecting the degree to which ice may be transported through the Canadian archipelago. Indeed, to a first order, the mass budget (Koerner 1973) may be considered to be a balance between outflow and growth, with higher outflow inducing thinner ice and thus more growth. The potential of partial "mechanical arching" across the Fram Strait to induce multiple equilibrium states in the arctic ice pack is investigated in the context of a relatively idealized model in section 16.7.

Arching is a statically indeterminate problem in that different arch shapes have different strengths and breaking limits. Nonetheless, the arching phenomenon can be relatively easily analyzed in planar systems. An idealized analysis (Richmond and Gardner 1962) of a free arch between two vertical walls can be made with some simplifying assumptions utilizing a coulombic rheology with an appreciable cohesive strength (i.e., $b \neq 0$, see section 16.5.2). Assuming that (1) in the vicinity of the static arch, there is no variation of stresses in the y-direction, and (2) a Mohr–Coulomb criterion with cohesive strength holds, Richmond and Gardner (1962) obtained the maximum channel width that could arch under a constant wind stress τ to be given by

$$\lambda_{max} = \frac{a\left(\cos\theta_c\right)b}{\tau} \qquad (16.31)$$

where a contains geometrical factors of order 1, whereas θ_c is defined in equations 16.24 and 16.25, and, as stated earlier, b is a measure of the cohesive strength of the material as specified by the relation $\tau_c = \mu\sigma + b$. For channel widths larger than this, the arch will fail at the boundaries and the ice will flow through the channel. This analysis also yields a minimum width through which no flow can occur for given τ and b. The key scaling here is the arch width, which is proportional to the cohesive strength and inversely proportional to the wind stress. Also, as the friction increases (wider coulombic yield with θ_c decreasing), a wider arch can be formed.

The arching problem has also been analyzed numerically by Ip (1993), who used a variety of channel configurations. These simulations were carried out with similar results both with and without thickness advection equations. Although there is some effect, the actual channel configuration was found not to be highly critical on the arching initiation. For a given arch configuration, the solutions were found to follow a nondimensional scaling very well, so that the numerical results can be used to examine the flow and arching through any size channel with given wind stresses and ice strengths.

When considering a tapered arch (Hibler and Hutchings 2003), the basic parameters are the width λ of the opening, the ice strength P, and the wind stress $\rho_i c_a u_g^2$. For this system, the x momentum equation in the absence of Coriolis force is given by

$$0 = \rho_a c_a u_g^2 - \rho_w c_w u^2 + \left(\frac{\partial \sigma_{xx}}{\partial x} + \frac{\partial \sigma_{xy}}{\partial y} \right). \tag{16.32}$$

Expressing x and y in terms of λ and stress in terms of P, we have, after dividing by the wind stress, the dimensionless equation

$$1 = \beta - \gamma \left(\frac{\partial \sigma'_{xx}}{\partial x'} + \frac{\partial \sigma'_{xy}}{\partial y'} \right) \tag{16.33}$$

where the primed values of stresses and x and y are dimensionless, and β and γ are dimensionless parameters given by

$$\beta = \frac{\rho_w c_w u^2}{\rho_a c_a u_g^2}, \quad \gamma = \frac{P}{\lambda \rho_a c_a u_g^2}. \tag{16.34}$$

Consequently, in dimensionless form, the solution for β, which is a measure of the ice velocity, should only depend on γ unless the geometry of the boundaries changes. Investigation of the solutions for different resolutions and different geostrophic wind speeds shows this scaling to hold.

The character of the outflow is illustrated in Figure 16.18, which has been taken from an investigation by Hibler and Hutchings (2003) of multiple equilibrium states induced by ice mechanics. Dimensional results for the area and volume outflow versus ice thickness (and hence ice strength) are shown. The results may be easily converted to nondimensional form for the area outflow as noted above. For the volume outflow the appropriate ordinate will scale as $\left(P/\tau^2 \right) \Delta$, where Δ is the volume outflow in dimensional form.

The basic character of the solutions is a gradual decrease of the ice velocity as the strength increases or the opening span (λ) decreases. At some point the velocity stops, and a static solution with an arch formed is obtained numerically, and the system is motionless with a free

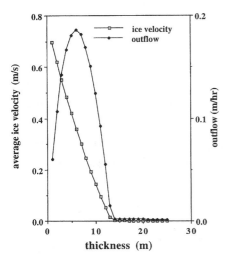

Figure 16.18. Results of a numerical "arching" study assuming a tapered channel with an opening with a width of λ. Shown are the dimensional average ice velocity and ice volume flow results versus ice thickness. A constant stress of $\tau = 0.4$ N/m^2 in the direction of the arrow is used for the body force. To simplify scaling, a linear water drag $(0.56 \, u)$ is used. For a plastic rheology, the modified coulombic yield curve of Hibler and Schulson (2000) is used and the ice strength is taken to scale linearly with thickness according to $P = 4 \times 10^4 h$ N/m, with h in meters. The ice area outflow may be put in nondimensional form by taking the ordinate to be $\beta = 0.56u/\tau$, with τ the wind stress ($= 0.4$ N/m^2) and the abcissa to be $\gamma = P/(\lambda\tau)$, where λ is the width of opening (Hibler and Hutchings 2003).

Figure 16.19. Dimensional plot of average ice velocity vs. wind stress for different ice strengths. These results were obtained from the nondimensional arching results of Figure 16.18. Note the similarity to observations shown in Figure 16.3.

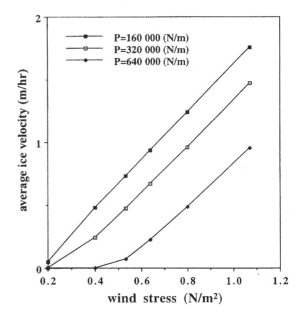

surface forming below an arch. Analysis of this system with a cavitating fluid has also been carried out by Ip (1993). In this case there is no arching, and, after a small decrease of ice velocity with h, the ice velocity becomes independent of h (i.e., ice strength). In the case of no ice interaction, there is some fixed velocity as a fraction of the wind speed. Analysis shows the estimated analytic strength limit to be slightly smaller $(\gamma_{crit} \approx 1.5)$ than the numerical result $(\gamma_{crit} \approx 2.1)$. This result is consistent with the fact that a constant channel width was assumed in the analytic calculation, whereas a tapered channel was used in the numerical experiments, so that arching was initiated at a slightly wider channel for the same forcing than the analytic limit.

Since the average ice velocity tends to scale with the areal outflow, it is instructive to plot average ice velocity in dimensional form for different ice strengths as a function of ice strength (Figure 16.19). This figure closely resembles observations of areal ice outflow through the Fram Strait shown in Figure 16.3 if we assume that during winter ice interaction effects are more pronounced. This figure also suggests that outflow results based only on correlations with atmospheric pressure gradients (Vinje 2001) may be somewhat high in winter months.

16.6 Ice Thickness Distribution Theory

A key coupling between sea ice thermodynamics (see Chapters 4 and 9) and ice dynamics is the ice thickness distribution. The basic physical notion here is that the deformation causes pressure ridging and open water creation. Thermodynamic processes, on the other hand, tend to ablate pressure ridges and remove thin ice by growth in winter, and create thinner ice and open water in summer. From a typical ice thickness distribution, we can think of deformation as affecting the higher (thick) and lower ends of the thickness distribution, thus causing a spreading of the distribution. Conversely, thermodynamic processes without deformation cause a concentration of the thickness toward a center value. The coupling of thermodynamics with deformation is critically affected by mechanical effects, which limit the character and amount of deformation. As a consequence there can be a significant feedback between mechanics and thickness because as the ice state changes the deformation can also change. Also, assumptions on how the ridging process occurs for a given deformation state can modify the temporal and spatial growth and thickness patterns of the ice pack and, as a result, can modify its response to climatic change. It should be noted that a major project (SEDNA; Sea Ice Experiment: Dynamic Nature of the Arctic) is currently underway (2007–2009) in the Beaufort Sea to provide data addressing different aspects of this general problem area (Hutchings et al. 2008). It is anticipated that project data will be publicly available in 2010. Initial data analysis indicates that during the winter there is coherence in pack ice deformation over a scale of at least 70 km, suggesting effective stress propagation through pack ice. However, after the early spring (May) this changes, with deformation on this scale becoming uncorrelated.

16.6.1 Evolution Equations for the Ice Thickness Distribution

A theory of ice thickness distribution can be formulated (Thorndike et al. 1975) by postulating an areal ice thickness distribution function and developing equations for both the dynamic and thermodynamic evolution of this distribution. An underlying assumption is that all thicknesses of the ice pack move with the same velocity field. Following Thorndike et al. (1975), $g(h)dh$ is defined to be the fraction of area (in a region centered at position x, y at time t) covered by ice with thickness between h and $h + dh$. Neglecting lateral melting effects, the governing equation for the thickness distribution is

$$\frac{\partial g}{\partial t} + \nabla \cdot (vg) + \frac{\partial (f_g g)}{\partial H} = \psi \qquad (16.35)$$

where f_g is the vertical growth (or decay) rate of ice of thickness H, and ψ is a redistribution function that depends on H and g and describes the creation of open water and the transfer of ice from one thickness to another by rafting and ridging. An underlying assumption here is that a region statistically large enough to describe a thickness distribution is used. However, this assumption is more a constraint on observations than on the model. Except for the last two terms, equation 16.35 is a normal continuity equation for g. The last term on the left-hand side can also be considered a continuity requirement in thickness space since it represents a transfer of ice from one thickness category to another by differences in the growth rates. An important feature of this theory is that growth occurs by rearranging the relative areal magnitudes of different thickness categories.

This theory was extended by Hibler (1980b) to include lateral melting by adding an additional term F_L to equation 16.35, where

$$\int_0^\infty F_L \, dh = 0. \qquad (16.36)$$

This constraint follows from the fact that lateral melting will be compensated for by a change in open water extent. The physical notion here is that thick ice will have a larger vertical interface with the ocean than thin ice, and hence assuming a constant value that reduces all categories proportional to their abundance is a good first approximation. This theory may also be extended explicitly to include ridged ice (Flato and Hibler 1995) by breaking the distribution into ridged $(g_r[h])$ and unridged $(g_u[h])$ components, each of which evolve according to an evolution equation of the form of equation 16.35.

The multilevel ice thickness distribution theory represents a very precise way of handling the thermodynamic evolution of a continuum comprising a number of ice thicknesses. However, the price paid for this precision is the introduction of a complex mechanical redistributor. The other issue is how to keep track of the stored energy and thermal properties of a variety of ice thicknesses as they are transferred into thicker ice by ridging. Also, as noted below, how the mechanical redistribution is carried out for arbitrary deformation states can significantly affect spatial patterns and ice thickness magnitudes.

To describe the redistribution, one must specify what portion of the ice distribution is removed by ridging, how the ridged ice is redistributed over the thick end of the thickness distribution, and how much ridging and open water creation occur for an arbitrary two-dimensional strain field, including shearing as well as convergence or divergence. In selecting a redistributor, one can be guided by the conservation conditions on ψ

$$\int_0^\infty \psi \, dh = \nabla \cdot v \tag{16.37}$$

$$\int_0^\infty h\psi \, dh = 0. \tag{16.38}$$

Equation 16.37 follows from the constraint that ψ renormalized the g distribution to unity due to changes in area. Equation 16.38 follows from conservation of mass, and basically states that ψ does not create or destroy ice but merely changes its distribution. An additional assumption in equation 16.38 is that the ice mass is related in a fixed manner to the thickness. Additional guidance on choosing redistributors maybe obtained from mechanical considerations as discussed below.

16.6.2 Consistency of Isotropic Plastic Models with Ridge Building

It is possible to base the smooth plastic descriptions of plastic flow, as, for example, the elliptical yield curve described above, on the physics of pressure ridging. Mechanistic pressure ridge models (Hopkins 1994; Parmerter and Coon 1972; for details see Chapter 12) show that the work needed to build ridges is relatively independent of how fast the ridge is formed, provided that the major work done in ridge building is due to potential energy changes caused by ice pileup. Rothrock (1975a) extended this concept to two dimensions by insisting that two-dimensional deformational work be explicitly related to the amount of work done by ridging. Here there is considerable uncertainty as to how to handle the amount of energy due to friction and deformation, which are largely caused by lateral sliding along leads or faults. If one assumes that the energy dissipated in pressure ridges is proportional to the change in potential energy, and that ridging is the only energy loss, then a constraint on ψ is

$$C \int_0^\infty h^2 \psi \, dh = \sum_{i,j=0}^{2} \sigma_{ij} \dot{\varepsilon}_{ij}, \tag{16.39}$$

where C is a constant related to the densities of water and ice and the frictional energy losses during ridging. In the case that only potential energy losses are considered, $C = C_b$ with

$$C_b = \frac{1}{2}\rho_i \left(\frac{\rho_w - \rho_i}{\rho_w} \right) g \tag{16.40}$$

where ρ_w is the density of water, ρ_i is the density of ice, and g is the acceleration due to gravity. In the special case of pure convergence, equation 16.39 can serve as a definition of ice strength. A key question is the magnitude of frictional losses in the ridging process. Early estimates of frictional losses put them at similar values to C_b, which have subsequently been shown to be low.

In order to explicitly ensure consistency of mechanical energy dissipation and ridging, the redistributor ψ may be written as $\psi = \delta\left(h\right)\left(M + \dot{\varepsilon}_{kk}\right) + M\omega_r$, where M is a normalized mechanical energy dissipation term and ω_r is the so-called ridging mode function. This function describes the transfer of thin ice into a distribution of thicker, ridged ice, acting simultaneously as a sink of thin ice and a source of thick ice in such a way as to conserve area and volume. If all of the energy losses result from ridging, then $M = p^{-1}\sum_{i,j=1}^{2}\sigma_{ij}\dot{\varepsilon}_{ij}$, where p is the hydrostatic compressive strength, σ_{ij} is the component of the two-dimensional stress rate tensor, and $\dot{\varepsilon}_{ij}$ is the component of the two-dimensional strain rate tensor. In the other extreme, $M = 0$. More details on these equations may be found in Hibler (1980b).

There are several major issues with the energetic consistency equations that make the mechanical aspects of the ice thickness distribution theory rather arbitrary and generally render the theory somewhat unwieldy. The most notable are: how to carry out the transfer of thin ice to thick ice by ridging, how much frictional energy losses occur during ridging, and how much ridging and concomitant open water creation occur under an arbitrary deformation state involving some shear. This latter feature can largely be dispensed with by using a fracture-based model at high resolution, rather than an aggregate model as discussed in section 16.5.2.

For the transfer of thin ice to thick ice by ridging, Thorndike et al. (1975) suggested a redistributor that transfers ice into categories that are fixed multiples of the initial thickness. However, ridge observations and theoretical considerations suggest that such a redistributor is unrealistic. Hibler (1980b) proposed a scaling law for ridge building with ice redistributed uniformly up to a maximum thickness, scaling as the square root of the thickness of ice being ridged. Limited data (see Chapter 12) on the block size of ice in ridges (Tucker and Govoni 1981) versus mean ridge height tend to support this scaling. Subsequent work (Babko et al. 2002) has added an explicit formulation of ice rafting.

With any of these redistributors, discrepancies remained for some time in that to obtain realistic ice drifts the local ridging losses were much smaller than needed in large-scale simulations (Flato and Hibler 1995; Hibler 1980b). A major contribution to resolving this inconsistency was made in the late 1980s through a series of large-scale simulations (Flato and Hibler 1995), as well as discrete element simulations of the ridging process by Hopkins and Hibler (1991a) and Hopkins (1994). The result was that frictional losses in ridging were found to be about an order of magnitude higher in the ridge building than potential energy losses. By increasing this energy loss in large-scale nonlinear plastic models (Flato and Hibler 1995), stresses large enough to yield realistic sea ice drifts were obtained, provided that the redistribution functions were realistic.

The other major uncertain aspect of the thickness distribution is the amount of ridging under pure shear. Within the conventional thickness distribution approach and an aggregate rheology (Rothrock 1975c; Thorndike et al. 1975), energetic consistency of a rheology requires that under shearing conditions both ridging and open water creation must occur. Variable thickness simulations (Flato and Hibler 1995) have shown that including or excluding this energetic consistency can lead to substantial changes in the simulated mean ice thicknesses and in the relative volume of ridged ice, even though the general shape of the thickness distribution is relatively unchanged. This consistency may be formulated in terms of sets of deforming oriented leads (Ukita and Moritz 1995) that provide a link to lead structure. However, such a procedure provides no unique way of choosing the lead orientation. Anisotropic models (Coon et al. 1998; Hibler and Schulson 2000) including continuity of stresses across the leads provide a unique way to make this choice based on dynamical constraints. These models typically include a great deal of shear energy losses due to sliding of leads so that additional open water creation is not usually required. Moreover, isotropic "fracture-based" rheologies (Hibler 2001; Hibler and Schulson 2000) provide a more precise theoretical determination of this partition, which ultimately derives from laboratory work provided that the "scale-invariant" assumption can be invoked.

16.6.3 Thickness Distribution Models Coupled to Specified Deformations

Although it does not consider mechanical feedback effects, coupling thickness distribution evolution models with thermodynamic models provides a useful way to estimate the role of a variable thickness sea ice cover. A number of "partially coupled" model studies of this type are reviewed by Steele and Flato (2000). One of the more comprehensive studies is that of Maykut (1986). Using relatively smoothed deformation from the AIDJEX experiment, Maykut investigated the role of thinner ice in the overall heat budget as well as the different heat budget components. These results show the dominance of thin ice in the overall heat budget. Basically, when variations in ice thickness were taken into account, the amount of heat conducted to the surface doubled. Maykut found that most of this additional heat was lost to the atmosphere via the turbulent fluxes. Notable was the fact that open water categories (0–0.1 m) did not dominate the balance; rather, it was the intermediate thicknesses (0.2–0.8 m) that were responsible for over half the heat loss through the ice. Maykut's study also emphasized the role of heat absorbed through leads, which may go to melting thick pressure ridges or possible lateral melting. Considerable experimental work has been performed on lateral melting (most notably by Perovich et al. 1997), which has allowed the development of formulations that can be used with the variable thickness models.

The degree to which inclusion of such improved thermodynamics might change variable thickness results that are subjected to a specified deformation has been examined by Zhang and Rothrock (2001) utilizing a three-level enthalpy-conserving model developed by Winton (2000). Each separate fixed thickness category was included in this formulation that made use of a fixed Eulerian 10-level thickness grid. Deformation and thermodynamic forcing near the North Pole were taken from a fully coupled dynamic thermodynamic model (Zhang et al. 2000; see also section 16.7). Their main result was that mirroring thermodynamic model results, including full enthalpy conservation, increased the mean ice thickness and substantially reduced the seasonal thickness swing. Specifically, they found a 5–10% effect on the ice thickness with much less reduction of thickness during the summer.

One difficulty with all specified strain comparisons is a realistic estimation of the deformation field. Portraying a typical deformation field with intersecting shearing and opening "faults" or "leads," the average strain field is very well defined, and can be obtained by a boundary integral using Green's theorem regardless of whether the faults are arbitrarily small. However, estimation of the strain by only a few velocity points on the ice can yield substantial errors, and one should minimally do a least squares analysis to estimate the error (e.g., Hibler, Ackley, et al. 1974). Moritz and Stern (2001) have effectively used the boundary integral method employing SAR data at three-day intervals. In that such data are not widely available, strain data are usually obtained from temporally and spatially smoothed velocity fields (see, e.g., Thorndike 1986), which underestimate high-frequency motion among other measures. Consequently, the resulting thickness distributions must be viewed with caution. In practice, aspects of the energetic consistency argument mentioned above (which are at best arbitrary even with remote-sensing estimates; e.g., Stern et al. 1995) must be invoked. Even with this inclusion, high-frequency variability is neglected. Heil and colleagues have shown, for example, that inclusion of this variability (from observations) yields a significant increase in the net sea ice production (Heil et al. 1998, 2001).

16.6.4 Two-Level Ice Thickness Distributions

Many features of the thickness distribution may be approximated by a two-level sea ice model (Hibler 1979), where the ice thickness distribution is taken to consist of two categories: thick and thin. Variations of this model have been used in a large number of investigations, most recently, for example, by Harder and Fischer (1999), Hilmer and Lemke (2000), Hollaway and Sou (2002), and Kreyscher et al. (2000) (see also Chapter 9). Within this two-level approach, the ice cover is broken down into an area A (often called the compactness), which is covered by thick ice, and a remaining area $1-A$, which is covered by thin ice, which, for computational convenience, is always taken to be of zero thickness (i.e., open water). The idea is to have the open water approximately represent the combined fraction of both open water and thin ice up to some cutoff thickness H_o. The thickness of the remainder of the ice is considered to be distributed between zero and $2H/A$.

For the overall mean thickness H and compactness A, the following continuity equations are used:

$$\frac{\partial H}{\partial x} = -\frac{\partial (uH)}{\partial x} - \frac{\partial (vH)}{\partial y} + S_H \tag{16.41}$$

$$\frac{\partial A}{\partial x} = -\frac{\partial (uA)}{\partial x} - \frac{\partial (vA)}{\partial y} + S_A \tag{16.42}$$

where $0 \leq A \leq 1$, u and v are the components of the ice velocity vector, and S_H and S_A are thermodynamic terms. Although the first equation is essentially a continuity equation for ice mass (characterized by the mean ice thickness or equivalently the ice mass per unit area) with a thermodynamic source term, the second equation for the compactness implicitly includes some treatment of the ridging process. By including the restriction that $A \leq 1$, a mechanical sink term for the areal fraction of ice has been added to a continuity equation for ice concentration. This sink term is considered to apply when $A = 1$, and under converging conditions removes enough ice area through ridging to prevent further increases in A. Note, however, that, independent of A, ice mass is rigidly conserved by the conservation equation for H. The thermodynamic growth and decay terms are typically determined by assuming a uniform distribution of ice thicknesses between zero and $2H/A$ (see, e.g., Hibler 1979; Hibler and Walsh 1982; Walsh et al. 1985).

To characterize the strength P in this two-level model, Hibler proposed that

$$P = P^* He^{-C(1-A)} \qquad\qquad (16.43)$$

where $C \sim 20$ and P^* is a constant. The most widely used value for the strength is $P^* = 2.75 \times 10^3$ N/m^2, which was obtained by Hibler and Walsh (1982) by fitting progressive vector plots of predicted ice drift to observations. This value is larger than the value initially proposed by Hibler (1979) owing to seven-day-averaged wind being used in the initial simulations. Similar values have been used by other investigators mentioned in this section.

16.6.5 Relative Characteristics of Two-Level and Multilevel Models

To allow either the multilevel or two-level sea ice model to be integrated over a seasonal cycle, it is necessary to include some type of oceanic boundary layer or ocean model. The simplest approach (e.g., Walsh et al. 1985) is to include a motionless fixed-depth mixed layer. Another approach is to use some type of one-dimensional mixed layer, such as the Kraus–Turner-like mixed layer used by Lemke et al. (1990) for the Weddell Sea. A third approach is to utilize a complete oceanic circulation model that allows lateral heat transport in the ocean (Hibler and Zhang 1994; Zhang, Rothrock, et al. 1998). This latter approach is discussed in more detail in section 16.7.1.

Of particular relevance to climatic variability is the inertia that a full thickness distribution inserts into the ice mass characteristics. This is illustrated in Figure 16.20, which shows autocorrelations of the mass anomalies for three different dynamic thermodynamic models of the arctic ice pack with different thickness distribution formulations. Note that with a full multilevel thickness distribution, it takes about eight years for the autocorrelation function of the thickness anomalies to fall off to $1/e$ of the variance. This is in sharp contrast to a two-level model, where there is a very rapid balancing of the strength and thickness on timescales of a few months, so much so that the autocorrelation time for a two-level model is less than for a thermodynamic model. The inertia of the multilevel model is very large because the thick ridged ice plays no direct role in the ice strength, and slowly reaches equilibrium by a combination of advection, growth, melt, and formation.

It should be emphasized that these temporal correlations occur even though a uniform distribution of thick ice is used in thermodynamic calculations in the two-level models. Without this thermodynamic correction, the two-level model (see, e.g., Hibler and Flato 1992) yields ice about 20% thinner than thermodynamic thicknesses for the Arctic Basin and about 35% thinner than a full variable-thickness simulation. This thermodynamic correction does improve the growth rates but still does not supply a precise treatment of

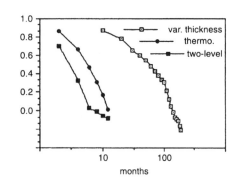

Figure 16.20. Monthly mass anomaly (deviation from the average seasonal value) autocorrelation of arctic ice mass for three different ice models used to simulate the arctic ice cover over several decades. The variable-thickness model results are from Flato (1995) and the thermodynamic and two-level model results are from Walsh et al. (1985).

thin ice, mechanical coupling, or a full thickness distribution for comparison with observations. For example, full thickness distribution models have been used by Flato and Hibler (1995), Hibler (1980b), Holland and Curry (1999), Polyakov et al. (1999), and Zhang et al. (2000).

16.6.6 Thickness Strength Coupling

High-frequency variability in ice motion and deformation at subdaily time scales can have a significant impact on the mass budget of sea ice (Heil et al. 1998). As noted above, the most pronounced form of high-frequency variability is inertial variability, which has some seasonal signature almost universally felt to be due to enhanced damping by ice interactions during the winter. The main explanation of these phenomena is a strong turbulent coupling between the ice and the oceanic boundary layer so that the ice motion is a signature of the larger inertia of the oceanic boundary layer. When coupled with the evolution of the thickness distribution (e.g., Hibler et al. 1998), significant variations in compactness can occur.

A separate, related mechanism that is probably related to inertial-scale variability in ice deformation are waves (loosely referred to as kinematic waves) arising from coupling between the momentum equations and ice thickness evolution equations. These waves have a more rapid propagation speed than the advection timescale, and hence at higher resolutions the use of longer time steps can result in instabilities in the coupled equations. In such cases, either coupled implicit solutions are utilized or smaller time steps are employed. We now briefly describe kinematic waves, the effects of high-frequency deformation on ice mechanics, and some ice arching issues.

16.6.6.1 Kinematic Waves

The basic idea of kinematic waves in sea ice is that the conservation equation in conjunction with the ice momentum equation sets up a wave owing to the velocity being dependent on quantities in the conservation equation (such as the compactness). The term *kinematic* arises from the fact that inertia plays no role in the wave.

To examine some analytic characteristics, consider ice piling up against the coast with a motion that is controlled by an ice pressure P. Considering the case where ice piles up continuously and smoothly so that the ice failure is always on the yield curve for appropriate forcing, we may take $P = f(A) = P_i e^{-C(1-A)}$, where A is the compactness. Then, with a linear water drag, the equilibrium equation of motion for a one-dimensional system is

$$\frac{\partial P}{\partial x} + \tau - cu = 0. \tag{16.44}$$

In the absence of thermodynamics, the conservation equation for A is

$$\frac{\partial A}{\partial t} + \frac{\partial(uA)}{\partial x} = 0. \tag{16.45}$$

Solving for u from the momentum equation, the conservation equation may be rewritten as

$$\frac{\partial A}{\partial t} - \frac{\partial}{\partial x} \left\{ \left[\frac{\partial P(A)}{\partial x} - \tau \right] A \right\} = 0. \tag{16.46}$$

Considering P to be dependent upon x through A and τ constant, this equation may be put in the form

$$\frac{\partial A}{\partial T} + Q(A)\frac{\partial A}{\partial X} = \eta(A)\frac{\partial^2 A}{\partial X^2} \qquad (16.47)$$

where $\eta(A) = A(\partial P/\partial A)$ and $Q(A)$ represents the remainder of the terms when the differentiation is carried out. This is basically a damped hyperbolic wave equation (Whitham 1974) with the essential nonlinearity inherent in the term $Q(A)(\partial A/\partial X)$. This equation may be thought of as a series of damped propagating waves with the wave speed dependent on the value of A (Whitham 1974). For example, an analytic simple case of equation 16.47 (still highly nonlinear) is Burgers' equation, where $Q(A) = A$ and $\eta(A)$ = constant. Burgers' equation is soluble analytically and tends to lead to sharp fronts propagating even though the initial conditions may be smooth.

A numerical example of ice buildup similar to the analytic example, together with attendant effects of nonlinear thickness strength coupling, was examined by Hibler et al. (1983). In this study, coupled momentum and conservation equations were numerically integrated in conjunction with an ice strength parameterization. The coupled equations were integrated for 18 hours at time steps of 15 minutes on a 9×9 grid. The initial conditions consisted of 10% open water together with a mean ice thickness of 0.5 m. To simplify analysis, the ice strength is taken to depend only on the compactness:

$$P = (0.5)P^* e^{-C(1-A)} \qquad (16.48)$$

In addition, the turning angles and Coriolis parameter were set equal to zero. A constant wind speed of 9.23 m/s in the positive x-direction was used.

The behavior of the velocity variability and ice buildup is shown in Figure 16.21. Even though the forcing is fixed, points nearest the boundary first slow down and then speed up. The initial slowing down of the ice near the coast is due to the ice becoming stronger as the compactness decreases. However, the region of low compactness eventually becomes large enough to accumulate an adequate wind fetch to overcome the differential of plastic stresses, and the nearshore ice begins to drift faster. In terms of the compactness, the wave is essentially a propagating front between high and low compactness propagating outward. The nonlinear coupling between the strength and the compactness (equation 16.48) can also lead to fluctuations in the ice velocity, as is apparent in the velocity plots in Figure 16.21 at ~ 6 hours. These secondary waves were removed when a linear compactness strength relationship was utilized. Overall it

Figure 16.21. Time series of x-velocity at grid points progressively further from right-hand boundary of an idealized ice buildup numerical experiment.

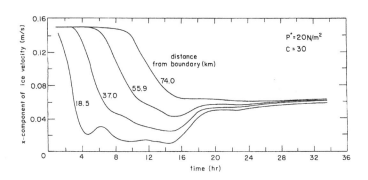

is possible that continual excitation of such waves may be responsible for observed mesoscale sea ice fluctuations which may affect the ice mass balance. Probably the main relevance of these waves to modeling the evolution of sea ice is their constraint on time step limitations in high-resolution models. Since the wave propagation time is probably an order of magnitude faster than the advection speed of the ice, typical explicit integrations of the coupled equations requires a Courant, Friedrichs, and Lewy (CFL) criterion (Richtmyer and Morton 1967) for overall stability, unless an implicit solution for the momentum and thickness equations is utilized.

16.6.6.2 Inertial Variability The basic character of simulated and observed inertial oscillations was briefly mentioned earlier in conjunction with the dynamical equations of motion. Here we focus on the mechanisms for this variability, and the role of ice mechanics in damping the oscillations. An idealized investigation of the effects of the inertial imbedding procedure proposed by McPhee (1978) was carried out by Hibler et al. (1998) with application to the antarctic ice pack. This simulation was similar to the kinematic wave study mentioned above. However, when a pulse of wind is applied to this system with appropriate strengths, the outward propagating front of high compactness couples with the inertial motion to form bands of oscillating deformation that are very slowly damped.

More relevant to the sea ice mass budget, however, are realistic two-dimensional simulations. Such simulations (Heil and Hibler 2002) demonstrate that with realistic temporally and spatially varying winds, nonlinear mechanical damping from plastic rheologies supplies about the correct mechanical damping with resultant deformation. The statistical characteristics of the resultant deformation are illustrated in Figure 16.22, which shows simulated and observed deformation spectra from two sets of three drifting buoys with the locations shown in Figure 16.23a. Basically, with imbedding the two deformation spectra agree very well with reasonable coherence. This improved coherence is also reflected in a very significant correlation coefficient ($r = 0.87$) between the simulated and observed divergence for the array corresponding to the left hand panels in Figure 16.22.

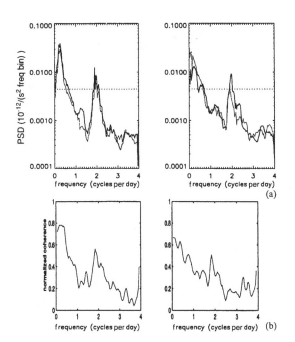

Inertial imbedding also greatly improves the variance of the simulated divergence compared with conventional simulations using, say, quadratic drag. Geiger et al. (1997), for example, found that while the viscous–plastic rheology performed best of all rheologies tested, modeled ice deformation components in the Weddell Sea showed a much lower variance than that yielded by the observed values, and a very low coherence with

Figure 16.22. (a) Power spectral densities of observed (darker curve) and simulated (with inertial imbedding) divergence rate for two different buoy arrays. (b) Normalized coherence between simulated and observed divergence rates.

Figure 16.23. (a) Outline of model domain used for an ice–ocean inertial oscillation study. The left-hand solid box indicates the region where ice thicknesses were analyzed. Trajectories of two arrays of three buoys each are shown: A1 near the Pole and A2 in the Beaufort Sea. Buoys shown traveled from the top toward the bottom of the graph. The trajectory of IABP 9357, which forms part of A1 (see Figure 16.18), is labeled. (b) Time series of ice thickness over the central Arctic Basin from a dynamic thermodynamic sea ice model with and without inclusion of "inertial embedding" and the resultant increased high-frequency deformation.

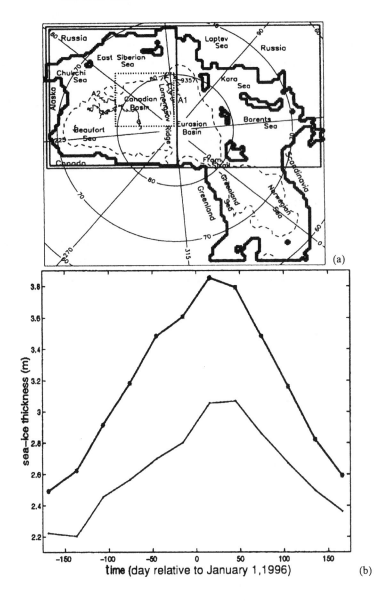

buoy deformation even at low frequencies. This was especially true of divergence, whose variance magnitude was only 10% of the observed divergence variance magnitude. Including the inertial energy of the ocean via inertial imbedding greatly ameliorates the simulation of divergence, and hence appears to be a key component in simulating realistic ice thickness characteristics.

Interestingly, much of the inertial energy in this deformation field arises from nonlinear ice mechanics. In particular, Heil and Hibler (2002) examined simulations with smoothed wind fields with inertial imbedding both with and without the effects of ice mechanics. The velocity magnitude spectra show that with an unsmoothed wind field the inertial imbedding alone supplies a resonance that will cause amplification of the small amount of high frequency noise in the wind, and hence create a spectral peak in the simulated buoy velocity. However, in the case of smoothed wind, the imbedding alone, which is essentially a linear operation on the wind field, produces no appreciable inertial power. These results show that the nonlinear ice mechanics coupled to the thickness evolution equations supplies a cascade of energy to higher frequencies that generates inertial power. Consequently, we may think of ice mechanics as generating energy through this cascade and then also damping this energy to realistic levels.

The effects of this variability on the mass balance are significant, resulting in up to 1 m of additional ice formation in the central arctic region (Figure 16.23b). Although utilizing only a two-level distribution simulation, Heil and Hibler (2002) showed that, in equilibrium, enhanced high-frequency variability substantially increases (Figure 16.23b) the ice volume averaged over a central portion of the ArcticBasin (as noted in Figure 16.23a). Basically, there is a greater increase in winter in the ice mass, with a somewhat reduced mass increase in summer. The likely physics here is an increase in open water in winter, which greatly increases ice production. However, during the decay season the additional radiation absorption provided by increased open water due to variability also causes more melt, hence the seasonal asymmetry.

16.7 Simulations of the Evolution of Sea Ice

To examine the interplay of dynamics and thermodynamics on the evolution of the mass balance of sea ice, a selected hierarchy of model simulations is presented here. Many of these model runs are coupled ice–ocean model studies. The focus is on simulations illustrating the role of dynamical processes. For this purpose we proceed from idealized models to models employing a full variable-thickness distribution. In practice there are a large number of contemporary dynamic thermodynamic model simulations of sea ice evolution, most emphasizing the thermodynamic characteristics. A review of different model simulations of the Arctic with an emphasis on thermodynamic response together with specified deformation may be found in Steele and Flato (2000). In addition, a more complete description of different simulations of contemporary change and future variability with emphasis on coupled climate simulations can be found in Bamber and Payne (2004, Chapter 9).

16.7.1 Ice–Ocean Circulation Models

It should be emphasized that if realistic ice margin simulations are desired in large-scale models, it is critical to utilize coupled ice–ocean models, otherwise the location of the ice margin cannot be realistically simulated. A useful mode for this purpose is a robust diagnostic model whereby a weak relaxation to observed temperature and salinity values is used everywhere except in the upper boundary layer where ice is present. This is particularly graphically illustrated in simulation results (Figure 16.24a) from such an ice–ocean model by Hibler and Bryan (1987). In this model a weak (three-year) relaxation to observed temperature and salinity was utilized. Sea ice was treated by a two-level dynamic thermodynamic model using a viscous–plastic rheology. The improved ice margin is due to the large oceanic heat flux from the deeper ocean into the upper mixed layer. Analysis shows that much of the heat flux occurs in winter and is the result of deep convection. This deep convection brings up warm water and prevents early winter ice formation.

Interannual variability of the ice margin in the Greenland and Barents Sea is, however, less critically affected by ocean variations. Results from a higher-resolution version (Hibler and Zhang 1994) of the Hibler–Bryan "robust" diagnostic ice–ocean model are shown in Figure 16.24(b). A typical stream function for this model showing ocean transport into the Barents

Figure 16.24. (a) February 50% concentration limits from a full coupled diagnostic ice–ocean model and for an ice-only model which includes a fixed-depth mixed layer. A one-degree resolution was used (Hibler and Bryan 1984, 1987). (b) Correlation between monthly simulated and observed ice edge anomalies using a 40-km-resolution diagnostic ice–ocean model. A two-level ice thickness distribution was used in both coupled models.

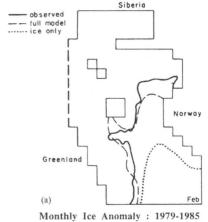

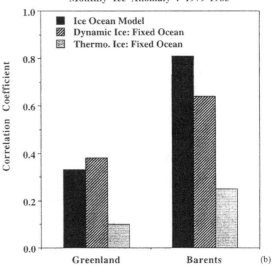

Sea and the model domain is given in Figure 16.25a. A five-year relaxation to observed temperature and salinity was used everywhere except in the upper two levels, which essentially comprise the oceanic boundary layer. Although there is some improvement in the ice margin variations with the inclusion of variable ocean circulation, the simulations with a fixed annual oceanic heat flux correlate essentially as well. Part of this is likely owing to the fact that ice transport effects and oceanic circulation effects tend to vary in a similar manner in these regions. For example, stronger northerly winds advecting ice further south into the Barents Sea also tend to cause a reduction of northward transport of heat in the ocean. The only model that works extremely poorly is a thermodynamics-only ice model where ice advection and ice deformation effects are not included.

As noted above, an important aspect of the ice–ocean model of the type used by Hibler and Bryan (1987) is the coupling of the ice to the ocean. This coupling consists of taking the stress into the ocean to be given by the wind stress minus the ice interaction. They argue that this procedure rigidly conserves momentum for the ice–ocean system and properly takes into account the combined Ekman flux of both the ice and oceanic boundary layer. The difference between this coupling and taking the stress into the ocean to be the water drag can be significant (see, e.g., Nazarenko et al. 1998; Wang and Ikeda 2000; the Community Climate System Model of the National Center for Atmospheric Research; and many others). Because the simulated ice drift correlates closely with the wind (see Figure 16.16), the water drag stress field closely resembles the wind.

With the full Hibler and Bryan coupling, the stream function from September 1983 (Zhang 1993) clearly shows an Alaskan countercurrent running along the arctic coast. Analysis of the vertical cross section of this current shows it to be similar to the Beaufort Sea undercurrent, based on hydrographic data and current measurements (Mountain 1974). The simulated cur-

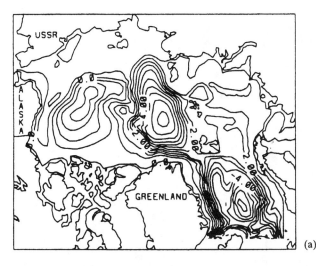

Figure 16.25. (a) Monthly mean stream function contours for September 1980 from a 40-km-resolution ice–ocean model of the Arctic, Greenland, and Norwegian Seas. Contour interval is 0.5 Sv (1 Sv = 106 m³/s). (b) Time series of vertically integrated flow ~50 km north of Point Barrow for 1979. Negative values represent eastward flow. The solid line is with the stress into the ocean modified by the ice interaction, and the dashed line is the current with wind stress transferred directly into the ocean without modification.

(a)

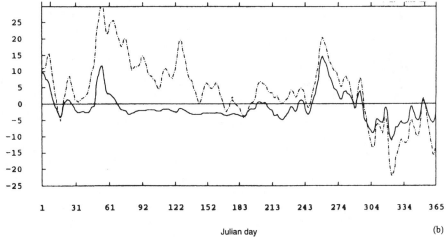

Julian day

(b)

rent was two to three grid cells wide (80 to 120 km), with a typical summer mean current speed of 15 cm/s. The current observed by Mountain was about 75 km wide with speeds up to 30 cm/s. However, in a simulation with the stress into the ocean taken to be the wind stress, the eastward character (Figure 16.26b) of the current in 1979 almost totally disappears except in the fall. Analysis (Zhang 1993) of the stress into the ocean system at this location shows that the modification of this stress by the ice interaction is very great, even though the ice is generally moving westward. Consequently, Zhang's explanation is that a portion of the inflow from the Bering Strait has a tendency to flow eastward along the Alaska coast toward the Beaufort Sea if ice is present to subdue the air stress. Otherwise the frequent easterly wind will prevent this from happening. It is, of course, possible that the ice drag is also reduced, but ice drift in this region is sometimes actually enhanced by the nonlinear character of the ice interaction. Given current interest in understanding the effects of fast ice and coastal processes on the evolution of sea ice, these coupling issues are important.

A particular problem with most ice–ocean circulation models is the tendency to produce too much salt in the central Arctic if the model is run without any "nudging" to observed temperatures and salinities (Häkkinen 1993; Häkkinen and Mellor 1992). This, in turn, tends to yield a weakened surface current circulation pattern compared with observations (see Zhang, Hibler,

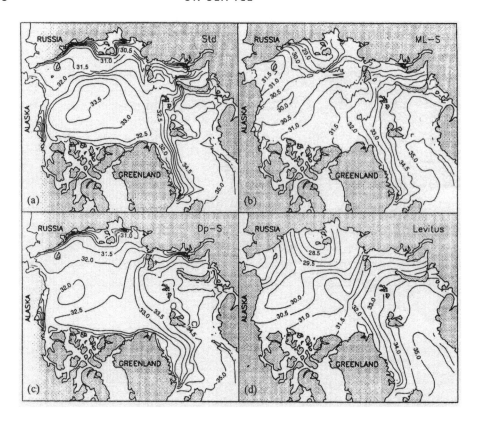

Figure 16.26. Seven-year (1979–1985) mean mixed-layer salinity distributions in parts per thousand predicted by three different ice–ocean models together with Levitus' (1982) data in (d). (a) The standard (Std) model is a fully prognostic model with imbedded mixed layer. (b) The specified mixed-layer salinity (ML-S) model is the same model with a diagnostic relaxation (30-day time constant) to the observed salinity only in the mixed layer. (c) The deep salinity (Dp-S) model has a five-year relaxation to observed salinity in the ocean everywhere below the mixed layer. For Dp-S there is no relaxation in the mixed layer. (A model with relaxation to both temperature and salinity was similar to Dp-S.) Contour interval is 0.5 ppt (Zhang, Hibler, et al. 1998).

et al. 1998). An examination of this problem was methodically carried out by Zhang, Rothrock, et al. (1998) by performing a series of seven-year simulations with and without various forms of "diagnostic nudging" to observed temperature and salinity. The results show that without relaxation of any kind the ice–ocean model tends to yield a dome of saltier surface water in the interior of the Arctic Basin (Figure 16.26a). This is in contrast to observations indicating that the surface salinity increases more or less uniformly as one approachs Fram Strait (Figure 16.26d). If, on the other hand, one utilizes either a strong (~30-day) relaxation to surface salinity (Figure 16.26b) or a weak relaxation (~ three years) to a salinity *below* the mixed layer (Figure 16.26c), then surface salinity fields are obtained in closer agreement with observation. These nudging effects also lead to more realistic ocean surface currents, and hence better simulations of ice drift. However, Zhang, Rothrock, et al. (1998) emphasize that diagnostic relaxation, especially when done below the mixed layer, can lead to somewhat unrealistic convection effects. The precise reason for these differences is an active area of research. Explanations range from indirectly hypothesizing the sinking of shelf water along the coasts (Aagaard and Carmack 1994; Aagaard et al. 1981); uncertainties with regard to the precipitation in the Arctic (Ranelli and Hibler 1991); and inadequate treatments of vertical boundary layer processes and horizontal eddies in models

(McPhee 1999; Zhang et al. 1999). Another possibility is inadequate simulation of ridged ice. This problem and related ice–ocean issues are currently active areas of investigation.

A final interesting feature of coupled ice–ocean models on the hemispheric scale is the presence of interannual and interdecadal oscillations (Hibler and Zhang 1995; Zhang et al. 1995) in the ice characteristics. These oscillations are related to the thermohaline overturning circulation and occur even with temporally constant atmospheric forcing in "sector models." They are not, however, present in the absence of the thermodynamic effects of a coupled ice cover, and hence appear to be an intrinsic ice–ocean phenomenon. Although the inclusion of ice transport can change the frequency of the oscillations, analysis shows these variations to be mainly caused by the thermal insulating effect of sea ice. Since the rate of heat transfer into the ocean depends nonlinearly on the ice thickness and the heat flux from the deeper ocean, a simple set of equations describing the system and admitting nonlinear oscillations can be formulated. The most notable variability found was in the ice thickness, which tended to vary in about a six-year cycle without ice transport, and at approximately four years with ice dynamics included. In addition, there are longer-term decadal variations in the northward water and heat transport in these models. The main relevance of these oscillations to the dynamic response of sea ice is that they naturally exist in the system, even in the absence of atmospheric feedback. Consequently, such variability is a candidate for variations of ice thickness in a natural climate sense as opposed to climatic warming.

16.7.2 Multiple Equilibrium States in Sea Ice Models

In the Arctic Basin, equilibrium ice thickness depends in a highly nonlinear manner upon residence time, ice growth, and ice outflow. Because of this, it is in principle possible for coupled models to have multiple equilibrium states for the same forcing, especially under cooler climate conditions. Whether or not this has happened in the past is an interesting question that has received little attention. The key mechanical component leading to multiple equilibrium states is the arching phenomenon, which can restrict ice flow. In addition there are albedo feedback effects that can, in principle, lead to multiple equilibrium states even under present climate conditions (Thorndike 1992a) owing to physics similar to the "small ice cap instability" inherent in Budyko–Sellers models (see, e.g., North 1988 and North et al. 1981).

The role of mechanical effects in inducing such multiple equilibrium states has been demonstrated by Hibler and Hutchings (2003) in a two-dimensional dynamic thermodynamic model of the arctic ice cover. In practice such states appear to be accessible to most nonlinear dynamic thermodynamic sea ice models. In order to consider a range of climate states, Hibler and Hutchings (2003) used the idealized thermodynamic ice model proposed by Thorndike (1992b). In addition to examining the overall qualitative sensitivity of sea ice to climatic perturbations, this model has also been used in conjunction with more traditional Budyko–Sellers energy balance models to examine multiple equilibrium states of a stationary sea ice cover (Thorndike 1992a). Although the seasonal response is unrealistic, the "Stefan"-like model of Thorndike reproduces salient aspects of seasonal thermodynamic models. The essential idea in this model is to divide the year into a warm and cold season each of length $Y = 0.5$ year. The climate is described by downwelling longwave radiation ($f\mathrm{lwc} = 180\,\mathrm{W/m^2}$) in the cold season and long- and shortwave radiation during the warm season of $f\mathrm{lww} = 270\,\mathrm{W/m^2}$ and $f\mathrm{sw} = 200\,\mathrm{W/m^2}$. The final variable is the heat supplied from the ocean, which is idealized as a constant of the order of 2 W/m^2 for pre-1990 arctic conditions (Maykut and Untersteiner 1971). Climatic change is introduced via a uniform perturbation δ to the longwave radiation fluxes $f$lwc and $f$lww. An important difference from the traditional "Stefan" models is the introduction of a heat capacity during the transition periods from warming to cooling and vice versa. To examine multiple equilibrium states, Hibler and Hutchings (2003) examined a fully coupled dynamic thermodynamic model of the Arctic Basin with a 100-km resolution and outflow only allowed at the Fram Strait. The ice strength P^* was related to the mean ice thickness h by $P^* = (4 \times 10^4)h$ N/m, with h given in meters. The mean ice thickness was the only thickness distribution variable, and evolved according to $\nabla \cdot (\mathbf{v}h) + f(h) = 0$, where $f(h)$ is the ice growth rate calculated from the idealized thermodynamic model. In the thermodynamic calculations the conductivity was increased to $k = 3.4$ W/m^2 to account for the modification to ice growth by deformation. Mean monthly wind forcing was used together with linear wind and water drag formulations. The growth period was considered to be from October to March.

The basic character of potential multiple equilibrium states is illustrated in Figure 16.27. Part (a) of this figure shows outflow and basin-averaged ice growth at the end of two years. To the degree that the initial thickness is a good indicator of the basin-averaged growth and outflow, three possible equilibrium states are identified by the intersection of the outflow and growth conditions in Figure 16.27a. In particular the rate of change of ice thickness dh/dt is given by

$$\frac{dh}{dt} = G(h) - O(h) \qquad (16.49)$$

where $G(h)$ is the growth rate and $O(h)$ is the outflow rate. Considering h_o to be a solution of this equation and $h_1 = h - h_o$ to be a small thickness perturbation relative to the solution, one obtains (to lowest order) the rate of change of the perturbation to be given by

$$\frac{dh_1}{dt} = \left[G'(h) - O'(h) \right] h_1. \qquad (16.50)$$

Clearly, the small perturbation will grow unless $\left[G'(h) - O'(h) \right] < 0$, a condition which, by inspection of Figure 16.27a, is met for solutions (1) and (3) but not for solution (2).

Because solution (2) is not stable, as one proceeds from a cold climate to a warm climate (or vice versa) there will be a rapid jump to a lower thickness state, which then changes less rapidly with warming as the outflow is much more significant and yields a lower ice thickness with higher growth rates. This is shown in Figure 16.27b, which shows mean ice thickness for actual multiple equilibrium states obtained by 300-year simulations initialized with either very thin or very thick ice. The change of slope is significant. If one were to interpolate linearly the trend before the drop, an approximately zero ice thickness under present conditions would be expected as compared with the ~2 m that is simulated. In addition, because of the multiple equilibrium states, the transition between thick and thin states will depend upon whether one is gradually warming or cooling.

Seasonal outflow and growth characteristics for two equilibrium states for $\delta = -9.5$ W/m^2 are shown in Figures 16.27c and d. Also shown is a control solution for $\delta = 0$. The main feature of the "thick" solution is a greatly reduced outflow, with the annual average outflow of the thick case being 16% of the thin solution and 12% of the standard simulation. In the thick solution the outflow occurs mainly during the melt season with little correlation with the wind stress in the Fram Strait region. In the standard case ($\delta = 0$), because of the weaker strengths the outflow tends to correlate strongly with the wind, causing the outflow to peak in late winter, even though the strength is largest then. However, with the thick solution, and, to a lesser degree, the thin multiple-equilibrium solution, the outflow tends to be seasonal, with much less outflow during the growth period. In terms of magnitude, the outflow is small compared with the growth rates, although outflow is, of course, comparable to the difference between the growth and melt rate. While the above model is very idealized, the fact that such multiple equilibrium states may exist in full sea ice circulation models indicates that such states may be latent in more complex thickness distribution simulations, especially under climatic cooling conditions. In the above analysis these states ultimately depend on shear strength in the ice rheology and should not, in principle, be present for a "cavitating fluid" rhe-

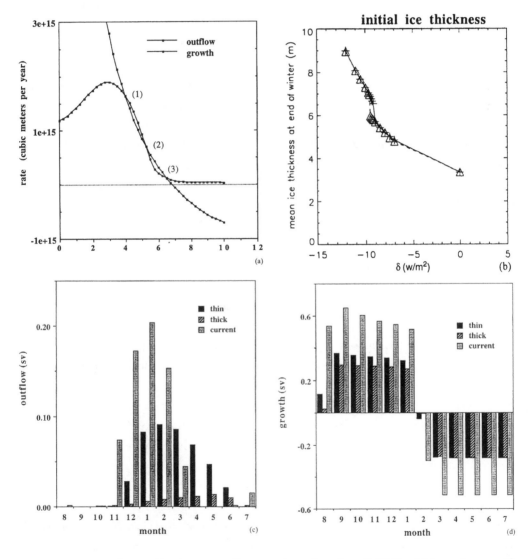

Figure 16.27. (a) Outflow and basin-averaged ice growth as a function of initial ice thickness for a specific value of $\delta = -11$ W/m^2. (b) Maximum ice thickness in seasonal equilibrium as a function of δ. States initialized with both thick and thin ice are shown. (c) Monthly basin outflow in Sverdrups where 1 Sv = 106 m^3/s. (d) Monthly basin-averaged ice growth (Sv) (Hibler and Hutchings 2003).

ology. In addition, in paleoclimate studies that include sea ice dynamics (e.g., Vavrus and Harrison 2003), the effect of nonlinear mechanics on the evolution of sea ice over the 20th century should be considered. Moreover, even if they are not possible for present climate conditions, the existence of such states can affect the response of the models to major changes in the wind fields related to interdecadal variability. Consequently, interpretations of the overall historical circulation of the arctic ice cover based only on correlations with winds (Vinje 2001) are questionable.

16.7.3 Arctic Basin Variable Thickness Simulations

Arctic Basin simulations employing a nonlinear plastic rheology together with full multiple-level thickness distributions and a heat-budget-based thermodynamic code provide a useful mechanism for examining the interplay between dynamics and thermodynamics in sea ice and ice–ocean models. In the highly selected simulations presented here, we focus on models with full thickness distributions and examine results using observed forcing, in some cases over 18-year periods. Natural variability in the forcing, or removal of variability, then provides the means to assess sensitivity.

The variable-thickness model simulations discussed below (Flato and Hibler 1995; Zhang and Rothrock 2001; Zhang et al. 2000) are largely based on the variable-thickness model solution framework of Hibler (1980b). This numerical framework represents the most direct solution of the equations of motion in that fixed Eulerian grids were utilized in both space and ice thickness space. These fixed thickness and space grids facilitate the maintenance of conservation properties of the thickness distribution, but for full resolution of the thickness distribution a large number of levels are typically required. Other formulations exist. For example, Bitz et al. (2001) have emphasized fewer categories, with both the mean thickness and ice area in each category as variables. Following Bratchie (1984) and Pritchard (1981), Polyakov et al. (1999) utilized fewer thickness categories up to some thick ice level where both thickness and ice area are utilized. Probably the most useful procedure, however, is that of Lipscomb (2001), who utilizes a fixed grid in thickness space together with a remapping procedure (Dukowicz and Baumgardner 2000) to treat the thickness advection term (df_g/dH) in equation 16.35. Using this procedure with five to seven fixed-thickness categories, Lipscomb was able to produce results commensurate with 20 to 30 categories (e.g., Hibler 1980b) without remapping. In most cases to date, when considering nonlinear ice dynamics, investigators have utilized a variation on the viscous–plastic constitutive law formulation with an elliptical yield curve of Hibler (1979), possibly with a numerical solution using the elastic–viscous–plastic method (Hunke and Dukowicz 1997).

Although it was a short simulation (five years) with relatively smooth wind fields (seven-day averages), the initial simulation by Hibler (1980b) demonstrated the characteristic ice buildup along the Canadian archipelago (Figure 16.28b). This buildup is largely caused by the ice velocity field advecting ice into this region as well as by episodic ice convergence. Because the ice convergence is episodic, it is difficult to tell whether there is a net convergence. This overall buildup is characteristic of variable-thickness sea ice models, and is in general agreement with contours of ice thickness from submarine sonar data (Bourke and Garrett 1987) as well as with ridge statistics (see Chapter 12). Analysis of actual ridge production over a year (Hibler (1989) shows much higher intensity of ridging near the outflow region. Consequently, the ice buildup is more

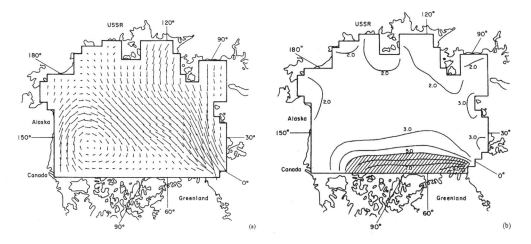

Figure 16.28. (a) Average annual ice velocity and (b) average April ice thickness contours (meters) at the end of a five-year variable-thickness dynamic thermodynamic sea ice model simulation. Seven-day averaged winds from May 1962 to May 1963 were used for dynamical forcing. Thermodynamic forcing similar to that used in Parkinson and Washington (1979) utilized monthly averaged climatological air temperatures, dew points, and daily long- and shortwave radiative forcing. In (a), a velocity vector one grid space long represents 0.02 m/s (Hibler 1980b).

of a complex combination of ice advection and local deformation. This is in contrast to results obtained with the two-level thickness distribution (sections 16.6.4 and 16.6.5), which tend to be more of a short-term balance of deformation and advection. Sensitivity studies (not shown; see Hibler 1980b) with greater ice strength do, however, show a decrease in the maximum thickness in the archipelago region, so that strength parameterizations in the variable-thickness model are important here. Strength sensitivity simulations carried out by Hibler (1980b) also demonstrate the effects of ice mechanics on ice outflow. In particular, increases of the strength by an order of magnitude over the standard case resulted in a total stoppage of flow for several of the winter months. The result supports the arching analysis carried out above and amplifies the importance of having as strong a variability in the wind forcing as possible. However, use of smoothed wind fields with the highly nonlinear rheologies employed in these models will likely result in sluggish ice flow and poorer renditions of ice deformation.

16.7.3.1 Ridged Ice and Mechanical Sensitivity   Using daily atmospheric forcing over the time period 1979 to 1985, Flato and Hibler (1995) carried out a series of variable-thickness simulations including an explicit treatment of ridging. The focus was largely on determining appropriate parameters for the mechanical redistributor and the buildup of ridged ice. In the standard version of this model, the energy dissipation during ridging is taken to be 17 times the potential energy of ridge building via the redistribution function. This value was obtained by insisting that the computed and average monthly drift magnitudes of several buoys were within about 2%. The value of 17 so obtained compares well with the range of 10–17 determined by discrete-element simulation of individual ridge-building events (Hopkins 1994).

   A notable characteristic (Figure 16.29a) of the buildup to equilibrium of this model is the long time (approximately 10 years) taken for the ridged ice fraction to come to equilibrium. This is opposed to the level ice fraction that comes into equilibrium more rapidly. Timescales

Figure 16.29. (a) Time series of total, ridged, and undeformed ice volume over the model domain for a multilevel sea ice simulation (Flato and Hibler 1995), with explicit accounting for the distribution of ridged ice. A 21-year simulation was obtained by repeating 1979–1985 forcing three times. (b) Sensitivity of ridged ice volume per unit area to energetic consistency assumption. Three cases were examined: convergence only, where ridges are only created by ice convergence proportional to the convergence rate; 50% consistency, where 50% of the mechanical energy is accounted for by additional ridging (and concomitant open water creation) under shearing conditions; and 100% consistency. The western arctic point was located at 76°N, 160°W and the data were compiled for October 31, 1982. The eastern arctic point was located at 84°N, 5°W, with data compiled for June 30, 1985 (Flato and Hibler 1995).

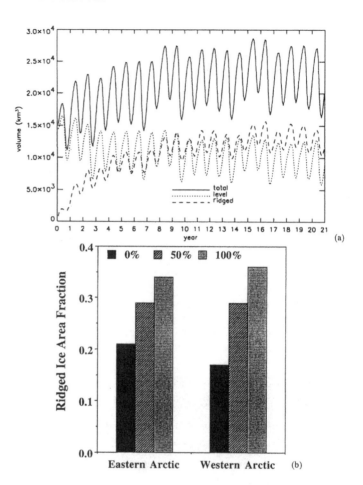

for different thickness categories were subsequently investigated more methodically by Holland and Curry (1999). It is also notable that in quasi-equilibrium the ridged ice accounts for more than half the ice volume, a feature that Flato and Hibler (1995) argue is qualitatively consistent with a wide variety of data. The long timescale (approximately seven years) is in contrast to a two-level model where the strength is directly coupled to ice thickness (see, e.g., Figure 16.20). This history effect means that wind patterns may take a number of years to affect the full thickness distribution. Basically, ridging is a complex balance between ridge formation, advection, and decay, with the actual ridging occurring in a given year being significantly smaller and differently distributed than the full distribution of ridged ice.

Investigation of different energetic consistency formulations by Flato and Hibler (1995) showed that much of the ridging was arising from the 50% energetic consistency assumption in the standard simulation. When either 100% or 0% energetic consistency was employed (in the latter case open water can only be created by divergence and ridged by convergence—not simultaneously by open water creation and ridging during shearing events), 10–14% changes in mean ice thickness were observed with the volume of ridged ice changing even more. Although not shown, these energetic consistency variations also change the spatial pattern of thickness, especially in the Beaufort Sea. Clearly, variations on how this feature is treated can significantly change results.

16.7.3.2 The Role of Dynamics and Thermodynamics in Historical Variability Changes in wind patterns can supply a very substantial change in the ice buildup. An example of how wind patterns can dominate, especially as one shifts between different arctic oscillation patterns, was provided by Zhang et al. (2000) using a multilevel-thickness sea ice model with atmospheric forcing over the time period 1979 to 1996. The formulation of this model was essentially the same as in Hibler and Flato (1992) and Hibler (1980b), and was coupled to an ocean circulation model (see, e.g., Zhang, Hibler, et al. 1998). Over this time period there was a general shift of the wind patterns and ice velocity patterns (see, e.g., Figure 16.2) from low–North Atlantic Oscillation (NAO) patterns with a strong gyrelike pattern to high-NAO patterns with a weakened and shrunken gyre. Some of the thickness patterns are shown in Figure 16.30. These thickness characteristics are qualitatively similar to most dynamic thermodynamic simulations. The main differences are the presence of significantly different thickness anomalies in response to changes in the wind forcing. Moreover, almost all the thickness changes were due to dynamical changes as opposed to thermodynamic changes (Figure 16.31). This figure shows simulations for the ice mass in the eastern

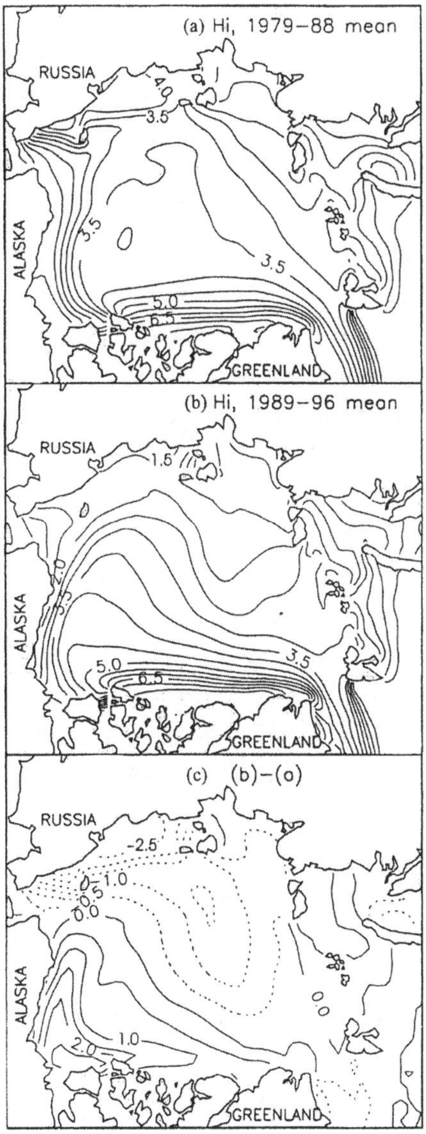

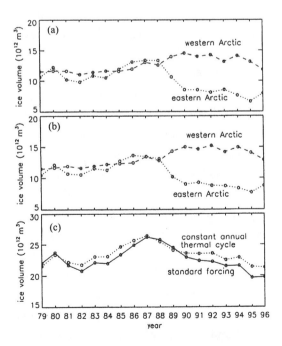

Figure 16.30. Simulated mean ice thickness fields (Zhang et al. 2000) for (a) 1979 to 1988 and (b) 1989 to 1996. Part (c) shows their difference field. The contour interval is 0.5 m.

Figure 16.31. Simulated annual mean ice volume in the eastern and western Arctic (a) for a standard forcing; (b) for a constant annual thermal cycle); and for the whole Arctic (c) for both types of forcing. The prime meridian divides the two regions (Zhang et al. 2000).

and western Arctic (taking the Greenwich meridian as the demarcation point) with fixed- and variable-thermal forcing.

The authors explain this difference to be largely caused by a reduced transport of ice from the western to the eastern Arctic under a low-NAO pattern. Conversely, during the high-NAO pattern, there is much less transport of thick ice into the eastern Arctic. When no variation in thermal forcing was employed, there was little change in the results. This occurred even though there was a substantial increase in mean air temperature over the Arctic Ocean of about 4°C between 1980 and 1996. It should be cautioned that the spatial pattern of the thickness increase in the western Arctic is not in agreement with recently reported submarine thickness observations by Tucker et al. (2001) occurring over the time period 1986 to 1994. In fact, Tucker et al. find a decrease in thickness along the transect in disagreement with the results of Zhang et al. (2000). Consequently, although the overall shifts are clearly consistent between time periods, there are some apparent discrepancies in the anomaly pattern. Clearly, there is motivation for model improvement.

16.7.4 The Effect of Ice Dynamics on the Response of Sea Ice to Climate Change

Early dynamic thermodynamic sea ice modeling studies (Hibler 1984) have suggested that sea ice motion might exert a negative climate feedback (Vavrus and Harrison 2003) on ice thickness changes. This negative feedback stems from two processes absent in a stationary ice pack: ice advection and local thickness variations due to ice deformation. Under this paradigm, regions characterized by sea ice convergence tend to resist thinning caused by atmospheric thermal perturbations. Local thickness variations resulting from deformation are important because they lead to differential growth rates. Since columnar ice growth is inversely propor- tional to ice thickness, sea ice tends to restore itself toward its original thickness, whereas ice motion causes thickness changes. Likewise, because the strength of ice is proportional to its thickness, mobile ice tends to resist thermodynamic perturbations through easier ridging as the floes thin and more difficult ridging as they thicken. Overall, these considerations suggest that models including dynamical sea ice processes should tend to be less sensitive to changes in forcing, at least with regard to sea ice thickness (Curry et al. 1995; Hibler 1989; Holland et al. 1993; Pollard and Thompson 1994).

In addition to ice-modifying ice thickness changes, ice dynamics and deformation can be argued to have a mitigating effect on high-latitude atmospheric warming. This characteristic was indirectly indicated by idealized warming experiments (Hibler 1984) using a two-level dynamic thermodynamic model of the Weddell Sea ice pack (Hibler and Ackley 1983). In these experiments, climatic change was crudely approximated by a 4°C temperature change, which affects incoming longwave radiation and sensible heat losses in both summer and win- ter. Warming was found to change thicknesses by similar factors in both the dynamic and ther- modynamic cases, with only a small negative feedback in the dynamic case. However, heat exchange into the atmosphere (Figure 16.32b) caused by ice growth and oceanic heat loss over the whole grid was appreciably less sensitive to warming in the dynamics case. Moreover, under the present climate both ice growth and heat gained by the atmosphere are substantially less for the thermodynamics-only model than for the full dynamics case. Based on these cur-

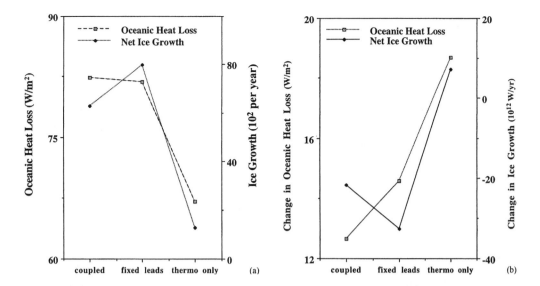

Figure 16.32. (a) Net sea ice growth and oceanic heat loss to the atmosphere over the growth period (Julian day 60–240) from a dynamic thermodynamic model (Hibler and Ackley 1983) of the Weddell Sea ice cover. Results are for a full dynamic thermodynamic model (coupled), a thermodynamic-only model (thermo only), and a thermodynamic model with a specified seasonal fraction of leads (fixed leads). (b) Change in net ice growth and oceanic heat loss over the same time period between a 5°C warming simulation and the control simulation for the same models as in part (a). Atmospheric forcing data from 1979 were used to drive the model (Hibler 1984).

rent climate characteristics, one would expect that with full ice dynamics the heat transfer into the atmosphere would be enhanced, causing a local warming anomaly compared to the thermodynamics-only situation. Moreover, because of the reduced change of sea-to-air heat exchange (Figure 16.32b), with dynamics one would also expect reduced polar temperature amplification under increased CO_2.

A methodical investigation of the effects of ice dynamics on the response to climatic change of a fully coupled atmosphere–sea ice model has been carried out by Vavrus (1999) and Vavrus and Harrison (2003). In addition to the reduced polar amplification mentioned above, Vavrus (1999) identified several new feedback mechanisms unique to coupled atmosphere–ice models. The atmosphere–ocean model used for this study was the GENESIS2 coupled atmosphere–mixed layer model that originated from the NCAR community climate model, version 1 (Thompson and Pollard 1997). Sea ice dynamics were treated using a cavitating fluid rheology together with a two-level ice thickness distribution. Ocean currents for sea ice drift were specified and the northward transport of heat by the ocean was parameterized by certain prognostic heat fluxes into the boundary layer.

Of particular interest are the different atmospheric characteristics simulated with and without sea ice dynamics. The results for the present climate show that even small numbers of leads cause a substantial warming at the Arctic Ocean surface and in the lower troposphere. This result is consistent with earlier GCM-based studies of both polar regions (Simmonds and Budd 1991; Vavrus 1995). In comparison to the thermodynamics-only (TI) simulation, Vavrus (1999) finds that inclusion of dynamics (DI) causes reduced ice coverage from the

Laptev Sea westward to Svalbard, caused by the mean offshore ice drift along the Eurasian Shelf and a much more diffuse ice margin in the North Atlantic (see also Maslanik 1997). These ice coverage differences cause higher temperatures in the lower troposphere in the vicinity of the reduced ice concentrations than does TI and allow generally warmer conditions to spread over the entire Arctic Ocean. The inclusion of dynamics also forces much greater ice coverage just east of Greenland, where only a minimal amount of sea ice grows thermodynamically, with these higher ice concentrations extending to the south along the east Greenland coast. As a result, annual air temperatures are up to 2°C colder than in TI along the east Greenland coast. These surface temperature differences propagate aloft into the middle troposphere, causing differences in geopotential height almost directly above the core surface anomalies. This results in anomalous higher geopotential heights in the Svalbard–Norwegian Sea region where the DI simulation produces warmer temperatures.

The sensitivity of the individual models with dynamics (DI) and without dynamics (TI) to various climatic warming scenarios is shown in Figure 16.33 (Vavrus and Harrison 2003). In the TI case, the spatial structure of the ice fraction changes (16.33b) consists of a roughly annular contraction of the ice pack, accompanied by a pronounced melt back along the east Greenland coast and at the boundary of the Arctic Ocean with the Greenland–Barents Seas. Conversely, the ice fraction anomaly pattern in DI features a dipole structure within the Arctic Ocean, with the largest decreases along and north of the Siberian coast and smallest decreases along and north of the Greenland–Canadian archipelago. Vavrus and Harrison (2003) note that this type of spatial anomaly resembles the trend of ice cover changes in recent decades. This general pattern in the DI simulations facilitates the melting of thin, less compact ice in the divergent Eurasian sector and hampers melting in the convergent Canadian sector, where sea ice is relatively thick and compact.

These differences in ice fraction cause corresponding heating anomalies in the lower troposphere. As in the case with the present climate, these heating anomalies also cause substantial pressure anomalies. In both pairs of experiments, the largest anomalies occur over regions of maximum retreat. Particularly notable is the muted and reduced warming over the central basin in the DI experiment. This basic result is that changes in oceanic heat loss in regions such as the Weddell Sea cause a reduced sensitivity of the temperatures over the basin as well as a reduced response in ice thickness in that region (Lemke et al. 1990, 1997). This reduced sensitivity in ice thickness persists over the whole range of warming and paleoclimate scenarios, as shown in Figure 16.33. Vavrus and Harrison (2003) attribute the change in thickness to two different processes. Smaller thickness anomalies under positive radiative perturbations appear to result from easier ridging partially compensating for the thermodynamic thinning. The muted thickness response under negative forcing, on the other hand, is caused by the equatorward spreading of the ice pack offsetting the thermally induced thickening. The differences in ice thickness between DI and TI constitute a negative feedback in every pair of simulations, all of which are statistically significant at the 99% confidence level, based on a student's t-test.

The changes in ice concentration (Figure 16.33b) resulting from dynamics-induced mobility demonstrate differing feedbacks under warming and cooling. Nonetheless, the results are amenable to consistent interpretation based on ice rectification effects. In particular, under positive radiative perturbations, the decrease in ice coverage is much smaller when dynam-

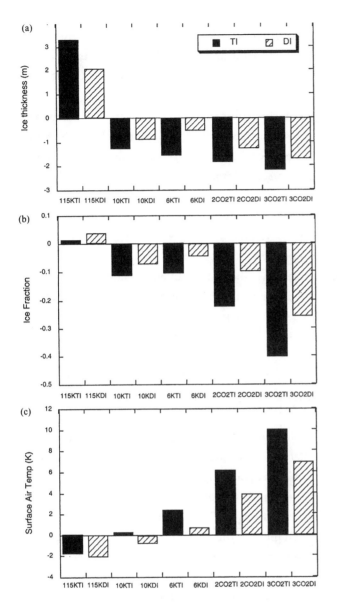

Figure 16.33. Mean annual anomalies in the central Arctic. (a) Sea ice thickness; (b) sea ice concentration; (c) surface air temperature for a variety of paleoclimate and warming experiments. The anomalies represent averages over the whole central portion of the Arctic Basin. The DI (TI) anomalies are computed as differences from the DI (TI) control simulation. TI anomalies are in black and DI anomalies are hatched. In addition to the 2 CO_2 and 3 CO_2 simulations, there are paleoclimate simulations using radiative and ice sheet reconstruction information from 6,000 (6K DI and 6K TI), 10,000 (10K DI and 10K TI). and 115,000 years (115K DI and 115K TI) before present. The 6K period represents a "climatic maximum" with warmer temperatures than today. The 10K period has stronger radiation forcing compared to today, but the Laurentide ice sheet was still present. The 115K period, which represents the beginning of the most recent glacial stage, has much lower radiative forcing, but no ice sheets were present. (From Vavrus and Harrison 2003.)

ics are included, ranging from 43% of the thermodynamic response at 6,000 years to 68% at 10,000 years. Conversely, the *increase* in ice fraction at 115,000 years is more than twice as large with ice dynamics than without. Sea ice dynamics thus represents a negative feedback on ice coverage when the arctic climate is warmer than present, but a positive feedback when the Arctic climate is colder. Vavrus and Harrison (2003) attribute this nonuniform response to the tendency of ice motion to spread ice equatorward. This feature opposes a contraction of the pack in a warm scenario, but reinforces an expansion in a cool scenario. The impact of dynamics on ice concentration is reflected strongly in the response of surface air temperature (Figure 16.33c). Hence, ice motion always promotes cooling in the interior Arctic relative to corresponding TI experiments. Although much of this effect can be attributed to the ice fraction feedback discussed above, further analysis by Vavrus and Harrison (2003) shows that modified circulation patterns play an important role. In particular, they note that the geostrophic surface wind anomalies associated with sea level pressure changes induced by the warming anomalies are such that the DI simulation produces enhanced outflow of cold arctic air and ice over the Norwegian and Barents Seas. The TI simulation, on the other hand, generates enhanced inflow of warm and moist Atlantic air into the central Arctic. Consequently, the shifting of the atmospheric circulation aids in the negative feedback effects under warming that are expected on the basis of air–sea heat exchange alone. As will be seen in Chapter 18, such changes are currently underway.

Although the details of the atmospheric circulation changes and different anomaly patterns are complex, the above sequences of simulations graphically emphasize the difference that sea ice dynamics can make to the response of the climate system to climate warming or cooling. Indeed, the base state of the circulation is itself modified. It is, of course, possible that some of this response is model-dependent. Simulations with the Canadian climate model, for example, yield little difference in sensitivity between dynamic and thermodynamic cases under climate warming (Flato et al. 2000). However, the other possibility is that the base state (Maslanik et al. 1996) of the GENESIS model used by Vavrus and Harrison (2003) is more realistic in the arctic region. It is also notable that shear strength effects not included in the cavitating fluid model may affect some of the sensitivity analysis, especially in cases of climatic cooling. Moreover, inclusion of full-thickness distribution formulations (see, e.g., Flato and Hibler (1995) and Lipscomb (2001)), and high-frequency variability (Heil and Hibler 2002) may modify aspects of the sensitivity.

16.8 Concluding Remarks

The main thrust of this chapter has been to elucidate the interplay between dynamics and thermodynamics in understanding and interpreting the dynamic response of sea ice. Developing the capability to model this response has necessitated the creation of sea ice circulation models that bear many resemblances to atmospheric and oceanic circulation models. Indeed, as these models become utilized at higher resolutions and as more coupled processes begin to be resolved, they will likely begin to take on more of the chaotic behavior that characterizes atmospheric circulation models and eddy-resolving ocean circulation models. Consequently, although a number of the dynamical processes such as "kinematic" waves examined in this chapter have received little attention in the climate community, as more detailed investigations of the role of sea ice in climatic change are explored, it will become more important to be aware of such phenomena.

It has recently become more commonplace to utilize some level of physically based sea ice dynamics models in numerical investigations of climate. With this fact in mind, this chapter has examined the major features of these models with extra emphasis on their dynamical and mechanical components. The hope here is that this information will aid in a more discerning interpretation of different model-based predictions of sea ice change. This detail also aids in identifying aspects of the dynamic response of sea ice requiring further attention or research.

Overall it is clear that although considerable progress has been made over the last several decades on modeling the dynamic response of sea ice, considerable uncertainties and research still remain. A recent discussion of the perceived inadequacies of several of the fundamental building blocks of many current sea ice models can be found in Coon et al. (2007). One current trend that will definitely continue into the near future is the increasingly detailed comparison of model estimates of the sea ice drift and deformation field with high-resolution observations such as those obtainable from buoy arrays and satellite-based synthetic aperture radar (SAR) systems. For a recently published example of such a study, see Kwok et al. (2008). Ultimately such studies should help us resolve current questions concerning the adequacy of different process parameterizations depending on the temporal and spatial resolution of the model and its ultimate use.

17 Underwater Ice

And ice mast-high came floating by,
As green as emerald.
Samuel Taylor Coleridge

17.1 Introduction

The ice types discussed earlier can all be considered to be part of the normal sea ice growth and decay cycle. However, there are several other, rather different, less common ice types that form from seawater and which, therefore, by definition are varieties of sea ice. These ice types all appear to share one "process" that, although active during the very initial frazil stages of normal sea ice growth and to a lesser extent during the freezing of subice melt layers, is largely missing during the growth of congelation sea ice. Here I refer to crystal growth in water that is supercooled: where the cold sink that is driving the freezing process is not heat loss through the overlying ice but instead heat loss into the water. The ice types involved are platelet ice, marine ice, and anchor ice. As will be seen, platelet ice and marine ice are believed to be different aspects of one general process that can occur at depth in water columns that are in contact with ice shelves. The process does not directly involve surface cooling. Anchor ice, on the other hand, although forming on the seafloor, is the result of cooling that occurs at the sea surface.

17.2 Platelet Ice

That there was a sea ice type, or at least a sea ice formation process, occurring in the McMurdo Sound region of the Antarctic that was different from the normal congelation ice cycle was first noted by members of the British Discovery Expedition of 1901–1904. This was followed by more detailed descriptions of this ice type by Wright and Priestly (1922) during their classic glaciological studies carried out as part of the British Terra Nova Antarctic Expedition. They noted that although the initial sea ice cover in the Sound appeared to be typical congelation ice, once the congelation had grown to a thickness of over a meter, a porous fragile layer of dendritic ice platelets started to form both at the growth interface as well as ahead of the interface. As the congelation ice continued to thicken during the growth season, ultimately many of the platelet crystals were incorporated into the congelation ice. Over 40 years later, further documentation of this phenomenon was provided by Paige's (1966) work also in the McMurdo Sound area and by Serikov's (1963) investigations along the east antarctic coast near Mirny, where he termed such ice *underwater ice*. Here I will refer to this ice type as *platelet ice* because of its characteristic crystal morphology. Certainly Serikov was quite correct in noting that such ice does indeed form underwater.

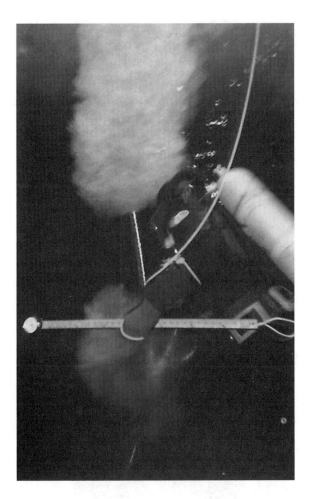

Figure 17.1. Photograph of platelet ice forming on a line hanging beneath congelation ice. (Photograph courtesy V. Grischenko, AARI Files.)

However, as will be seen, both marine ice and anchor ice also form underwater. Therefore, applying the term *underwater ice* only to platelet ice types could result in some confusion. In my view, it is better to use the term *underwater ice* in a more general sense by applying it to the group of ice types discussed in this chapter. Here the expression *platelet ice* will be used to describe both the mesh of fragile crystals that can develop below a growing sheet of congelation ice and the plumes of crystals occasionally found at depth within the water column at locations invariably in the vicinity of ice shelves.

Photographs of platelet ice as it develops on objects such as ropes hanging beneath an ice sheet can be found in a number of papers (Dayton et al. 1969; Leonard et al. 2006; Lewis and Perkin 1986) as well as in Figure 17.1. However, long-term photographic time series of the platelet layer as it develops ahead of the lower congelation ice interface appear to be rare. Nevertheless the general characteristics of this ice can be surmised from core hole observations and by examining thin sections showing the platelets after they have been incorporated into the thickening congelation. Fortunately, a number of photos of such thin sections have been published (Gow et al. 1982, 1998; Jeffries et al. 1993; Paige 1966). The Jeffries reference is particularly recommended as it contains color photographs of the thin sections. The platelet layer frequently appears to be both quite open and fragile in that disturbances caused while obtaining core samples invariably result in disrupting the layer, with the core hole filling with loose platelets. The individual platelets are bladelike, with the plane of the blade being in the (0001) direction. The blades frequently show sharp tips and relatively even edges. Their sizes in the plane of the plate can be several centimeters while thicknesses measured parallel to the c-axis are frequently 4 to 5 mm. These later values are significantly larger than brine layer spacings (λ) measured parallel to the c-axis in congelation ice. Also, the platelets do not show the development of the substructure with the entrapped brine and gas inclusions that invariably occur in congelation crystals. Another difference is that although congelation ice crystals in the columnar zone invariably show c-axis horizontal orientations and frequently show strong alignments in the horizontal plane, platelet orientations, although variable, are quite different. To date random,

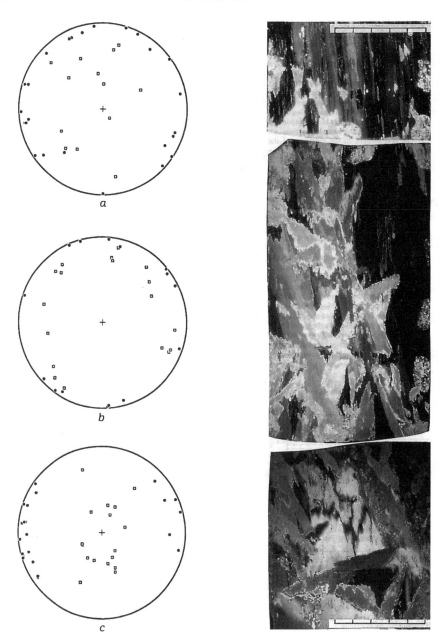

Figure 17.2. Fabric diagrams and vertical thin sections of platelet ice collected at McMurdo Sound, Antarctica (Jeffries et al. 1993).

nonhorizontal girdles and c-axis vertical patterns have been reported (Figure 17.2; Jeffries et al. 1993). Furthermore, congelation ice crystals growing through a developed platelet layer show weaker horizontal c-axis alignments than occur in comparable situations where the formation of platelets is not observed. The assumed reason for this difference is that the formation of a well-developed platelet layer undoubtedly results in both a reduction of the effective current speed at the congelation–seawater interface and an increase in the directional variability of the currents as observed at the growth interface.

I believe that the above structural differences clearly show that platelet ice is not a member of the congelation ice family but is quite different. When present, it can also make a significant contribution to ice sheet growth. For example, Jeffries et al. (1993) have compared the growth of the composite congelation/platelet ice sheet at McMurdo with the growth of a pure congelation ice sheet at Station Alert located on the northern coast of Ellesmere Island in the high Canadian Arctic. It was found that although McMurdo is warmer and has a shorter ice growth season than Alert, its maximum reported ice thickness is larger (236 cm vs. 205 cm), a nontrivial difference.

In that McMurdo Sound platelet ice usually forms below continuous fast ice and, in most cases, without direct contact with the fast ice, the cooling source driving the process clearly cannot be the atmosphere, or at least not the local atmosphere. Somewhere in the near vicinity there must be a process occurring that produces the supercooled water essential to the formation of the platelets.

Increased interest in platelet ice was sparked when Foldvik and Kvinge (1974) suggested a formation mechanism. Although they were carrying out oceanographic studies in the Weddell Sea near the Filchner Ice Shelf, their interest in the platelet ice problem was apparently not the result of direct experience with this ice type. Instead, it was the result of reading Russian reports of impressive accumulations of platelet ice near Mirny (Buynitskiy 1967) where platelet ice thicknesses exceeded congelation ice thickness by a factor of five or more. Moreover, they realized that a theory advanced earlier for certain arctic phenomena by Lewis and Lake (1971) was not adequate to explain the large quantities of platelets that were being reported at the antarctic sites. Its difficulty was that it relied on heat loss through the congelation ice and was therefore limited by both increasing ice thickness and the typically thick snow covers reported at these antarctic locations. What was needed was a theory that was independent of the thickness of the local sea ice and, if possible, utilized cryological elements known to exist at or near the platelet sites. Foldvik and Kvinge realized that

1. all the sites where platelet ice had been reported are located in the near vicinity of ice shelves;
2. glacier ice at the base of the shelves is not only fresh but is presumably at or near its pressure melting point;
3. water's pressure dependence is atypical in that it is negative; i.e., as pressure increases the freezing point decreases;
4. seawater, at any temperature above its freezing point, is capable of melting freshwater ice at its freezing point if they are placed in contact;
5. when ice melts into seawater the result is both cooling and more importantly, freshening; and finally,
6. as seawater freshens (becomes less saline), it also becomes less dense.

Considering the above facts, they suggested that the origin of the supercooling assumed to be driving the platelet growth is as follows. Let high-salinity shelf water (HSSW) that forms the upper layer of the ocean in the Ross and Weddell Seas come in contact with freshwater glacier ice at locations near the ice shelf grounding line. The expected result would be melting, leading to the development of a freshened water layer at the ice–seawater interface. Although this might appear to be stable in that the least dense fluid layer is on top, it is not, because the base of the ice shelf is frequently not a flat surface. Instead, it slopes upward as the shelf gradually thins towards its seaward edge. As a result the low-density water resulting from the melting will start to flow upward along the inclined base of the shelf. If this occurs rapidly, it can be approximated as an adiabatic decompression process. As can be seen from examining the last term in equation 5.2, every decibar of pressure decrease results in a rise in the freezing point of 0.00075°C (1 dbar = 10^{-1} bar $\approx$ 1 m depth in the ocean). If the rising water maintains contact with the ice, supercooling cannot occur and ice crystals will start to form. If the contact with the ice is lost, supercooling becomes possible. Probably a combination of the two situations will occur where some ice crystals form but not enough to completely eliminate the supercooling.

A formal development of the appropriate equations can be found in Foldvik and Kvinge's (1974) original paper. There they considered what would happen to a "mass particle" of seawater located in contact with the base of the ice shelf if it were to move upward toward the surface while continuously maintaining thermodynamic equilibrium (i.e., if it were to rise adiabatically toward the sea surface without supercooling occurring). They referred to this postulated process as an example of conditional instability. Here I will refer to the process as an ice pump, using the term suggested somewhat later by Lewis and Perkin (1983, 1986) in their studies of different applications of this process. Introducing appropriate property values into their equations, Foldvik and Kvinge obtained the following results. If the crystallization of ice were to start at a depth of 400 m and a packet of water were then raised to the surface, the salinity of the packet would increase by 0.13‰, the macroscopic density would decrease by 0.37 σ_t, and the density of the residual water would become 0.1 σ_t-units larger than its original value. This last increase is a result of the increase in salinity resulting from the growth of the ice crystals. Note that it is not necessary for the seawater to be exactly at the freezing point when it initially contacts the ice. If the seawater is above the freezing temperature, when it comes into contact with the base of the shelf, it will melt ice until the freezing temperature is reached. They also suggested that it is not necessary for the crystallization process to take place in the near vicinity of an ice shelf or iceberg in that, once the water has been appropriately conditioned at depth, it may move laterally before starting to move to the surface. Finally they suggested that "possibly this underwater ice crystal production may contribute to the formation of the relatively porous underwater sea ice which in the Antarctic is underlying a layer of more massive congelation sea ice."

The question then remained, was this actually occurring? There was never any doubt about the analysis of the proposed ice pump process as it was solidly built on the thermodynamics of seawater and ice. Gradually the facts began to come in. These were as follows: in 1980 Gow et al. (1982, 1998) and in 1990 Jeffries and others (Jeffries and Weeks 1992a; Jeffries, Weeks, et al. 1993) undertook studies of the variations of sea ice structure in the McMurdo Sound region. They found the following:

1. well-developed *c*-axis horizontal alignments are common in the congelation-type fast ice reflecting consistent current patterns within the Sound;

2. although platelet ice was commonly observed in the southern part of the Sound, there was no clear pattern to its occurrence, other than the fact that it invariably occurred in the near vicinity of the ice shelf;

3. further to the north, platelet ice was not observed, except in the near vicinity of ice shelves or ice tongues.

The reader might well inquire at this point, "do you really know that platelet ice forms from supercooled water, or is this just a guess assumed to be necessary to explain the observed crystal morphology?" The answer is that it has now been definitely established that supercooled water is flowing out from beneath the McMurdo Ice Shelf, which is a local offshoot of the much larger Ross Ice Shelf. However, there are still differences in the estimates of the degree of supercooling. The first such measurements were by Lewis and Perkin (1985), who observed cold, relatively fresh supercooled water exiting from beneath the McMurdo Shelf south of the New Zealand station at Scott Base and then flowing to the northwest in the Sound. Supercooling values ranged from 0.010 to a maximum of 0.047°C. More recently, Smith et al. (2001) reported average values of supercooling measured ~50 cm beneath the platelet layer in McMurdo Sound to be between 0.008 to 0.011 ± 0.004 °C. At another site where only columnar ice was present, the measured values (0.004 ± 0.004 °C) did not definitively indicate the presence of supercooling.

The most detailed study of platelet ice formation to date is the work of Leonard et al. (2006), carried out during the austral winter (March to September) of 2003 at two different sea ice sites located in front of the McMurdo Ice Shelf. Measurements included current, temperature, salinity, and the occurrence of platelets in the water column and on cables and ropes located beneath the ice cover. Also included were the thicknesses, temperature profiles, and the structure of the FY fast ice. They found that the presence of platelets within the sea ice was linked to the time history of the occurrence of ice crystals in the water column (suggested by an increase in the backscatter profile as determined by an acoustic profiler). When the amount of supercooling was compared to the amount of platelet growth on a wire suspended beneath the fast ice, it was found that platelet growth only occurred when supercooling was present, a conclusion supporting the earlier results of Smith et al. (2001). The largest supercooling observed was 0.010°C, with values of 0.002 to 0.006 °C being the most common.

With one exception for a brief period around 11 April when platelets were incorporated into the ice cover, prior to mid-May the ice forming at the study sites was normal platelet-free congelation ice with a c-axis alignment parallel to the primary tidal current directions. During this overall period, the water within 0.15 m of the ice–water interface was essentially at the freezing point and exhibited $\delta^{18}O$ values characteristic of antarctic surface water.

On ~15 May a change started that, over a week or two, converted the water column down to a depth of 50 m to temperatures close to the surface freezing temperature. At this time the thickness of the near-surface mixed layer immediately under the ice was less than 50 m. Below the mixed layer the water column conditions were dynamic with appreciable temperature and salinity fluctuations, including the warmest and coldest temperatures observed during the complete measurement program. From this time, a band of very cold water was observed beneath the mixed layer, although it was initially unable to reach the ice–water interface because of the more buoyant water in the mixed layer. About the same time, the water within meters of the interface became supercooled in situ. Once this change occurred, the sea ice cover started to incorporate platelet ice, and platelets also formed on wires beneath the sea ice. With time this supercooled layer not only thickened but also became additionally supercooled, persisting through the remainder of the measurement season. By late winter, possibly as the result of brine rejection from the sea ice as initially suggested by Lewis and Perkin (1985), it was possible for the very cold water formed beneath the ice shelf to move into position beneath the sea ice. I note that at least in McMurdo Sound, most of the ice crystals are either fixed in the mesh that is attached to the bottom of the congelation ice or fixed to a static wire or rope. This is clearly not crystal growth in a static system. The growth is fed by the continuous flow of supercooled water past crystals that are relatively fixed in space. Using variations in the strength of the backscattered signal measured by an acoustic Doppler current profiler as an index of the volume of ice crystal scatterers present in the water column, it was found that in the late winter there is a significant zero-lag correlation between decreases in water temperature, typically associated with supercooling, and the rise in backscatter strength, suggesting that the scatterers are indeed ice crystals. The origins of the water producing the platelets can be inferred from the fact that the observed potential water temperatures are such as to result in below-freezing temperatures if the water is adiabatically raised to the surface. In order to achieve such low temperatures, the water must have been in contact with ice at depth. Such water has been termed Ice Shelf Water (ISW) (Jacobs et al. 1979), and in this case the contact point has been assumed to be the bottom of the nearby McMurdo Ice Shelf (Lewis and Perkin 1985).

I have used the McMurdo observations as a sort of case study of platelet ice formation as it is the location that has received the attention of the largest number of investigators. However, quite similar observations have been noted by Kipfstuhl (1991) at Atka Bay located at ~8°W near the Ekström Ice Shelf. There the congelation ice (2 m) and subice platelet layer (4 m) thicknesses, when combined, would result in a total ice column of 3 to 4 m. This is in striking contrast to the situation in the nearby Weddell Sea, where the average thickness of the sea ice is commonly slightly less than 1 m. Particularly interesting aspects of the Kipfstuhl paper are the isotopic measurements (δD and $\delta^{18}O$) that indicate that the platelet ice originates from a water mass that is strongly depleted in both D and $\delta^{18}O$. As will be seen in the next section, such depletion is a characteristic of ice shelf water (ISW), which is believed to form by interaction between the seawater in the subshelf cavity and the ice of the shelf as envisioned by the Foldvik and Kvinge mechanism.

Other field observations clearly support Foldvik and Kvinge's suggestion that it is not necessary for the crystallization process to take place in the near vicinity of an ice shelf or iceberg, as once the water has been appropriately conditioned at depth it may move laterally before starting to move to the surface. For instance, Dieckmann et al. (1986), working approximately 25 km from the edge of the Filchner Ice Shelf, observed returns on a 30-kHz echo sounder from depths of up to 250 m that, they were able to show by trawl sampling, were produced by clouds of frazil crystals. The average diameter and thickness of the crystals were 20 mm and 0.5 mm, respectively. Similar results have been obtained on the other side of the continent in the general area of Prydz Bay near the Amery Ice Shelf by Penrose et al. (1994), who were carrying out a biological survey.

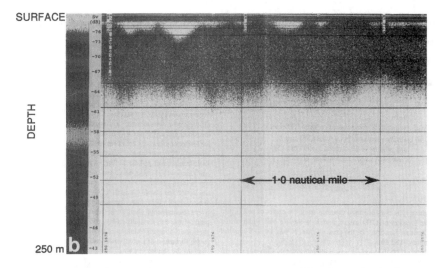

Figure 17.3. An echogram showing characteristic return patterns arising from platelet ice crystals dispersed in the water column (Penrose et al. 1994).

Several different classes of echo returns were observed, some caused by krill and other biota as revealed by net trawls. They noted that there was one type of echo that was of relatively uniform intensity that frequently occurred in the upper 50 m of the water column over much of the bay (Figure 17.3). Again, via the use of trawls they were able to verify that the scatterers were crystals of platelet ice. The crystals, which were roughly disk-shaped with diameters in the range of 10–25 mm and thicknesses of ~1mm, were similar to those reported earlier by Dieckmann and his coinvestigators. The Prydz Bay trawl results suggested in-situ occurrence values in the range of 0.1 to 1.0 crystal per 1 m^{-3} of seawater. Furthermore, the authors were able to obtain very precise conductivity–temperature–depth (CTD) measurements at their sampling sites. Figure 17.4 shows an example of such a record. Indicated on the record is the freezing temperature (dotted line) profile, the salinity (dashed line) profile, and the approximate vertical extent of the echo trace as observed on the echogram. Note that the vertical extent of the echo trace is from 20 to 75 m, whereas the area of supercooling extends from ~40 to 240 m. The interpretation of these observations is that although the nucleation and growth of platelets is possible at depths of up to 240 m, it is only at ~75 m that

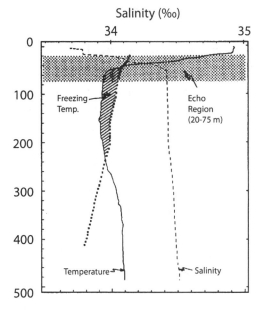

Figure 17.4. A CTD record taken at Prydz Bay off the edge of the Amery Ice Shelf, East Antarctica. Also indicated are the depths where the observed water temperatures are colder than the theoretical freezing temperatures (dotted line), the salinity profile (dashed line), and the approximate vertical extent of the echo trace for the volume reverberation greater than −76 dB assumed to indicate the presence of ice crystals because of the presence of ice in trawls (Penrose et al. 1994).

the number of platelets per m³ becomes large enough to produce a measurable echo. As platelets rise above a depth of 40 m they start to melt, and at a depth of ~20 m the number of platelets per m³ again becomes small enough to cause a disappearance of the echo return. Note that during this cruise platelets were never noted on the sea surface.

I now suggest the following conclusions:

1. Although platelet ice has been reported in the Arctic (Jeffries et al. 1995), extensive development of platelets has only been reported in the Antarctic and there only in the near vicinity of ice shelves.
2. Platelets crystallize from low-salinity water that has been supercooled and appears to be depleted in both D and ^{18}O.
3. The ice pump mechanism as proposed by Foldvik and Kvinge (1974) is both soundly based on the thermodynamic properties of seawater and ice and appears to adequately explain the current field observations on platelet ice.

17.3 Marine Ice

17.3.1 Green Icebergs

There is an obvious question that follows from the previous discussion: is platelet ice only the tail end of a larger process underway beneath the large antarctic ice shelves? Clues that this was an investigation well worth pursuing came in an unlikely form: green icebergs. Now I do not mean greenish icebergs. I mean "as green as emerald," as Mr. Coleridge so aptly put it in "The Rime of the Ancient Mariner." Now I have often wondered whether Coleridge had heard reports of the existence of such icebergs or whether this was purely a flight of fantasy. In that the poem was written in 1797, it is conceivable that he had heard of the sighting of such an iceberg on one of James Cook's voyages. However, if I were a betting man I would put my money on a flight of fancy. After all, it was Coleridge's purpose at the time that he wrote it to treat supernatural subjects to illustrate the common emotions of humanity. Besides, I doubt that "as blue as azurite" would have served his poetic purposes.

In the scientific literature reports of green icebergs can be found in scattered papers dating back to 1921 (Amos 1978; Betts 1988; Binder 1972; Dieckmann et al. 1987; Lee 1990; Moulton and Cameron 1976; Von Drygalski 1921; Wordie and Kemp 1933). Now, as I mentioned, these icebergs are striking. Also, their sightings are rare. I have only had the good fortune to see one. Unfortunately, it was while I was lecturing on a cruise down the west side of the Antarctic Peninsula. We had just dropped anchor for the evening and I went up to the bridge and there, in the distance, was the iceberg. It was strikingly different. Unfortunately, I was unable to examine it closely. Even so, I could see that it contained both green ice and normal iceberg ice of the white variety and that the boundaries between these two ice types appeared sharp.

Two questions immediately come to mind:

1. What is the origin of the green ice, which is now generally referred to as marine ice to distinguish it from glacier ice as well as from normal sea ice?
2. Why does it appear green?

A clear answer to the first query can be found in a paper by Kipfstuhl et al. (1992), building upon Dieckmann's et al. (1987) earlier work. In suggesting an answer, the Kipfstuhl group had a significant advantage over previous workers in that not only were they able to obtain samples from a green iceberg, they also were in possession of a 215-m core from the Filchner Ice Shelf in which the lower portion was comprised of marine ice. As a result they were able to make a direct comparison of the two sets of samples. They found that the crystal structure, isotopic composition, amount and type of incorporated sediment particles, and electrical conductivity were essentially the same in both sample sets. The conductivities corresponded to salinities below 0.1‰. The fact that these values lie above those of meteoric ice and far below those of normal sea ice provides further evidence for the assumption that marine ice is indeed a distinct ice type. The Kipfstuhl et al. paper is well worth examining in that it contains a color photograph of a green iceberg as well as photographs of thin sections of both the marine ice from the iceberg and the core. Additional work on the microstructure and electrical properties of this core can be found in Moore et al. (1994).

Figure 17.5 summarizes their conceptual model for the processes leading to the formation of green icebergs. In that marine ice forms the bottom layer of the shelf, it only becomes visible after a shelf iceberg has capsized. In that shelf icebergs are invariably tabular, they only capsize after their horizontal dimensions become less than their thickness. This only occurs when the icebergs are small with a limited life expectancy. As a result, although in some regions the occurrence of marine ice may be common, green icebergs are rarely seen.

However, knowing the origin of green icebergs does not explain why they are green. Several speculative explanations have been advanced without reinforcement from associated optical measurements. For instance Lee (1990) suggested, on the basis of examining color photographs, that inherently blue icebergs could appear green under direct illumination when the sun was near the horizon (reddened). The trouble with this explanation is the observational fact that green icebergs appear green and are easily distinguished from both white and bluish icebergs under highly varied lighting conditions including illumination other than by the late afternoon sun.

Fortunately, both optical absorption measurements and spectrophotometric analyses have now been completed on specimens collected from a green iceberg grounded near Mawson

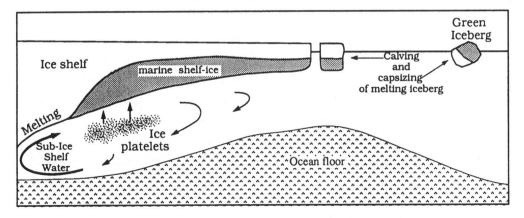

Figure 17.5. A conceptual model of processes leading to the formation of green icebergs as advanced by Kipfstuhl et al. (1992).

Figure 17.6. A plot of absorption coefficient vs. wavelength based on measurements on pure ice as well as on several types of ice collected from green icebergs (Warren et al. 1993).

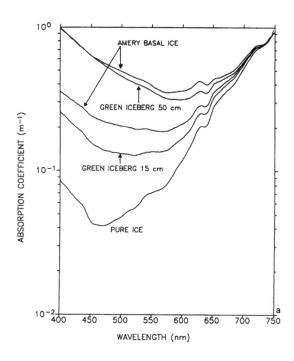

Station, which is located somewhat to the west of the Amery Ice Shelf (Warren et al. 1993). When the marine ice sample that they used was examined closely (both in hand specimen and in thin section), it was found to be free of both air bubbles and brine pockets. Moreover, it appeared to be colorless. This fact alone would tend to lead one to guess that the green color is somehow inherent to the ice and not an artifact of the illumination. For example, a similar situation occurs when one closely examines ice samples from blue icebergs: they are colorless. The blue color results from the selective transmission of light in the blue frequency range as it passes through the ice while the ice attenuates light at other wavelengths. This effect only becomes visible when the path through the ice is sufficiently long. The results of the measurements on marine ice from the green iceberg are shown in Figure 17.6. As can be seen, the absorption minimum has shifted from the 460-to-480-nm range exhibited by pure ice (Grenfell and Perovich 1981) to somewhat higher values in the range of 500 to 610 nm. Also, the locations of the absorption minimums vary from sample to sample and are generally broader than for pure ice. Now it is known that the color of natural ocean water is shifted from blue to green by organic material. This material can be present in particulate form as either phytoplankton or nonphotosynthetic particles or in a degraded dissolved form referred to as gelbstoff. It has also been found that most organic components found in seawater absorb strongly in the blue region of the spectrum and that the absorption minimum gradually shifts from 470 nm to higher values as the concentration of organic material increases (Warren et al. 1993). As both seawater off the Amery Ice Shelf and antarctic seawater in general are known to be rich in organic matter, an examination of the iceberg samples was carried out by electron microscopy. Although phytoplankton frustules and detrital particles were found, their concentrations were too low to explain the observed shifts in the absorption frequency. Therefore, the possibility that the absorption shift resulted from the presence of dissolved organic matter was examined. It was found, via spectrophotometric measurements, that samples from basal ice from the Amery, the green iceberg, and seawater collected in Prydz Bay during the bloom all showed the enhanced blue absorption characteristic of particulate or dissolved organic matter. Finally, it was possible to confirm the presence of dissolved organic matter in all the samples except one that exhibited the optical properties of pure ice obtained from the green marine ice cliffs located near Casey Station. This latter result was not believed to be the result of observational errors but to reflect real variations in the optical properties of the ice samples. In short, marine ice does not have to appear green. Warren and his coauthors concluded their study by suggesting four reasons why green icebergs are so rare. These are as follows:

1. The seawater being frozen in the marine ice forming process must contain adequate concentrations of blue absorbing components.
2. The iceberg must originate from an ice shelf with a thick basal layer of marine ice or with bottom crevasses that contain such material.
3. The marine ice has to survive to the calving front without melting.
4. The iceberg has to capsize before the marine ice melts off the bottom.

17.3.2 Structure and Composition

Very little work has been completed on the structure of marine ice as it is difficult to obtain suitable specimens. This is the result of the fact that the lower portions of ice shelves are inaccessible, requiring either major coring programs or rotated icebergs for sampling. At first glance there might appear to be two types of marine ice: one that forms on the base of an ice shelf as the result of heat loss into the ice shelf, and one that is produced by the adiabatic cooling of upwelling seawater that has previously reached its equilibrium freezing point at some depth while in contact with ice.

This first ice type was briefly discussed in Chapter 7 (section 7.2.4.4, Cell Size Variations) and is similar to normal congelation sea ice in that it shows c-axis horizontal crystal orientations and a well-developed substructure. It differs in that the plate width λ of the substructure is comparatively large and the salinity of the ice is lower. Both of these differences are associated with the extremely low growth rates that occur at the base of thick ice shelves as the result of the insulating effect of the shelves. The most striking example (Figure 7.45) to date is from the core obtained at Site J-9 on the Ross Ice Shelf, where the annual growth rate has been estimated to be 2 cm (Zotikov et al. 1980). There λ values of 5 mm were observed as contrasted with a fraction of a millimeter in normal sea ice. As this type of ice is simply an extremely slow-growth example of congelation ice, it should not be considered as a type of marine ice in the sense that the term is being used here. The term *marine ice* is assumed to only apply to the ice produced by the ice pump mechanism forming as suggested by Foldvik and Kvinge. As their mechanism is not limited by the insulating capabilities of large thicknesses of overlying shelf ice, it is capable of producing large volumes of ice.

The first published photographs of thin sections of marine ice samples obtained from a central location in an ice shelf (Ronne) were those presented by Kipfstuhl et al. (1992). The views of the two vertical sections photographed under crossed polaroids are informative. An appreciable number of platelike crystals are oriented with their wide surfaces in the horizontal plane. This orientation would be preferred by tabular, platelike crystals that float upward and attach themselves to a generally horizontal ice "roof." In that the c-axes of platelike ice crystals are invariably oriented perpendicular to the plane of the plate, one would guess that such samples would exhibit a preference for c-axis vertical orientations. The fact that these thin sections also occasionally show thin horizontal layers of dark material can be taken to suggest that the upward deposition at this site did not occur in a turbulent environment. More detailed study of this core by Eicken, Oerter,

Figure 17.7. (a) A horizontal thin section of marine ice at 153.76 m depth taken between crossed polaroids. A mm scale is located along the lower flat surface. (b) A *c*-axis fabric diagram (Schmidt-net plot) of the same sample. The sample size is 100 crystals (Eicken, Oerter, et al. 1994).

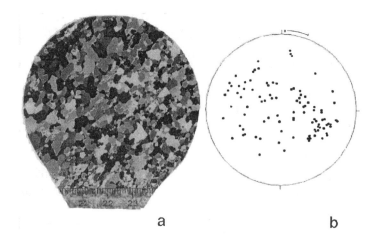

a   b

et al. (1994) verifies the above speculation. There is definitely a tendency for *c*-axis vertical orientations in the marine ice (Figure 17.7). Note the complete absence of crystals with *c*-axes horizontal or nearly horizontal. However, the *c*-axis vertical tendency is not nearly as pronounced as in samples from the lower part (137 m) of the overlying meteoric ice. Eicken et al. suggest that the tendency for vertical *c*-axis orientations in both the lower meteoric ice as well as in the marine ice are possibly the result of *c*-axis rotation through basal glide as suggested by Alley (1988) and Lipenkov et al. (1989). Perhaps. Examination of a set of cores obtained down a flow line should answer this question.

The most detailed studies currently available of ice presumed to have been produced via the ice pump mechanism are based on samples from a small ice shelf, a floating glacier, and also from the nearby FY ice located in Terra Nova Bay on the coast of Victoria Land (Souchez et al. 1995; Tison et al. 1993, 1998). These studies provide a very different view of marine ice as the samples are extremely complex. Perhaps this should not be surprising as the general geometry of this area is complex. The Tison et al. (1998) paper is of particular interest in that it provides quite detailed descriptions of the observed textural variations. A large range of grain sizes and textures were observed, including four different textural types: columnar (2.5%), platelet (4.8%), orbicular granular (49.4%), and banded granular (43.3%). Some of the ice was both bubble and debris free and was believed to form by the intrusion of brackish water into basal crevasses. Other ice was comprised of thin clear layers alternating with layers of debris. This ice was believed to have formed when a subglacial freshwater-filled sediment came into contact with seawater and froze via a double diffusion process. Particularly interesting is the fact that although the coarser varieties of orbicular frazil show random *c*-axes, as might be expected, the fine-grained (0.1–0.2 cm) orbicular as well as the two varieties of banded frazil show strongly developed horizontal alignments. To add confusion the two strongest alignments are not only horizontal but are also oriented at right angles to each other. Tison et al. (1993) have explained these alignments as resulting from frazil crystals aligning themselves parallel to the local ice–ocean interface as the result of differential fluid shear. I believe that another explanation may be possible. Note that the *c*-axis directions generally follow the boundary of the central sector (see their Fig. 1). If it is assumed that these sector boundaries tend to fix the local current direction, then we find that the *c*-axes are also parallel to the current direction, as was the case in normal congelation ice. One difficulty with this explanation is that it would require that the alignments develop by selective growth after the frazil layer has formed. Needless to say, whatever the real explanation of the origins of the structural relations in this ice may be, they are clearly complex and worthy of further study. It certainly would be useful to have additional studies completed on marine ice samples obtained from more central locations on all of the large ice shelves. This would allow comparisons to be made with the results of the studies of the Ronne and the Terra Nova Bay cores and would provide a better picture of which is most representative of marine ice. The difficulty in obtaining such samples has been the fact that the majority of the holes punched through ice shelves are produced by systems designed only to produce holes to gain oceanographic access to the subice cavity. As hot-water drill systems are fast and relatively inexpensive whereas coring is slow and expensive, I expect that cores of marine ice will continue to be as rare as green icebergs.

One other unusual aspect of many of the marine ice samples, and apparently of all the marine ice samples obtained to date from green icebergs, is the complete lack of gas bubbles and brine inclusions in the ice. One might be tempted to explain the lack of bubbles as the result of very low freezing rates, which Warren et al. (1993) estimate to be in the range of ~30 cm yr^{-1}. However, they note that this rate is similar to that occurring in thick MY ice, which invariably does contain bubbles. They then note what would appear to be a more plausible explanation: the fact that if the ice they examined initially formed at a depth of ~400 m (near the base of the nearby Amery Ice Shelf), the increased pressure would result in a greater solubility of air in seawater than at the sea surface. Another possible factor could be that as the marine ice consolidates on the roof of the shelf cavity, the initial compact is quite porous. There possibly could be a continuous flux of supercooled water through the compact. The effect of this would be to continuously remove gas-enriched water at the growing crystal interfaces, replacing it with water containing air quantities below saturation values. If I recall correctly, a related process is occasionally used to produce bubble-free ice cubes.

One final, poorly understood aspect of marine ice has to do with its composition. Current data indicate that it commonly has salinities in the range of 0.015 to 0.1‰, with values occasionally reaching 0.2‰ (Figure 17.8). These values are well above typical values for meteoritic ice and well below values for old sea ice. As I noted earlier, this in itself is reason to suspect that such ice should be considered a different ice type. Currently postulated desalination mechanisms as reviewed in Chapter 8 will not result in an ice of this composition. Futhermore, we do not even know exactly what this composition is in that it is inferred from an electrical conductivity measurement rather than based on a more direct chemical analysis. Tison et al. (2001) suggest that the chemical development of marine ice can be considered as a two-phase process. The first of these is the frazil ice stage, in which the free-floating crystals nucleate and grow in a liquid that is believed to be a mixture of primarily HSSW with a small amount of added meteoric water from the base of the ice shelf. Note that the range of $\delta^{18}O$ values obtained by the Tison group on marine ice from the Nansen Ice Shelf is similar to the range of $\delta^{18}O$ values (+1.6 to 2.2) obtained from free-floating platelets collected by trawl off the Filchner Shelf (Kipfstuhl et al. 1992). The fact that the $\delta^{18}O$ values are positive supports the conclusion that these crystals form from a liquid that was essentially seawater, the so-called parent water. This also suggests that the observed isotopic composition was already fixed before deposition and that any evolution with time has been

Figure 17.8. The $\delta^{18}O$/salinity relationship for 99 samples collected from the Nansen Ice Shelf, Victoria Land, East Antarctica (open circles; Tison et al. 2001). Also included are samples from the B13 core, Filchner Shelf (solid triangles; H. Oertner and W. Graff, pers. comm.), and from the G1 core, Amery Ice Shelf (black circles; V. Morgan, pers. comm.).

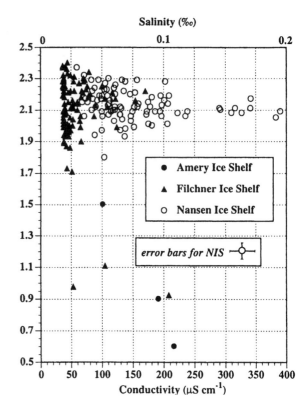

relatively minor. The surprising observation, at least to me, is that these primary crystals do not contain either gas or brine inclusions.

The second phase of the process, termed the *consolidation phase*, is even more poorly understood. In it the accumulated crystals are welded into a compact that also contains no gas or brine inclusions. Certain elements of this process appear relatively certain. The upper layers in a thick marine ice sequence will definitely be under stress as the result of the accumulation of platelet layers below. Also, the initial compact can contain an appreciable amount of what Tison et al. refer to as host water. At present one can only speculate on the initial volume of the host water. Although one is tempted to assume values that would be expected from random packing (~40%), in that one is dealing with platelets instead of spheres, it might be possible to start with much lower values. Processes that Tison et al. suggest as possibly influencing the further densification of the ice are (1) loss of heat to the overlying shelf, leading to the freezing of the host water, and (2) local melt regulation within the densifying compact. Process 2 will clearly occur. It is difficult for me to believe that process 1 is of much importance in that it would result in classic one-dimensional freezing as was observed at site J-9 on the Ross Shelf and should reveal itself in the structure of the resulting ice. There is no indication of such structure. Any such process would be extremely slow as it would be limited by the heat loss through the thick overlying shelf. However densification works it should ultimately reach a limit that has been termed the "Rule of Fives" in the sea ice literature and "close-off" in glaciology. This refers to the brine or gas volume at which the remaining inclusions are sealed and no longer communicate with one another. If there is either dissolved gas or salt in these sealed inclusions, when this material freezes brine pockets and "air bubbles" should appear. They apparently do not. Tison et al. circumvent this problem by suggesting that the stresses during compaction could result in melting at contact points thereby feeding freshwater into the interstitial host water. They note, however, that such a process would terminate at close-off. Ultimately they suggest that any remaining gas and salt impurities will remain located at grain boundaries, presumably as air bubbles and brine pockets. If this is the case, it should be observable. I guess that all one can say at present is that no one has looked.

By now readers will have guessed that I have been floundering around hoping that something would click in my brain allowing me to say, "Eureka, now I see it, the real answer is simple and it is—." Unfortunately, instead of clicking my brain has clunked. I invite the reader to examine Eicken, Oerter, et al. (1994) and Tison et al. (2001) as they are both very interesting papers. I would bet that there is a "Eureka" out there somewhere. At the very least you will be able to produce your own list of desirable further investigations.

17.3.3 Distribution

Assuming that marine ice occurs at many locations beneath large ice shelves, is there a technique that would allow its distribution to be mapped without repeatedly coring through the shelves? There is, it is surprisingly simple, and different variations of it have been applied to the Filchner-Ronne Ice Shelf (Grosfeld et al 1998; Thyssen 1988) and to the Amery Ice Shelf (Fricker et al. 2001). In that an ice shelf is free-floating at locations more than a few ice thicknesses seaward of the grounding line, the relationship between the surface height H (relative to sea level) and the thickness Z at any point can be calculated from the hydrostatic equation

$$H = \frac{Z\left(\rho_w - \rho_i\right)}{\rho_w} \tag{17.1}$$

where ρ_w and ρ_i are the column-averaged densities of seawater and ice, respectively. Then consider the variations in the hydrostatic height anomaly $(\delta h')$, taken to be the difference between the observed surface height and the surface height as calculated from the measured ice thickness using equation 17.1. Values for $\delta h'$ will be significant when there are errors in the total thickness or density values or when the ice is grounded, in which case equation 17.1 would not hold. In cases when marine ice is present, there will be an error proportional to its thickness. The reason for this error is the fact that the radio echo sounding (RES) systems that have been used to determine ice thickness typically do not penetrate through the marine ice layer to sense the marine ice–ocean interface (Blindow 1994; Robin et al. 1983; Thyssen 1988). RES does, however, detect the glacier ice–marine ice boundary quite well. Figure 17.9 presents a plot of ice surface elevation as a function of meteoric ice thickness as determined by RES. Note the large number of data points that lie significantly off the hydrostatic line, indicating the presumed presence of marine ice. Discussions of possible errors are provided in the papers mentioned above. In the Amery, a combination of the possible errors gives a marine ice layer uncertainty of ~30 m. Fortunately there was one checkpoint in the study area where a 315-m core had been obtained in 1968 (Morgan 1972). There the thickness of the marine layer was inferred to be 158 m. The Fricker et al. estimate at the drill hole location was 141 ± 30 m consistent with the more direct measure.

An example of the difference between the RES returns for a meteoric ice–ocean boundary and a meteoric ice–marine ice boundary can be found in Fig. 1 in the Grosfeld et al. (1998) paper. If strong basal returns are taken to indicate bottom melting, the distribution of melting and freezing on the bottom of the shelf appears to be in reasonable agreement with results from models of the three-dimensional ocean circulation under the shelf. A detailed map showing current

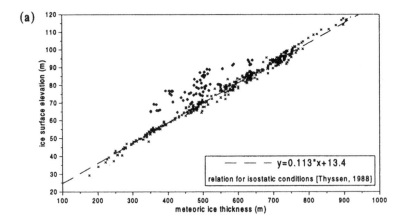

Figure 17.9. The thickness of meteoric ice as determined by radio echo sounding (RES) measurements vs. the elevation of the ice surface as determined by coincident altimeter measurements on the Filchner Ice Shelf (Grosfeld 1998). Note that numerous data points depart significantly from isostatic balance, an imbalance indicating the presence of marine ice.

estimates of the magnitude of basal melting and freezing for the Filchner Ice Shelf can be found in Grosfeld et al. (1998, Fig. 4). Maps showing variations in the thickness of the meteoric, marine, and total ice thickness of the Filchner-Ronne can be found in Sandhäger et al. (2004). Although most of the freezing occurs along the western boundary of the Shelf, the overall pattern is quite complex. In the Amery the thickest marine ice occurs in two longitudinal bands that are oriented in the direction of ice flow positioned on either side of where a major ice stream enters the Shelf. This distribution is in keeping with the idea that marine ice should accumulate at locales where the glacier ice is thinner on the flanks of ice streams, i.e., in domes or troughs in the glacier ceiling. The maximum marine ice thickness obtained on the Amery Shelf was 190 m. This compares with 140 m for the Filchner (Grosfeld et al. 1998), 350 m for the Ronne (Thyssen et al. 1993), and less than 10 m for the Ross. Fricker et al. suggest that these variations are presumably the result of a combination of factors, including different cavity geometries, ice shelf drafts, seawater properties, and circulation differences. For instance, the Amery has a long, narrow, sub-ice-shelf cavity with a maximum draft of approximately 2200 m. The Filchner and Ronne also have deep drafts of about 1400 m. This is in contrast to the Ross, which only has a draft of about 800 m plus a smaller length-to-width ratio. An important difference between the marine ice in the Amery and that in these other shelves is that in the Amery the marine ice persists all the way to the ice front, resulting in the possible occurrence of green icebergs. The marine ice layer of the Filchner and Ronne Ice Shelves does not persist until the shelf edge is reached, presumably as the result of warm currents (Thyssen et al. 1993). However, in 1986 a major calving event occurred between Berkner Island and Coats Land so that as last reported the new ice front contained ~100 m of marine ice, with a total thickness of ~500 m. It is worth noting that although the principles of the techniques used here are simple, obtaining and merging the necessary data sets is far from simple.

More recently a less direct remote sensing technique has been applied to evaluating the spatial distribution of melt beneath the Filchner-Ronne Ice Shelf (FRIS) by Joughin and Padman (2003). The technique used standard glaciological techniques first applied to the FRIS by Jenkins and Doake (1991). These procedures require measurements of ice thickness, surface accumulation, and ice flow velocity. As data sets of ice thickness and surface accumulation were available, the specific contribution of Joughin and Padman was to compile a greatly improved ice velocity map using interferometric synthetic aperture radar (InSAR) data obtained by RADAR-SAT. If it is then assumed that a given point on the ice shelf is in a steady state, then the horizontal divergence of the volume flux equals the combined surface and basal accumulation. As a result, if velocity, thickness, and surface accumulation are known, then the basal accumulation can be obtained by difference. The procedure was carried out two different ways. In the first the inflow and outflow were evaluated on the ice shelf perimeter. This was considered to be the most accurate method as there was some flexibility in choosing the location of the perimeter. In the second procedure the flux divergence was integrated over the complete ice shelf using gridded thicknesses, a procedure that yields the melt/freeze distribution. The first method yielded a melt rate of 83.4 ± 24.8 Gtons/yr, a rate that is 2.5 to 5 times less than estimates made earlier using a similar approach (Jenkins and Doake 1991). However, the Joughin and Padman value is in the range of earlier oceanographic estimates that ranged between 45 and 90 Gtons/yr. I think that everyone would feel better if these estimates were better resolved.

17.3.4 Oceanographic Aspects

The oceanographic aspects of marine ice are complex. First one might ask, "Are there oceanographic observations that support the contention that marine ice is formed as the result of the ice pump?" In fact there are several. Perhaps the most striking is the occurrence of ice shelf water (ISW). This water type, which is found beneath and in the near vicinity of ice shelves, has several unique characteristics that can only result from ocean–ice shelf interactions. For example (Nicholls 2001), consider a unit mass of seawater at temperature T_0 and salinity S_0 that comes into direct contact with the base of an ice shelf where the in-situ freezing point is T_f. First, the water warms m kg of ice to the freezing point. Then it supplies the latent heat of melting. The resulting mixture of melt and seawater has a temperature T and a salinity S. Taking the initial ice temperature to be T_i, the latent heat to be L, and the specific heat capacities of seawater and ice to be c_w and c_i, respectively, then heat and salt conservation requires that

$$(T\text{-}T_f)(1+m)c_w + m\left(c_i(T_f\text{-}T_i)+L\right) = (T_0\text{-}T_f)c_w \qquad S(1+m)=S_0. \qquad (17.2)$$

Then eliminating m and expressing T as a function of S shows the trajectory of the resulting mixture in T–S space as a straight line that passes through (S_0, T_0) with the gradient

$$\frac{dT}{dS} = \frac{L}{S_0 c_w} + \frac{(T_f\text{-}T_i)c_i}{S_0 c_w} + \frac{(T_0\text{-}T_f)}{S_0}. \qquad (17.3)$$

Equation 17.3 is dominated by the first term, which evaluates to ~2.4°C‰$^{-1}$. The second term is roughly one-tenth the size of the first term and the third term is two orders of magnitude less (Nicholls 2001). Therefore ISW should reveal itself as a straight line with the appropriate slope on a potential temperature–salinity plot. Figure 17.10 shows an example of such water as observed by CTD measurements taken through a drill hole at a site on the Ronne Ice Shelf (Nicholls and Makinson 1998). Note the deviation from a straight line connecting the

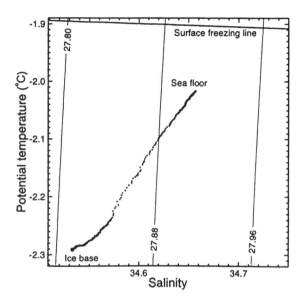

Figure 17.10. A potential temperature–salinity plot for seawater observed at SITE 1 beneath the Ronne Ice Shelf. Also shown is a line indicating the surface freezing point and lines of constant potential density referenced to sea level (Nicholls and Makinson 1998).

upper and lower layers occurring between the salinity values of 34.54 and 34.58. This bulge, which occurs at a pressure of approximately 610 to 700 dBar, can be explained by intrusions of anomalously cold water. As the authors note, this is the only depth interval in which the (θ, S) variability cannot be ascribed to vertical motion associated with the ice pump. Another characteristic of ISW as seen in Figure 17.10 is that it is always colder than the surface freezing temperature. To the best of my knowledge, ISW is the only water mass in the World Ocean with these characteristics. There are also geochemical clues to the origin of ISW. For instance, antarctic glacier ice derived from snow of coastal origin has δD and $\delta^{18}O$ values more negative than –130‰ and –17‰, respectively. If the ice comes from the interior of the continent, the δ values are even more negative. Figure 17.11 shows the $\delta D \delta^{18}O$ relationship observed in basal marine ice collected from the Campbell glacier tongue (Souchez et al. 1995). The open symbols represent marine ice samples whose plot can be represented by a straight line ($\delta D = 7.86$ $\delta^{18}O + 0.29$) with a correlation coefficient of 0.998 ($n = 71$). Type 1 ice comes from the upper part of the marine ice sequence whereas type 2 comes from the lower part. Note that all the basal ice has δD and $\delta^{18}O$ values that are less negative than –130‰ and –17‰, indicating that their source material cannot have been unmodified glacier ice. In that the local seawater has δD and $\delta^{18}O$ values of –3.23‰ and –0.69‰, the linear compositional variation can be taken as an indicator of mixing as might be expected to occur during the ice pump process. Note that the seawater values are similar to the values obtained by extrapolating the least-squares line. The black symbols are part of a computational simulation carried out assuming different mixing ratios. See Souchez et al. (1995) for details. Another useful indicator is based on the fact that helium contained in the air trapped in the bubbles in the glacier ice is introduced into seawater when the glacier ice melts. In that helium's solubility in seawater increases as pressure increases, high concentrations of helium in ISW indicate that the water equilibrated at depth, as would have occurred in an ice pump initiating near the grounding line.

The water masses that are believed to initiate the ice pump process are those that are encountered over the antarctic continental shelves. There appear to be two different types. The

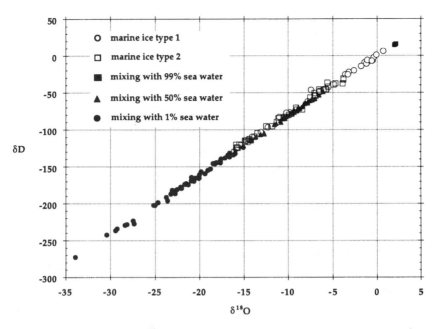

Figure 17.11. The δD–$\delta^{18}O$ relationship observed in basal marine ice collected from the Campbell glacier tongue (Souchez et al. 1995). The open symbols represent the marine ice samples, whose plot can be represented by a straight line with a correlation coefficient of 0.998 (71 samples). Type 1 ice comes from the upper part of the marine ice sequence and type 2 comes from the lower part.

first and most frequent type, at least at locations off the Filchner-Ronne and Amery Ice Shelves, is high-salinity shelf water (HSSW). This water has been conditioned via the action of polynyas north of the shelf margins. It is both very cold and saline, at times reaching salinities of more than 34.8‰ and temperatures near the surface freezing point (–1.9°C). Moreover, this water type characteristically fills the complete water column at the shelf edge. The other extreme is encountered over the shelf in the Bellingshausen and Amundsen Seas. There the shelf break appears to be far less effective in isolating the shelf from water masses located farther to the north, allowing circumpolar deep water (CDW) with temperatures as high as +1°C to move onto the shelf and gain access to the sub-ice-shelf cavity. These two quite different extremes are referred to as the cold and warm regimes.

In the cold regime, HSSW flows into the sub-ice cavity and on down to the grounding line, where the depths may reach values up to 2000 m. There, because of the increased pressure, HSSW may be as much as 1.5°C warmer than the freezing point, and when it comes into contact with glacier ice, rapid melting and cooling occurs. The result is ice shelf water that is defined as water that has a temperature below the surface freezing point. As the resulting meltwater is less saline, it is less dense and relatively buoyant and will start to flow up the inclined base of the shelf, resulting in a gradual increase in its freezing point as the pressure decreases. There are then two possibilities. If the rising plume only entrains small amounts of HSSW, which is warm relative to the plume, the decreasing pressure will cause the plume to become supercooled resulting in the nucleation and growth of frazil crystals, which in turn are buoyant and float upward, ultimately creating a layer of marine ice at the base of the ice shelf. If the entrainment of HSSW and the formation of crystals results in the plume density reaching the density of the surrounding water, then it is possible for the plume to detach from the base of the ice shelf and emerge from the shelf at some intermediate depth.

Model calculations suggest that at least on the Ronne Ice Shelf an internal recirculation may also be possible. The mechanism is as follows. If there is intense nucleation of frazil crystals in the plume and the separation of many of the crystals upward to the accumulating marine ice layer, the remaining water in the plume may become sufficiently dense from the salt rejected during crystal growth to drain backward toward the grounding line. This again is an ice pump, but now one acting between the grounding line and the central part of the Ronne Shelf. Such a pump works purely as a transfer mechanism moving ice from one part of the shelf to another. As pointed out by Nicholls (2001), such a recirculation does not result in a gain or loss of ice. Also the heat required to melt the ice at the grounding line is later recovered in the freezing region. The external heat required to drive such a process is only that needed to warm the ice at the grounding line to the freezing point. As this value is less than 20% of that required for melting, such a recirculation requires only a small fraction of the external heat necessary to melt and remove ice from the system. Nevertheless, direct evidence for such a recirculation is still sparse. However, its occurrence does explain one curious observation: although the Ronne Shelf is believed to be underlain by a thick layer of marine ice based on ice thickness measurements, to date no sizable plume of ISW has been observed exiting the front of the shelf.

The warm regime, as observed along the Bellingshausen and Amundsen coasts, is quite different. Field observations indicate an inflow of warm circumpolar deep water (CDW) at a temperature of +1.0°C into the subshelf cavity and an outflow of CDW mixed with meltwater from the base of the shelf, an interpretation supported by the helium and oxygen isotope contents of the water. The Pine Island Glacier located on the coast of the Amundsen Sea is estimated to have an annual melt rate at the grounding line of ca. 12 m a^{-1}. This very high value is believed to be the result of a variety of factors, in particular the extreme depth of the grounding line (> 1100 m) resulting in a lower in-situ freezing point. Also of importance is a steep upward slope away from the grounding line, a factor favoring more effective turbulent exchange between the CDW and the glacier base.

17.3.5 Models

17.3.5.1 Plume Models   Significant progress has been made in applying both plume models and more general circulation models to the problem of ascertaining oceanic behavior in the sub-ice-shelf cavity. The problem here is not with the models, which in my view appear to be very physically realistic, but in the general dearth of observational data required to verify them. The current plume models have been developed and applied in a series of papers (Bombosch and Jenkins 1995; Jenkins 1991; Jenkins and Bombosch 1995; Khazendar and Jenkins 2003; Smedsrud and Jenkins 2004). The first two papers develop the theory and discuss earlier work on inclined plumes; the third discusses a number of possible applications of the theory; and the fourth expands some of the crystal growth aspects of the theory and explores the specific problem of the infilling of rifts in ice shelves by marine ice. Here I will only try to give the reader a general sense of this effort. The model is one-dimensional and tracks the initiation and behavior of the plume from the grounding line until it exits the shelf. The initial seawater is assumed to be HSSW as this not only is the obvious choice, it also is the coldest and most saline water available and is therefore able to interact with the shelf at the deepest point of the grounding line, thus providing the largest temperature difference between the seawater and the glacier ice. The theory is based on the conservation equations for the mass, momentum, and concentration of the ice crystals and the residual liquid in the plume. Included are considerations of the thermodynamics of the interaction between the plume and the base of the shelf, the growth of the suspended crystals, and the process of crystal deposition. The motion of the plume is considered to be the result of the density difference between the water in the plume and the stationary ambient water mass. During melting the water produced must be warmed from the temperature of the water–ice crystal contact to the temperature of the plume. During freezing the temperature of the water lost must be raised from the (supercooled) temperature of the plume to the temperature of the water–ice crystal contact. The salinity of the ice formed is assumed to be zero. During the freezing and melting process only molecular diffusion is considered in that turbulent transport must be negligible at the crystal–seawater boundary. In addition, the temperature and salinity of the water in contact with the ice is constrained by the pressure–freezing point relationship. The analysis is further complicated by the fact that the size of the growing ice crystals influences the process of heat and salt transfer and that at any one time there are a range of crystal sizes involved. To simplify this, all the crystals are assumed to have the same dimensions following the earlier work of Omstedt (1985a). The buoyant drift velocity of the crystals is assumed to equal that in calm water and is calculated using relations suggested by Gosink and Osterkamp (1983) based on experimental as well as field observations on frazil growth in rivers. Although the plume model could hardly be called simple, I think that it is fair to say that when some aspect of the model could be formulated in a variety of different ways, Jenkins and Bombosch have consistently chosen to start with the most straightforward, reasonable relations.

So what does the model tell us? In considering crystal diameters in the range of 0.5 to 4.5 mm and disc thicknesses of between 5 and 45 μm (the aspect ratio was taken to be constant at 0.01), the 4.5 mm crystals settled out (up) so rapidly that all the seeds were lost before the concentration could get an opportunity to grow. When the crystal size is smaller, the plume is able to hold the frazil in suspension, allowing the concentration to grow to higher values. On the other hand, the smallest crystals never settle out and the concentration continues to grow all the way to the ice front. Between these two extremes, although growth and precipitation occur at varying rates, the total amount of ice generated is similar because the freezing rate adjusts to a value that is almost independent of the dimensions of the crystals. The biggest difference appears to be the fact that the response of the larger crystals to changes in supercooling is small in comparison to their response to the gravitational forces driving precipitation. As a result both freezing and precipitation start more slowly and then undergo a damped oscillation. For the smaller crystals, the oscillations are not apparent as the result of the more rapid thermal response and the slower drift velocity. Perhaps more importantly, in that all the plumes deposit roughly the same amount of ice, they all achieve neutral buoyancy at nearly the same point. Furthermore, when a significant number of crystals is present in a plume, thereby presenting a larger surface area available for growth, supercooling is more readily converted to ice. This in turn keeps the level of supercooling low and minimizes the possibility that freezing will take place directly on the base of the shelf. As a result the deposition of crystals becomes the major mechanism of basal ice growth. The presence of ice crystals in the plume also gives it added buoyancy, resulting in a positive feedback between ice concentration and plume velocity. As a result, if the ice concentration is increasing, the increase in momentum will make the deposition of crystals less likely. On the other hand, once precipitation has started the loss of buoyancy will result in the plume slowing and more rapid precipitation of crystals. Therefore, supercooling resulting from the ascent of the plume may result in suspended ice crystals that are deposited over a restricted region. The authors also note that under some conditions precipitation of crystals can lead to a density inversion. These are cases where the rejection of salt resulting from crystal growth has caused the residual liquid in the plume to become denser than the underlying unaffected seawater. Although the plume may still remain buoyant as the result of the effect of the suspended frazil crystals, once the frazil is deposited this is no longer the case and the residual liquid will sink into the underlying fluid, modifying it and perhaps driving a recirculation.

Now let us examine how the predictions of the model compare with observational facts for the Filchner-Ronne Shelf. Bombosch and Jenkins consider several different plume paths on the shelf. As the results are generally similar, I will use the results for the Foundation Ice Stream as an example. Figure 17.12 shows a cross section of the Ronne Ice Shelf along the presumed path of the plume, starting at the grounding line for the Foundation Ice Stream. Shown are the surface and basal elevations as well as the extent of the marine ice (stippling). Also indicated are modeled rates of basil melting (positive) and accretion (negative) for three different

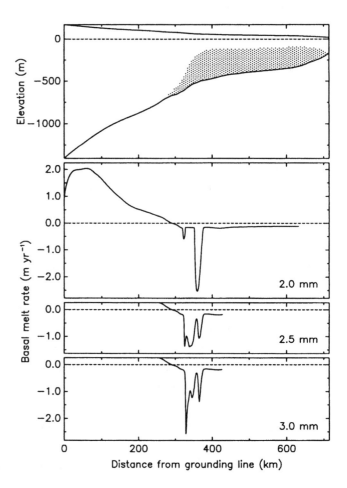

Figure 17.12. A cross-section of the Ronne Ice Shelf along the presumed path of the plume starting at the grounding line for the Foundation Ice Stream (Bombosch and Jenkins 1995). Also shown are the surface and basal elevations of the shelf as well as the extent of the marine ice (stippling). Indicated on separate plots are modeled rates of basil melting (positive) and accretion (negative) for three different crystal diameters. In that basal melting is essentially the same in all cases, it is only shown once. Melting and freezing rates are presented as solid ice equivalent thickness per unit time.

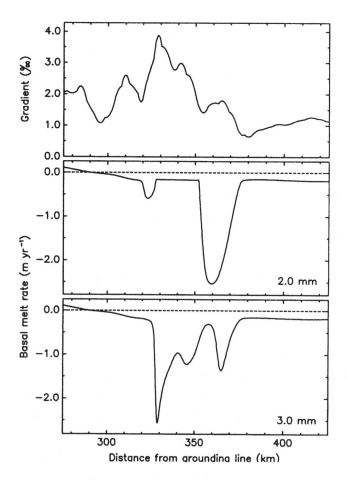

Figure 17.13. The slope of the ice shelf base along a section of the path followed by the Foundation Ice Stream as well as the modeled rates of basal melting (positive) and accretion (negative) for frazil ice crystals of two different diameters (Bombosh and Jenkins 1995).

crystal diameters. As melting is similar in all three cases, it is only shown once. Peak melt rates occur a short distance downstream from the grounding line and reach values of ~2 m yr[-1]. Although the peak of the accumulation rate depends somewhat on crystal size, in all cases there is a peak of between 1.5 to 2.5 m yr[-1] located between 300 and 400 km downstream from the grounding line. This is the same region where observation shows that marine ice is growing rapidly. Figure 17.13 shows a detailed plot of the freezing rate for the 2.0- and 3.0-mm-diameter ice crystals with a plot of the basal slope of the ice shelf. For the 3.0-mm crystals there is a direct correlation between the gradient and the accumulation of ice in that steep basal slopes are associated with peaks in the ice concentration. This is the result of the fact that a rapid rise in the plume produces a rapid increase in the supercooling and an associated increase in ice concentration. This in turn leads to a rapid increase in deposition rate, provided that the plume velocity is not high enough to appreciably reduce the deposition rate of the larger crystals. For the smaller crystals there is an inverse relation between gradient and accumulation. Although the crystal deposition begins earlier associated with the more rapid increase in ice concentration, it is stopped by the rise in the basal gradient, which results in a plume velocity too high to allow deposition. It is only after the plume reaches a local minimum in basal slope that precipitation recommences. As Bombosch and Jenkins note, the varying slope of the base of the ice shelf influences the behavior of the ISW in two different ways. First, through its effect on the upward component of the plume velocity, it determines how rapidly the ice concentration can grow. Second, through its effect on the forward component of the plume velocity, it determines where deposition is likely. Once marine ice is deposited on the shelf base, the flow of the glacier will carry it down the flow line to the terminus unless it is removed by melting at the shelf base, as is the case on the Foundation Ice Stream as well as on all of the flow lines examined.

A later paper in this series (Khazendar and Jenkins 2003) applies the plume model to the specific problem of the infilling of rifts in ice shelves from below by the generation of marine ice resulting from melt-driven convection at the sides of the rifts. The proposed mechanism is believed to be capable of generating tens of meters of marine ice in a rift as well as a thick layer of supercooled water. These results are shown to be in good agreement with field observations made over a 23-month period on a rift in the Fimbulisen Shelf in Dronning Maud Land, east Antarctica (Østerhus and Orheim 1992).

The most recent application of a plume model to the distribution of marine ice beneath the Filchner-Ronne Ice Shelf is by Holland et al. (2007). The model predicts ISW plumes that exit the shelf in the observed locations. In addition, the estimated basal freezing rates account for marine ice thicknesses as observed in the western part of the shelf. Furthermore, both the freezing rate and the plume properties in this area appear to be influenced by the joining of plumes from different meltwater sources. Elsewhere on the shelf, the match between modeling and observations is less successful. The authors suggest that this inadequacy is probably due to their model neglecting ocean circulations outside of the plume.

17.3.5.2 General Ocean Models   The application of more typical ocean circulation models to ice shelf cavity studies would appear to be a natural extension of the use of such models in the open ocean. The differences, however, are significant. First, the ice shelf provides a "fixed" roof possessing an ill-defined surface roughness. Furthermore, no longer is the pressure at the upper liquid boundary essentially fixed at one atmosphere. Instead, it varies according to variations in the ice shelf thickness. Fortunately, these values can be assumed to be fixed if the simulation is only concerned with a period of a few years. In addition, the overall geometry of the cavity is frequently quite complex; the Filchner-Ronne Shelf is a good example. An excellent place to start an exploration of this literature is the paper by Williams et al. (1998) and its associated reference list. As is pointed out, over the last 15 to 20 years there has been a gradual increase in both the number and diversity of the ocean models that have been applied to the problem of the sub-ice-shelf circulation. Gradually the models have progressed from two-dimensional to three-dimensional, including the open ocean seaward of the shelf front. Considerable attention has been given to tidal models in that tides are believed to possibly represent the most important process driving mixing in the water column and melting at the shelf base. More recent related papers include Jenkins and Holland (2002), Jenkins et al. (2004), Makinson and Nicholls (1999), and Williams et al. (2001).

One problem that has always interested me, and that will require both modeling and expanded focused data collection to solve, returns us to where we started this chapter: McMurdo Sound. If you will recall, in that area the flux of ISW from beneath the McMurdo Shelf does not occur year-round. At least, we know that it is not occurring in the fall when freeze-up occurs, as platelet ice does not form until congelation ice has reached an appreciable thickness. For instance, in 1980 Gow et al. (1998) found that roughly 1.5 m of congelation ice was present before platelet ice started to form. Similar observations have been made in 1990 (Jeffries et al. 1993). When platelet ice stopped growing is unknown but it had obviously stopped prior to freeze-up. Whether or not this occurs at other ice shelf locations is unknown. One simple guess would be that the flux of ISW from beneath the McMurdo Shelf only starts after the water in front of the nearby (and connected) Ross Ice Shelf is appropriately conditioned by lowering its temperature and increasing its salinity. What would cause this? Clearly the development of a polynya along the edge of the Ross Shelf during the fall and early winter is a strong possibility. Unfortunately, suggesting possibilities is easy; nailing them down is another matter. Recent work on such timing on the Filchner-Ronne Shelf results in a more complex picture (Jenkins et al. 2004). There it was found that peak inflows into the subshelf cavity occur in midwinter in association with pronounced convective activity north of the shelf. However a significant, although slightly lesser, peak occurs in the summer. Moreover, these two inflows do not appear to ventilate the same parts of the cavity.

To conclude this section a few words should be added concerning the interactions between the outflow from beneath the antarctic ice shelves, namely ISW, and the bottom water on the lowest level of the World Ocean, specifically, Antarctic Bottom Water (AABW). Here the unusual properties of ISW are important. Not only is ISW characterized by potential temperatures that are lower than the surface freezing point, but it also has a high compressibility resulting from its extremely low temperature. These characteristics favor its ability to sink to the ocean floor in regions where it spills off the continental shelf. Let us examine what is known about these matters in the Weddell Sea, a region where more than half of AABW is believed to originate (Orsi et al. 1999). Fortunately, this is also the region bordered by the Filchner and Ronne Ice Shelves which, to date, are the most thoroughly studied of the antarctic shelves. Much of this work has been recently summarized by Foldvik et al. (2004), who note that there are a variety of pathways by which HSSW can arrive in the deep Weddell Sea as Weddell Sea Bottom Water (WSBW), the Weddell Sea's precursor to AABW. However, only one of these pathways involves the circulation under the Filchner-Ronne Shelf that generates ISW. That being said, it is found that the ISW pathway is quite complex. Data from 20 different moorings deployed between 1968 and 1998 are combined in the study and indicate that ISW and its mixing products are commonly denser than Weddell Deep Water (WDW) and are therefore capable of sinking further down the continental slope. It was found that the cold shelf water sometimes reaches depths of more than 2000 m with little mixing with the overlying water. There appear to be two main reasons that the ISW plumes show so little interaction with the surrounding waters. First, any decrease in buoyancy that occurs as the result of mixing during sinking is offset by what has been termed the *thermobaric effect*. This is that in the cold ISW plumes the increase in density with increasing pressure (i.e., depth) is greater than for the surrounding WDW. Furthermore, the flow within the plume frequently appears to be supercritical; that is, the plume speed is larger than the internal phase speed. At one of the moorings discussed by Foldvik et al. (2004), the average current speed was 45 cm s^{-1}. For such flows, resonance waves do not form and associated mesoscale mixing does not occur. As a result, these dense plumes can reach great depths with very little mixing and reduced friction. Once the ISW plume slows so that a transition to subcritical flow occurs, there is associated mixing in the hydraulic jump. At a depth of 2000 m, the data indicate that the plume has disappeared and been replaced by a new boundary-layer water mass: newly formed WSBW resulting from the mixing of WDW and ISW from the plume. Foldvik et al. (2004) estimate that the production of WSBW at the Filchner outflow resulting these processes amounts to 4.3 ± 1.4 Sv.

I will leave the problem of converting WSBW to AABW to the oceanographers. The point that I hope to have made here is that the processes involved are far from simple. Personally, I have trouble just keeping the water type acronyms straight. For individuals wishing to explore these matters further, the Foldvik et al. (2004) paper is a good place to start in that it incorporates data from a number of earlier works and has an extensive reference list.

17.4 Anchor Ice

As its name implies, this ice type characteristically forms on the seafloor. It can form if vertical turbulence is sufficiently strong to rapidly transport seawater that has been supercooled at the sea surface to the seafloor. There, objects on the bottom can serve as nucleation sites for ice growth. The supercooled water may also contain small frazil ice crystals that stick to objects on the seafloor and serve as the nucleation sites for further ice growth. This general process has been termed *suspension freezing* by Campbell and Collin (1958), and the conditions necessary for its occurrence are believed to be strong winds, extreme subfreezing temperatures, and open water that is sufficiently shallow that the entire water column can become supercooled. As noted earlier in Chapter 11, these are conditions that one might expect to encounter in the polynya regions located along arctic and antarctic coasts.

Anchor ice appears to form selectively on the coarser seafloor material. There appear to be several reasons for this. For one, as the coarser clasts project further into the flow, they cool more quickly to the subfreezing temperatures required for anchor ice formation. Also, clasts projecting into the flow will collect more frazil crystals. The ice that forms is typically comprised of spongy ice masses, with the lower part of the mass composed of large ice platelets and the upper part of the mass of smaller frazil crystals. There do not appear to be any detailed studies of crystal size distributions or of orientations. In regions where there is a consistent directional flow of water, ice buildup appears to be heaviest on the sides of objects facing into the flow. Recent interest in this subject has been focused on the ability of anchor ice to transport material from the seafloor upward to the base of the overlying congelation ice where it can become incorporated into the thickening ice sheet.

To date, the efficiency of this process in sea ice has primarily been studied in the Arctic, where in 1979 it was found that the sea ice between the Colville and the Sagavanirktok Rivers contained 16 times more sediment than the annual sediment supply to the same region (Kempema et al. 1989; Reimnitz and Kempema 1987). Significant sediment entrapment occurred in water depths up to at least 25 to 30 m. Reimnitz has suggested that in extreme cases the affected depths may be as large as 50 m. A particularly interesting description of the results of suspension freezing in the Northwest Passage can be found in Reimnitz, Marincovich, et al. (1992). In rivers the formation of anchor ice is well established as an important process, as large ice masses can form and lift large objects such as lost anchors and rocks weighing as much as 30 kg off the bottom (Ashton 1986; Martin 1981). Here the obvious application relates to the possibility of sea ice incorporating pollutants and, in particular radioactive materials that were initially deposited on the seafloor, and then transporting them long distances (Pfirman et al. 1989, 1995; Weeks 1994).

In the Antarctic it is well established that ice can form on the seafloor at significantly greater depths. For instance, at McMurdo Sound its presence has been directly observed at depths of up to 33 m (Dayton et al. 1969), and at nearby White Island ice buildup has occurred on fish traps at depths of over 70 m. Evidence also suggests that in the vicinity of the Filchner Ice Shelf organisms have been raised from the bottom in water up to 250 m deep (Dieckmann et al. 1986). Admittedly such ice could be considered anchor ice in that it forms both on the bottom and from supercooled water. Nevertheless, I believe that this ice should be considered to be platelet ice because, as discussed earlier in this chapter, the mechanism involved in producing the supercooling is quite different from a suspension freezing process applied at the sea surface.

17.5 Conclusions

I have found that trying to understand the different processes and interactions involved in the world of underwater ice is quite fascinating. Moreover, the amazingly rapid pace of discovery in this research area has been impressive. Thirty years ago all that existed was a theory that might explain how platelet ice forms. The fact that large volumes of marine ice existed under the antarctic ice shelves and that the process that produced this ice was a factor in the generation of AABW had not yet surfaced. Impressive!

18 Trends

18.1 Introduction

One of my wiser associates warned me some time ago to avoid the subject of trends like the plague. "It's simply too much of a moving target," he counseled. Of course, he was, and still is, correct. Nevertheless, if I were reading this book I would feel shortchanged if some space was not devoted to this subject. If nothing were changing, then perhaps the subject could be dismissed in a few words. However, this clearly is not the case as anyone knows who is reasonably acquainted with events of the day as reported in the popular press. Moreover, with some exceptions the current trends point in the same direction. Besides, this is a good subject for a final chapter in a book of this type.

First, before we discuss what is happening in the polar regions, let us consider what is known about global average temperature trends. Recently, it was announced that 2005 was the warmest year since at least the 1860s and that 1998 and 2007 are tied for second place. Furthermore, the eight warmest years have occurred since 1998 and the fourteenth since 1990. Now one does not have to be a statistician to realize that the probability of this being a random sequence is minuscule. In fact, a recent analysis places the probability of such an occurrence at $p = 0.001$ for a global record and even lower for some regional records (Zorita et al. 2008). In addition, based on reconstructions of temperatures for times prior to 1860, several studies have suggested that in the northern hemisphere, the current surface temperatures are warmer than they have been at any time during at least the last thousand years. Available data indicate that since 1900 global surface temperatures have increased by approximately 0.75°C, with land areas warming the most.

When indications of global warming were first announced and the suggestion was made that this change was being driven by increases in so-called greenhouse gases, many individuals within the scientific community were very skeptical. In particular, it was pointed out that temperatures in the lower atmosphere appeared to show little or no warming, a state that appeared to contradict the postulated upward temperature trend in the surface data. However, improved temperature observations and a better understanding of both radiosonde data and of upper atmosphere temperature measurements obtained via the use of satellite-based sensor systems have recently resolved these problems. It is now clear that at least since 1979, the starting date for effective satellite monitoring, both the lower atmosphere and the surface have been warming and in a similar manner (Mears and Wentz 2005). Moreover, the vertical pattern of the observed temperature change is in agreement with simulated predictions made using global climate models. With this improved knowledge, skepticism concerning the reality of current global warming has largely disappeared, although questions still remain concerning the separation of naturally occurring thermal cycles from temperature changes resulting from the increasingly modified state of the Earth's surface and the rising level of greenhouse gases resulting from human activities. The current assessment of these contributing factors by the Intergovernmental Panel on Climate Change (IPCC) concludes that human activities are a significant contributing factor. However, as yet a hard number has not been placed on the term *significant*.

What else do we know? For one, the observed increases in CO_2 content of the atmosphere are in excellent agreement with the projections made in 1990 by the IPCC. However, the high quality of the agreement is to some degree believed to be partially the result of compensating errors in the estimates of the industrial emissions and the carbon sinks. Current observations of trends in the global mean surface temperature (land and ocean combined) suggest an increase of +0.33°C over 16 years of observations starting in 1990. Although this number is higher than predicted, it is nevertheless within the upper part of the range suggested by the IPCC's scenarios (Rahmstorf et al. 2007). Global sea level is rising at a current (1993–2002) rate of 3.4 mm yr^{-1} based on TOPEX/Poseidon satellite altimeter measurements. This is also a value higher than predicted by the IPCC models (~2 mm yr^{-1}). A recent reconstruction (Jevrejeva et al. 2008) based on observational data since 1700 indicates that since the end of the 18th century, the rate of global sea level rise has been accelerating at about 0.01 mm yr^{-1}. These authors also note that if the conditions that established the acceleration continue, sea level will rise an additional 34 cm between 1990 and 2090, a value that is also higher than the IPCC projections. The most recent assessment of global glacier mass balance and sea level changes with particular emphasis on the possibility of abrupt changes is by Steffen et al. (2008).

Let us now look at what is currently happening in the polar regions. There, if our current understanding of the positive feedbacks associated with decreases in the amount of sea ice and snow cover are correct as discussed in Chapter 9, we would expect to observe appreciable changes in temperature in excess of those observed in more temperate regions as well as measurable changes in the extent, thickness, and types of sea ice (Christensen et al. 2007).

18.2 Arctic

18.2.1 General

The fact that during recent years the Arctic has been undergoing an appreciable change in climatic regime is becoming increasingly well documented (Dickson 1999; Serreze et al. 2000). For instance, in the stratosphere (10 to 40 km above the surface) the arctic polar vortex, the circumpolar wind in the stratosphere that isolates cold air within it, has both strengthened and become more persistent (Shindell 2003). One noticeable effect of this stratospheric cooling is the recently observed increase in stratospheric ozone during the spring, an increase that requires both very cold air temperatures plus the presence of sunlight. Associated with this upper-level change, arctic surface pressure has decreased, resulting in a significant winter warming. During the period 1950–1999 there has been an overall winter temperature increase in the lower troposphere. In the middle troposphere, temperatures also increased during the first 10 years and then remained steady whereas in the lower stratosphere there was a slight cooling (Kahl et al. 2001). These results are based on all available observations made north of 80°N, including the observations made on the Russian drifting NP stations (Environmental Working Group 2000a). There has also been a weakening of the blocking effect of the typically high pressure over the North Pole as well as a shift in the mean location of the high. Other coupled effects associated with the weakening of the polar high in the troposphere are the more frequent summer penetration of cyclonic storms into regions such as the Arctic Ocean and the Canadian Archipelago, where cumulonimbus clouds and the associated thunder and lightning have not previously been observed.

Other well-documented recent trends include a general decrease of glacier extent and thickness in the northern hemisphere (Kaser et al. 2006). For example, a recent study of ice loss in the British Columbian glaciers indicates an average thinning of 0.78 m a^{-1}, an amount sufficient to account for a 0.67 mm sea level rise since 1985 (Schiefer et al. 2007). Moreover, the recent rate of glacier loss (17.0 $km^3 a^{-1}$) from this geographic area is approximately double that observed for the two previous decades. Recent analyses of glacier mass wastage in this same general area (including Alaska) suggest that a primary driver in this retreat and thinning is, for the glaciers that terminate in the ocean, a nonlinear instability resulting from the fact that their beds are below sea level (Meier et al. 2007). For instance, the Columbia Glacier in Alaska, which has been extensively studied, has not only retreated over 15 km since ~1980, it has also thinned extensively. On Svalbard, located at ~80°N, the thinning rates for 2003–2005 are four times the average for the earliest measurement period, 1936–1962 (Kohler et al. 2007). It is presently estimated that (a) ice loss to the sea accounts for virtually all of the observed sea level rise that cannot be attributed to steric effects (i.e., expansion of seawater upon heating), and (b) that approximately 60% of this loss comes from such "small" glaciers and ice caps as the Columbia, as contrasted with the giant ice masses in Greenland and the Antarctic.

Current observations of Greenland document pronounced ablation and ice discharge from the southern portions of the ice cap. In fact, the 2007 melt season for Greenland has proven to be the most intense since satellite observations were started in the late 1970s, with large inland areas of south Greenland showing pronounced melting (Mote 2007; Tedesco 2007). Specifically, melting occurred at areas above 2000 m for over 25 to 30 days longer in 2007 than the average number of days calculated for the period 1988 to 2006. In addition, the 2007 melting index—defined as the melting area times the number of melting days—was 153% greater for elevations above 2000 m than the average, setting a new record. Although the 2007 value for elevations below 2000 m did not set a new record, it also exceeded the average by about 30%. Limited data from the northern parts of the cap suggest that there, at least at higher elevations, the mass balance has been slightly positive through 2007. This is not unexpected in that warmer air masses are capable of transporting larger amounts of moisture that would result in increased snowfall in the high interior. Preliminary analysis of satellite data indicate extreme snowmelt at lower elevations in northern Greenland during the summer of 2008, with the number of melting days lasting 18 days longer than the previous maximum value (Tedesco et al. 2008). The most recent estimate of overall ice volume loss from Greenland averaged over the 2002–2005 time period is 108 $km^3 yr^{-1}$ (Howat et al. 2008). In weighing the above information, one should also consider the fact that the present Greenland warming is not unprecedented in recent Greenland history; a similar warming occurred between 1920 and 1930. At least through 2005, the 1920–1930 warming was of a similar magnitude to that presently observed (Chylek et al. 2006).

Other changes associated with global warming include greatly enhanced coastal erosion resulting from the increased length of the ice-free season, the greatly extended wind fetch resulting from the recession of the ice pack, and gradually rising sea level. There are also shifts in the distribution patterns of marine fisheries and the migration of species of vegetation further north to locations where, in the past, they were unable to flourish. Particularly interesting temperature observations were recently reported for Svalbard (Isaksen et al. 2007) in that its location (74° to 81°N and 10° to 35°E) places it in the Atlantic sector of the Arctic Ocean just slightly to the east of the southerly flowing East Greenland Drift Stream and the northerly flowing West Spitzbergen Current. There near-surface permafrost temperatures indicate a mean warming of 1–2°C over the last six to eight decades, with an accelerating warming occurring since 1999 of the order of 0.6–0.7°C/decade. A rather singular temperature anomaly occurred during the winter of 2005–2006, with air temperatures running 4–9°C above the 1961–1990 average. This is the warmest period for this time of the year since record keeping began in 1911 and was 2.8°C higher than the previous extreme (1954), amounting to an offset of 3.7 standard deviations from the mean. The time period coincided with open water in many of the fjords, unusually early and extreme snowmelt on the islands, and long periods of strong mild southerly winds. The authors note that although extreme, these high temperatures fall well within the predicted warming scenarios for the 21st century.

Because many areas of interest to this book such as the Arctic Ocean are still singularly lacking in long-term meteorological stations, alternate means of estimating temperature trends have been developed. For instance, monthly surface temperature trends can be obtained by using surface temperatures measured using NOAA's Advanced Very High Resolution Radiometer (AVHRR) during clear-sky periods. Although the first satellites were launched in the late 1950s, it was not until the late 1970s that multichannel sensor systems were included, allowing surface temperature measurements to be made. This means that the total time period of such measurements is short: slightly over 30 years. Even so, analysis of the AVHRR data is very informative. Comiso and Parkinson (2004) report that between 1981 and 2003 mean annual surface temperatures at locations north of 60°N have increased at a rate of ca. +0.5°C per decade. More specifically, over the sea ice a warming of +0.54°C per decade was observed; over Greenland the rate was +0.85°C; and over North America, +0.79°C. In contrast, there was a slight cooling of –0.14°C per decade over Eurasia, with an uncertainty of ±0.2°C. Standard meteorological observations for North America over the same time period give a trend of only +0.39°C per decade, as opposed to the +0.79°C value. However, if the AVHRR data are restricted to the same points as the meteorological data, Comiso and Parkinson report that a similar underestimate results. As might be expected, warming trends varied considerably with season in the polar areas, showing an increase of +0.84°C per decade in the spring and only +0.25°C in the summer. This is hardly surprising, considering the damping effect of the ice–water transition during the summer. The negative trend in Eurasia was largely the result of cooling during the winters.

However, the general atmospheric warming is only part of the problem, as changes are also occurring in the Arctic Ocean. The reason for the concern here is related to the rather unusual hydrography of the Arctic Ocean, as discussed in Chapter 3. Here I refer to the fact that the sea ice cover drifts in a cold, less saline, and therefore less dense surface layer. This layer isolates the ice from the underlying more saline and warmer (above-freezing) Atlantic Water (AW) that initially entered the Arctic in the form of the West Spitzbergen Current (WSC) (see Chapter 3). The stable vertical density distribution in the halocline at the base of the surface layer is the result of the salinity profile insofar as, at near-freezing temperatures, the density of seawater is primarily a function of its salinity. It is this stable density distribution that suppresses vertical convective motions and limits the upward heat flux from the warmer underlying water.

Now, it has been known for some time that the arctic halocline was not horizontally uniform. For instance, in the Eurasian Basin salinity increases rapidly with depth, reaching 34.9 to 35.0‰ at approximately 200 m. In contrast, in the Canadian Basin the halocline is deeper and the salinity increase with depth is gradual. Until the 1990s, although it was clear that the hydrography of the Polar Basin was not simple, there were no firm observations indicating that the halocline might become unstable. In the early 1990s this began to change insofar as a slight warming was observed near the Pole during a cruise of the *Odin* (Anderson et al. 1994; Rudels et al. 1994). In addition, observations indicated that the temperatures of the Atlantic inflow into the Basin at depths of 150 to 900 m appeared to be slightly warmer than usual (Quadfasel 1991). Also, the influence of the AW in the Arctic Ocean was becoming both more widespread and important (Carmack et al. 1997). Moreover, the front separating the more saline AW from the less saline Pacific water, as well as the associated circulation and ice drift patterns, moved counterclockwise from the pre-1990 position over the Lomonosov Ridge to a position that is in general alignment with the Alpha–Mendelyev ridge system. The result has been a more cyclonic circulation and a weakening of the cold halocline layer that isolates the ice (Steele and Boyd 1998). Specific changes at the North Pole and the Makarov Basin included a 2‰ salinity increase within 200 m of the surface as well as a 1 to 2°C increase in the AW temperature, changes estimated to produce a 30 to 40% increase in the annual average upward heat flux to the base of the sea ice. As pointed out by Morison et al. (2006), these changes occurred in tandem with a strengthening of the Arctic Oscillation (AO), as discussed by Thompson and Wallace (1998), corresponding to a drop in the surface pressure over the Arctic and a cyclonic spinup of the polar vortex. These changes were relative to pre-1990 observations that include the combined Russian–U.S.–Canadian data sets as compiled by the Environmental Working Group (1997, 2000a). These new 1990s conditions continued through 2003, arousing considerable interest in the arctic research community. In 2004 and 2005 the strength of the AO decreased and the hydrographic conditions within the Basin appear to have started to return to the pre-1990 EWG climatology (Morison et al. 2006).

However, at about the same time (February and late August 2004) additional pulses of unusually warm AW with water temperatures reaching +0.8°C were observed entering the Basin (Polyakov et al. 2005). These are temperature values comparable to the anomalies observed in the 1990s. As the result of these events there has now been a significant increase in both the thickness of the AW layer within the Basin and its heat content. Moreover AW temperatures measured east of Svalbard in 2004 reached 4.2°C, the highest temperature recorded to date in this area. At present these warm anomalies are in the process of moving across the Basin. Current observations indicate that the depth range (near-surface to almost 1000 m) and horizontal extent of the present anomaly (over most of the Barents Sea slope) are exceptional (Polyakov et al. 2007). Clearly, these oceanographic changes are the result of interactions between the polar and the subpolar basins in that the driving factor is the increase in the temperature of the AW as it enters the Arctic Basin. A related piece of information is the fact that current profiling buoy data indicate that the upper 1500 m of the North Atlantic has been warming during the time period 1999 to at least 2005 (Ivchenko et al. 2006). Interestingly, this increase in the heat content occurred north of 50°N. South of this latitude a slight cooling was observed.

Although it is easy to focus one's attention on the warming of the AW residing at depths of 150 to 800 m in the Arctic Ocean, what is known about changes in the temperature of the near-surface layer that actually contains the ice? This subject has been examined in two different ways. For example, Steele et al. (2008) examined sea surface temperature (SST) obtained between 60 and 90°N for the time period 1805 to 2007. By the nature of these regions, the data (240,851 profiles) are from the summer (considered to be July, August, and September) and are essentially derived from the marginal shelf seas. The measures examined are the mean temperature of the upper 10 m and the upper ocean heat content (OHC), that is, the integrated ocean heat content, relative to –2°C from the surface to 100 m or to the bottom, whichever is shallower. It was found that during periods such as 1930–1965 when the Arctic Oscillation Index fell (AO–), the SST generally declined. Conversely, during periods such as 1965–1995 when the AO increased (AO+), the SST increased. As might be anticipated, there are exceptions. For instance, the Beaufort Sea showed an anticorrelation until recent years, when an SST increase was associated with AO+ conditions. Since the late 1990s most regions have been warming, a trend that is particularly strong in the Bering and Chukchi Seas and in the eastern portion of the Beaufort Sea. During 2005 the maximum temperature anomaly in the northern Chukchi Sea was ~3°C above freezing representing an OHC of 240 MJ m^{-2} assuming a summer mixed-layer depth of 20 m. This is a value roughly half of estimated solar heating (400 MJ m^{-2}) as calculated for that year by Perovich et al. (2007). If it is assumed that all this heat is expended in melting sea ice, this amounts to 56–75 cm less ice growth in 1995, relative to 1965. This is a significant percentage of the average annual ice growth in this region. In 2007 the SST anomalies were as large as +5°C. More on this problem later.

The second approach examines the magnitude of the contribution of the ice–albedo feedback to the observed warming of the upper layer of the Arctic Ocean (Perovich et al. 2007). In this study satellite-derived ice concentrations (C), shortwave flux estimates based on meteorological reanalysis products F_r, and field observations of ocean albedo (a) over the Arctic Ocean are combined to calculate the solar heat flux directly into the ocean (F_{rw}) by using

$$F_{rw} = F_r \left(1 - a\right)\left(1 - C\right). \tag{18.1}$$

Calculations were made daily for every grid cell, starting in 1979 and concluding in 2005. As the amount of radiation penetrating to the ocean through the ice was assumed to be zero, the above approach should provide a lower bound on the total solar heating of the upper ocean. The results were interesting, with the annual cumulative heat input ranging from a few hundred MJ m^{-2} at high latitudes to a few thousand MJ m^{-2} at lower latitudes. As might be expected, the highest values occurred in the seasonal sea ice zone at locations such as the Bering Sea, Davis Strait, and Greenland Sea. It was found that although the annual trends in solar heat input were modest, the cumulative effects were significant. For instance, the median increase of 0.64% yr^{-1} accumulates to a total increase of 17% by 2005. If one integrates over the whole study area from 1979 to 2005, the additional heat input is 2.9×10^{15} MJ, enough heat to melt 9.3×10^{12} m^3 of ice. The time series of the input is particularly interesting in that it indicates that the thermal input was fairly constant from 1979 until 1992. Then it started to steadily increase through 2005 from an annual value of ~200 MJ m^{-2} to about 400 MJ m^{-2}.

18.2.2 Sea Ice Extent

Now that we have some general idea of the recent changes in the meteorology of the Arctic Basin as well as of the changes in the thermal structure of the Arctic Ocean, it is natural to inquire as to what is happening to the material in the middle, the sea ice.

There are two aspects of sea ice extent that need to be examined. These are the maximum extent of the pack ice in the late winter–early spring, and the minimum extent of the pack in the late summer–early fall. This latter value is particularly important in that it specifies the amount of ice that survives the melt season. Such ice has recently been referred to as *perennial*. I personally think that perennial is a very poor descriptor for such ice. In my dictionary the term *perennial* implies enduring, perpetual, and everlasting. As Weyprecht pointed out in the late 1800s, there is no perpetual, everlasting sea ice. Furthermore, if current trends continue, the amount of perennial ice may become essentially zero in the foreseeable future. I find the idea of a quantity termed perennial having a value of zero to be more than a little perplexing. This said, I will rarely use the term perennial in the remainder of this chapter. One should also note that the fact that ice has survived the summer does not provide one with specific information relative to the ages of the surviving floes. Admittedly, most of this ice is presumably comprised of thicker MY ice. Nevertheless, in most cases some FY ice will also survive the melt season, perhaps only as mechanically thickened pressure ridges.

One should also realize that although there are a few exceptions, such as the long historical record of ice conditions in the vicinity of Iceland, consistent high-quality observations of ice extent and type only date back to 1972, with the launch of the Nimbus-5 satellite carrying a scanning passive microwave (μ-wave) system (Massom 1991). The important factor here was that this system and its successors provided ice observations irrespective of weather, time of day, or season. The importance of this cannot be overemphasized, considering that the polar winters are periods of darkness and the polar summers are frequently periods of extensive cloud cover. The reader will probably not be surprised to hear that there are still questions remaining concerning some of the finer points of extracting sea ice information from passive μ-wave data (Andersen et al. 2007; Carsey 1992). However, from the point of view of examining trends in sea ice extent and type, such differences, though interesting, are both small and unimportant, in that the trends revealed are quite consistent independent of procedural differences, provided that the same operational procedures are used in processing the data sets. In the following discussion the total ice

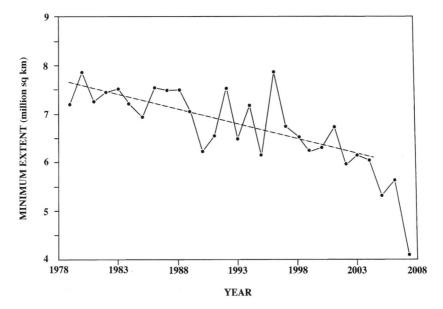

Figure 18.1. Yearly minimum sea ice extents in the Arctic in millions of square kilometers for the time period 1979–2007 (National Snow and Ice Data Center 2007). The linear trend indicated by the dashed straight line is based on data from 1979–2004 and indicates a decline in area of slightly over 8% per decade.

area reported is, unless specifically noted otherwise, obtained by summing the number of pixels that indicate ice concentrations equal to or greater than 15%.

The fact that the Arctic's sea ice cover has recently exhibited signs of both decreasing extent and duration has been known since the mid-1980s (Gloersen and Campbell 1991; Johannessen et al. 1995, 1996; Maslanik et al. 1996; Parkinson 1992). The following are the results of more recent studies containing longer time series. Minimum sea ice extent at the end of summer (September) has shown an appreciable downward trend of ~8% per decade since the late 1970s (Meier et al. 2005), resulting in a 22% loss of ice area through 2004. During the last five years, end-of-summer values have consistently been at or near record lows that have been sufficiently striking to gain the attention of the popular press. For instance, the 2002 low was 4% lower than during any previous September since 1978 and 14% lower than the long-term mean (1979–2000; Stroeve et al. 2005). Moreover, analysis of extended μ-wave, aircraft, and ship reports indicate that the 2002 end-of-summer value was the lowest in at least 50 years (Rayner et al. 2003). In 2003 and 2004, although the minimum values were slightly higher than the 2002 value, they still did not exceed the linear least-squares line fitted to the data from 1979 to 2004 (Figure 18.1). In 2005 the minimum ice area dropped to below the 2002 value, setting a new record low of 5.32×10^6 km². Although the minimum ice area of 5.7×10^6 km² increased slightly in 2006, it was still the fourth lowest value, as of that date. Moreover, if the average extent for the different Septembers was compared, the 2006 value of 5.9×10^6 km² was found to be the second lowest, only exceeded by that for 2005.

The most recent examination of these trends examines the 28.2-year record between 1979 and 2006 (Parkinson and Cavalieri 2008). It is a particularly useful paper in that it documents trends in both ice extent and ice area in specific oceanic regions. The overall negative trend in sea ice extent for the northern hemisphere, based on the yearly averages, was $-45,100 \pm 4600$ km² a⁻¹

or $-3.7 \pm 0.4\%$/decade, with negative trends occurring for the four seasons and for each of the 12 months. The largest decreases occurred in the Kara and Barents Seas $(-10,600 \pm 2800 \text{ km}^2\text{a}^{-1})$ and in the Arctic Ocean $(-10,100 \pm 2,200 \text{ km}^2\text{a}^{-1})$. These regional decreases were followed by similar decreases in the Baffin Bay/Labrador Sea $(-8,000 \pm 2,000 \text{ km}^2\text{a}^{-1})$, the Greenland Sea $(-7000 \pm 1400 \text{ km}^2\text{a}^{-1})$, and Hudson Bay $(-4500 \pm 900 \text{ km}^2\text{a}^{-1})$. All the above negative slopes were significant at the 99% confidence level. The Okhotsk and Japan Seas also showed significant decreases at the 95% confidence level. The remaining three areas examined (Bering Sea, Canadian Archipelago, Gulf of St. Lawrence) also exhibited negative slopes, but the trends were not statistically significant. Similar trends in ice area are also documented. As Parkinson and Cavalieri point out, these negative trends in sea ice extent and area are almost certainly the result of the widespread warming observed in the Arctic. It is also interesting to note that in general the strongest negative trends occur in the more northerly locations, as would be expected if a classic ice–albedo feedback effect is a significant contributor to the observed changes.

In 2007 the minimum ice extent dropped even further to $4.13 \times 10^6 \text{km}^2$. As pointed out by NSIDC, this is a decrease of $2.61 \times 10^6 \text{km}^2$ over the average minimum area, as calculated by averaging the observations from 1979 to 2000. This is not a trivial change, being roughly equal to the areas of Texas and California combined. In fact, the ice extent and area covered by ice were 24% and 27% lower than the previous record lows, which were both reached during September 2005. They are also 37% and 38% less than the climatological averages (Comiso et al. 2008). Furthermore, the 2007 minimum is now only slightly over one-half of the generally accepted minimum value ($\sim 8 \times 10^6 \text{ km}^2$) for ice area during the 1950s through the 1970s (see Table 1.1 in Chapter 1). Figure 18.1, which is an extension of a similar plot published online by NSIDC, shows this gradual decline followed by the more rapid dropoff since 2004. Not only are minimum values of both ice extent and area declining, the decline rate is increasing. For instance, the values per decade changed from -2.2 and -3.0%, during the time period 1979–1996, to -10.1 and 10.7% during the last 10 years (1997–2007). As Comiso et al. note, these are values that are now comparable to the high negative trends of -10.2 and 11.4% per decade of the extent and area of the old ice within the pack.

Figure 18.2 shows a map of the ice-covered regions at the minima for both 2005 and 2007 as also presented by NSIDC. The lines on the two figures show the mean ice extent during September, based on data from 1979 to 2000. The 2005 and 2007 distributions show both certain similarities and interesting differences. For one, the 2007 image shows the most northerly ice edge ever reported, at 85.5°N near the 160°E longitude line. It also shows a completely ice-free direct NW Passage. On the other hand the NE Passage is blocked by ice extending all the way to the Russian coast at Cape Chelyuskin on the north end of the Taymyr Pennisula, a frequent choke point. In 2005 the ice distribution was just the opposite, with Cape Chelyuskin clear of ice and the west end of the NW Passage blocked. During both years the ice can be said to be crowded toward Fram Strait, the only deepwater exit from the Basin, and also pressed against the north coasts of Greenland and the Canadian Arctic Islands. In contrast, the shelf areas of the Beaufort, Chukchi, East Siberian, Laptev, and Barents Seas were wide open.

A number of authors have discussed different aspects of the fact that a continued decrease in the amount of sea ice present in the Arctic Basin at the end of the summer will result in a sig-

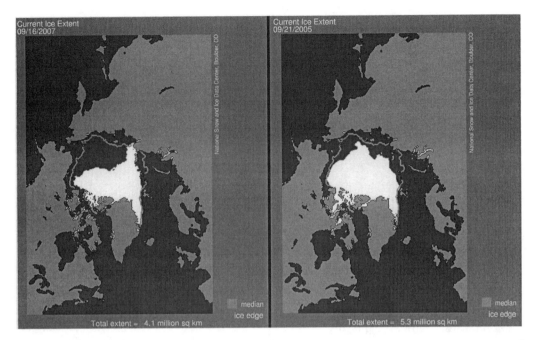

Figure 18.2. Map presentation of the distribution of sea ice during the September minimum extents of 2007 and 2005 (NSIDC, published online).

nificant decrease in the amount of older ice types present during the ice growth season (Comiso 2002, 2006; Francis et al. 2005; Rigor and Wallace 2004; Stroeve et al. 2007; Sturm et al. 2003). For example, a recent paper on this subject by Nghiem et al. (2007) uses backscatter data collected since 1999 by the QuickSCAT satellite (QSAT) to map on a daily basis the distributions of ice types during the minimum extent periods. Ice types distinguished were referred to as seasonal (i.e., FY), perennial (old, using the WMO terminology), and mixed. This latter ice type is thought to consist of seasonal ice that has been thickened by compression and desalinated. The authors then combined these results with a sequence of model runs using a buoy-driven Drift-Age (DM) simulation (Rigor and Wallace 2004). Comparisons were also made with the ice charts prepared by the National Ice Center (NIC). The QSAT data also show large reductions in old ice during recent years, as discussed earlier. Figure 18.3 shows the results of the DM model calculations from 1957 to 2007, combined with the QSAT observations from 1999 to 2007. During the period between 1957 and 1972 the amount of old ice remained comparatively stable at just under 5.6×10^{-6} km^2, although there was considerable year-to-year variation. After 1973 a gradual, somewhat variable decline set in, resulting in the area of old ice decreasing to 4.0×10^{-6} km^2 by 2002. After ~2002 the decline rate drastically increased. In short, although the techniques used are very different, the results—that the amount of ice surviving the summer is rapidly decreasing—are similar to those described earlier based on passive μ-wave observations.

A related study that also uses both satellite and drifting buoy observations and similar analysis techniques in estimating the changes in the age of the ice in the Arctic Basin is by Maslanik, Fowler et al. (2007). The time period studied is also similar (1979 to the summer of 2007). The results add considerable detail to the earlier results of Johannessen et al. (1999), to the effect that the amount of old ice in the basin is undergoing a steady and significant reduction. For instance, the area where at least half of the ice in March is a minimum of 5 years old decreased by 56%

Figure 18.3. Time series of the area of perennial (old) sea ice extent in March of each year as estimated by the Drift–Age model (fitted with a fifth-order regression) and as observed by the QuikSCAT satellite scatterometer within the model domain (Nghiem et al. 2007). In each year the model result was an average over March, and the satellite observation was on the spring equinox (21 March).

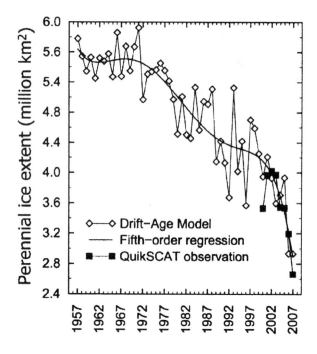

between 1985 and 2007. Furthermore, most of the old ice is now only 2 to 3 years old. Most importantly, the oldest and presumably the thickest ice types have essentially disappeared. In the central portion of the basin, 57% of the ice was 5 or more years old in 1987 with 25% at least 9 years old. In 2007 the coverage of the +5-year-old ice decreased to 7% and no very old (+9) ice appears to have survived. Although one might quibble over some of the fine details in these studies, the trends are striking. As the authors point out, the trends are also consistent with the thickness changes observed in ULS data by Tucker et al. (2001) and Yu et al. (2004); the losses that are occurring are clearly in the oldest, thickest ice types.

An additional associated trend that has been documented only recently (Häkkinen et al. 2008) is that since the 1950s there has been a gradual acceleration in both the observed ice drift rate and the wind stress in the region of the Transpolar Drift Stream, with significant positive trends occurring in both winter and summer. As noted earlier these trends are believed to be the result of both more intense storm activity in this region of the Arctic and the general shift of storm tracks toward higher latitudes noted earlier.

How well do the observed changes in minimum sea ice extent agree with computer simulations? This question has been examined by Stroeve et al. (2007), who compared the satellite observations with trends obtained by combining the 1900 to 2100 results of 13 of the IPCC Fourth Assessment Report (IPCC AR4) models. Figure 18.4 shows the results. The 13 individual model results are shown as faint dotted lines. Also shown are the multimodel ensemble means and standard deviations. As can clearly be seen, the slope of the observational data (the heavy irregular line) is appreciably more negative than the slope suggested by the model results. Note also that if the 2007 value of 4.28×10^6 km^2 (Stroeve, pers. comm.) is added to this figure, not only are the slopes very different, but the 2007 value would appear to be almost two standard deviations from the modeled mean.

It appears reasonable to suggest that there may be factors affecting the ice extent minima that are not currently considered in the IPCC models. In that these GCMs (Global Climate Models) were specifically selected because they appear to be in reasonably good agreement with observed sea ice variations, what could be wrong? Some possibilities are discussed by Eisenman et

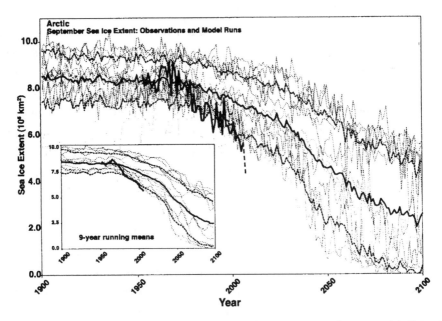

Figure 18.4. Arctic September sea ice extent based on 13 different IPCC AR4 climate models (faint dotted lines). Also shown is the ensemble mean (solid black line) and the standard deviations (dashed black lines) based on the model results and the observational results (1953 to 2006) (Stroeve et al. 2007). The inset shows the same data presented using a nine-year running mean. The original of this figure also contains an inset identifying the specific models used. I have also added the 2007 value of 4.28 million km².

al. (2007), who point out that intermodel differences in simulated cloud cover can be significant, resulting in appreciable differences in downwelling longwave radiation, differences large enough to result in major changes in the equilibrium ice thickness (1 to 10 m) of MY ice floes. In that equilibrium thickness values are also sensitive functions of the albedo of the ice, unrealistic values here resulting from errors in simulated cloud cover can be compensated for by tuning the values of the ice albedo. Therefore, the reader is cautioned to examine current GCM results with some skepticism when they are used in long-term predictions. For further discussion of these comments, see DeWeaver, Hunke, et al. (2008) and Eisenmann et al. (2008).

(The reader should note that the observational results plotted in Figure 18.4 are not identical to the results plotted in Figure 18.1. In Figure 18.4, the values are the average ice areas during the month of September. In Figure 18.1, the values are the sequence of minimum values observed during September.)

As the 2007 ice extent minimum appears to be different from both the observed and the modeled trends, it seems fair to inquire whether there was anything unusual about that specific year. A number of investigators have considered this question. Although the long-term negative trend in arctic sea ice extent is clearly associated with the general warming trend in the Arctic (Arctic Climate Impact Assessment 2005; Johannessen et al. 2004), in 2007 a number of other factors have been suggested as contributing. These are as follows. First, there was persistent high pressure over the Beaufort Sea through most of the 2007 summer. Skies remained clear, allowing large amounts of downwelling shortwave radiation to reach the surface, thereby accelerating the melting of the ice (Kay et al. 2008). In fact, during late June and July the area covered by ice decreased by an average of 200,000 km² per day. This is the most rapid decline noted in the satellite record to date. Second, the atmospheric pressure pattern resulted in an ice circulation pattern that pushed the ice to the north and west out of the western Arctic (Kwok 2008; Maslanik, Drobot, et al. 2007; Ogi et al. 2008; Rigor and Wallace 2004) and through Fram Strait (Nghiem et al. 2007). Third, ocean temperatures north of the Bering Strait were warmer than usual (Woodgate et al. 2006), a fact that presumably enhanced ice melt from both the bottoms and the sides of the floes. Fourth, previous low-ice-extent years had, in a sense, preconditioned the ice cover, leaving it much thinner and younger than in the past (Maslanik, Fowler, et al. 2007). Such ice is more susceptible both to melt and to being pushed around and through Fram Strait by the wind. Finally, the increased amount of open water, coupled with the clear skies, resulted in additional heating of the surface layer of the ocean, further contributing to the loss of ice (Perovich et al. 2007). An additional possible factor is the unusually high temperature of the AW entering the Arctic Basin in the vicinity of Svalbard. Although this factor ultimately must result in a decrease in the amount of sea ice, whether this is a contributing factor at the present time is not clear.

It must be added here that not everyone has been equally enthusiastic about all of the above suggestions. For instance, although Schweiger et al. (2008) concur with the suggestion that the summer of 2007 was unusually clear, they point out that the geographic match between the cloud anomalies and the ice anomalies is poor. As a result they conclude that the shortwave fluxes of the summer of 2007 appear to have contributed little to the extreme minimum in ice extent. Further work, by a rearranged but otherwise identical group of authors (J. L. Zhang et al. 2008), favors a combination of preconditioning, anomalous winds, and ice–albedo feedback. Of the additional 10% of the total ice mass lost during the summer of 2007, they conclude that 70% was lost due to amplified melting and 30% to the unusual ice advection.

I find the most interesting paper to date focused on the 2007 ice loss to be that of Perovich et al. (2008). This study had the advantage of including the results of measurements made by autonomous ice mass balance buoys (Richter-Menge et al. 2006) deployed in both the Beaufort Sea and near the North Pole. Not surprisingly, they found that the average annual surface melt in the Beaufort Sea (0.64 m) was larger than near the Pole (0.26 m) as the result of differences in shortwave solar radiation. They also note that despite the extreme ice retreat of 2007, the amount of surface melt in both regions was not significantly different from that of previous years. The same was the case for bottom melt near the Pole: no significant change over previous years. What was different in 2007 was the amount of ice lost off the bottoms of the floes in the Beaufort Sea: more than six times the annual average value of 0.34 m for the 1990s and two and a half times the 2006 value. The solar heat input into the upper ocean was estimated using

$$F_{rw} = F_r \left(1 - \alpha_w \right) A_w \qquad\qquad (18.2)$$

where F_r is the incident solar radiation, α_w is the albedo of the ocean, and A_w is the fractional area of ice-free ocean (see also Perovich et al. 2007). In that equation 18.2 does not consider the light that penetrates through the ice cover to warm the upper ocean, it provides a lower bound on the warming estimate. Even so, these estimates indicate that the large areas of open water present during the summer of 2007 absorbed twice the amount of heat required to account for the observed melting. The authors suggest that what is occurring is a classic ice–albedo feedback signature: more open water leads to more solar heat absorbed, resulting in more melting, which in turn results in more open water. I believe that the Perovich et al. (2007) and Perovich et al. (2008) papers have conclusively answered the question concerning whether or not the ice–albedo feedback is currently playing a significant role in the observed decrease in the amount of old ice in the Arctic Basin. The answer is, "It is." In fairness it should also be mentioned that general circulation models (GCMs) have for some time indicated that any reduction in ice area driven by external forcing such as a general global greenhouse gas–induced warming would result in a further reduction in ice area as the result of solar heat input into the newly accessible open water (Holland and Bitz 2003; Zhang and Walsh 2006).

In the past, after a low summer minimum it has been common for the following winter to rebound with near-normal ice extent values. Although the trend of the maximum ice extent has been negative prior to 2005, at about 1.5% per decade (Comiso 2006), this change is clearly less striking than that exhibited by minimum ice extent values. Moreover, the winter trend also appears to be changing in that it is now approaching –3% per decade. A figure similar to Figure 18.4 but based on March values can be found in Stroeve et al. (2007). It shows that although the slope of the observational data is also more negative than the climate model estimates, the disagreement between the observations and the simulations is not as striking as was the case for the September values. When looking at this figure, the reader should note that the March 2007 value was 14.66×10^6 km². Moreover, in the past when one area of the Arctic showed an unusually low ice extent, it was typically countered by another area showing higher than normal values, a sort of regional seesaw (Parkinson 1991). This also appears to be changing, with observed reductions now occurring on both the Atlantic and Pacific sides of the Arctic.

A slightly different approach to measuring the variations in the maximum ice extent can be found in Francis and Hunter (2007), who note that at most arctic locations the maximum ice extent southwards is limited by the presence of a coastline. However, there are two principal areas where this is not the case: the Bering Sea and the Barents Sea. As a result they studied the variations in the southernmost latitude of 50% ice concentration as determined by the passive μ-wave observations located along 4-degree-wide longitude bands centered on 30°E (Barents) and 170°W (Bering). The locations are then expressed as anomalies by subtracting the position for each year from the 27-year mean position in that region (1979–2006). The monthly northern hemisphere anomalies for ice extent during March are shown in Figure 18.5. As can be seen, there is a gradual decrease in the anomaly values from 1979 until 2003, followed by a sharp decline between 2004 and 2006. Anomaly time series were also prepared for the forcing parameters using daily mean values in order to investigate the factors influencing the location of the ice edge. The authors found an appreciable difference in the relative importance of the different factors affecting the ice edge position. In the Barents Sea the major factor controlling the position was the presence of warm surface winds near the ice edge as well as anomalous meridional winds, which can both mechanically move the ice and advect warmer surface air from the south. The Bering was different because zonal wind anomalies accounted for the majority of the ice edge variability, with winds from the east corresponding to retreating ice extent. The authors note that this result is supported by both theory and observations, in that sea ice moves to the right of the geostrophic wind in the northern hemisphere (Colony and Thorndike 1985; Zubov 1945). Similar

Figure 18.5. Monthly anomalies in northern hemisphere sea ice extent during March, expressed in percentage of the mean extent during 1979 to 2000 (Francis and Hunter 2007). The data source is NSIDC. Here the edge of the compact ice is defined as the southernmost latitude with an ice concentration of 50%.

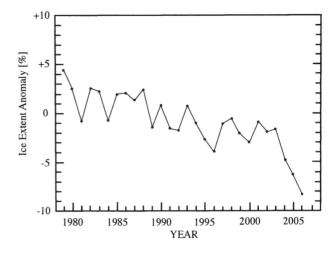

conclusions have been reached by Stabeno and Overland (2001) and Ukita et al. (2007). The authors particularly note the sharp contrast between the factors influencing the winter ice edge and those influencing the summer minimum ice extent, where the downwelling longwave radiation flux dominated the anomalies (Francis and Hunter 2006).

Other recent studies of the variation of ice conditions with time in a specific location or region in the Arctic include two that focus on the variability of ice extent in the shelf seas to the north of Russia (Mahoney et al. 2008; Rodrigues 2008). The first of these is a slightly more focused version of Parkinson and Cavalieri (2008) in that the study is also based on the passive microwave data set. As a result, the oldest data are again from 1979. Over this period a large reduction in ice extent was noted in all the seas and straits in the Russian Arctic during the summer. In the more southerly Barents Sea a significant reduction also occurred during the winter months. In all the locations examined, the length of the ice-free season increased. The rates of change in the Barents and the Chukchi Seas were particularly high. On the other hand, the Mahoney et al. paper is very different in that its primary data sources are the AARI ice reconnaissance charts. In that these visual observations were started as part of the second International Polar Year in 1932, they offer a glimpse of sea ice variability over a 74-year period. Data sources include observations from polar meteorological stations, from ships, from aerial reconnaissance flights, and, during more recent times, from drifting buoys and satellites. As might be expected, coverage varied from year to year and there are missing years. Nevertheless, over the complete time period examined there has been a gradual decrease in ice extent. However, the retreat has been far from continuous, with a partial recovery occurring between the mid-1950s and the mid-1980s. The data suggest that the mid-1980s were the time of significant transition in that after this time the ice along the Russian coast began a retreat that was similar to that occurring in the rest of the Arctic.

18.2.3 Sea Ice Thickness

The fact that both the maximum extent and the amount of sea ice surviving the summer melt period is decreasing brings to mind a related question: is the thickness of the ice also decreasing? Until very recently, the only data sets that could adequately address this question have been ice draft (~ice thickness) measurements made by upward-looking sonar (ULS) systems deployed on submarines transiting the Arctic Ocean. Additional information on the use and interpretation of ULS data can be found in Chapter 12, section 12.5.3. A general discussion of the advantages and disadvantages of the different possible techniques currently available for measuring the thickness of sea ice can be found in Wadhams (1994, 2000).

Individuals studying data sets from different submarine cruises were well aware of the fact that there was considerable scientific interest in possible thickness change trends in arctic sea ice. However, the problem of piecing together a data set adequate to address this problem was far from simple. First, data from a number of different cruises were required in order to obtain observations over an appreciable time span. Moreover, to make meaningful comparisons, the observations had to be made from the same general geographic region and at the same time of the year. Otherwise, it would be impossible to separate temporal variations occurring both within and between years from spatial variations. Such a data set was first assembled by McLaren et al. (1994) by utilizing the fact that U.S. submarines on arctic cruises invariably included the North Pole in their itinerary. As a result they were able to combine the ULS observations obtained on 12 different cruises conducted between 1958 and 1992. Track lengths varied from 50 to 338 km, with the majority of the cruises occurring in April or May (i.e., late winter–early spring). An additional advantage of the North Pole location was that seasonal variations would be expected to be small. The study found a mean ice thickness of 3.6 m and a large year-to-year variability ranging from 2.8 m in 1986 to 4.4 m in 1970. Large changes in the mean ice draft and in the amount of open water/new ice were observed occurring on timescales as short as one year. A linear regression on a plot of the mean ice thickness vs. time gave a slope (-0.0126 m yr^{-1}) that was not statistically significant. The scatter was appreciable. Portions of this data set were also examined by Shy and Walsh (1996) who, although able to explain some of the thickness changes through ice dynamics, also concluded that there was no clear temporal trend.

In the same volume as the McLaren et al. (1994) paper, Wadhams (1994) contributed a less North Pole–focused look at the same problem, discussing ULS results obtained from the Canadian Basin, the area north of Greenland, the Transpolar Drift Stream, the Eurasian Basin, and the North Pole. Again, appreciable interannual variations were noted. In most of these regions there was no convincing evidence for a significant decrease in ice thickness. The exception occurred in the triangular region that extended north of Greenland to the Pole. There he found that when data obtained during the fall of 1976 were compared with results from May 1987, there was a 15% decrease in mean draft (Wadhams 1990). He suggested that the primary difference resulting in the change in draft was the replacement of old and ridged ice by young and FY ice. He ultimately concluded that although he believed the 15% decrease to be statistically significant, there was still no conclusive evidence for the systematic thermodynamic thinning of the ice cover. However, he felt that his results were sufficiently suggestive to justify further thickness profiling.

Clearly what was needed was systematic ice thickness sampling over a period of several years. In that the only viable technique available was the use of ULS deployed on a nuclear submarine, arranging such a program was far from simple. The cruises would have to provide repeat sampling of selected critical areas and the analyzed results would have to be made available to the scientific community. Fortunately, such arrangements proved possible (Gossett 1996). The resulting U.S. Navy program, termed the Scientific Ice Expeditions or SCICEX, consisted of three end-of-summer cruises (September 1993, September–October 1996, and September 1997). To date there have been three different studies using the SCICEX data set. In the first of these, the SCICEX results

Figure 18.6. (a) Mean ice drafts at the crossings of earlier cruises (1958–1976) with SCICEX cruises made in the 1990s (Rothrock et al. 1999). The early data are indicated by open triangles and the 1990s data by solid squares. Both data sets are adjusted to 15 September, with the small dots showing the original data point before the adjustment. The crossings are grouped into six regions and named appropriately. (b) Changes in the mean draft at cruise crossings (dots) from the early data to the 1990s. The change in the mean draft for all crossings in each region is indicated by a large diamond.

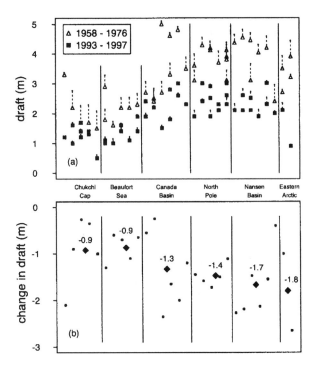

were compared with draft measurements taken on cruises made between 1958 and 1976 at approximately the same time of the year (August 1958, August 1960, July 1962, August 1970, and October 1976; Rothrock et al. 1999). Sampling points were selected where the older cruises either crossed or ran in close proximity to cruise tracks made in the 1990s. All drafts were adjusted downward based on the difference between the data collection date and the assumed minimum date of 15 September. The adjustment was made based on a modeled annual thickness cycle derived via the use of a 12-category ice thickness ice–ocean model (Zhang, Rothrock, et al. 1998). This adjustment reduced all measurements, with the largest adjustment being 0.6 m. As the dates of the pre-1977 data sets were in general temporally located further from the 15 September minimum, this meant that the early data were reduced the most. This in turn meant that if the 1990 thicknesses were less than those from the earlier years, the correction would tend to minimize the differences. As can be seen in Figure 18.6, the differences in the data are striking both in magnitude and in the consistency of their sign. Overall the mean draft has decreased by 1.3 m, or 40% from that of the earlier data set. There were regional differences, with the largest change occurring in the eastern Arctic and the Nansen Basin (−1.7 to −1.8 m) and the smallest in the Beaufort Sea and Chukchi Cap areas (−0.9 m). Because the mean thickness calculations include portions of the tracks that are open water, the authors note that the thinning appears to be minimally related to changes in the open-water fraction and are primarily the result of changes in draft.

This paper was followed in 2000 by a September–October study by Wadhams and Davis (2000), who compared ULS data obtained during 1976 and then during 1996 on British submarine cruises across the Eurasian Basin between Fram Strait and the North Pole. After adjustments were made to a common date of 15 September, it was found that the mean draft had decreased from 3.1 m to 1.8 m during the 20-year period separating the cruises. In that this was a decrease of 42%, a figure essentially identical with the Rothrock et al. (1999) value of

40%, these two studies served to verify each other, indicating that these changes were not only surprisingly large but real.

The third study (Tucker et al. 2001), and the second using the SCICEX data set, was different in that it contrasted the results of nine spring cruises, one of which occurred in 1976 and the remainder between 1986 and 1994. All the cruises occurred in April except one that occurred in May 1988. The advantage of using spring data is that during this period of maximum thickness, thermodynamic changes in thickness are small. Therefore, the thickness adjustments used by Rothrock et al. and by Wadhams and Davis were not required. In addition, the fraction of open water has little influence on mean ice draft during this time period. The area sampled was a swath extending from off the Beaufort coast of Alaska to, and slightly over, the Pole. The ice proved to be appreciably thinner in the early 1990s than in 1976, 1986, and 1987, with a mean decrease of −1.5 m. During the 1980s the larger mean drafts (> 3.5 m) appeared to be primarily the result of a larger fraction of deformed ice. As also reported by several other authors, ice draft was generally found to increase poleward. During some years. the north/south variability was large. The proportions of deformed ice observed in the 1980s (> 64%) were found to be at least 25% larger than the proportions of the same ice type observed during the 1990s. At the Pole itself, the mean thicknesses did not suggest the presence of an appreciable trend in agreement with the earlier observations of McLaren et al. (1994). The question then became why the ice was thicker in the 1980s in the western Arctic, and why this difference was not apparent at the Pole. The authors suggest the following possibility. It is known that during the 1976 and early 1980 cruises the AO index was generally negative, indicating an anticyclonic circulation regime in the Arctic (Proshutinsky and Johnson 1997). During 1988–1989 the index shifted to more positive values and remained positive through the mid-1990s. Such a shift is linked to a weakened anticyclone and a reduced or nonexistent Beaufort Gyre. It is also known that anticyclonic wind forcing favors convergent ice motion and increased deformation (Walsh et al. 1996). Longer residence times in the gyre also allow more deformed ice to accumulate, a result supported by the correspondence between mean ice age as determined from buoy motions (Colony and Thorndike 1985) and mean ice draft (Bourke and McLaren 1992). During the alternate situation in the 1990s, because ice is more rapidly advected out of the Basin and less deformation occurs, large amounts of deformed ice do not form. Buoy motions suggest (Tucker et al. 2001, Fig. 5) that the North Pole lies in a strongly advective regime during both periods. When the gyre is strong, much of the ice crossing the Pole originates in the Laptev Sea; when it is weak more of the ice originates in the East Siberian Sea. In that positive AOs are also associated with warmer arctic air temperatures, the above ice dynamics mechanism is also supported by the atmospheric thermodynamics.

The next study using the SCICEX data set (Yu et al. 2004) was a more detailed analysis of the same set of cruises examined by the Rothrock et al. (1999) study. However, the two studies differ in that the 1999 paper focused on changes in average ice thickness whereas the 2004 paper examined changes in the thickness distributions, as a decrease in mean thickness can result either from an increase in the percentage of thin ice or from a decrease in the amount of thicker ridged ice, or both. The 2004 paper also included an expanded treatment of procedural details. In their analysis its authors explored the temporal changes in the ice thickness distributions as observed in four different

Figure 18.7. (a) Ice volume as a function of ice thickness averaged over the four regions studied (Chukchi Cap, Canadian Basin, eastern Arctic, and central Arctic) during each of the two time periods. (b) Corresponding cumulative volume as a function of ice thickness (Yu et al. 2004).

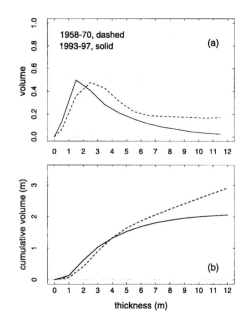

regions of the Arctic Ocean. Although there were distinct regional differences, here I will only summarize their results using overall averages for the whole Basin. The volume of 0–1 m ice was about twice as large during the 1990s as during the earlier time period. The volume of 1–2 m ice also increased by 39%. In addition, the fractional area covered by open water and FY ice increased from 0.19 to 0.30. There was also a net volume loss from all categories of ice thicker than 2 m, as shown in Figure 18.7a.

The total net loss of volume between 1958–1970 and 1993–1997 was 32%. (The 8% difference between this value and the 40% value reported earlier by Rothrock et al. [1999] is the result of additional Chukchi Sea tracks being included in the Yu et al. paper.) As the volume of ice thinner than 4 m remained essentially unchanged, this implies that all the changes in ice volume are primarily due to losses of ridged and MY ice. Figure 18.7b also shows that volume reduction increases with increasing ice thickness. As the authors point out, this pattern clearly implies a long-term depletion of ridged ice through either bottom melting and/or decreased deformation. The Yu et al. paper also includes detailed discussions of how each of the following possible drivers could contribute to the observed loss of ice thickness within the Basin. Processes considered include changes in ice advection patterns, in ice export, in growth and melting, and in ice deformation. They conclude that, at present, the ice thickness data in themselves are insufficient to establish the relative importance of changes in the thermodynamic and dynamic forcings. They also note that this distinction is important in that variations in ice export can potentially reverse even if the climate continues to warm.

The most recent paper in this series is by Rothrock et al (2008), who use data from 34 U.S. submarine cruises that span the years between 1975 and 2000. Again the area covered is the so-called Gore box, an irregular polygon within the Arctic Ocean outside of the exclusive economic zone of non-U.S. countries. The data was extracted using digital processing techniques (Wensnahan and Rothrock 2005). Earlier data extracted using manual digitization were not used because of a possible positive bias resulting from the difficulty of resolving the troughs between ridges. The final data set consisted of 2203 50-km mean draft values, with values ranging from 0 to 6.09 m. In calculating the means, open-water values were included. The great majority of the data collection occurred during the spring and the fall. Although this data set spans a shorter period than those of some previous studies, the resulting data set is larger and is believed to be of higher quality. The overall mean draft was 2.97 m, with the mean declining from a peak of 3.42 m in 1980 to a minimum of 2.29 m in 2000. This corresponds to a thickness change of 1.25 m. Although this thickness decrease of 36% from the maximum is less than the 43% decline reported earlier by Rothrock et al. (1999),

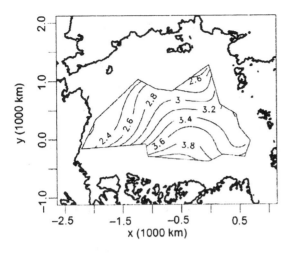

Figure 18.8. The spatial field of mean ice draft in meters as determined from ULS observations averaged over both the 26 years between 1975–2000 and the annual cycle (Rothrock et al. 2008).

that value was a comparison made using data from an earlier period (1958–1976). The annual thickness cycle was 1.12 m. A contour map of mean draft averaged over the 26 years covered by the study shows values ranging from a maximum of 4 m north of Ellesmere to lows of 2.2 m off the Alaskan Beaufort coast (Figure 18.8). Although the general trends shown in this map are similar to trends suggested by previous authors (see Figures 16.28b and 16.30), a number of specific differences are noted and discussed.

In addition to the ULS measurements, there are a few observations that compare ice thicknesses at the same approximate geographic regions during periods separated by a number of years. The most detailed of these is the paper by McPhee et al. (1998), who contrasted ice conditions at the SHEBA site in the Beaufort Sea during 1997 with observations made in the same general area during 1975–1976 as part of AIDJEX. This location, which is within the Beaufort Gyre, has historically been ideal as a site for establishing drift stations in that it traditionally has included some of the oldest and most compact concentrations of MY ice found in the Arctic Ocean. Based on the AIDJEX observations, MY ice would be expected to be common, with mean floe thicknesses near 3 m. In fact, however, the above authors were impressed by the lack of MY ice in the vicinity of the SHEBA site and found difficulty in locating floes thicker than 1.5 m. The relatively thick floe on which the initial SHEBA site was located only rarely exceeded 1.8 m in thickness. An interesting aspect of this study was the comparison of the changes in the upper (0 to 500 m) ocean temperature with salinity profiles that occurred over the 22-year span separating the observations, in that the upper ocean tends to be more spatially uniform than the ice. During SHEBA the upper ocean was both warmer and less saline than would be expected based on the AIDJEX observations. The salinity deficit in the upper 100 m of the water column was equivalent to a freshwater input of 2.4 m, whereas the heat content of the same layer increased by approximately 67 MJm^{-2}. Assuming that the freshening was the result of the melting of local sea ice, these values translate into ~0.8 m of melt during AIDJEX and 2.0 m during SHEBA, an increase of 2 1/2 times. These are large values given that a 2.5 m melt would eliminate much of the pack ice in this region at the end of the summer melt season. Although these results could be looked at cynically as a trend based on two data points separated by 22 years, the suggested trend is clearly in line with the other observations discussed here.

The trends suggested by the above ULS observations have recently been strongly supported by helicopter-borne electromagnetic thickness measurements carried out over the Transpolar Drift Stream since 1991 by Haas et al. (Haas 2004; Haas et al. 2008). These essentially ground-based observations indicate an ongoing reduction of the mean ice thickness by as much as 53% between

2001 and 2007. Furthermore, the results in conjunction with buoy-based ice age model calculations again suggest that this change is predominantly the result of SY and MY ice being replaced by FY ice that has a mean summer thickness of only 1.27 m. Observations near the North Pole indicate a mean thickness of the surviving SY ice during the late summer to be only 1.81 m. The authors suggest that these trends will result in the near future in an ice-free North Pole during the late summer.

An examination of ice thickness variations up to 81.5°N during the time period between the winters of 2002–2003 and 2007–2008 has recently been completed by combining radar altimetry observations from several different satellites (Giles et al. 2008). The results indicate that after the melt season of 2007 the mean ice thickness was 0.26 m below the 2002–2003 to 2007–2008 average. Furthermore, in the western Arctic the decrease was 0.49 m below the six-year mean. More surprising was the fact that there was no suggestion of short-term preconditioning between 2002 and 2007. However, after both of the ice extent minima in 2005 and 2007 a thinning was noted, particularly in the western Arctic. I find this to be a bit odd in view of the trends shown in Figure 18.3. Perhaps some of these differences are related to differences in the maximum latitude covered.

If the Arctic Ocean becomes seasonally ice free during this century, how long has it been since it last exhibited a similar state? This question has until recently been unanswerable owing to a lack of long cores from bottom sediments obtained at central locations where heavy pack ice has been present throughout the year. Fortunately, recently (2004) the Arctic Coring Expedition (ACEX), by using two icebreakers, has been able to obtain cores up to 410 m in length at sites within 220 km of the North Pole. Preliminary analysis of the cores suggests that the age of the oldest material obtained was 80 M yr BP. The most interesting portion of the core to date is 55 M yr old, spans the Paleocene–Eocene Thermal Maximum (PETM), and contains plant and animal microfossils characteristic of +20°C subtropical waters in contrast to the 0 to −1.8°C temperatures experienced today. It is believed that during the PETM deep ocean temperatures were as much as 10°C warmer than at present and that at the start of this period the high-latitude oceans warmed by 8 to 10°C in less than 10,000 years (Zachos et al. 2001). Clearly the Arctic Ocean was ice free during this period. These are temperatures that are appreciably higher than estimates using current climate models, even if extreme CO_2 levels are assumed. The current leading hypothesis for this event is the release of methane gas from clathrate material on the seafloor that had become unstable as the result of the increase in the water temperature (Dickens et al. 1995), although other possibilities have been suggested. However, Schrag and Alley (2004) note that although the release of methane may have acted as a trigger, the duration of the PETM (50,000 to 200,000 years) makes methane, because of its short atmospheric half-life, unlikely as a sole agent. They suggest that the warming was mainly caused by an increase in atmospheric CO_2. Clearly, this issue is far from resolved. Preliminary examination of the ACEX cores indicates that the first clear signs of the presence of ice after the PETM (bits of sand that were apparently ice-rafted to these mid-ocean locations) occurred in 46 M-yr-old sediments. Somewhat later, at between 13 and 14 Ma BP, the delivery of iceborne material increased significantly, indicating a heavier, thicker year-round ice cover (Kerr 2004b). Considering the above, it is clear that the Arctic Ocean has had an extensive sea ice cover for at least ~13 M yrs. If the current trend continues, the Arctic will enter a very different regime during the lifetime of many readers of this book with essentially ice-free periods during the summer.

18.3 Antarctic

18.3.1 *General*

In discussing climatic changes in the Arctic, I pointed out that although the great majority of the Arctic is clearly exhibiting a pronounced warming trend, there are some areas, specifically northern Russia, where a slight cooling trend has occurred during the last 10 years (Comiso and Parkinson 2004). A similar but more pronounced situation prevails in the Antarctic. In this case the contrast is between the Antarctic Peninsula/West Antarctica, where the largest positive temperature changes recorded anywhere in the southern hemisphere have been observed, and East Antarctica, where a slight cooling trend has prevailed. The changes on the Peninsula are the most striking. Its climate has warmed by +3.4°C during the last century and by +2°C in the last 30 years. This is a rate of more than five times the global mean. As will be seen, the climate trends affecting antarctic sea ice are even more interesting than those occurring in the Arctic.

Perhaps the most striking change on the Peninsula is the continuing retreat of the ice shelves. Cook et al. (2005), who have summarized these retreats, have found that 87% of the 244 marine glacier fronts studied have retreated, and that the boundary between mean advance and retreat has steadily moved southward during the recent half-century. Furthermore, the scale of the retreats is generally large compared to the scale of the advances. Some of the retreats, such as the disintegration of the different portions of the Larsen Ice Shelf, have been both rapid and spectacular (Scambos et al. 2000). The reason for the particular interest in the stability of such shelves was the speculation, based on fundamental ice mechanics (Thomas 1973; Thomas and MacAyeal 1982), that ice shelves exert a back-pressure, restraining the flow of ice from the interior onto the shelf. This would be particularly true if there were significant friction between the ice shelf and its bounding walls, and/or if the ice shelf contained an ice rise (i.e., was locally grounded). In short, once the effect of the shelf is removed, the flow rate of the glacier that feeds the shelf increases, potentially leading to the depletion of the upstream accumulation basin. Recent satellite observations of glacier flow velocities in the Peninsula have shown that in the few glaciers studied, this has occurred (De Angelis and Skvarca 2003). Although the recent spectacular retreat of the Larsen B Ice Shelf appears to have been unique to the Holocene, resulting from long-term thinning and increased surface melting, shelves on the west and northeast parts of the Peninsula have experienced earlier retreat events associated with warm periods in the early to middle Holocene (Hodgson et al. 2006). Current research suggests that ocean temperatures are also a major factor affecting shelf stability. Recent radar data indicate that as much as 2.78 ± 0.08 m of ice is ablated every year from the base of the George VI Ice Shelf. This shelf, on the western side of the peninsula, is located both further south than the Larsen Shelf and south of what would appear to be the present-day limit of ice shelf stability. Yet it is known to have retreated in the early Holocene at a time when the Larsen remained stable. There are currently several hypotheses that have attempted to explain these differences, including an intensification of the climatic contrast between the two sides of the peninsula. Both atmospheric and oceanic forcing mechanisms would appear to be operating at different times and scales in order to produce the retreat patterns both at the present and in the recent geologic past. I also note that as I am in process of editing this chapter (July 2008), The National Snow and Ice Data Center (2008) has just announced that the Wilkins Ice Shelf has started to break up. This shelf, which is located on the west side of the Peninsula, is located at approximately the same latitude as the George VI Ice Shelf, although in a somewhat less protected location. It is easy to suggest some possible contributing factors: thinning in a manner similar to that in the George VI Shelf, plus flexural failures associated with increased exposure to wave action, resulting from the fact that regional sea ice has now essentially vanished during the austral summer (see Chapter 14).

18.3.2 Sea Ice Extent

In light of the above, what is happening to sea ice in the Antarctic? Our knowledge of the antarctic pack, based on passive microwave observations between 1979 and 2006, is as follows (Cavalieri and Parkinson 2008). (Incidentally, this paper is the antarctic flip side of Parkinson and Cavalieri's 2008 study of sea ice extent variations in the Arctic. As such the two papers comprise a self-consistent bipolar data set.)

Sea ice extent varies significantly seasonally from a maximum value in September of $18.9 \times 10^6 \, km^2$ in 2006 to a minimum value of $2.46 \times 10^6 \, km^2$ in 1993. This amounts to a maximum difference of ~87% decrease in the area covered by sea ice between austral winter and austral summer. It also means that only a small amount of ice (< 20%) survives the summer melt period to become old ice. This is a larger annual variation than observed in the Arctic where, even in the extreme minimum year of 2007, 28% of the March maximum extent remained at the end of the summer ($14.66 \times 10^6 \, km^2$ in March and $4.13 \times 10^6 \, km^2$ in September). There is also an appreciable variability in sea ice extent both between years in the same sea and between seas in the same year. The trends are as follows. The total antarctic sea ice extent trend increased slightly over the 28-year period studied (+1.0 ± 0.4% per decade) from the value obtained earlier by Zwally et al. (2002) of +0.96 ± 0.61% per decade) for the 20-year period between 1979 and 1998. Not only is the trend positive, in contrast to the consistently negative trends observed in the Arctic, it increases slightly as the more recent data are added. The changes in the trends, as shown by the different sectors of the Southern Ocean as designated in Figure 18.9, are particularly interesting. For example, the eight additional years of data resulted in smaller positive yearly trends in ice extent for the Weddell Sea (+0.80 ± 1.4%/decade), the Western Pacific Ocean (+1.4 ± 1.9% per decade), and the Ross Sea (+4.4 ± 1.7% per decade), a shift from a negative to a positive trend for the Indian Ocean (+1.9 ± 1.4% per decade), and a lessening of the negative trend for the Bellingshausen/Amundsen Seas (5.4 ± 1.9% per decade). The positive trends for the southern hemisphere as a whole and the Ross Sea sector proved to be significant at the 95% level. On the other hand, the negative trend for the Bellingshausen/Amundsen sector was significant at the 99% level. The seasonal trends in the Weddell Sea are particularly interesting, with winter and spring sharing negative values whereas summer and autumn values are positive.

Finally, additional documentation of the contrast between sea ice advances and retreats in the Peninsular/Bellingshausen Sea (P/BS) region and that in the Western Ross Sea (WRS) can be found in Stammerjohn et al. (2008). Over the time span 1979–2004, in the P/BS region sea ice retreated 31 ± 10 days earlier and advanced 54 ± 9 days later, resulting in a loss of 85 ± 20 ice season days. In contrast, in the WRS area just the opposite occurred, with the retreat occurring 29 ± 6 days later and advancing 31 ± 6 days earlier, resulting in an increase of 60 ± 10 ice season days. Clearly

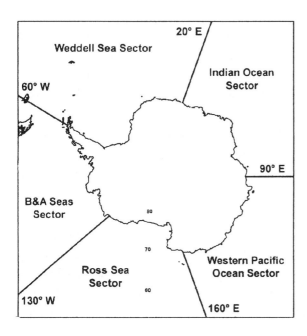

Figure 18.9. Southern hemisphere sector map utilized by Cavalieri and Parkinson (2008).

current changes in sea ice extent (and also in area) are both very different and possibly more complex than what is occurring in the Arctic.

18.3.3 Sea Ice Thickness

There are no submarine-based ULS data sets on sea ice draft available for the Antarctic. There are several reasons for this absence. First and most important, the Antarctic Treaty prohibits the presence of nuclear systems in the Antarctic. Furthermore, even if nuclear-powered submarines were allowed, operating them as ice draft profiling systems would be extremely difficult owing to the frequent presence of deep-draft tabular icebergs. As a result our current knowledge of ice thickness variations in the antarctic pack is far less exact than in the Arctic.

Our present knowledge is primarily based on two types of information. The first type of thickness observation is obtained by either drilling or coring. Although such observations are locally quite precise, obtaining them is slow and labor-intensive. Therefore, such measurements are invariably limited both spatially and temporally. Another possibility would be to let a ship assist in the making of observations. When a breaker is moving through pack ice, it is frequently possible to obtain thickness estimates by observing the thickness of blocks that have been rotated 90 degrees during the breaking process. Although such observations are admittedly a bit rough-and-ready, they do provide a general picture of the variations in ice thickness as the breaker moves ahead. Based on my experience in assisting with such observations, this procedure works well when a breaker is moving through ice that is less than a meter thick. In thicker ice the procedure is less successful as both the speed and direction of the motion through the ice become highly variable. Also, the breaking process can produce horizontal splitting failures in thicker ice that result in thickness estimates that are artificially low. Finally, FY deformation features are frequently poorly "cemented" together. As a result, the thicknesses that are observed are frequently those of the blocks in the ridge instead of the thickness of the ridges.

The results have just been published of a major multinational program (ASPeCt) to collect and analyze this type of ship-based ice and snow thickness data collected between 1981 and 2005 (Worby et al. 2008). In addition to useful data summaries this paper contains detailed descriptions of both observational and analysis procedures. The overall mean thickness for undeformed sea ice was found to be 0.62 m whereas the long-term mean and standard deviation for all observations, including ridged ice, were 0.87 and 0.91 m, respectively. The extreme variability of the observations is attested to by the fact that the standard deviation is larger than the mean. As expected, the thickest mean ice thickness (1.33 ± 1.13 m) occurred in the western Weddell Sea. Annual mean snow thick-ness values, which were only taken on level ice, ranged between 0.11 and 0.24 m.

Of course, there are other techniques for measuring ice thickness (see for example Wadhams 1994, 2000). Of these, the only other technique that has been used successfully in the Antarctic is bottom-mounted ULS. Unfortunately at present there appears to be only one example: be-tween 1990 and 1994 Strass and Fahrbach (1998) were able to obtain up to two-year-long ice coverage and thickness profiles using bottom-mounted ULS systems at six sites in the Weddell Sea. As might be expected there were appreciable differences in the resulting draft profiles at the different instrumented sites. Also, one should note that the Weddell is hardly typical of the seas that surround the Antarctic. Nevertheless, these results support the suggestions of other authors discussed above that the maximum thickness of undeformed FY ice at locations off the antarctic coast is usually less than 1 m. In addition, all of the draft probability distributions show pronounced positive skews similar to those observed in the Arctic and that are believed to be the result of deformation. The authors also suggest that the monotonic increase in ice draft during the freezing season is indicative of rafting while the ice is still quite thin. Although I do not find this argument to be definitive, the suggestion is in agreement with other field observations (see Chapter 12). Current knowledge of age and ice thickness variations in older ice types is unfortu-nately still poor, making it difficult to arrive at any generalizations.

What does all the above tell us about trends in sea ice thickness? The answer is very little. Considering the very large variability, the data are clearly too sparse and temporally limited to establish a trend. The locations where there would appear to be the best chance for a decrease in ice thickness are the Amundsen and Bellingshausen (A&B) Seas. At present, all we can say is that because the ice extent there is decreasing, it would not be at all surprising to see an associated decrease in thickness. As for hard data supporting this conjecture, at present there appears to be none. Perhaps the offbeam lidar technique currently being developed by Varnai and Cahalan (2007) will prove to be useful in filling this data gap.

18.4 Causes and Predictions

So what is to be made of all of the above? Let us start up north, where we have a better obser-vational data set. What is occurring there is obviously complex and, as yet, only partially under-stood. In the atmosphere, during the period between the 1950s and the 1970s, the Arctic Oscil-lation (AO) index, a descriptor of the atmospheric pressure field and a measure of the strength of the circumpolar winds, was, with rare exception, consistently negative. These are values indica-tive of an anticyclonic (clockwise) atmospheric pattern and a strongly developed Beaufort Gyre, a situation favoring the development of MY ice. Between 1970 and 1989, the AO weakened and appeared to oscillate between positive and negative values until 1989, when the values became strongly positive. This indicated an atmospheric circulation favoring a weakened Beaufort Gyre, a shift of the Transpolar Drift Stream towards the Pole, and more rapid exit of ice from the Basin. The strong positive values continued until 1997, when low AO values that oscillated between positive and negative returned, a situation similar to that observed between 1970 and 1989. This roughly neutral pattern has continued to the present (i.e., December 2008). The most recent study of these changes (Zhang, Sorteberg, et al. 2008) concludes, based on the use of involved statistical techniques, that during the last 20 years there has been a systematic shift toward the northeast in the locations of the center of maximum variance in the pressure field from over the Icelandic Sea to the Eurasian arctic coast. This finding provides further support for the observations discussed earlier, to the effect that there is an increasingly efficient transport of heat into the Arctic from more southerly locations. Furthermore, the nature of the pressure pattern appears to have changed during the last five years (2003–2008) from a tripolar pattern characteristic of the AO to a dipole pattern that the authors refer to as the Arctic Rapid change Pattern (ARP).

As noted earlier, in the ocean the 1990s were a period when the seawater temperature and salinity fields appeared to change dramatically relative to earlier values as reported by the Environmental Working Group (EWG). These were changes that generally favored a thinner, less extensive, sea ice cover. However, measurements made during 2000 to 2006 indicate that at least at some regions within the Arctic Basin such as the North Pole, the oceanographic conditions favored during the 1990s appear to have relaxed to a climatology similar to that observed pre-1990. This change has been interpreted as a first-order response to the changes in the AO with a five-year time constant and a three-year time delay (Morison et al. 2006). During the same time period, the hydrography of the Canadian Basin also changed, but in an opposite direction to the changes observed at the Pole, as the salinity of the upper layer has decreased a bit. In fact, since 2000, the temperatures of both the Pacific and the Atlantic waters in the Canadian Basin have been higher than during the 1990s and 0.8–1.0°C higher than the EWG climatology (Proshutinsky and Morison 2007). Also, although the total freshwater content of the Beaufort Gyre, the major freshwater reservoir in the Arctic Ocean, has not changed drastically relative to the EWG climatology, the center of the Gyre has shifted towards Canada and intensified. The heat content of the Gyre has also increased as the result of the increase in the temperature of the Atlantic water layer. Not surprisingly, it has been suggested that the pronounced reduction in sea ice in this region in 2005 (and also in 2007) is, at least partially, the result of the release of heat from this layer (Shimada et al. 2004).

In that the meteorological circulation in the Arctic Basin as indicated by the AO appears to have returned to its pre-1990s pattern, one might expect that the oceanography would follow suit as soon as any excess heat acquired during this period was dissipated. However, as noted, this does not appear to be the case in that pulses of warm Atlantic water have continued, at least through 2004, to enter the Basin in oceanic events similar to those observed during the 1990s. Even if the change in the temperature of the Atlantic water entering the Basin were to respond instantaneously to AO changes, transport times within the Basin are measured in years (Polyakov et al. 2005). For example, Kikuchi et al. (2005) have reported a seven-year lag between water temperature changes measured at the Pole and related changes as observed in the Makarov Basin. An initial analysis of baroclinic response, transport lags, and AO declines can be found in Morison et al. (2006), who conclude that the slow recovery of the central Arctic Ocean hydrography to climatological EWG values with the decline of the AO appears to be consistent with the response of the hydrography to the rapid rise in the AO in observed in 1989.

As far as sea ice is concerned, the long-term consequences of the above events are, as yet, not clear. One might expect the return of AO values to those observed before 1990 to at least slow the current precipitous decrease in the amount of old sea ice in the Basin. The effect of the ocean on all this seems to be particularly unclear. One uncertainty is how much effect, if any, the additional AW heat has had to date on the sea ice cover. Another uncertainty is how long and at what level will additional heat continue to enter the Basin from the North Atlantic. To add to the confusion, current data indicate that the Bering Sea is currently cooling from the temperatures observed during 2000–2005, when values were +2°C warmer than earlier (Stabeno et al. 2002), with the winter of 2006–2007 being a relative extensive ice-year (Overland et al. 2007).

My prognostication is as follows. Considering the general worldwide atmospheric warming trend plus the fact that unusually warm AW is continuing to flow into the Arctic Basin via the WSC, the amount of sea ice surviving the summer will continue to decrease. For the thick MY ice pack to start to reestablish itself would require the AO to remain negative and the temperature of the AW entering the Basin to return to pre-1990 values. Although such a situation remains possible, considering current warming trends it hardly appears probable. If the North Atlantic continues to pump heat into the Arctic, there is currently no consensus as to the rate that this will affect the ice cover. In the short term there may be little, if any, effect. However, it is clear that in the long term such a situation will not favor either the expansion or the thickening of the ice pack. Although this is true, based on oceanographic data collected over the last hundred years there is little basis for the expectation that oceanic fluxes from the Atlantic will remain relatively constant. In fact, during the 20th century AW variability has been dominated by low-frequency oscillations occurring on timescales of 50–80 years with warm periods recently and during two periods in the 1930s, and two cold periods early in the century and in the 1960s–1970s (Polyakov et al. 2004). That said, when this hundred-year period is looked at as a whole, there does appear to be a general warming trend, particularly if data from the most recent decades are included.

In addition, we now have direct observational confirmation of the long-assumed "fact" that the ice–albedo feedback is currently adding enough heat to the near-surface environment of the Arctic Ocean to result in a significant decrease in the amount of old ice surviving the summer melt season. The summer of 2007 has also shown us that if the meteorological and oceanographic factors are appropriately aligned, a drastic decrease in the amount of old ice present in the Arctic Basin can occur during one year. Does the above result in an inevitable path to a sea ice–free summer, as suggested by some authors? I would suggest that it does, unless there is a drastic and currently unanticipated change in the trends. Certainly more oceanographic detail is needed. I also would have to agree with the opinions of Perovich et al. (2007) and Yu et al. (2004) that there is still not enough information available to confidently partition the different thermal contributions to the present sea ice decline. However, at present this information lack seems to be immaterial when the end result of current trends is concerned. The current choice as to when the Arctic Basin becomes essentially ice free by the end of the summer would appear to be between soon—2070 (Holland et al. 2007)—and even sooner—2030 (Stroeve et al. 2007). As of December 2008, NSIDC has announced (http://www.nsidc.org) that the 2008 minimum sea ice extent for the Arctic was 4.52×10^6 km². This is the second-lowest value reported since the time series based on satellite observations was initiated in 1979. Furthermore the 2008 minimum is 15.0% less than the next-lowest minimum (2005) and 33.1% less than the average minimum during the time period from 1979 to 2000. In that environmental conditions during 2008 did not appear to favor a low ice extent, the 2008 value can be taken as support for the contention that we are now on the downside of a tipping point leading to an essentially ice-free Arctic Ocean during the late summer–early fall. I also would not be surprised if further analysis indicates that the end-of-the-summer ice volume in 2008 is less than in 2007, in spite of the fact that the 2008 end-of-the-summer ice extent is slightly larger.

As far as the Antarctic is concerned, there is a pronounced but lessening decrease in sea ice extent in the A&B Seas, both locations to the west of the Antarctic Peninsula. Elsewhere, the sea ice area appears be expanding slightly. In the A&B region there would appear to be two factors at work. As noted above, the Peninsula is the one area in the Antarctic where there has been a pronounced atmospheric warming. In fact, recently a very interesting hypothesis has been advanced as to why the peninsular region might be warming while the rest of the Antarctic is cooling slightly (Thompson and Solomon 2002). This explanation posits that stratospheric cooling from springtime ozone depletion in the upper atmosphere favors the positive phase of the Southern Annular Mode (SAM) characteristic of a cooler climate over most of the antarctic coastline and a warming over the peninsula. The observed sea ice decrease in the vicinity of the Peninsula is believed to be the result of this atmospheric warming, plus the fact that there is recent evidence that warmer, deeper water comes to the surface to the west of the Peninsula (Martinson 2005). To date, the relative contributions of these two factors have not been worked out. Based purely on thermodynamic considerations I would be very surprised if there has not also been a commensurate thinning of the sea ice in the western portion of the Weddell Sea. Clearly the reason that this area remains a region of heavy pack ice during the austral summer owes more to the local dynamics, including the blocking effect of the Antarctic Peninsula, than to the thermodynamics.

Assuming that the Thompson and Solomon theory is correct, it implies that the reason that the East Antarctic sea ice extent is slightly positive is to some extent the result of human activity in that the ozone hole is definitely the result of the atmospheric effects of CFCs used primarily as propellants in spray cans. As CFC introduction into the atmosphere has been greatly reduced since the Montreal Protocol was signed in 1989, it is currently estimated that the austral ozone hole will essentially disappear by ~2050. When this occurs, one possible consequence could be a reversal in the positive trend of the SAM index, presumably resulting in a reduction in the warming rate as observed in the Peninsula and more direct greenhouse gas impact on air temperatures (increases) and on sea ice extent (decreases) in the seas off of East Antarctica (Overland et al. 2008). This possibility is also suggested by recent computer simulations that include consideration of changes in stratospheric chemistry, including the recovery of the ozone layer (Son et al. 2008). However, the magnitudes of these changes remain to be determined, as do the exact mechanisms involved. Whatever happens, one can be certain that considering the thinness of antarctic pack ice, its response will be essentially instantaneous.

One fundamental question posed by the behavior of the non-A&B portion of the Southern Ocean's sea ice cover is this: considering the general warming atmospheric and oceanic conditions, why is there an expansion in sea ice extent (and area)? One possibility suggested by a modeling study carried out by Zhang (2007) involves a reduction in convective overturning beneath the ice, resulting in a reduced ocean heat flux. Such a change could lead to an increase in sea ice extent and volume. Whether this is actually what is occurring remains to be determined.

Although I was willing to speculate a bit about the future of the Arctic's sea ice cover, I believe that the future of the antarctic ice pack is far less clear. That said, it is hard to believe that the expansion of the non-A&B portion of the Southern Ocean's sea ice cover can be maintained if climatic warming continues. I also note that the current expansion rates are small and in several regions appear to be decreasing.

I feel that the one encouraging piece of data in this whole picture is that the glacial ice cover of East Antarctica has, to date, appeared to be oblivious to the effects of global warming. It is bad enough that global temperatures have warmed enough to affect the behavior of alpine glaciers worldwide, the southern portion of the Greenland Ice Cap, and the ice shelves of the Antarctic Peninsula. Once it becomes clear that the activities of the human race have affected the behavior of the East Antarctic ice sheet, it will also be clear that the world as we know it is about to undergo a drastic change. The reader should also note that the state of current GCMs in the Antarctic appears to be less adequate than in the Arctic (Connolley and Bracegirdle 2007). At present, there is a wide scatter in simulated temperature trends. A particularly disturbing fact is that the large trend during the austral winter observed in the general area of the Peninsula is not well represented by the simulations. The fact that this is currently the most striking feature in the antarctic climate is not encouraging.

Hopefully the data being collected during the current International Polar Year, in concert with increasingly focused modeling, will assist in resolving these varied questions. In the Arctic, the next three to four years should be very informative. Unless there is a significant shift in the current trend, old, thick sea ice in the Arctic will become a historical curiosity sooner rather than later. To keep abreast of these matters one should check monthly with the National Snow and Ice Data Center (http://www.nsidc.org) and for a yearly perspective (with references) with the NOAA Arctic Report Card (http://www.arctic.noaa.gov/reportcard/). Individuals wishing even more detail concerning the recent sea ice decline in the Arctic should examine DeWeaver, Bitz, et al. (2008).

I guess that my feelings concerning the current trends in sea ice specifically, and in glaciology in general, can be summed up in a two-word quote from the comedian Arte Johnson of *Rowan & Martin's Laugh-In* fame, who frequently proclaimed in a voice reminiscent of Peter Lorre, "Veeeeery interesting."

May you live in interesting times!
Anon, Ancient Chinese curse

19 Conclusions

It is fatal to wait for perfection.
Gen. George Patton

It has taken me a long time to complete this book, considering the fact that I had essentially completed the chapter on the history of sea ice research by the time that I retired in 1996 and it is now 2008. Not very impressive considering that I was told that Peter Hobbs wrote all 836 pages of *Ice Physics* in two years. Now that's impressive! However, in retirement I was singularly lacking in his degree of focus and dedication. Besides, it is a lot more fun to play contrabass and to travel than to sit in one's basement looking at a computer screen and trying to comprehend seemingly endless numbers of reprints. To make certain that I was under no pressure from an editor on a deadline, I also deliberately avoided talking to potential publishers, assuming that there were some. At times I was certain that I was never going to finish. Ultimately it dawned on me that there might be some daylight at the end of the tunnel that I seemed to be crawling down. Moreover, I realized that if I focused a bit I could finish the book during the data collection period of the 2007–2008 International Polar Year. In that the IPY presented a natural break point in the continuum of research on sea ice, it was clearly time to wrap up my literary effort. Besides I have always been a better finisher that a starter.

It is my hope that readers will find the book to be useful. In particular that they will find approaches and references to papers that relate to problems that they are currently confronting. After glancing at the bibliography, the uninitiated might conclude that I have referenced every paper ever written on sea ice. As I noted in the introduction, this is far from the case. There are important active research areas that I have barely mentioned in passing. However, with a little luck the material that I have referenced should aid in bringing at least some of these additional papers to light. So good luck. Besides, digging around in a good library is never a waste of time as you invariably learn something old that is new to you.

When I started this book I had two goals: to close out my career with a broad overarching view of sea ice science and to provide a document that could prove useful to individuals interested in this general subject area. I am sorry to say that I have failed to achieve the first goal. As the book neared completion I realized that although I was learning new things about the subject of the particular chapter on which I was working, I simultaneously was forgetting material from all the other chapters. Unfortunately I seemed to be forgetting faster than I was learning. I can't make a judgment on the second goal. You will have to decide that.

In closing, may I recommend that you follow the advice of the esteemed Mr. Marx as stated below and devote a little thought to the 1920 insight of Professor Eddington.

From the moment I picked your book up until
I laid it down, I was convulsed with laughter.
Someday I intend reading it.
Groucho Marx

We have found a strange footprint on the shores of the unknown.
We have devised profound theories, one after the other, to
account for its origin. At last we have succeeded in
reconstructing the creature that made the footprint.
And lo! It is our own.
Arthur Eddington

Appendix A Symbols

Because there is no symbology convention used in the sea ice literature and because the scientists who work in this field come from a wide variety of backgrounds, a given symbol may be used to represent more than one entity. The following is a listing of most, but not all, of the symbols used in the present book, along with, in many cases, a sample equation where a given symbol appears. In all cases symbols are defined when used for the first time. "(Chapter 6)" indicates that this usage of the symbol only occurs in Chapter 6.

In the text, subscripts are frequently used to indicate the state of parameters as follows: a (air or gas); b (brine); i (ice); l (liquid); si (sea ice); ss (solid salt); w (water, usually seawater).

A	compactness
a	cutoff value in gouge sampling (13.9b)
b	fitted constant (9.21)
C	number of components; concentration (7.8, 8.4); proportionality constant (6.14)
C	cloud fraction (9.8); constant related to the frictional energy losses during ridging (16.39)
C_t	averaged surface heat transfer coefficient (4.8, 9.16, 9.17)
c	speed of light (10.51)
D	diffusion coefficient (7.8); number of gouges with depths greater or equal to d (13.6)
d	sea ice draft (7.21); grain size or diameter (8.23); potential open water (11.8)
E	elastic modulus
e	near-surface partial vapor pressure of water (9.12); emissivity
F	degrees of freedom (Chapter 6); flux of the type designated by the subscripts as follows: F_c conductive (9.1); F_r incoming shortwave radiation (9.3); emitted radiation (9.3); $F_{L\downarrow}$ incoming longwave (9.3); $F_{L\uparrow}$ emitted longwave (9.3); F_s sensible heat (9.3); F_e latent heat (9.3); F_w oceanic heat; F_{clr} clear sky downwelling shortwave (9.6); F_{cld} cloudy sky downwelling shortwave (9.7)
F_g	average area of a brine pocket in the BG plane (10.28)
F_L	lateral melting term
F_{wr}	wave force per unit width (14.1)
$F(x)$	function of the parameter x (e.g., 6.12, 6.18)
f_b	freeboard (10.4)
f_g	vertical growth or decay rate of ice of thickness H (16.35)
G	temperature gradient (7.13)
G_t	fraction of the seafloor gouged since time T_o (13.7)
$\bar{g}$	average number of gouges per km per year (13.8)
H, h	ice thickness
I_o	net influx of radiant energy (9.3)
J	solute flux (8.12)
K	permeability (8.23); brine compressibility (10.43)
k, k_o	solute partition coefficient (6.1, 6.7, 8.3)
k	thermal conductivity (4.6); i pure ice (9.23); si sea ice (9.23)
L	latent heat of fusion (4.6); latent heat of vaporization (9.17)
L_0	lead number (11.14)

ℓ_* characteristic length of the finger rafting pattern (12.2)

M bulk mass

M_w integrated mass transport of the oceanic boundary layer (16.9)

m mass of a component; slope of the liquidus line from the phase diagram (7.12)

N Total number of gouges that will occur during a proposed lifetime of a pipeline

P number of phases; atmospheric pressure (9.20); ice strength

p pressure; probability distribution for ridge heights (12.4)

Q_s surface heat flux

q specific humidity (9.20)

r radius of curvature (7.5)

r_a, r_b brine pocket dimensions as used in the Assur model (10.33)

r correlation coefficient; the value $r^2 \times 100$ gives the percentage of the scatter that is treated by the regression line

Ra Rayleigh number (8.29)

S salinity, usually in parts per thousand ‰ (8.1); Stefan number (8.27)

s entropy (7.1)

T_a Air temperature

T_f freezing temperature of water (usually refers to the freezing point of seawater)

T_o surface temperature of ice; also T_i (9.11b)

t Time

U wind velocity normal to a lead (11.11)

u_g geostrophic wind

V volume (6.2)

V_i advection rate of solidified ice (11.1)

v mole volume; v flow velocity (7.8); growth velocity (7.18)

v_t total porosity (10.42)

W lead width

W weight of sample in air (10.3) or weight of sample submerged in a liquid

X_p polynya width (11.1)

Z ice thickness at a given point calculated from the hydrostatic equation 17.1

Z_p predicted plume height (11.11)

z level in the ice (i.e., $z = 0$ is the upper ice surface and $z = H$ is the lower ice surface)

a albedo

a_1 and a_2 experimental constants (4.3, 6.3)

γ ridging intensity (12.10–12.12)

Γ base-state stratification (11.11)

δ interface perturbation (8.6)

$\delta h'$ hydrostatic height anomaly

δ_T potential open water (11.8)

ε_{ij} component of a two-dimensional strain rate tensor (16.18)

ε' and ε'' real and imaginary permittivity (10.6.2)

η porosity (10.16)

θ cumulative number of freezing degree-days (4.1)

k_λ spectral extinction coefficient at wavelength λ (9.31)

λ wavelength of a perturbation (cell size, platelet width, brine layer spacing); wavelength (9.5); mean lead width (11.13); wavelength of the finger-rafting pattern (12.1)

λ_{max} maximum channel width that can arch (16.31)

λ_1 distribution shape parameter (12.4)

$\hat{\lambda}$ maximum likelihood estimate of the free parameter of the exponental distribution (13.2)

μ chemical potential (7.1); local friction coefficient (16.23)

$\overline{\mu}_h$ average ridge height (12.5)

μ Poisson's ratio (12.13); dry friction in a ridge sail

ξ_{sn} thermal conductivity of snow (chapter 15)

$\xi_{sn(e)}$ effective thermal conductivity of snow (10.13)

Π isotropic permeability

ρ density

ρ_L lead density

σ surface tension (7.4); Stefan–Boltzman constant (9.4); strength (as specified by the subscript 10.25) c compressive; f flexural; s shear; t tensile

σ_o the σ-axis intercept at zero brine volume (10.37)

σ_{ij} component of a two-dimensional stress rate tensor (16.18)

σ_t sigma t (see equation 5.1)

φ local mean solid fraction (see equation 8.24)

φ', φ floe edge angles (Figure 12.51)

ψ plane porosity (10.26); redistribution function (16.35)

AABW	Antarctic Bottom Water
AACC	Antarctic Circumpolar Current
AARI	Arctic and Antarctic Research Institute. This Russian institute, located in St. Petersburg, is the largest laboratory in the world specializing in polar research.
AASW	Antarctic Surface Water
A&B	Amundsen and Bellingshausen (Seas)
ACEX	Arctic Coring Expedition
AIDJEX	Arctic Ice Dynamics Joint Experiment. A multidisciplinary research project focused on the drift and dynamics of arctic pack ice. The project operated between 1970 and 1976.
ANARE	Australian National Antarctic Research Expedition
AO	Arctic Oscillation. The leading empirical orthogonal function of the winter sea level pressure field in the Arctic, interpreted as the surface signature of modulations of the strength of the polar vortex aloft. High AO values are associated with below-normal Arctic sea level pressure, enhanced surface westerlies in the North Atlantic, and wetter than normal conditions in northern Europe. Low AO values are associated with higher than normal atmospheric pressure over the Arctic, the development of a strong clockwise atmospheric gyre in the western Arctic, and weak westerlies. Until recently low AO values were considered to be the normal state of the Arctic.
ARLIS	Arctic Research Laboratory Ice Station. One of several small research stations established on drifting ice islands by ONR. Support was provided by NARL, Barrow, Alaska (Sater 1968).
ARPA	Advanced Projects Research Agency. A U.S. Defense Department agency that funded speculative research projects. It has now changed its name to DARPA (Defense Advanced Projects Research Agency).
ASEV	Arctic Surface Effect Vehicle. An ARPA program ostensibly to develop a large hovercraft capable of transiting the Arctic Ocean at high speeds.
ASW	Arctic Surface Water
AVHRR	Advanced Very High Resolution Radiometer
AW	Atlantic Water. The water that enters the Arctic Basin from the North Atlantic via the West Spitzbergen Current.
BLP	Boundary layer parameterization
CDW	Circumpolar Deep Water
CRREL	Cold Regions Research and Engineering Laboratory. A U.S. Army Corps of Engineers laboratory located in Hanover, NH, that specializes in winter and polar physical science and engineering.
CS	Constitutional supercooling
CTD	Conductivity–temperature–density. Typically refers to a type of oceanographic sensor system that is capable of measuring these three parameters.
D	Deuterium
DI	Diffusive instability
DMSP	Defense Meteorological Satellite Program
ECMWF	European Centre for Medium-Range Weather Forecasting
EGC	East Greenland Current

ESA	European Space Agency
EWG	Environmental Working Group. A group of U.S. and Russian scientists that compiled a large database of previously unavailable environmental data concerning the Arctic Ocean. The data sets are available on CD from NSIDC.
FY	First-year sea ice. Sea ice of not more than one winter's growth.
G	Gitterman (chapter 6). An early student of the phase relations in cold brine.
GCM	Global circulation model. A computational model based on physical principles that simulates the global circulation of the atmosphere and/or the ocean. Such models that include factors such as sea ice or glacier ice extent and that examine long periods of time are frequently referred to as global climate models.
GPS	Global positioning system. This system uses a number of polar-orbiting satellites to determine the position of a transceiver located on or near the Earth's surface.
HSSW	High-salinity shelf water
IABP	International Arctic Buoy Programme. This effort, which has its roots in the AIDJEX project, has developed and maintained sets of data buoys on the ice pack of the Arctic Ocean from the 1970s to the present.
IBCAO	International Bathymetric Chart of the Arctic Ocean. An effort of the IASC.
ICESat	Ice, Cloud, and land Elevation Satellite. A part of NASA's Earth Observing System. The satellite carries one instrument called GLAS (Geoscience Laser Altimeter System) that serves both as a precision lidar for providing surface elevation data and a dual-frequency lidar for cloud and aerosol studies.
IGY	International Geophysical Year. This important multidisciplinary international scientific effort occurred during 1957–1958. Under its banner important research expeditions were mounted to both the Arctic and the Antarctic.
IHT	Instantaneous heat Transfer
InSAR	Interferometric SAR (synthetic aperture radar)
IPCC	Intergovernmental Panel on Climate Change. A multinational scientific panel charged with accessing the reality and consequences of and potential remedies for climate change.
IPY	International Polar Year. A multinational research program (2007–2009) focused on the polar regions.
IR	Infrared
ISW	Ice shelf water. A distinctive type of sea water that forms beneath antarctic ice shelves.
LEADEX	Lead Experiment
LES	Large eddy simulation
Lidar	Light detection and ranging. Based on the details of the system, lidar are effective for measuring distance, speed, rotation, and chemical composition and concentration.
MIZ	Marginal ice zone
MY	Multiyear sea ice. In the WMO ice code, the term *multiyear* refers to sea ice that has survived at least two summer melt periods. Other related expressions are *second-year (SY) ice*, which denotes ice that has been through one summer but not two, and *old ice*, which denotes ice that has survived at least one summer. However, in the scientific and engineering literature MY is frequently used to describe ice whose exact age is unknown other than it has been through at least one melt season.
NAO	North Atlantic Oscillation. An index useful in describing atmospheric pressure patterns in the Arctic. Similar to the AO.
NASA	National Aeronautics and Space Administration
NATO	North Atlantic Treaty Organization
NAVOCEANO	Navy Oceanographic Office

NCAR	National Center for Atmospheric Research
NMR	Nuclear magnetic resonance
NOAA	National Oceanic and Atmospheric Administration. A U.S. government agency that is concerned with meteorological and oceanographic matters.
NP-1, 2, 3, etc.	North Pole-1, 2, 3, etc. The English designation commonly given to the Soviet manned drifting stations in the Arctic Basin. They are often also referred to as the *SP* stations for *Severny Polius* (Russian for North Pole).
NSF	National Science Foundation (U.S.)
NSIDC	National Snow and Ice Data Center. University of Colorado, Boulder, Colorado
OCSEAP	Outer Continental Shelf Environmental Assessment Program. A multidisciplinary program operated by NOAA that focused on the environment of the outer continental shelf of Arctic Alaska.
OMAE	Offshore Mechanics and Arctic Engineering
ONR	Office of Naval Research. The organization that funds research programs of interest to the U.S. Navy.
PDF	Probability density function
PW	Polar Water
RAC	Return Atlantic Current
RADARSAT	Radar Satellite. A Canadian satellite designed to collect SAR data on sea ice.
RES	Radio echo sounding
SAC	Strategic Air Command. A branch of the U.S. Air Force.
SAR	Synthetic aperture radar
SCICEX	Submarine Science Experiment. A series of semiannual submarine cruises made in the Arctic during the late 1980s and 1990s by the U.S. Navy, primarily to collect unclassified environmental observations.
SHEBA	Surface Heat Budget of the Arctic. A multidisciplinary field experiment carried out in the Beaufort Sea between October 1997 and October 1998. As the field observations continued through the summer melt season, an ice-strengthened ship was locked in the ice and used as a support platform.
SIPRE	Snow, Ice and Permafrost Research Establishment. One of the two organizations that in 1961 merged to form CRREL.
SLS	Side-looking sonar
SP-1, 2, 3, etc.	The Russian *Severny Polius* (North Pole) stations. See NP-1, 2, 3, etc. above.
SSIZ	Seasonal sea ice zone
SY	Second-year. Refers to sea ice that has survived one but not two summer melt periods.
T-3	A portion of the Ellesmere Ice Shelf that broke off in 1950s and drifted around the Arctic Ocean until the 1970s when it exited the Arctic via the East Greenland Drift Stream and ultimately melted in the North Atlantic. The T refers to Target, as T-3 was the third of several ice islands initially identified as unusual targets by aircraft radar in 1951. T-3 is also frequently referred to as Fletcher's Ice Island in honor of Col. J. O. Fletcher, who spearheaded the initial investigation of the nature of the island. A manned research station was maintained on T-3 for many years.
ULS	Upward-looking sonar. In this book ULS refers to sonar systems deployed on submarines transiting under the arctic pack ice for the purpose of measuring the draft of the ice.
WDW	Weddell Deep Water
WMO	World Meteorological Organization, Geneva, Switzerland
WSC	West Spitzbergen Current

Appendix C Terminology & Glossary

*Explain the meaning of words and
you will help the world get rid of half of its mistakes.*
Descartes

The individual exploring the subject of sea ice for the first time has undoubtedly already encountered in this book a number of words and phrases concerning ice in general and sea ice in particular whose exact meanings were not obvious. Therefore a brief discussion of sea ice terminology is included here.

Sea ice terminology appears to have gradually developed over time to describe the wide variety of ice features that can be observed while traveling on or through sea ice areas. The origin of a number of the more unusual-appearing terms (to the English speaker) comes from the Russian Pomor people who settled in the coastal regions of the White Sea during the 11th century (Borodachev et al. 2000). In fact, the word *Pomor* is derived from *"po moryu,"* which means "on the sea" in Russian. They gained experience on sea ice while hunting from the ice and sailing through the ice during the summer. Their terminology was gradually appropriated by other Russians working on or studying sea ice. Ultimately a number of their phrases became accepted internationally. Examples include the terms *stamukha, nilas, shuga, sastruga,* and *polynya*. Fortunately a lexicon of Pomor terminology has been published by Badigin (1956), who was the captain during the drift of the icebreaker *Georgii Sedov* during 1937–1940.

Many of the other sea ice expressions presumably were developed by whalers to describe the ice features that they saw while in pursuit of their quarry. Whalers were fundamentally very practical people and much of sea ice terminology reflects this in that the phrase used catches the essence of the feature or phenomenon being described. Examples include terms such as pancake ice, glass ice, ice keels, frost smoke, old ice, and growlers.

In the present book many of these expressions are explained in some detail where they are first used. In addition this appendix offers a number of definitions of many of the more commonly used sea ice terms. For individuals interested in exploring terminology in more detail several sources are to be recommended. The international standard is the World Meteorological Organization's (WMO) Publication No. 259, which gives sea ice phrases in English, Spanish, German, Russian, and French. Publication 259 also includes a description of the symbols used in preparing ice charts (a subject not discussed in the present book). Also to be recommended is the Scott Polar Research Institute glossary (Armstrong et al. 1966), which also includes English, Russian, Danish, Finnish, French Canadian, and Icelandic terms, as well as photographs of many of the ice features. Perhaps the most detailed glossary of sea ice terminology is the English translation of a Russian sea ice glossary (Borodachev et al. 1994) which was initially prepared at AARI. The English version also contains a large number of photographs of different sea ice features (Borodachev et al. 2000).

As should be clear by now, there are still terminological problems related to sea ice. One difficulty is that the terminology is frequently not carefully applied. Perhaps the best examples of this are the expressions old ice, second-year (SY) ice, and multiyear (MY) ice. The WMO definitions are quite clear. All three of these terms refer to ice that has at least survived a summer melt season, thereby separating such ice from the various first-year (FY) ice types. Furthermore, SY ice has survived one, but not two, melt seasons, whereas MY ice has survived at least two melt seasons. The least specific term is old ice, which refers to ice that has survived at least one melt season. This means that both SY and MY ice are types of old ice.

The problem comes from the fact that in much of the recent technical literature, the term MY is often applied to ice whose age is unknown except for the fact that it has been through at least one melt season and, therefore, should properly be termed old. As a result, if the SY/MY distinction is important in the problem that you are considering, don't accept the label MY on blind faith. This is not to imply that a careful observer cannot tell these ice "varieties" apart. In particular, observers making sequential aerial reconnaissance flights over a region frequently can distinguish the differences between SY and MY ice as they manifest themselves in that region. Even if one is flown into a site where there is no history on the ice that is about to be sampled except that it shows melt features indicating that it has experienced at least one summer, there are ways to distinguish SY from MY ice. However, such matters are rarely explored as they may involve time-consuming petrographic procedures.

Sea Ice Glossary

The following is a brief glossary of many of the sea ice terms that are used in the present book. Definitions marked with an asterisk either are not included in, or are appreciably modified from, the definitions in the WMO Glossary. **Boldface** terms are defined elsewhere in the list.

Anchor ice: Submerged ice that is or has been attached or anchored to the bottom, irrespective of the nature of its formation.

Belt: A term describing a large feature of pack ice arrangement, longer than it is wide, and with dimensions of from 1 to more than 100 km in width (cf. **strip**).

Beset: Situation of a vessel surrounded by ice and unable to move.

Brash ice: An accumulation of floating ice made up of fragments not more than 2 m across (small **ice cakes**), the wreckage of other forms of ice.

Breakup*: A general expression applied to the formation of a large number of **fractures** through a compact ice cover, followed by a rapid diverging motion of the separate fragments.

Buckling*: The flexure of a floating ice sheet into a series of open folds as the result of the elastic instability of the sheet under lateral pressure. Buckling is usually observed only in thin ice.

Bummock: The underside of a **hummock** that projects down below the lower surface of the surrounding ice. This term, coined by the submarine community, has not been extensively used in the scientific literature.

Candling*: The separation of the elongate ice crystals that comprise the columnar zone in fresh- and brackish-water ice into individual crystals (candles) as the result of differential melting along grain boundaries caused by the absorption of solar radiation.

Compact ice edge*: An **ice edge** in which the transition between **open water** and **pack ice** is sharply defined. Compact ice edges characteristically occur when the wind direction as well as the wave field direction are oriented toward the area of sea ice.

Compacting*: Pieces of sea ice are said to be compacting when they are subjected to a converging motion that increases ice **concentration** and **compactness** and/or produces stresses that may result in ice deformation.

Compactness*: The ratio of the area of the sea surface actually covered by ice to the total area of the sea surface under consideration. Therefore a compactness of 0 indicates an ice-free area and a compactness of 1 refers to pack ice containing no open water.

Concentration: The ratio, in tenths, of the sea surface covered by ice to the total area of the sea surface, both ice-covered and ice-free at a specific location or over a defined area. Concentration may be expressed in the following terms:

Compact pack ice: concentration 10/10, no water visible.

Consolidated pack ice: concentration 10/10, **floes** *frozen together.*

Very close pack ice: concentration 9/10 to less than 10/10.

Close pack ice: concentration 7/10 to 8/10, **floes** *mostly in contact.*

Open pack ice: concentration 4/10 to 6/10, many **leads** *and* **polynyas,** *floes generally not in contact.*

Very open pack ice: concentration 1/10 to 3/10.

Convergence*: A term describing the condition when *div* v_i is negative (cf. **divergence**).

Converging*: **Ice fields** and **floes** are said to be converging when they are subjected to a convergent motion that increases the **concentration** and **compactness** of the ice or increases the stresses in the ice.

Crack*: Any **fracture** which has not yet parted. In **pack ice** a parted fracture is still commonly referred to as a crack provided it is sufficiently narrow to jump across.

Deformed ice: A general term applied to ice that has been squeezed together and in places forced upward (and downward). Forms of deformation include **rafting**, **ridging**, and **hummocking**.

Divergence*: Formally defined as $div\ v_i = \dfrac{\partial v_x}{\partial x} + \dfrac{\partial v_y}{\partial y}$ where v_i is the ice drift velocity. The divergence can be considered as the change in area per unit area at a given point. The word is also used to indicate a generally **diverging** motion in the ice.

Diverging*: **Ice fields** and **ice floes** are said to be diverging when they are subjected to a divergent or dispersive motion, thus reducing the ice **compactness** and **concentration**, or relieving stresses in the ice (cf. **converging**).

Draft*: The distance at a given point, measured normal to the sea surface, between the lower surface of the ice and sea level.

Fast ice*: Sea ice of any origin that remains fast (attached with little horizontal motion) along a coast or to some other fixed object.

Finger rafting: A type of **rafting** in which interlocking overthrusts and underthrusts are formed. Common in **nilas** and **gray ice**.

Firn: Old snow that has recrystallized into a dense material. Unlike old snow the grains are joined together, but unlike ice the voids are interconnected, causing it to be permeable to air.

First-year (FY) ice: Sea ice of not more than one winter's growth, developing from **young ice**, with a thickness of 30 cm to ~3 m. FY may be subdivided into thin FY ice/white ice (30–70 cm), medium FY ice (70–120 cm), and thick FY ice (>120 cm).

Flaw: A narrow separation zone between **pack ice** and **fast ice** that forms when pack ice shears under the effect of a strong wind or current along the fast ice boundary.

Flaw lead: A **lead** occurring between **pack ice** and **fast ice**.

Floe: Any relatively flat piece of sea ice 20 m or more across. Floes are subdivided according to horizontal extent:

Small: 20 to 100 m across

Medium: 100 to 500 m across

Big: 500 to 2000 m across

Vast: 2 to 10 km across

Giant: >10 km across

Floeberg: A massive piece of sea ice composed of deformed ice (a **hummock**, a group of hummocks, or a **rubble field**) that has been frozen together and is separable from any surrounding ice. Floebergs may have freeboards of up to 5 m.

Flooded ice: Sea ice that has been flooded by meltwater, seawater, or river water and that is heavily loaded with water and wet snow.

Fracture: Any break or rupture through very close, compact, or consolidated **pack ice** (see **concentration**), **fast ice**, or a single **floe** that results from deformation processes. Fractures may contain **brash ice** and be covered with **nilas** or **young ice**. The length may be a few meters or many kilometers.

Fracture zone: An area of ice that has a great many fractures.

Fracturing: The process whereby the ice is permanently deformed and rupture occurs.

Frazil ice: Fine spicules or platelets of ice suspended in seawater.

Freeboard*: The distance, measured normal to the sea surface, between the upper surface of the ice and sea level.

Frost smoke: Foglike clouds that form by the contact of very cold air with relatively warm water. Frost smoke can appear over newly formed openings in the ice or to the leeward of the **ice edge** and may persist while thin ice is forming.

Grease ice: A stage of freezing, later than that of **frazil ice**, in which the crystals have coagulated to form a soupy layer on the sea surface. As grease ice reflects little light, its presence gives the upper surface of the sea a dull matte appearance.

Gray ice: Young ice, 10–15 cm thick. Less elastic than **nilas**, it breaks on swell. It usually rafts when under pressure.

Gray-white ice: Young ice, 15–30 cm thick, that under pressure is more likely to ridge than to raft.

Grounded ice*: Floating ice (e.g., **ridge**, **hummock**, **ice island**) which is aground (stranded) in shoal water.

Hummock: A hillock of broken ice that has been forced upward by pressure. Hummocks may be either fresh (FY hummock) or weathered (MY or old hummock). The submerged volume of ice under the hummock, forced downward by pressure, has been called a **bummock** by the submarine community.

Hummock field*: An area of sea ice that has essentially all been deformed into a series of **hummocks** (cf. **rubble field**).

Hummocking: The process whereby sea ice is forced into **hummocks**.

Iceberg: A massive piece of ice of greatly varying shape with a freeboard of more than 5 m that has broken away from a glacier. Icebergs can either be afloat or aground.

Ice cover: An expression referring to the general properties of the ice within some large geographic locale such as the Weddell Sea or Baffin Bay.

Ice edge: The demarcation at any given time between the open sea and sea ice of any kind, whether fast or drifting.

Ice field: An area of **pack ice** greater than 10 km across (cf. **ice patch**), consisting of **floes** of any size. Ice fields are subdivided as follows: *Small ice fields:* 10 to 15 km across; *Medium ice fields:* 15 to 20 km across; *Large ice fields:* > 20 km across.

Ice-free: No sea ice is present. However there may be **icebergs** present.

Ice island: A large piece of floating ice with a freeboard of ca. 5 m that has broken away from an arctic ice shelf, most commonly the **ice shelf** located along the north coast of Ellesmere Island. Ice islands typically have thicknesses of ca. 30–50 m, areas of from a few thousand square meters to several hundred square kilometers, and a regularly undulating upper surface.

Ice limit: A climatological term referring to the extreme minimum or extreme maximum extent of the **ice edge** in any given month or period, based on observations made over a number of years.

Ice masssif: A significant region of an ocean or sea that contains heavy, compact **pack ice** during all seasons of the year. Ice massifs typically have a **compactness** of over 7 points and an area in excess of a thousand km².

Ice patch: An area of **pack ice** less than 10 km across (cf. **ice field**).

Ice rind: A brittle shiny crust of ice formed on a quiet surface by direct freezing or from **grease ice**, usually in water of low salinity. Thicknesses are usually less than 5 cm. Ice rind is easily broken by wind or swell and commonly breaks into rectangular pieces (cf. **nilas**).

Ice sheet*: A general expression for a laterally continuous, relatively undeformed piece of sea ice with lateral dimensions of 10 m or larger.

Ice shelf: A floating ice sheet of considerable thickness, showing freeboards of between 2 and 50 m, that is attached to the coast. They are usually of great horizontal extent and possess a level to undulating upper surface. Usually only a small part of an ice shelf is comprised of sea ice in that they are both nourished by the annual accumulation of snow and often are the seaward extension of large glaciers. Parts of an ice shelf can be aground. The seaward edge is called an ice front. Although ice shelves are fairly rare in the Arctic, they make up a significant percentage of the coastline of Antarctica.

Katabatic*: An expression referring to the downslope, topographically steered drainage winds that frequently form near the margins of large glacial ice sheets such as Greenland and the Antarctic. They frequently are the primary factor controlling the locations of **polynyas**.

Keel*: The underside of a **pressure ridge** that projects downward below the lower surface of the surrounding sea ice.

Lead*: Any **fracture** or passage through sea ice that is generally too wide to jump across. A lead may contain **open water** (open lead) or be ice covered (refrozen lead).

Level ice: Sea ice that has been unaffected by deformation.

Melt hummock*: A round hillock-shaped raised portion of the surface of an ice cover that is caused by differential ablation during a summer melt period, or over several summers.

Melt pond*: An accumulation of meltwater on the surface of sea ice that because of the appreciable melting of the ice surface exceeds 20 cm in depth, is embedded in the ice (has distinct banks of ice), and may reach several tens of meters in diameter. In extreme cases melt ponds may perforate completely through the ice sheet (cf. **thaw hole**).

Multiyear (MY) ice*: **Old ice** 3 m or more thick that has survived at least two summer melt seasons. The **hummocks** are even smoother than in **second-year ice** and the near-surface portion of the ice is almost salt free. The color when bare is usually a shade of blue. The melt pattern consists of large interconnecting irregular **puddles** and **melt ponds** and the drainage system is well integrated. (Frequently, in the recent literature, the term MY has been used in a more general sense to describe ice that has survived at least one melt season. When used in this sense, the term MY replaces the more specific **second-year** and **old** descriptors.)

New ice: A general term for recently formed ice, which includes **frazil ice**, **grease ice**, **slush**, and **shuga**. These ice types are composed of ice crystals that are only weakly frozen together (if at all) and have a definite form only while afloat.

Nilas: A thin elastic crust of ice up to 10 cm thick possessing a matte surface. Nilas bends easily under lateral pressure, thrusting into an interlocking pattern of "fingers" (**finger rafting**). Dark nilas, up to 5 cm thick, is very dark in color whereas light nilas, 5–10 cm thick, is slightly lighter in color.

Nip: Ice is said to nip when it presses forcibly against a ship. A vessel so caught, even though undamaged, is said to have been nipped.

Old ice: Sea ice that has survived at least one summer's melt. Most topographic features are smoother than on first-year ice. Old ice may be subdivided into **second-year ice** and **multiyear ice**. (For a definitive field guide to identifying the different varieties of old sea ice during the arctic summer, refer to Johnston and Timco 2008.)

Open water: A large area of freely navigable water in which sea ice is present at a concentration of < 1/10 concentration (see also **ice-free**).

Pack ice: Any accumulation of sea ice, other than **fast ice**, no matter what form it takes or how it is disposed.

Pancake ice: Predominately circular pieces of ice from 30 cm to 3 m in diameter, and up to about 10 cm in thickness, with raised rims caused by the pieces striking against one another. It may be formed on a slight swell from **grease ice**, **shuga**, or **slush**, or as a result of the breaking up of **ice rind**, **nilas**, or, under severe conditions of swell or waves, **gray ice**.

Perennial ice*: A term that has appeared in recent papers using remote-sensing techniques to study sea ice extent. I know of no place where it is actually defined. It appears to be synonymous with the WMO term **old ice** in that it refers to the ice that on a given year has survived the melt season.

Podsov*: The Russian expression for the pieces and plates of ice that are pushed under an ice sheet as the result of **rafting**, **ridging**, and **hummocking**.

Polynya*: An area of significant size with lateral dimensions < 5 km that is comprised of **open water** and primary ice types that are appreciably thinner than would be expected considering the regional climatology and the thicknesses of the ice in the areas surrounding the polynya.

Pressure ridge: A general expression for any elongated (in plan view) ridgelike accumulation of broken ice resulting from ice deformation.

Puddle*: An accumulation of meltwater on the surface of sea ice. Puddles are usually only a few meters across and less than 20 cm deep. As puddles deepen as melting progresses, they become **melt ponds**.

Rafting*: A deformation process in which one piece of ice overrides another. Rafting is most obvious in new and young ice (cf. **finger rafting**) but occurs in ice of all thicknesses.

Ridging: The process whereby ice is deformed into the different types of **pressure ridges**.

Rotten ice: Sea ice that has become honeycombed and is in an advanced state of disintegration.

Rubble field*: An area of sea ice that has essentially all been deformed. Unlike the expression **hummock field**, it does not imply any specific form of the upper or lower surface of the deformed ice.

Sail*: The upper portion of a **pressure ridge** that projects above the upper surface of the surrounding ice.

Sastrugi: Sharp, irregular, parallel ridges formed on a snow surface by wind erosion and deposition. On mobile pack ice the ridges parallel the direction of the prevailing wind at the time when the sastrugi formed.

Second-year (SY) ice: Old ice that has survived only one summer's melt. Because it is thicker and less dense than **first-year ice**, it has a generally higher **freeboard**. In contrast to **multiyear ice**, **second-year ice** during the summer melt shows a regular pattern of numerous small **puddles**. Bare patches and puddles are generally greenish blue.

Shear ridge: A type of pressure ridge in which the primary motion between the interacting ice sheets is one of shear. Such ridges are frequently straight in plan view and commonly show a vertical side in profile.

Shear zone: An area in which a large amount of shearing deformation has occurred.

Shore lead: A **lead** between the **pack ice** and the shore or between pack ice and the edge of an **ice shelf** or glacier.

Shuga: An accumulation of spongy white ice lumps a few centimeters across. Shuga usually forms from **grease ice** or **slush** and sometimes from **anchor ice** rising to the surface.

Slush: Snow that is saturated and mixed with water on land or ice surfaces, or that forms as a viscous mass floating in water after a heavy snowfall.

Snow ice: The equigranular ice that forms when **slush** freezes completely.

Stamukha: The Russian expression for a large **hummocky** pile of ice grounded in water depths of more than 20 m. The height of these features can be 10 m or more. Chains of stamukhi often form along coastal shoals.

Strip: A long narrow area of **pack ice**, about 1 km or less in width, usually composed of small fragments detached from the main mass of ice and run together under the influence of wind, current, and swell.

Thaw hole: A vertical hole in sea ice formed when a **melt pond** melts through to the underlying water.

Weathering: Processes of ablation and accumulation that gradually eliminate irregularities on an ice surface.

Young ice*: Ice in the transition stage between **nilas** and **first-year ice**, 10 to 30 cm in thickness. It is subdivided into **gray ice** and **gray-white** ice. The expression young ice is also commonly used in a more general way to indicate the complete range of ice thicknesses between 0 and 30 cm (as in "the formation and growth of young ice"). Usually these differences in meaning are clear from the context of the discussion.

Appendix D Sampling

When sea ice is quite thin it is fairly easy to sample as one can cut through it with a small saw and remove a piece without great difficulty. However, once the ice exceeds ~25 cm in thickness, the difficulty in obtaining a sample increases rapidly and specialized equipment becomes necessary. Usually on arriving at a sampling site, the first item of business is to discover the local ice thickness. The best way to do this is with a hand drill. There are several different versions available. The most commonly used one is configured in the shape of a continuous spiral that has a diameter of ~5 cm and is fitted with a sharp cutter at the bottom. The drill segments usually come in 1-m lengths. The cutting fragments are passed up the spiral to the surface of the ice. Another version of an ice drill has a somewhat larger diameter and has a radial spiral flight passing up the outside of the drill for the purpose of removing the cuttings. If one only has a few holes to drill, then a simple brace driven by the person requiring the thickness measurement is perfectly adequate. Besides, the work entailed keeps one warm. If there are an appreciable number of holes required, the brace can be replaced by a motorized power head. Another option, which I have found to be preferable, is to drive the drill with a heavy-duty, variable-speed (< 650 rpm), electric drill motor and power the drill motor with a small portable generator which can be moved from hole to hole on a sled.

If there are large numbers of holes to drill in thick ice, then one should consider the possibility of using one of the several versions of a hot-water drill. The advantage of this technique is that once the setup is completed, a large number of holes can be drilled rapidly with comparatively little effort. The problems are that the "setup" and "takedown" procedures are more involved than with purely mechanical drilling systems and a considerable amount of fuel can be required. However, it should be noted that recently the effectiveness of portable hot-water systems has increased thanks to the development of improved methods for drilling holes in glaciers. Therefore, it is recommended that attention be paid to developments in this area that may significantly increase the attractiveness of such procedures as used on sea ice.

Unfortunately, the above procedures only allow one to obtain ice thicknesses, possibly sample the underlying seawater, and implant small sensors such as thermistors in the ice. If one wishes to obtain a continuous ice sample for study, a corer should be used. For many years the most commonly used corer was the so-called SIPRE corer, which obtained a ~7.6-cm core. The corer was a hollow steel tube with radial flights around the outside to pass the cuttings upward. The core barrel was ~1 m in length and, if deeper sampling was required, was attached to a series of 1-m rods that were driven at the surface by a T-bar. The SIPRE corer was a tough, durable, and heavy piece of equipment, developed initially in the late 1940s. It inevitably accomplished the job if the glaciologist applied sufficient profanity and perspiration. Ironically, the fact that the SIPRE corer usually obtained the desired samples slowed the development of improved corers. Ultimately the sound level of the complaints grew sufficiently loud to cause John Rand, a CRREL engineer, to extensively redesign the SIPRE corer, making a number of changes. These were as follows: the barrel diameter was increased to give a core diameter of ~10 cm; the barrel and the flights were now made of plastic; the cutting head was changed; and a pair of spring-loaded "dogs" were added to facilitate core retrieval. Recently, Rand's design has been further fine-tuned by Austin Kovacs. The end result is a better cutting, lighter corer that although still a bit temperamental, is a significant improvement over the earlier SIPRE version. Several well-made versions of this corer (with core diameters of 7.5, 9, and 14 cm) as well as a durable and effective ice drill and a hot-water drill are available from Kovacs Enterprises (info@kovacsicedrillingequipment.com). Although these items would not be described as inexpensive, they are extremely well made, and based on my experience it has invariably proven to be more expensive to produce similar items on an individual basis. As was the case with the drills, corers can also be modified to be driven by either power heads or electric motors.

A few comments are in order regarding the process of drilling and coring. For individuals who have never participated in this "fun" activity, you should start by taking several cores by hand in order to gain a rough feel for the drill or corer. With a little practice you will be able to tell by the level of the resistance and by small vibrations associated with cutting whether you are cutting or whether the drill has spun out and is no longer in effective contact with the ice. Once you have started to drill using a power system, remember that faster rotation does not necessarily result in better cutting rates. Also, pay particular attention to any rapid increase in the resistance to turning as it is an indication of one of two different situations, both of them undesirable. Usually it is the result of the increased friction produced by the buildup of cuttings along the penetrated length of the drill or corer. The solution is to withdraw from the hole and clean off the cuttings. I have found it useful to do this frequently. Occasionally increased resistance is the result of the drill actually screwing itself into the ice. Again the remedy is to quit attempting to penetrate and to come out of the hole to clean off the equipment and to clean out the hole. Sometimes the buildup of resistance can occur rapidly, resulting in the equipment becoming stuck. Commonly, a period of pulling, pushing, twisting, and expressive language will result in freeing the drill. In a worst case, it can be necessary to pour antifreeze down the hole to free the gear. If you sense that freeze-in is imminent, be careful to ascertain that your thumb can readily release the deadman throttle or speed control. I know of several cases where the drill froze in rapidly but the drill and the attached driller, who was unable to release the throttle, started turning instead. In the worst case, the result was a broken arm. When pulling equipment out of a hole, do it smartly in order to maximize the amount of cuttings that will come out of the hole with the equipment. Also, when encountering large resistance it may be necessary to frequently go down the hole, not for further drilling, but simply to spin the cuttings up the flight so that they can be removed from the hole.

For sawing ice, experience has shown that wide-tooth spacings give smooth, even cuts without the blade binding. Saws with fine-tooth spacings do not show good consistent clearing of the cuttings from between the teeth, resulting in the teeth sliding on their own cuttings and diminished contact with the ice. For a small handsaw, the coarse side of a standard pruning saw has been quite successful. On band saws, we have successfully used between 8 and 16 teeth per 10 cm (2, 3, and 4 teeth/inch), depending upon available spacings. It is also desirable to use wider ($\sim$1/2 inch) blades as they do not distort as much during cutting, resulting in straighter cuts and more exactly dimensioned samples. Always take several extra spare band saw blades along on field operations as breakage is common.

On thicker ice, most sampling is performed using chain saws: electric chain saws are excellent for smaller jobs requiring more precise cutting, and larger gasoline-engine-driven models are required for larger jobs such as working on thick ice. It is generally believed that the differences in power between different brands of chain saws is not particularly important and that if a particular model is adequate to drive a blade length suitable for the ice thickness that you wish to sample, the power will prove to be adequate to deal with the ice. There seem to be several different views as to how to deal with the fine-tuning of the chains. One is to grind off the rakers (the small raised points ahead of the cutting surface on the teeth that limit the depth of the cut). Another approach that is used by *Ice Alaska* in sampling over 1-m-thick cold lake ice is not to grind the rakers, but instead grind the teeth so that the leading edges make an angle of 20° from the normal to the line of the chain when viewed from above (rather than the 30 to 35° that the manufacturers initially provide). These angles are marked on the guides for chain saw sharpeners. The third approach that is offered by chain manufacturers is to use "skip-tooth" blades with every other blade missing but supplied with the standard cutting angles and with the rakers intact. I know of no comparative study of the relative merits of these different approaches.

There are several other matters that one should keep in mind when cutting samples in thick ice. If one does not cut too close to the bottom of the ice, a dry quarry can be cut, in which sampling can proceed fairly rapidly. Once the saw fully penetrates the ice the quarry will instantly flood, causing sampling to become more difficult. When one cuts with one end of the blade in the water, there is an appreciable spray of −1.8°C seawater on the lower portions of the individual holding the saw. The effect can best be described as chilling and very antiromantic even if rubberized clothing is worn.

If one wishes to produce a large hole in the ice for the purpose of diving or oceanographic sampling, several other approaches have been used. Although explosives are fast and fun, the end result is usually a very irregular and messy hole that is not convenient as a working location. One solution to this problem is to use a rectangular frame whose bottom is composed of a thermally conductive pipe through which hot water is circulated. Once a rectangular column of ice is melted free, it is either hauled to the surface or pushed downward and disposed of under the ice sheet. The fact that sea ice weighs roughly 900 kg/m³, plus a few simple calculations, will prove to one that the extraction of even fairly modest-sized (in a horizontal sense) samples that contain the complete thickness of the ice will require special lifting equipment.

One particular problem that should be noted occurs when sampling sea ice for studies in which the brine that is included in the ice is of importance. The problem is simple: if there is an appreciable amount of brine in the ice, it will tend to drain out while the specimen is being removed from the ice sheet. This problem is particularly vexing during the summer when sea ice is extremely porous with very large internal brine volumes. Although various schemes have been tried to circumvent this difficulty, I know of no particularly successful procedure. I believe that the best strategy is, if at all possible, to sample while the ice is still quite cold. I have found that if sea ice is colder than ca. −10°C brine drainage is not a significant problem, as long as the samples are processed immediately and placed in sealed containers for melting. However, even at these temperatures, brine loss is still a problem in samples taken from the lower portions of the ice sheet. The difficulty here is, of course, the result of the fact that this ice is always at near-freezing temperatures in that it is only a few centimeters away from the ice–seawater interface, which is fixed by the freezing point of seawater at −1.8°C (in the summer this boundary can be at 0°C). Unfortunately, brine loss occurs quite rapidly whereas the core or ice block sample cools comparatively slowly. One way to minimize brine loss a bit is to cut the samples from the core starting at the bottom.

If samples must be shipped to warmer climes it may be necessary to use insulated containers and place dry ice (solid CO_2) with the ice samples. When undertaking such an operation such as shipping large numbers of cores from Prudhoe Bay, Alaska, to Hanover, New Hampshire, during April and May, one should recall that most locales that are near to sea ice do not stock large quantities of dry ice and that shipments containing large quantities of dry ice can only be carried on special cargo flights. It is best to contact the airlines to update yourself on the current regulations. One should also remember when storing boxes containing dry ice in small rooms such as the typical cold room, that breathing air containing high levels of CO_2 can be fatal even though the amount of O_2 in the air is still adequate to support life. The CO_2 causes a cessation of the breathing reflex. This is particularly dangerous in that CO_2 is both odorless and colorless. A more theoretical discussion of the effect of changes in sea ice salinity and porosity during shipping and storage can be found in Cox and Weeks (1986).

Appendix E   Thin Sections

The most frequently used method for examining the internal structure of sea ice is to prepare thin slices of the ice (i.e., thin sections) and then to examine the sections using a polarizing microscope. Although in principle the technique is similar to procedures used by geologists in studying thin sections of rocks, the details of the preparation procedures are quite different. It should also be noted that although sections of glacier, lake, and river ice can be thinned in the field by melting (Bader 1951; Rigsby 1951), such procedures should particularly be avoided in studying sea ice because of the large changes in the volume of the included brine at near-melting temperatures. This caution is critical if one is interested in studying the geometry of the brine inclusions in the sea ice.

It is frequently convenient to remove a rectangular block from the sheet. Before removing the block a mark, commonly a groove cut by a chain saw or a handsaw along the side of the block, should be referenced relative to true north. Then a smaller square, vertically oriented column measuring ~15 × 15 cm (in the horizontal) by the vertical thickness of the block should be cut from the initial block removed from the sheet. Electric chain saws are very useful for making such cuts in that they both weigh less and vibrate less than gasoline-powered saws. The block should then be placed on a band saw and the 15 × 15 cm horizontal cross section reduced to dimensions slightly less than those of the glass slide on which the ice is to be mounted (typically ~8 × 8 cm). Next, ~1-cm-thick horizontal slices should be cut from the 8 × 8 cm block with the lower surface of each slice located ~1 to 2 mm below the vertical level that you wish to examine. In making these as well as the subsequent cuts, it is important that the tension on the band saw blade be carefully adjusted so that the blade makes straight cuts. A piece of wire mesh "sandpaper" is then placed on a flat table surface and the lower surface of each horizontal ice slice is "sanded" on it until a flat surface is obtained. Although conventional sandpaper can be used in this step, the wire mesh variety has one significant advantage. In both regular and wire mesh sandpaper, the sanding process produces a fine-grained powder of ice crystals that rapidly reduces the polishing effectiveness until it becomes necessary to replace either the paper or the wire screen. As the powder can be knocked out of the wire screen, it has a much longer life before it must be replaced. However, even more important is the fact that at the end of a day's work, the wire screen can be placed in a warm room where the ice melts. After drying and cooling back to a below-freezing temperature, it is essentially as good as new. This is not the case for normal sandpaper.

Next, the ice is mounted on a glass slide which should have a thickness of ~2 mm. If the glass is too thin, it will break easily. If the glass is too thick, its higher effective heat capacity will cause problems during mounting. The mounting process is the trickiest step in thin-section preparation. First, an electric hot plate is set on a suitably low setting and allowed to warm to a steady-state temperature. The glass slide is then placed on the plate and allowed to remain until it becomes warm to the touch. Next, the glass is removed and the ice slice is pressed tightly onto the surface of the slide. The goal here is to melt only a thin layer of the ice at the ice–glass interface without heating the ice any more than necessary. After some practice you will begin to sense how much you must warm the glass slide under your working conditions to achieve just the right amount of melting. Next, the slide is immediately placed on a table or, even better, on a block of metal, which causes the slide to cool rapidly to ambient temperature and the thin liquid film of water at the ice–glass interface to freeze. When making thin sections of freshwater ice, this newly established glass–ice bond is usually sufficient to hold the ice on to the glass during subsequent processing. When dealing with sea ice samples this frequently is not the case in that the salt in the sea ice weakens the glass–ice bond, ultimately resulting in the ice separating from the glass. To prevent this from happening the following has proven to be useful. Partially fill a thermos bottle with a slurry composed of crushed ice and freshwater and then, using an eyedropper, transfer water from the thermos onto the surface of the thin section, building up a dam of freshwater ice completely around the edge of the sea ice sample. During this process some of the water invariably works its way under the outer edges of the thin section. Fortunately this ice can readily be distinguished from sea ice by its distribution pattern as well as by its smaller grain size and lack of substructure. Therefore its occurrence does not result in a significant problem.

Next the mounted ice sample is returned to the band saw and turned on its side so that a cut can be made parallel to the ice–glass interface. After this is completed, if a well-adjusted band saw has been used, the remaining ice specimen should be 2 to 5 mm in thickness. Depending on your purposes this can be the final step in thin section preparation. If you are interested in brine pocket distributions within the ice, thicker sections are desirable. If you are undertaking crystallographic studies of finer-grained ice types, very thin sections are needed. In most cases an additional procedure is added: the ice samples are thinned further using a sledge microtome. This is a desirable additional procedure even if a thinner section is not required, as it removes the pattern produced by the band saw on the upper surface of the thin section. A microtome is a device that was initially developed to prepare thin slices of frozen tissue for medical research. Although there are several varieties of microtome available, the version that has invariably been used in snow and ice studies is produced by Leitz. The glass slide containing the ice sample on its upper surface is mounted on a metal plate on the microtome by either freezing the glass to the metal by using a small amount of water or by using a vacuum pump to hold the glass onto the metal by suction. The metal plate holding the slide is then manually pushed back and forth and sequentially raised between 1 and 20 microns each stroke, causing a fixed overlying blade to remove ice from the upper surface of the thin section. As the microtoming progresses the cuttings can readily be removed from the surface of the thin section by use of a small brush. As the forces during this cutting process tend to shear the ice sample along the surface of the glass slide, the existence of a well-made ice dam around the sample edges as described in the previous paragraph is essential. With care, thicknesses of a few tenths of a millimeter can be reached. The thinner the section becomes, the greater the chance of having it disintegrate. Fortunately it is rarely necessary to produce sections that are this thin, as most sea ice specimens possess a columnar structure oriented in the vertical direction, and equiaxed grains are not particularly common. An exception is, of course, frazil ice, in which the crystals are both equiaxed and small. It should be noted here that to date detailed crystallographic studies have not been undertaken on frazil ice because of its typically small grain sizes. However, if one were to use a Fedorov-type stage similar to those used in studies of the fabrics of fine-grained rocks, studies of frazil ice should not prove to be too difficult.

Subsequent to thin section preparation the sections are typically photographed using both polarized and unpolarized light. This can be done using either an SLR camera with a macrolens or a large-format camera. It is also useful to obtain photographs using a Polaroid camera in that they provide you with "instant" working prints on which you can mark locations of particular interest as you examine a thin section. This is particularly useful in studying crystal orientations in order to avoid measuring the same crystal more than once.

In many types of studies, the next step is to determine the fabric of the ice. The theory behind the examination of hexagonal materials such as ice under polarized light is discussed in books on optical mineralogy and will not be described here. Two things to remember are: (1) when you observe an ice crystal that remains at or near extinction (dark) under crossed polaroids during a 360° rotation of the thin section, this indicates that you are looking down the c-axis of the ice crystal; and (2) the inclusions in sea ice are primarily located along the basal or (0001) planes. The fact that ice is hexagonal greatly simplifies the measurement of crystal orientations in that a 3-axis universal stage is sufficient in contrast to the more complex 5-axis stages that are used to study monoclinic and triclinic materials. Incidentally, when one examines a thin section of sea ice, one can only observe the nature of the ice and the shapes and distributions of the brine inclusions. The salt cryohydrates that may also be present in the ice at colder temperatures ($< -8.7°C$)

Figure E.1. Schematic plan view of a Rigsby 3-axis universal stage.

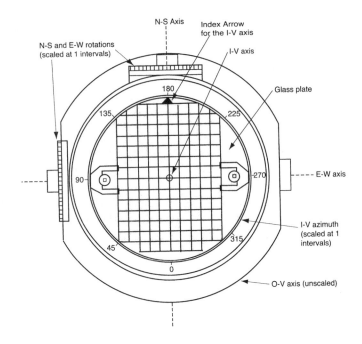

are fine-grained and cannot easily be observed unless a fairly high-powered microscope is used (Light et al. 2003a).

The instrument commonly used in crystallographic studies of sea ice is the so-called Rigsby stage, which is a large 3-axis universal stage initially developed by George Rigsby to accommodate large thin sections (~8 × 8 cm) of glacier ice. The procedures for determining the crystal orientations are described in detail in Langway (1958). As this classic reference is not easy to obtain, the essence of this procedure follows.

A schematic drawing of a Rigsby universal stage is shown in Figure E.1. There are actually four axes of possible rotation. These are as follows:

I-V = inner vertical axis

N-S = north–south axis

E-W = east–west axis

O-V = outer vertical axis. (On a Rigsby stage a rotation on the O-V is achieved by rotating the complete stage on its uncalibrated mounting ring.)

In the initial or rest position with all rotations set at zero, I-V and O-V are parallel to the line of sight as one looks vertically downward at the stage, and N-S and E-W are mutually perpendicular and lie in the horizontal plane. The goals of the optical procedures that follow are to either orient the c-axis of the selected ice crystal vertically parallel to the I-V axis or to orient it horizontally parallel to the E-W axis. Once these orientations are obtained, the rotation readings will allow one to plot the spatial orientation of the c-axis of the ice crystal. In that the great majority of congelation sea ice crystals are oriented with their c-axes approximately oriented in the horizontal plane, the first operation is the most frequently utilized.

In each of the following three cases, initially set the N-S and E-W axes to zero and select a crystal to be measured.

Case 1: When the crystal's defect structure reveals the direction of the (0001) plane

Examine the crystal to see if the orientation of the arrays of brine pockets within the crystal indicates the direction of the trace of the basal plane. In most cases this direction will be quite clear.

1. Rotate on the I-V axis to orient the basal plane of the crystal in the N-S direction while bringing the grain to extinction. The c-axis is now located in the E-W plane.
2. Depress on the E-W axis until the crystal becomes illuminated.
3. Then rotate on the N-S axis until extinction is achieved.
4. Finally, return the E-W axis to its zero position and note the readings on the I-V and N-S axes. They give the orientation of the c-axis that is oriented in an equatorial position.

Case 2: When the *c*-axis is approximately vertical

Crystals with a nearly vertical (polar) *c*-axis orientation will exhibit extinction or near-extinction during a complete rotation on the I-V axis. Such crystals also frequently exhibit a generally rounded shape; this distinguishes them from crystals oriented with their *c*-axes in an equatorial position, which typically have very irregular outlines.

1. Rotate on N-S until the grain is illuminated.
2. Then rotate on I-V until extinction is reached.
3. Return N-S to zero while retaining the rotation on the I-V axis.
4. Then rotate on the E-W axis until the grain becomes illuminated.
5. Next rotate on the N-S axis until extinction is achieved.
6. Finally return the E-W axis to its zero position and note the readings. They give the orientation of the *c*-axis which is now oriented in a near-vertical polar position.

Case 3: When the *c*-axis is not approximately vertical (does not remain at extinction during a complete rotation on the I-V axis) and the crystal does not exhibit a substructure that reveals the orientation of the (0001) plane

1. Rotate on I-V to bring the grain to extinction.
2. Rotate on N-S axis to test the extinction. If extinction is maintained, skip to step 3. If the grain becomes illuminated, return N-S to zero and rotate the crystal 90° to extinction on the I-V. Again rotate on the N-S axis to ascertain that the extinction is maintained. The *c*-axis should now be oriented in the E-W plane. *Note:* if the above procedures maintain extinction when tested in both I-V positions, this indicates that the *c*-axis is normal to the line of sight. In such cases Langway (1958) recommends the following procedure to be certain that the *c*-axis is located in the E-W direction. Rotate the E-W axis an arbitrary amount and depress the N-S axis. If extinction is lost, the *c*-axis lies in the E-W plane. If the grain remains dark, rotate 90° on the I-V axis so that the *c*-axis lies in the E-W plane. Although this procedure is not essential, having all the equatorial measurements made with the *c*-axis located in the E-W plane simplifies plotting of the data.
3. Then rotate on the E-W axis until the grain becomes illuminated.
4. Next rotate on the N-S axis until extinction is achieved.
5. Finally return the E-W axis to its zero position and note the readings on the I-V and N-S axes. They give the orientation of the *c*-axis which is oriented in an equatorial position.

During mineralogical investigations using a Fedorov stage, glass hemispheres are typically used to increase the possible measurement angles before total reflection occurs and to reduce the need for angular corrections. Such hemispheres do not exist for the Rigsby stage. As a result, Rigsby has experimentally determined the appropriate corrections by producing thin sections of a single ice crystal in a variety of known orientations. A table of these corrections can be found in Langway (1958). The corrections become larger as the orientation angle deviates farther from either the polar or the equatorial position. As most crystallographic orientations observed in sea ice are within 10° of either the polar or equatorial positions, these corrections are commonly not used.

These types of observations are typically presented as plots of the orientations of individual crystals as shown on a Schmidt equal-area net, which is constructed so that a unit area in any position on the net corresponds to a unit area on the spherical projection from which the net was derived. Detailed descriptions of the procedures can be found in Langway (1958) or in papers on structural petrology such as those by Fairbairn (1954) or Haff (1938). In sea ice studies crystal orientations are typically plotted on the lower hemisphere following the practice used in structural petrology. A fast and easy way to complete this step is to utilize a computer program that carries out these procedures for you. One such program, which runs on a Macintosh computer, is available from Dr. Richard Allmendinger of the Department of Geology at Cornell University (allmendin@geology.cornell.edu). Although this program can provide contour plots based on the data, I believe that the simpler approach of plotting each measurement as an individual point is much preferable in that it allows the viewer to make a better assessment of the adequacy of the data relative to the problem under study. Besides, contour plots can occasionally give misleading impressions depending on the choice of the contour intervals used. If one wishes to numerically summarize the results of typical sea ice fabrics in the columnar zone, where c-axes are all essentially horizontal, then one can assume that the data set is axial and compute a circular mean and circular variance. Details of these proceedures can be found in Mardia (1972), and applications to sea ice are discussed in Weeks and Gow (1978). If there is significant nonhorizontality to the data, then one may wish to consider using the more involved vector procedures described by Ferrick and Claffey (1992). As is clear, the above procedures are time-consuming. Therefore, the possibility of utilizing automated techniques is attractive. To date this has not occurred in work on sea ice but possible techniques have been explored in studies of freshwater ice. The most promising approach to date would appear to be that of Wilen (2000), who utilized computerized photographic procedures. Once his procedures are finalized they would appear to be particularly useful in studying orientations in frazil ice where the c-axes commonly exhibit large angular variations from the horizontal. Whether his procedures will prove to be sufficiently accurate to apply to congelation ice, where c-axis deviations from horizontal are commonly less than 5°, remains to be determined. In that setting up an automated procedure is time-consuming and requires special equipment, such approaches would appear to be most attractive when one is faced with performing large numbers of fabric analyses. It is also possible to utilize automated procedures to extract additional textural information such as the distribution of grain and pore sizes and shapes from sea ice thin sections (Eicken 1991, 1993; Eicken and Lange 1991; Eicken et al. 1990; Perovich and Gow 1991, 1992). Although such procedures are not routine, they should be considered by anyone confronted with the task of analyzing large numbers of thin sections. If properly utilized they also have the advantage of removing some of the subjectivity commonly inherent in thin-section studies.

In closing it should be mentioned that there are other interesting techniques available for studying the internal structure of sea ice. For instance, a replicate technique can be used to examine the grain and subgrain structure of the ice via optical and scanning electron microscopy (Sinha 1977). Other possibilities are X-ray computer tomography and nuclear magnetic resonance imaging, which can be used to examine both brine layers and brine drainage channels and offers the possibility of three-dimensional measurements of these features (Eicken et al 2000; Kawamura 1988). Although the potential of both of these techniques has been demonstrated, to date they have not received the attention they would appear to deserve and are not in routine use.

Appendix F   Remote Sensing

A quick glance at recent treatments of remote sensing should convince the reader that this is a very large topic and that even a cursory discussion of this subject would significantly increase the length of the present long book: for instance Carsey (1992, 462 pp.); Elachi and van Zyl (2006, 584 pp.); Martin (2004, 426 pp.); Rees (2001, 343 pp.); Tsatsoulis and Kwok (1998, 290 pp.); and, finally, the two-volume set by Massom and Lubin (2006, 1201 pp.). In that satellite-based remote sensing is an essential tool in many types of sea ice studies, see for instance Chapter 18, this brief appendix attempts to provide the reader with a sense of the capabilities of some of the more important remote sensing techniques that have been used in such studies. For a slightly longer discussion of the results obtained from the application of passive microwave remote sensing to variations in the extent of the global sea ice cover, Comiso (1991) is to be recommended.

Orbits

Most artificial satellites utilized as remote sensing platforms are placed into orbits that are nominally circular with very small eccentricities (< 0.01). The angle between the plane of the orbit and the plane of the equator is referred to as the inclination of the orbit (i). The value of i is always taken as positive. If i is 0°, the orbit is in the plane of the equator. In that we are interested in polar regions, orbits that place a satellite into equatorial or near-equatorial orbits are of little use to us. If i < 90° the orbit is referred to as *prograde* and is in the same direction as the Earth's rotation; if i > 90° the orbit is termed *retrograde* and is counter to the Earth's rotation; if i = 90° the orbit is exactly polar. Although all the orbits that are of interest to the sea ice community are of the near-polar variety, I am not aware of any satellite orbits where i = 90°. Interestingly, although near-polar orbits provide the best coverage of the Earth's surface, they are more expensive to launch as less advantage can be taken of the Earth's rotation. Because of the Earth's rotation the subsatellite track would not trace a great circle on the Earth's surface even if the Earth were a perfect sphere and the satellite orbit were perfectly spherical. When the satellite has completed one orbit, the earth's rotation will cause the satellite track to appear to drift to the west. This is clearly shown in Figure F.1 and is true for both prograde and retrograde orbits. The figure also illustrates the fact that a prograde orbit of inclination i reaches maximum north and south latitudes of i whereas a retrograde orbit reaches a maximum latitude of 180° – i. Additional factors that must be considered are the fact that the Earth is not a sphere but is instead an oblate spheroid and orbits are never exactly circular. One of the effects of this is that the elliptical orbit will precess in its own plane, as shown in Figure F.2.

This figure also illustrates another important point: invariably there is a hole near the pole that is not imaged. The reason for this is that such tilted orbits are sun-synchronous in that they cross a given latitude at an identical solar time. As far as sea ice is concerned this is not a particular problem in the Antarctic as most areas of interest are north of 75° S. The Arctic is another matter as essentially the whole Arctic Ocean is north of 70° N and the majority of the MY ice is north of 75 to 80° N. For example, the inclinations of the radar satellites ERS-1 and Radarsat, which were specifically focused on sea ice research and operations, particularly in the Arctic, were 98.52° and 98.6°, respectively. This means that their nadir points are never north of 81.48° and 81.4°. Even considering the fact that the 100-km swath imaged by the Synthetic Aperture Radar (SAR) system on ERS-1 is offset to the right of the orbital track, there is no coverage of the ice to the north of 85° N. To remove this problem, Radarsat has a ScanSAR mode that offers either a 300- or a 500-km-wide imaged swath allowing daily coverage all the way to the pole. Of coarse, there is a price to be paid for the additional coverage in that the resolution changes from 25 m to 75–100 m.

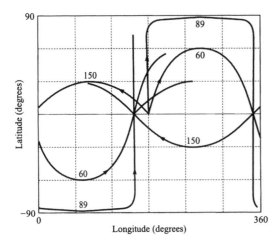

Figure F.1. Subsatellite tracks for circular orbits of inclination 60°, 89°, and 150°. All tracks begin at the equator and at longitude 180°. The orbital period is 100 min and just over one complete orbit has been plotted (Rees 2001).

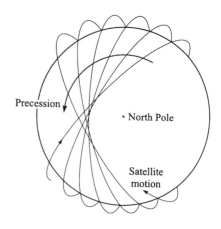

Figure F.2. A view from space looking down on the North Pole showing the precession of a satellite around the Earth's polar axis (Rees 2001).

Orbits are often designed to exactly repeat in a fixed amount of time. There are advantages to this in that sequential images have identical viewing angles, and referencing image locations is greatly simplified. In addition, if a satellite is carrying optical or infrared sensors it is desirable to have the repeats at the same time of day. Such orbits are referred to as sun-synchronous. What I am trying to convey here is that although there are many possible orbits, there is no perfect orbit. Every one is a series of compromises depending on the region of the Earth that is of interest and the specific remote-sensing instruments that are on board.

Techniques

The different types of remote-sensing techniques are separated based on the portion of the electromagnetic spectrum that is used.

VIR

For many types of applications the visible and near-infrared (VIR) portions of the electromagnetic spectrum are very useful (0.4 μm to about 20 μm). The advantage here is that the images that are acquired are readily interpretable in that they appear to be identical to imagery obtained by the use of conventional aerial photography. The illumination is provided by sunlight and at these short wavelengths the absorption and scattering processes are concentrated in the upper few millimeters of the surface layer of the material being sensed. The problem with obtaining data gathered in this portion of the EM spectrum is that like the eye, such systems cannot collect sea ice data if it is either dark or cloudy. This is a severe limitation in that the polar regions are dark during the winter and cloudy during the summer. In short, if you are lucky you can occasionally obtain excellent VIR imagery of sea ice for a short period of time. What you cannot do is obtain a lengthy time series of sea ice images because of the light and cloud limitations. As a result VIR imagery is typically used to provide snapshots of specific features or of the results of certain processes. Examples of studies using Landsat imagery are Hibler, Ackley, et al. (1974; coastal pack ice motions in the vicinity of Barrow, Alaska) and Toyota et al. (2006; floe size distributions in the Sea of Okhotsk).

TIR

In the thermal and infrared (TIR) portion of the EM spectrum (3.5 to 20 μm), the important characteristics of the material being sensed are its emissivity and the penetration depth. For naturally occurring snow and ice, these values are typically small. As a result the signals detected are derived from surface and near-surface layers. In that the emissivities are typically high (0.97–1.00), most of the variation in the observed signal is the result of temperature changes as opposed to differences in the targeted material such as grain size variations. In that the energy measured by a TIR sensor is a mixture of solar reflected and thermally emitted energy, TIR observations are typically made at night. As with VIR observations, TIR observations are cloud-limited. However, as noted, they are not limited by darkness. One obvious potential application of TIR to sea ice studies would be the estimation of ice thicknesses via the determination of the surface temperature of the ice. Unfortunately such studies have, to date, not proved to be generally successful, in that they require independent measures of snow cover thickness and its associated thermal properties. An exception here would be the estimation of ice thicknesses in recently formed leads as these would not have been in existence long enough to accumulate an appreciable snow cover. Interesting examples of the use of TIR imagery in characterizing arctic leads can be found in Lindsay and Rothrock (1995) and Miles and Barry (1998).

Passive Microwave

Although passive microwave (μ-wave) sensors detect radiation that originates by the same basic thermal mechanisms as TIR radiation, the wavelengths involved are much longer (30 cm to 1 mm or 1 to 100 GHz). As a result there is more penetration into the media and the nature of the target material has a greater effect. The advantage possessed by μ-wave systems is that observations are not limited by either darkness or clouds, thereby making time series possible. The fact that data collection is limited by heavy rain is rarely a problem in the polar regions. A second advantage is that the brightness temperatures of the Earth's surface, atmosphere, and other extraterrestrial radiation sources are of the order of 300°K or less. In the microwave at these temperatures, the Rayleigh–Jeans approximation to Planck's law applies permitting the radiative transfer equation to be written in terms of brightness temperatures instead of radiances. The disadvantages result from the fact that the longer μ-wave wavelengths necessitate an appreciably larger antenna to obtain the same spatial resolution as VIR and TIR systems. Unfortunately, current μ-wave antennas have diameters in the range of 1 to 4 m, a fact resulting in μ-wave resolutions ranging between 5 and 100 km. For instance, the Special Sensor Microwave/Imager (SSM/I) that flew aboard the satellites of the Defense Meteorological Satellite Program, and has been used extensively to observe sea ice, monitored four μ-wave channels (19.35, 22.235, 37.00, and 85.50 GHz). The effective resolutions were respectively 45×70, 40×60, 30×38, and 14×16 km. As most sea ice elements, such as floes and leads, have lateral dimensions less than 14 km, this means that values represented by individual pixels will, in most cases, be determined by a combination of the emissions from different types of sea ice and seawater. The different passive μ-wave sea ice algorithms, such as the Team and the Bootstrap, represent different approaches to this problem (Comiso et al. 1997). Both approaches take advantage of the differences in the frequency-dependent emissivities to estimate the percentages of open water, FY ice, and MY ice within each pixel element. As the reader has undoubtedly anticipated, the differences between FY and MY ice are the result of the fact that FY ice is saline, containing brine, thereby limiting the μ-wave emission to the near-surface portion of the ice. On the other hand, the upper portion of MY ice is both desalinated and contains appreciable numbers of air bubbles. In the Antarctic, although it is possible to distinguish two different sea ice types, the physical differences between these are less well documented. Fortunately the large emissivity differences between open water and either FY or MY sea ice simplify data interpretation in both hemispheres. The most recent reports on variations on sea ice extent based on the analysis of μ-wave imagery are Cavalieri and Parkinson (2008) and Parkinson and Cavalieri (2008). For additional details refer to Carsey (1992).

Active Microwave

For many sea ice studies, the active μ-wave system of primary interest is synthetic aperture radar or SAR. As was the case for the passive μ-wave systems discussed above, SAR is not darkness- or cloud-limited.

Instead of sensing the microwave emissions of the material of interest, SAR systems emit pulses of microwave energy and measure the strength of the return. As with passive microwave systems, to obtain the high resolutions needed to "see" surface details a very large antenna would be necessary, a requirement that is not realistic for a satellite-based system. The "synthetic aperture" part of SAR refers to an ingenious way of removing the large antenna restriction. SAR antennas are in fact small and emit a very short pulse whose return is then resolved in time, providing good across-track spatial resolution. The along-track resolution is where the synthetic aperture comes in. Assuming a pulse interval of T and a satellite velocity of v, the antenna is carried through a distance vT between each pulse. Now, assuming that your system has the capability to store the returns from n sequential pulses, the system has now acquired return information similar to that obtained by an antenna nvT long. With each new pulse, the system adds a return at the forward end of the track and loses a return at the back end of the track. However, it is not sufficient just to collect these data; at each time step the data must be numerically processed, thus creating the synthetic aperture. What has been accomplished is a way around the fact that one cannot currently launch a giant antenna on a spacecraft. Unfortunately, the solution results in a formidable data-processing problem. Fortunately the necessary data processing can be dealt with on the ground.

The end result is a high-resolution image in which the gray scale is determined by the strength of the radar return. The question now becomes, "What determines the return strength?" Part of the answer is the roughness of the surface. Consider undeformed FY sea ice. The ice is typically salty, resulting in higher brine volumes. This limits the penetration of the radar pulse into the ice. In addition, the upper ice surface is flat. Such ice acts as a specular (forward) reflector and no energy is reflected back to the radar antenna. Another good example would be a flat calm sea. As the surface becomes rougher more energy is reflected back to the radar system and the intensity of the return increases. A useful criterion for whether or not a particular surface will appear rough is the so-called Rayleigh criterion, which considers a surface rough if $\Delta h \cos \theta > \dfrac{\lambda}{8}$.

Here Δh is the average vertical displacement of the surface, θ is the incidence angle of the radar beam, and λ is the wavelength of the radar pulse. In short, just because a surface appears rough to you and me does not necessarily mean that it will provide a strong radar return. For example, during the first year of the ERS-1 mission, I compared the radar imagery with ice features located in the FY fast ice at Barrow, Alaska. The situation was static in that the fast ice was horizontally fixed. Also, identifiable buildings provided definite reference points. The highest feature (~2.5 m) in the study area was a rounded pressure ridge. There was no indication of the presence of this feature in the radar image. The rounded shape of the ridge apparently resulted in it serving as a forward scatterer. Even more surprising was the fact that the strongest return in the area was from a location that at first glance appeared to be perfectly flat. However, a little digging in the ~20-cm-thick snow cover revealed that the surface of the underlying ice had, during the formation of the initial ice skim, been broken into a number of irregularly oriented plates whose dimensions were such as to satisfy the Rayleigh criterion. Clearly, the interpretation of SAR imagery can, at times, be counterintuitive. For instance, one might think that SAR would prove to be useful in providing a quantitative measure of the intensity of ice deformation as evidenced by the degree of ridging. This does not appear to be the case, at least for observations made at C-band frequencies (5.3 GHz, $\lambda = 5.7$ cm).

As sea ice ages, and particularly after it passes through a summer melt period, the strength of the radar return changes. The sequence of these changes is as follows (Gogineni et al. 1992):

1. In winter and spring, the return from MY ice is much stronger than that from FY ice.
2. When the snow starts to melt in the early summer, the contrast between MY and FY signatures disappears.
3. When the FY ice is snow free the winter contrast is reversed, with FY returns exceeding MY returns by a few dB.
4. The reversal vanishes as soon as the superimposed ice is melted.

5.Changes 3 and 4 occur over short periods of time (one to two weeks).

6. As melt ponds increase, MY returns again exceed FY returns, a state that continues for the remainder of the summer.

In short, the sequence in backscatter changes exhibited by sea ice is far from simple. The positive aspect of this is that these changes can be used to separate different ice types and also to provide information relative to the onset of surface melt and freeze-up. For instance, in the spring the formation of surface meltwater converts relatively transparent dry snow into highly attenuating wet snow that masks the scattering from the bubble-rich upper layer of the underlying ice. In the fall, as the dielectrically lossy liquid water freezes to low-loss, relatively fresh ice, the volume scattering from the bubbles in the upper portion of the sea ice once again contributes to the overall return (Winebrenner et al. 1998).

In my view, the interpretation of SAR imagery of sea ice is every bit as complex as the interpretation of passive microwave imagery. Yet SAR has one saving grace: its high resolution allows viewers as well as automated systems to track specific ice features for appreciable periods of time (weeks to months), thereby allowing a determination of the ice motion field (Kwok 1999). Furthermore, many of the trackable features can be associated with features such as leads or particularly shaped MY floes. Two recent papers demonstrating interesting applications of SAR to sea ice studies are Kwok (2006) and Kwok and Cunningham (2008).

In designing SAR satellites such as ERS-1 and RADARSAT, the study of sea ice was a primary consideration. Occasionally systems designed for other purposes also prove to be useful for sea ice studies. One example is the scatterometer system on board the QuickSCAT satellite. Although this active microwave system's main purpose is to provide estimates of the near-surface wind field over the ocean, its polar orbit, all-weather capabilities, and sensitivity to changes in the backscatter from the underlying ice have provided useful information on the distribution of large icebergs in the Antarctic as well as MY ice in the Arctic (Nghiem et al. 2007; see also Figure 18.3).

Lidar

Lidar (light detection and ranging) is a relatively new addition to the suite of satellite-based systems useful in the study of sea ice. The system of particular interest here is the lidar launched in 2003 as the sole instrument on ICESat, the Ice, Cloud and Land Elevation Satellite that is one part of NASA's Earth Observing System. The ICESat mission was designed to provide elevation data required for the determination of the mass balance of the world's large glacial ice sheets, in particular Greenland and the Antarctic. Other aspects of the mission include cloud and aerosol studies. As was the case with QuickSCAT, sea ice studies were not a primary focus. However, as noted above, when a remote-sensing system is launched to study X and Y, the ensuing data often prove to be useful in studies of Z. This has clearly been the case with lidar and sea ice.

The Geoscience Laser Altimetry System (GLAS) on ICESat samples the Earth's surface from an orbit of 94° with a footprint of ~70 m in diameter spaced at ~170 m intervals. The Arctic Ocean is covered to a latitude of 86°N. Assuming the necessary clear-sky conditions, the elevation accuracy for low slope surfaces is estimated to be ~14 cm. The question then becomes, "What useful information can be extracted from such a data set?" The answer is a great deal. The two obvious targets are ice thickness and snow depth. As the reader will recall, current ice thickness estimates come from determining the draft of the ice from ULS data sets. Unfortunately, such data are limited both spatially and temporally. The advantage of ULS data is that the accuracy requirements for determining thickness from draft are much less stringent than when one attempts to determine thickness from freeboard. Also, the resolution is much finer. As a result, the first step in obtaining ice thickness estimates from GLAS data is to obtain the best possible freeboard estimates. Kwok et al. (2007) have accomplished this by using three different procedures to obtain tie points that specify the local surface of the ocean. In order of decreasing quality, these use samples (1) of new openings identified in ICESat profiles and SAR imagery, (2) where ICESat reflectivities are below the background snow-covered sea ice and their elevations exceed an expected deviation below that of the local mean surface, and 3) where the only condition is that their elevations exceed an expected deviation below that of the local mean surface. The advantage of the second and third approaches is that they do not depend on the availability of SAR imagery and offer a larger and more spatially complete data set. Via the use of these procedures, the final uncertainty estimate of the freeboard can be reduced to better than 7 cm. In that freeboard here is considered to be the air–ice surface, to obtain the elevation of the snow–ice surface and in turn the actual ice thickness it is necessary to obtain estimates of the snow loading from a knowledge of the snow thickness and density. This is accomplished starting with the ECMWF snowfall estimates combined with results from field observations on snow properties on arctic sea ice. See, for instance, Warren et al. (1999). Ultimately, snow thickness and density can be combined with freeboard values and sea ice density measurements to estimate ice thicknesses (Kwok and Cunningham 2008). The different assumptions and correlations used in making these transitions are clearly identified. The end results of the above exercise are ice volume estimates for the Arctic Basin that are within the bounds of other available estimates. Six data gaps are identified, of which the first four involve snow and its properties.

I think that the Kwok and Cunningham paper is important in that it presents a clear route leading to verifiable ice thickness measurements from space. This work, in turn, leads to another aspect of lidar remote sensing. Offbeam lidars are capable of determining the thickness of highly opaque media by observing the horizontal spread of lidar pulses. The bright halo observed around the illuminated spot extends further out in thicker layers because photons can travel further without escaping through the bottom of the layer. This measurement approach has been used to determine thickness in several different disciplines varying from dentistry (tooth enamel) to clouds. That the approach could provide accurate ground-based thickness measurements for snow and sea ice was first shown by the field experiments of Haines et al. (1997). That the approach could possibly be extended to airborne systems has recently been supported by a theoretical analysis of the various associated uncertainties by Varnai and Cahalan (2007), who point out that snow and sea ice present quite different problems in that, although sea ice is commonly much thicker, snow contains a higher concentration of scatterers. As a result, although sea ice halos are larger, snow halos are brighter. They conclude that a properly designed airborne lidar system should be capable of obtaining sea ice thickness retrievals, but only at night, whereas snow retrievals could occur during either day or night. Although a satellite-based lidar system capable of direct snow and sea ice thickness measurements would be ideal, an airborne system would also be very useful in providing detailed snow thickness distribution measurements that could be combined with lidar-determined freeboard determinations à la Kwok and associates.

Bibliography

There are certain journals and report series that through the years have contained large numbers of papers concerning sea ice. To avoid needless repetition, the following compact abreviations will be used in the reference list. The titles of most papers will be given in the language used in the paper. However, in the cases of papers written in Russian and Japanese, the letters *R* or *J* will appear in parentheses after the title to indicate the language of the paper. Recent papers written in Japanese have lengthy English abstracts. Many Russian papers do not have English abstracts. This is particularly true of the older Russian literature. Less frequently cited reference sources will be given in more detail.

AIDJEX Bull: The gray literature publication of the *Arctic Ice Dynamics Joint Experiment*, Polar Research Center, Applied Physics Laboratory, University of Washington, Seattle, WA. Fortunately these papers, many of which contain considerably more detail than the formally published versions, are now available online at http://psc.apl.washington.edu/aidjex.

AINA: Arctic Institute of North America. Over the years AINA has published the journal *Arctic* as well as a number of symposia volumes focused on the Arctic Ocean. Although in the past AINA has maintained offices in both Canada and the United States, presently their offices are located in Calgary, Alberta.

Ann Glaciol: *Annals of Glaciology*. The conference proceedings publication of the International Glaciological Society, Cambridge, England.

CRREL: Cold Regions Research and Engineering Laboratory, Hanover, NH, a research facility operated by the U.S. Army Corps of Engineers. There are several varieties of CRREL reports (Research, Technical, Special).

CRST: *Cold Regions Science and Technology*. A journal published by Elsevier.

GRL: *Geophysical Research Letters*, American Geophysical Union (AGU), Washington, D.C.

IASH: *International Association of Scientific Hydrology*. An organization that operates as part of the IUGG (International Union of Geodesy and Geophysics).

JGR: *Journal of Geophysical Research*, American Geophysical Union (AGU), Washington, D.C. (Most sea ice–related papers are published in the C Series, which contains oceanography and is commonly referred to as *JGR Green* from the color of the cover.)

J Glaciol: *Journal of Glaciology*. The journal of the International Glaciological Society, Cambridge, England.

LTS-A: *Low Temperature Science, Series A*. The physical science journal published by the Institute of Low Temperature Science (ILTS), Hokkaido University, Hokkaido, Japan.

PA: *Problemy Arktiki*. This Russian journal changed its name to *PAiA* in 1958.

PAiA: *Problemy Arktiki i Antarktiki*. The house journal of the Arctic and Antarctic Research Institute (AARI), St. Petersburg, Russia.

POAC: *Port and Ocean Engineering under Arctic Conditions*. An organization that has held a number of conferences relating to sea ice engineering. Unfortunately, the fact that each conference is published by the organization hosting the conference makes obtaining specific conferences difficult. A listing of the different sources can be found in Timco (2005).

SIPRE: *Snow, Ice, and Permafrost Research Establishment*. In 1961 SIPRE was merged with another Corps of Engineers laboratory and its name was changed to CRREL.

Trudy AANII: *Trudy Arkticheskogo i Antarkticheskogo Nauchno-issledovatel'skogo Instituta*. A monograph series published by AARI, St. Petersburg, Russia.

Aagaard, K., and E. Carmack. 1989. The role of sea ice and fresh water in the Arctic circulation. *JGR* 94(C10):14485–98.

———. 1994. The Arctic Ocean and climate: A perspective. In *The polar oceans and their role in shaping the global environment,* ed. O. M. Johannessen, R. D. Muench, and J. E. Overland, 5–20. Geophys. Monograph Series 85. Washington, DC: AGU.

Aagaard, K., L. K. Coachman, and E. C. Carmack. 1981. On the halocline of the Arctic Ocean. *Deep-Sea Res* 28A:529–45.

Aagaard, K., and P. Greisman. 1975. Toward new mass and heat budgets for the Arctic Ocean. *JGR* 80:3821–27.

Abel, G. 1983. Observations of daily temperature variations in a snow cover (G). *Rept. Meteorol. Herausgegeben Kaiserlichen Akad. Wiss.* 16:1–53.

Ackley, S. F. 1986. Sea ice pressure ridge microbial communities. *Antarct J US* 21:172–74.

Ackley, S. F., W. D. Hibler III, F. K. Kugzruk, A. Kovacs, and W. F. Weeks. 1976. Thickness and roughness variations of Arctic multiyear sea ice. *CRREL Report* 76-18.

Ackley, S. F., M. A. Lange, and P. Wadhams. 1990. Snow cover effects on Antarctic sea ice thickness. *CRREL Monograph* 90-1:16–21.

Ackley, S. F., and C. W. Sullivan. 1994. Physical controls on the development and characteristics of Antarctic sea ice biological communities: A review and synthesis. *Deep-Sea Res* 41(10):1583–1604.

Adams, C. M., Jr., D. N. French, and W. D. Kingery. 1960. Solidification of sea ice. *J Glaciol* 3(28):745–61.

Addison, J. R. 1970. Electrical relaxation in saline ice. *J App Phys* 41(1):54–63.

———. 1977. Impurity concentrations in sea ice. *J Glaciol* 18(78):117–27.

Aksenov, Ye., and W. D. Hibler III. 2001. Failure propagation effects in an anisotropic sea ice dynamics model. In *IUTAM Conf on Scaling Laws in Ice Mechanics and Ice Dynamics,* ed. J. P. Dempsey and H. H. Shen, 363–72. Dordrecht: Kluwer.

Alam, A., and J. A. Curry. 1995. Lead-induced atmospheric circulations. *JGR* 100(C3):4643–51.

———. 1998. Evolution of new ice and turbulent fluxes over freezing winter leads. *JGR* 103:15783–802.

Alfultis, M. A., and S. Martin. 1987. Satellite passive microwave studies of the Sea of Okhotsk ice cover and its relation to oceanic processes, 1978–1982. *JGR* 92(C12):13013–28.

Alley, R. B. 1988. Fabrics in polar ice sheets: Development and prediction. *Science* 240(129):493–95.

Allison, I., and S. Qian. 1985. Characteristics of sea ice in the Casey region. *ANARE Res Notes* 28:47–56.

Amos, A. F. 1978. Green iceberg sampled in the Weddell Sea. *Antarct J US* 13:63–64.

Andersen, S., R. Tonboe, L. Kaleschke, G. Heygster, and L. T. Pedersen. 2007. Intercomparison of passive microwave sea ice concentration retrievals over the high-concentration Arctic sea ice. *JGR* 112:C08004, doi:10.1029/2006JC003543.

Anderson, D. L. 1958a. A model for determining sea ice properties. In *Arctic sea ice,* ed. W. Thurston, 148–52. Pub. No. 598, US Nat. Acad. Sci., Nat. Res. Council.

———. 1958b. Preliminary results and review of sea ice elasticity and related studies. *Trans Eng Inst Canada* 2:116–22.

———. 1960. The physical constants of sea ice. *Research* 13(8):310–18.

———. 1961. Growth rate of sea ice. *J Glaciol* 3(30):1170–72.

Anderson, D. L., and W. F. Weeks. 1958. A theoretical analysis of sea ice strength. *Trans Am Geophys Union* 39:632–40.

Anderson, L. G. 1998. Chemical oceanography in polar oceans. In *Physics of ice-covered seas,* ed. M. Lepparanta, 2:787–809. Helsinki: Univ Helsinki Press.

Anderson, L. G., and E. P. Jones. 1985. Measurements of the total alkalinity, calcium, and sulfate in natural sea ice. *JGR* 90(C5):9194–98.

Anderson, L. G., and 9 others. 1994. Water masses and circulation in the Eurasian Basin: Results from the *Odin* 91 expedition. *JGR* 99(C2):3273–83.

Andreas, E. L. 1980. Estimation of heat and mass fluxes over Arctic leads. *Mon Weather Rev* 108:2057–63.

———. 1986. A theory for the scalar roughness and the scalar transfer coefficients over snow and sea ice. *CRREL Report* 86-9.

———. 1987. Comment on Bennett and Hunkins, Atmospheric boundary layer modification in the Marginal Ice Zone. *JGR* 92:3965–68.

———. 1996. The atmospheric boundary layer over polar marine surfaces. *CRREL Monograph* 96-2.

———. 1998. The atmospheric boundary layer over polar marine surfaces. In *Physics of ice-covered seas*, ed. M. Leppäranta, 715–71. Helsinki: Univ Helsinki Press.

Andreas, E. L., and S. F. Ackley. 1981. On the differences in ablation seasons of the Arctic and Antarctic sea ice. *J Atmos Sci* 39:440–47.

Andreas, E. L, P. S. Guest, P. O. G. Persson, C. W. Fairall, T. W. Horst, R. E. Moritz, and S. R. Semmer. 2002. Near-surface water vapor over polar sea ice is always near ice saturation. *JGR* 107:C10, 10.1029/2000JC000411.

Andreas, E. L., M. A. Lange, S. F. Ackley, and P. Wadhams. 1993. Roughness of Weddell Sea ice and estimates of the air–ice drag coefficient. *JGR* 98(C7):12439–52.

Andreas, E. L., and B. Murphy. 1986. Bulk transfer coefficients for heat and momentum over leads and polynyas. *J Phys Oceanog* 16:1875–83.

Andreas, E. L., C. A. Paulson, R. M. Williams, R. W. Lindsey, and J. A. Businger. 1979. The turbulent heat flux from Arctic leads. *Boundary-Layer Meteorol* 17(1):57–91.

Arakawa, K. 1954. Studies of the freezing of water, II. *J Faculty Sci Hokkaido Univ* Ser. II (4):311–39.

Arakawa, K., and K. Higuchi. 1954. On the freezing process of aqueous solutions. *LTS-A* 12:73–86.

Arbetter, T. E., J. Curry, and J. A. Maslanik. 1999. Effects of rheology and ice thickness distribution in a dynamic–thermodynamic sea ice model. *J Phys Oceanog* 29:2656–70.

Arcone, S. A., A. J. Gow, and S. McGrew. 1986. Structure and dielectric properties at 4.8 and 9.5 GHz of saline ice. *JGR* 91(C12):14281–303.

Arctic Climate Impact Assessment. 2005. *Scientific report*. Cambridge Univ Press.

Arctowski, H. 1903. Die antarktischen Eisverhältnisse. *Petermanns Mitteilung* XXX(144):1–121.

———. 1908. Les glaces. Glace de mer et banquises. In *Resultats du voyage de S. Y. Belgica en 1897–99, Rapports scientifiques*. Anvers: J. E. Buschmann.

———. 1909. La congelation de l'eau de mer. *Bull Soc Belge Astron* 14:182–95.

Armstrong, R. L. 1980. An analysis of compressive strain in adjacent temperature-gradient and equi-temperature layers in a natural snow cover. *J Glaciol* 26:283–89.

Armstrong, T. E. 1952. *The northern sea route. Soviet exploitation of the northeast passage*. Cambridge Univ Press.

———. 1955. Soviet work on sea ice forecasting. *Polar Record* 7(49):302–11.

———. 1996. *The historical and current uses of the Northern Sea Route, part I*. INSROP Working Paper No. 28–1996, IV.1.1, Lysaker, Norway: Fridtjof Nansen Institute.

Armstrong, T. E., B. Roberts, and C. Swithinbank. 1966. *Illustrated glossary of snow and ice*. Scott Polar Research Institute. Special Publication Number 4.

Arnol'd-Aliab'ev, V. I. 1929a. The determination of air bubbles in ice and instruments for measuring the same (R). *Izvest Glav Geofiz Observat, Leningrad* 4:34–36.

———. 1929b. Investigation of ice strength of the Gulf of Finland in 1923, 1927, and 1928 (R). *Izvest Glav Geofiz Observat, Leningrad*,2:15–28.

———. 1934. Determination of density and porosity of ice under field conditions (R). *Trudy Komissii po izucheniiu vechnoi merzloty. Academy of Sciences USSR*, 3:127–50.

———. 1939. Strength of ice in the Barents and Kara Seas (R). *PA* 6:21–30.

Arrigo, K. R., A. Weiss, and W. O. Smith. 1998. Physical forcing of phytoplankton dynamics in the southwestern Ross Sea. *JGR* 103:1007–21.

Ashby, M. F., and S. D. Hallam. 1986. The failure of brittle solids containing small cracks under compressive stress states. *Acta Metall* 34(3):497–510.

Ashton, G. D., ed. 1986. *River and lake ice engineering*. Littleton, CO: Water Resources Publications.

Ashworth, T. 1972. Measurement of the thermal properties of ice. *Proc 4th Internat Cryogenic Engineering Conf* 148–52.

Assur, A. 1958. Composition of sea ice and its tensile strength. In *Arctic sea ice*, ed. W. Thurston, 106–38. Pub. No. 598, US Nat Acad Sci, Nat Res Council.

———. 1967. Flexural and other properties of sea ice sheets. In *Proceed Internat Conf on Low Temperature Science, Physics of Snow and Ice*, ed. H. Oura, I(1):557–67. Sapporo: Hokkaido Univ, ILTS.

Assur, A., and W. F. Weeks. 1963. Growth, structure and strength of sea ice. *Internat Union Geodesy and Geophysics* (Commission on Snow and Ice), Berkeley Meeting. Pub. 61:95–108.

Auty, R. P., and R. H. Cole. 1952. Dielectric properties of ice and solid D_2O. *J Chem Phys* 20(8):1309–14.

Babko, O., D. A. Rothrock, and G. A. Maykut. 2002. Role of rafting in the mechanical redistribution of sea ice thickness. *JGR* 107(C8):10.1029/1999JC000190.

Backstrom, L. G. E., and H. Eicken. 2006. Capacitance probe measurements of brine volume and bulk salinity in first-year sea ice. *CRST* 46:167–80.

Bader, H. 1951. Introduction to ice petrofabrics. *J Geol* 59(6):519–36.

Badgley, F. I. 1966. Heat balance at the surface of the Arctic Ocean. In *Proc Symposium on the Arctic Heat Budget and Atmospheric Circulation*, ed. J. O. Fletcher, 215–46. Santa Monica, CA: RAND Corporation. RM-5233-NSF.

Badigin, K. S. 1956. *In cold seas: Sketches on the history of the ice navigation of the Russian Pomors* (R). Izd Geograf Literatury.

Bailey, D. A., A. H. Lynch, and T. E. Arbetter. 2004. Relationship between syntopic forcing and polynya formation in the Cosmonaut Sea: Regional climate model simulations. *JGR* 109(C04023):doi:10.10 29/2003JC001838.

Baines, P. G., and S. Condie. 1998. Observations and modeling of Antarctic downslope flows: A review. In *Ocean, ice and atmosphere interactions at the antarctic continental margin*, ed. S. S. Jacobs and R. F. Weiss, 29–49. Antarctic Research Series 75. Washington, DC: AGU.

Bamber, J. L., and A. J. Payne, eds. 2004. *Mass balance of the cryosphere: Observations and modelling of contemporary and future changes*. Cambridge Univ Press.

Baranov, G. I., Yu. L. Nazintsev, and N. V. Cherepanov. 1968. Formation conditions and certain properties of Antarctic sea ice (according to observations of 1963). *Trudy AANII* 38. (In *Collected Papers of Soviet Antarctic Expeditions*, ed. V. G. Aver'yanov, 74–86. Israel Program for Scientific Translations, 1969.)

Bari, S. A., and J. Hallett. 1974. Nucleation and growth of bubbles at an ice–water interface. *J Glaciol* 13(69):489–520.

Barker, K., G. Timco, and B. Wright. 2006. Traversing grounded rubble fields by foot: Implications for evacuation. *CRST* 46(2):79–99.

Barnes, P. W., D. McDowell, and E. Reimnitz. 1978. Ice gouging characteristics: Their changing patterns from 1975–1977, Beaufort Sea, Alaska. *USGS Open-File Report* 78-730.

Barnes, P. W., D. M. Rearic, and E. Reimnitz. 1984. Ice gouging characteristics and processes. In *The Alaskan Beaufort Sea: Ecosystems and environment*, ed. P. W. Barnes, D. M. Schell, and E. Reimnitz, 185–212. Orlando: Academic Press.

Barnes, P. W., and E. Reimnitz. 1974. Sedimentary processes on arctic shelves off the northern coast of Alaska. In *The coast and shelf of the Beaufort Sea*, eds. J. C. Reed and J. C. Sater, 439–76. AINA.

———. 1976. Flooding of sea ice by the rivers of northern Alaska. *USGS Professional Paper* 929:356–59.

———. 1979. Ice gouge obliteration and sediment redistribution event, 1977–1978, Beaufort Sea, Alaska. *USGS Open-File Report* 79-351.

Barnes, P. W., D. M. Schell, and E. Reimnitz, eds. 1984. *The Alaskan Beaufort Sea: Ecosystems and environments*. Orlando: Academic Press.

Barr, W. 1991. The Arctic Ocean in Russian history to 1945. In *The Soviet maritime Arctic*, ed. L. W. Brigham, 11–32. Annapolis: Naval Institute Press.

Barrow, John. 1846. *Voyages of discovery and research within the arctic regions, from the year 1818 to the present time*. London: John Murray.

Barry, R. G. 1979. Study of climatic effects on fast ice extent and its seasonal decay along the Beaufort–Chukchi coasts. In *Environmental assessmant of the Alaska continental shelf, physical science studies*, 2:272–375. Boulder, CO: Environ Res Lab.

———. 1986. The sea ice data base. In *Geophysics of sea ice*, ed. N. Untersteiner, 1099–1134. NATO Advanced Sciences Institutes Series, Series B: Physics. New York: Plenum.

Barry, R. G., R. E. Moritz, and J. C. Rodgers. 1979. The fast ice regimes of the Beaufort and Chukchi sea coasts, Alaska. *CRST* 1:129–52.

Bauer, J., and S. Martin. 1980. Field observations of the Bering Sea ice edge properties during March 1979. *Mon Weather Rev* 108(12):2045–56.

———. 1983. A model of grease ice growth in small leads. *JGR* 88(C5):2917–25.

Bazant, Z. P. 1992. Large-scale thermal bending fracture of sea ice plates. *JGR* 97(11):17739–51.

———. 2001. Scaling laws for sea ice fracture. In *IUTAM Conf on Scaling Laws in Ice Mechanics and Ice Dynamics*, ed. J. P. Dempsey and H. H. Shen, 195–206. Dordrecht: Kluwer.

Behounek, F. 1928. *Männen på isflaket. Med Italia till Nordpolen*. Uppsala: Lindblads.

Beketsky, S. P., V. N. Astafiev, and P. A. Truskov. 1996. Structure of hummocks offshore of northern Sakhalin. *Proceed 6th Internat Offshore and Polar Engineering Conf*, II:398–400. Los Angeles.

Bennington, K. O. 1963a. Some crystal growth features of sea ice. *J Glaciol* 4(36):669–88.

———. 1963b. Some chemical composition studies on arctic sea ice. In *Ice and snow: Processes, properties, and applications*, ed. W. D. Kingery, 248–57. Cambridge, MA: MIT Press.

Bennon, W. D., and F. P. Incropera. 1987a. A continuum model for momentum, heat and species transport in binary solid–liquid phase change systems. I. Model formulation. *Int J Heat Mass Transfer* 30(10):2161–70.

———. 1987b. A continuum model for momentum, heat and species transport in binary solid–liquid phase change systems. II. Application to solidification in a rectangular cavity. *Int J Heat Mass Transfer* 30(10):2171–87.

Berdennikov, V. P. 1948. Investigation of the elastic modulus of ice (R). *Proceed State Hydrological Inst* N7(61).

Bergdahl, L. 1977. *Physics of ice and snow as it affects thermal pressure*. Chalmers Univ Technol, Dept Hydraulics, Rept. Ser. A:1.

Berton, P. 1988. *The arctic grail*. New York: Viking Press.

Betts, M. 1988. Why are some icebergs green? *ANARE News*, 16. Australian Antarctic Division, Kingston, Tasmania.

Bilello, M. A. 1961. Formation, growth and decay of sea ice. *Arctic* 14(1):3–24.

———. 1980. Decay patterns of fast sea ice in Canada and Alaska. In *Sea ice processes and models*, ed. R. S. Pritchard, 313–26. Seattle: Univ Washington Press.

Binder, A. R. 1972. Black and white icebergs, Southern Ocean. *Mar Observ* 42:15–16.

Bindoff, N. L., G. D. Williams and I. Allison. 2001. Sea-ice growth and water-mass modification in the Mertz Glacier polynya, East Antarctica. *Ann Glaciol* 33:399–406.

Birkeland, K. W. 1998. Terminology and predominant processes associated with the formation of weak layers of near-surface faceted crystals in the mountain snow pack. *Arct Alp Res* 30:193–99.

Bitz, C. C., M. M. Holland, A. J. Weaver, and M. Eby. 2001. Simulating the ice thickness distribution in a coupled climate model. *JGR* 106:2441–64.

Bitz, C. C., and W. H. Lipscomb. 1999. An energy-conserving thermodynamic model of sea ice. *JGR* 104:15669–77.

Blindow, N. 1994. The central part of the Filchner-Ronne Ice Shelf, Antarctica: Internal structures revealed by 40 MHz minopulse RES. *Ann Glaciol* 20:365–71.

Blinov, L. K. 1965. The salt content of seawater and sea ice (R). *Trudy Gos Okeanograf Inst* 81:5–55.

Bogorodskii, V. V., and V. P. Gavrilo. 1980. *Ice: Its physical properties: Contemporary methods of glaciology* (R). Leningrad: Gidrometeoizdat.

Bohren, C. 1983. Colours of snow, frozen waterfalls and icebergs. *J Opt Soc Amer* 73:1646.

Bolling, G. F., and W. A. Tiller. 1960. Growth from the melt. Part II. Cellular interface morphology. *J Appl Phys* 31(11):2040–45.

Bombosch, A., and A. Jenkins. 1995. Modeling the formation and deposition of frazil ice beneath Filchner-Ronne Ice Shelf. *JGR* 100(C4):6983–92.

Borodachev, V. E., V. P. Gavrilo, and M. M. Kazanskii. 1994. *Glossary of marine ice terminology* (R). St. Petersburg: AARI.

Borodachev, V. E., V. P. Gavrilo, M. M. Kazanskii, and W. F. Weeks. 2000. Glossary of sea ice terminology. In *The Environmental Working Group Arctic Sea Ice Atlas*, CD-ROM No. 1. Boulder, CO: National Snow and Ice Data Center.

Borodkin, V. A., V. P. Gavrilo, S. M. Kovalev, and G. A. Lebedev. 1992. Influence of structural anisotropy of sea ice on its mechanical and electrical properties. In *Proc of the Second (1992) Internat Offshore and Polar Engineering Conf*, ed. M. S. Triantafyllou, J. S. Chung, K. Karal, and A. L. Tunik, 670–74. Internat Soc of Offshore and Polar Engineers, San Francisco.

Bourke, R. H., and R. P. Garrett. 1987. Sea ice thickness distribution in the Arctic Ocean. *CRST* 13:259–80.

Bourke, R. H., and A. S. McLaren. 1992. Contour mapping of Arctic Basin ice draft and roughness parameters. *JGR* 97(C11):17715–28.

Brace, W. F., and E. G. Bombolakis. 1963. A note on brittle crack growth in compression. *JGR* 68:3709.

Bragg, W. H. 1922. The crystal structure of ice. *Proc Phys Soc* 34:98–103.

Bratchie, I. 1984. Rheology of an ice-floe field. *Ann Glaciol* 5:23–28.

Bretterbauer, K. 1983. J. Payer, C. Weyprecht, and H. Wilczek: The promoters of international polar research. In *Historical events and people in geosciences*, ed. W. Schröder, 49–57. Selected papers from the symposia of the interdivisional commission on history of IAGA during the IUGG General Assembly, Hamburg. Peter Lang: Frankfurt am Main.

Brigham, L. W. 1991. *The Soviet maritime Arctic*. Naval Institute Press.

————. 2000. Sea ice variability in the Russian Arctic coastal seas: Influences on the Northern Sea Route. Ph.D. Dissertation. SPRI, Univ Cambridge.

Bromwich, D. H., and D. D. Kurtz. 1984. Katabatic wind forcing of the Terra Nova Bay polynya. *JGR* 89(C3):3561–72.

Bromwich, D. H., T. R. Parish, and C. A. Zorman. 1990. The confluence zone of the intense katabatic winds at Terra Nova Bay, Antarctica, as derived from airborne sastrugi surveys and mesoscale numerical modeling. *JGR* 95(D5):5495–5509.

Brown, R. A. 1980. Planetary boundary layer modeling for AIDJEX. In *Sea ice processes and models*, ed. R. S. Pritchard, 387–401. Seattle: Univ Washington Press.

Brown, R. D., and P. Cole. 1992. Interannual variability of landfast ice thickness in the Canadian High Arctic, 1950–1989. *Arctic* 45(3):273–84.

Bruns, B. P. 1937. Observations on the processes of formation and melting of ice (R). *Trudy Arkticheskogo Instituta* 83:75–97.

Brunt, D. 1932. Notes on radiation in the atmosphere. *Q J R Meteorol Soc* 58:389–420.

Buchanan, J. Y. 1874. Some observations on sea-water ice. *Proceed Roy Soc (Edinburgh)* 11:431.

————. 1887. On ice and brines. *Proceed Roy Soc (Edinburgh)* 14:129–47.

————. 1911. Ice and its natural history. *Proceed Roy Inst Gt Brit* 19:243–76.

Buckley, J. R., T. Gammelsrød, J. A. Johannessen, O. M. Johannessen, and L.-P. Røed. 1979. Upwelling: Oceanic structure at the edge of the arctic icepack in winter. *Science* 203:165–67.

Bulatov, V. 1997. The historical and current uses of the Northern Sea Route. Part IV. The administration of the Northern Sea Route (1917–1991). INSROP Working Paper No. 84-1997, IV.1.1. Norway: Fridtjof Nansen Inst.

Burden, R. P., and G. W Timco. 1995. A catalogue of sea ice ridges. Nat Res Council of Canada Report TR-1995-27. Ottawa.

Burk, S. D., R. W. Fett, and R. E. Englebretson. 1997. Numerical simulation of cloud plumes emanating from Arctic leads. *JGR* 102:16529–44.

Burke, A. K. 1940. *Sea ice* (R). Leningrad: Izd Glasevmorputi.

Burton, J. A., R. C. Prim, and W. P. Slichter. 1953. The distribution of solute in crystals grown from the melt. Part I. Theoretical. *J Chem Phys* 21:1987–91.

Butiagin, I. P. 1966a. Scale factor in the evaluation of ice strength (R). *Mater. VIII Vsesoiuz. mezhduvedom soveshchan geokriolog (merzlot.)* 5:73–78. Yakutsk.

————. 1966b. *Strength of ice and ice cover* (R). Novosibirsk: Izd "Nauka" Sibirskoe Otdelenie.

Butkovich, T. R. 1956. Strength studies of sea ice. *SIPRE* Research Report 20.

————. 1959. On the mechanical properties of sea ice, Thule, Greenland, 1957. *SIPRE* Research Report 54.

Buynitskiy, V. Kh. 1967. Structure, principal properties, and strength of Antarctic sea ice. *Sov Antarct Exped Inform Bull* 6(6):504–10.

Buzuyev, A. Ya., I. P. Romanov, and V. Ye. Fedyakov. 1997. Variability in snow distribution on ice in the Arctic Ocean (R). *Meteorologiya i Gidrologiya* 9:76–85.

Campbell, K. J., and A. S. Orange. 1974. The electrical anisotropy of sea ice in the horizontal plane. *JGR* 79(33):5059–63.

Campbell, N. J., and A. E. Collin. 1958. The discoloration of Foxe Basin ice. *J Fish Res Bd Can* 15:1175–88.

Campbell, W. J. 1965. The wind-driven circulation of ice and water in a polar ocean. *J Geophys Res* 70(14):3279–3301.

———. 1968. Sea ice dynamics. In *Arctic drifting stations*, ed. J. E. Sater, 189–96. Washington, DC: AINA.

Campbell, W. J., E. G. Jossberger, and N. M. Mognard. 1994. Southern Ocean wave fields during the austral winters, 1985–1988. In *The polar oceans and their role in shaping the global environment*, ed. O. M. Johannessen, R. D. Muench, and J. E. Overland, 421–34. Geophys Monogr 85, AGU.

Carmack, E. C. 1986. Circulation and mixing in ice-covered waters. In *Geophysics of sea ice*, ed. N. Untersteiner, 641–712. NATO Advanced Sciences Institutes Series, Series B: Physics. New York: Plenum.

———. 1990. Large-scale physical oceanography of polar oceans. In *Polar oceanography. Part A: Physical science*, 171–222. San Diego: Academic Press.

Carmack, E. C., K. Aagaard, J. H. Swift, R. W. Macdonald, F. A. McLaughlin, E. P. Jones, R. G. Perkin, J. N. Smith, K. M. Ellis, and L. R. Kilius. 1997. Changes in temperature and tracer distributions within the Arctic Ocean: Results from the 1994 Arctic Ocean section. *Deep-sea Res II* 44:1487–1502.

Carsey, F. D. 1980. Microwave observation of the Weddell Polynya. *Mon Wea Rev* 108(12):2032–44.

———, ed. 1992. *Microwave remote sensing of sea ice*. Monograph 68. Washington, DC: AGU.

Carslaw, H. S., and J. C. Jaeger. 1959. *Conduction of heat in solids*. Oxford: Clarendon.

Carsola, A. J. 1954. Microrelief on the Arctic sea floor. *Bull Amer Assoc Petrol Geol* 38:1587–1601.

Carte, A. E. 1961. Air bubbles in ice. *Proc Phys Soc (London)* 77(495):757–68.

Cavalieri, D. J., P. Gloersen, C. L. Parkinson, J. C. Comiso, and H. J. Zwally. 1997. Observed hemispheric asymmetry in global sea ice changes. *Science* 278:1104–6.

Cavalieri, D. J., and S. Martin. 1985. A passive microwave study of polynyas along the Antarctic Wilkes Land coast. In *Oceanology of the antarctic continental shelf*, ed. S. S. Jacobs, 227–52. Washington, DC: AGU (Antarctic Research Series).

———. 1994. The contribution of Alaskan, Siberian, and Canadian coastal polynyas to the cold halocline layer of the Arctic Ocean. *JGR* 99(C9):18343–62.

Cavalieri, D. J., and C. L. Parkinson. 2008. Antarctic sea ice variability and trends, 1979–2006. *JGR* 113:C07004, doi:10.1029/2007JC004564.

Chalmers, B. 1964. *Principles of solidification*. New York: Wiley.

Chalmers, B., and R. V. Williamson. 1965. Crystal multiplication without nucleation. *Science* 148(3676): 1717–18.

Chao, S. Y. 1999. Ocean feedback to wind-driven coastal polynyas. *JGR* 104(C8):18073–85.

Chapman, W. L., and G. Gawarkiewicz. 1997. Shallow convection and buoyancy equilibration in an idealized coastal polynya. *J Phys Oceanog* 27:555–66.

Chapman, W. L., W. J. Welch, K. P. Bowman, J. Sacks, and J. E. Walsh. 1994. Arctic sea ice variability: Model sensitivities and a multidecadal simulation. *JGR* 99(C1):919–35.

Chavanne, J. 1875. Die Eisverhältnisse im Arktischen Polarmeere und ihre periodischen Veränderungen. *Petermanns Geograph Mitt* 21:134–43, 245–80.

Chen, A. C. T., and J. Lee. 1986. Large-scale ice strength tests at slow strain rates. *Proc OMAE*: 374–78.

Cherepanov, N. V. 1957. Using the methods of crystal optics for determining the age of drift ice (R). *PA* 2:179–84.

———. 1964a. Structure of ice of great thickness (R). *Trudy AANII* 267:13–18.

———. 1964b. Determination of periods of movement of ice fields based on the formation characteristics of their crystalline structure (R). *Trudy AANII* 267:48–53.

———. 1968. The role of the thermal regime of the water reservoir in the formation of the ice structure (R). *PAiA* 29:55–63.

———. 1971a. Spatial arrangement of sea ice crystal structure (R). *PAiA* 38:137–40.

———. 1971b. Classification of the crystalline structure of Arctic ice (R). *PAiA* 40:75–80.

———. 1974. Classification of ice of natural water bodies. *IEEE Ocean '74*: 97–101.

———. 1976. A classification of the ice of natural bodies of water (R). *Trudy AANII* 11(331):77–99.

Cherepanov, N. V., V. I. Fodotov, and K. P. Tyshko. 1997. Sea ice structure (R). In *Sea ice*, ed. I. E. Frolov and V. P. Gavrilo, 36–67. St. Petersburg: Gidrometeoizdat.

Cherepanov, N. V., and A. Kamyshnikova. 1973. Sizes and shapes of congealed-ice crystals. In *Studies in ice physics and ice engineering* 300:170–6. Jerusalem: Israel Program for Scientific Translations Ltd.

Cherepanov, N. V., and A. M. Kozlovskiy. 1973a. Autumn formation of underwater ice in the Lazarev Ice Shelf area. *Sov Antarct Exped Inform Bull* 8(8):438–40.

———. 1973b. Structure and some properties of old Antarctic fast ice. *Sov Antarct Exped Inform Bull* 8(7):389–92.

———. 1973c. Underwater ice in the coastal waters of Antarctica. *Sov Antarct Exped Inform Bull* 8(6):335–38.

———. 1973d. Classification of Antarctic sea ice by the conditions of its formation (R). *PAiA* 42:49–58.

Cherepanov, N. V., and M. V. Strakhov. 1989. Laboratory studies of oriented growth of ice crystals in a moving stream of water. In *Electro-physical and physical-mechanical properties of ice*, ed. V. V. Bogorodskii and V. P. Gavrilo, 125–33. Leningrad: Gidrometeoizdat.

Chernigovskii, N. T. 1963. Radiational properties of the Central Arctic ice cover (R). *Trudy AANII* 253: 249–60.

Cho, C. H., S. Singh, and G. W. Robinson. 1996. An explanation of the density maximum in water. *Phys Rev Lett* 76(10):1651–54.

Cho, H., P. R. Stepson, L. A. Barrie, J. P. Cowin, and R. Zaveri. 2002. NMR investigation of the quasi-brine layer in ice/brine mixtures. *J Phys Chem B* 106:11226–32.

Christensen, J. H., et al. 2007. Regional climate projections. In *Climate change 2007: The physical science basis. Contribution of Working Group I to the Fourth Assessment Report of the Intergovernmental Panel on Climate Change*, ed. S. Solomon et al., 849–940. New York: Cambridge Univ Press.

Chu, P. C. 1987. An instability theory of ice–air interaction for the formation of ice edge bands. *JGR* 92(C7):6966–70.

Church, J. A., and N. J. White. 2006. A 20th century acceleration in global sea-level rise. *GRL* 33: L01602, doi:10.1029/2005GL024826.

Chylek, P., M. K. Dubney, and G. Lesins. 2006. Greenland warming of 1920–1930 and 1995–2005. *GRL* 33: L11707, doi:10.1029/2006GL026510.

Coachman, L. K., and K. Aagaard. 1988. Transports through Bering Strait: Annual and interannual variability. *JGR* 93(C12):15535–39.

Coachman, L. K., K. Aagaard, and R. B. Tripp. 1975. *Bering Strait: The regional physical oceanography.* Seattle: Univ Washington Press.

Coates, D. E., S. V. Subramanian, and G. R. Purdy. 1968. Solid-liquid interface stability during solidification of dilute ternary alloys. *Trans Mettalurgical Soc AIME* 242:800–9.

Colbeck, S. C. 1973. Theory of metamorphism of wet snow. *CRREL Research Rept* 313.

———. 1979. Grain clusters in wet snow. *J Colloid Interface Sci* 72(3):371–84.

———. 1982. An overview of seasonal snow metamorphism. *Rev Geophys Space Phys* 20(1):45–61.

Cole, D. M., L. D. Gould, and W. D. Burch. 1985. A system for mounting end caps on ice specimens. *J Glaciol* 31(109):362–65.

Cole, D. M., H. Eicken, K. Frey, and L. H. Shapiro. 2004. Observations of banding in first-year Arctic sea ice. *JGR* 109:C08012, doi:10: 1029/2003JC001993.

Cole, D. M., and L. M. Shapiro. 1998. Observations of brine drainage networks and microstructure in first-year sea ice. *JGR* 103(C10):21739–50.

Cole, D. M., L. H. Shapiro, W. F. Weeks, C. Byers, J. P. Dempsey, R. M. Adamson, V. F. Petrenko, and O. V. Gluschenkov. 1995. In-situ and laboratory measurements of the physical and mechanical properties of first-year sea ice. In *Ice Mechanics–95*, 161–78. UCLA: ASME.

Colony, R. 1990. Seasonal mean ice motion in the Arctic Basin. In *Proceed Internat Conf on the Role of the Polar Regions in Global Change*, 290. Fairbanks: Univ Alaska.

Colony, R. L., I. Rigor, and K. Runciman-Moore. 1991. A summary of observed ice motion and analyzed atmospheric pressure in the Arctic Basin, 1979–1990. Appl Phys Lab, Univ Washington, Seattle, Tech Rept, APL-UW TR 9112.

Colony, R., and A. S. Thorndike. 1985. Sea ice motion as a drunkard's walk. *JGR* 90(C1):965–74.

Comiso, J. C. 1991. Satellite remote sensing of the Polar Oceans. *J Mar Syst* 2(3–4):395–434.

———. 1993. Ice regimes in the Antarctic and Arctic using satellite multichannel passive microwave data. In *IGARSS '93*, 2:843–46. Tokyo: IEEE.

———. 2002. A rapidly declining perennial sea ice cover in the Arctic. *GRL* 29(20):doi:10.1029/2002GL015650.

———. 2003. Large-scale characteristics and variability of the global sea ice cover. In *Sea ice: An introduction to its physics, chemistry, biology and geology*, ed. D. N. Thomas and G. S. Dieckmann, 112–42. Oxford: Blackwell.

———. 2006. Abrupt decline in the Arctic sea ice cover. *GRL* 33(L18504):doi:10.1029/2002GL015650.

Comiso, J. C., D. J. Cavalieri, C. P. Parkinson, and P. Gloersen. 1997. Passive microwave algorithms for sea ice concentration: A comparison of two techniques. *Remote Sens Environ* 60:357–84.

Comiso, J. C., and A. L. Gordon. 1987. Recurring polynyas over the Cosmonaut Sea and the Maud Rise. *JGR* 92(C3):2819–33.

———. 1996. Cosmonaut Polynya in the Southern Ocean: Structure and variability. *JGR* 101(C8):18297–313.

———. 1998. Interannual variability in summer sea ice minimum, coastal polynyas and bottom water formation in the Weddell Sea. In *Antarctic sea ice: Physical processes, interactions and variability*, ed. M. O. Jeffries, 293–315. AGU, Antarctic Research Series, vol. 74.

Comiso, J. C., and C. L. Parkinson. 2004. Satellite-observed changes in the Arctic. *Physics Today* 57(8):38–44.

Comiso, J. C., C. L. Parkinson, R. Gersten, and L. Stock. 2008. Accelerated decline in the Arctic sea ice cover. *GRL* 35: L01703, doi:10.1029/2007GL031972.

Comiso, J. C., P. Wadhams, W. B. Krabill, R. N. Swift, J. P. Crawford, and W. B. Tucker III. 1991. Top/bottom multisensor remote sensing of Arctic sea ice. *JGR* 96(C2):2693–2709.

Connolley, W. M., and T. J. Bracegirdle. 2007. An Antarctic assessment of IPCC AR4 coupled models. *GRL* 34:L22505, doi:10.1029/2007GL031648.

Cook, A. J., A. J. Fox, D. G. Vaughan, and J. G. Ferrigno. 2005. Retreating glacier fronts on the Antarctic Peninsula over the past half-century. *Science* 308(22 April):541–44.

Coon, M. D. 1974. Mechanical behavior of compacted Arctic floes. *J Petrol Technol* 26:466–70.

———. 1980. A review of AIDJEX modeling. In *Sea ice processes and models*, ed. R. S. Pritchard, 12–27. Seattle: Univ Washington Press.

Coon, M. D., G. S. Knoke, D. C. Echert, and R. Pritchard. 1998. The architecture of an elastic-plastic sea ice mechanics constitutive law. *JGR* 103:21915–25.

Coon, M. D., R. Kwok, G. Levy, M. Pruis, H. Schreyer, and D. Sulsky. 2007. Arctic Ice Dynamics Joint Experiment (AIDJEX) assumptions revisited and found inadequate. *JGR* 112:C11S90, doi:10.1029/2005JC003393.

Coon, M. D., and R. Pritchard. 1974. Application of an elastic-plastic model of Arctic pack ice. In *The coast and shelf of the Beaufort Sea*, ed. J. C. Reed and J. E. Slater, 173–94. AINA.

Coriell, S. R., G. B. McFadden, P. W. Voorhees, and R. F. Sekerka. 1987. Stability of a planar interface during solidification of a multicomponent system. *J Crystal Growth* 82:295–302.

Cottier, F. H., H. Eicken, and P. Wadhams. 1999. Linkages between salinity and brine channel distribution in young sea ice. *JGR* 104(C7):15859–71.

Cox, G. F. N. 1983. Thermal expansion of saline ice. *J Glaciol* 29(103):425–32.

Cox, G. F. N., and J. B. Johnson. 1983. Stress measurements in ice. *CRREL Rept* 83-23.

Cox, G. F. N., and J. A. Richter-Menge. 1985a. Triaxial compression testing of ice. *Conf Arctic* 85:476–88. San Francisco: ASCE.

———. 1985b. Tensile strength of multiyear pressure ridge sea ice samples. *J Energy Resources Technol* 107(3):375–80.

Cox, G. F. N., J. A. Richter-Menge, W. F. Weeks, H. W. Bosworth, N. Perron, M. Mellor, and G. Durell. 1985. Mechanical properties of multiyear sea ice, Phase II: Test results. *CRREL Rept* 85-16.

Cox, G. F. N., J. A. Richter, W. F. Weeks, and M. Mellor. 1984. A summary of the strength and modulus of ice samples from multiyear pressure ridges. In *Proceed 3rd Internat Offshore Mechanics and Arctic Engineering Sympos* III:126–33. New York: ASME.

———. 1985. A summary of the strength and modulus of ice samples from multi-year pressure ridges. *J Energy Resources Technol* 107(3):93–98.

Cox, G. F. N., J. A. Richter, W. F. Weeks, M. Mellor, and H. Bosworth. 1984. The mechanical properties of multi-year sea ice, Phase I: Test results. *CRREL Rept* 84-9.

Cox, G. F. N., and W. F. Weeks. 1974. Salinity variations in sea ice. *J Glaciol* 13:109–20.

———. 1975. Brine drainage and initial salt entrapment in sodium chloride ice. *CRREL Research Rept* 345.

———. 1983. Equations for determining the gas and brine volumes in sea ice samples. *J Glaciol* 29(102):306–16. (*CRREL Rept* 82-30.)

———. 1986. Changes in the salinity and porosity of sea-ice samples during shipping and storage. *J Glaciol* 32(112):371–75.

———. 1988a. Numerical simulations of the profile properties of undeformed first-year sea ice during the growth season. *JGR* 93(C10):12449–60.

———. 1988b. Profile properties of undeformed first-year sea ice. *CRREL Rept* 88-13.

Cramer, H. 1946. *Mathematical methods of statistics*. Sweden: Hugo Gebers Förlag.

Crary, A. P. 1958. Arctic ice island and shelf ice studies. Part 1. *Arctic* 11(1):3–42.

———. 1960. Arctic ice island and shelf ice studies. Part 2. *Arctic* 13(1):32–50.

Croasdale, K., G. Comfort, and K. Been. 2005. Investigations of ice limits to ice gouging. In *18ᵗʰ POAC Conference* 1:23–32. Potsdam, NY: Clarkson University.

Crocker, G. B., and P. Wadhams. 1989. Modelling Antarctic fast ice growth. *J Glaciol* 35(119):3–8.

Cross, W. 2000. *Disaster at the pole*. New York: Lyons Press.

Cumming, W. 1952. The dielectric properties of ice and snow at 3.2 cm. *J App Phys* 23:768–73.

Cunningham, F. F., R. Kwok, and J. Banfield. 1994. Ice lead orientation characteristics in the winter Beaufort Sea. Internat Geoscience and Remote Sensing Symp, IEEE. Pasadena, CA.

Currier, J. H., and E. M. Schulson. 1982. The tensile strength of ice as a function of grain size. *Acta Metall* 30:1511–14.

Curry, J. A., and E. E. Ebert. 2006. Annual cycle of radiation fluxes over the Arctic Ocean: Sensitivity to cloud optical properties. *J Climatology* 5:1267–80.

Curry, J. A., J. L. Schramm, and E. E. Ebert. 1995. Sea ice-albedo climate feedback mechanism. *J Climate* 8:240–47.

Curry, J. A., and P. J. Webster. 1999. *Thermodynamics of atmospheres and oceans*. International Geophysics Series vol 65. San Diego: Academic Press.

Daly, S. F. 1984. Frazil ice dynamics. *CRREL Monograph* 84-1.

Damiano, V., and M. Herman. 1959. Cellular substructure in Zn crystals grown from the melt. *Trans Metal Soc AIME* 215:136–38.

Darby, D. A. 2003. Sources of sediment found in sea ice from the western Arctic Ocean, new insights into processes on entrainment and drift patterns. *JGR* 108(C8):3257, doi:10.1029/2002JC001350.

Darby, M. S., A. J. Wilmott, and T. A. Sommerville. 1995. On the influence of coastal orientation on the steady state width of a latent heat polynya. *JGR* 100(C7):13625–33.

Darwin, C. R. 1855. On the power of icebergs to make rectilinear, uniformly-directed grooves across a submarine undulatory surface. *London, Edinburgh, and Dublin Phil Mag, J Sci* 10:96–98.

Dash, J. G., H. Fu, and J. S. Wettlaufer. 1995. The premelting of ice and its environmental consequences. *Rep Prog Phys* 58:115–67.

Davis, N. R., and P. Wadhams. 1995. A statistical analysis of Arctic pressure ridge morphology. *JGR* 100(C6):10915–25.

Davis, S. H., H. E. Huppert, U. Müller, and M. G. Worster. 1992. *Interactive dynamics of convection and solidification*. NATO ASI E219. Kluwer.

Dayton, P. K., and S. Martin. 1971. Observations of ice stalactites in McMurdo Sound, Antarctica. *JGR* 76:1595–99.

Dayton, P. K., G. A. Robillard, and A. L. DeVries. 1969. Anchor ice formation in McMurdo Sound, Antarctica, and its biological effects. *Science* 163:273–74.

De Angelis, H., and P. Skvarca. 2003. Glacier surge after ice shelf collapse. *Science* 299(5612):1560–62.

DeBondt, S., L. Froyen, and A. Deruyttere. 1992. Electrical conductivity of composites: A percolation approach. *J Mater Sci* 27:1983–88.

Debye, P. 1929. *Polar molecules*. New York: Chemical Catalog Company.

de LaChapelle, S., P. Duval, and B. Baudelet. 1995. Compressive creep of polycrystalline ice containing a liquid phase. *Scripta Met et Mat* 33(3):447–50.

Delgado, J. P. 1999. *Across the top of the world: The quest for the Northwest Passage*. New York: Facts on File.

Dempsey, J. P. 1989. The fracture toughness of ice. *Proc. IUTAM/IAHR Symp on Ice/Structure Interaction*. St. John's, Newfoundland, 109–145.

Dempsey, J. P., R. M. Adamson, and S. V. Mulmule. 1999. Scale effects on the in situ tensile strength and fracture of ice. Part II: first-year sea ice at Resolute, NWT. *Internat J Fracture* 95:347–366.

Dempsey, J. P., A. C. Palmer, and D. S Sodhi. 2001. High pressure zone formation during compressive ice failure. *Eng Fracture Mech* 68:1961–74.

Dempsey, J. P., and Y. D. S. Rajapakse, eds. 1995. *Ice mechanics-95*. Vol AMD-207. ASME.

Dempsey, J. P., and H. H Shen, eds. 2001. IUTAM symposium on scaling laws in ice mechanics and ice dynamics. In *Solid mechanics and its applications*, vol 94, ed. G. M. L. Gladwell. Dordrecht: Kluwer.

Den Hartog, G., S. D. Smith, R. J. Anderson, D. R. Topham, and R. G. Perkin. 1983. An investigation of a polynya in the Canadian archipelago. III: Surface heat flux. *JGR* 88(C5):2911–16.

Deser, C., and H. Teng. 2008. Evolution of Arctic sea ice concentration trends and the role of atmospheric circulation forcing. *GRL* 35:L02504, doi:10.1029/2007 GL032023.

DeWeaver, E. T., C. M. Bitz, and L.-B. Tremblay, eds. 2008. *Arctic sea ice decline: Observations, projections, mechanisms, and implications*. Washington, DC: AGU.

DeWeaver, E. T., E. C. Hunke, and M. M. Holland. 2008. Comment on the reliability of simulated Arctic sea ice in global climate models by I. Eisenman, N. Untersteiner, and J. S. Wettlaufer. *GRL* 35:L04501, doi:10.1029/2007GL031325.

Dickens, G. R., J. R. O'Neil, D. K. Rea, and R. M. Owen. 1995. Dissociation of oceanic methane hydrate as a cause of the carbon isotope excursion at the end of the Paleocene. *Paleoceanog* 10(6):965–71.

Dickins, D. F., and V. F. Wetzel. 1981. Multiyear pressure ridge study, Queen Elizabeth Islands. Proceedings of the Sixth International Conference on Port and Ocean Engineering under Arctic Conditions (POAC), II, 765–75.

Dickson, B. 1999. All change in the Arctic. *Nature* 397:389–91.

Dieckmann, G. S., G. Rohardt, H. Hellmer, and J. Kipfstuhl. 1986. The occurrence of ice platelets at 250 m depth near the Filchner Ice Shelf and its significance for sea ice biology. *Deep-Sea Res* A(33):141–48.

Dieckmann, G. S., C. Hemleben, and M. Spindler. 1987. Biogenic and mineral inclusions in a green iceberg from the Weddell Sea, Antarctica. *Polar Biol* 7:31–33.

Dieckmann, G. S., G. Nehrke, S. Papadimitriou, J. Göttlicher, R. Steininger, H. Kennedy, D. Wolf-Gladrow, and D. Thomas. 2008. Calcium carbonate as ikaite crystals in Antarctic sea ice. *GRL* 35:L08501. doi:10.1029/2008GL033540.

Dierking, W. 1995. Laser profiling of the ice surface topography during the Winter Weddell Gyre Study 1992. *JGR* 100(C3):4807–20.

Divine, D. V., and C. Dick. 2006. Historical variability of sea ice edge position in the Nordic Seas. *JGR* 111:C01001, doi:10, 1029/2004JC002851.

Doble, M. J., M. D. Coon, and P. Wadhams. 2003. Pancake ice formation in the Weddell Sea. *JGR* 108:C73209, doi: 10.1029/2002JC001373.

Dobrowolski, A. B. 1923. *Historja Naturalna Lodu*. Warsaw: Edition de la Kaswa pompcy dla asob pracujacych na polu nakowem inuemia D–ra Mianowskiego.

———. 1933. Sur la necessité de la fondation d'une institution internationale pour l'étude de la glace. Leningrad: *IV Conference on the Hydrology of the Baltic States*.

Doronin, Yu. P. 1963. On the heat balance of the central Arctic. *Trudy AANII* 253:178–84.

———. 1969. *Thermal interaction of the atmosphere and the hydrosphere in the Arctic* (R). Leningrad: Gidrometeozidat.

———. 1997. The growth and melting of sea ice (R). In *Sea ice*, ed. I. E. Frolov and V. P. Gavrilo, 107–125. St. Petersburg: Gidrometeoizdat.

Doronin, Yu. P., and D. E. Kheisin. 1977. *Sea ice*. Translation for the National Science Foundation, TT–75–52088. New Delhi: Amerind Publishing Co.

Drake, M. J. 2005. Origin of water in the terrestrial planets. *Meteoritics & Planetary Science* 40(4):1.

Drinkwater, M. R., and G. B. Crocker. 1988. Modelling changes in the dielectric and scattering properties of young snow-covered ice at Ghz frequencies. *J Glaciol* 34:274–82.

Drobot, S. D., J. A. Maslanik, and C. Fowler. 2006. A long-range forecast of Arctic summer sea-ice minimum extent. *GRL* 33:L10501, doi:10.1029/2006GL0262216.

Drucker R., S. Martin, and R. Moritz. 2003. Observations of ice thickness and frazil ice in the St. Lawrence Island polynya from satellite imagery, upward looking sonar, and salinity/temperature moorings. *JGR* 108(C5):3149, doi:10.1029/2001JC001213.

Dukowicz, J. K., and J. R. Baumgardner. 2000. Incremental remapping as a transport/advection algorithm. *J Comput Phys* 160:318–35.

Dunbar, M. 1960. Thrust structures in young sea ice. *J Glaciol* 3(28):724–32.

———. 1962. Note on the formation process of thrust structures in young sea ice. *J Glaciol* 4(32):147–50.

Dunbar, M., and W. F. Weeks. 1975. Interpretation of young sea ice in the Gulf of St. Lawrence using side-looking radar and infrared imagery. *CRREL Res Rept* 337.

Dykins, J. E. 1963. Construction of sea ice platforms. In *Ice and snow: Properties, processes, and applications*, ed. W. D. Kingery, 289–301. Cambridge, MA: MIT Press.

———. 1967. Tensile properties of sea ice grown in a confined system. In *Proc Internat Conf on Low Temperature Science. Physics of Snow and Ice*, ed. H. Oura, I(1):523–37. Sapporo: ILTS.

———. 1971. *Ice engineering: Material properties of saline ice for a limited range of conditions.* Tech Rept R720. Port Hueneme, CA: Naval Civ Eng Lab.

Ebert, E. E., and J. A. Curry. 1993. An intermediate one-dimensional thermodynamic sea ice model for investigating ice-atmosphere interactions. *JGR* 98(C6):10085–109.

Ebert, E. E., J. L. Schramm, and J. A. Curry. 1995. Disposition of solar radiation in sea ice and the upper ocean. *JGR* 100(C8):15965–75.

Edie, D. D, and D. J. Kirwan. 1973. Impurity trapping during crystallization from melts. *I&EC Fundamentals* 12:100–6.

Efimova, N. A. 1961. On methods of calculating monthly values of net longwave radiation (R). *Meteorol Gidrol* 10:28–33.

Eicken, H. 1991. Quantifizierung von Meereiseigenschaften: Automatische Bildanalyse von Dünnschnitten und Paramatrisierung von Chlorophyll- und Salzgehaltsverteilungen. Alfred-Wegener-Institut für Polar- und Meeresforschung *Berichte zur Polarforschung* 82:1–105.

———. 1992. Salinity profiles of Antarctic sea ice: Field data and model results. *JGR* 97(C10):15545–57.

———. 1993. Automated image analysis of ice thin sections: Instrumentation, methods and extraction of stereological parameters. *J Glaciol* 39(132):341–52.

———. 1994. Structure of under-ice melt ponds in the Central Arctic and their effect on the sea ice cover. *Limnol Oceanog* 39(3):682–94.

———. 1998. Deriving modes and rates of ice growth in the Weddell Sea from microstructural, salinity and stable-isotope data. In *Antarctic sea ice: Physical processes, interactions and variability*, ed. M. O. Jeffries, 89–122. AGU, Antarctic Research Series, vol. 74.

———. 2003. From the microscopic to the macroscopic to the regional scale: Growth, microstructure and properties of sea ice. In *Sea ice: An introduction to its physics, biology, chemistry, and geology*, ed. D. N. Thomas and G. S. Dieckmann, 22–81. London: Blackwell.

Eicken, H., C. Bock, R. Wittig, H. Miller, and H.-O. Poertner. 2000. Magnetic resonance imaging of sea ice pore fluids: Methods and thermal evolution of pore microstructure. *CRST* 31(3):207–25.

Eicken, H., H. Fischer, and P. Lemke. 1995. Effects of the snow cover on Antarctic sea ice and potential modulation of its response to climate change. *Ann Glaciol* 21:369–76.

Eicken, H., T. C. Grenfell, D. K. Perovich, J. A. Richter-Menge, and K. Frey. 2004. Hydraulic controls of summer Arctic pack ice albedo. *JGR* 109:C08007, doi.10.1029.2003JC001989.

Eicken, H., H. R. Krouse, D. Kadko, and D. K. Perovich. 2002. Tracer studies of pathways and rates of meltwater transport through Arctic summer sea ice. *JGR* 107:C10, 8046, doi.10.1029/2000 JC000583.

Eicken, H., and M. A. Lange. 1989a. Sea ice thickness: The many vs. the few. *Geophys Res Lett* 16(6):495–98.

———. 1989b. Development and properties of sea ice in the coastal regime of the southeastern Weddell Sea. *JGR* 94(C6):8193–8206.

———. 1991. Image analysis of sea ice thin sections: A step towards automated texture classification. *Ann Glaciol* 15:204–9.

Eicken, H., M. A. Lange, and S. F. Ackley. 1990. Quantification of sea ice textures through automated digital analysis. In *Sea ice properties and processes*, ed. S. F. Ackley and W. F. Weeks, 28–32. CRREL Report 90-1.

Eicken, H., M. A. Lange, H.-W. Hubberten, and P. Wadhams. 1994. Characteristics and distribution patterns of snow and meteoric ice in the Weddell Sea and their contributions to the mass balance of sea ice. *Ann Geophys* 12:80–93.

Eicken, H., M. Lensu, M. Leppäranta, W. B. Tucker III, A. J. Gow, and O. Salmela. 1995. Thickness, structure and properties of level summer multiyear sea ice in the Eurasian sector of the Arctic Ocean. *JGR* 100(11):22697–710.

Eicken, H., H. Oerter, H. Miller, W. Graf, and J. Kipfstuhl. 1994. Textural characteristics and impurity content of meteoric and marine ice in the Ronne Ice Shelf, Antarctica. *J Glaciol* 40(135):386–98.

Eide, L. I., and S. Martin. 1975. The formation of brine drainage features in young sea ice. *J Glaciol* 14(70):137–54.

Eisenberg, D., and W. Kauzmann. 1969. *The structure and properties of water*. Oxford: Clarendon.

Eisenman, I., N. Untersteiner, and J. S. Wettlaufer. 2007. On the reliability of simulated Arctic sea ice in global climate models. *GRL* 34:L10501, doi:10.1029/ 2007GL 029914.

———. 2008. Reply to comment by E. T. DeWeaver et al. On the reliability of simulated Arctic sea ice in global climate models. *GRL* 35:L04502, doi:10.1029/2007GL032173.

Ekman, W. V. 1905. On the influence of the Earth's rotation on ocean currents. *Arkiv f Mat Astrom och Fysik* 2(11):1–52.

Elachi, C., and J. van Zyl. 2006. *Introduction to the physics and techniques of remote sensing*. New York: Wiley-Interscience.

Elbaum, C. 1959. Substructures in crystals grown from the melt. *Prog Met Phys* 8:203–53.

Enkvist, E. 1972. On the ice resistance encountered by ships operating in the continuous mode of ice breaking. *Svenska Tekniska Vetenskapsakademien i Finland* Report 24.

Environmental Working Group. 1997. Joint U.S.–Russian atlas of the Arctic Ocean: Oceanography atlas for the winter period (CD-ROM). Silver Springs, MD: Nat Oceanog Data Center.

———. 2000a. Joint U.S.–Russian arctic atlas: The arctic climatology project, arctic meteorology and climate atlas (CD-ROM). Boulder, CO: National Snow and Ice Data Center, Univ of Colorado.

———. 2000b. Joint U.S.–Russian sea ice atlas: Sea ice atlas for the arctic ocean (CD-ROM). Boulder, CO: National Snow and Ice Data Center, Univ of Colorado.

Epstein, S. 1956. Variation of the $^{18}O/^{16}O$ ratio of fresh water and ice. In *Nuclear Processes in Geologic Settings*, 20–24. US Nat Acad Sci Pub 400.

Epstein, S., and T. Mayeda. 1953. Variation of the ^{18}O content of water from natural sources. *Geochem Cosmochem Acta* 4: 213–24.

Erlingsson, B. 1988. Two-dimensional deformation patterns in sea ice. *J Glaciol* 34(118):301–8.

———. 1991. The propagation of characteristics in sea-ice deformation fields. *Ann Glaciol* 15:73–80.

Evans, R. J. 1971. Cracks in perennial sea ice due to thermally induced stress. *JGR* 76(33):8153–55.

Evans, R. J., and D. A. Rothrock. 1975. Stress fields in pack ice. In *Third Internat Symp on Ice Problems*, 527–39. Hanover, NH: IAHR.

Evans, R. J., and N. Untersteiner, N. 1971. Thermal cracks in floating ice sheets. *JGR* 76(3):694–703.

Evans, S. 1965. Dielectric properties of ice and snow: A review. *J Glaciol* 5(42):773–92.

Ewing, M., and A. P. Crary. 1934. Propagation of elastic waves in ice. II. *Physics* 5:181–84.

Ewing, M., A. P. Crary, and A. M. Thorne. 1934. Propagation of elastic waves in ice. I. *Physics* 5:165–68.

Fahrbach, E., M. Knoche, and G. Rohardt. 1991. An estimate of water mass transformation in the southern Weddell Sea. *Mar Chem* 35:25–44.

Fairbairn, H. W. 1954. *Structural petrology of deformed rocks*. Cambridge, MA: Addison-Wesley.

Farhadieh, R., and R. S. Tankin. 1972. Interferometric study of freezing of sea water. *JGR* 77(9):1647–57.

Fedotov, V. I., and N. V. Cherepanov. 1991. The initial ice layer and its role in the formation of the crystalline structure of the ice cover (R). In *Physical and technical problems of ice research*, ed. V. P. Gavrilo and G. A. Lebedev, 7–17. Leningrad: AARI, Gidrometeoizdat.

Fedotov, V. I., N. V. Cherepanov, and K. P. Tyshko. 1998. Some features of the growth, structure and metamorphism of East Antarctic landfast sea ice. In *Antarctic sea ice: Physical processes, interactions and variability*, ed. M. O. Jeffries, 343–54. AGU, Antarctic Research Series, vol. 74.

Feltham, D. L., N. Untersteiner, J. S. Wettlaufer, and M. G. Worster. 2006. Sea ice as a mushy layer. *GRL* 33:L14501, doi:10.1029/2006GL026290.

Fernandez, R., and J. Barduhn. 1967. The growth rate of ice crystals. *Desalination* 3:330–42.

Ferrick, M. G., and K. J. Claffey. 1992. Vector analysis of ice fabric data. *J Glaciol* 39(132):292–302 (also CRREL Report 92-1).

Feruck, L., J. W. Spyker, and W. H. W. Husband. 1972. Computing salinity profiles in ice. *Canad J Phys* 50(2):264–67.

Fetterer, F., and N. Untersteiner. 1998. Observations of melt ponds on Arctic sea ice. *JGR* 103(C11):24812–35.

Flato, G. M. 1995. Spatial and temporal variability of Arctic ice thickness. *Ann Glaciol* 21:323–29.

Flato, G. M., and R. D. Brown. 1996. Variability and climate sensitivity of land-fast Arctic sea ice. *JGR* 101:25767–77.

Flato, G. M., and W. D. Hibler III. 1990. On a simple sea-ice dynamics model for climate studies. *Ann Glaciol* 14:72–77.

———. 1992. On modeling pack ice as a cavitating fluid. *J Phys Oceanog* 22:626–51.

———. 1995. Ridging and strength in modelling the thickness distribution of Arctic sea ice. *JGR* 100:C9,18611–26.

Fleagle, R. G., and J. Businger. 1980. *An introduction to atmospheric physics*. New York: Academic Press.

Fleming, F. 1998. *Barrow's boys*. London: Granata Books.

Fleming, G. H., and A. J. Semtner. 1991. A numerical study of interannual ocean forcing on Arctic ice. *JGR* 96(C3):4589–4603.

Flemings, M. C. 1974. *Solidification processing*. New York: McGraw-Hill.

Fletcher, J. O. 1965. *The heat budget of the Arctic Basin and its relation to climate*. RAND Corporation, R-444-PR.

Fletcher, N. H. 1970. *The chemical physics of ice*. Cambridge: Cambridge Univ Press.

Fletcher, R. J. 1990. Military radar defense lines of northern North America: A historical geography. *Polar Record* 26(159):265–76.

Flocco, D., and D. L. Feltham. 2007. A continuum model of melt pond evolution on Arctic sea ice. *JGR* 112:C08016, doi:10.1029/2006JC003836.

Foldvik, A., T. Gammelsrød, S. Østerhus, E. Fahrbach, G. Rohardt, M. Schröder, K. W. Nicholls, L. Padman, and R. A. Woodgate. 2004. Ice shelf water overflow and bottom water formation in the southern Weddell Sea. *JGR* 109:C02015, doi: 10.1029/2003JC002008.

Foldvik, A., and T. Kvinge. 1974. Conditional instability of sea water at the freezing point. *Deep-Sea Res* 21(3):169–74.

———. 1977. Thermohaline convection in the vicinity of an ice shelf. In *Polar oceans*, ed. M. J. Dunbar, 247–55. Calgary, Alberta: AINA.

Forsberg, R., and H. Skourup. 2005. Arctic Ocean gravity, geoid and sea-ice freeboard heights from ICESat and GRACE. *GRL* 32:L21502, doi:10.1029/2005GL023711.

Fox, C., and V. A. Squire. 1990. Reflection and transmission characteristics at the edge of shore fast sea ice. *JGR* 95(C7):11629–39.

———. 1994. On the oblique reflexion and transmission of ocean waves from shore fast sea ice. *Phil Trans Roy Soc* A347:1682, 185–218.

Francis, J. A., and E. Hunter. 2006. New insight into the disappearing Arctic sea ice. *Eos Trans AGU* 87:509.

———. 2007. Drivers of declining sea ice in the Arctic winter: A tale of two seas. *GRL* 34:L17503, doi:10.1029/2007GL030995.

Francis, J. A., E. Hunter, E., J. R. Key, and X. Wang. 2005. Clues to variability in Arctic minimum sea ice extent. *GRL* 32:L21501, doi:10.1029/2005GL024376.

Frankenstein, G. E. 1969. Ring tensile strength studies of ice. *CRREL Technical Report* 172.

———. 1970. The flexural strength of sea ice as determined from salinity and temperature profiles. In *Tech Memo No. 98*, 66–73. Ottawa: Nat Research Council of Canada (Associate Committee on Geotechnical Research).

Frankenstein, G. E., and R. Garner. 1967. Equations for determining the brine volume of sea ice from –0.5° to –22.9°C. *J Glaciol* 6(48):943–44.

Frederking, R. M. W. 1977. Plane-strain compressive strength of columnar-grained and granular snow-ice. *J Glaciol* 18:505–16.

Frederking, R. M. W., and F.-U. Häusler. 1978. The flexural behavior of ice from *in situ* cantilever beam tests. In *Proc IAHR Symposium on Ice Problems* vol 1, 197–215. Luleå, Sweden: Internat Assoc for Hydraulic Res.

Frederking, R. M. W., and O. J. Svec. 1985. Stress-relieving techniques for cantilever beam tests in an ice cover. *CRST* 11:247–53.

Frederking, R. M. W. and G. W. Timco. 1984a. Compressive behavior of Beaufort sea ice under vertical and horizontal loading. *Proc 3rd OMAE*, vol. III:145–49.

———. 1984b. Measurement of shear strength of granular/discontinuous-columnar sea ice. *CRST* 9:215–20.

———. 1986. Field measurements of the shear strength of columnar-grained sea ice. *IAHR Ice Symp*, 279–92. Iowa City.

Fredsoe, J. 1979. Natural backfilling of pipeline trenches. *J Petroleum Technology* October:1223–30.

Freitag, J., and H. Eicken. 2003. Meltwater circulation and permeability of Arctic summer sea ice derived from hydrological field experiments. *J Glaciol* 49(166):349–56.

Fricker, H. A., S. Popov, I. Allison, and N. Young. 2001. Distribution of marine ice beneath the Amery Ice Shelf. *GRL* 28(11):2241–44.

Friedman, I., B. Schoen, and J. Harris. 1961. The deuterium concentration in arctic sea ice. *JGR* 66:1861–64.

Frolov, I. E., and V. P. Gavrilo, eds. 1997. *Morskoi led* (Sea ice). St. Petersburg: Gidrometeoizdat.

Fujino, K. 1967. Electrical properties of sea ice. In *Proc of the Internat Conf on Low Temperature Science. Physics of Snow and Ice*, ed. H. Oura, I(1):633–60. Sapporo: ILTS.

Fujino, K., and Y. Suzuki. 1959. Observations on the process of ice rind formation on the surface of still water. *LTS-A* 18:149–55.

Fujioka, T., and R. F. Sekerka. 1974. Morphological stability of disc crystals. *J Cryst Grow* 24–25:84–93.

Fukutomi, T., and K. Kusunoki. 1951. On the form and formation of hummocky ice ranges (J). *LTS-A* 8:59–88.

Fukutomi, T., M. Saito, and Y. Kudo. 1953. Study of sea ice (16th report): On the structure of ice rind, especially on the structure of thin ice sheet and ice-sheet block. *LTS* 9:113–23.

Fung, Y. C. 1997. *A first course in continuum mechanics*. 2nd ed. Englewood Cliffs, NJ: Prentice Hall.

Gakkel', Y. Y. 1959. Natural modeling of tectonic deformation in an ice sheet (R). *Izvest Vsesoyuznogo Geograficheskogo Obshchestva* 91(1):27–41. Canad Defense Res Board Trans T328R.

Garabedian, H., and R. F. Strickland-Constable. 1974. Collision breeding of ice crystals. *J Cryst Growth* 22:188–92.

Gargett, A., J. Wells, A. E. Tejada-Martinez, and C. E. Grosch. 2004. Langmuir supercells: A mechanism for sediment resuspension and transport in shallow seas. *Science* 306 (10 December):1925–28.

Gavrilo, V. P. 1997. The mechanical properties of ice: Terminology and experimental characteristics (R). In *Sea ice*, ed. I. E. Frolov and V. P. Gavrilo, 126–70. St. Petersburg: Gidrometeoizdat.

Gavrilo, V. P., and B. Y. Gaitskhoki. 1971. The statistics of air inclusions in ice. In *The physics of ice*, ed. V. V. Bogorodskii, 125–28. Washington, DC: Nat Science Foundation.

Gavrilo, V. P., S. M. Kovalev, G. A. Lebedev, and O. A. Nedoshivin. 1995. A method of estimating sea ice flexural strength based on hydrometeorological data. In *Proceed 5th Internat Offshore and Polar Eng Conf*, vol II:521–28. The Hague: Internat Soc Offshore and Polar Engineers.

Gavrilo, V. P., G. A. Lebedev, V. I. Fedotov, and N. V. Cherepanov. 1991. Seasonal variability of physical-mechanical characteristics of sea ice. *Internat J Offshore Polar Engineering* 1(1):53–57.

Geiger, C. A., W. D. Hibler III, and S. F. Ackley. 1997. Large-scale sea ice drift and deformation: Comparison between models and observations in the western Weddell Sea during 1992. *JGR* 103:21893–913.

Gibbs, J. W. 1957. *The collected works of J. Willard Gibbs*, vol 1. New Haven, CT: Yale Univ Press.

Giles, K. A., S. W. Laxon, and A. L. Ridout. 2008. Circumpolar thinning of Arctic sea ice following the 2007 record ice extent minimum. *GRL* 35:L22502, doi:10.1029/2008GL035710.

Gill, A. E. 1982. *Atmosphere–ocean dynamics*. New York: Academic Press.

Gitterman, K. E. 1937. Thermal analysis of sea water (R). *Trudy Solyanoy Lab Akad Nauk SSSR* 15(1):5–23. (CRREL Translation TL 287.)

Glen, J. W. 1955a. Comments on the paper of Professor Arakawa on the growth of ice crystals in water. *J Glaciol* 2:483.

———. 1955b. The creep of polycrystalline ice. *Proc R Soc London* Ser. A228:519–38.

———. 1968. The effect of hydrogen disorder on dislocation movement and plastic deformation in ice. *Phys Condens Mater* 7:43–51.

———. 1974. The physics of ice. *CRREL Monograph* II-C2a.

———. 1975.The mechanics of ice. *CRREL Monograph* II-C2b.

Glendening, J. W. 1994a. Dependence of plume heat budget upon lateral advection. *J Atmos Sci* 51:3517–30.

———. Dependence of boundary-layer structure near an ice-edge coastal front upon geostrophic wind direction. *JGR* 99(D3):5569–81.

———. 1995. Horizontally integrated atmospheric heat flux from an Arctic lead. *JGR* 100(C3):4613–20.

Glendening, J. W., and S. D. Burk. 1992. Turbulent transport from an Arctic lead: A large-eddy simulation. *Boundary-Layer Meteorol* 59:315–39.

Gloersen, P., and W. J. Campbell. 1991. Recent variations in Arctic and Antarctic sea ice covers. *Nature* 352:33–36

Gloersen, P., W. J. Campbell, D. J. Cavalieri, J. C. Comiso, C. L. Parkinson, and H. J. Zwally. 1992. *Arctic and antarctic sea ice, 1978–1987: Satellite passive-microwave observations and analysis.* NASA Rept. SP-511.

Gogineni, S. P., R. K. Moore, T. C. Grenfell, D. G. Barber, S. Digby, and M. Drinkwater. 1992. The effects of freeze-up and melt processes on microwave signatures. In *Microwave remote sensing of sea ice*, ed. F. Carsey, 329–41. Geophysical Monograph 68, AGU.

Gold, L. W. 1965. The initial creep of columnar grained ice. Part I. Observed behavior. *Canad J Physics* 43(8):1414–22.

———. 1972. The process of failure in columnar-grained ice. *Phil Mag* eighth ser 26(2):311–28.

———. 1977. Engineering properties of freshwater ice. *J Glaciol* 19(81):197–212.

———. 1983. Creep of columnar grained ice at low stress. *Ann Glaciol* 4:73–78.

Golden, K. M. 1986. Bounds on the complex permittivity of a multicomponent material. *J Mech Phys Solids* 34(4):333–58.

———. 1995. Bounds on the complex permittivity of sea ice. *JGR* 100(C7):13699–711.

———. 1997a. Critical behavior of transport in lattice and continuum percolation models. *Phys Rev Lett* 78:3935–38.

———. 1997b. Percolation models for porous materials. In *Homogenization and porous media*, ed. U. Hornung, 27–43. Springer-Verlag.

———. 1997c. The interaction of microwaves with sea ice. In *Wave propagation in complex media*, ed. G. Papanicolaou, 75–94. IMA Volumes in Mathematics and Uts Applications, 96. Springer-Verlag.

———. 2001. Brine percolation and the transport properties of sea ice. *Ann Glaciol* 33:28–36.

———. 2003. Critical behavior of transport in sea ice. *Physica B* 338:274–83.

Golden, K. M., and S. F. Ackley. 1981. Modeling of anisotropic electromagnetic reflection from sea ice. *JGR* 86(C9):8107–16.

Golden, K. M., S. F. Ackley, and V. I. Lytle. 1998. The percolation phase transition in sea ice. *Science* 282:2238–41.

Golden, K. M., H. Eicken, A. L. Heaton, J. Miner, D. J. Pringle, and J. Zhu. 2007. Thermal evolution of permeability and microstructure in sea ice. *JGR* 34:L16501, doi:10.1029/2007GL030447.

Golovkov, M. P. 1936. The petrography of Kara Sea ice (R). *Trudy Vsesoiuznogo Arkticheskogo Instituta* 60:7–40.

———. 1951. On the structure of natural ice-masses of various origins. *Dok Akad Nauk* 78(3):573–75 (trans E. R. Hope, Defense Research Board of Canada, T111R, 1953).

Goncharov, A. F., V. V. Struzhkin, M. S. Somayazulu, R. J. Hemley, and H. K. Mao. 1996. Compression of ice to 210 Giga pascals: Infrared evidence for a symmetric hydrogen-bonded phase. *Science* 273(5272):218–20.

Gonin, G. B. 1960. Calculation of the hummockiness of ice by statistical treatment of air photographic data (R). *PAiA* 3:93–100.

Goodwin, C. R., J. C. Finley, and L. M. Howard. 1985. *Ice scour bibliography*. Environmental Studies Revolving Funds Rept No. 010, Ottawa.

Gordeev, V. 1940. The Sedov's ice cushion (R). *Sovetskaia Arktika* 8:54–56.

Gordon, A. L. 1978. Deep Antarctic convection west of Maud Rise. *J Phys Oceanog* 8:600–12.

Gordon, A. L., and B. A. Huber. 1990. Southern Ocean winter mixed layer. *JGR* 95(C7):11655–72.

Gordon, A. L., and W. B. Owens. 1987. Polar Oceans. *Rev Geophys* 25:227–33.

Gosink, J. P., and T. E. Osterkamp. 1983. Measurements and analyses of velocity profiles and frazil ice-crystal rise velocities during periods of frazil-ice formation in rivers. *Ann Glaciol* 4:79–84.

Gosink, T. A., and J. J. Kelley. 1982. Annual sea ice: An air–sea gas exchange moderator. Unpublished manuscript.

Gottstein, G., and L. S. Shvindlerman. 1999. *Grain boundary migration in metals*. Boca Raton: CRC Press.

Gow, A. J. 1970. The hydrology and compositional structure of the Koettlitz Glacier tongue, McMurdo Sound, Antarctica. In *Symp on the Hydrology of Glaciers*. Cambridge, England: IASH 257.

———. 1986. Orientation textures in ice sheets of quietly frozen lakes. *J Cryst Growth* 74:247–58.

Gow, A. J., S. F. Ackley, K. R. Buck, and K. M. Golden. 1987. Physical and structural characteristics of Weddell Sea pack ice. *CRREL Report* 87-14.

Gow, A. J., S. F. Ackley, J. W. Govoni, and W. F. Weeks. 1998. Physical and structural properties of landfast sea ice in McMurdo Sound, Antarctica. In *Antarctic sea ice: Physical processes, interactions and variability*, ed. M. O. Jeffries, 355–74. AGU, Antarctic Research Series, vol. 74.

Gow, A. J., S. F. Ackley, W. F. Weeks, and J. W. Govoni. 1982. Physical and structural characteristics of Antarctic sea ice. *Ann Glaciol* 3:113–17.

Gow, A. J., and S. Epstein. 1972. The use of stable isotopes to trace the origins of ice in a floating ice tongue. *JGR* 77:6552–57.

Gow, A. J., and J. W. Govoni. 1994. An 80 year record of retreat of the Koettlitz Ice Tongue, McMurdo Sound, Antarctica. *Ann Glaciol* 20:237–41.

Gow, A. J., and D. Langston. 1977. Growth history of lake ice in relation to its stratigraphic, crystalline and mechanical structure. *CRREL Report* 77-1.

Gow, A. J., D. A. Meese, D. K. Perovich, and W. B. Tucker III. 1990. The anatomy of a freezing lead. *JGR* 95(C10):18221–32.

Gow, A. J., W. B. Tucker III, and W. F. Weeks. 1987. Physical properties of summer sea ice in the Fram Strait, June–July, 1984. *CRREL Report* 87-16.

Gow, A. J., H. T. Ueda, and J. A. Ricard. 1978. Flexural strength of ice on temperate lakes; comparative tests of large cantilever and simply supported beams. *CRREL Report* 78-9.

Gow, A. J., and W. F. Weeks. 1977. The internal structure of fast ice near Narwhal Island, Beaufort Sea, Alaska. *CRREL Report* 77-29.

Gow, A. J., W. F. Weeks, G. Hendrickson, and R. Rowland. 1965. New light on the mode of uplift of the fish and fossiliferous moraines of the McMurdo Ice Shelf, Antarctica. *J Glaciol* 5(42):813–28.

Gow, A. J., W. F. Weeks, P. Kosloff, and S. Carsey. 1992. Petrographic and salinity characteristics of brackish water ice in the Bay of Bothnia. *CRREL Report* 92-13.

Granberg, H. B., and M. Leppäranta. 1999. Observations of sea ice ridging in the Weddell Sea. *JGR* 104(C11):25735–46.

Gratz, E. T., and E. M. Schulson. 1994. Preliminary observations of brittle compressive failure of columnar saline ice under triaxial loading. *Ann Glaciol* 19:33–38.

Graystone, P., and M. P. Langleben. 1962. Ring tensile strength of sea ice. In *Ice and snow: Properties, processes, and applications*, ed. W. D. Kingery, 114–23. Cambridge, MA: MIT Press.

Green, J. C. 1970. Finger-rafting in fresh-water ice. *J Glaciol* 9(57):401–4.

Grenfell, T. C. 1983. A theoretical model of the optical properties of sea ice in the visible and near infrared. *JGR* 88:9723–35.

———. 1991. A radiative transfer model of sea ice with vertical structure variations. *JGR* 98(C9):16991–17002.

Grenfell, T. C., and G. A. Maykut. 1977. The optical properties of snow and ice in the Arctic Basin. *J Glaciol* 18:445–63.

Grenfell, T. C., and D. K. Perovich. 1981. Radiation absorption coefficients of polycrystalline ice from 400 to 1400 nm. *JGR* 86(C8):7447–50.

Grigg, S. B., and N. J. Holbrook. 2001. The impact of polynyas on the stability of the thermohaline circulation as simulated in a coupled ocean-atmosphere-sea ice box model. *GRL* 28(5):767–70.

Grigor'yev, A. A., and M. V. Strakhov. 1991. Anisotropy of the elastic properties of an ice cover with spatially organized structure. In *Physical and technical problems of ice research*, ed. V. P. Gavrilo and G. A. Lebedev, 18–25. Leningrad: AARI, Gidrometeoizdat.

Grosfeld, K., H. H. Hellmer, M. Jonas, H. Sandhäger, M. Schulte, and D. G. Vaughan. 1998. Marine ice beneath the Filchner Ice Shelf: Evidence from a multi-disciplinary approach. In *Ocean, ice, and atmosphere: Interactions at the antarctic continental margin*, ed. S. S. Jacobs and R. F. Weiss, 319–39. AGU, Antarctic Research Series vol 75.

Grosfeld, K., M. Schröder, E. Fahrbach, R. Gerdes, and A. Mackensen. 2001. How iceberg calving and grounding change the circulation and hydrography in the Filchner Ice Shelf–Ocean system. *JGR* 106(C5):9039–55.

Guest, P. S., and K. L. Davidson. 1991. The aerodynamic roughness length of different types of sea ice. *JGR* 96(C3):4709–21.

Guest, P. S., K. L. Davidson, J. E. Overland, and P. A. Frederickson. 1994. Atmosphere–ocean interactions in the marginal ice zones of the Nordic Seas. In *Arctic oceanography: Marginal ice zones and continental shelves*, ed. W. O. Smith and J. Grebmeier, 51–95. AGU.

Haarpaintner, J., J-C. Gascard, and P. M. Haugen. 2001. Ice production and brine formation in Storfjorden, Svalbard. *GJR* 106(C7):14001–13.

Haas, C. 2004. Late-summer sea ice thickness variability in the Arctic Transpolar Drift 1991–2001 derived from ground-based electromagnetic sounding. *GRL* 31:L09402, doi:10.1029/203GL019394.

Haas, C., A. Pfaffling, S. Hendricks, L. Rabenstein, J.-L Etienne and I. Rigor. 2008. Reduced ice thickness in the Arctic Transpolar Drift favors rapid ice retreat. *GRL* 35:L17501, doi:10/1029/2008GL034457.

Haas, C., D. N. Thomas and J. Bareiss, J. 2001. Surface properties and processes of perennial Antarctic sea ice in summer. *J Glaciol* 47(159):613–25.

Haff, J. C. 1938. Preparation of petrofabric diagrams. *Amer Mineralogist* 23(9):543–74.

Haines, E. M., R. G. Buckley, and H. J. Trodahl. 1997. Determination of the depth dependent scattering coefficient of sea ice. *JGR* 102:1141–51.

Häkkinen, S. 1986. Ice banding as a response of the coupled ice-ocean system to temporally varying winds. *JGR* 91(C4):5047–53.

———. 1990. Models and their application to polar oceanography. In *Polar oceanography*, ed. W. O. Smith, 335–84. San Diego: Academic Press.

———. 1993. An arctic source for the great salinity anomaly: A simulation of the Arctic ice–ocean system for 1955–1975. *JGR* 98(C9):16397–410.

Häkkinen, S., and G. L. Mellor. 1992. Modeling the seasonal variability of a coupled arctic ice–ocean system. *JGR* 97(C12):20285–304.

Häkkinen, S., A. Proshutinsky, and I. Ashik. 2008. Sea ice drift in the Arctic since the 1950s. *GRL* 35:L19704, doi:10.1029/2008GL034791.

Hallam, S. D., N. Jones, and M. W. Howard. 1987. The effect of sub-surface irregularities on the strength of multi-year ice. In *6th Internat Symp Offshore Mech Arctic Eng*, 235–37. Houston, TX: ASME.

Hallett, J. 1960. Crystal growth and the formation of spikes in the surface of supercooled water. *J Glaciol* 3(28):698–702.

Hallikainen, M. T. 1983. A new low salinity sea ice model for UHF radiometry. *Internat J Remote Sens* 4:655–81.

Hallikainen, M. T., F. T. Ulaby, and M. Abelrazik. 1986. Dielectric properties of snow in the 3 to 37 GHz range. *IEEE Trans Anten Propag* AP-34:1329–45.

Hallikainen, M. T., and D. P. Winebrenner. 1992. The physical basis for sea ice remote sensing. In *Microwave remote sensing of sea ice*, ed. F. D. Carsey, 29–46. AGU Geophysical Monograph 68.

Hansen, D. P., and L. A. Wilen. 2002. Performance and applications of an automated *c*-axis ice fabric analyzer. *J Glaciol* 48(160):159–70.

Hanson, A. M. 1965. Studies of the mass balance of arctic pack ice floes. *J Glaciol* 5:701–9.

———. 1980. The snow cover of sea ice during the Arctic Ice Dynamics Joint Experiment, 1975 to 1976. *Arctic and Alpine Research* 12(2):215–26.

Harder, M., and H. Fischer. 1999. Sea ice dynamics in the Weddell Sea simulated with an optimized model. *JGR* 104:11151–62.

Hardy, S. C., and S. R. Coriell. 1969. Morphological stability of cylindrical ice crystals. *J Crystal Growth* 5:329–37.

———. 1973. Surface tension and interface kinetics of ice crystals freezing and melting in sodium chloride solutions. *J Crystal Growth* 20:292–300.

Harned, H. S., and B. B. Owen. 1950. *The physical chemistry of electrolytic solutions.* New York, Reinhold.

Harrison, J. D. 1965. Measurement of brine droplet migration in ice. *J App Phys* 36(12):3811–15.

Harrison, J. D., and W. A. Tiller. 1963a. Controlled freezing of water. In *Ice and snow: Processes, properties, and applications,* ed. W. D. Kingery, 215–25. Cambridge, MA: MIT Press.

———. 1963b. Ice interface morphology and texture developed during freezing. *J Appl Phys* 34(11):3349–55.

Harvey, L., and D. Danny. 1990. Testing alternative parameterizations of lateral melting and upward basal heat flux in a thermodynamic sea ice model. *JGR* 95(C5):7359–65.

Hastings, A. D., Jr. 1971. Surface climate of the Arctic Basin. Selected climatic elements related to surface-effects vehicles. Fort Belvoir, VA: US Army Topographic Laboratory, ETL-TR-71-5.

Haüsler, F.-U. 1981. Multi-axial compressive strength tests on saline ice with brush-type loading platens. *IAHR Internat Symp on Ice* 2:526–39.

Heil, P., I. Allison, and V. I. Lytle. 1996. Seasonal and interannual variations of the oceanic heat flux under a landfast Antarctic sea ice cover. *JGR* 101(C11):25741–52.

———. 2001. Effect of high-frequency deformation on sea-ice thickness. In *IUTAM symp on scaling laws in ice mechanics and ice dynamics,* ed. J. P. Dempsey and H. H. Shen, 417–26. Dordrecht: Kluwer.

Heil, P., and W. D. Hibler III. 2002. Modelling the high frequency component of Arctic sea-ice drift and deformation. *J Phys Oceanog* 32(11):3039–57.

Heil, P., V. I. Lytle, and I. Allison. 1998. Enhanced thermodynamic ice growth by sea-ice deformation. *Ann Glaciol* 27:493–97.

Hellawell, A., and P. M. Herbert. 1962. The development of preferred orientations during the freezing of metals and alloys. *Proc Roy Soc* A269:560–73.

Hendrickson, G., and R. Rowland. 1965. Strength studies on Antarctic sea ice. *CRREL Tech Rept* 157.

Hertz, H. 1884. Über das Gleichgewicht schimmender elasticher Platten. *Wiedmanns Ann Phys Chem* 22:449–55.

Hetenyi, M. 1946. *Beams on elastic foundation.* Ann Arbor: Univ Michigan Press.

Hibler, W. D., III. 1972. Removal of aircraft altitude variation from laser profiles of the Arctic ice pack. *JGR* 77(36):7190–95.

———. 1974. Differential sea ice drift II. Comparison of mesoscale strain measurements to linear drift theory predictions. *J Glaciol* 13(69):457–71.

———. 1975. Characterization of cold-regions terrain using airborne laser profilometry. *J Glaciol* 15(73):329–47.

———. 1977. A viscous sea ice law as a stochastic average of plasticity. *JGR* 82(27):3932–38.

———. 1979. A dynamic thermodynamic sea ice model. *J Phys Ocean* 9(4):815–46.

———. 1980a. Documentation for a two-level dynamic thermodynamic sea ice model. *CRREL Special Report* 80-8.

———. 1980b. Modeling a variable thickness sea ice cover. *Mon Weath Rev* 108(12):1943–73.

———. 1980c. Modelling pack ice as a viscous-plastic continuum. In *Sea ice processes and models,* ed. R. S. Pritchard, 163–76. Seattle: Univ Washington Press.

———. 1980d. Sea ice growth, drift, and decay. In *Dynamics of snow and ice masses,* ed. S. C. Colbeck, 141–209. Academic Press.

———. 1984. Role of sea ice dynamics in modeling CO_2 increases. In *Climate processes and climate sensitivity,* 238–53. AGU Geophysical Monograph 29.

———. 1985. Modeling sea ice dynamics. In *Issues in atmospheric and ocean modelling, Part A: Climate dynamics* (S. Manabe, ed.) *Advances in Geophysics* 28:549–578.

———. 1986. Ice dynamics. In *Geophysics of sea ice,* ed. N. Untersteiner, 577–640. New York: Plenum.

———. 1988. Modeling sea ice thermodynamics and dynamics in climate studies. In *NATO Advanced Study Institute on Physically-Based Modelling and Simulation of Climate and Climate Change, Part 1, Proceedings,* ed. M. E. Schlesinger, 509–63. Dordrecht: Kluwer.

———. 1989. Arctic ice–ocean dynamics. In *The arctic seas: Climatology, oceanography, geology, and biology*, ed. Y. Hermann, 47–91. New York: Van Nostrand Reinhold.

———. 2001. Sea ice fracturing on the large scale. *Engrng Fracture Mech* 68:2013–43.

———. 2004. Modelling the dynamic response of sea ice. In *The mass balance of the cryosphere: Observations and modeling of contemporary and future changes*, ed. J. Bamberg and T. Payne, 227–334. Cambridge: Cambridge Univ Press.

Hibler, W. D., III, and S. F. Ackley. 1982. On modeling the Weddell Sea pack ice. *Ann Glaciol* 3:125–30.

———. 1983. Numerical simulation of the Weddell Sea pack ice. *JGR* 88:2873–87.

Hibler, W. D., III, S. F. Ackley, W. K. Crowder, H. W. McKim, and D. M. Anderson. 1974. Analysis of shear zone ice deformation in the Beaufort Sea using satellite imagery. In *The coast and shelf of the Beaufort Sea*, ed. J. C. Reed and J. E. Sater, 285–96. AINA.

Hibler, W. D., III, and K. Bryan. 1984. Ocean circulation: Its effect on seasonal sea-ice simulations. *Science* 224(4648):489–92.

———. 1987. A diagnostic ice-ocean model. *J Phys Oceanogr* 17(7):987–1015.

Hibler, W. D., III, and G. M. Flato. 1992. Sea ice modeling. In *Climate systems modeling*, ed. K. E. Trenberth. Cambridge: Cambridge University Press.

Hibler, W. D., III, P. Heil, and V. I. Lytle. 1998. On simulating high frequency variability in Antarctic sea ice dynamics. *Ann Glaciol* 27:443–48.

Hibler, W. D., III, and J. Hutchings. 2003. Multiple equilibrium ice cover states induced by ice mechanics. *16th IAHR Conf on Sea Ice Processes*, Dunedin, New Zealand.

Hibler, W. D., III, and C. F. Ip. 1995. The effect of sea ice rheology on Arctic buoy drift. In *Ice Mechanics*, ed. J. Dempsey and Y. D. S. Rajapakse, 255–63. ASME-AMD 207. New York: Amer Soc Mech Eng.

Hibler, W. D., III, and L. A. Le Schack. 1972. Power spectrum analysis of undersea and surface sea ice profiles. *J Glaciol* 11(63):345–56.

Hibler, W. D., III, S. J. Mock, and W. B. Tucker III. 1974. Classification and variation of sea ice ridging in the western Arctic Basin. *JGR* 79(18):2735–43.

Hibler, W. D., III, and E. Schulson. 2000. On modeling the isotropic failure and flow of flawed sea ice. *JGR* 105(C7):17105–19.

Hibler, W. D., III, and W. B. Tucker III. 1977. Seasonal variations in apparent sea ice viscosity on the geophysical scale. *GRL* 4(2):87–90.

———. 1979. Some results from a linear viscous model of the Arctic ice cover. *J Glaciol* 22(87):293–304.

Hibler, W. D., III, I. Udin, and A. Ullerstig. 1983. On forecasting mesoscale ice dynamics and buildup. *Ann Glaciol* 4:110–15.

Hibler, W. D., III, and J. E. Walsh. 1982. On modeling seasonal and interannual fluctuations of Arctic sea ice. *J Phys Oceanog* 12:1514–23.

Hibler, W. D., III, W. F. Weeks, A. Kovacs, and S. F. Ackley. 1974. Differential sea-ice drift. I: Spatial and temporal variations in sea-ice deformation. *J Glaciol* 13(69):437–55.

Hibler, W. D., III, W. F. Weeks, and S. J. Mock. 1972. Statistical aspects of sea ice ridge distributions. *JGR* 70(30):5954–70.

Hibler, W. D., III, and J. Zhang. 1994. On the effect of ocean circulation on Arctic ice-margin variations. In *The polar oceans and their role in shaping the global environment*, ed. O. M. Johannessen, R. D. Muench, and J. E. Overland. Geophysical Monograph, AGU.

———. 1995. On the effect of sea ice dynamics on oceanic thermohaline circulation. *Ann Glaciol* 21:361–68.

Hillig, W. B. 1958. The kinetics of freezing of ice in the direction perpendicular to the basal plane. *Growth and perfection of crystals*, ed. R. H. Doremus, 350–59. New York: Wiley.

Hilmer, M., and P. Lemke. 2000. On the decrease of Arctic sea ice volume. *GRL* 27:3751–54.

Hilmer, M., M. Harder, and P. Lemke. 1998. Sea ice transport: A highly variable link between Arctic and North Atlantic. *GRL* 25:3359–62.

Hobbs, P. V. 1974. *Ice physics*. Oxford: Clarendon Press.

Hochstein, M. P. 1967. Pressure ridges of the McMurdo Ice Shelf near Scott base, Antarctica. *New Zealand J of Geology and Geophysics* 10(4):1165–68.

Hodgson, D. A., M. J. Bentley, S. J. Roberts, J. A. Smith, D. E. Sugden, and E. W. Domack. 2006. Examining Holocene stability of Antarctic ice shelves. *Eos* 97(31):305, 308.

Hoekstra, P., and P. Cappellino. 1971. Dielectric properties of sodium chloride ice at UHF and microwave frequencies. *JGR* 76(20):4922–31.

Hoekstra, P., T. E. Osterkamp, and W. F. Weeks. 1965. The migration of liquid inclusions in single ice crystals. *JGR* 70(20):5035–41.

Hoerner, R. A. 1989. Arctic sea-ice biota. In *The arctic seas: Climatology, oceanography, geology, and biology*, ed. Y. Herman, 123–46. New York: Van Nostrand Reinhold.

Holland, C. 1994. *Arctic exploration and development: c. 500 B.C. to 1915*. New York: Garland Publishing.

———. 1999. *Farthest north: Endurance and adventure in the quest for the North Pole*. New York: Carrol and Graf.

Holland, D. M. 2001. Explaining the Weddell Polynya: A large ocean eddy shed at Maud Rise. *Science* 292(1 June):1697–1700.

———. 2002. Computing marine-ice thickness at an ice-shelf base. *J Glaciol* 48(160):9–19.

Holland, D. M., L. A. Mysak, D. K. Manak, and J. M. Oberhuber. 1993. Sensitivity study of a dynamic–thermodynamic sea ice model. *JGR* 98(C2):2561–68.

Holland, M. M., and C. M. Bitz. 2003. Polar amplification of climate change in coupled models. *Clim Dyn* 21:221–32.

Holland, M. M., and J. A. Curry. 1999. The role of physical processes in determining the interdecadal variability of Central Arctic Sea Ice. *J Climate* 12:3319–30.

Holland, P. R., D. L. Feltham, and A. Jenkins. 2007. Ice shelf water plume flow beneath Filchner-Ronne Ice Shelf, Antarctica. *JGR* 112:C05044, doi;10.1029/2006JC003915.

Hollaway, G., and T. Sou. 2002. Has Arctic sea ice rapidly thinned. *J Climate* 15:1691–1701.

Holmes, J. F., and L. V. Worthington. 1953. Oceanographic studies on project Skijump II. *Woods Hole Oceanographic Institution Technical Rept* 53-23.

Holt, B., and S. A. Digby. 1985. Processes and imagery of first-year fast sea ice during the melting season. *JGR* 90(C3):5045–62.

Holtsmark, B. E. 1955. Insulating effect of a snow cover on the growth of young sea ice. *Arctic* 8:60–65.

Hopkins, M. A. 1992. Numerical simulation of systems of multitudinous polygonal blocks. *CRREL Rept* 92-22.

Hopkins, M. A. 1994. On the ridging of intact lead ice. *JGR* 99(C8):16351–60.

———. 1996. On the mesoscale interaction of lead ice and floes. *J Geophys Res* 101(C8):18315–26, 10.1029/96JC01689.

———. 1998. Four stages of pressure ridging. *JGR* 103:21883–91.

Hopkins M. A., S. Frankenstein, and A. S. Thorndike. 2004. Formation of an aggregate scale in Arctic sea ice, *JGR* 109:C01032, doi:10.1029/2003JC001855.

Hopkins, M. A., and W. D. Hibler III. 1991a. On the ridging of a thin sheet of lead ice. *Ann Glaciol* 15:81–86.

———. 1991b. Numerical simulations of a compact convergent system of ice floes. *Ann Glaciol* 15:26–30.

Hopkins, M. A., W. D. Hibler III, and G. M. Flato. 1991. On the numerical simulation of the sea ice ridging process. *JGR* 96(3):4809–20.

Hopkins, M. A., and A. S. Thorndike. 2006. Floe formation in Arctic sea ice. *JGR* 111:C11S23, doi:10.1029/2005JC003352.

Hopkins, M. A., and J. Tuhkuri. 1998. Simulation of ridging and rafting in first-year ice. In *Ice in surface waters*, ed. H. T. Shen, 623–30. Proceed 14th Internat Symp on Ice. Rotterdam: Balkema.

Hopkins, M. A., J. Tuhkuri, and M. Lensu. 1999. Rafting and ridging in thin ice sheets. *JGR* 104(C6):13605–13.

Horii, H., and S. Nemat-Nasser. 1985. Compression-induced microcrack growth in brittle solids: Axial splitting and shear failure. *JGR* 90:3105–25.

Howat, I. M., B. E. Smith, I. Joughin, and T. A. Scambos. 2008. Rates of southeast Greenland ice volume loss from combined ICESat and ASTER observations. *GRL* 35(L17505):doi:10.1029/2008GL 034496.

Høyland, K. V. 2002a. Consolidation of first-year sea ice ridges. *JGR* 107(C6):10.1029/2000JC000526.

———. 2002b. Simulations of the consolidation process in first-year sea ice ridges. *CRST* 34(3):143–58.

———. 2007. Morphology and small-scale strength of ridges in the North-western Barents Sea. *CRST* 48(3):169–87.

Hudier, E. J., R. G. Ingram, and K. Shirasawa. 1995. Upward flushing of seawater through first year ice. *Atmos.-Ocean* 33(3):569–80.

Hughes, B. A. 1991. On the use of lognormal statistics to simulate one- and two-dimensional under-ice draft profiles. *JGR* 96(C12):22101–11.

Hunke, E. C., and S. F. Ackley. 2001. A numerical investigation of the 1997–1998 Ronne polynya. *JGR* 106(C10):22373–82.

Hunke, E. C., and J. K. Dukowicz. 1997. An elastic viscous plastic model for sea ice dynamics. *J Phys Oceanog* 27:1840–67.

———. 2002. The elastic-viscous-plastic sea ice dynamics model in general orthogonal curvilinear coordinates on a sphere: Incorporation of metric terms. *Month Weather Rev* 130(7):1848–65.

Hunke, E. C., and Y. Zhang. 1999. A comparison of sea ice dynamics models at high resolution. *Month Weather Rev* 127:396–408.

Hunkins, K. 1960. Seismic studies of sea ice. *JGR* 65:3459–72.

———. 1962. Waves on the Arctic Ocean. *JGR* 67:2477–89.

Huppert, H. E., and M. G. Worster. 1985. Dynamic solidification of a binary melt. *Nature* 314(25 April):703–7.

Hurrell, J. W. 1995. Decadal trends in the North Atlantic Oscillation: Regional temperatures and precipitation. *Science* 269:676–79.

Hutchings, J., and W. D. Hibler III. 2002. Modelling sea ice deformation with a viscous-plastic isotropic rheology. *Proceed 16th IAHR Internat Symp on Ice* 2:358–66.

Hutchings, J., and 15 others. 2008. Role of ice dynamics in the sea ice mass balance. *Eos* 89(50):515–16.

Hutter, K. 1975. Floating sea ice plates and the significance of the dependence of the Poisson ratio on the brine volume. *Proc Royal Soc, London A* 343(1962):85–108.

Hvidegaard, S. M., and R. Forsberg. 2002. Sea-ice thickness from airborne laser altimetry over the Arctic Ocean north of Greenland. *GRL* 29(20):1952, doi:10.1029/2001GL014474.

Iliescu, D., and E. M. Schulson. 2004. The brittle compressive failure of fresh-water columnar ice loaded biaxially. *Acta Materialia* 52(20):5723–35.

Ip, C. F. 1993. Numerical investigation of the effect of rheology on sea ice drift. Ph.D. thesis, Dartmouth College.

Ip, C. F., W. D. Hibler III, and G. M. Flato. 1991. The effect of rheology on seasonal sea ice simulations. *Ann Glaciol* 15:17–25.

Isaksen, K., R. E. Benestad, C. Harris, and J. L. Sollid. 2007. Recent extreme near-surface permafrost temperatures on Svalbard in relation to future climate scenarios. *GRL* 34:L17502, doi:10/1029/2007 GL031002.

Ivchenko, V. O., N. C. Wells, and D. L. Aleynek. 2006. Anomaly of heat content in the northern Atlantic in the last 7 years: Is the ocean warning or cooling? *GRL* 33:L22606, doi:10.1029/2006GL027691.

Jacka, T. H., I. Allison, R. Thwaites, and J. C. Wilson. 1988. Characteristics of the seasonal sea ice of East Antarctica and comparisons with satellite observations. *Ann Glaciol* 9:85–91.

Jacobs, S. S. 1985. Oceanographic evidence for land ice/ocean interactions in the Southern Ocean. In *Glaciers, ice sheets and sea level: Effects of a CO_2 induced climate change*, 116–28. Report of a Workshop, Seattle, 13–15 Sept 1984. Washington, DC: Nat Acad Press.

Jacobs, S. S., A. L. Gordon, and A. F. Amos. 1979. Effect of glacial ice melting on the Antarctic surface water. *Nature (London)* 277:469–71.

Jacobs, S. S., and R. F. Weiss, eds. 1998. *Ocean, ice, and atmosphere: Interactions at the antarctic continental margin*. Antarctic Research Series vol. 75. Washington, DC: AGU.

Jakobsson, M., R. Macnab, M. Mayer, R. Anderson, M. Edwards, J. Hatzky, H-W. Schenke, and P. Johnson. 2008. An improved bathymetric portrayal of the Arctic Ocean: Implications for ocean modeling and geological, geophysical and oceanographic analyses. *GRL* 35:L07602, doi:10.1029/2008GL033520.

Jeffries, M. O. 1991. Massive, ancient sea-ice strata and preserved physical-structural characteristics in the Ward Hunt ice shelf. *Ann Glaciol* 15:125–31.

———. 1992. Arctic ice shelves and ice islands: Origin, growth and disintegration, physical characteristics, structural-stratigraphic variability, and dynamics, *Rev Geophys* 30(3):245–67.

———, ed. 1998. *Antarctic sea ice: Physical processes, interactions and variability*, ed. M. O. Jeffries. AGU, Antarctic Research Series, vol. 74.

Jeffries, M. O., and U. Adolphs. 1997. Early winter snow and ice thickness distribution, ice structure and development of the western Ross Sea pack ice between the ice edge and the Ross Ice Shelf. *Antarct Sci* 9(2):188–200.

Jeffries, M. O., B. Hurst-Cushing, H. R. Krouse, and T. Maksym. 1998. *The role of snow in the thickening and mass budget of first-year floes in the eastern Pacific sector of the Antarctic pack*. Univ Alaska Fairbanks, Geophysical Institute Rept UAG R–327.

Jeffries, M. O., S. Li, R. Jaña, H. R. Krouse, and B. Hurst-Cushing. 1998. Late winter first-year ice floe thickness variability, seawater flooding and snow ice formation in the Amundsen and Ross Seas. In *Antarctic sea ice: Physical processes, interactions and variability*, ed. M. O. Jeffries, 69–87. AGU, Antarctic Research Series vol 74.

Jeffries, M. O., and W. M. Sackinger. 1989. Some measurements and observations of very old sea ice and brackish ice, Ward Hunt ice shelf, N.W.T. *Atmosphere-Ocean* 27(3):553–64.

Jeffries, M. O., K. Schwartz, K. Morris, A. D. Veazy, H. R. Krouse, and S. Cushing. 1995. Evidence for platelet ice accretion in Arctic sea ice development. *JGR* 100(C6):10905–14

Jeffries, M. O., R. A. Shaw, K. Morris, A. L. Veazy, and H. R. Krouse. 1994. Crystal structure, stable isotopes ($d^{18}O$), and development of sea ice in the Ross, Amundsen, and Bellingshausen seas, Antarctica. *JGR* 99(C1):985–95.

Jeffries, M. O., A. L. Veazey, K. Morris, and H. R. Krouse. 1994. Depositional environment of the snow cover on West Antarctic pack-ice floes. *Ann Glaciol* 20:33–38.

Jeffries, M. O., and W. F. Weeks. 1992a. Fast ice properties and structure in McMurdo Sound. *Antarctic J U S* 26(5):94–95.

———. 1992b. Summer pack ice properties and structure in the western Ross Sea. *Antarctic J US* 26(5):95–97.

———. 1993. Structural, characteristics and development of sea ice in the western Ross Sea. *Antarctic Science* 5(1):63–75.

Jeffries, M. O., W. F. Weeks, R. Shaw, and K. Morris. 1993. Structural characteristics of congelation and platelet ice and their role in the development of Antarctic land–fast sea ice. *J Glaciol* 39(132):223–38.

Jeffries, M. O., A. P. Worby, K. Morris, and W. F. Weeks. 1997. Seasonal variations and structural composition of sea ice and snow cover in the Bellingshausen and Amundsen Seas, Antarctica. *J Glaciol* 43(143):138–51.

Jenkins, A. 1991. A one-dimensional model of ice shelf–ocean interaction. *JGR* 96(C11):20671–77.

Jenkins, A., and A. Bombosch. 1995. Modeling the effects of frazil ice crystals on the dynamics and thermodynamics of Ice Shelf Water plumes. *JGR* 100(C4):6967–81.

Jenkins, A., and C. S. M. Doake. 1991. Ice–ocean interaction on Ronne Ice Shelf, Antarctica. *JGR* 96(C1): 791–813.

Jenkins, A., and D. M. Holland. 2002. A model study of ocean circulation beneath Filchner-Ronne Ice Shelf, Antarctica: Implications for bottom water formation. *GRL* 29(8):10.1029/2001GL014589.

Jenkins, A., D. M. Holland, K. W. Nicholls, M. Schröder, and S. Østerhus. 2004. Seasonal ventilation of the cavity beneath Filchner-Ronne Ice Shelf simulated with an isopycnic coordinate model. *JGR* 109:C01024, doi:10.1029/2001JC001086.

Jesse, R. E., and H. Giller. 1970. Cellular growth: The relation between growth velocity and cell size of some alloys of cadmium and zinc. *J Cryst Growth* 7:348–52.

Jevrejeva, S., J. C. Moore, A. Grinsted, and P. L. Woodworth. 2008. Recent global sea level acceleration started over 200 years ago. *GRL* 35:08715, doi:10.1029/2008 GL033611.

Jin, Z., K. Stamnes, W. F. Weeks, and S.-C. Tsay. 1994. The effect of sea ice on the solar energy budget in the atmosphere–sea ice system: A model study. *JGR* 99(C12):25281–94.

Johannessen, O. M., J. A. Johannessen, E. Svendsen, R. A. Shuchman, W. J. Campbell, and E. Josberger. 1987. Ice edge eddies in the Fram Strait marginal ice zone. *Science* 236(4800):427–29.

Johannessen, O. M., M. Miles, and E. Bjørgo. 1995. The Arctic's shrinking sea ice. *Nature* 376:126–27.

———. 1996. Global sea ice monitoring from microwave satellites. *Proc 1996 IGARSS*:932–34.

Johannessen, O. M., S. Sandven, W. P. Budgell, J. A. Johannessen, and R. A. Shuchman. 1994. Observation and simulation of ice tongues and vortex pairs in the Marginal Ice Zone. In *The Polar oceans and their role in shaping the global environment*, ed. O. M. Johannessen, R. D. Muench, and J. E. Overland, 109–36. AGU Geophysical Monograph 85.

Johannessen, O. M., E. V. Shalina, and M. Miles. 1999. Satellite evidence for an Arctic sea ice cover in transformation. *Science* 286(5446):1937–39.

Johannessen, O. M., and 11 others. 2004. Arctic climate change: Observed and modelled temperature and sea ice. *Tellus* 56A(4):328–41.

Johnson, H. L. 1989. The specific heat of sea ice. B.Sc. honour's thesis, Victoria Univ of Wellington, NZ.

Johnson, J. B., and R. C. Metzner. 1990. Thermal expansion coefficients for sea ice. *J Glaciol* 36(124):343–49.

Johnston, M. 2006. A comparison of physical properties and strength of decaying first-year ice in the Arctic and sub-Arctic. *Ann Glaciol* 44:154–62.

Johnston, M., and A. Barker. 2000. *Microstructure of first year sea ice ridges.* Canadian Hydraulics Centre Tech Rept HYD-TR-043.

Johnson, N. G. 1943. Studier av isen i Gullmarfjorden. *Svenska Hydrog-Biol Komm Skr, Ny Serie; Hydrografi* 18:1–21.

Johnston, M., R. Frederking, and G. W. Timco. 2001. Decay induced changes in the physical and mechanical properties of first-year sea ice. *Proc. POAC'01*, Vol. 3, pp. 1395–1404, Ottawa.

———. 2002. Properties of decaying first-year sea ice: two seasons of field measurements. *Proc 17th Int Symp on Okhotsk Sea and Sea Ice*, pp. 303–11, ILTS, Hokkaido, Japan.

———.2003a. Properties of decaying first-year sea ice at five sites in Parry Channel. *Proc. 17th Int. POAC*, Vol. 1, pp. 131–140, Trondheim, Norway.

———. 2003b. Property changes of first-year ice and old ice during summer melt. NRC Canadian Hydraulics Centre Report CHC-TR-010, TP14098E, Ottawa.

Johnston, M. E., and G. W. Timco. 2008. *Understanding and identifying old ice in summer.* Canadian Hydraulics Centre, NRC, Ottawa, CHC-TR-055.

Jones, D. R. H. 1973. The temperature-gradient migration of liquid droplets through ice. *J Cryst Growth* 20:145–51.

———. 1974. Determination of the kinetics of ice–brine interfaces from the shapes of migrating droplets. *J Cryst Growth* 26:177–79.

Jones, E. P., and L. G. Anderson. 1986. On the origin of the chemical properties of the Arctic Ocean halocline. *JGR* 91(C9):10759–67.

Jones, E. P., and A. R. Coote. 1981. Oceanic CO_2 produced by the precipitation of $CaCO_3$ from brines in sea ice. *JGR* 86(C11):11041–43.

Jones, S. J. 2007. A review of the strength of iceberg and other freshwater ice and the effect of temperature. *CRST* 47(3):256–62.

Jones, S. J., and B. T. Hill. 2001. Structure of sea ice in McMurdo Sound, Antarctica. *Ann Glaciol* 33:5–12.

Jordan, R. E. 1991. A one-dimensional temperature model for a snow cover: Technical documentation for SNTHERM.89. *CRREL Special Rept* 91-16.

Jordan, R. E., E. L. Andreas, and A. P. Makshtas. 1999. Heat budget of snow-covered sea ice at North Pole-4. *JGR* 104(C4):7785–7806.

———. 2002. Modeling the surface energy budget and the temperature structure of snow and brinesnow at Ice Station Weddell. In *Sixth Conf on Polar Meteorology and Oceanography*, 129–32. Boston: Amer Meteorol Soc.

Josberger, E. G. 1979. Laminar and turbulent boundary layers adjacent to melting vertical ice walls in salt water. Ph.D. thesis, Dept Oceanog, Univ Washington, Seattle.

Joughin, I., and L. Padman. 2003. Melting and freezing beneath the Filchner-Ronne Ice Shelf, Antarctica. *GRL* 30(9):1477, doi:10.1029/2003GL016941.

Kahl, J. D. W., M. Jansen, and M. A. Pulrang. 2001. Fifty-year record of North Polar temperatures shows warming. *Eos* 82(1):1, 5.

Kankaanpää, P. 1991. *Morphology of sea ice pressure ridges in the Baltic Sea.* Lic Phil Dept Geography, University of Helsinki.

Kantha, L. H. 1995. A numerical model of arctic leads. *JGR* 100(C3):4653–72.

Kantha, L. H., and G. L. Mellor. 1989. A numerical model of the atmospheric boundary layer over a marginal ice zone. *JGR* 94(C4):4959–70.

Karelin, D. B. 1937. Some data on the arctic ice cover (R). *Priroda* 26:32–38.

———. 1940. Melting of ice in the Arctic (R). *Probl Arktiki* 5:118–19.

Kaser, G., J. C. Cogley, M. B. Dyergerov, M. F. Meier, and A. Ohmura. 2006. Mass balance of glaciers and ice caps: Consensus estimates for 1961–2004. *GRL* 33:L19501, doi:10.1029/2006GL027511.

Kassner, K. 1996. *Pattern formation in diffusion-limited crystal growth.* Singapore: World Scientific.

Katsaros, K. B. 1973. Supercooling at the surface of an arctic lead. *J Phys Oceanog* 3:482–86.

Kawamura, T. 1986. A method for growing large single crystals of sea ice. *J Glaciol* 32(111):302–3.

———. 1987. Studies on preferred growth of sea ice grain. *ILTS Contributions*, Series A No 36.

———. 1988. Observations of the internal structure of sea ice by X ray computed tomography. *JGR* 93(C3):2343–50.

Kawamura, T., M. O. Jeffries, J. L. Tison, and H. R. Krouse. 2004. Superimposed ice formation in summer on Ross Sea pack-ice floes, Antarctica. *Ann Glaciol* 39:563–68.

Kawamura, T., and N. Ono. 1980. Freezing phenomenon at seawater surface opening in polar winter. III. Measurement of crystallographic orientation in newly grown sea ice (J). *LTS-A* 39:175–80.

Kay, J. E., T. L'Ecuyer, A. Gettleman, G. Stephens, and C. O'Dell. 2008. The contribution of cloud and radiation anomalies to the 2007 Arctic sea ice extent minimum. *GRL* 35:L08503, doi:10.1029/2008 GL033451.

Keller, J. 1997. Gravity waves on ice covered-water. *JGR* 103(C4):7663–69.

Keller, J. B., and M. Weitz. 1953. Reflection and transmission coefficients for waves entering or leaving an icefield. *Communs Pure Appl Math* 6(3):415–17.

Kempema, E. W., E. Reimnitz, and P. W. Barnes. 1989. Sea ice sediment entrainment and rafting in the Arctic. *J Sed Petrol* 59:308–17.

Kennedy, K. P., K. J. Mamer, J. P. Dempsey, R. M. Adamson, P. A. Spenser and D. M. Masterson. 1994. Large scale fracture experiments: Phase 2. *Proc 12th IAHR Ice Symp* 1:315–324. Trondheim, Norway.

Kerr, A. D., and W. T. Palmer. 1972. The deformations and stresses in floating ice plates. *Acta Mech* 15:57–72.

Kerr, R. A. 2004a. Sea change in the Atlantic. *Science* 303(2 January):35.

———. 2004b. Signs of a warm, ice-free Arctic. *Science* 305(17 September):1693.

Ketchum, R. D., Jr. 1971. Airborne laser profiling of the Arctic pack ice. *Remote Sensing of the Environment* 2:41–52.

Ketchum, W. M., and P. V. Hobbs. 1967. The preferred orientation in the growth of ice from the melt. *J Cryst Growth* 1:263–70.

Key, J. R., and M. Haefliger. 1992. Arctic ice surface temperature retrieval from AVHRR thermal channels. *JGR* 97:5885–93.

Key, J. R., and A. S. McLaren. 1989. Periodicities and keel spacings in the under-ice draft distribution of the Canada Basin. *CRST* 16:1–10.

———. 1991. Fractal nature of the sea ice draft profile. *GRL* 18(8):1437–40.

Key, J. R., and S. Peckham. 1991. Probable errors in width distributions of sea ice leads measured along a transect. *JGR* 96:18417–23.

Key, J. R., R. A. Silcox, and R. S. Stone. 1996. Evaluation of surface radiative flux parameterizations for use in sea ice models. *JGR* 101(C2):3839–49.

Kharitonov, V. V. 2008. Internal structure of ice ridges and stamukhas based on thermal drilling data. *CRST* 52(3):302–25.

Khazendar, A., and A. Jenkins. 2003. A model of marine ice formation within Antarctic ice shelf rifts. *JGR* 108(C7):3235, doi:10.1029/2002JC001673.

Kikuchi, T., J. Inoue, and J. Morison. 2005. Temperature difference across the Lomonosov Ridge: Implications for the Atlantic water circulation in the Arctic Ocean. *GRL*:L20604, doi:10.1029/2005GL023982.

Killawee, J. A., I. J. Fairchild, J. L. Tison, L. Janssens, and R. Lorrain. 1998. Segregation of solutes and gasses in experimental freezing of dilute solutions: Implications for natural glacier systems. *Geochem Cosmochim Acta* 62(23–24):3637–55.

Kindle, E. M. 1924. Observations on ice-borne sediments by the Canadian and other arctic expeditions. *Amer J Sci* 7:251–86.

Kingery, W. D., and W. H. Goodnow. 1963. Brine migration in salt ice. In *Ice and snow: Properties, processes and applications*, ed. W. D. Kingery, 35–47. Cambridge, MA: MIT Press.

Kipfstuhl, J. 1991. Zur Entstehung von Unterwassereis und das Wachstum und die Energiebilanz des Meereises in der Atka Bucht, Antarktis. *Alfred-Wegener-Institut für Polar- und Meeresforschung* 85:1–88.

Kipfstuhl, J., G. Dieckmann, H. Oerter, H. Hellmer, and W. Graf. 1992. The origin of green icebergs in Antarctica. *JGR* 97(C12):20319–24.

Kirschfield, R. A., and D. K. Perovich. 2005. Spatial and temporal variability of oceanic heat flux to the Arctic ice pack. *JGR* (110):C07021, doi:10.1029/2004JC002293.

Klein, L. A., and C. T. Swift. 1977. An improved model for the dielectric constant of seawater at microwave frequencies. *IEEE Trans Antennas Propagat* AP-25:104–11.

Knauss, J. A. 1978. *Introduction to physical oceanography*. Englewood Cliffs, NJ: Prentice-Hall.

Knight, C. A. 1962a. Polygonization of aged sea ice. *J Geol* 70:240–46.

———. 1962b. Studies of arctic lake ice. *J Glaciol* 4(33):319–35.

———. 1962c. Curved growth of ice on surfaces. *J Appl Phys* 33:1808–15.

Koch, L. 1945. The East Greenland ice. *Meddelelser om Gronland* 130(3).

Koerner, R. M. 1963. The Devon Island Expedition, 1960–64. *Arctic* 16:57–62.

———. 1973. The mass balance of sea ice of the Arctic Ocean. *J Glaciol* 12:173–85.

Kohler, J., and 7 others. 2007. Acceleration in thinning rate on western Svalbard glaciers. *GRL* 34:L18502, doi:10.1029/2007GL030681.

Kohnen, H. 1976. On the DC-resistivity of sea ice. *Z für Gletscherkunde Glazialgeol* 11(2):143–54.

Kohout, A. L., and M. H. Meylan. 2008. An elastic plate model for wave attenuation and ice floe breaking in the marginal ice zone. *JGR* 113(C09016), doi:10.1029/2007JC004434.

Kolchak, A. V. 1909. The ice of the Kara and Siberian Seas (R). *Russkaia Poliarnaia Ekspeditsiia 1900–1903 gg.*

Kolesnikov, A. G. 1958. On the growth rate of sea ice. In *Arctic sea ice*, ed. W. Thurston, 157–61. Pub. No. 598, US Nat Acad Sci Nat Res Council.

Konuk, I., S. Yu, and R. Gracie. 2005a. A 3-dimensional continuum ALE model for ice scour-study of trench effects. *24th Internat Conf on Offshore Mechanics and Arctic Engineering* (OMAE 2005-67547). Halkidiki, Greece.

———. 2005b. An ALE FEM model of ice scour. *Proceed 11th Internat Conf of the Internat Assoc Computer Methods and Advances in Geomechanics*, vol 3:63–70. Turin, Italy.

Koo, K-K., R. Ananth, and W. N. Gill. 1991. Tip splitting in dendritic growth of ice crystals. *Phys Rev A44*(6):3782–90.

Kovacs, A. 1971. On pressured sea ice. In *Sea ice*, ed. T. Karlsson, 276–95. Proc Internat Sea Ice Conf. Reykjavik.

———. 1972. Ice scoring marks floor of the Arctic shelf. *Oil and Gas Journal* 70:92–106.

———. 1976. Grounded ice in the fast ice zone along the Beaufort Sea coast of Alaska. *CRREL Report* 76-32.

———. 1983a. Characteristics of multiyear pressure ridges. *POAC 83*, VTT, Finland, vol. III.

———. 1983b. Shore ice ride-up and pile-up features, Part I: Alaska's Beaufort Sea coast. *CRREL Report* 83-9.

———. 1984. Shore ice ride-up and pile-up features, Part II: Alaska's Beaufort Sea Coast: 1983 and 1984. *CRREL Report* 84-26.

———. 1993. Axial double-ball test versus the uniaxial unconfined compression test for measuring the compressive strength of freshwater and sea ice. *CRREL Report* 93-25.

———. 1996. Sea ice part I. Bulk salinity vs. ice floe thickness. *CRREL Report* 96-7.

———. 1997. Estimating the full scale flexural, and compressive strength of first-year sea ice. *JGR* 102(C4):8681–89. (*CRREL Report* 96-11.)

Kovacs, A., and A. J. Gow. 1976. Some characteristics of grounded floebergs near Prudhoe Bay, Alaska. *CRREL Report* 76-34.

Kovacs, A., A. J. Gow, and W. F. Dehn. 1976. Islands of grounded sea ice. *CRREL Report* 76-4.

Kovacs, A., and M. Mellor. 1974. Sea ice morphology and ice as a geologic agent in the southern Beaufort Sea. In *The coast and shelf of the Beaufort Sea*, ed. J. C. Reed and J. C. Sater, 113–61. AINA.

Kovacs, A., and R. M. Morey. 1978. Radar anisotropy of sea ice due to preferred azimuthal orientation of the horizontal *c*-axes of ice crystals. *JGR* 83(C12):6037–46.

———. 1980. Investigations of sea ice anisotropy, electromagnetic properties, strength, and under-ice current orientation. *CRREL Report* 80-20.

Kovacs, A., and D. S. Sodhi. 1980. Shore ice pile-up and ride-up: Field observations, models, theoretical analyses. *CRST* 2:211–88.

———. 1988. Onshore ice pile-up and ride-up. In *Arctic coastal processes and slope protection design*, ed. A. T. Chen and C. B. Leidersdorf, 108–42. Technical Council on Cold Regions Engineering Monograph. New York: Amer Soc Civ Eng.

Kovacs A., W. F. Weeks, S. F. Ackley, and W. D. Hibler III. 1973. Structure of a multiyear pressure ridge. *Arctic* 26(1):22–32.

Kowalik, Z., and I. Polyakov. 1999. Diurnal tides over Kashevarov Bank, Okhotsk Sea. *JGR* 104:1389–1409.

Kozlovskii, A. M. 1976a. Some peculiarities of the formation and structure of shore ice along the Amery Ice Shelf. *Bull Sov Antark Exped* 92:40–45.

———. 1976b. Some peculiarities and causes of multi-year sea ice formation in the Cosmonaut Sea. *Bull Sov Antark Exped* 92:46–49.

Kozo, T. L. 1983. Initial model results for Arctic mixed layer circulation under a refreezing lead. *JGR* 88(C5):2926–34.

Kraus, E. 1930. Ice pressure ridges. In *Third Hydrological Conference of the Baltic States*. Warsaw, May 1930. (An English translation of this paper may possibly be available as CRREL Draft Translation 721.)

Kreyscher, M., M. Harder, P. Lemke, and G. M. Flato. 2000. Results of the sea ice model intercomparison project: Evaluation of sea ice rheology schemes for use in climate simulations. *JGR* 105:11299–320.

Krishfield, R. A., and D. K. Perovich. 2005. Spatial and temporal variability of oceanic heat flux to the Arctic ice pack. *JGR* 110:C07021, doi:10.1029/2004JC002293.

Kry, P. R. 1980a. Ice forces on wide structures. *Canad Geotech J* 17(1):97–113.

———. 1980b. A statistical prediction of effective ice crushing stresses on wide structures. In *Proceed ASCE Convention and Exposition* Preprint No. 80-084, Portland.

Krylov, A. N. 1901. Observations of ice strength (R). In *The Ermak in the ice fields: A description of the construction and voyages of the ice breaker Ermak and a summary of scientific results from the voyages*, ed. S. O. Makarov, 418–35. St. Petersburg: Tipografiia Sanktpetersburgskogo aktsionnogo obshchestva pechatnogo dela v Rosii E. Evdokimova.

Kuehn, G. A., and E. M. Schulson. 1994. The mechanical properties of saline ice under uniaxial compression. *Ann Glaciol* 19:39–48.

Kumai, M., and Itagaki, K. 1953. Cinematographic study of ice crystal formation in water. *J Faculty Sci, Hokkaido Univ* Ser 2(4):235–46.

Kusunoki, K. 1955. Observations on the horizontal and vertical distribution of chlorinity of sea ice. *J Oceanog Soc Japan* 11:1–5.

———. 1958. Measurement of gas bubble content in sea ice (J). *LTS-A* 17:123–34.

Kusy, R. P. 1977. Influence of particle size ratio on the continuity of aggregrates. *J Appl Phys* 48(12):5301–3.

Kuz'mina, N. P., and V. E. Sklyarov. 1984. Drifting ice as a tracer for the general circulation of marginal seas (R). *Issled Zemli Kosm* 1:16–25.

Kvajic, G., V. Brajovic, M. Milosevic-Kvajic, and E. R. Pounder. 1973. Selective growth of ice crystals from dilute water solution. In *Physics and chemistry of ice*, ed. E. Whalley, S. J. Jones, and L. W. Gold, 251–55. Ottawa: Royal Soc Canada.

Kvajic, G., E. R. Pounder, and V. Brajovic. 1971. Instability of a smooth-planar solid–liquid interface on an ice crystal growing from a melt. *Canad J Phys* 49(21):2636–45.

Kwok, R. 1999. The RADARSAT geophysical processor system. In *Analysis of SAR data of the polar oceans*, ed. C. Tsatsoulis and R. Kwok, 235–57. Heidelberg: Springer-Verlag.

———. 2001. Deformation of the Arctic Ocean sea ice cover between November 1996 and April 1997: A qualitative survey. In *Proc IUTAM Conf on Scaling Laws in Ice Mechanics and Ice Dynamics*, ed. J. P. Dempsey and H. H. Shen, 315–22. Dordrecht: Kluwer.

———. 2004. Annual cycles of multiyear sea ice coverage of the Arctic Ocean: 1999–2003. *JGR* 109: C11004, doi:10.1029/2003JC002238.

———. 2006. Contrasts in sea ice deformation and production in the Arctic seasonal and perennial ice zones. *JGR* 111:C11S22, doi:10.1029/2005JC003246.

———. 2008. Summer sea ice motion from the 18 GHz channel of AMSR-E and the exchange of sea ice between the Pacific and Atlantic sectors. *GRL* 35:L03504, doi:10.1029/2008GL032692.

Kwok, R., and G. F. Cunningham. 2008. ICESat over Arctic sea ice: Estimation of snow depth and ice thickness. *JGR* 113:C08010, doi:10.1029/2008JC004753.

Kwok, R., G. F. Cunningham, N. LaBelle-Hamer, B. Holt, and D. A. Rothrock. 1999. Ice thickness derived from high-resolution radar imagery. *Eos* 80(42):495, 497.

Kwok, R., G. F. Cunningham, H. J. Zwally, and D. Yi. 2007. ICESat over Arctic sea ice: Retrieval of free-board. *JGR* 112:C12013, doi:10.1029/2006JC003978.

Kwok, R., E. C. Hunke, W. Maslowski, D. Menemenlis, and J. Zhang. 2008. Variability of sea ice simulations assessed with RGPS kinematics. *JGR* 113:C11012, doi:10.1029/2008JC004783.

Kwok, R., and D. A. Rothrock. 1999. Variability of Fram Strait ice flux and North Atlantic Oscillation. *JGR* 104(C3):5177–89.

Lainey, L., and R. Tinawi. 1981. Parametric studies of sea ice beams under short and long term loadings. *IAHR Internat Symp on Ice* 2:607–13, Quebec City.

Lake, R. A., and E. L. Lewis. 1970. Salt rejection by sea ice during growth. *JGR* 75(3):583–97.

Lamb, H. 1932. *Hydrodynamics*. Dover.

Landauer, J. K., and H. Plumb. 1956. Measurements on anisotropy of the thermal conductivity of ice. *SIPRE* Research Paper 16.

Lane, J., and J. Saxton. 1953. Dielectric dispersion in pure polar liquids at very high radio frequencies. *Proceed Royal Soc* 214A:531–45.

Lange, M. A. 1988. Basic properties of Antarctic sea ice as revealed by textural analyses of ice cores. *Ann Glaciol* 10:95–101.

Lange, M. A., S. F. Ackley, P. Wadhams, G. S. Dieckmann and H. Eicken. 1989. Development of sea ice in the Weddell Sea. *Ann Glaciol* 12:92–96.

Lange, M. A., and H. Eicken. 1991. Textural characteristics of sea ice and the major mechanisms of ice growth in the Weddell Sea. *Ann Glaciol* 15:210–15.

Lange, M. A., P. Schlosser, S. F. Ackley, P. Wadhams, and G. S. Dieckmann. 1990. ^{18}O concentrations in sea ice of the Weddell Sea, Antarctica. *J Glaciol* 36(124):315–23.

Langhorne, P. J. 1980. Crystal anisotropy in sea ice in the Beaufort Sea. *Proc Workshop on Remote Estimation of Sea Ice Thickness*, 189–224. C-CORE Pub. 80-5.

———. 1982. Crystal alignment in sea ice. Ph.D. thesis, Cambridge University.

———. 1983. Laboratory experiments on crystal orientation in NaCl ice. *Ann Glaciol* 4:163–69.

Langhorne, P. J., and W. H. Robinson. 1983. Effect of acceleration on sea ice growth. *Nature* 305:695–98.

———. 1986. Alignment of crystals in sea ice due to fluid motion. *CRST* 12(2):197–214.

Langleben, M. P. 1959. Some physical properties of sea ice. II. *Canadian J Physics* 37:1438–54.

———. 1966. On the factors affecting the rate of ablation of sea ice. *Can J Earth Sci* 3:431–39.

———. 1972. The decay of an annual cover of sea ice. *J Glaciol* 11:337–44.

Langleben, M. P., and E. R. Pounder. 1963. Elastic parameters of sea ice. In *Ice and snow: Properties, processes and applications*, ed. W. D. Kingery, 67–78. Cambridge, MA: MIT Press.

———. 1964. Arctic sea ice of various ages. I. Ultimate strength. *J Glaciol* 5(37):93–98.

Langway, C., Jr. 1958. Ice fabrics and the universal stage. *SIPRE* Technical Report 62.

Lasca, N. P. 1971. River ice fabrics: Preliminary results. *J Glaciol* 10(58):151–52.

Laws, R. M. 1985. The ecology of the Southern Ocean. *American Scientist* 1:26–40.

Leavitt, E. 1980. Surface-based air stress measurements made during AIDJEX. In *Sea ice processes and models*, ed. R. S. Pritchard, 419–29. Seattle: Univ Washington Press.

Lebedev, V. L. 1968. Maximum size of a wind-generated lead during sea freezing. *Oceanology* (English translation) 8:313–18.

Lebedev, V. V. 1938. Growth of ice in Arctic rivers and seas in relation to negative air temperatures (R). *PA* 5-6:9–25.

Lee, J., T. D. Ralston, and D. H. Petrie. 1986. Full-thickness sea ice strength tests. *Proc IAHR Ice Symp*, vol 1:293–306. Iowa City.

Lee, R. L., Jr. 1990. Green icebergs and remote sensing. *J Opt Soc Amer* 7:1862–74.

Lee, R. W. 1986. A procedure for testing cored ice under uniaxial tension *J Glaciol* 32(112):540–41.

Legen'kov, A. P. 1963. Dispersal, concentration and compression of ice as a result of tidal action (R). *TAANII* 248:52–61.

Lemke, P. 1986. Stochastic description of atmosphere–sea ice–ocean interaction. In *Geophysics of sea ice*, ed. N. Untersteiner, 785–824. NATO Advanced Sciences Institutes Series, Series B, Physics. New York: Plenum Press.

———. 1987. Coupled one-dimensional sea ice–ocean model. *JGR* 92(C12):13164–72.

————. 1988. Results from sea ice–ocean modeling experiments in the Weddell Sea. In *Working Group on Sea Ice and Climate (Report of the Third Session)*, 13. WMO/TD-272 (Annex B). Oslo, Norway: World Climate Program (World Meteorological Organization).

————. 2001. Open windows to the polar oceans. *Science* 292(1 June):1670–71.

Lemke, P., W. D. Hibler III, G. Flato, M. Harder, and M. Kreyscher. 1997. On the improvement of sea ice models for climate simulations: The sea ice model inter-comparison project. *Ann Glaciol* 25:183–87.

Lemke, P., W. B. Owens, and W. D. Hibler III. 1990. A coupled sea ice-mixed layer-pycnocline model for the Weddell Sea. *JGR* 95(C6):9513–25.

Lensu, M., J. Tuhkuri, and M. Hopkins. 1998. Measurements of curvilinear ridges in the Bay of Bothnia during the ZIP-97 Experiment. Rept. M-231. Otaniemi, Finland: Helsinki Univ Technology, Dept Mechanical Engineering, Ship Laboratory.

Leonard, G. H., C. R. Purdie, P. J. Langhorne, T. G. Haskell, M. J. M. Williams, and R. D. Few. 2006. Observations of platelet ice growth and oceanic conditions during the winter of 2003 in McMurdo Sound, Antarctica. *JGR* 111:C04012, doi:10.1029/ 2005JC002952.

Leppäranta, M. 1980. On the drift and deformation of sea ice fields in the Bothian Bay. Winter Navigation Board, Helsinki, Finland. *Research Report* 29.

————. 1993. A review of analytical models of sea ice growth. *Atmosphere-Ocean* 31(1):123–38.

————, ed. 1998. *Physics of ice covered seas*. Vols. I and II. Helsinki Univ Pub House.

————. 2005. *The drift of sea ice*. Chichester, UK: Springer/Praxis.

Leppäranta, M., and R. Hakala. 1992. The structure and strength of first-year ice ridges in the Baltic Sea. *CRST* 20:295–311.

Leppäranta, M., and W. D. Hibler III. 1985. The role of plastic ice interaction in marginal ice zone dynamics. *JGR* 90:11899–909.

————. 1987. Mesoscale sea ice deformation in the east Greenland marginal ice zone. *JGR* 92:7060–70.

Leppäranta, M., M. Lensu, P. Kosloff, and B. Veitch. 1995. The life story of a first-year sea ice ridge. *CRST* 23:279–90.

Leppäranta, M., and T. Manninen. 1988. The brine and gas content of sea ice with attention to low salinities and high temperatures, Merentutkimuslaitos Sisäinen raportti [Finnish Institute of Marine Research] Internal Report 88-2:1–10.

Levitus, S. 1982. Climatological Atlas of the World. NOAA Publ. 13, Washington, DC: US Dept Commerce.

Lewis, C. F. M. 1977a. *Bottom scour by sea ice in the southern Beaufort Sea*. Beaufort Sea Technical Report 23 (draft). Victoria, BC: Department of Fisheries and the Environment.

————. 1977b. The frequency and magnitude of drift ice groundings from ice-scour tracks in the Canadian Beaufort Sea. In *POAC '77*, vol 1, 567–76. Newfoundland: Memorial University.

Lewis, E. L. 1967. Heat flow through winter ice. In *Physics of snow and ice*, ed. H. Oura, 611–31. Hokkaido: ILTS.

————. 1982. The Arctic Ocean: Water masses and energy exchanges. In *The Arctic Ocean, the hydrographic environment and the fate of pollutants*, ed. L. Ray, 43–68. Monaco: Comité Arctique International.

Lewis, E. L., and R. A. Lake. 1971. Sea ice and supercooled water. *JGR* 76(24):5836–41.

Lewis, E. L., and A. R. Milne. 1977. Underwater ice formation. In *Polar oceans*, ed. M. J. Dunbar, 239–45. Calgary: AINA.

Lewis, E. L., and R. G. Perkin. 1983. Supercooling and energy exchange near the Arctic Ocean surface. *JGR* 88(C12):7681–85.

————. 1985. The winter oceanography of McMurdo Sound, Antarctica. In *Oceanology of the Antarctic Continental Shelf*, 145–64. AGU Antarctic Research Series, No. 43.

————. 1986. Ice pumps and their rates. *JGR* 91:11756–62.

Lewis, J. E., M. Leppäranta, and H. B. Granberg. 1993. Statistical properties of sea ice surface topography in the Baltic Sea. *Tellus, Ser. A* 45:127–42.

Lewis, J. K. 1993. A model for thermally-induced stresses in multi-year sea ice. *CRST* 21(4):337–48.

————. 1995. A conceptual model of the impact of flaws on the stress state of sea ice. *JGR* 100(C5):8829–35.

————. 1998. Thermomechanics of pack ice. *JGR* 103(C10):21869–82.

Lewis, J. K., and J. A. Richter-Menge. 1998. Motion induced stresses in pack ice. *JGR* 103(C10):21831–44.

Lewis, J. K., W. B. Tucker III, and P. J. Stein. 1994. Observations and modeling of thermally induced stresses in first-year sea ice. *JGR* 99(C8):16361–71.

Liferov, P., and K. V. Høyland. 2004. In-situ ridge scour tests: Experimental set up and basic results. *CRST* 40(1–2):97–110.

Light, B., R. E. Brandt, and S. G. Warren. 2009. Hydrohalite in cold sea ice: Laboratory observations of single crystals, surface accumulations, and migration rates under a temperature gradient with application to "Snowball Earth." *JGR* 114(C07018), doi:10.1029/2008JC005211.

Light, B., T. C. Grenfell, and D. K. Perovich. 2008. Transmission and absorption of solar radiation by Arctic sea ice during the melt season. *JGR* 113(C7):C03023, doi:10.1029/2006JC003977.

Light, B., H. Eicken, G. A. Maykut, and T. C. Grenfell. 1998. The effect of included particulates on the spectral albedo of sea ice. *JGR* 103(C12):27739–52.

Light, B., G. A. Maykut, and T. C. Grenfell. 2003a. Effects of temperature on the microstructure of first-year Arctic sea ice. *JGR* 108(C2):3051, doi:10.1029/2001JC000887.

———. 2003b. A two-dimensional Monte Carlo model of radiative transfer in sea ice. *JGR* 108(C7):3219, doi:10.1029/ 2002JC001513.

———. 2004. A temperature-dependent, structural-optical model of first-year sea ice. *JGR* 109:C06013, doi:10.1029/2003JC002146.

Lin, F. C., J. A. Kong, R. T. Shin, A. J. Gow, and S. A. Arcone. 1988. Correlation function study for sea ice. *JGR* 93(C11):14055–63.

Lindsay, R. W. 1976. Wind and temperature profiles taken during the Arctic Lead Experiment. Master's thesis, Univ Washington, Seattle.

Lindsay, R. W. 1998. Temporal variability of the energy balance of thick Arctic pack ice. *J Climate* 11(3):313–33.

Lindsay, R. W., R. Kwok, L. deSteur, and W. Meier. 2008. Halo of ice deformation observed over the Maud Rise seamount. *GRL* 35:L15501, doi:10.1029/2008GL034629.

Lindsay, R. W., and D. Rothrock. 1994. Arctic sea surface temperature from AVHRR. *J Climate* 7(1):174–83.

———. 1995. Arctic sea ice leads from advanced very high resolution radiometer images. *JGR* 100(C3): 4533–44.

Lindsay, R. W., and J. Zhang. 2005. The thinning of Arctic sea ice, 1988–2003: Have we passed a tipping point? *J Climate* 18(22):4879–94.

Lindsay, R. W., J. Zhang, A. J. Schweiger, and M. A. Steele. 2008. Seasonal predictions of ice extent in the Arctic Ocean. *JGR* 113:C02023, doi:10.1029/2007JC004259.

Lipenkov, Y. Ya., N. I. Barkov, P. Duval, and P. Pimienta. 1989. Crystalline texture of the 2083 m ice core at Vostok Station, Antarctica. *J Glaciol* 35(121):392–98.

Lipscomb, W. H. 2001. Remapping the thickness distribution in sea ice models. *JGR* 106(C7):13,989–14,000.

Liu, A. K., S. Häkkinen, and C. Y. Peng. 1993. Wave effects in ocean–ice interaction in the marginal ice zone. *JGR* 98(C6):10025–36.

Liu, A. K., B. Holt, and P. Vachon. 1991. Wave propagation in the marginal ice zone: Model predictions and comparisons with buoy and synthetic aperture radar data. *JGR* 96:4605–21.

Liu, A. K., and E. Mollo-Christensen. 1988. Wave propagation in a solid ice pack. *J Phys Oceanogr* 18(11): 1702–12.

Liu, A. K., and C. Y. Peng. 1992. Waves and mesoscale features in the marginal ice zone. *Proc 1st ERS-1 Symp: Space at the Service of Our Environment*, 343–48. Cannes, France.

Liu, A. K., P. W. Vachon, and C. Y. Peng. 1991. Observation of wave refraction at an ice edge by synthetic aperture radar. *JGR* 96(C3):4803–8.

Liu, A. K., P. W. Vachon, C. Y. Peng, and A. S. Bhogal. 1992. Wave attenuation in the marginal ice zone during LIMEX. *Atm–Ocean* 30(2):192–206.

Liu, J., J. A. Curry, and Y. Hu. 2004. Recent Arctic sea ice variability: Connections to the Arctic Oscillation and the ENSO. *GRL* 31:L09211, doi:10.1029/2004GL019858.

Lizotte, M. P., and K. R. Arrigo, eds. 1998. *Antarctic sea ice: Biological processes, interactions and variability*. AGU, Antarctic Research Series, vol. 74.

Lofgren, G., and W. F. Weeks. 1969. Effect of growth parameters on substructure spacing in NaCl ice crystals. *J Glaciol* 8(52):153–64.

Long, D. G., J. Ballantyne, and C. Bertoia. 2002. Is the number of Antarctic icebergs really increasing? *Eos* 83(42):469, 474.

Longuet-Higgins, M. S. 1977. The mean forces exerted by waves on floating or submerged bodies with applications to sand bars and wave power machines. *Proc Roy Soc* A352:463–80.

Loring, S. 1996. Snow by many other names. *Washington Post* (9 November):A25.

Loshchilov, V. S. 1964. Snow cover on the ice of the Central Arctic. *PAiA* 17:36–45.

Lowry, R. T., and P. Wadhams. 1979. On the statistical distribution of pressure ridges in sea ice. *JGR* 84(C5): 2487–94.

Lu, S.-Z., and J. D. Hunt. 1992. A numerical analysis of dendrite and cellular array growth: The spacing adjustment mechanisms. *J Cryst Growth* 123:17–34.

Lubin, D., and R. A. Massom. 2006. *Polar remote sensing. Vol I: Atmospheres and oceans.* Springer.

Lüpkes, C., V. M. Gryanik, B. Witha, M. Gryschka, S. Raasch, and T. Gollnik. 2008. Modeling convection over arctic leads with LES and a non-eddy-resolving microscale model. *JGR* 113:C09028, doi:10.1029/2007JC004099.

Lüthje, M., D. L. Feltham, P. D. Taylor, and M. G. Worster. 2006. Modeling the summertime evolution of melt ponds. *JGR* 111:C02001, doi:10.1029/2004JC002818.

Lyon, W. K. 1961. Ocean and sea-ice research in the Arctic Ocean via submarine. *Trans NY Acad Sci* (srs. 2) 23:662–74.

Lyons, J. B., S. M. Savin, and A. J. Tamburi. 1971. Basement ice, Ward Hunt Ice Shelf, Ellesmere Island, Canada. *J Glaciol* 10(58):43–52.

Lytle, V. I., and S. F. Ackley. 1991. Sea ice ridging in the eastern Weddell Sea. *JGR* 96:18411–16.

———. 1996. Heat flux through sea ice in the western Weddell Sea: Conductive and convective transfer processes. *JGR* 101(C4):8853–68.

Lytle, V. I., A. P. Worby, R. Massom, M. J. Paget, I. Allison, X. Wu, and A. Roberts. 2001. Ice formation in the Mertz Glacier polynya, East Antarctica, during winter. *Ann Glaciol* 33:368–72.

MacDonald, W. J. P., and T. Hatherton. 1961. Movement of the Ross Ice Shelf near Scott Base. *J Glaciol* 3(29):859–66.

Macklin, W. C., and B. F. Ryan. 1966. Habits of ice grown in supercooled water and aqueous solutions. *Phil Mag* 14:847–60.

Magono, C., and C. W. Lee. 1966. Meteorological classification of natural snow crystals. *J Fac Sci Hokkaido Univ* ser 2:321–35.

Mahoney, A., R. G. Barry, V. Smolyanitsky, and F. Fetterer. 2008. Observed sea ice extent in the Russian Arctic. *JGR* 113:C11005, doi:10.1029/2008JC004830.

Mahoney, A., H. Eicken, A. Graves Gaylord, and L. Shapiro. 2007. Alaska landfast sea ice: Links with bathymetry and atmospheric circulation. *JGR* 112:C02001, doi:10.1029/2006JC003559.

Mahoney, A., H. Eicken, and L. Shapiro. 2007. How fast is landfast sea ice? A study of the attachment and detachment of near shore ice at Barrow, Alaska. *CRST* 47(3):233–55.

Mailhot, J., A. Tremblay, S. Belair, I. Gultepe, and G. A. Isaac. 2002. Mesoscale simulation of surface fluxes and boundary layer clouds associated with a Beaufort Sea polynya. *JGR* 107:C10, 10.1029/2000JC000429.

Makarov, S. O. 1901. *The Ermak in the ice fields: A description of the construction and voyages of the ice breaker Ermak and a summary of scientific results from the voyages* (R). St. Petersburg: Tipografiia Sanktpetersburgskogo aktsionnogo obshchestva pechatnogo dela v Rosii E. Evdokimova.

Makinson, K., and K. W. Nicholls. 1999. Modelling tidal currents beneath Filchner-Ronne Ice Shelf and on the adjacent continental shelf: Their effect on mixing and transport. *JGR* 104(C6):13449–65.

Makshtas, A. P. 1991. *The heat budget of arctic ice in the winter.* Cambridge: International Glaciological Society.

Maksym, T., and M. O. Jeffries. 2000. A one dimensional percolation model of flooding and snow ice formation on Antarctic sea ice. *JGR* 105(C11):26313–31.

Malmgren, F. 1927. On the properties of sea ice. In *Scientific results of the Norwegian north pole expedition with the* Maud, *1918–1925,* 1a(5).

Malvern, L. E. 1969. *Introduction to the mechanics of a continuous media.* Englewood Cliffs, NJ: Prentice Hall.

Marchenko, A. 2008. Thermodynamic consolidation and melting of sea ice ridges. *CRST* 52:278–301.

Mardia, K. V. 1972. *Statistics of directional data*. New York: Academic Press.

Marion, G. M., and R. E. Farren. 1999. Mineral solubilities in the Na-K-Mg-Ca-Cl-SO_4-H_2O system: A re-evaluation of the sulfate chemistry in the Spencer-Møller-Weare model. *Geochim Cosmochim Acta* 63:1305–18.

Marion, G. M., R. E. Farren, and A. J. Komrowski. 1999. Alternative pathways for seawater freezing. *CRST* 29(3):259–66.

Marion, G. M., and S. A. Grant. 1994. FREZCHEM: A chemical-thermodynamic model for aqueous solutions at subzero temperatures. *CRREL Special Repprtt* 94-18.

Marko, J. R., and R. E. Thompson. 1975. Spatially periodic lead patterns in the Canadian Basin sea ice: A possible relationship to planetary waves. *JRL* 2:431–34.

———. 1977. Rectolinear leads and internal motions in the ice pack of the western Arctic Ocean. *GRL* 82:979–87.

Markus, T., and D. J. Cavalieri. 1998. Snow depth distribution over sea ice in the Southern Ocean from satellite passive microwave data. In *Antarctic sea ice: Physical processes, interactions and variability*, ed. M. O. Jeffries, 9–39. AGU Antarctic Res Series, vol. 74.

Markus, T., C. Kottmeier, and E. Fahrbach. 1998. Ice formation in coastal polynyas in the Weddell Sea and their impact on oceanic salinity. In *Antarctic sea ice: Physical processes, interactions and variability*, ed. M. O. Jeffries, 273–92. AGU Antarctic Res Series, vol. 74.

Marshunova, M. S. 1961. Principal characteristics of the radiation balance of the underlying surface and of the atmosphere in the Arctic. *Trudy Arkt Antarkt Nauch Issle Inst* 229 (Eng. trans. by RAND Corp., Santa Monica, rept. RM-5003-PR, 1966).

Martin, S. 1970. A hydrodynamic curiosity: The salt oscillator. *Geophys Fluid Dyn* 1:143–60.

———. 1972. Ice stalactites and the desalination of sea ice. *Naval Research Reviews* 25(9):17–26.

———. 1974. Ice stalactites: Comparison of a laminar flow theory with experiment. *J Fluid Mech* 63(1):51–79.

———. 1981. Frazil ice in rivers and oceans. *Ann Rev Fluid Mech* 13:379–97.

———. 2001. Polynyas in the Arctic and Antarctic. In *Encyclopedia of ocean sciences*, ed. J. Steele, S. Thorpe, and K. Turekian, 2241–47. London: Academic Press.

———. 2004. *An Introduction to ocean remote sensing*. Cambridge: Cambridge Univ Press.

Martin, S., and J. Bauer. 1981. Bering Sea ice-edge phenomena. In *The eastern Bering Sea shelf: Oceanography and resources*, 189–212. Seattle: Univ Washington Press.

Martin, S., and D. J. Cavalieri. 1989. Contributions of the Siberian shelf polynyas to the Arctic Ocean intermediate and deep water. *JGR* 94(C9):12725–38.

Martin, S., R. Drucker, and M. Fort. 1995. A laboratory study of frost flower growth on the surface of young sea ice. *JGR* 100(C4):7027–36.

Martin, S., R. Drucker, R. Kwok, and B. Holt. 2004. Estimation of the thin ice thickness and heat flux for the Chukchi Sea Alaskan coast polynya from SSM/I data, 1990–2001. *JGR* 109:C10012, doi:10.1029/2004JC002428.

Martin, S., and P. Kauffman. 1974. The evolution of under-ice melt ponds, or double diffusion at the freezing point. *J Fluid Mech* 64:507–27.

———. 1977. An experimental and theoretical study of the turbulent and laminar convection generated under a horizontal ice sheet floating on warm salty water. *J Phys Oceanogr* 7(2):272–83.

———. 1981. A field and laboratory study of wave damping by grease ice. *J Glaciol* 27(96):283–313.

Martin, S., P. Kauffman, and C. Parkinson. 1983. The movement and decay of ice edge bands in the winter in the Bering Sea. *JGR* 88(C5):2803–12.

Martin, S., and A. S. Thorndike. 1985. Dispersion of sea ice in the Bering Sea. *JGR* 90(C4):7223–26.

Martin, S., Y. Yu, and R. Drucker. 1996. The temperature dependence of frost flower growth on laboratory sea ice and the effects of the flowers on infrared observations of the surface. *JGR* 101(C5):12111–25.

Martinson, D. 1990. Evolution of the Southern Ocean winter mixed layer and sea ice: Open ocean deep-water formation and ventilation. *JGR* 95(C7):11641–54.

———. 1991. Role of the Southern Ocean/sea ice interaction in global climate change. In *Internat Conf on the Role of the Polar Regions in Global Change*, 269–74. Fairbanks (June 1990): Univ of Alaska Fairbanks.

———. 2005. Distribution and ventilation of ocean heat on the western Antarctic Peninsula continental shelf. *Geophys Res Absts* 7:10293.

Martinson, D. G., P. D. Killworth, and A.L. Gordon. 1981. A convective model for the Weddell Polynya. *J Phys Oceanogr* 11:466–87.

Martinson, D. G., and C. Wamser. 1990. Ice drift and momentum exchange in winter antarctic pack ice. *JGR* 95(C2):1741–56.

Maslanik, J. 1997. On the role of sea ice transport in modifying polar responses to global climate change. *Ann Glaciol* 25:102–6.

Maslanik, J. A., S. Drobot, C. Fowler, W. Emery, and R. Barry. 2007. On the Arctic climate paradox and the continuing role of atmospheric circulation in affecting sea ice conditions. *GRL* 34:L03711 doi:10.1029/2006GL128269.

Maslanik, J. A., C. Fowler, J. Strove, S. Drobot, J. Zwally, D. Yi, and W. Emery. 2007. A younger, thinner Arctic ice cover: Increased potential for rapid sea ice loss. *GRL* 34:L24501, doi:10.1029/2007GL032043.

Maslanik, J. A., and J. Key. 1995. On treatments of fetch and stability sensitivity in large-area estimates of sensible heat flux over sea ice. *JGR* 100(C3):4573–84.

Maslanik, J. A., M. C. Serreze, and R. G. Barry. 1996. Recent decreases in Arctic summer ice cover and linkages to atmospheric circulation anomalies. *GRL* 23:1677–80.

Maslowski, W., B. Newton, P. Schlosser, A. Sempter, and D. Martinson. 2000. Modelling recent climate variability in the Arctic Ocean. *GRL* 27:3743–46.

Massom, R. A. 1991. *Satellite remote sensing of polar regions*. Boca Raton, FL: Lewis Publishers.

Massom, R. A., J. C. Comiso, A. P. Worby, V. I. Lytle, and L. Stock. 1999. Regional classes of sea ice cover in the East Antarctic pack observed by satellite and *in situ* data during a winter time period. *Remote Sens Environ* 68(1):61–76.

Massom, R. A., K. L. Hill, V. I. Lytle, A. P. Worby, M. J. Paget, and I. Allison. 2001. Effects of regional fast-ice and iceberg distributions on the behaviour of the Mertz Glacier polynya, East Antarctica. *Ann Glaciol* 33(3):391–98.

Massom, R. A., K. Jacka, M. J. Pook, C. Fowler, N. Adams, and N. Bindoff. 2003. An anomalous late-season change in the regional sea ice regime in the vicinity of the Mertz Glacier Polynya, East Antarctica. *JGR* 108(C7):3212, doi:10.1029/ 2002JC001354.

Massom, R. A., and D. Lubin. 2006. *Polar remote sensing. Vol. II: Ice sheets*. Springer.

Massom, R. A., V. I. Lytle, A. P. Worby, and I. Allison. 1998. Winter snow variability on East Antarctic sea ice. *JGR* 103(C11):24837–55.

Massom, R. A., H. Eicken, and 12 others. 2001. Snow on Antarctic sea ice. *Rev Geophys* 39(3):413–45.

Masterson, D. M. 1996. Interpretation of in situ borehole strength measurement tests. *Can J Civ Eng* 23(1):165–79.

———. 2009. Ice strength: In situ measurement. In *Sea ice handbook*, ed. H. Eicken, chapter 3.11. Fairbanks: Univ Alaska Press.

Masterson, D. M., W. P. Graham, S. J. Jones, and G. R. Childs. 1997. A comparison of uniaxial and borehole jack tests at Fort Providence Ice Crossing. *Can Geotech J* 34:471–75.

Matsuo, S., and Y. Miyake. 1966. Gas composition in ice samples from Antarctica. *JGR* 71(22):5235–41.

Mätzler, C., and U. Wegmüller. 1987. Dielectric properties of freshwater ice at microwave frequencies. *J Physics D: Applied Physics* 20:1623–30.

Maxwell, J. C. 1891. *Electricity and magnetism*, 3rd ed. Vol. I. New York: Dover Publications.

Maykut, G. A. 1978. Energy exchange over young sea ice in the Central Arctic. *JGR* 83(C7):3646–58.

———. 1986. The surface heat and mass balance. In *The geophysics of sea ice*, ed. N. Untersteiner, 395–463. New York: Plenum Press.

Maykut, G. A., and P. E. Church. 1973. Radiation climate of Barrow, Alaska, 1962–66. *J Appl Met* 12:620–28.

Maykut, G. A., T. C. Grenfell, and W. F. Weeks. 1992. On estimating spatial and temporal variations in the properties of ice in the polar oceans. *J Marine Sys* 3:41–72.

Maykut, G. A., and B. Light. 1995. Refractive index measurements in freezing sea-ice and sodium chloride brines. *Appl Optics* 34(6):950–61.

Maykut, G. A., and M. G. McPhee. 1995. Solar heating of the Arctic mixed layer. *JGR* 100(C12):24691–703.

Maykut, G. A., and D. K. Perovich. 1987. The role of shortwave radiation in the summer decay of a sea ice cover. *JGR* 92(C7):7043–44.

Maykut, G. A., and N. Untersteiner. 1969. Numerical prediction of the thermodynamic response of the Arctic sea ice to environmental change. *Rand Corp Memo* RM-6093-PR.

———. 1971. Some results from a time dependent thermodynamic model of sea ice. *JGR* 76(6):1550–75.

McGuinness, M. J., H. J. Trodahl, K. Collins, and T. G. Haskell. 1998. Non-linear thermal transport and brine convection in first-year sea ice. *Ann Glaciol* 27:471–76.

McLaren, A. S. 1989. The under-ice thickness distribution of the Arctic basin as recorded in 1958 and 1970. *JGR* 94(C4):4971–83.

McLaren, A. S., R. H. Bourke, J. E. Walsh, and R. L. Weaver. 1994. Variability in sea-ice thickness over the North Pole from 1958 to 1992. In *The polar oceans and their role in shaping the global environment,* ed. O. M. Johannessen, R. D. Muench, and J. E. Overland, 363–71. AGU Geophys Monograph 85.

McLaren, A. S., P. Wadhams, and R. Weintraub. 1984. The sea ice topography of M'Clure Strait in winter and summer of 1960 from submarine profiles. *Arctic* 37(2):110–20.

McPhee, M. G. 1978. A simulation of inertial oscillation in drifting pack ice. *Dyn Atmos & Oceans* 2:107–22.

———. 1979. The effect of the oceanic boundary layer on the mean drift of pack ice: Application of a simple model. *J Phys Oceanog* 9(2):388–400.

———. 1980. An analysis of pack ice drift in summer. In *Sea ice processes and models,* ed. R. S. Pritchard, 62–75. Seattle: Univ Washington Press.

———. 1981. An analytic similarity theory for the planetary boundary layer stabilized by surface buoyancy. *Bound-Layer Met* 21:325–339.

———. 1982. Sea ice drag laws and simple boundary layer concepts, including application to rapid melting. *CRREL Report* 82-4.

———. 1986. The upper ocean. In *The geophysics of sea ice,* ed. N. Untersteiner, 339–94. New York: Plenum Press.

———. 1987. Time-dependent model for turbulent transfer in a stratified oceanic boundary layer. *JGR* 92(C7):6977–86.

———. 1988. Inferring ice/ocean surface roughness from horizontal current measurements. In *Proc 7th Internat Conf on Offshore Mechanics and Offshore Engineering,* ed. D. S. Sodhi, C. H. Luk, and N. K. Sinha, 91–98. Vol. 4: Arctic Engineering and Technology. Houston: American Society of Mechanical Engineers.

———. 1990a. Small-scale processes. In *Polar oceanography. Part B: Chemistry, biology and geology,* ed. W. O. Smith, Jr., 287–334. San Diego: Academic Press.

———. 1990b. Using the free-drift force balance to estimate drag coefficients and ice thickness. In *Sea ice properties and processes,* ed. S. F. Ackley and W. F. Weeks, 281–84. CRREL Monograph 90-1.

———. 1992. Turbulent heat flux in the upper ocean under sea ice. *JGR* 97:5365–79.

———. 1999. Parameterizing mixing in the oceanic boundary layer. *J Marine Syst* 21:55–65.

———. 2008. *Air–ice–ocean interaction: Turbulent ocean boundary layer exchange processes.* New York: Springer.

McPhee, M. G., G. A. Maykut, and J. H. Morison. 1987. Dynamics and thermodynamics of the ice/upper ocean system in the marginal ice zone of the Greenland Sea. *JGR* 92(C7):7017–31.

McPhee, M. G., T. P. Stanton, J. H. Morison, and D. G. Martinson. 1998. Freshening of the upper ocean in the Arctic: Is the perennial sea ice disappearing? *GRL* 25(10):1729–32.

McPhee, M. G., and N. Untersteiner. 1982. Using sea ice to measure vertical heat flux in the ocean. *JGR* 87(C3):2071–74.

McPhee, M. G., and 8 others. 1996. The antarctic flux zone experiment. *Bull Amer Meteorol Soc* 77:1221–32.

Mears, C. A., and F. J. Wentz. 2005. The effect of diurnal correction on satellite-derived lower tropospheric temperature. *Science* 309(2 September):1548–51.

Meese, D. A. 1989. The chemical and structural properties of sea ice in the southern Beaufort Sea. *CRREL Report* 89-25.

Meier, M. F., M. B. Dyurgerov, and 6 others. 2007. Glaciers dominate eustatic sea-level rise in the 21st century. *Science* 317(24 August):1064–67.

Meier, W., J. Stroeve, F. Fetterer, and K. Knowles. 2005. Reductions in Arctic sea icc cover no longer limited to summer. *Eos* 86(36):326.

Melling, H., and D. A. Riedel. 1995. The underside topography of sea ice over the continental shelf of the Beaufort Sea in the winter of 1990. *JGR* 100(C7):13641–53.

———. 1996. Development of seasonal pack ice in the Beaufort Sea during the winter of 1991–1992: A view from below. *JGR* 101:11975–92.

Melling, H., D. A. Riedel, and Z. Gedalof. 2005. Trends in the draft and extent of seasonal pack ice, Canadian Beaufort Sea. *GRL* 32:L24501, doi:1029/2005 GL024483.

Mellor, G. L., and L. H. Kantha. 1989. An ice–ocean coupled model. *JGR* 94(C8):10937–54.

Mellor, M. 1964. Properties of snow. *CRREL Monograph* III-A1.

———. 1964b. Optical measurements on snow. *CRREL Research Report* 169: 17 pp.

———. 1977. Engineering properties of snow. *J Glaciol* 19(81):15–65.

———. 1986. Mechanical behavior of sea ice. In *Geophysics of sea ice*, ed. N. Untersteiner, 165–281. New York: Plenum Press.

Mellor, M., and I. Hawkes. 1971. Measurement of tensile strength by diametral compression of discs and annuli. *Engineer Geol* 5(3):173–225.

Mel'nichenko, N. A., V. I. Mikhaylov, and V. I. Chizhik. 1979. Studies of the temperature dependence of the brine content of sea ice by the pulse NMR method (R). *Okeanologiya* 19(5):811–14.

Melnikov, I. A. 1997. *The arctic sea ice ecosystem.* Gordon and Breach.

Mendelson, A. 1968. *Plasticity: Theory and applications.* Malabar, FL: Krieger Pub Co.

Meyers, R., S. Blasco, G. Gilbert, and G. Shearer. 1996. *1990 Beaufort Sea ice scour repetitive mapping program.* Environmental Studies Research Funds Report No. 129.

Meylan, M., and V. A. Squire. 1993a. A novel analytic model for the motion and bending of an ice floe in ocean waves. *Proc 3rd Internat Offshore and Polar Eng Conf* vol. II, 718–23. Singapore, 6–11 June 1993.

———. 1993b. A model for the motion and bending of an ice floe in ocean waves. *Int J Offshore Polar Eng* 3(4):322–23.

———. 1993c. Finite-floe wave reflection and transmission coefficients from a semi-infinite model. *JGR* 98(C7):12537–42.

———. 1994. The response of ice floes to ocean waves. *JGR* 99(C1):891–900.

Michel, B., and R. O. Ramseier. 1971. Classification of river and lake ice. *Canad Geotech J* 8(1):36–45.

Miles, M. W., and R. G. Barry. 1998. A 5-year satellite climatology of winter sea ice leads in the western Arctic. *JGR* 103(C10):2172–34.

Milne, A. R. 1972. Thermal tension cracking in sea ice: A source of under-ice noise, *JGR* 77(12):2177–92.

Mironenko, M. V., S. A. Grant, G. M. Marion, and R. E. Farren. 1997. FREZCHEM2: A chemical-thermodynamic model for aqueous solutions at subzero temperatures. *CRREL Report* 97-5.

Mock, S. J., S. F. Ackley, and W. D. Hibler III. 1973. Surface effect vehicle program: Interim sea ice statistics. *CRREL* 2.

Mock S. J., A. D. Hartwell, and W. D. Hibler. 1972. Spatial aspects of pressure ridge statistics. *JGR* 77(30):5945–53.

Monaghan, A. J., D. H. Bromwich, and D. P. Schneider. 2008. Twentieth century Antarctic air temperature and snowfall simulations by IPCC climate models. *GRL* 35:L07502, doi:10/1029/2007GL032630.

Moore, J. C., A. P. Reid, and J. Kipfstuhl. 1994. Microstructure and electric properties of marine ice and its relationship to meteoric ice and sea ice. *JGR* 99(C3):5171–80.

Morales-Maqueda, M. A., A. J. Willmott, and N. R. T. Biggs. 2004. Polynya dynamics: A review of observations and modeling. *Rev Geophys* 42:RG1004, doi:10.1029/2002 RG000116.

Morey, R. M., A. Kovacs, and G. F. N. Cox. 1984. Electromagnetic properties of sea ice. *CRREL Rept* 84-2.

Morgan, V. I. 1972. Oxygen isotope evidence for bottom freezing on the Amery Ice Shelf. *Nature* 238:393–94.

Morison, J. H., M. G. McPhee, T. B. Curtin, and C. A. Paulson. 1992. The oceanography of winter leads. *JGR* 97(C7):11199–218.

Morison, J. H., M. G. McPhee, and G. A. Maykut. 1987. Boundary layer, upper ocean, and ice observations in the Greenland Sea marginal ice zone. *JGR* 92(C7):6987–7011.

Morison, J., M. Steele, T. Kikuchi, K. Falkner, and W. Smethie. 2006. Relaxation of central Arctic Ocean hydrography to pre-1990s climatology. *GRL* 33:L17604, doi:10.1029/2006GL026826.

Moritz, R. E., and H. L. Stern. 2001. Relationships between geostrophic winds, ice strain rates and the piecewise rigid motions of pack ice. In *Scaling laws in ice mechanics and ice dynamics*, ed. J. P. Dempsey and H. H. Shen, 335–48. Dordrecht: Kluwer.

Moslet, P. O. 2007. Field testing of uniaxial compression strength of columnar sea ice. *CRST* 48(1):1–14.

Mote, T. L. 2007. Greeland surface melt trends 1973–2007: Evidence in a large increase in 2007. *GRL* 34:L22507, doi:10.1029/2007GL031976.

Moulton, K. N., and R. L. Cameron. 1976. Bottle green iceberg near the South Shetland Islands. *Antarct J US* 11:94–95.

Mountain, D. G. 1974. Preliminary analysis of Beaufort shelf circulation in summer. In *The coast and shelf of the Beaufort Sea*, ed. J. C. Reed and J. E. Sater, 27–48. AINA.

Mueller, D. R., W. F. Vincent, and M. O. Jeffries. 2003. Break-up of the largest Arctic ice shelf and associated loss of an epishelf lake. *GRL* 30:2031, doi:10,1029/2003GL017931.

Muench, R. D., ed. 1983. Marginal Ice Zone. *JGR* 88(C5):2715–2972.

Muench, R. D., and A. L. Gordon. 1995. Circulation and transport of water along the western Weddell Sea margin. *JGR* 100:18503–15.

Muench, R. D., P. H. LeBlond, and L. E. Hachmeister. 1983. On some possible interactions between internal waves and sea ice in the marginal ice zone. *JGR* 88(C5):2819–26.

Muench, R. D., S. Martin, and J. E. Overland, eds. 1987. Marginal Ice Zones. *JGR* 92(C7):6716–7225.

Muench, R. D., J. H. Morison, L. Padman, D. Martinson, P. Schlosser, B. Huber, and R. Hohmann. 2001. Maude rise revisited. *JGR* 106(C2):2423–40.

Muench, R. D., D. C. Smith, and C. A. Paulson. 1995. Convection beneath freezing leads: New observations compared with numerical model results. *JGR* 100(C3):4681–92.

Müller-Stoffels, M. 2006. Preferred crystal orientations in fresh-water ice sheets. Master's thesis, Dunedin University, New Zealand.

Müller-Stoffels, M., P. J. Langhorne, C. Petrich, and E. W. Kempema. 2009. Preferred crystal orientation in fresh water ice. *CRST*: in press.

Mullins, W. W., and R. F. Sekerka. 1963. Morphological stability of a particle growing by diffusion or heat flow. *J Appl Phys* 34:323–29.

———. 1964. Stability of a planar interface during solidification of a dilute binary alloy. *J Appl Phys* 35:444–51.

Murat, J. R., and L. M. Lainey. 1982. Some experimental observations on the Poisson's ratio of sea ice. *CRST* 6:105–13.

Murray, B. J., and A. K. Bertram. 2007. Strong dependence of cubic ice formation on droplet ammonium to sulfate ratio. *GRL* 34:L16810, doi:10/1029/2007GL030471.

Myerson, A. S., and D. J. Kirwan. 1977a. Impurity trapping during dendritic crystal growth. 1. Computer simulation. *I&EC Fundamentals* 16:414–20.

———. 1977b. Impurity trapping during dendritic crystal growth. 2. Experimental results and correlation. *I&EC Fundamentals* 16:420–25.

Nada, H., and Y. Furukawa. 1996. Anisotropic growth kinetics of ice crystals from water studied by molecular dynamics simulation. *J Cryst Growth* 169:587–97.

Nagashima, K., and Y. Furukawa. 1997. Solute distribution in front of an ice/water interface during directional growth of ice crystals and its relationship to interfacial patterns. *J Phys Chem B* 101:6174–76.

Nagle, J. F. 1966. Lattice statistics of hydrogen bonded crystals. I. The residual entropy of ice. *J Math Phys* 7:1484–91.

Nakawo, M. 1983. Measurements on air porosity of sea ice. *Ann Glaciol* 4:204–8.

Nakawo, M., and N. K. Sinha. 1981. Growth rate and salinity profile of first-year sea ice in the high Arctic. *J Glaciol* 27(96):315–30.

———. 1984. A note on brine layer spacing in first-year sea ice. *Atmos-Ocean* 22(2):193–206.

Nakaya, U. 1954. *Snow crystals: Natural and artificial.* Cambridge, MA: Harvard Univ Press.

———. 1956. Properties of single crystals of ice, revealed by internal melting. *SIPRE Res Rept* 13.

Nansen, F. 1897. *Farthest North; being the record of a voyage of exploration of the ship* Fram *1893–96.* London: Archibald Constable and Co.

———. 1900. Ice drift in the North Polar Basin. In *The Norwegian north polar expedition 1893–1896. Scientific results*, ed. F. Nansen, 365–93. New York: Longmans, Green and Co.

———. 1900–1906. *The Norwegian north pole expedition 1893–1896. Scientific results.* London: Longmans, Green and Co. 6 vols.

————. 1902. The oceanography of the North Polar Basin. In *The Norwegian north polar expedition 1893–1896. Scientific results*, ed. F. Nansen, vol. 3(9):357–86. New York: Longmans, Green and Co.

————. 1911. *Nebelheim. Entdeckung und Erforschung der nördlichen Länder und Meere* (2 vols.). Leipzig: F. A. Brockhaus (published in English in 1911 as *Northern Mists: Arctic exploration in early times*, London: W. Heinemann).

National Aeronautics and Space Administration. 1986. *Earth system science: Overview* (1988). A section of *Earth system science: A program for global change*. Washington, DC: NASA.

National Snow and Ice Data Center. 2007. NSIDC track record shattering summer for Arctic sea ice. *NSIDC Notes* 61:1.

————. 2008. Antarctic ice shelf disintegration underscores a warming world. *NSIDC Notes* 63:1.

Nayar, H. S., F. V. Lenel, and G. S. Ansell. 1971. Creep of dispersions of ultrafine amorphous silica in ice. *J Appl Phys* 42(10):3786–89.

Nazarenko, L., G. Holloway, and N. Tausner. 1998. Dynamics of transport of "Atlantic signature" in the Arctic Ocean. *JGR* 103(13):31003–15.

Nazarov, V. S. 1962. *Ice of the Antarctic waters* (R). Results of Researches on the Program of the International Geophysical Year, Oceanology Section X(6):1–81. Moscow: Izdatel'stvo Akademii Nauk SSSR.

————. 1963. Amount of ice in the world's oceans and its variation (R). *Okeanologiia* 3(2):243–49.

Nazintsev, Iu. L. 1963. On the role of thermal processes in the ice in melting and modifying the relief of multiyear ice floes in the Central Arctic (R). *PAiA* 12:69–75.

————. 1964a. Thermal balance of the surface of the perennial ice cover in the central Arctic (R). *Trudy AANII* 267:110–26.

————. 1964b. Some data on the calculation of thermal properties of sea ice (R). *Trudy AANII* 267:31–47.

————. 1997. The phase relations and thermal properties of sea ice (R). In *Morskoi led (Sea ice)*, ed. I. Frolov and V. L. Gavrilo, 68–106. St. Petersburg: Gidrometeoizdat.

Nazintsev, Iu. L., Zh. A. Dmitrash, and V. I. Moiseev. 1988. *Thermophysical properties of sea ice* (R). Leningrad: Leningradskii Gosudarstvennyi Universitet.

Nelson, K. H. 1953. A study of the freezing of sea water. Ph.D. thesis, Department of Oceanography, Univ Washington, Seattle.

Nelson, K. H., and T. G. Thompson. 1954. Deposition of salts from sea water by frigid concentration. *J Mar Res* 13:166–82.

Newyear, K., and S. Martin. 1997. A comparison of theory and laboratory measurements of wave propagation and attenuation in grease ice. *JGR* 102(C11):25091–99.

————. 1999. Comparison of laboratory data with a viscous two-layer model of wave propagation in grease ice. *JGR* 104(C4):7837–40.

Nghiem, S. V., Y. Chao, G. Neumann, P. Li, D. K. Perovich, T. Street, and P. Clemente-Colon. 2006. Depletion of perennial sea ice in the East Arctic Ocean. *GRL* 33:L17501, doi:10.1029/2006GL027198.

Nghiem, S. V., I. G. Rigor, D. K. Perovich, P. Clemente-Colon, J. W. Weatherly, and G. Neumann. 2007. Rapid reduction of Arctic perennial sea ice. *GRL* 34:L19504, doi:10.1029/2007GL031138.

Nicholls, K. W. 2001. Sub ice shelf circulation and processes. In *Encyclopedia of ocean sciences*, ed. J. Steele, S. Thorpe, and K. Turekian. Academic Press.

Nicholls, K. W., and K. Makinson. 1998. Ocean circulation beneath the western Ronne Ice Shelf, as derived from *in situ* measurements of water currents and properties. In *Ocean, ice and atmosphere: Interactions at the antarctic continental margin*, ed. S. S. Jacobs and R. F. Weiss, 301–18. AGU Antarctic Research Series 75.

Niebauer, H. J., and R. H. Day. 1989. Causes of interannual variability in the sea ice cover of the eastern Bering Sea. *Geo J* 18(1):45–59.

Niedrauer, T. M., and S. Martin. 1979. An experimental study of brine drainage and convection in young sea ice. *JGR* 84(C3):1176–86.

Nielsen, J. P. 1996. *Historical and current uses of the Northern Sea Route. Part III: The Period 1855–1917.* INSROP Working Paper No. 61-1996, IV.1.1, Lysaker, Norway: The Fridtjof Nansen Institute.

Nihashi, S., K. I. Ohshima, M. O. Jeffries, and T. Kawamura. 2005. Sea-ice melting processes inferred from ice–upper ocean relationships in the Ross Sea, Antarctica. *JGR* 110:C02002, doi:10.1029/2003JC002235.

North, G. R. 1988. Lessons from energy balance models. In *Physically-based modelling and simulation of climate and climate change*, ed. M. E. Schlesinger, 627–52. Dordrecht: Kluwer.

North, G. R., R. F. Cahalan, and J. A. Coakley. 1981. Energy balance climate models. *Rev Geophys Space Phys* 19(1):91–121.

Notz, D., M. G. McPhee, M. G. Wooster, G. A. Maykut, K. H. Schlünzen, and H. Eicken. 2003. Impact of underwater-ice evolution on Arctic summer sea ice. *JGR* 108(C7):3223, doi:10.1029/2001JC001173, 2003.

Notz, D., J. S. Wettlaufer, and M. G. Worster. 2005. A non-destructive method for measuring the salinity and solid fraction of growing sea ice in situ. *J Glaciol* 51(172):159–66.

Notz, D., and M. G. Worster. 2008. In situ measurements of the evolution of young sea ice. *JGR* 113:C03001, doi:10.1029/2007JC994333.

Oertner, H., J. Kipfstuhl, J. Determann, H. Miller, D. Wagenbach, A. Minikin, and W. Graf. 1992. Ice core evidence for basal marine shelf ice in the Filchner-Ronne Ice Shelf. *Nature* 358:399–401.

Oertling, A. B., and R. G. Watts. 2004. Growth and brine drainage from NaCl-H$_2$O freezing: A simulation of young sea ice. *JGR* 109:C04013, doi:10.1029/2001 JC001109.

Ogi, M., I. G. Rigor, M. G. McPhee, and J. M. Wallace. 2008. Summer retreat of Arctic sea ice: Role of summer winds. *GRL* 35:L24701, doi:10.1029/2008GL035672.

Okamoto, T., and K. Kishitake. 1975. Dendritic structure in unidirectional solidified aluminum, tin, and zinc base binary alloys. *J Crystal Growth* 29:137–46.

Okamoto, T., K. Kishitake, and I. Bessho. 1975. Dendritic structure in unidirectional solidified cyclohexanol. *J Crystal Growth* 29:131–36.

Omstedt, A. 1984. A forecasting model for water cooling in the Gulf of Bothnia and Lake Vänern. *SMHI Report*. RHO 36.

———. 1985. Modelling frazil ice and grease ice formation in the upper layers of the ocean. *CRST* 11:87–98.

———. 1985b. On supercooling and ice formation in turbulent sea water. *J Glaciol* 31(109):263–71.

———. 1986. Modelling initial ice formation in rivers and oceans. In *IAHR Ice Symp*, 559–68. Iowa City, Iowa.

———. 1990a. A coupled one-dimensional sea ice-model applied to a semi-enclosed basin. *Tellus* 42A(5): 568–82.

———. 1990b. Modelling the ice/water boundary layer. In *IAHR, The 10th symp on ice*, 386–95. 1 Espoo, Finland: Helsinki Univ Technology.

———. 1994. Numerical simulation of frazil ice. In *IAHR working group on thermal regimes report on frazil ice*, ed. S. F. Daly, 25–43. CRREL Special Report 94-23.

Omstedt, A., and J. Sahlberg. 1977. Some results from a joint Finnish–Swedish sea ice experiment, March 1977. Research Rept 26. Norrkoping, Sweden: Winter Navagation Research Board.

Omstedt, A., and U. Svensson. 1984. Modelling supercooling and ice formation in a turbulent Ekman layer. *JGR* 89(C1):735–44.

———. 1992. On the melt rate of drifting ice heated from below. *CRST* 21(1):91–100.

Omstedt, A., and J. S. Wettlaufer. 1992. Ice growth and oceanic heat flux: Models and measurements. *JGR* 97(C6):9383–90.

O'Neil, R. V., D. L. DeAngelis, J. B. Waide, and T. F. H. Allen. 1986. *A hierarchical concept of ecosystems*. Princeton Univ Press.

Ono, N. 1967. Specific heat and heat of fusion of sea ice. In *Physics of snow and ice*, ed. H. Oura, 1:599–610. Sapporo: ILTS, Hokkaido Univ.

———. 1968. Thermal properties of sea ice. IV. Thermal constants of sea ice. *LTS-A* 26:329–49.

Ono, N., and T. Kasai. 1985. Surface layer salinity of young sea ice. *Ann Glaciol* 6:298–99.

Orsi, A. H., G. C. Johnson, and J. L. Bullister. 1999. Circulation, mixing and production of Antarctic bottom water. *Prog Oceanog* 43:55–109.

Østerhus, S., and O. Orheim. 1992. Studies through Jutulgryta, Fimbulisen in the 1991/92 season. In *Filchner-Ronne ice shelf programme*, ed. H. Oertner, 103–9. Rep. 6. Bremmerhaven: Alfred Wegener Inst for Polar and Marine Research.

Osterkamp, T. E. 1977. Frazil-ice nucleation by mass-exchange process at the air–water interface. *J Glaciol* 19(81):619–25.

———. 1978. Frazil ice formation: A review. *J Hydraul Div, ASCE* 104(HY9):1239–55.

Osterkamp, T. E., T. Ohtake, and D. C. Warniment. 1974. Detection of airborne ice crystals near a supercooled stream. *J Atmos Sci* 31(5):1464–65.

Ostrogorsky, A. G., and G. Müller. 1992. A model of effective segregation coefficient, accounting for convection in the solute layer at the growth interface. *J Crystal Growth* 121:587–98.

Ou, H. W. 1988. A time-dependent model of a coastal polynya. *J Phys Oceanogr* 18(4):584–90.

———. 1991. Some effects of a seamount on oceanic flows. *J Phys Oceanogr* 21:1835–45.

Ou, H. W., and A. L. Gordon. 1986. Spin-down of baroclinic eddies under sea ice. *JGR* 91(C6):7623–30.

Overgaard, S., P. Wadhams, and M. Leppäranta. 1983. Ice properties in the Greenland and Barents Sea during summer. *J Glaciol* 29(101):142–64.

Overland, J. E. 2007. Could Mother Nature give the warming Arctic a reprieve? *Science* 315(5 January):36.

Overland, J. E., L. Bengtsson, R. Przybylak, and J. Walsh. 2007. Atmosphere. In *Arctic Report Card 2007*. http://www.arctic.noaa.gov/reportcard.

Overland, J. E., and K. L. Davidson. 1992. Geostrophic drag coefficients over sea ice. *Tellus* 44A(1):54–66.

Overland, J. E., and P. S. Guest. 1991. The arctic snow and air temperature budget over sea ice during winter. *JGR* 96(C3):4651–62.

Overland, J. E., S. L. McNutt, J. Groves, S. Salo, E. L. Andreas, and P. O. G. Person. 2000. Regional sensible and radiative heat flux estimates for the winter Arctic during the Surface Heat Budget of the Arctic Ocean (SHEBA) experiment. *JGR* 105(C6):14093–102.

Overland, J. E., S. L. McNutt, S. Salo, J. Groves, and S. Li. 1998. Arctic sea ice as a granular plastic. *JGR* 103:21845–68.

Overland, J. E., H. O. Mofjeld, and C. H. Pease. 1984. Wind-driven ice drift in a shallow sea. *JGR* 89(C4):6525–31.

Overland, J. E., and C. H. Pease. 1988. Modeling ice dynamics of coastal seas. *JGR* 93(C12):15619–37.

Overland, J. E., R. M. Reynolds, and C. H. Pease. 1983. A model of the atmospheric boundary layer over the marginal ice zone. *JGR* 88(C5):2836–40.

Overland, J., J. Turner, J. Francis, N. Gillett, G. Marshall, and M. Tjernström. 2008. The Arctic and Antarctic: Two faces of climate change. *Eos* 89(19):177–78.

Overland, J. E., B. A. Walter, T. B. Curtin, and P. Turet. 1995. Hierachy and sea ice mechanics: A case study from the Beaufort Sea. *JGR* 100(C3):4559–71.

Overland, J. E., and M. Wang. 2007. Future regional Arctic sea ice declines. *GRL* 34:L17705, doi:10.1029/2007GL030808.

Overpeck, J. T., and 20 others. 2005. Arctic system on trajectory to new, seasonally ice-free state. *Eos* 86(34): 309, 312–13.

Oxtoby, D. W. 1990. New perspectives on freezing and melting. *Nature* 347:725–30.

Ozüm, B., and D. J. Kirwan. 1976. Impurities in ice crystals grown from stirred solutions. *AIChE Symp Ser* 153(72):1–10.

Padman, L. 1995. Small-scale physical processes in the Arctic Ocean. In *Arctic Oceanography: Marginal ice zones and continental shelves*, ed. W. O. Smith Jr. and J. M. Grebmeier, 97–129. Coastal and Estuarine Studies No. 49. Washington: AGU.

Paige, R. A. 1966. Crystallographic studies of sea ice in McMurdo Sound, Antarctica. *US Naval Civil Eng Lab Tech Rept* R-494.

———. 1970. Stalactite growth beneath sea ice. *Science* 167:171.

Paige, R. A., and C. W. Lee. 1967. Preliminary studies on sea ice in McMurdo Sound, Antarctica, during Deep Freeze 65. *J Glaciol* 6(46):515–28.

Palacio, H. A., M. Solari, and H. Biloni. 1985. Microsegregation in cellular dendritic growth in binary alloys of Al–Cu. *J Cryst Growth* 73:369–78.

Palmer, A. C. 1997. Geotechnical evidence of ice scour as a guide to pipeline burial depth. *Canadian Geotech J* 34(6):1002–3.

Palmer, A. C., I. Konuk, G. Comfort, and K. Been. 1990. Ice gouging and the safety of marine pipelines. *Proc 22nd Offshore Technol Conf* (Paper 6371) 3:235–44.

Palmer, A. C., and A. Niedoroda. 2005. Ice gouging and pipelines: Unresolved questions. *Proc 18th POAC Conf*, vol 1:11–22. Potsdam, NY: Clarkson University.

Palmisano, A. C., and D. L. Garrison. 1993. Microorganisms in Antarctic sea ice. In *Antarctic microbiology*, ed. E. I. Friedmann, 210–18. New York: Wiley.

Palosuo, E. 1974. Formation and structure of ice ridges in the Baltic. Res Rept 12. Helsinki: Winter Navigation Research Board.

Panov, V. V., and A. O. Shpaikher. 1964. Influence of Atlantic waters on some features of the hydrology of the Arctic Basin and adjacent seas. *Deep-Sea Res* 11:275–85.

Papadimitriou, S., H. Kennedy, G. Kattner, G. S. Dieckmann, and D. N. Thomas. 2003. Experimental evidence for carbonate precipitation and CO_2 degassing during sea ice formation. *Geochim Cosmochim Acta* 68(8):1749–61.

Parkinson, C. 1991. Interannual variability of the spatial distribution of sea ice in the north polar region. *JGR* 96(C3):4791–4801.

———. 1992. Spatial patterns of increases and decreases in the length of the sea-ice season in the north polar region, 1979–1986. *JGR* 97:9401–10.

———. 1998. Length of the sea ice season in the Southern Ocean, 1988–1994. In *Antarctic sea ice: Physical processes, interactions and variability*, ed. M. O. Jeffries, 173–86. AGU Antarctic Res Series, vol. 74.

———. 2002. Trends in the length of the Southern Ocean sea-ice season, 1979–1999. *Ann Glaciol* 34:435–40.

Parkinson, C. L., and D. J. Cavalieri. 1989. Arctic sea ice 1973–1987: Seasonal, regional, and interannual variability. *JGR* 94(C10):14499–523.

———. 2008. Arctic sea ice variability and trends, 1979–2006. *JGR* 113:C07003, doi:10.1029/2007JC004558.

Parkinson, C. L., D. J. Cavalieri, P. Gloersen, H. J. Zwally, and J. C. Comiso. 1999. Arctic sea ice extents, areas and trends, 1978–1996. *JGR* 104(C9):20837–56.

Parkinson, C. L., J. C. Comiso, H. J. Zwally, D. J. Cavalieri, P. Gloersen, and W. J. Campbell. 1987a. *Arctic sea ice, 1973–1976: Satellite passive-microwave observations*. Publication SP-489. Washington, DC: NASA.

———. 1987b. Seasonal and regional variations of Northern Hemisphere sea ice as illustrated with satellite passive-microwave data for 1974. *Ann Glaciol* 9:119–26.

Parkinson, C. L., K. Y. Vinnikov, and D. J. Cavalieri. 2006. Evaluation of the simulation of the annual cycle of Arctic and Antarctic sea ice coverages by 11 major global climate models. *JGR* 111:C07012, doi:10.1029/2005JC003408.

Parkinson, C. L., and W. M. Washington. 1979. A large-scale numerical model of sea ice. *JGR* 84(C1):311–37.

Parmerter, R. R. 1975. A model of simple rafting in sea ice. *JGR* 80(15):1948–52.

Parmerter, R. R., and M. Coon. 1972. Model of pressure ridge formation in sea ice. *JGR* 77(33):6565–75.

———. 1973. Mechanical models of ridging in the Arctic sea ice cover. *AIDJEX Bull* 19:59–112.

Parsons, B. L., M. Lal, F. M. Williams, J. P. Dempsey, J. B. Snellen, J. Everhard, T. Slade, and J. Williams. 1992. The influence of beam size on the flexural strength of sea ice, freshwater ice and iceberg ice. *Phil Mag* 66(6):1017–36.

Pauling, L. 1935. Structure and entropy of ice and of other crystals with randomness of atomic arrangements. *J Amer Chem Soc* 57:2608–84.

Pease, C. H. 1980. Eastern Bering Sea ice processes. *Monthly Weather Rev* 108(12):2015–23.

———. 1987. Size of wind-driven coastal polynyas. *JGR* 92(C7):7049–59.

Pease, C. H., and J. E. Overland. 1984. An atmospherically driven sea-ice drift model for the Bering Sea. *Ann Glaciol* 5:111–14.

Pease, C. H., and P. Turet. 1989. Sea ice drift and deformation in the western Arctic. In *Oceans '89 Proc*, 1276–81. Seattle, WA: MTS/IEEE.

Penrose, J. D., M. Conde, and T. J. Pauly. 1994. Acoustic detection of ice crystals in Antarctic waters. *JGR* 99(C6):12573–80.

Percival, D. B., D. A. Rothrock, A. S. Thorndike, and T. Gneiting. 2008. The variance of mean sea-ice thickness: Effect of long range dependence. *JGR* 113:C01004, doi:10.1029/2007JC004391.

Perepezko, J. H., and G. J. Shiflet. 1984. Symposium on the establishment of microstructural spacing during dendritic and cooperative growth. *Metallurgical Trans* A(15A):960–1036.

Perey, F. G. J., and E. R. Pounder. 1958. Crystal orientation in ice sheets. *Canad J Phys* 36:494–502.

Perkin, R. G., and E. L. Lewis. 1984. Mixing in the West Spitsbergen Current. *J Phys Oceanog* 14:1315–25.

Perovich, D. K. 1983. *On the summer decay of a sea ice cover*. Ph.D. thesis, Geophys Program. Seattle: Univ Washington.

———. 1996. The optical properties of sea ice. *CRREL Monograph* 96-1.

———. 1998. The optical properties of sea ice. In *Physics of ice covered seas*, ed. M. Leppäranta, 195–230. Univ Helsinki Printing House.

———. 2005. On the aggregate-scale partitioning of solar radiation in Arctic sea ice during the SHEBA field experiment. *JGR* 110:C3002, doi:10.1029/2004JC 002512.

Perovich, D. K., and B. Elder. 2002. Estimates of ocean heat flux at SHEBA. *GRL* 29(9):10.1029/2001 GL014171.

Perovich, D. K., B. Elder, and J. A. Richter-Menge. 1997. Observations of the annual cycle of sea-ice temperature and mass balance. *GRL* 24(5):555–58.

Perovich, D. K., and A. J. Gow. 1991. A statistical description of the microstructure of young sea ice. *JGR* 96(C9):16943–954.

———. 1992. Towards a quantitative characterization of sea ice microstructure. In *12th International Geoscience and Remote Sensing Symposium (IGARSS '92)*, 2:1249–52. Houston: IEEE.

Perovich, D. K., T. C. Grenfell, B. Light, and P. V. Hobbs. 2002. Seasonal evolution of the albedo of multiyear Arctic sea ice. *JGR* 107(C10):8044, doi:10.1029/2000 JC000438.

Perovich, D., T. C. Grenfell, J. A. Richter-Menge, B. Light, W. B. Tucker III, and H. Eicken. 2003. Thin and thinner: Sea ice mass balance measurements during SHEBA. *JGR* 108(C3):8050, doi:10.1029/2001JC001079.

Perovich, D. K., B. Light, H. Eicken, K. F. Jones, K. Runciman, and S. V. Nghiem. 2007. Increasing solar heating of the Arctic Ocean and adjacent seas, 1979–2005: Attribution and role in the ice–albedo feedback. *GRL* 34:L19505, doi:10.1029/ 2007GL031480.

Perovich, D. K., and G. A. Maykut. 1990. Solar heating of a stratified ocean in the presence of a static ice cover. *JGR* 95(C10):18233–45.

Perovich, D. K., and J. A. Richter-Menge. 1994. Surface characteristics of lead ice. *JGR* 99(C8):16341–50.

———. 2000. Ice growth and solar heating in springtime leads. *JGR* 105(C3):6541–48.

Perovich, D. K., J. A. Richter-Menge, K. F. Jones, and B. Light. 2008. Sunlight, water, and ice: Extreme arctic sea ice melt during the summer of 2007. *GRL* 35:L11501, doi:1029/2008GL034007.

Perovich, D. K., J. A. Richter-Menge, and J. A. Morison. 1995. The formation and morphology of ice stalactites observed under deforming lead ice. *J Glaciol* 41(138):305–12.

Perovich, D. K., W. B. Tucker III, and K. A. Ligett. 2002. Aerial observations of the evolution of ice surface conditions during summer. *JGR* 107(C10):8048, doi:1029/2000JC000449.

Peschanskii, I. S. 1960. Arctic and Antarctic sea ice (R). *PAiA* 4:111–29.

Peters, A. S. 1950. The effect of a floating mat on water waves. *Communs Pure Appl Math* 3:319–54.

Peterson, S. W., and H. A. Levy. 1957. A single-crystal neutron diffraction study of heavy ice. *Acta Crystallographia* 10:70–76.

Petrenko, V. F. 1993a. Structure of ordinary ice I$_h$, Part I: Ideal structure of ice. *CRREL Special Report* 93-25.

———. 1993b. Electrical properties of ice. *CRREL Special Report* 93-20.

———. 1994. The surface of ice. *CRREL Special Report* 94-22.

———. 1996. Electromechanical phenomena in ice. *CRREL Special Report* 96-2.

Petrenko, V. F., and O. V. Gluschenkov. 1995. Measurements of crack velocity and electromagnetic emission from cracks in sea ice. In *Ice Mechanics-95*, vol. AMD-207, ed. J. P. Dempsey and Y. D. S. Rajapakse, 201–9. ASME.

———. 1996. Crack velocities in freshwater and saline ice. *JGR* 101(B5):11541–51.

Petrenko, V. F., and R. W. Whitworth. 1994a. Structure of Ordinary Ice Ih, Part II: Defects in ice. Volume 1: Point defects. *CRREL Special Report* 94-4.

———. 1994b. Structure of ordinary Ice Ih, Part II: Defects in ice. Volume 2: Dislocations and plane defects. *CRREL Special Report* 94-12.

———. 1999. *Physics of ice*. Oxford Univ Press.

Petrich, C., P. J. Langhorne, and T. G. Haskell. 2007. Formation and structure of refrozen cracks in land-fast first-year sea ice. *JGR* 112:C04006, doi:10.1029/2006JC003466.

Petrich, C., P. J. Langhorne, and Z. F. Sun. 2006. Modelling the interlationships between permeability, effective porosity and total porosity in sea ice. *CRST* 44:131–44.

Petrov, I. G. 1954–1955. Physical and mechanical properties and thickness of ice cover. In *Observations of the drifting research station of 1950–1951*, ed. M. M. Somov, 2:103–65. Leningrad: ANII.

Pettersson, O. 1883. On the properties of water and ice. In *Vega Expeditions Vetenskapliga Iakttagelser*, ed. A. E. Nordenskiöld, 2:247–323. Stockholm: F. and G. Beijers Förlag.

———. 1904. On the influence of ice-melting upon oceanic circulation. I. *Geograph J* 24:285–333.

———. 1907. On the influence of ice-melting upon oceanic circulation. II. *Geograph J* 30:339–44.

Peyton, H. R. 1963. Some mechanical properties of sea ice. In *Ice and snow: Processes, properties and applications*, ed. W. D. Kingery, 107–13. Cambridge, MA: MIT Press.

———. 1966. *Sea ice strength*. University of Alaska Report UAG-182. Fairbanks: Geophysical Institute.

———. 1968. Sea ice strength-effects of load rates and salt reinforcement. In *Arctic drifting stations*, ed. J. E. Sater, 197–217. AINA.

Pfann, W. G. 1958. *Zone melting*. New York: Wiley.

Pfirman, S. L., H. Eicken, D. Bauch, and W. F. Weeks. 1995. The potential transport of pollutants by Arctic sea ice. *The science of the total environment* 159:129–46.

Pfirman, S. L., J.-C. Gasacard, I. Wollenburg, et al. 1989. Particle-laden Eurasian Arctic sea ice: Observations from July and August 1987. *Polar Research* 7:59–66.

Pharr, G. M., and P. S. Godaverti. 1987. Comparison of the creep behavior of saline ice and frozen saline Ottawa sand at −8°C. *CRST* 14(3):273–79.

Phillips, F. C. 1946. *Introduction to crystallography*. London and New York: Longman.

Phipps, C. J. 1774. *A voyage toward the North Pole, undertaken by His Majesty's command 1773*. London: Nourse.

Pilkington, G. R., and R. W. Marcellus. 1981. Methods of determining pipeline trench depths in the Canadian Beaufort Sea. In *POAC '84 Proc* 2:674–87.

Pinto, J. O., A. Alam, J. A. Maslanik, J. A. Curry, and R. S. Stone. 2003. Surface characteristics and atmospheric footprint of springtime Arctic leads at SHEBA. *JGR* 108(C4):8051, doi:10.1029/ 2000JC000473.

Pinto, J. O., and J. A. Curry. 1995. Atmospheric convective plumes emanating from leads. 2. Microphysical and radiative processes. *JGR* 100(C3):4633–42.

Pinto, J. O., J. A. Curry, and K. L. McInnes. 1995. Atmospheric convective plumes emanating from leads. 1. Thermodynamic structure. *JGR* 100(C3):4621–31.

Pollard, D., and S. L. Thompson. 1994. Sea-ice dynamics and CO_2 sensitivity in a global climate model. *Atmos–Ocean* 32:449–67.

Polonichko, V. 1997. Generation of Langmuir circulation for nonaligned wind stress and the Stokes drift. *JGR* 102(C7):15773–80.

Polyakov, I. V., G. V. Alekseev, L. A. Timokhov, U. S. Bhatt, R. L. Colony, H. L. Simmons, D. Walsh, and V. F. Zakharov. 2004. Variability of the intermediate water of the Arctic Ocean over the last 100 years. *J Climate* 17(23):4485–97.

Polyakov, I., and S. Martin. 2000. Interaction of the Okhotsk Sea diurnal tides with the Kashevarov bank polynya. *JGR* 105(C2):3281–94.

Polyakov, I. V., A. Y. Proshutinsky, and M. A. Johnson. 1999. Seasonal cycles in two regimes of Arctic climate. *JGR* 104:25761–88.

Polyakov, I. V., D. Walsh, I. Dmitrenko, R. L. Colony, and L. A. Timokhov. 2003. Arctic Ocean variability derived from historical observations. *GRL* 30(6):1298, doi:1029/2002GL016441.

Polyakov, I. V., and 22 others. 2005. One more step toward a warmer Arctic. *GRL* 32:L17605, doi:10.1029/ 2005GL023740.

Polyakov, I. V., and 18 others. 2007. Observational program tracks Arctic Ocean transition to a warmer state. *Eos* 88(40):398–99. doi:10.1029/2007EO400002.

Pounder, E. R. 1965. *Physics of ice*. Oxford: Pergamon Press.

Pounder, E. R., and M. P. Langleben. 1964. Arctic sea ice of various ages. II. Elastic properties. *J Glaciol* 5(37):99–105.

Pounder, E. R., and E. M. Little. 1959. Some physical properties of sea ice. *Canad J Phys* 37:443–73.

Pringle, D. J., H. Eicken, H. J. Trodahl, and L. G. E. Backstrom. 2007. Thermal conductivity of landfast Antarctic and Arctic sea ice. *JGR* 112:C04017, doi:10.1029/2006JC003641.

Pringle, D. J., H. J. Trodahl, and T. J. Haskell. 2006. Direct measurement of sea ice thermal conductivity: No surface reduction. *JGR* 111:C05020, doi:10.1029/2005 JC002990.

Pritchard, R. S. 1975. An elastic-plastic law for sea ice. *J Appl Mech* 42(E2):379–84.

———. 1977. An estimate of the strength of Arctic pack ice. *AIDJEX Bull* 34:94–113.

———. 1980a. Simulation of near-shore winter ice dynamics in the Beaufort Sea. In *Sea ice processes and models*, ed. R. S. Pritchard, 49–61. Seattle: Univ Washington Press.

———, ed. 1980b. *Sea ice processes and models*. Seattle: Univ Washington Press.

———. 1981. Mechanical behavior of pack ice. In *Mechanics of Structured Media*, ed. A. P. S. Selvadurai, 371–405. Amsterdam: Elsevier.

———. 1988a. Chukchi Sea ice motion 1981–1982. In *POAC '87*, ed. W. M. Sackinger and M. O. Jeffries, 3:255–70. Fairbanks, AK: Geophysical Institute.

———. 1988b. Mathematical characteristics of sea ice dynamics models. *JGR* 93(C12):15609–18.

———. 2001. Sea ice dynamics models. In *IUTAM Symp on Scaling Laws in Ice Mechanics and Ice Dynamics*, ed. J. P. Dempsey and H. H. Shen, 265–88. Dordrecht: Kluwer.

Pritchard, R. S., M. D. Coon, and M. G. McPhee. 1977. Simulation of sea ice dynamics during AIDJEX. *J Pressure Vessel Technol* 99J:491–97.

Proshutinsky, A., and M. Johnson. 1997. Two circulation regimes of the wind-driven Arctic Ocean. *J Geophys Res* 102(C6):12493–514.

Proshutinsky, A., and J. Morison. 2007. *Arctic report card 2007: Ocean*. http://www.arctic.noaa.gov/reportcard/ocean.html.

Pruppacher, H. R. 1967. Growth modes of ice crystals in supercooled water and aqueous solutions. *J Glaciol* 6(47):651–62.

Quadfasel, D. 1991. Warming in the Arctic. *Nature* 350:385.

Radionov, V. F., N. N. Bryazgin, and E. I. Alexandrov. 1997. *The snow cover of the Arctic Basin*. Rep. APL-UW-TR 9701. Appl Phys Lab, Univ Washington.

Ragle, R. H. 1962. The formation of lake ice in a temperate climate. *CRREL Research Rept* 107.

Rahmstorf, S., A. Cazenave, and 5 others. 2007. Recent climate observations compared to projections. *Science* 316(4 May):709.

Rampal, P., J. Weiss, D. Marsden, R. Lindsay, and H. Stern. 2008. Scaling properties of sea ice deformation from buoy dispersion analysis. *JGR* 113:C03002, doi:10.1029/2007JC004143.

Randall, D. A., R. A. Wood, and 11 others. 2007. Climate models and their evaluation. In *Climate change 2007: The physical science basis*. Contribution of Working Group I to the Fourth Assessment Report of the IPCC. Cambridge Univ Press.

Ranelli, P. H., and W. D. Hibler III. 1991. Seasonal Arctic sea-ice simulations with a prognostic ice–ocean model. *Ann Glaciol* 15:45–53.

Raymond, C. F., and Tusima, K. 1979. Grain coarsening in water-saturated snow. *J Glaciol* 22(86):83–105.

Rayner, N. A., and 7 others. 2003. Global analyses of sea surface temperature, sea ice, and night marine air temperature since the late nineteenth century. *JGR* 108(D14):4407, doi:10.1029/2002JD002670.

Rearic, D. M. 1986. Temporal and spatial character of newly formed ice gouges in eastern Harrison Bay, Alaska, 1977–1982. *US Geological Survey Open-File Report* 86-391.

Rearic, D. M., P. W. Barnes, and E. Reimnitz. 1990. Bulldozing and resuspension of shallow-shelf sediments by ice keels: Implications for Arctic sediment transport trajectories. *Mar Geol* 91(1–2):133–47.

Reddy, T. E., K. R. Arrigo, and D. M. Holland. 2007. The role of thermal and mechanical processes in the formation of the Ross Sea summer polynya. *JGR* 112:C07027, doi:10.1029/2006JC003874.

Reeburgh, W. S., and M. Springer-Young. 1983. New measurements of sulfate and chlorinity in natural sea ice. *JGR* 88(C5):2959–66.

Reed, J. C., and Sater, J. E., eds. 1974. *The coast and shelf of the Beaufort Sea*. AINA.

Reed, R. J., and W. J. Campbell. 1960. Theory and observations of the drift of Ice Station ALPHA. *JGR* 67:281–97.

Rees, W. G. 2001. *Physical principles of remote sensing*. Cambridge Univ Press.

Reimnitz, E., and P. W. Barnes. 1974. Sea ice as a geologic agent on the Beaufort Shelf of Alaska. In *The coast and shelf of the Beaufort Sea*, ed. J. C. Reed and J. E. Sater, 301–53. AINA.

Reimnitz, E., P. W. Barnes, and T. R. Alpha. 1973. Bottom features and processes related to drifting ice. *US Geological Survey Misc Field Studies*, Map MF-532.

Reimnitz, E., P. W. Barnes, T. Forgatsch, and C. Rodeick. 1972. Influence of grounding ice on the Arctic Shelf of Alaska. *Mar Geol* 13:323–34.

Reimnitz, E., P. W. Barnes, L. J. Toimil, and J. Melchior. 1977. Ice gouge recurrence and rates of sediment reworking, Beaufort Sea, Alaska. *Geology* 5(7):405–8.

Reimnitz, E., D. Dethleff, and D. Nürnberg. 1994. Contrasts in Arctic shelf sea-ice regimes and some implications: Beaufort Sea versus Laptev Sea. *Mar Geol* 119(3–4):215–25.

Reimnitz, E., D. Dethleff, D. Nürnberg, and Y. P. Savchenko. 1992. The Laptev Sea as sediment source for the Transpolar Drift. In *4th Internat Conf on Paleoceanography, IPC IV; Short- and long-term global change: Records and modeling*, 237–38. Kiel, Germany.

Reimnitz, E., and E. W. Kempema. 1982. Dynamic ice-wallow relief of northern Alaska's nearshore. *J Sediment Petrol* 52:451–61.

———. 1984. Pack ice interaction with Stamukhi Shoal, Beaufort Sea, Alaska. In *The Alaskan Beaufort Sea: Ecosystems and environment*, ed. P. W. Barnes, D. M. Schell, and E. Reimnitz, 159–83. Orlando: Academic Press.

———. 1987. Field observations of slush ice generated during freeze-up in arctic coastal waters. *Mar Geol* 77:219–31.

Reimnitz, E., E. W. Kempema, and P. W. Barnes. 1987. Anchor ice, seabed freezing and sediment dynamics in shallow Arctic seas. *J Geophys Res* 92(C13):14671–78.

Reimnitz, E., L. Marincovich Jr., M. McCormick, and W. M. Briggs. 1992. Suspension freezing of bottom sediment and biota in the Northwest Passage and implications for Arctic Ocean sedimentation. *Can J Earth Sci* 29:693–703.

Reimnitz, E., and D. K. Maurer. 1978. Storm surges in the Alaskan Beaufort Sea. *US Geological Survey Open File Report* 78-593.

Reimnitz, E., M. McCormick, K. McDougall, and E. Brouwers. 1993. Sediment export by ice rafting from a coastal polynya, Arctic Alaska, USA. *Arct Alpine Res* 25(2):83–98.

Reimnitz, E., C. A. Rodeick, and S. C. Wolf. 1974. Strudel scour: A unique Arctic marine geologic phenomenon. *J Sed Petrol* 44(2):409–20.

Reimnitz, E., L. Toimil, and P. W. Barnes. 1978. Arctic continental shelf morphology related to sea ice zonation, Beaufort Sea, Alaska. *Marine Geol* 28:179–210.

Rex, R. W. 1955. Microrelief produced by sea ice grounding in the Chukchi Sea near Barrow, Alaska. *Arctic* 8:177–86.

Richardson, C. 1976. Phase relations in sea ice as a function of temperature. *J Glaciol* 17(77):507–19.

Richardson, C., and E. E. Keller. 1966. The brine content of sea ice measured with a nuclear magnetic resonance spectrometer. *J Glaciol* 6(43):89–100.

Richmond, O., and G. C. Gardner. 1962. Limiting spans for arching of bulk materials in vertical channels. *Chem Eng Sci* 17:1071–78.

Richter-Menge, J. A. 1997. Towards improving the physical basis for ice dynamics models. *Ann Glaciol* 25:177–82.

Richter-Menge, J. A., K. J. Claffey, and M. R. Walsh. 1993. End-capping procedure for cored ice samples used in tension tests. *J Glaciol* 39(133):698–700.

Richter-Menge, J. A., and G. F. N. Cox. 1985. A preliminary examination of the effect of structure on the compressive strength of ice samples from multi-year pressure ridges. *J Energy Resources Technol* 107:99–102.

Richter-Menge, J. A., G. F. N. Cox, N. Perron, G. Durell, and H. W. Bosworth. 1986. Triaxial testing of first-year sea ice. *CRREL Report* 88-16.

Richter-Menge, J. A., and B. C. Elder. 1998. Characteristics of pack ice stress in the Alaskan Beaufort Sea. *JGR* 103:21817–29.

Richter-Menge, J. A., and K. F. Jones. 1993. The tensile strength of first-year sea ice. *J Glaciol* 39(133):609–18.

Richter-Menge, J. A., D. K. Perovich, B. C. Elder, K. J. Claffey, I. Rigor, and M. Ortmeyer. 2006. Ice mass balance buoys: A tool for measuring and attributing changes in the thickness of the Arctic sea ice cover. *Ann Glaciol* 44(1):205–10.

Richtmyer, R. D., and K. W. Morton. 1967. *Difference methods for initial value problems*. 2nd ed. New York: Wiley Interscience.

Ridley, J., J. Lowe, C. Brierley, and G. Harris. 2007. Uncertainty in the sensitivity of Arctic sea ice to global warming in a perturbed parameter climate model ensemble. *GRL* 34:L19704, doi:10.1029/2007GL031209.

Rigby, F. A., and A. Hanson. 1976. Evolution of a large Arctic pressure ridge. *AIDJEX Bull* 34:43–71.

Rignot, E., and S. S. Jacobs. 2002. Rapid bottom melting widespread near Antarctic ice sheet grounding lines. *Science* 296(14 June):2020–23.

Rigor, I. G., and J. M. Wallace. 2004. Variations in the age of Arctic sea ice and summer sea ice extent. *GRL* 31:L09401, doi:10.1029/2004GL019492.

Rigsby, G. P. 1951. Crystal fabric studies on Emmons Glacier, Mount Rainier, Washington. *J Geol* 59(6): 590–98.

Ringer, W. E. 1906. Über die Veränderungen in der Zusammensetzung des Meereswassersalzes beim Ausfrieren (Changes in the composition of sea-water salt during freezing). *Verh Rijksinst Onderz Zee* 3:1–55.

———. 1928. Über die Veränderungen in der Zusammensetzung des Meereswassersalzes beim Ausfrieren. *J Conseil Permanent Intern, Exploration Mer* Rapp. 47:226–32.

Rintoul, S. R. 1998. On the origin and influence of Adelie Land bottom water. In *Ocean, ice and atmosphere: Interactions at the antarctic continental margin*, ed. S. S. Jacobs and R. F. Weiss, 151–172. Antarct Res Ser 74. Washington, DC: AGU.

Rintoul, S. R., S. Sokolov, and R. A. Massom. 2008. Rapid development and persistence of a massive sea ice tongue. *JGR* 113:C07045, doi:10.1029/2007JC004541.

Ripperger, A. E., and N. Davids. 1947. Critical stresses in a circular ring. *Amer Soc Civil Eng Trans* 12:619–35.

Roberts, A., I. Allison, and V. I. Lytle. 2001. Sensible- and latent-heat flux estimates over the Mertz Glacier Polynya, East Antarctica, from in-flight measurements. *Ann Glaciol* 33:377–84.

Roberts, B., and T. Armstrong. 1964. Proposed new terms and definitions of sea ice for the use of submariners. *Polar Record* 12(77):197–98.

Robertson, R., L. Padman, and M. D. Levine. 1995. Fine structure, microstructure, and vertical mixing processes in the upper ocean in the western Weddell Sea. *JGR* 100(C9):18517–35.

Robin, G. de Q. 1963. Wave propagation through fields of pack ice. *Phil Trans R Soc* A255:313–39.

Rodrigues, J. 2008. The rapid decline of sea ice in the Russian Arctic. *CRST* 54(2):124–42.

Røed, L-P., and J. J. O'Brien. 1983. A coupled ice-ocean model of upwelling in the marginal ice zone. *JGR* 88(C5):2863–72.

Rohatgi, P. K., and C. M. Adams. 1967a. Freezing rate distributions during unidirectional solidification of solutions. *Trans Metall Soc AIME* 239(6):850–57.

———. 1967b. Ice-brine dendritic aggregate formed on freezing of aqueous solutions. *J Glaciol* 6(47):663–79.

———. 1967c. Effect of freezing rates on dendritic solidification of ice from aqueous solutions. *Trans Metall Soc AIME* 239(11):1729–36.

Rohatgi, P. K., S. M. Jain, D. N. French, and C. M. Adams. 1969. Effect of solute concentration and diffusivity on dendritic solidification of dilute binary aqueous solutions. *Trans Metall Soc AIME* 245:267–73.

Romanov, I. P. 1993. *Ice cover of the Arctic Basin*. Privately published monograph, St. Petersburg.

Rossby, C. G., and R. B. Montgomery. 1935. The layer of frictional influence in wind and watercurrents. *Papers Phys Oceanog and Meteorol* 3. MIT and Woods Hole Oceanog Inst.

Röthlisberger, H. 1972. Water pressure in intra- and subglacial channels. *J Glaciol* 11(62):177–203.

Rothrock, D. A. 1975a. The mechanical behavior of pack ice. *Ann Rev Earth Planet Sci* 3:317–42.

———. 1975b. The steady drift of an incompressible arctic ice cover. *JGR* 80(3):387–97.

———. 1975c. The energetics of the plastic deformation of pack ice by ridging. *JGR* 80:4514–49.

———. 1986. Ice thickness distribution-measurement and theory. In *The geophysics of sea ice*, ed. N. Untersteiner, 551–75. New York: Plenum Press.

Rothrock, D. A., R. Colony, and A. S. Thorndike. 1980. Testing pack ice constitutive laws with stress divergence measurements. In *Sea ice processes and* models, ed. R. S. Pritchard, 102–23. Seattle: Univ Washington Press.

Rothrock, D. A., D. B. Percival, and M. Wensnahan. 2008. The decline in arctic sea-ice thickness: Separating the spatial, annual, and interannual variability in a quarter century of submarine data. *JGR* 113:C05003, doi:10/1029/2007JC004252.

Rothrock, D. A., and D. Thomas. 1989. Estimating sea ice concentration from satellite passive microwave data and a physical model. In *Internat Geoscience and Remote Sensing Symp IGARRS '88*, 1–7. Edinburgh, Scotland.

Rothrock, D. A., and A. S. Thorndike. 1980. Geometric properties of the underside of sea ice. *JGR* 85(C7):3955–63.

Rothrock, D. A., Y. Yu, and G. A. Maykut. 1999. Thinning of the Arctic sea-ice cover. *GRL* 26(23):3469–72.

Rothrock, D. A., and J. Zhang. 2005. Arctic Ocean sea ice volume: What explains its recent depletion. *JGR* 110:C01002, doi:10.1029/2004JC002282.

Rothrock, D. A., J. Zhang, and Y. Yu. 2003. The Arctic ice thickness anomaly of the 1990s: A consistent view from observations and models. *JGR* 108:3083, doi:10.1029/2001JC001208.

Rowe, C., K. C. Kuivinen, and R. Jordan. 1995. Heat budget of snow-covered sea ice at North Pole-4. *JGR* 104(C4):7785–7806.

Rudels, B. 1987. On the mass balance of the polar ocean with special emphasis on the Fram Strait. *Skr Norsk Polarinst* 188:1–53.

Rudels, B., E. P. Jones, L. G. Anderson, and G. Kattner. 1994. On the intermediate depth waters of the Arctic Ocean. In *The polar oceans and their role in shaping the global environment*, ed. O. M. Johannessen, R. D. Muench, and J. E. Overland, 33–46. The Nansen Centennial Volume 85. Washington, DC: AGU.

Ruffieux, D., P. Persson, G. Ola, C. W. Fairall, and D. E. Wolfe. 1995. Ice pack and lead surface energy budgets during LEADEX 1992. *JGR* 100(C3):4593–4612.

Rutter, J. W., and B. Chalmers. 1953. A prismatic substructure formed during solidification of metals. *Can J Phys* 31:15–39.

Ryan, B. F. 1969. The growth of ice parallel to the basal plane in supercooled water and supercooled metal fluoride solutions. *J Cryst Growth* 5:284–88.

Ryan, B. F., and W. C. Macklin. 1968. The growth of ice in supercooled aqueous solutions. *J Cryst Growth* 2:337–40.

Saito, T., and N. Ono. 1978. Percolation in sea ice. I. Measurements of kerosene permeability of NaCl ice. *LTS-A* 37:55–62.

———. 1980. Percolation in sea ice. II. Brine drainage channels in young sea ice. *LTS-A* 39:127–32.

Salzmann, C. G., P. G. Radarlli, A. Hallbrucker, E. Mayer, and J. L. Finney. 2006. The preparation and structures of hydrogen ordered phases of ice. *Science* 311(5768):1758–61.

Sammonds, P. R., S. A. F. Murrell, and M. A. Rist. 1998. Fracture of multiyear sea ice. *JGR* 103(C10):21795–815.

Sanderson, T. J. O. 1988. *Ice mechanics: Risks to offshore structures*. Boston: Graham and Trottman.

Sandhäger, H., D. G. Vaughan, and A. Lambrecht. 2004. Meteoric, marine and total ice thickness maps of Filchner-Ronne-Schelfeis, Antarctica. *Forum for Research into Ice Shelf Processes (FRISP) Report* No 15:23–29.

Sauliev, V. K. 1957a. A method of numerical solution for the diffusion equation (R). *Dokl Akad Nauk SSSR(NS)* 115:1077–79.

———. 1957b. Numerical integration of parabolic equations (R). *Dokl Akad Nauk SSSR(NS)* 117:36–39.

Savel'yev, B. A. 1954. Procedural notes on determination of content of solid, liquid and gaseous phases in saline ice (R). *Materialy po Laboratornym Issledovaniyam Merzlykh Gruntov*. Collection no 2, 176–92 (also available as an English translation by E. R. Hope published by the Defense Research Board of Canada, T344R, 1960).

———. 1958. Study of ice in the region of the drift of the station SP-4 during melting and breakup in 1955 (R). *Problemy Severa* 2:47–79.

———. 1963. *Structure, composition and properties of the ice cover on marine and fresh water bodies* (R). Moscow: Izd Moskovskogo Universiteta.

Scambos, T. A., C. Hulbe, M. Fahnestock, and J. Bohlander. 2000. The link between climate warming and break-up of ice shelves in the Antarctic Peninsula. *J Glaciol* 46(154):516–30.

Schanda, E. 1986. *Physical fundamentals of remote sensing*. Springer-Verlag.

Schaus, R. H., and J. A. Galt. 1973. A thermodynamic model of an arctic lead. *Arctic* 26:208–21.

Schiefer, E., B. Menounos, and R. Wheate. 2007. Recent volume loss of British Columbian glaciers, Canada. *GRL* 34:L16503, doi:10.1029/2007GL030780.

Schledermann, P. 1980. Polynyas and prehistoric settlement patterns. *Arctic* 33(2):292–302.

Schneider, W., and G. Budeus. 1994. The north-east water polynya (Greenland Sea). 1. A physical concept of its generation. *Polar Biology* 14(1):1–9.

———. 1995. On the generation of the northeast water polynya. *JGR* 100(C3):4269–86.

Schell, R. C., R. G. Barry, M. W. Miles, E. L. Andreas, L. F. Radke, C. A. Brock, M. P. McCormick, and J. L. Moore. 1989. Lidar detection of leads in Arctic sea ice. *Nature* 339:530–32.

Schrag, D. P., and R. B. Alley. 2004. Ancient lessons for our future climate. *Science* 306:821–22.

Schramm, J. L., G. M. Flato, and J. A. Curry. 1999. Towards the modeling of enhanced basal melting in ridge keels. *JGR* 105(C6):14081–92.

Schroder, D., and W. M. Connolley. 2007. Impact of instantaneous sea ice removal in a coupled general circulation model. *GRL* 34:L14502, doi:10.1029/2007GL030253.

Schulkes, R. M. S. M. 1996. Asymptotic stability of the viscous–plastic sea ice rheology. *J Phys Oceanog* 26:279–83.

Schulson, E. M. 2001a. Fracture of ice on scales large and small. In *IUTAM symp on scaling laws in ice mechanics and ice dynamics*, ed. J. P. Dempsey and H. H. Shen, 161–70. Dordrecht: Kluwer.

———. 2001b. Brittle failure in ice. *Engineering Fracture Mech* 68:1839–87.

———. 2002. Compressive shear faults in ice: Plastic vs. Coulombic faults. *Acta Mater* 50:3415–24.

———. 2004. Compressive shear faults within arctic sea ice: Fractures on scales large and small. *JGR* 109:C07016, doi:10.1029/2003JC002108.

Schulson, E. M., and P. Duval. 2009. *Creep and fracture of ice*. Cambridge University Press.

Schulson, E. M., A. L. Fortt, D. Iliescu, and C. E. Renshaw. 2006. Failure envelope of first-year Arctic sea ice: The role of friction in compressive fracture. *JGR* 111:C11S25, doi:10.1029/2005JC003235.

Schulson, E. M., and W. D. Hibler III. 1991. The fracture of ice on scales large and small: Arctic leads and wing cracks. *J Glaciol* 37:319–22.

———. 2004. Fracture of the winter sea ice cover on the Arctic Ocean. *Comptes Rendus Physique* 5:753–67.

Schulson, E. M., and O. Y. Nickolayev. 1995. Failure of columnar saline ice under biaxial compression: Failure envelopes and the brittle to ductile transition. *JGR* 100:22383–400.

Schumaker, J. D., K. Aagaard, C. H. Pease, and R. B. Tripp. 1983. Effects of a shelf polynya on flow and water properties in the northern Bering Sea. *JGR* 88(C5):2723–32.

Schwarz, J., and W. F. Weeks. 1977. Engineering properties of sea ice. *J Glaciol* 19(81):499–531.

Schwarzacher, W. 1959. Pack ice studies in the Arctic Ocean. *JGR* 64:2357–67.

Schweiger, A. J., J. Zhang, R. W. Lindsay, and M. Steele. 2008. Did unusually sunny skies help drive the record sea ice minimum of 2007. *GRL* 35:L10503, doi:10.1029/ 2008GL033463.

Schwerdtfeger, P. 1963a. The thermal properties of sea ice. *J Glaciol* 4(36):789–807.

———. 1963b. Theoretical derivation of the thermal conductivity and diffusivity of snow. *IASH* Pub. 61, 75–81. Berkeley.

Scoresby, W. 1818. On the Greenland or polar ice. *Mem Wernerian Nat Hist Soc* 2:261–336 (reprinted 1980 by Caedmon of Whitby Press).

———. 1820. *An account of the Arctic Regions with a history and description of the Northern Whale-Fishery*. 2 vols. Edinburgh: Constable (reprinted 1969, David and Charles, Newton Abbot).

Seidensticker, R. G. 1965. Comment on paper by P. Hoekstra, T. E. Osterkamp, and W. F. Weeks, The migration of liquid inclusions in single ice crystals. *JGR* 71(8):2180–81.

Sekerka, R. F., R. G. Seidensticker, D. R. Hamilton, and J. D. Harrison. 1967. *Investigation of desalination by freezing*. Office of Saline Water, Contract No. 14-01-0001-605, Final Report. Pittsburgh: Westinghouse Research Laboratories.

Seligman, G. 1949. Growth of glacier crystal. *J Glaciol* 1(5):254–68.

Semtner, A. J. 1976. A model for the thermodynamic growth of sea ice in numerical investigations of climate. *J Phys Oceanogr* 6(3):379–89.

Serikov, M. I. 1963. Structure of Antarctic sea ice. *Info Bull Sov Ant Exped* 4(5):265–66.

Serreze, M. C., R. G. Barry, and J. E. Walsh. 1995. Atmospheric water vapor characteristics at 70°N. *J Climate* 8:719–31.

Serreze, M. C., J. D. Kahl, E. L. Andreas, J. A. Maslanik, M. C. Rehder, and R. C. Schnell. 1992. Theoretical heights of buoyant convection above open leads in the winter Arctic pack ice cover. *JGR* 97:9411–22.

Serreze, M. C., and 9 others. 2000. Observational evidence of recent change in the northern high-latitude environment. *Climate Change* 46:159–207.

Serson, H. V. 1972. *Investigation of a plug of multi-year sea ice in the mouth of Nansen Sound.* Defense Res Establ Tech. Note 72-6. Ottawa.

Shames, I. H., and F. A. Cozzarelli. 1992. *Elastic and inelastic stress analysis.* Englewood Cliffs, NJ: Prentice-Hall.

Shen, H. H., and S. F. Ackley. 1991. One-dimensional model for wave-induced ice–floe collisions. *Ann Glaciol* 15:87–95.

Shen, H. H., W. D. Hibler III, and M. Leppäranta. 1987. The role of ice floe collisions in sea ice rheology. *JGR* 92(C7):7085–96.

Shapiro, A., and L. S. Simpson. 1953. The effect of a broken icefield on water waves. *Trans Am Geophys Union* 34(1):36–42.

Shapiro, L. H. 1987. Mechanical processes of sea ice deformation in the near Shore Zone. In *OCSEAP Final Reports,* 72:357–584. US Dept Commerce, NOAA.

Shapiro, L. H., and W. F. Weeks. 1993. The influence of crystallographic and structural properties on the flexural strength of small sea ice beams. In *Ice Mechanics '93,* ed. J. P. Dempsey, Z. P. Bazant, Y. D. S. Rajapakse, and S. S. Sunder, 177–88. ASME.

———. 1995. Controls on the flexural strength of small plates and beams of first-year sea ice. In *Ice Mechanics '95,* 179–88. ASME, UCLA.

Shapiro, M. A., T. Hampel, and L. S. Fedor. 1989. Research aircraft observations of an arctic front over the Barents Sea. In *Polar and arctic lows,* ed. P. F. Twitchell, E. A. Rasmussen, and K. L. Davidson, 279–89. Hampton, VA: Deepak.

Shearer, J. M., and S. M. Blasco. 1975. Further observations of the scouring phenomena in the Beaufort Sea. In *Report of activities, part A,* 483–93. Paper 75-1A. Geological Survey of Canada.

Shearer, J. M., E. F. MacNab, B. R. Pelletier, and T. B. Smith. 1971. Submarine pingos in the Beaufort Sea. *Science* 174:816–18.

Shearwood, C., and R. W. Whitworth. 1991. The velocity of dislocations in ice. *Philos Mag* A64(2):289–302.

Shilling, J. E., M. A. Tolbert, O. B. Toon, E. J. Jensen, B. J. Murray, and A. K. Bertram. 2006. Measurements of the vapor pressure of cubic ice and their implications for atmospheric ice clouds. *GRL* 33:L17801, doi:10.1029/2006GL026671.

Shimada, K., F. McLaughlin, E. Carmack, A. Proshutinsky, S. Nishino, and M. Itoh. 2004. Penetration of the 1990s warm temperature anomaly of Atlantic Water in the Canada Basin. *GRL* 31:L20301, doi:10.1029/2004GL020860.

Shindell, D. 2003. Whither Arctic climate? *Science* 299(10 January):215–16.

Shine, K. P. 1984. Parameterization of shortwave flux over albedo surfaces as a function of cloud thickness and surface albedo. *QJR Meteorol Soc* 110:747–64.

Shirasawa, K., and R. G. Ingram. 1991. Characteristics of the turbulent oceanic boundary layer under sea ice. Part 1: A review of the ice–ocean boundary layer. *J Mar Sys* 2(1–2):153–60.

Shirasawa, K., and M. P. Langleben. 1976. Water drag on arctic sea ice. *JGR* 6:6451–54.

Shreve, R. L. 1967. Migration of air bubbles, vapor figures and brine pockets in ice under a temperature gradient. *JGR* 72(16):4093–4100.

———. 1972. Movement of water in glaciers. *J Glaciol* 11(62):205–14.

Shuleikin, V. V. 1938. The drift of ice fields (R). *Compt Rend (Doklady)* Acad Sci USSR 19(8):589–94.

Shumskiy, P. A. 1955a. A study of ice in the Arctic Ocean (R). *Vest Akad Nauk SSSR* 25(2):33–38.

———. 1955b. *Principles of structural glaciology* (R). Publishing House of the Academy of Sciences of the USSR (English translation by D. Krauss published in 1964 by Dover Press).

Shy, T. L., and J. E. Walsh. 1996. North Pole ice thickness and association with ice motion history 1977–1992, 1979–1986. *GRL* 23:2975–78.

Signorini, S. R., and D. J. Cavalieri. 2002. Modeling dense water production and salt transport from Alaskan coastal polynyas. *JGR* 107(C6):10.1029/2000JC000491.

Sillars, R. W. 1937. The properties of a dielectric containing semi-conducting particles of various shapes. *J Inst Elect Engrs* 80:378–94.

Simmonds, I., and W. F. Budd. 1991. Sensitivity of the southern hemisphere circulation to leads in the Antarctic pack ice. *QJR Met Soc* 117:1003–24.

Simpson, H. C., G. C. Beggs, J. Deans, and J. Nakamura. 1973. The growth of ice crystals. In *4th Internat Symp Fresh Water From the Sea* 3:395–407.

Singarayer, J. S., and J. L. Bamber. 2003. EOF analysis of three records of sea ice concentration spanning the last 30 years. *GRL* 30(5):doi.10.1029/2002GL016640.

Sinha, N. K. 1977. Technique for studying structure of sea ice. *J Glaciol* 18(79):315–23.

———. 1983. Does the strength of ice depend on grain size at high temperatures? *Scripta Metall* 17:1269–73.

———. 1984. Uniaxial compressive strength of first-year and multi-year sea ice. *Canad J Civ Eng* 11:82–91.

———. 1985. Confined strength and deformation of second-year columnar-grained sea ice in Mould Bay. *Proc 4th Internat Offshore Mechanics and Arctic Engineering Symp* (OMAE '85), 2:209–19. Dallas, TX.

———. 1986. The borehole jack: Is it a useful tool? *Proc 5th Internat Offshore Mechanics and Arctic Engineering Symp* (OMAE '85), 4:328–35. Tokyo.

———. 1991. In situ multi-year ice strength using NRCC Borehole Indenter. *Proc 10th Internat Offshore Mechanics and Arctic Engineering Symp* (OMAE), 4:229–36. Stavanger, Norway.

Sinha, N. K., and R. M. W. Frederking. 1979. Effect of test system stiffness on strength of ice. *POAC '79*, 1:708–17.

Sinha, N. K., and M. Nakawo. 1981. Growth of first year sea ice, Eclipse Sound, Baffin Island. *Can J Geotech J* 18(1):17–23.

Skyllingstad, E. D., and C. A. Paulson. 2007. A numerical study of melt ponds. *JGR* 112:C08015, doi:10.1029/2006JC003729.

Skyllingstad, E. D., C. A. Paulson, and W. S. Pegau. 2005. Simulation of turbulent exchange processes in summertime leads. *JGR* 110:C05021, doi:10.1029/2004 JC002502.

Slack, G. A. 1980. Thermal conductivity of ice. *Phys Rev B* 22(6):3065–71.

Slesarenko, Y., and A. D. Frolov. 1974. Comparison of elasticity and strength characteristics of salt-water ice. *IAHR Symp on Ice and Its Action on Hydraulic Structures*, 2:85–87. Leningrad.

Smedsrud, L. H. 2002. A model for entrainment of sediment into sea ice by aggregation between frazil-ice crystals and sediment grains. *J Glaciol* 48(160):51–61.

Smedsrud, L. H., and A. Jenkins. 2004. Frazil ice formation in an ice shelf plume. *JGR* 109:C03025, doi:10.1029/2003JC001851.

Smedsrud, L. H., T. M. Saloranta, P. M. Haugan, and T. Kangas. 2003. Sea ice formation on a very cold surface. *JGR* 30(6):1284, doi:10.1029/2002GL016786.

Smedsrud, L. H., and R. Skogseth. 2006. Field measurements of Arctic grease ice properties and processes. *CRST* 44(3):171–83.

Smirnov, V. I. 1961. Quantitative characteristics of ice as a material (R). *Trudy AANII* 256:40–46.

Smith, C. L. 1971. A comparison of Soviet and American drifting ice stations. *Polar Record* 15(99):877–85.

Smith, D. C., IV, and A. A. Bird. 1991. The interaction of an ocean eddy with an ice edge ocean jet in the marginal ice zone. *JGR* 96(C3):4675–89.

Smith, D. C. IV, and J. Morison. 1993. A numerical study of haline convection beneath leads in sea ice. *JGR* 98(C6):10069–83.

Smith, D. D. 1964. Ice lithologies and structures of ice island ARLIS II. *J Glaciol* 5(37):17–38.

Smith, D. R., and E. M. Schulson. 1994. Brittle compressive failure of salt-water columnar ice under biaxial loading. *J Glaciol* 40(135):265–76.

Smith, I. J., P. J. Langhorne, T. G. Haskell, H. J. Trodahl, R. Frew, and M. R. Vennel. 2001. Platelet ice and the land-fast sea ice of McMurdoe Sound, Antarctica. *Ann Glaciol* 33:21–27.

Smith, J. A. 1992. Observed growth of Langmuir circulation. *JGR* 97(C4):5651–64.

———. 1998. Evolution of Langmuir circulation during a storm. *JGR* 103(C6):12649–68.

Smith, R. B. 1983. A note on the constitutive law for sea ice. *J Glaciol* 29:191–95.

Smith, S. D., R. D. Muench, and C. H. Pease. 1990. Polynyas and leads: An overview of physical processes and environment. *JGR* 95(C6):9461–79.

Smith, V. G., W. A. Tiller, and J. W. Rutter. 1955. A mathematical analysis of solute redistribution during solidification. *Canad J Phys* 33:723–45.

Smith, W. O., Jr., ed. 1990. *Polar oceanography. Part A: Physical science.* San Diego: Academic Press.

Sodhi, D. S. 1977. Ice arching and the drift of pack ice through restricted channels. *CRREL Rept* 77-18.

Sodhi, D. S., and W. D. Hibler III. 1980. Non-steady ice drift in the Strait of Belle Isle. In *Sea ice processes and models*, ed. R. S. Pritchard, 177–86. Seattle: Univ Washington Press.

Solomon, S., and 7 others, eds. 2007. *Climate change 2007: The physical science basis.* New York: Cambridge Univ Press.

Son, S.-W., and 9 others. 2008. The impact of stratospheric ozone recovery on the southern hemisphere westerly jet. *Science* 320(13 June):1486–89.

Soper, A. K. 2002. Water and ice. *Science* 297(23 August):1288–89.

Souchez, R., and J. Jouzel. 1984. On the isotopic composition in δD and δ^{18}O of water and ice during freezing. *J Glaciol* 30(106):369–72.

Souchez, R., J. L. Tison, and J. Jouzel. 1988. Deuterium concentration and growth rate of Antarctic first-year sea ice. *GRL* 15:1385–88.

Souchez, R., J. L. Tison, R. Lorrain, C. Fléhoc, M. Stiévenard, J. Jouzel, and V. Maggi. 1995. Investigating processes of marine ice formation in a floating ice tongue by high-resolution isotopic study. *JGR* 100(C4):7019–25.

Spencer, R. J., N. Møller, and J. H. Weare. 1990. The prediction of mineral solubility in natural waters: A chemical equilibrium model for the Na-K-Ca-Mg-Cl-SO$_4$-H$_2$O system at temperatures below 25°C. *Geochim Cosmochim Acta* 54:575–90.

Spichkin, V. A. 1963. Roll of melting in forming the relief of the upper surface of old ice (R). *PAiA* 14:71–73.

———. 1966. Sea water accumulation on the Antarctic fast ice. *Sov Antarct Exped Inform Bull* 6(3):235–36.

Squire, V. A. 1983. Numerical modelling of realistic ice floes in ocean waves. *Ann Glaciol* 4:277–82.

———. 1991. Atmosphere–ice–ocean: Do we really understand what is going on. In *Antarctica and global climatic change*, ed. C. M. Harris and B. Stonehouse, 82–89. Boca Raton, FL: Belhaven Press.

———. 1993. The breakup of shore fast sea ice. *CRST* 21(3):211–18.

———. 1993. A comparison of the mass–loading and elastic plate models of an ice field. *CRST* 21(3):219–29.

———. 1994. Marginal ice zone rigidity parameterization from ocean wave refraction. *CRST* 22(3):235–41.

———. 1998. The marginal ice zone. In *Physics of ice covered seas*, ed. M. Leppäranta, I:381–446. Helsinki Univ Pub House.

———. 2007. Of ocean waves and sea ice revisited. *CRST* 49(2):110–33.

Squire, V., J. Dugan, P. Wadhams, P. Rottier, and A. Liu. 1995. Of ocean waves and sea ice. *Ann Rev Fluid Mech* 27:115–68.

Squire, V. A., and C. Fox. 1992. On ice-coupled waves: A comparison of data and theory. In *3rd Internat Conf on Ice Technology*, ed. T. K. S. Murthy, W. M. Sackinger, and P. Wadhams, 269–80. Computational Mechanics Publications.

Squire, V. A., and S. C. Moore. 1980. Direct measurement of the attenuation of surface waves by pack ice. *Nature* 283(5745):365–68.

Stabeno, P. J., N. B. Kachel, M. Sullivan, M., and Whitledge, T. E. 2002. Variability of physical and chemical characteristics along the 70-m isobath in the SE Bering Sea. *Deep-Sea Res* Pt II 49:5931–43.

Stabeno, P. J., and J. E. Overland. 2001. The Bering Sea shifts toward an earlier spring transition. *Eos Trans AGU* 82:317.

Stammerjohn, S. E., D. G. Martinson, R. C. Smith, X. Yuan, and D. Rind. 2008. Trends in Antarctic annual sea ice retreat and advance and their relation to El Niño–Southern Oscillation and Southern Annular Mode variability. *JGR* 113:C03S90, doi:10.1029/ 2007JC004269.

Stamp, T., and C. Stamp. 1976. *William Scoresby: Arctic scientist*. London: Caedmon of Whitby Press.

Stander, E., and G. A. Gidney. 1980. The measurement of finite strain in sea ice by impulse radar techniques. In *Proc of the Workshop on Sea Ice Field Measurement*, 127–64. C-CORE Pub No 80-21. St. John's, Newfoundland: Memorial University.

Stander, E., and B. Michel. 1989a. The effect of fluid flow on the development of preferred orientations in sea ice: Laboratory experiments. *CRST* 17(2):153–61.

———. 1989b. The development of aligned columnar sea ice: A field investigation. *J Glaciol* 35(120):217–23.

———. 1991. The effect of deformation on fabric development in columnar sea ice. In *POAC '91*, ed. D. B. Muggeridge, D. B. Colbourne, and H. M. Muggeridge, 527–40. Memorial University of Newfoundland, Ocean Engineering Research Centre.

Stanley, H. E., and N. Ostrowsky, eds. 1990. *Correlations and connectivity, geometric aspects of physics, chemistry and biology.* Dordrecht: Kluwer.

Starokadomskiy, L. M. 1976. *Charting the Russian northern sea route.* Montreal: McGill–Queen's Univ Press.

Steele, M., and T. Boyd. 1998. Retreat of the cold halocline layer in the Arctic Ocean. *JGR* 103:10419–35.

Steele, M., W. Ermold, and J. Zhang. 2008. Arctic Ocean surface warming trends over the past 100 years. *GRL* 35:L02614, doi:10.1029/2007GL031651.

Steele, M., and G. M. Flato. 2000. Sea ice growth melt and modeling: A survey. In *The freshwater budget of the Arctic Ocean*, ed. E. L. Lewis et al., 549–87. Dordrecht: Kluwer.

Steele, M., and J. H. Morison. 1993. Hydrography and vertical fluxes of heat and salt northeast of Svalbard in autumn. *JGR* 98:10013–24.

Steele, M., J. Zhang, D. Rothrock, and H. Stern. 1997. The force balance of sea ice in a numerical model of the Arctic Ocean. *JGR* 102(C9):21061–79.

Stefan, J. 1889. Über einige Probleme der Theorie der Warmeleitung. *Sitzungsberichte der Math-Natur Classe der Kaiserlichen, Akad der Wissenschaften* 98(2a):473–84.

———. 1891. Über die Theorie der Eisbildung, inbesondere über die Eisbildung im Polarmeere. *Wiedemann's Ann Phys Chem, Neue Folge* 42:269–86.

Steffen, K., and 6 others. 2008. *Abrupt climate change: Final report, synthesis and assessment Product 3.4.* US Geological Survey.

Stern, H. L., D. A. Rothrock, and R. Kwok. 1995. Open water production in Arctic sea ice: Satellite measurements and model parameterizations. *JGR* 100:20601–12.

Stewart, M. J., and F. Weinberger. 1972. Fluid flow through a solid–dendritic interface. *Metallurgical Trans* 3(1):333–37.

Stirling, I. 1980. The biological importance of polynyas in the Canadian Arctic. *Arctic* 33(2):303–15.

———. 1997. The importance of polynyas, ice edges, and leads to marine mammals and birds. *J Mar Syst* 10:9–21.

Stogryn, A. 1971. Equations for calculating the dielectric constant of saline water. *IEEE Trans Microwave Theory and Techniques* MTT-19:733–36.

———. 1987. An analysis of the tensor dielectric constant of saline water. *IEEE Trans Microwave Theory and Techniques* GE-25(2):147–58.

Stogryn, A., and G. J. Desargant. 1985. The dielectric properties of brine in sea ice at microwave frequencies. *IEEE Trans Antennas Propagation* AP-33(5):523–32.

Stokes, G. 1851. On the effect of internal friction in causing the motion of a fluid to subside: Application to the case of oscillatory waves. *Trans Camb Phil Soc* 9:57-62.

Strakhov, M. V. 1987. The effect of currents on the alignment of the ice structure near the station Mirnyi (R). *Trudy Sovetskoi Antarkticheskoi Ekspeditsii* 82:118–24.

———. 1989a. On spatial arrangement of crystalline structure of ice in Antarctic seas. *Sov Antarkt Eksped Inform Biull* 111:67–70.

———. 1989b. A study of spatially oriented crystal structures of arctic sea ice. In *The electrophysical and physicomechanical properties of ice*, ed. V. V. Bogorodskiy and V. P. Gavrilo, 134–45. Leningrad: AARI, Gidrometeoizdat.

———. 1991. A study of the influence of currents on the formation of an ice cover with an aligned crystal structure in polar seas. In *Glaciers–ocean–atmosphere interactions*, ed. V. M. Kotliakov, A. Ushakov, and A. Glazovski, 95–103. IAHS 208.

Strass, V. H., and E. Fahrbach. 1998. Temporal and regional variation of sea ice draft and coverage in the Weddell Sea obtained from upward looking sonars. In *Antarctic sea ice: Physical processes, interactions and variability*, ed. M. O. Jeffries, 123–39. AGU, Antarctic Research Series, vol. 74.

Stringer, W. J. 1974. Sea ice morphology of the Beaufort shorefast ice. In *The coast and shelf of the Beaufort Sea*, ed. J. C. Reed and J. A. Sater, 165–72. Arlington: AINA.

———. 1978. *Morphology of Beaufort, Chukchi and Bering Seas nearshore ice conditions by means of satellite and aerial remote sensing.* Geophysical Institute, University of Alaska Fairbanks report 1.

Stringer, W. J., and S. A. Barrett. 1976. Ice motion in the vicinity of a grounded floeberg. *3rd Internat POAC Conf* 1:527–51. Fairbanks, Alaska.

Stroeve, J. C., M. M. Holland, W. Meier, T. Scambos, and M. Serreze. 2007. Arctic sea ice decline: Faster than forecast. *GRL* 34:L09501, doi:10.1029/2007GL029703.

Stroeve, J. C., M. Serreze, S. Drobot, S. Gearheard, M. Holland, J. Maslanic, W. Meier, and T. Scambos. 2008. Arctic sea ice extent plummets in 2007. *Eos* 89(2):13–14.

Stroeve, J. C., M. C. Serreze, F. Fetterer, T. Arbetter, W. N. Meier, J. Maslanik, and K. Knowles. 2005. Tracking the Arctic shrinking ice cover: Another extreme minimum in 2004. *GRL* 32:L04501, doi:10.1029/2004GL021810.

Sturm, M., and C. S. Benson. 1997. Vapor transport, grain growth and depth hoar development in subarctic snow. *J Glaciol* 43:42–59.

Sturm, M., J. Holmgren, M. König, and K. Morris. 1997. The thermal conductivity of seasonal snow. *J Glaciol* 43(143):26–41.

Sturm, M., J. Holmgren, and D. K. Perovich. 2002. Winter snow cover on the sea ice of the Arctic Ocean at the Surface Heat Budget of the Arctic Ocean (SHEBA): Temporal evolution and spatial variability. *JGR* 107:C10, 8047, doi:10.1029/ 2000JC000400.

Sturm, M., K. Morris, and R. A. Massom. 1998. The winter snow cover of the West Antarctic pack ice: Its spatial and temporal variability. In *Antarctic sea ice: Physical processes, interactions and variability*, ed. M. O. Jeffries, 1–18. AGU, Antarctic Research Series, vol. 74.

Sturm, M., D. K. Perovich, and J. Holmgren. 2002. Thermal conductivity and heat transfer through the snow on the ice of the Beaufort Sea. *JGR* 107:C21, 8043, doi:10.1029/2000JC000409.

Sturm, M., D. K. Perovich, and M. C. Serreze. 2003. Meltdown in the north. *Sci Amer* 288:60–67.

Sullivan, C. W., A. C. Palmisano, and J. B. SooHoo. 1984. Influence of sea ice and ice biota on downwelling irradiance and spectral composition of light at McMurdo Sound. In *Proc Internat Soc for Optical Engineering*, 159–65. Vol 489 (Ocean Optics VII).

Sunder, S. S. 1986a. An integrated constitutive theory for the mechanical behavior of sea ice: Experimental verification. In *IAHR '86*, 1:253–64.

———. 1986b. An integrated constitutive theory for the mechanical behavior of sea ice: Micromechanical interpretation. In *Ice technology*, ed. T. K. S. Murthy, J. J. Connor, and C. A. Brebbia, 87–102. Springer-Verlag.

Suzuki, Y. 1967. On disorder entropy of ice. In *Physics of snow and ice*, 1(1):21–41. ILTS, Hokkaido University.

Svec, O. J., and R. M. W. Frederking. 1981. Cantilever beam tests in an ice cover: Influence of plate effects at the root. *CRST* 4:93–101.

Svec, O. J., J. C. Thompson, and R. M. W. Frederking. 1985. Stress concentrations in the root of an ice cover cantilever: Model tests and theory. *CRST* 11:63–73.

Sverdrup, H. U. 1928. The wind-drift of the ice on the North Siberian Shelf. In *Scientific results of the Norwegian north pole expedition with the* Maud, *1918–1925*, 4(1).

Sverdrup, H. U., M. W. Johnson, and R. H. Fleming. 1942. *The oceans*. New York: Prentice-Hall.

Svishchev, I. M., and P. G. Kusalik. 1996. Quartzlike polymorph of ice. *Phys Rev* B(53)(1 April):721.

Swift, J. H., and K. Aagaard. 1981. Seasonal transitions and water mass formation in the Iceland and Greenland Seas. *Deep Sea Res* 28:1107–29.

Swift, J. H., T. Takahashi, and H. Livingston. 1983. The contribution of the Greenland and Barents Seas to the deep water of the Arctic Ocean. *JGR* 88(C10):5981–86.

Swinzow, G. K. 1966. Ice cover of an arctic pro-glacial lake. *CRREL Research Rept* 155.

Szpir, M. 1996. Bits of ice XI and ice XII. *American Scientist* 84(5):437–38.

Tabata, T. 1960. Studies on mechanical properties of sea ice. V. Measurement of flexural strength (J). *LTS-A* 19:187–201 (Am Met Soc trans T-J-17).

———. 1966. Studies on the mechanical properties of sea ice. IX. Measurement of the flexural strength *in situ* (J). *LTS-A* 24:259–68.

———. 1967. The flexural strength of small sea ice beams. In *Physics of snow and ice*, I:1:481–97. ILTS, Hokkaido University.

Tabata, T., and K. Fujino. 1964. Studies on the mechanical properties of sea ice. VIII. Measurement of the flexural strength *in situ* (J). *LTS-A* 23:157–66.

———. 1965. Studies on the mechanical properties of sea ice. VII. Measurement of the flexural strength *in situ* (J). *LTS-A*.

Tabata, T., K. Fujino, and M. Aota. 1967. Studies on the mechanical properties of sea ice: The flexural strength of sea ice *in situ*. In *Physics of snow and ice*, I(1):539–50. ILTS, Hokkaido University.

Tabata, T., and N. Ono. 1957. On the structure of sea ice. *LTS-A* 16:197–210.

———. 1962. On the crystallographic study of several kinds of ice (J). *LTS-A* 20:199–214.

Tabata, T., Y. Suzuki, and M. Aota. 1975. Ice study in the Gulf of Bothnia. II. Measurements of flexural strength. *LTS-A* 33:199–206.

Takizawa, T. 1984. Characteristics of snow cover on sea ice and the formation of snow ice. II. *LTS-A* 43:157–61.

Tamura, T., K. I. Ohshima, and S. Nihashi. 2008. Mapping of sea ice production for Antarctic coastal polynyas. *GRL* 34:L07606, doi:10/1029/2007GL032903.

Tauber, G. M. 1957. On the katabatic winds of Antarctica and their influence on the formation of sea ice. *Dokl Gos Okeanogr In-ta Glav Upr Gidrometeorol Sluzhby SSSR* (referenced in Cherepanov and Kozlovskii 1973d).

Taylor, L. D., and J. B. Lyons. 1959. *Ice structures, Angiussaq Lake, Northwest Greenland.* Air Force Cambridge Research Center, Geophysics Research Directorate TN-59-461.

Taylor, R. B. 1978. The occurrence of grounded ice ridges and the shore ice piling along the northern coast of Somerset Island, NWT. *Arctic* 31(2):133–49.

Tedesco, M. 2007. A new record in 2007 for melting in Greenland. *Eos* 88(39):383.

Tedesco, M., X. Fettweiss, M. van den Broeke, R. van de Wal, and P. Smeets. 2008. Extreme snowmelt in Northern Greenland during summer 2008. *Eos* 89(41):391.

Thomas, D. N., and G. S. Dieckmann. 2002. Antarctic sea ice: A habitat for extremophiles. *Science* 295(5555):641–44.

———, eds. 2003. *Sea ice: An introduction to its physics, biology, chemistry, and geology.* London: Blackwell.

Thomas, D. N., and S. Papadimitriou. 2003. Biogeochemistry of sea ice. In *Sea ice: An introduction to its physics, biology, chemistry, and geology*, ed. D. N. Thomas and G. S. Dieckmann, 267–302. London: Blackwell.

Thomas, R. H. 1973. The creep of ice shelves: Theory. *J Glaciol* 12(64):45–53.

Thomas, R. H., and D. R. MacAyeal. 1982. Derived characteristics of the Ross Ice Shelf, Antarctica. *J Glaciol* 28(100):397–412.

Thompson, D. W. J., and S. Solomon. 2002. Interpretation of recent Southern Hemisphere climate change. *Science* 296(3 May):895–99.

Thompson, D. W. J., and J. M. Wallace. 1998. The Arctic Oscillation signature in the wintertime geopotential height and temperature fields. *GRL* 25:1297–1300.

Thompson, S. L., and D. Pollard. 1997. Greenland and Antarctic mass balances for present and doubled atmospheric CO_2 from the GENESIS version 2 global climate model. *J Climate* 10:871–900.

Thorndike, A. S. 1986a. Diffusion of sea ice. *JGR* 91(C6):7691–96.

———. 1986. Kinematics of sea ice. In *Geophysics of sea ice*, ed. N. Untersteiner, 489–550. New York: Plenum.

———. 1987. A random discontinuous model of sea ice motion. *JGR* 92(C6):6515–20.

———. 1988. A naive zero-dimensional sea ice model. *JGR* 93(C5):5093–99.

———. 1992a. Estimates of sea ice thickness distribution using observations and theory. *JGR* 97(C8): 12601–5.

———. 1992b. A toy model linking atmospheric thermal radiation and sea ice growth. *JGR* 97(C6):9401–10.

———. 2000. Sea ice thickness as a stochastic process. *JGR* 105(C1):1311–13.

Thorndike, A. S., and R. Colony. 1980. Large-scale ice motion in the Beaufort Sea during AIDJEX, April 1975–April 1976. In *Sea ice processes and models*, ed. R. S. Pritchard, 249–60. Seattle: Univ Washington Press.

———. 1982. Sea ice motion in response to geostrophic winds. *JGR* 87(C8):5845–52.

Thorndike, A. S., D. A. Rothrock, G. A. Maykut, and R. Colony. 1975. The thickness distribution of sea ice. *JGR* 80(33):4501–13.

Thyssen, F. 1988. Special effects of the central part of Filchner-Ronne Ice Shelf, Antarctica. *Ann Glaciol* 11:173–79.

Thyssen, F., A. Bombosch, and H. Sandhager. 1993. Elevation, ice thickness and structure mark maps of the central part of the Filchner-Ronne Ice Shelf. *Polarforschung* 62(1):17–26.

Tiller, W. A. 1957. Preferred growth direction of metals. *J Metals* 9:847–55.

———. 1962. Effect of grain boundaries on solute partitioning during progressive solidification. *Acta Met* 1(4):428–37.

———. 1963. Migration of a liquid zone through a solid. *J Appl Phys* 34(9):2757–62.

———. 1991. *The science of crystallization: Macroscopic phenomenon and defect generation.* Cambridge Univ Press.

Tiller, W. A., K. A. Jackson, J. W. Rutter, and B. Chalmers. 1953. The redistribution of solute atoms during the solidification of metals. *Acta Met* 1:428–37.

Timco, G. W. 2005. The history of the POAC conference. *18th POAC Conf,* 1:1–10. Potsdam, NY: Clarkson Univ.

Timco, G. W., and R. P. Burden. 1997. An analysis of the shapes of sea ice ridges, *CRST* 25:65–77.

Timco, G. W., K. Croasdale, and B. Wright. 2000. *An overview of first-year sea ice ridges.* Canadian Hydraulics Centre Tech Rept HYD-TR-047 (PERD/CHC Report 5-112).

Timco, G. W., and R. M. W. Frederking. 1983. Confined compressive strength of sea ice. *Proc POAC '83* (VTT Symp 27), 1:243–53.

———. 1986. Confined compression tests: Outlining the failure envelope of columnar sea ice. *CRST* 12:13–28.

———. 1990. Compressive strength of sea ice sheets. *CRST* 17:227–40.

———. 1991. Seasonal compressive strength of Beaufort Sea ice sheets. In *Ice-Structure Interaction,* ed. S. Jones, R. F. McKenna, J. Tillotson, and I. Jordaan, 267–80. Springer-Verlag.

———. 1996. A review of sea ice density. *CRST* 24(1):1–6.

Timco, G. W., and S. O'Brien. 1994. Flexural strength equation for sea ice. *CRST* 22(3):285–98.

Timco, G. W., and W. F. Weeks. 2009. A review of the engineering properties of sea ice. *CRST* 60:107–29.

Tin, T., and M. O. Jeffries. 2003. Morphology of deformed first-year sea ice features in the Southern Ocean. *CRST* 36:141–63.

Tinga, W. R., W. A. G. Voss, and D. F. Blossey. 1973. Generalized approach to multiphase mixture theory. *J App Phys* 44:3897–3902.

Tirmizi, S. H., and W. N. Gill. 1989. Experimental investigation of the dynamics of spontaneous pattern formation during dendritic ice crystal formation. *J Crystal Growth* 96:277–92.

Tison, J.-L., C. Haas, M. M. Gowing, S. Sleewaegen, and A. Bernard. 2002. Tank study of physico-chemical controls on gas content and composition during growth of young sea ice. *J Glaciol* 48(161):177–91.

Tison, J.-L., and J. Haren. 1989. Isotopic, chemical and crystallographic characteristics of first-year sea ice from Breid Bay (Princess Ragnhild Coast–Antarctica). *Antarctic Science* 1(3):261–68.

Tison, J.-L., A. Khazendar, and E. Roulin. 2001. A two-phase approach to the simulation of the combined isotope/salinity signal of marine ice. *JGR* 106(C12):31387–401.

Tison, J. L., R. D. Lorrain, A. Bouzette, M. Dini, A. Bondesan, and M. Stiévenard. 1998. Linking landfast sea ice variability to marine ice accretion at Hells Gate Ice Shelf, Ross Sea. In *Antarctic sea ice: Physical processes, interactions and variability,* ed. M. O. Jeffries, 375–407. AGU, Antarctic Research Series, vol. 74.

Tison, J.-L., E. M. Morris, R. Souchez, and J. Jouzel. 1991. Stratigraphy, stable isotopes and salinity in multi-year sea ice from the rift area, south George VI Ice Shelf, Antarctic Peninsula. *J Glaciol* 37(127):357–67.

Tison, J.-L., D. Ronveaux, and R. D. Lorrain. 1993. Low salinity frazil ice generation at the base of a small Antarctic ice shelf. *Antarct Sci* 5(3):309–21.

Tokmakoff, A. 2007. Shining light on the rapidly evolving structure of water. *Science* 317(5834):54–55.

Tomlinson, C. 1871. *Frozen stream: Formation and properties of ice.* London: SPCK.

Tooma, S. G., and W. B. Tucker III. 1973. Statistical comparisons of airborne laser and stereophotogrammetric sea ice profiles. *Remote Sensing of the Environment* 2:261–72.

Topham, D. R., R. G. Perkin, S. D. Smith, R. J. Anderson, and G. Den Hartog. 1983. An investigation of a polynya in the Canadian archipelago. 1. Introduction and oceanography. *JGR* 88:2888–99.

Toyota, T., S, Takatsuji, and M. Nakayama. 2006. Characteristics of sea ice floe size distribution in the seasonal ice zone. *GRL* 33:L02616, doi:10.1029/2005GL024556.

Transehe, N. A. 1928. The ice cover of the arctic sea, with a genetic classification of sea ice. In *Problems of polar research,* 91–123. New York: American Geographical Society.

Tratteberg, A., L. W. Gold, and R. Frederking. 1975. The strain rate and temperature dependence of Young's modulus of ice. In *IAHR Symp on Ice Problems*, 479–86. Hanover, NH.

Tremblay, L. B., and L. A. Mysak. 1997. Modelling sea ice as a granular material including the dilatancy effect. *J Phys Oceanog* 27:2342–60.

Trodahl, H. J., M. McGuinness, P. Langhorne, K. Collins, A. Pantoja, I. Smith, and T. Haskell 2000. Heat transport in McMurdo Sound first-year fast ice. *JGR* 28:1279–82.

Trodahl, H. J., S. O. F. Wilkinson, M. J. McGuinness, and T. G. Haskell. 2001. Thermal conductivity of sea ice: Dependence on temperature and depth. *GRL* 28(7):1279–82.

Tsang, G. 1982. *Frazil and anchor ice: A monograph*. National Research Council of Canada, Subcommittee on the Hydraulics of Ice Covered Rivers.

Tsang, G., and T. O'D. Hanley. 1985. Frazil formation in water of different salinities and supercoolings. *J Glaciol* 31(108):74–85.

Tsatsoulis, C., and R. Kwok, eds. 1998. *Analysis of SAR data of the polar oceans. Recent advances*. Berlin: Springer.

Tsurikov, V. L. 1940. The problem of ice strength (R). *Severnyi Morskoi Put'* 16:45–74.

———. 1947a. The influence of the salinity of sea ice on its strength (R). *Trudy Gosudarstvennogo Okeanograficheskogo Instituta* 2(14):89–108.

———. 1947b. The problem of the influence of the cavities in ice on its strength (R). *Trudy Gosudarstvennogo Okeanograficheskogo Instituta* 2(14):66–88.

———. 1974. Statistics of salt composition in sea ice. *Oceanology* 14(3):360–67.

———. 1976. *Liquid phase in sea ice* (R). Moscow: Nauka.

———. 1979. The formation and composition of the gas content of sea ice. *J Glaciol* 22(86):67–81.

Tsurikov, V. L., and V. Tsurikova. 1972. The brine content of sea ice (Statement of the problem). *Oceanology* 12(5):663–72.

Tucker, W. B., III, and J. W. Govoni. 1981. Morphological investigations of first-year pressure ridge sea ice sails. *CRST* 5:1–12.

Tucker, W. B., III, A. J. Gow, and J. A. Richter. 1984. On small-scale horizontal variations of salinity in first-year sea ice. *JGR* 89(C4):6505–14.

Tucker, W. B., III, A. J. Gow, and W. F. Weeks. 1987. Physical properties of summer sea ice in the Fram Strait. *JGR* 92:6787–6803.

Tucker, W. B., III, and D. K. Perovich. 1992. Stress measurements in drifting pack ice. *CRST* 20:119–39.

Tucker, W. B. ,III, D. K. Perovich, A. J. Gow, W. F. Weeks, and M. R. Drinkwater. 1992. Physical properties of sea ice relevant to remote sensing. In *Microwave remote sensing of sea ice*, ed. F. D. Carsey, 9–28. Geophysical Monograph Series. Washington, DC: AGU.

Tucker, W. B., III, D. K. Perovich, M. A. Hopkins, and W. D. Hibler III. 1991. On the relationship between local stresses and strains in Arctic pack ice. *Ann Glaciol* 15:265–70.

Tucker, W. B., III, D. S. Sodhi, and J. W. Govoni. 1984. Structure of first-year pressure ridge sails in the Prudhoe Bay region. In *The Alaskan Beaufort Sea: Ecosystems and environments*, ed. P. W. Barnes, D. M. Schell, and E. Reimnitz, 115–35. New York: Academic Press.

Tucker, W. B., III, J. W. Weatherly, D. T. Eppler, D. Farmer, and D. L. Bentley. 2001. Evidence for the rapid thinning of sea ice in the western Arctic Ocean at the end of the 1980s. *GRL* 28(14):2851–54.

Tucker, W. B., III, W. F. Weeks, and M. Frank. 1979. Sea ice ridging over the Alaskan continental shelf. *JGR* 84(C8):4885–97. (See also CRREL Rept 79-8 which contains data summaries.)

Tucker, W. B., III, W. F. Weeks, A. Kovacs, and A. J. Gow. 1980. Nearshore ice motion at Prudhoe Bay, Alaska. In *Sea ice processes and models*, ed. R. S. Pritchard, 261–72. Seattle: Univ Washington Press.

Tuhkuri, J., M. Lensu, and M. A. Hopkins. 1998. Laboratory and field studies on ridging of an ice sheet. In *Ice in surface waters*, ed. H. T. Shen, 397–404. Proc 14th Internat Symp on Ice. Rotterdam: Balkema.

Tuhkuri, J., M. Lensu, and S. Saarinen. 1999. Laboratory and field studies on the mechanics of ice ridge formation. *Proc 15th Internat Conf on POAC* 3:1118–29. Helsinki.

Tulk, C. A., C. J. Benmore, J. Urquidi, D. D. Klug, J. Neuefeind, J., B. Tomberli, and P. A. Egelstaff. 2002. Structural studies of several distinct metastable forms of amorphous ice. *Science* 297(23 August):1320–24.

Ukita, J., M. Honda, H. Nakamura, Y. Tachibana, D. J. Cavalieri, C. L. Parkinson, H. Koide, and K. Yamamoto. 2007. Northern hemisphere sea ice variability: Lag structure and its implications. *Tellus* A59:261–72.

Ukita, J., and R. E. Moritz. 1995. Yield curves and flow rules of pack ice. *JGR* 100:4545–57.

Ulaby, F. T., R. K. Moore, and A. K. Fung. 1982–1986. *Microwave remote sensing: Active and passive.* Reading, MA: Addison-Wesley.

United States Navy. 1952. *A functional glossary of ice terminology.* US Navy Hydrographic Office. HO Pub 609:1–88.

Untersteiner, N. 1961. On the mass and heat budget of Arctic sea ice. *Arch Meteorol Geophys Bioklimatol* A(12):151–82.

———. 1964. Calculations of temperature regime and heat budget of sea ice in the Central Arctic. *JGR* 69(22):4755–66.

———. 1968. Natural desalination and equilibrium salinity profile of perennial sea ice. *JGR* 73(4):1251–57.

———, ed. 1986. *The geophysics of sea ice.* NATO Advanced Sciences Institutes Series, Series B, Physics 146, Proc of NATO Advanced Study Institute on Air–Sea–Ice Interaction. New York: Plenum.

———. 1990. Some problems of sea ice and climate modeling. In *Festbuch commemorating the 100 year anniversary of the Institut für Meteorology und Geophysik.* Veröffentlichungen der Universität Innsbruck 178.

Untersteiner, N., and F. I. Badgley. 1958. Preliminary results of thermal budget studies of arctic pack ice during summer and autumn. In *Arctic sea ice,* ed. W. Thurston, 85–92. Publication No. 598, National Academy of Sciences–Nat Res Council.

Untersteiner, N., and R. Sommerfeld. 1964. Supercooled water and the bottom topography of floating ice. *JGR* 69(6):1057–62.

Untersteiner, N., A. S. Thorndike, D. A. Rothrock, and K. L. Hunkins. 2007. AIDJEX revisited: A look back at the U.S.–Canadian Arctic Ice Dynamics Joint Experiment 1970–78. *Arctic* 60(3):327–36.

Urabe, N., and M. Inoue. 1986. Mechanical properties of Antarctic sea ice. In *Proc 5th OMAE Symp* 4:303–9. Tokyo.

Ushio, S., and M. Wakatsuchi. 1989. Rapid frazil ice production in coastal polynya: Laboratory experiments. In *Proc NIPR Symp on Polar Meteorology and Glaciology* 2:117–26. Tokyo: Nat Inst of Polar Res.

Vali, G. 1971. Quantitative evaluation of experimental results on the heterogeneous freezing nucleation of supercooled liquids. *J Atmos Sci* 28:402–9.

Vant, M. R. 1976. *A combined empirical and theoretical study of the dielectric properties of sea ice over the frequency range 100 MHz to 40 GHz.* Technical Report. Ottawa: Carleton Univ.

Vant, M. R., R. Gray, R. O. Ramseier, and V. Makios. 1974. Dielectric properties of fresh and sea ice at 10 and 35 GHz. *J App Phys* 45(11):4712–17.

Vant, M. R., R. O. Ramseier, and V. Makios. 1978. The complex-dielectric constant of sea ice at frequencies in the range 0.1–40 GHz. *J App Phys* 49(3–1):1264–80.

Van Woert, M. L. 1999. Wintertime dynamics of the Terra Nova Bay polynya. *JGR* 104(C4):7753–69.

Varnai, T., and Cahalan, R. F. 2007. Potential for airborne offbeam lidar measurements of snow and sea ice thickness. *JGR* 112:C12S90, doi:10.1029/2007JC004091.

Vaudrey, K. D. 1977. Ice engineering: Study of related properties of floating sea-ice sheets and summary of elastic and viscoelastic analysis. *US Navy Civil Engineering Lab Rept* no TR-860. Port Hueneme, CA.

Vaughn, D. G., G. J. Marshall, W. M. Connolley, J. C. King, and R. Mulvaney. 2001. Devil in the detail. *Science* 293(7 September):1777–79.

Vavrus, S. J. 1995. Sensitivity of the Arctic climate to leads in a coupled atmosphere/mixed-layer ocean model. *J Climate* 8:158–71.

———. 1999. The response of the coupled Arctic sea ice–atmosphere system to orbital forcing and ice motion at 6 kyr and 115 kyr BP. *J Climate* 12:873–96.

Vavrus, S. J., and S. P. Harrison. 2003. The impact of sea ice dynamics on the Arctic climate system. *Climate Dyn* doi:10.1007/s00382-003-0309-5.

Velasco, E., L. Mederos, and G. Navascués. 1997. Comment on "An explanation of the density maximum in water." *Phys Rev Lett* 79(1):179.

Vella, D., and J. S. Wettlaufer. 2007. Finger rafting: A generic instability in floating elastic sheets. *Phys Rev Lett* 98(23 February):088303.

———. 2008. Explaining the patterns formed by ice floe interactions. *JGR* 113:C11011, doi:10.1029/2008JC004781.

Vinje, T. E. 1976. Sea ice conditions in the European sector of the marginal seas of the Arctic, 1966–1975. In *Norsk Polar-Instititt Arbok 1975*, 163–74. Oslo.

————. 2001. Fram Strait ice fluxes and atmospheric circulation. *J Climate* 14(16):3508–17.

Vinje, T. E., and O. Finnekasa. 1986. The ice transport through Fram Strait. *Norsk Polarinstitutt Rept* 186:1–39.

Vinje, T. E., N. Nordlund, and A. Kvambekk. 1998. Monitoring ice thickness in Fram Strait. *J Geophys Res* 103(C5):10437–49.

Vittoratos, E. S. 1979. Existence of oriented sea ice by the McKenzie Delta. In *POAC 79*, 643–50. Trondheim.

Vize, V. Iu. 1923. On the possibility of forecasting the state of the ice in the Barents Sea (R). *Izv Tsen Gidro Byuro* 1:1–45.

————. 1924. Polareis und atmosphärische Schwankungen. *Geograf Ann* 6:273–99.

————. 1944. Principles of long-range forecasts for arctic seas (R). *Trudy Arkt Inst* 190.

Vlahakis, J. G., and A. J. Barduhn. 1974. Growth rate of an ice crystal in flowing water and salt solutions. *Amer Inst Chem Eng J* 20(3):581–91.

Vogt, P. R., K. Crane, and E. Sundvor. 1994. Deep Pleistocene iceberg plowmarks on the Yermak Prareau: Sidescan and 3.5 kHz evidence for thick calving ice fronts and a possible marine ice sheet in the Arctic Ocean. *Geology* 22:403–6.

Voitkovskii, K. F. 1960. *The mechanical properties of ice.* Moscow: Izd Akad Nauk SSSR. (Trans US Air Force Cambridge Research Laboratory, 62-838.)

Volkov, N. A., and P. C. Voronov. 1967. Studies of the cryotectonics of sea ice for glaciological and geological use (R). *PAiA* 27:152–68.

von Drygalski, E. 1921. *Deutsche südpolar Expedition 1901–1903: Band I(4). Das Eis der Antarktis und sudantarktischen Meere.* Berlin und Leipzig: Vereinigung Wissenschaftlicher Verleger, 365–709.

von Wrangell, F. 1840. *Narrative of an expedition to the polar sea in the years 1820, 1821, 1822, & 1823* (E. Sabine, ed.). London: James Madden and Co.

Wadhams, P. 1972. Measurement of wave attenuation in pack ice by inverted echo sounding. In *Sea ice*, ed. T. Karlsson, 255–60. Reykjavik: Nat Res Council of Iceland.

————. 1973. Attenuation of swell by sea ice. *JGR* 78(18):3552–63.

————. 1975. Airborne laser profiling of swell in an open ice field. *JGR* 80(33):4520–28.

————. 1976a. Sea ice topography in the Beaufort Sea and its effect on oil containment. *AIDJEX Bull* 33:1–52. Seattle: Div Marine Resources, Univ Washington.

————. 1976b. Oil and ice in the Beaufort Sea. *Polar Record* 18(114):237–50.

————. 1978a. Characteristics of deep pressure ridges in the Arctic Ocean. *POAC 77*, Proc 4th Internat Conf on Port and Ocean Engineering under Arctic Conditions, 1:544–55. St. John's, Newfoundland: St. John's Memorial Univ.

————. 1978b. Wave decay in the marginal ice zone measured from a submarine. *Deep-Sea Res* 25:23–40.

————. 1979. Field experiments on wave–ice interaction in the Labrador and East Greenland Currents, 1978. *Polar Rec* 19(121):373–76.

————. 1980. A comparison of sonar and laser profiles along corresponding tracks in the Arctic Ocean. In *Sea ice processes and models*, ed. R. S. Pritchard, 283–99. Seattle: Univ Washington Press.

————. 1981a. Sea ice topography of the Arctic Ocean in the region 70°W to 25°E. *Phil Trans Roy Soc, London* A302(1464):45–85.

————. 1981b. The ice cover in the Greenland and Norwegian Seas. *Rev Geophys Space Phys* 19(3):345–93.

————. 1983a. The prediction of extreme keel depths from sea ice profiles. *CRST* 6(3):257–66.

————. 1983b. Sea ice thickness distribution in Fram Strait. *Nature* 305:108–11.

————. 1983c. The prediction of extreme keel depths from sea ice profiles. *CRST* 6:257–66.

————. 1983d. A mechanism for the formation of ice edge bands. *JGR* 88(C5):2813–18.

————. 1984. Arctic sea ice morphology and its measurement. In *Arctic technology and policy*, ed. I. Dyer and C. Chryssostomidis, 179–95. New York: Hemisphere Pub Corp.

————. 1986. The seasonal ice zone. In *The geophysics of sea ice*, ed. N. Untersteiner, 825–991. New York: Plenum.

————. 1990. Evidence for thinning of the Arctic ice cover north of Greenland. *Nature* 345:795–97.

————. 1992. Sea ice thickness distribution in the Greenland Sea and Eurasian Basin, May 1987. *JGR* 97(C4):5331–48.

———. 1994. Sea ice thickness changes and their relation to climate. In *The polar oceans and their role in shaping the global environment*, ed. O. M. Johannessen, R. D. Muench, and J. E. Overland, 337–62. AGU Geophys Monograph 85.

———. 1997. Ice thickness in the Arctic Ocean: The statistical reliability of experimental data. *J Geophys Res* 102(C13):27951–59.

———. 2000. *Ice in the ocean*. Amsterdam: Gordon and Breach.

Wadhams, P., and J. C. Comiso. 1992. The ice thickness distribution inferred using remote sensing techniques. In *Microwave remote sensing of sea ice*, ed. F. Carsey, 375–83. Geophysical Monograph 68. Washington: AGU.

Wadhams, P., J. C. Comiso, J. Crawford, and 8 others. 1991. Concurrent remote sensing of Arctic sea ice from submarine and aircraft. *Internat J Remote Sensing* 12(9):1829–40.

Wadhams, P., J. C. Comiso, E. Prussen, and 6 others. 1996. The development of the Odden ice tongue in the Greenland Sea during winter 1993 from remote sensing and field observations. *JGR* 101(C8):18213–35.

Wadhams, P., and N. R. Davis. 2000. Further evidence of ice thinning in the Arctic Ocean. *GRL* 27(24):3973–75.

Wadhams, P., and T. Davy. 1986. On the spacings and draft distributions for pressure ridge keels. *JGR* 91(C9):10697–708.

Wadhams, P., and B. Holt. 1991. Waves in frazil and pancake ice and their detection in Seasat synthetic aperture radar imagery. *JGR* 96(C5):8835–52.

Wadhams, P., and R. J. Horne. 1980. An analysis of ice profiles obtained by submarine sonar in the Beaufort Sea. *J Glaciol* 15(93):401–24.

Wadhams, P., M. A. Lange, and S. F. Ackley. 1987. The ice thickness distribution across the Atlantic sector of the Antarctic Ocean in midwinter. *JGR* 92(C13):14535–52.

Wadhams, P., and S. Martin. 1990. Processes determining the bottom topography of multiyear Arctic sea ice. In *Sea ice properties and processes*, ed. S. F. Ackley and W. F. Weeks, 136–41. CRREL Monograph 90-1.

Wadhams, P., A. S. McLaren, and R. Weintraub. 1985. Ice thickness distribution in Davis Strait in February from submarine sonar profiles. *JGR* 90(C1):1069–77.

Wadhams, P., V. A. Squire, J. A. Ewing, and R. W. Pascal. 1986. The effect of the marginal ice zone on the directional wave spectrum of the ocean. *J Phys Oceanogr* 6(2):358–76.

Wadhams, P., V. A. Squire, D. J. Goodman, A. M. Cowan, and S. C. Moore. 1988. The attenuation rates of ocean waves in the marginal ice zone. *J Geophys Res* 93(C6):6799–6818.

Wadhams, P., W. B. Tucker III, W. B. Krabill, R. N. Swift, J. C. Comiso, and N. R. Davis. 1992. Relationship between sea ice freeboard and draft in the Arctic Basin, and implications for ice thickness monitoring. *JGR* 97(C12):20325–34.

Wadhams, P., and J. P. Wilkinson. 1999. The physical properties of sea ice in the Odden Ice Tongue. *Deep-Sea Res* II 46(6–7):1275–1300.

Wadhams, P., J. P. Wilkinson, and S. D. McPhail. 2006. A new view of the underside of Arctic sea ice. *GRL* 33:L04501, doi:10.1029/2005GL025131.

Wakatsuchi, M. 1977. Experiments on haline convection induced by freezing of seawater (J). *LTS-A* 35:249–58.

———. 1983. Brine exclusion process from growing sea ice. *ILTS Science Contributions* 33:29–65.

Wakatsuchi, M., and T. Kawamura. 1987. Formation processes of brine drainage channels in sea ice. *JGR* 92(C7):7195–97.

Wakatsuchi, M., and N. Ono. 1983. Measurements of salinity and volume of brine excluded from growing sea ice. *JGR* 88(C5):2943–51.

Walczowski, W., and J. Piechura. 2006. New evidence of warming propagating toward the Arctic Ocean. *GRL* 33:L12601, doi:10.1029.2006GL025872.

Walker, E. R., and P. Wadhams. 1979. Thick sea ice floes. *Arctic* 32(2):140–47.

Waller, W. H., and P. W. Hodge. 2003. *Galaxies and the cosmic frontier*. Cambridge, MA: Harvard Univ Press.

Walsh, J. E. 1986. Diagnostic studies of large-scale air–sea–ice interactions. In *Geophysics of sea ice*, ed. N. Untersteiner, 755–84. New York: Plenum.

———. 1993. The elusive Arctic warming. *Nature* 361(28 January):300–1.

Walsh, J. E. , W. L. Chapman, and T. L. Shy. 1996. Recent decrease in sea level pressure in the Central Arctic. *J Climate* 9:480–86.

Walsh, J. E., W. D. Hibler III, and B. Ross. 1985. Numerical simulation of northern hemisphere sea ice variability 1951–1980. *JGR* 90(C3):4847–65.

Walsh, J. E., V. Kattsov, D. Portis, and V. Meleshko. 1998. Arctic precipitation and evaporation: Model results and observational estimates. *J Climate* 11:72–87.

Walsh, J. E., and H. J. Zwally. 1990. Multiyear sea ice in the Arctic: Model- and satellite-derived. *JGR* 95(C7):11613–28.

Walsh, J. E., H. J. Zwally, and J. W. Weatherly. 1991. Antarctic sea ice and temperature variations. In *Internat Conf on the Role of the Polar Regions in Global Change*, 1:263–68. Fairbanks: Univ Alaska Fairbanks.

Walter, B. A., J. E. Overland, and R. L. Gilmer. 1984. Air–ice drag coefficients for first-year sea ice derived from aircraft measurements. *JGR* 89(C3):3550–60.

Walter, B. A., J. E. Overland, and P. Turet. 1995. A comparison of satellite-derived and aircraft-measured regional surface sensible heat fluxes over the Beaufort Sea. *JGR* 100(C3):4585–91.

Walton, D., and B. Chalmers. 1959. The origin of preferred orientation in the columnar zone of metal ingots. *Trans AIME* 215:447–57.

Walton, D., W. A. Tiller, J. W. Rutter, and W. C. Winegard. 1955. Instability of a smooth solid–liquid interface during solidification. *Trans AIME* 203 (*J Metals* 7):1023.

Wang, A. T. 1991. Numerical simulations for rare ice gouge depths. *CRST* 19(1):19–32.

Wang, J., and M. Ikeda. 2000. Arctic oscillation and Arctic sea ice oscillation. *GRL* 27(9):1287–90.

Wang, L., and R. A. Schapery. 1995. Prediction of elastic and viscoelastic properties of anisotropic columnar ice. In *Ice Mechanics-95*, ed. J. P. Dempsey and Y. D. S. Rajapakse, 33–47. ASME.

Wang, Y. S. 1979. Crystallographic studies and strength tests in the Alaskan Beaufort Sea. *Proc POAC* 79, I:651–65. Trondheim, Norway.

Warren, S. G., I. G. Rigor, N. Untersteiner, V. F. Radionov, N. N. Bryazgin, Y. I. Aleksandrov, and R. Colony. 1999. Snow depth on Arctic sea ice. *J Climate* 12(6):1814–29.

Warren, S. G., C. S. Roesler, V. I. Morgan, R. E. Brandt, I. D. Goodwin, and I. Allison. 1993. Green icebergs formed by freezing of organic-rich seawater to the base of Antarctic ice shelves. *JGR* 98(C4):6921–28. (See also correction *JGR* 98(C10):18309.)

Weber, J. E. 1987. Wave attenuation and wave drift in the marginal ice zone. *J Phys Oceanog* 17(12):2351–61.

Weeks, J. D., W. van Saarloos, and M. Grant. 1991. Stability and shapes of cellular profiles in directional solidification: Expansion and matching methods. *J Cryst Growth* 112:244–82.

Weeks, W. F. 1958. The structure of sea ice: A progress report. In *Arctic sea ice*, ed. W. R. Thurston, 82–83, 93, 96–99, 156, 187, 238. Pub. 598. Washington, DC: US Nat Academy of Sciences/Nat Res Council.

———. 1962. Tensile strength of NaCl ice. *J Glaciol* 4(31):25–52.

———. 1985. The variation of sea ice strength within and between multiyear pressure ridges in the Beaufort Sea. *J Energy Resources Technol* 107(2):167–72.

———. 1994. Possible roles of sea ice in the transport of hazardous material in the Arctic Ocean. In *Workshop on Arctic Contamination*, 34–52. US Interagency Arctic Research Policy Committee, 8.

———. 1998a. On the history of research on sea ice. In *Physics of ice covered seas*, ed. M. Leppäranta, 1:1–24. Helsinki Univ Printing House.

———. 1998b. Growth conditions and the structure and properties of sea ice. In *Physics of ice-covered seas*, ed. M. Leppäranta, 1:25–104. Helsinki Univ Printing House.

Weeks, W. F., and S. F. Ackley. 1986. The growth, structure, and properties of sea ice. In *The geophysics of sea ice*, ed. N. Untersteiner, 9–164. New York: Plenum.

Weeks, W. F., S. F. Ackley, and J. Govoni. 1989. Sea ice ridging in the Ross Sea, Antarctica, as compared with sites in the Arctic. *JGR* 94(C4):4984–88.

Weeks, W. F., and D. L. Anderson. 1958a. Sea ice thrust structures. *J Glaciol* 3(23):173–75.

———. 1958b. An experimental study of the strength of young sea ice. *Trans AGU* 39(4):641–47.

Weeks, W. F., and A. Assur. 1963. Structural control of the vertical variation of the strength of sea and salt ice. In *Ice and snow: Properties, processes, and applications*, ed. W. D. Kingery, 258–76. Cambridge, MA: MIT Press.

———. 1968. The mechanical properties of sea ice. In *Proc Conf on Ice Pressures Against Structures* Tech Memo no 92, Quebec, Laval University, Assoc Comm on Geotech Res, Nat Res Council of Canada: 25–78. (CRREL Monograph II-C3.)

———. 1972. Fracture of lake and sea ice. In *Fracture, an advanced treatise. Vol 7: Fracture of nonmetals and composites*, ed. H. Liebowitz, 879–978. New York: Academic Press. (CRREL Research Report 269.)

Weeks, W. F., P. W. Barnes, D. M. Rearic, and E. Reimnitz. 1983. Statistical aspects of ice gouging on the Alaskan shelf of the Beaufort Sea. *CRREL Report* 83-21.

———. 1984. Some probabilistic aspects of ice gouging on the Alaskan shelf of the Beaufort Sea. In *The Alaskan Beaufort Sea: Ecosystems and environment*, ed. P. W. Barnes, E. Reimnitz, and D. W. Schell, 213–36. Orlando: Academic Press.

Weeks, W. F., and G. F. N. Cox. 1984. The mechanical properties of sea ice: A status report. *Ocean Sci Eng* 9(2):135–98.

Weeks, W. F., and A. J. Gow. 1978. Preferred crystal orientations in the fast ice along the margins of the Arctic Ocean. *JGR* 83(C10):5105–21.

———. 1980. Crystal alignments in the fast ice of arctic Alaska. *JGR* 85(C2):1137–46.

Weeks, W. F., A. J. Gow, P. Kosloff, and S. Digby-Argus. 1989. The internal structure, composition and properties of brackish ice in the Bay of Bothnia during the BEPERS-88 experiment. In *POAC '89*, 3:1318–33. Luleå, Sweden.

Weeks, W. F., and W. L. Hamilton. 1962. Petrographic characteristics of young sea ice, Point Barrow, Alaska. *Amer Mineralogist* 47:945–61.

Weeks, W. F., and A. Kovacs. 1970. On pressure ridges. *CRREL* Tech. Rept. IR505.

Weeks, W. F., A. Kovacs, and W. D. Hibler III. 1971. Pressure ridge characteristics in the Arctic coastal environment. *POAC* 1:152–83. Technical Univ Norway.

Weeks, W. F., A. Kovacs, S. J. Mock, W. D. Hibler III, and A. J. Gow. 1977. Studies of the movement of coastal sea ice near Prudhoe Bay, Alaska, USA. *J Glaciol* 19(81):533–46.

Weeks, W. F., and O. S. Lee. 1958. Observations on the physical properties of sea ice at Hopedale, Labrador. *Arctic* 11(3):134–55.

———. 1962. The salinity distribution in young sea ice. *Arctic* 15(2):92–108.

Weeks, W. F., and G. Lofgren. 1967. The effective solute distribution coefficient during the freezing of NaCl solutions. In *Proc Internat Conf on Low Temperature Science. Physics of Snow and Ice*, ed. H. Oura, I(1):579–97. Sapporo: ILTS.

Weeks, W. F., and M. Mellor. 1984. Mechanical properties of ice in the arctic seas. In *Arctic technology and policy*, ed. I. Dyer and C. Chryssostomidis, 235–359. New York: Hemisphere Pub Co.

Weeks, W. F., W. B. Tucker III, and A. W. Niedoroda. 1985. A numerical simulation of ice gouge formation and infilling on the shelf of the Beaufort Sea. In *POAC '85* 1:393–407. Narssarssuaq, Greenland.

Weeks, W. F., and G. Weller. 1980. Offshore oil in the Alaskan Arctic. *Science* 225(4660):371–78.

Weeks, W. F., and J. Wettlaufer. 1996. Crystal orientations in floating ice sheets. In *The Johannes Weertman Symposium*, ed. R. J. Arsenault, D. Cole, T. Gross, G. Kostorz, P. K. Liaw, S. Parameswaran and H. Sizek, 337–50. The Minerals, Metals & Materials Society.

Weinberg, B. P., V. Ia. Al'tberg, V. I. Arnol'd-Aliab'ev, and M. P. Golovkov. 1940. Ice, its properties, appearance and disappearance (R). Moscow: Gosudarstvennoe Izdatel'stvo Tekhniko-Teoreticheskoi Literatury.

Weinberg, F. 1963. Growth substructure in rapidly solidified Zn–2 pct Au alloys. *Trans Metal Soc AIME* 227:276–77.

Weiss, J., E. M. Schulson, and H. L. Stern. 2007. Sea ice rheology from in-situ, satellite and laboratory observations; fracture and friction. *Earth and Planetary Sci Lett* 255:1–8.

Weissenberger, J., G. S. Dieckmann, R. Gradinger, and M. Spindler. 1992. A cast technique to examine and analyze brine pocket and channel structure. *Limnol Oceanogr* 37(1):179–83.

Weitz, M., and J. Keller. 1950. Reflection of water waves from floating ice in water of finite depth. *Comm Pure Appl Math* 3(3):305–18.

Weller, G. 1968a. Heat-energy transfer through a four-layer system: Air, snow, sea ice, sea water. *JGR* 73(4):1209–20.

———. The heat budget and heat transfer processes in Antarctic plateau ice and sea ice. ANARE Sci Repts, Srs A(IV), Glaciology, Publn 102.

———. 1972. Radiation flux investigation. *AIDJEX Bull* 14:28–30.

Wensnahan, M. R., and D. A. Rothrock. 2005. Sea-ice draft from submarine-based sonar: Establishing a consistent record from analog and digitally recorded data. *GRL* 32:L11502, doi:10.1029/2005GL022507.

Wensnahan, M. R., D. A. Rothrock, and P. Hezel. 2007. New Arctic sea ice draft data from submarines. *Eos* 88(5):55–56.

Wernick, J. H. 1956. Determining the diffusivities in liquid metals by means of temperature-gradient zone melting. *J Chem Phys* 25(1):47–49.

Wettlaufer, J. S. 1991. Heat flux at the ice–ocean interface. *JGR* 96(C4):7215–36.

———. 1992a. Directional solidification of salt water: Deep and shallow cells. *Europhys Lett* 19(4):337–42.

———. 1992b. Singular behavior of the neutral modes during directional solidification. *Physics Rev* A46(10):6568–77b.

———. 1998. Introduction to crystallization phenomena in natural and artificial sea ice. In *Physics of ice-covered seas*, ed. M. Leppäranta, 1:105–94. Helsinki Univ Printing House.

———. 2001. The Stefan Problem: Polar exploration and the mathematics of moving boundaries. *Festschrift 150 Jahre Institut für Met und Geophysik, Univ Wien*.

Wettlaufer, J. S., M. Jackson, and M. Elbaum. 1994. A geometric model for anisotropic crystal growth. *J Phys* A27:5957–68.

Wettlaufer, J. S., N. Untersteiner, and R. Colony. 1990. Estimating oceanic heat flux from sea-ice thickness and temperature data. *Ann Glaciol* 14:315–18.

Wettlaufer, J. S., and M. G. Worster. 1995. Dynamics of premelted films: Frost heave in a capillary. *Phys Rev E Stat Phys Plasmas Fluids Relat Interdiscip Top* 51(5):4679–89.

Wettlaufer, J. S., M. G. Worster, and H. E. Huppert. 1997a. The phase evolution of young sea ice. *GRL* 24(10):1251–54.

———. 1997b. Natural convection during solidification of an alloy from above with application to the evolution of sea ice. *J Fluid Mech* 344:291–316.

———. 2000. The solidification of leads: Theory, experiment and field observations. *JGR* 105(C1):1123–34.

Weyl, P. 1964. On the change in the electrical conductance of seawater with temperature. *Limnol Oceanog* 9:75–78.

Weyprecht, K. 1872. Über die Eisverhältnisse im Arktischen Norden. *Ausland* LVX(2).

———. 1875. Die 2. Österr.-Hungarische Nordpolar-Expedition unter Weyprecht und Payer. *Petermanns Geographische Mitt* 21:65–72.

———. 1879. *Die Metamorphosen des Polareises*. Wien: Moritz Perles.

Whitham, G. B. 1974. *Linear and nonlinear waves*. New York: Wiley.

Whitman, W. G. 1926. Elimination of salt from seawater ice. *Amer J Sci* 211:126–32.

Wilen, L. A. 2000. A new technique for ice-fabric analysis. *J Glaciol* 46(152):129–39.

Williams, E., C. Swithinbank, and G. Robin. 1975. A submarine sonar study of Arctic pack ice. *J Glaciol* 15:349–62.

Williams, M. J. M., K. Grosfeld, R. C. Warner, R. Gerdes, and J. Determann. 2001. Ocean circulation and ice–ocean interaction beneath the Amery Ice Shelf, Antarctica. *JGR* 106(C10):22383–99.

Williams, M. J. M., A. Jenkins, and J. Determann. 1998. Physical controls on ocean circulation beneath ice shelves revealed by numerical models. In *Ocean, ice and atmosphere: Interactions at the Antarctic Continental Margin*, ed. S. S. Jacobs and R. F. Weiss, 285–99. AGU Antarctic Research Series, vol 75.

Wilson, J. T., J. H. Zumberge, and E. W. Marshall. 1954. A study of ice on an inland lake. *SIPRE Report* 5.

Winebrenner, D. P., D. G. Long, and B. Holt. 1998. Mapping the progression of melt onset and freeze-up on Arctic sea ice using SAR and scatterometry. In *Analysis of SAR data of the polar oceans*, ed. C. Tsatsoulis and R. Kwok, 129–44. Heidelberg: Springer-Verlag.

Winsor, P., and G. Björk. 2000. Polynya activity in the Arctic Ocean from 1958 to 1997. *JGR* 105(C4):8789–8803.

Winsor, P., and D. C. Chapman. 2002. Distribution and interannual variability of dense water production from coastal polynyas on the Chukchi Shelf. *JGR* 107(C7):10.1029/2001JC000984.

Winton, M. 2000. A reformulated three-layer sea ice model. *J Atmos Oceanic Technol* 17:525–31.

Wittmann, W. I., and J. J. Schule. 1966. Comments on the mass budget of Arctic pack ice. In *Proc Symp on the Arctic Heat Budget and Atmospheric Circulation*, 215–46. Santa Monica, CA: RAND Corp.

Woodgate, R. A., K. Aagaard, and T. J. Weingartner. 2006. Interannual changes in the Bering Strait fluxes of volume, heat and freshwater between 1991 and 2004. *GRL* 33:L15609, doi:10.1029/2006GL026931.

Woods, A. W. 1992. Melting and dissolving. *J Fluid Mech* 239:429–48.

Woods, A. W., and S. J. Linz. 1992. Natural convection and dispersion in a tilted fracture. *J Fluid Mech* 241:59–64.

Worby, A. P., C. A. Geiger, M. J. Paget, M. L. Van Woert, S. F. Ackley, and T. L. DeLiberty. 2008. Thickness distribution of Antarctic sea ice. *JGR* 113:C05S92, doi:10.1029/2007JC004254.

Worby, A. P., M. O. Jeffries, W. F. Weeks, K. Morris, and R. Jaña. 1996. The thickness distribution of sea ice and snow cover during late winter in the Bellingshausen and Amundsen Seas, Antarctica. *JGR* 101(C12):28441–55.

Worby, A. P., and R. A. Massom. 1995. The structure and properties of sea ice and snow cover in East Antarctic pack ice. Res Rept no 7. Hobart: Antarctic CRC.

Worby, A. P., R. A. Massom, I. Allison, V. I. Lytle, and P. Heil. 1998. East Antarctic sea ice: A review of its structure, properties and drift. In *Antarctic sea ice: Physical processes, interactions and* variability, ed. M. O. Jeffries, 41–68. AGU Antarctic Research Series, vol. 74.

Wordie, C. S., and S. Kemp. 1933. Observations on certain Antarctic icebergs. *Geogr J* 81:428–34.

Wordie, J. M. 1921. Shackleton Antarctic Expedition, 1914–1917: The natural history of pack ice as observed in the Weddell Sea. *Trans Roy Soc Edinburgh* 52:795–829.

World Meteorological Organization (WMO). 1989. *WMO Sea-Ice Nomenclature.* Rept. No. 259, Supplement No. 5. Geneva, Switzerland: World Meteorological Organization.

Worster, M. G. 1997. Convection in mushy layers. *Ann Rev Fluid Mech* 29:91–122.

Worster, M. G., and J. S. Wettlaufer. 1997. Natural convection, solute trapping, and channel formation during solidification of seawater. *J Phys Chem B* 101(32):6132–36.

Wright, B. D., J. Hnatiuk, and A. Kovacs. 1978. Sea ice pressure ridges in the Beaufort Sea. In *Proc IAHR Symp on Ice Problems*, 1:249–71.

Wright, C. S., and R. E. Priestley. 1922. *Glaciology.* British (Terra Nova) Expedition, 1910–1913. London: Harrison and Sons.

Yakovlev, G. N. 1955. Visual observations on the drifting ice cover and study of the morphology of the ice cover by surveying (R). In *Materialy nabliudenii nauchno-issledovatel'skoi Dreifuiushchei Stantsii 1949/51 goda*, ed. M. M. Somov, 6–51, 52–102. Leningrad: Izd "Morskoi Transport" (trans Amer Met Soc ASTIA Doc. AD117136).

———. 1960. The snow cover on the drifting ice of the Central Arctic (R). *PAiA* 3:65–76.

———. 1963. On the forms of relief of the surface of drifting ice (R). *PAiA* 12:61–68.

Yang, J., J. Comiso, R. Krishfield, and S. Honjo. 2001. Synoptic storms and the development of the 1997 warming and freshening event in the Beaufort Sea. *GRL* 28:799–802.

Yen, Y. C. 1981. Review of thermal properties of snow, ice and sea ice. *CRREL Rept* 81-10.

Yu, Y., G. A. Maykut, and D. A. Rothrock. 2004. Changes in the thickness distribution of Arctic sea ice between 1958–1970 and 1993–1997. *JGR* 109:C08004, doi:10.1029/ 2003JC001982.

Yu, Y., D. A. Rothrock, and R. W. Lindsay. 1995. Accuracy of sea ice temperature derived from the advanced very high resolution radiometer. *JGR* 100(C3):4525–32.

Zachos, J., M. Pagani, L. Skoan, E. Thomas, and K. Billups. 2001. Trends, rhythms, and aberrations in global climate 65 Ma to present. *Science* 292(5517):686–93.

Zhang, J. 1993. A high resolution ice–ocean model with imbedded mixed layer. Ph.D. thesis, Thayer School of Enginering, Dartmouth College.

———. 2007. Increasing Antarctic sea ice under warming atmospheric and oceanic conditions. *J Climate* 20:2515–29.

Zhang, J., and W. D. Hibler III. 1997. On an efficient numerical method for modeling sea ice dynamics. *J Geophys Res* 102(C4):8691–8702.

Zhang, J. L., W. D. Hibler III, M. Steele, and D. A. Rothrock. 1998. Arctic ice–ocean modeling with and without climate restoring. *J Phys Oceanog* 28:191–217.

Zhang, J. L., C. A. Lin, and R. J. Greatbatch. 1995. A decadal oscillation due to the coupling between an ocean circulation model and a thermodynamic sea-ice model. *J Marine Res* 53:79–106.

Zhang, J. L., W. Maslowski, and A. J. Semtner. 1999. Impact of mesoscale ocean currents on sea ice in high-resolution Arctic ice and ocean simulations. *JGR* 104(C8):18409–29.

Zhang, J. L., and D. A. Rothrock. 2001. A thickness and enthalpy distribution sea ice model. *J Phys Oceanog* 31:2986–3001.

Zhang, J. L., D. A. Rothrock, and M. Steele. 1998. Warming of the Arctic Ocean by a strengthened Atlantic inflow: Model results. *GRL* 25:1745–48.

———. 2000. Recent changes in Arctic sea ice: The interplay between ice dynamics and thermodynamics. *J Climate* 13:3099–3114.

Zhang, J. L., M. Steele, R. Lindsay, A. Schweiger, and J. Morison. 2008. Ensemble 1-year predictions of Arctic sea ice for the spring and summer of 2008. *GRL* 35:L08502, doi:10.1029/2008GL033244.

Zhang, J. L., and J. E. Walsh. 2006. Toward a seasonally ice-covered Arctic Ocean: Scenarios from the IPCC AR4 model simulations. *J Climat* 19:1730–47.

Zhang, X., A. Sorteberg, J. Zhang, R. Gerdes, and J. C. Comiso. 2008. Recent radical shifts of atmospheric circulations and rapid changes in Arctic climate system. *GRL* 35:L22701, doi:10.1029/2008GL 035607.

Zillman, J. W. 1972. A study of some aspects of the radiation and heat budget of the Southern hemisphere oceans. In *Meteorological Studies 26*. Canberra, Australia: Bureau of Meteorology, Dept of the Interior.

Zorita, E., T. F. Stocker, and H. von Storch. 2008. How unusual is the recent series of warm years? *GRL* 35:L24706, doi:10.1029/2008GL036228.

Zotikov, I. A., V. S. Zagorodnov, and J. V. Raikovski. 1980. Core drilling through the Ross Ice Shelf (Antarctica) confirmed basal freezing. *Science* 207(4438):1463–65.

Zubavicus, Y., and M. Grunze. 2004. New insights into the structure of water with ultrafast probes. *Science* 304(14 May):974–76.

Zubov, N. N. 1938. On the maximum thickness of perennial sea ice (R). *Meteorol Gidrol* 4:123–31.

———. 1945. *Arctic ice* (US Navy Electronics Laboratory translation, NTIS no. AD 426972; originally published in Russian by Izdatel'stvo Glasevmorputi in 1943).

Zubov, N. N., and M. M. Somov. 1940. Drift of the ice in the central part of the Polar Basin. *Prob Artk* 2:51–68.

Zukriegel, J. 1935. *Cryologia maris (Study of sea ice)*. Prague: Geographical Institute of the Charles IV Univ.

Zulauf, M. A., and S. K. Krueger. 2003. Two-dimensional simulations of Arctic leads: Plume penetration height. *JGR* 108:C2, 8050, doi:10.1029/2000JC000495.

Zwally, H. J., J. C. Comiso, and A. L. Gordon. 1985. Antarctic offshore leads and polynyas and oceanographic effects. In *Oceanology of the antarctic continental shelf*, ed. S. S. Jacobs, 203–26. Antarctic Res Srs 43. Washington, DC: AGU.

Zwally, H. J., J. C. Comiso, C. L. Parkinson, W. J. Campbell, F. D. Carsey, and P. Gloersen. 1983. In *Antarctic sea ice, 1973–1976: Satellite passive-microwave observations*. National Aeronautics and Space Administration Special Publication SP-459.

Zwally, H. J., J. C. Comiso, C. L. Parkinson, D. J. Cavalieri, and P. Gloersen. 2002. Variability of Antarctic sea ice 1979–1998. *JGR* 107:C5, 3041, 10.1029/2000 JC000733.

Index

Note: Entries followed by "f" indicate figures; "n" notes; "t" tables.